Solid State Industrial Electronics

Solid State Industrial Electronics

Richard A. Pearman

Reston Publishing Company, Inc.
A Prentice-Hall Company
Reston, Virginia

Library of Congress Cataloging in Publication Data

Pearman, Richard A.
Solid state industrial electronics.

1. Industrial electronics. 2. Solid state electronics. I. Title.
TK7881.P43 1983 621.3815 82-21472
ISBN 0-8359-7041-8

Reston Publishing Company, Inc.
A Prentice-Hall Company
Reston, Virginia

10 9 8 7 6 5 4 3 2 1

Printed in the United States of America

Contents

Preface

In this text the author has departed from the traditional approach of studying industrial electronics from a device point of view and instead has approached the subject from a systems approach.

The text assumes that the reader possesses a sound understanding of algebra and trigonometry, as well as a good background in electrical and electronic fundamentals combined with a good foundation of the theory and application of ac and dc rotating machines.

The purpose of the text is to provide a sound understanding of the application of solid-state industrial electronics in industry as required to meet the needs of two-year technician and three-year technologist level programs at community colleges, and in-house industrial training programs.

Every effort has been made to present the material in a logical manner by giving the background theory before going on to specific applications. The mathematical level has been deliberately limited to ensure that the students approaching the subject for the first time do not find the material obscured by abstract theory, since in applying their knowledge they are most concerned with the installation and maintenance of closed-loop and process control systems rather than with their design.

Much of the information in this text has been obtained from industrial sources. The author particularly wishes to express his appreciation to the following companies for providing material and permissions: The Algoma Steel Corporation Limited; Brown, Boveri & Cie. AG; Lovejoy Electronics; Micro Switch; Motorola Semiconductor Products Inc.; Texas Instruments Inc.; and Westinghouse Electric Corporation, Semiconductor Division.

The author wishes to acknowledge his indebtedness to his wife for her patience and encouragement during the many hours of work involved in the preparation of this text.

Richard A. Pearman

1

Semiconductor Physics

1-1 INTRODUCTION

Before embarking upon a study of **semiconductor devices** and their application, it is necessary to understand the basis of their operation. The knowledge of their electrical properties must be built upon the physics defining the behavior of semiconductors.

1-2 MATTER

Matter by definition is anything that has mass and occupies space. In turn, matter is classified chemically as simple or complex. **Simple matter** consists of just one element and **complex matter** consists of a combination of two or more elements. The smallest portion of simple matter that retains all the properties of an element is the **atom.** By contrast, a **molecule** is the smallest portion of complex matter.

Matter in turn is classified into one of three forms: liquid, gaseous, or solid. Electrically, matter is classified as being an **insulator, conductor,** or **semiconductor.** It can only exist in one of the three electrical forms at any given temperature.

1-3 THE STRUCTURE OF MATTER

In 1913 Dr. Niels Bohr put forward a theory on which our understanding of the structure of matter is based. The Bohr atomic theory suggested that the atom consisted of a minute central positively charged **nucleus** where the major portion of the mass of the atom was concentrated. Distributed around the nucleus at relatively large distances there are negatively charged particles called **electrons,** which form a relatively small fraction of the atom's **total mass.** The negative charge possessed by each electron is 1.602×10^{-19} coulombs. Since the atom in its normal state is electrically neutral, the negative charges possessed by the atom must be neutralized by an equal number of positive charges. The positively charged particles are called **protons** and together with neutrally charged particles called **neutrons** form the nucleus of the atom. Each proton and neutron present in the nucleus has a mass of 1.67×10^{-27} kg. The mass of the electron is 9.107×10^{-31} kg; from this it can be seen that each proton and neutron is approximately 1840 times as heavy as the electron. As a result the number of protons and neutrons present in the nucleus determine the mass of the atom. The number of electrons present in a neutral atom is called the **atomic number.** The **atomic mass** of an atom is the total number of protons and neutrons present in the nucleus. It should be noted that although all the atoms of a given element have the same number of electrons, that is, they have the same atomic number, their atomic masses may differ. Atoms having different masses but with the same charge or atomic number are called **isotopes.** Most elements can exist in the form of several isotopes, for example, hydrogen and its isotopes, deuterium and tritium.

Each element differs from all other elements only by the number of protons in the nucleus and the number of orbiting electrons. However, all electrons are identical to each other, as are all protons and neutrons.

1-3-1 Electron Orbits

The Bohr Atomic Theory stated that the atom consisted of a small positively charged nucleus consisting of protons and neutrons, surrounded by electrons orbiting in concentric **shells** somewhat in the same manner as a satellite orbits the earth. The orbiting electrons are free to move anywhere within their own shell, but they may not move from one shell to another. The shells have been designated K to Q inclusively, with the K shell being the innermost shell. The number of electrons permitted in each shell is given by the following rule:

$$\text{Maximum number of electrons} = 2n^2, \tag{1-1}$$

where n = shell number ($K = 1, L = 2, M = 3, N = 4, O = 5,$
$P = 6,$ and $Q = 7$).

As a result of this rule then

K shell may have a maximum of 2 electrons
L shell may have a maximum of 8 electrons

M shell may have a maximum of 18 electrons, etc.

There are two modifications to this rule:

1. The outermost or **valence** shell may not contain more than 8 electrons.

2. The next to the outermost shell cannot contain more than 18 electrons, irrespective of the number permitted by the rule.

Some examples of the distribution of electrons in their orbits may be instructive at this point (see Figure 1-1).

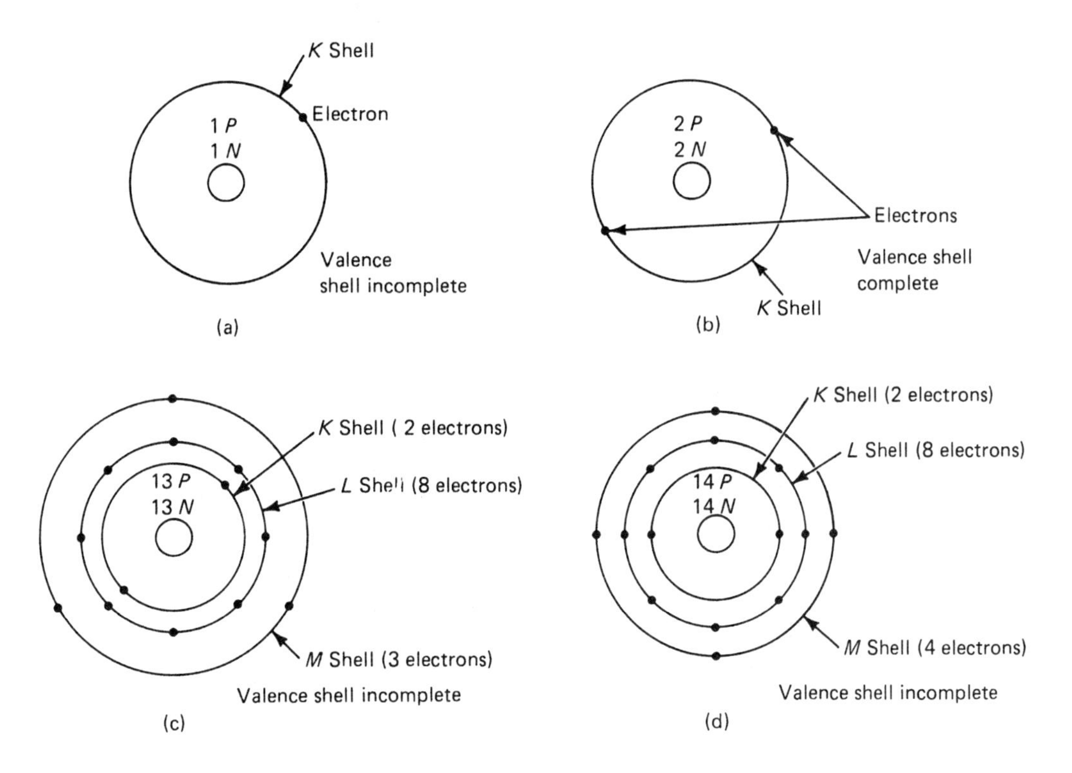

Figure 1-1. Atomic structures: (a) hydrogen atom, (b) helium atom, (c) aluminum atom, (d) silicon atom.

Examples:

1. Hydrogen (H), atomic number 1, has an electron grouping of 1. The *K* shell is the outermost shell and is partially filled.

2. Helium (He), atomic number 2, has an electron grouping of 2. The *K* shell is completely filled.

3. Aluminum (Al), atomic number 13, has an electron grouping of 2, 8, 3. The *K* and *L* shells are completely filled, and the valence shell contains 3 electrons; that is, it is partially filled.

4. Silicon (Si), atomic number 14, has an electron grouping of 2, 8, 4. The *K* and *L* shells are completely filled, and the valence shell contains 4 electrons.

5. Silver (Ag), atomic number 47, has an electron grouping of 2, 8, 18, 18, 1. The *K, L, M,* and *N* shells are completely filled, and the valence shell contains 1 electron.

6. Argon (A), atomic number 18, has an electron grouping of 2, 8, 8. All shells are completely filled.

The electrical (and chemical) properties of an element are determined by the electrons in the valence shell. These properties may be summarized as follows:

1. Atoms with less than four electrons in the valence shell tend to give up electrons with increasing ease as the number of electrons decreases; that is, they release electrons and thus decrease the resistance of the material. Elements with less than four electrons in the valence shell are **conductors.**

2. Atoms with more than four electrons in the valence shell tend to acquire one or more electrons and thus remove electrons as the conducting mechanism. This group of elements is called **resistors.**

3. Atoms with exactly eight electrons in the valence shell neither accept or release electrons and are classified as **insulators.**

4. Atoms with exactly four electrons in the valence shell share electrons with similar atoms to fill the valence shell. This is known as **covalent bonding,** and it is the peculiar property of materials such as silicon and germanium, members of a class called semiconductors.

1-3-2 Energy Levels and Energy Gaps

The behavior of electrons can be analyzed in terms of **energy.** The electrons in the shells at different radii from the nucleus are considered to have different amounts of energy. Each electron is considered to exist at an **energy level;** each energy level is separated from the next energy level by an **energy gap.** The assumption that an electron possesses energy is based on the fact that it must possess energy to counteract the electrostatic attraction to the nucleus in order to stay in its orbit. The electron must move at such a velocity that the centrifugal force acting on it is equal and opposite to the attractive force, and since it is in motion it must also have **kinetic energy** ($\frac{1}{2} mv^2$). It also posesses **potential energy** or the **energy of position,** since work would be done if the electron fell back into the nucleus. As the radius of orbit of the electron increases, the potential energy increases; but since the attractive force decreases as the radius increases, the velocity of the electron decreases. The net result is that as the radius of the electron orbit increases the total net energy of the electron has increased.

The Pauli Exclusion Principle states that two electrons cannot occupy the same energy state in a closed system. Each energy level contains exactly two energy states; therefore, only one electron can exist in any energy state and only two electrons can exist in a given energy level. To accommodate this principle the shells K to Q inclusive must in turn have **subshells.** These are shown in Table 1-1.

Table 1-1 Distribution of Shells, Subshells and Electrons

Shell	Number of Subshells	Electron Distribution
K	1	2
L	2	2,6
M	3	2,6,10
N	4	2,6,10,14
O	4	2,6,10,14
P	3	2,6,10
Q	2	2,6

The energy levels associated with each subshell are best illustrated by means of an energy level diagram. Figure 1-2 is the energy level diagram for germanium. In this diagram the horizontal lines represent the diameter of the shells, and the vertical positioning of the horizontal lines represents the energy levels of the shells. Note that there are minor energy gaps between subshells and major energy gaps between shells. The energy gaps are sometimes called **forbidden gaps** because no electrons exist in these regions. It should also be noted that the valence electrons exist in the highest energy level shell.

We have previously noted that the energy possessed by an electron increases as the distance from the nucleus increases. An electron may change its energy level by gaining or losing energy. To move to a higher energy level two conditions must be met: (1) there is a vacancy at the

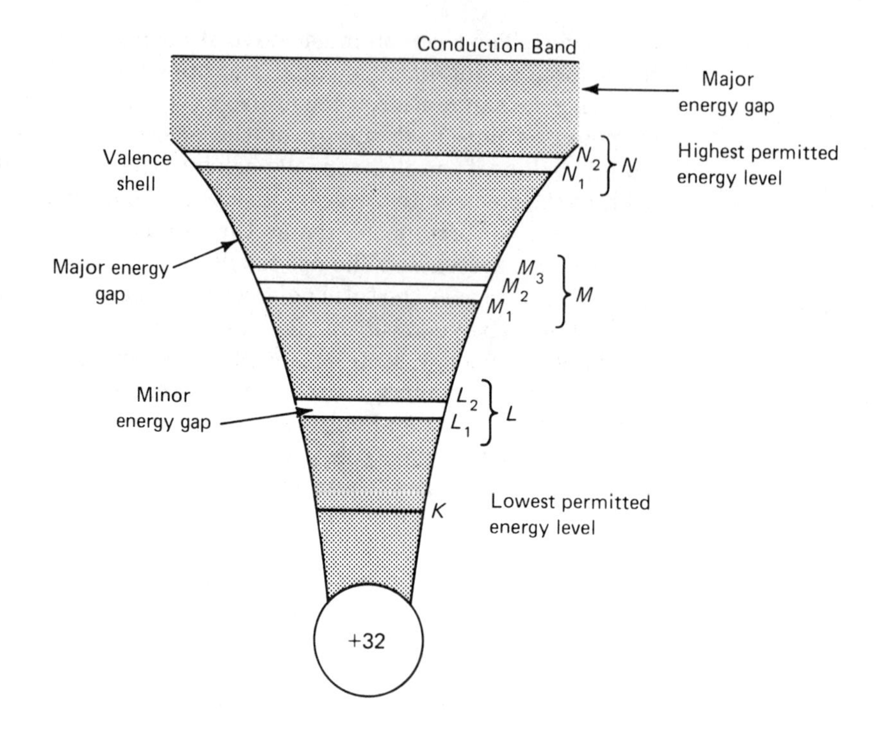

Figure 1-2. Energy level diagram of the germanium atom.

higher level and (2) the electron must gain sufficient energy to overcome the energy gap. If insuffient energy is given to the electron it will not move from its original position. On the other hand, if an electron loses energy it will fall through the energy gap to the next lower energy level provided there is a vacancy. The electrons in the **valence band** are the least tightly held electrons, and as a result are most easily affected by the application of energy from external sources.

The **conduction band** is considered to be at the next permissible energy level above the valence band. When valence electrons have sufficient energy applied to them, they then move up into the conduction band where they become free electrons separated from their parent atoms.

1-4 MONOCRYSTALLINE STRUCTURE

Germanium and silicon atoms each have four electrons in the valence shell; this is an unstable state. Since each atom of germanium or silicon requires four more electrons to completely fill the valence shell, they must share the four electrons with four adjacent atoms.

When two valence electrons are shared by two atoms they are said to be covalently bonded (see Figure 1-3). The covalent bonding forms a crystalline structure.

In pure or intrinsic germanium or silicon crystals, each atom is connected to four adjacent atoms by covalent bonds. The whole arrangement of covalently bonded atoms forms a three-dimensional arrangement known as a **monocrystalline lattice structure** which does not have any grain or grain boundaries. The characteristics of the monocrystalline structure can be developed from the energy level diagrams. When two identical atoms are brought close together, there is an interaction between the various energy levels which causes a shift of such levels (see Figure 1-4). Since by the Pauli Exclusion Principle it is only possible for two electrons to exist at the same energy level, the electrons in the *K* shell of the second atom cannot exist at the same energy level as those of the first atom. When the two atoms are in close proximity, two energy levels must now exist where previously there was only one. Applying the same principle to each of the shells and subshells, it can be seen that there will be twice as many energy levels as there were in the single atoms. With the addition of an ever increasing number of atoms, there are more increases to the energy levels until these energy levels become so packed and numerous that they are regarded as **energy bands.**

It should be noted that the first shell that becomes common between the atoms is the *N* shell, the valence shell in the case of germanium. As a result there is a continuous valence and conduction band throughout the whole monocrystalline structure, the two being separated by a continuous energy gap.

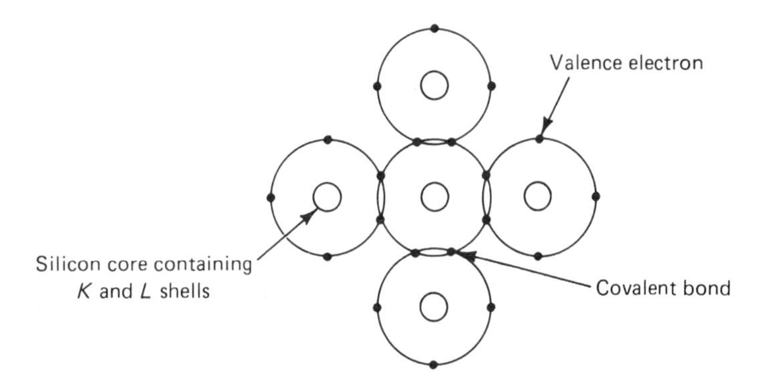

Figure 1-3. Silicon crystal structure.

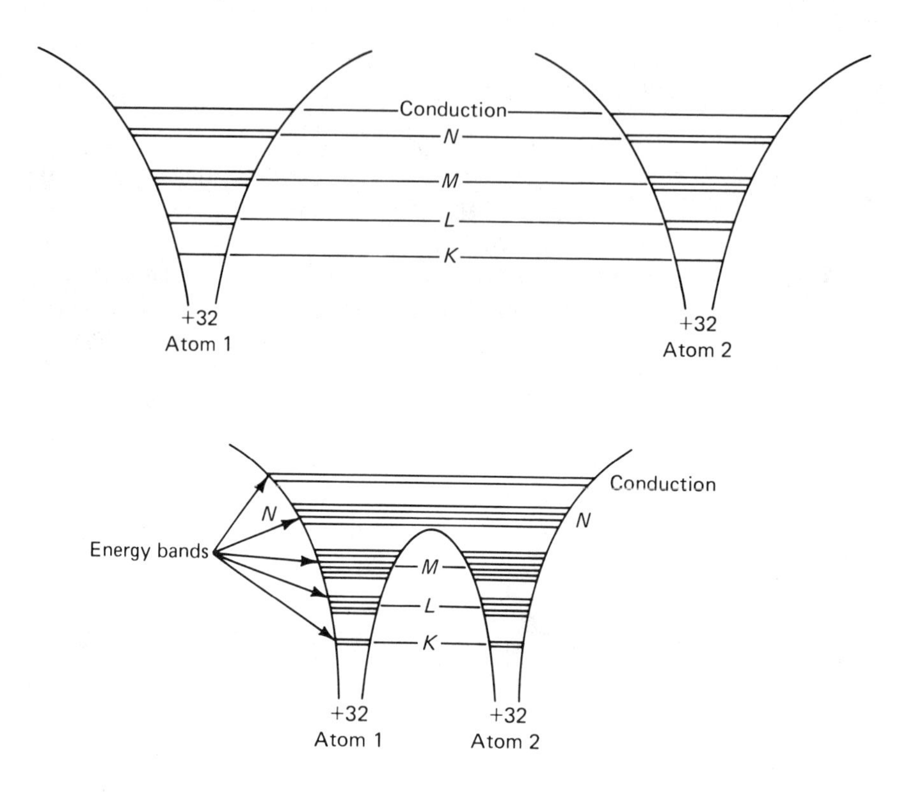

Figure 1-4. Energy band diagram of germanium: (a) energy levels of two separate atoms, (b) energy levels of two atoms in close proximity.

1-5 THE PRODUCTION OF HOLE AND ELECTRON CARRIERS

At absolute zero temperature, −273.15°C (−459.7°F), no external energy is being supplied; each energy band including the valence is filled with electrons at their lowest energy state. Electrons are not present in the conduction band. Therefore, at absolute zero the valence shell of semiconductor materials such as germanium and silicon are filled with eight electrons. Thus the semiconductor material is acting as a perfect insulator, since electrons are not available for the conduction process in the conduction band.

As the temperature increases from absolute zero, **phonons** of heat energy are imparted to the electrons in the monocrystalline structure. When a sufficient amount of energy has been applied to a valence electron, it crosses the energy gap between the valence and conduction bands, where, since it is now separated from the parent atom, it is free to take part in the conduction process. When the electron leaves the valence band, it creates a **hole**, or a deficiency of an electron in the covalent bond. The hole acts as a mobile positive charge carrier in the valence band (see Figure 1-5).

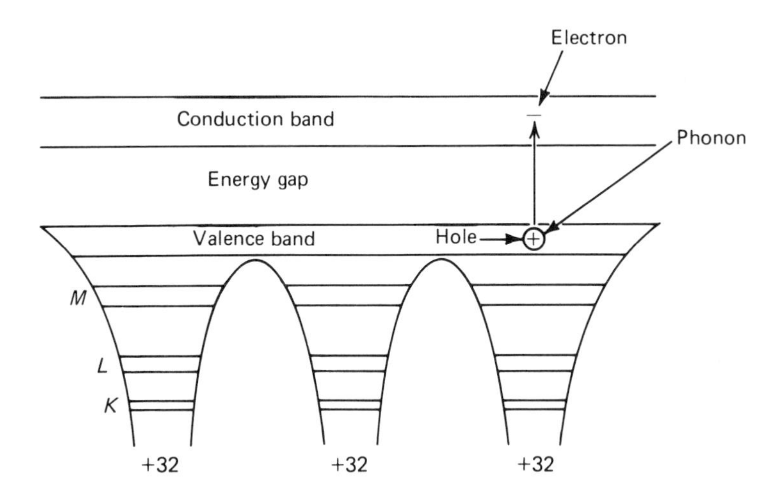

Figure 1-5. Electron-hole generation in intrinsic germanium.

As the temperature of the monocrystalline structure continues increasing because of an increase in the ambient temperature, a greater number of valence band electrons gain sufficient energy to cross the energy gap into the conduction band. At the same time as the electron carriers in the conduction band are increasing, there is a similar increase in the number of hole carriers in the valence band. The simultaneous production of the two types of carriers is called **thermal generation of electron-hole pairs.** The carriers are in continuous random motion in their respective bands, but conduction through the crystalline structure does not occur because the carrier movements cancel each other.

Electron and hole carriers can also be generated by the visible portion of the electromagnetic spectrum, called **photon energy,** as well as by higher frequency radiations such as X-ray and gamma radiations.

Decreasing the temperature of the crystalline structure will cause the electrons in the conduction band to lose energy, and if the loss of energy is great enough the electrons will fall back into the valence band and fill vacant holes. This process is called **recombination.** The lifetime of these carriers is usually between 10^{-4} and 10^{-2} sec.

1-6 INTRINSIC CONDUCTION

When an external emf is applied to an intrinsic crystal, conduction will take place; equal numbers of electrons and holes will be produced to act as current carriers (see Figure 1-6). Electron flow in the conduction band is relatively easy since the majority of the permissible energy states are vacant, and the electrons move toward the positive terminal of the source of external emf. The situation in the valence band is a little more complicated since an electron can only fill an empty energy state and most of the energy states are filled. However, the holes which are empty energy states do permit limited valence electron conduction. Whenever an electron frees itself from a covalent bond, it will fill a hole as it attempts to move toward the positive terminal. But in the process of breaking loose from the covalent bond it creates a hole, and this hole is further away from the positive terminal. In turn this hole will be filled by an electron breaking loose from a covalent bond. Effectively the holes are moving toward the negative terminal while the electrons are being attracted to the positive terminal.

There are two factors affecting conductivity in an intrinsic crystal. First, the temperature, since there will be no conduction until the temperature is great enough to cause large numbers of electron-hole pairs to be formed. Second, the mobility of the current carriers is proportional to the applied emf and electron energy. As a result, the electrons in the conduction band which are at higher energy levels than those in the valence band will move at a faster rate. Electron mobility at room temperature in the conduction band for germanium is about 3900 cm^2/V sec and in the valence band the hole mobility is about 1900 cm^2/V sec. Similar figures for silicon are about 1300 cm^2/V sec and 500 cm^2/V sec, respectively. The reason for the lower mobilities of the carriers in silicon is because the valence shell of silicon is at a lower energy level than that of germanium. Thus there are two current carriers in semiconductors moving at different velocities, and the total current is the sum of electron current and hole current.

1-7 CONDUCTORS, SEMICONDUCTORS, AND INSULATORS

Conductors, semiconductors, and insulators may be classified from energy band diagrams in accordance with their intrinsic conductivity. The width of the energy gap is measured in electron volts, where 1 eV is the amount of kinetic energy gained when an electron falls through a potential difference of 1 V, that is,

$$1 \text{ eV} = 1.602 \times 10^{-19} \text{ J}. \quad (1\text{-}2)$$

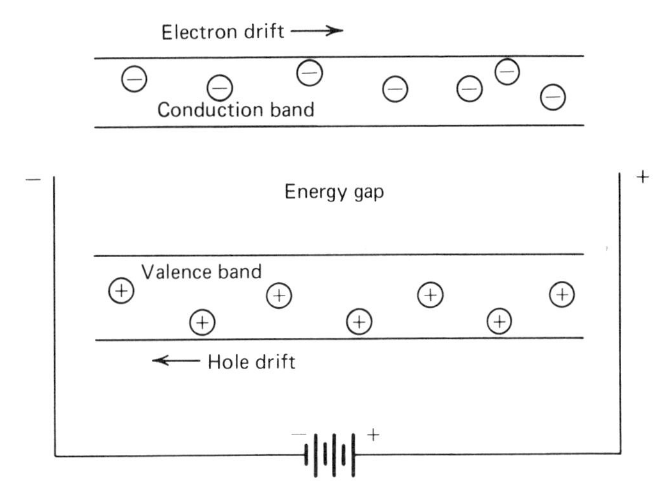

Figure 1-6. Conduction in an intrinsic crystalline structure when an external emf is applied.

Figure 1-7 shows the energy band diagrams for a conductor, a semiconductor, and an insulator. The energy band diagram for a conductor shows that both the valence band and the conduction band overlap, which implies very little energy is required for electrons to move into the conduction band and be available as current carriers. Figure 1-7 (b) shows that, while there is an energy gap for a semiconductor material, it is relatively small (in the case of silicon 1.1 eV), and it is still relatively easy to cause valence electrons to move up to the conduction band. Since there are eight tightly held electrons in the valence band, the insulator in Figure 1-7 (c) requires a considerable energy input before an electron moves up to the conduction band.

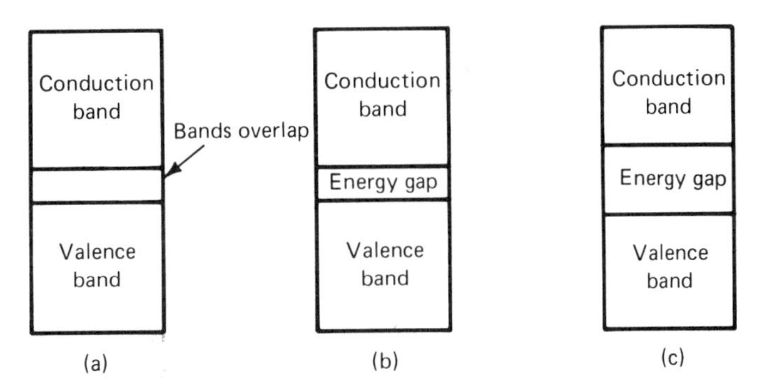

Figure 1-7. Energy band diagrams representing (a) conductors, (b) semiconductors, and (c) insulators.

1-8 DOPING

Because of its poor conductivity the intrinsic semiconductor is not effective for use in semiconductor devices. The conductivity may be improved by the deliberate addition of controlled amounts of suitable additives (usually about 1 part in 100 million). This process is called **doping.** There are two groups of elements that are used for doping. The first group consists of **pentavalent elements**, that is, elements that have five electrons in the valence shell. Typical pentavalent additives are phosphorous, arsenic, and antimony. The second group consists of **trivalent elements** or elements with three valence electrons, such as aluminum, gallium, boron, and indium. These additives are introduced into the intrinsic and semiconductor material while it is in the molten state. Doped semiconductor materials are called extrinsic semiconductors.

1-8-1 n-Type Materials

When a pentavalent additive is introduced into the intrinsic crystalline structure of germanium or silicon there will be an excess electron left over after the covalent bonds are completed (see Figure 1-8). The fifth valence electron of the pentavalent additive has no place in the valence bands of germanium or silicon, and it possesses an energy level just below the conduction band of either the germanium or silicon. However, a slight increase in thermal energy will cause the electrons to move up into the conduction band where they become free electrons available for the conduction process.

The loss of the electron from the valence shell of the pentavalent additive creates a hole, but unlike the holes formed in the valence band during intrinsic conduction, the hole is not free to move. When each pentavalent atom has given up its electron to the conduction band, it becomes a positive ion. The fixed positive charges shown in Figure 1-8(b) represent the location of the donated electrons prior to their movement to the conduction band. The majority of the electrons in the conduction band have been donated by the donor atoms, and the conduction caused by these carriers is called **extrinsic conduction.**

It should be realized that intrinsic conduction is also taking place at the same time as a result of thermal electron-hole pair generation. In an *n*-type material the number of electrons greatly exceeds the number of holes. As a result, the electrons are called **majority carriers** and the holes are called **minority carriers.** The addition of pentavalent impurities, such as phosphorous, arsenic, and antimony to intrinsic germanium or silicon will always form *n*-type materials.

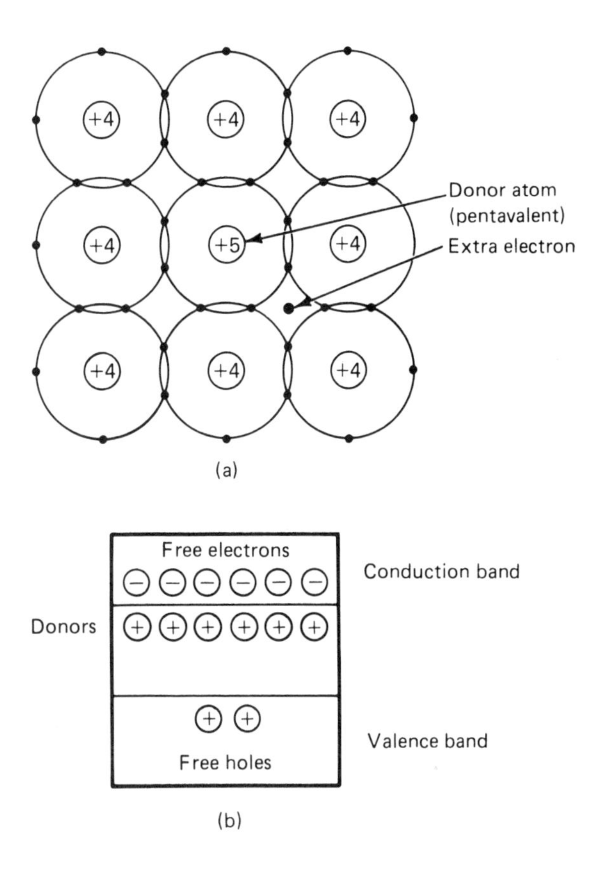

Figure 1-8. *n*-type material: (a) crystalline structure, (b) charge carrier distribution.

1-8-2 p-Type Materials

When an atom of germanium or silicon in a monocrystalline structure is replaced by a trivalent atom, that is, one with three valence electrons such as aluminum, gallium, or indium, covalent bonds will be formed with three adjacent germanium or silicon atoms. See Figure 1-9(a). As a result one covalent bond will not be made so that a hole or vacant energy state will exist. At temperatures slightly in excess of absolute zero, valence electrons from silicon or germanium will acquire sufficient thermal energy to move into the holes in the trivalent atoms, that is, the trivalent atoms act as acceptors. The trivalent atoms are ionized to become negative ions, these charges remaining fixed in position in the energy gap as in the case of the *n*-type material. However, the holes created in the valence band of the semiconductor are mobile and are responsible for the extrinsic conductivity. Holes in the valence band of the semiconductor material are also created by thermal generation of electron-hole pairs as well as those created by extrinsic conduction, but the electrons in the conduction band are caused strictly by the thermally generated electron-hole pairs. In an acceptor doped material the number of holes far exceeds the number of electrons, and thus the majority carrier in a *p*-type material is the hole, and the electrons are the minority carriers.

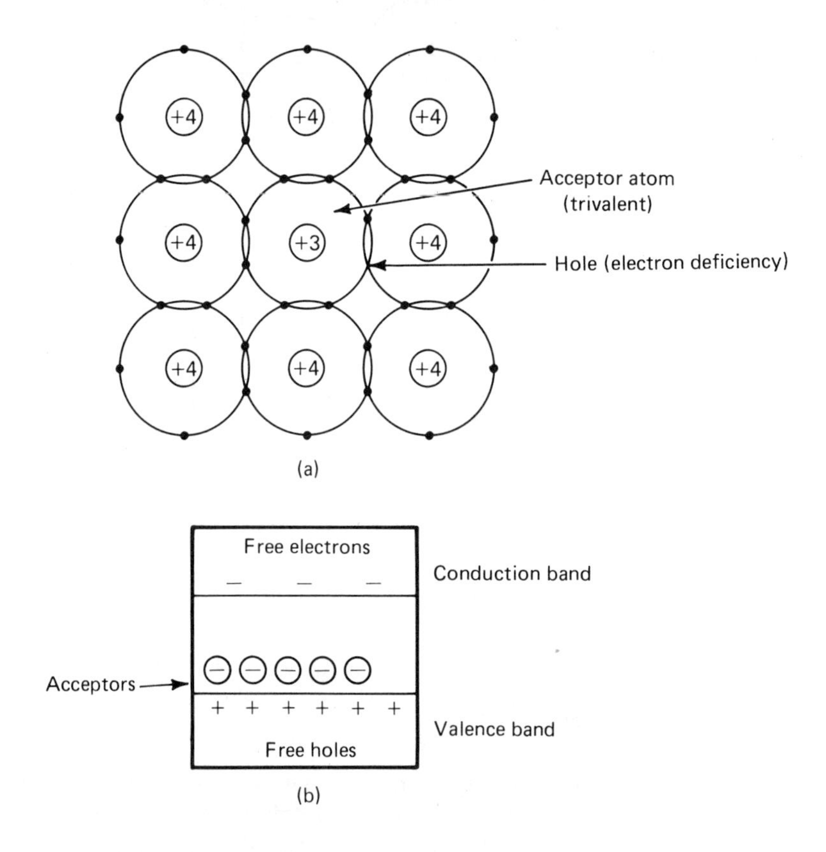

Figure 1-9. *p*-type material: (a) crystalline structure, (b) charge carrier distribution.

1-9 FERMI LEVEL

The **Fermi level** is a convenient method of indicating the relative densities of the charge carriers in the conduction and valence bands. In an intrinsic conductor, since all the charge carriers are the result of thermal electron-hole generation, the number of electron carriers in the conduction band is equal to the number of hole carriers in the valence band, and the Fermi level is assumed to be midway between the valence and conduction bands. See Figure 1-10(a). In the case of an *n*-type material, because of the excess of free electrons in the conduction band as compared to the holes in the valence band, the Fermi level is assumed to be closer to the conduction band. See Figure 1-10(b). Similarly, in the case of a *p*-type material the number of holes in the valence band is in excess of the number of electrons in the conduction band, and as a result the Fermi level is assumed to lie very close to the valence band. See Figure 1-10(c). In general, it can be said that the Fermi level lies closest to the band which contains the majority of carriers, and the degree of closeness is dependent upon the density of the carriers in that band as compared to the other band.

The availability of carriers depends upon both the degree of doping and the temperature. Therefore, as the amount of doping increases, for example, in an *n*-type carrier, the Fermi level will move closer to the conduction band. But as the temperature increases, the number of electron-hole pairs increases; however, the hole density increases at a faster rate than the electron density, and as a result the Fermi level will move toward the center of the energy gap. Similarly, in the case of

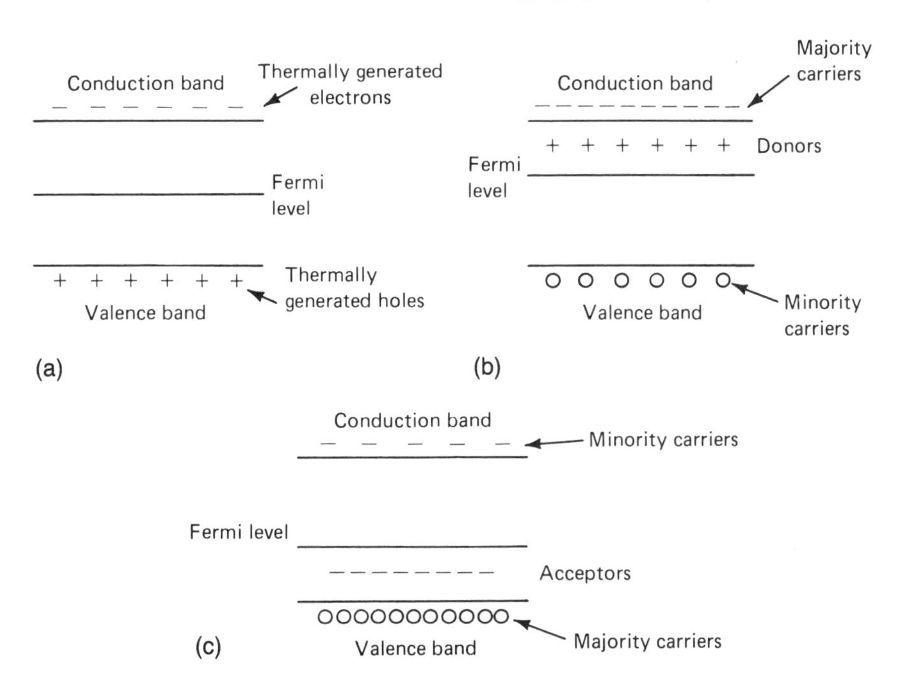

Figure 1-10. Fermi levels: (a) intrinsic material, (b) *n*-type material, (c) *p*-type material.

the *p*-type material, the electron density increases at a faster rate than the hole density and the Fermi level moves up toward the center of the energy gap. The precise position of the Fermi level in both the *n*-type and *p*-type semiconductor materials is therefore dependent upon the degree of doping and the operating temperature.

1-10 P-N JUNCTIONS

So far we have considered *n*- and *p*-type materials separately. However, in semiconductor devices there is at least one junction between *n*- and *p*-type materials. These junctions are formed by fusing the appropriate donor and acceptor impurities into intrinsic semiconductor materials such as germanium and silicon. The resistance of the *n*- and *p*-type materials is dependent upon the degree of doping. In our discussion we will assume that the two materials are equally doped, that is, there are as many donors as there are acceptors. This is called **symmetrical doping.**

From the Fermi level diagrams it was noted that the Fermi level was close to the conduction band in the *n*-type material, while in the *p*-type material it was closer to the valence band. As a result, at the *p-n* junction there will initially be attraction of the electrons in the conduction band of the *n*-type to the conduction band of the *p*-type material. There will also be a migration of holes from the valence band of the *p*-type to the valence band of the *n*-type material and recombinations will take place, thus removing charge carriers.

However, when an electron leaves the *n*-type material the pentavalent impurity atom is deficient one electron and becomes a **positive ion.** Similarly, when the electron recombines with a hole in the *p*-type material the trivalent impurity atom has one excess electron and becomes a **negative ion.** As a result, holes in the *p*-type material cannot cross the junction because of the repulsion offered by the positive ions in the *n*-type material. Similarly, electrons attempting to cross from the *n*-type material are repelled by the negative ions in the *p*-type material. This is called the **space charge effect.** The current carriers in the immediate vicinity of the *p-n* junction have been depleted by recombination; the area in the immediate vicinity of the junction is called the **depletion region** (see Figure 1-11). As can be seen, a difference of potential exists across the depletion region, and is approximately 0.3 V for a germanium *p-n* junction and approximately 0.7 V for a silicon *p-n* junction. As soon as this potential has been established, conduction across the *p-n* junction ceases until an external emf is applied.

Another way of representing the inherent emf across the junction is

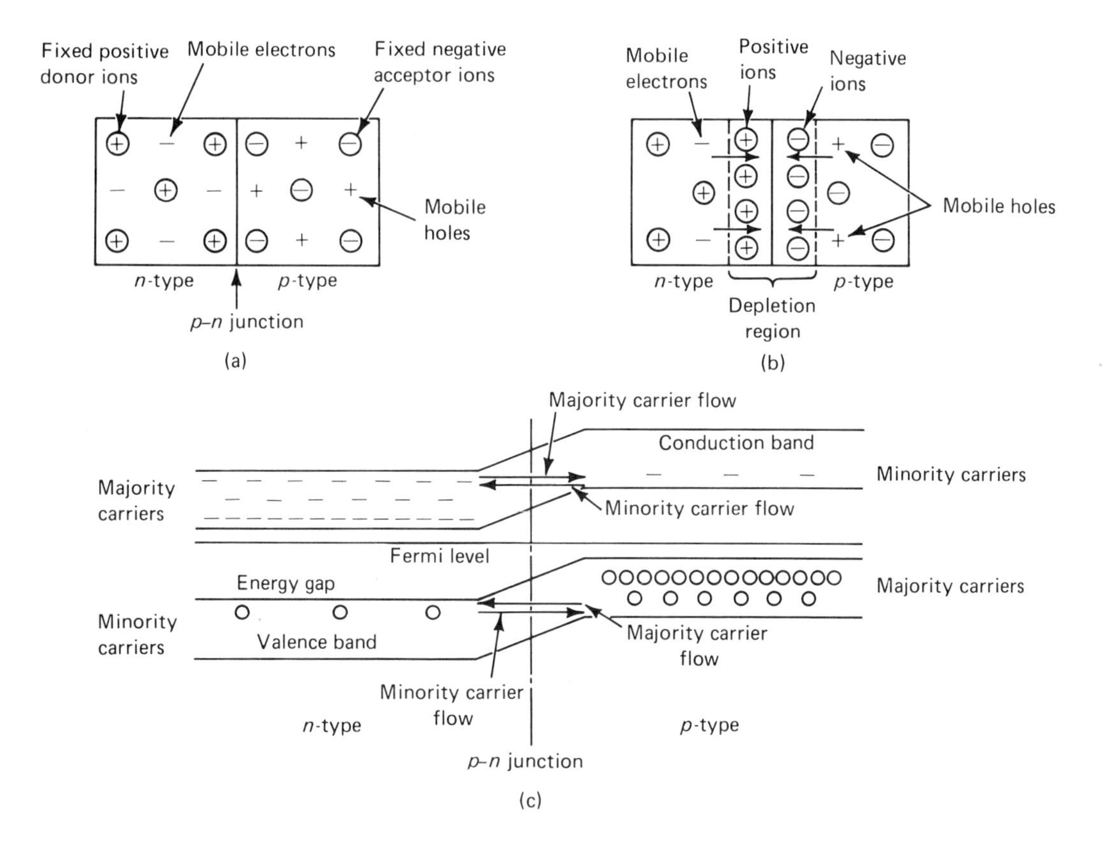

Figure 1-11. *p-n* junctions: (a) symmetrically doped junction, (b) depletion region and electric field, (c) energy band diagram of junction at equilibrium.

by an energy band diagram of the *p-n* junction [Figure 1-11(c)]. When the *p-* and *n*-type materials are in electrical contact the Fermi levels align, since the recombination process has left the relative densities of the electrons and holes at the junction in balance, and as a result the Fermi level will be in the center of the energy gap at the junction.

The alignment of the Fermi levels causes the conduction and valence bands to be warped at the junction. The magnitude of the warp is a measure of the inherent emf across the depletion region. From our discussion on Fermi levels, it should be recalled that an increase in temperature causes the Fermi level to move toward the center of the energy gap, which in our case means a reduction in the inherent emf across the junction. The Fermi level will be moved away from the center of the energy gap with increased doping, resulting in an increase of the inherent emf across the depletion region.

From Figure 1-11(c) it can be seen that most of the majority carriers in the conduction band of the *n*-type material will require an increase in energy before they can cross the junction into the *p*-type material. However, there are a few majority carriers in the conduction band of the *n*-type material that have received sufficient thermal energy so that they are at a higher energy level than the lowest energy level in the conduction band of the *p*-type material. These carriers can cross the junction freely. In the conduction band of the *p*-type material there are also a few thermally generated minority carrier electrons, which, since they exist at energy levels comparable to the conduction band of the *n*-type material, can also cross the junction freely.

Most of the majority carriers (holes) in the valence band of the *p*-type material are at energy levels in excess of those in the valence band of the *n*-type material. There are a few majority carriers in the *p*-type material at low enough energy levels so that they can cross the junction freely. At the same time there are a few thermally generated holes in the valence band of the *n*-type material that exist at high enough energy levels to cross the junction into the valence band of the *p*-type material.

Thus it is apparent that the barrier effect offered by the inherent emf is effective against the majority carriers but has no effect on the minority carriers. It should also be emphasized that the net current flows in both the conduction band and the valence band are zero in an unbiased *p-n* junction.

1-10-1 The Forward-Biased p-n Junction

The inherent emf produced by the donor and acceptor ions across the junction forms a barrier to majority carriers. When the *p-n* junction is forward biased by an external source of emf (see Figure 1-12), the effect of the inherent emf is reduced and, as a result, more carriers can cross the junction and more recombinations occur. The effect of the forward-bias voltage is that the positive side of the external voltage repels holes in the *p*-type material and causes them to drift across the junction. Similarly, the negative terminal of the external emf source repels the free electrons in the *n*-type material and causes them to also drift across the junction. Under the effect of the forward-bias voltage, the ion charges offer less opposition to the movement of majority carriers across the junction and to their subsequent recombination.

Whenever a recombination occurs in the junction area, a new hole carrier in the *p*-type material and a new electron carrier in the *n*-type material must be created. The new electron carriers are supplied from the negative terminal of the external voltage source, and at the same time electrons move from the *p*-type material into the external voltage source. Hole movement only takes place in the semiconductor

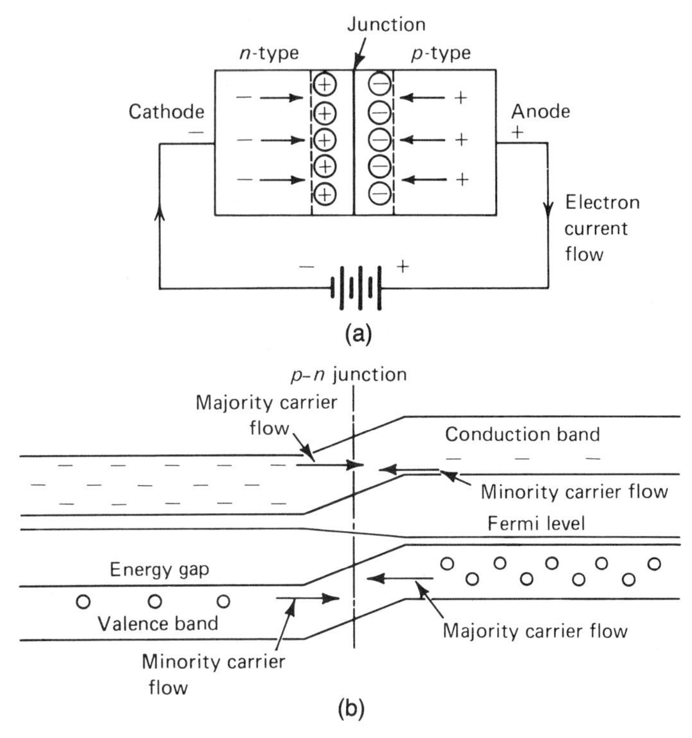

Figure 1-12. Forward-biased *p-n* junction: (a) junction, (b) energy band diagram.

material. Current flow in the external circuit is solely due to electrons, but this external current is the sum of the electron and hole currents in the semiconductor material.

The effect of the external voltage is to decrease the inherent emf, not to eliminate it, since most of the applied voltage is used in overcoming the resistances of the *n*- and *p*-type materials. As can be seen from Figure 1-12(b), the Fermi level becomes bent to accommodate the reduction in the opposing voltage across the junction. The majority carrier currents are controlled by the forward-bias voltage; however, the number of minority carriers is strictly determined by the temperature of the semiconductor material.

1-10-2 The Reverse-Biased p-n Junction

When the polarity of the external emf applied to a *p-n* junction is reversed, it is said to be reverse biased. The effect of the reverse bias is to aid the effects of the inherent emf and reduce recombinations.

Additionally, holes in the p-type material and electrons in the n-type material are attracted away from the junction by the reverse-bias voltage, which results in a widening of the depletion region and leads to a further increase in the inherent emf in the junction area (see Figure 1-13).

From Figure 1-13(b) it can be seen that the effect of the reverse bias increases the difference in energy levels between the conduction bands and the valence bands; there are no majority carriers in the conduction band of the n-type material with sufficient energy to cross the junction to the conduction band of the p-type material. Similarly, the majority carriers in the valence band of the p-type material do not exist at a low enough energy level to be able to cross the junction to the valence band of the n-type material. However, thermally produced minority carriers will flow, their magnitude being strictly dependent upon the temperature. This reverse current is called the **reverse saturation current.**

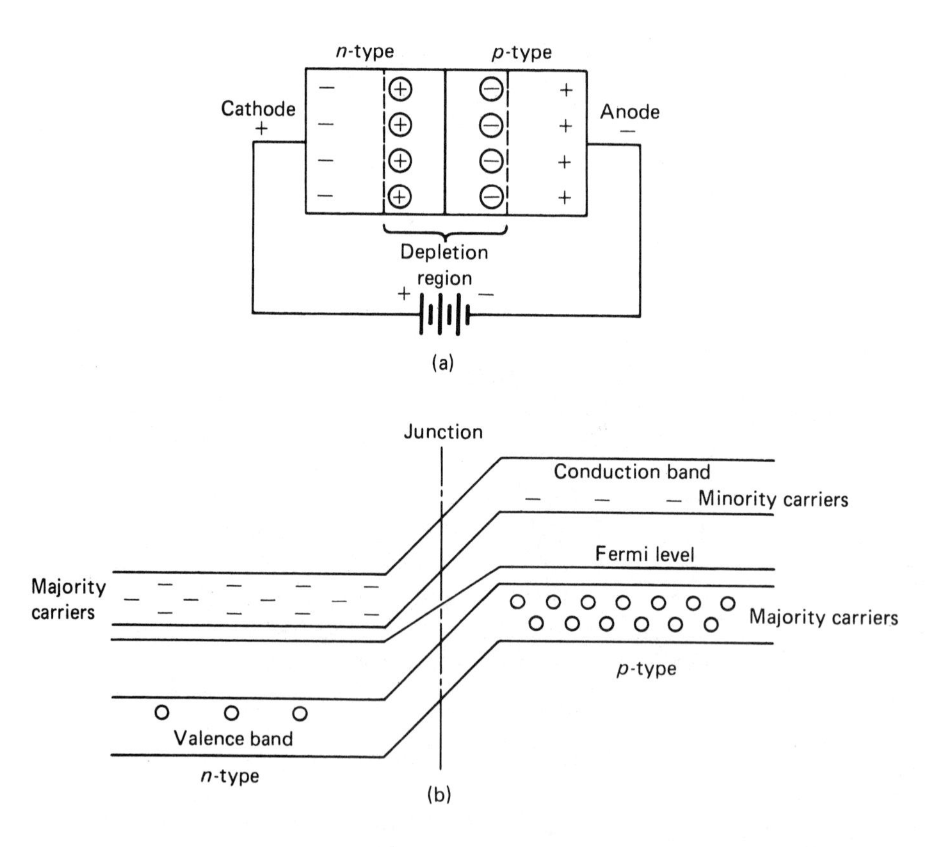

Figure 1-13. Reverse-biased p-n junction: (a) junction, (b) energy band diagram.

1-11 CARRIER INJECTION EFFICIENCY

So far we have assumed that the *p-n* junctions have been symmetrically doped; however, in actual practice this condition is very difficult to obtain. In fact there are times when it is desirable to have **unbalanced** or **asymmetrically doped** junctions, such as in transistors.

Let us assume that the *n*-type material is more heavily doped than the *p*-type material. The depletion region will extend further into the *p*-type material than in the *n*-type material because the density of the donor atoms is greater. Under conditions of forward-bias, and with the correct selection of the doping ratio, the total current flow can be made to consist almost exclusively of electrons.The ratio of the electron current to the hole current is called **carrier injection efficiency** γ. The carrier injection efficiency is 50% for a symmetrically doped junction and is greater than 50% for an asymmetrical junction. A similar reasoning applies if the *p*-type material is heavily doped.

Under reverse-bias conditions the majority of the reverse current consists of electrons. The minority carrier (electron) lifetime is relatively long in the *p*-type material since the shortage of holes restricts the number of recombinations that can take place. However, in the *n*-type material the lifetime of the mimority carriers (holes) is very short because of the excess number of electrons available for recombinations. As a result, since there are more minority electron carriers than minority hole carriers, the reverse current like the forward current is almost exclusively electrons.

1-12 JUNCTION CAPACITANCES

The *p-n* junction can be considered as a paralled resistance-capacitance combination. The magnitude of the junction resistance R_J is determined by the degree of doping, the temperature of the semiconductor material, the cross-sectional area of the junction, and the amplitude and type of bias applied. The junction capacitance C_J is due to the absence of charge carriers in the depletion region and is proportional to the cross-sectional area of the junction and inversely proportional to the width of the depletion region. The width of the depletion region is decreased by additional doping or by applying a forward bias, which in turn increases the junction capacitance. On the other hand, the application of a reverse bias increases the width of the depletion region and reduces the junction capacitance.

When used in an ac application such as rectification, the junction capacitance will vary in a cyclic manner, since the applied ac voltage will cause the width of the depletion region to vary in a cyclic manner.

In rectifier applications, R_J is low when forward biased and high when reverse biased. Normally the junction capacitance C_J is relatively small, the capacitive reactance $1/\omega\ C_J$ will be high, and effectively C_J acts as an open circuit at low to medium frequencies. However, at high frequencies the capacitive reactance is small and will effectively short out the junction resistance and eliminate the reverse blocking capability of the *p-n* junction (see Figure 1-14).

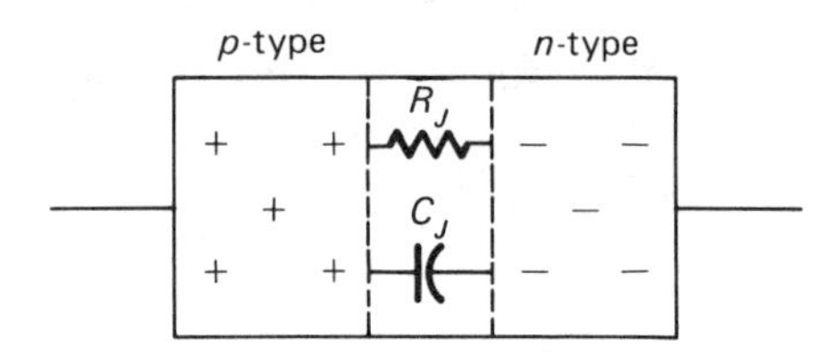

Figure 1-14. Electrical equivalent of a *p-n* junction.

GLOSSARY OF IMPORTANT TERMS

Matter: Anything that has mass and occupies space. It exists at any time in one of three forms: liquid, gas, or solid. In the electrical sense matter is either a conductor, semiconductor, or insulator and can exist only in one of these forms at a given temperature.

Atom: The smallest indivisible unit of an element.

Nucleus: The central core of the atom.

Electron: Very small negatively charged particles with a charge of 1.602×10^{-19}C and a mass of 9.107×10^{-31}kg.

Proton: Positively charged particle contained in the nucleus of an atom with a charge of 1.602×10^{-19}C and a mass of 1.67×10^{-27}kg.

Neutron: Neutrally charged particles with a mass of 1.67×10^{-27}kg which together with the protons form the nucleus of the atom.

Atomic number: The number of electrons or protons present in a neutral atom.

Atomic weight: Approximately equal to the total number of protons and neutrons in the nucleus of an atom.

Isotopes: Atoms of an element with the same atomic number but different atomic masses.

Shells: The orbital paths of electrons about the nucleus.

Valence shell: The outermost electron orbit.

Semiconductors: Elements with exactly four electrons in the valence shell.

Covalent bonding: Type of atomic bonding where atoms with exactly four electrons in the valence shell share electrons with similar adjacent atoms to form a complete interlinking of atoms in a structure.

Energy level: The amount of energy contained by an orbital electron and which is proportional to the radius of the orbit about the nucleus.

Energy bands: The grouping of energy levels in solid materials and which is a range of energies in which an electron may be found.

Energy gaps: Regions between permissible energy bands in which electrons cannot exist.

Valence band: Energy band of the valence electrons.

Conduction band: Energy band of electrons that have escaped from the valence band.

Hole carrier: A deficiency of an electron in the valence band. The hole acts as a mobile positive charge carrier.

Electron carrier: A valence electron to which sufficient energy has been imparted to move it up to the conduction band. The electron acts as a mobile negative charge carrier.

Recombination: The recombining of holes and electrons as a result of a conduction band electron losing energy.

Charge carriers: Holes or electrons.

Intrinsic semiconductor: An undoped semiconductor.

Doping: The controlled addition of impurity atoms to improve the conductivity of an intrinsic semiconductor.

Donor impurity: Doping with a pentavalent additive creates one electron in excess of the covalent bonding requirement. The excess electron is donated to the conduction band.

Acceptor impurity: Doping with a trivalent additive results in a lack of sufficient valence electrons to complete a covalent bond. An electron from an adjacent atom is accepted to complete the bond; as a result a hole is created.

Extrinsic semiconductor: A doped seimiconductor whose properties are determined by the doping additives.

***n*-type materials:** An extrinsic semiconductor doped with donor atoms, in which the majority carriers are electrons.

***p*-type materials:** An extrinsic semiconductor doped with acceptor atoms, in which the majority carriers are holes.

***p-n* junction:** The junction between *p*-type and *n*-type materials in a single semiconductor crystal structure.

Space charge region (depletion region): The region in the immediate vicinity of the *p-n* junction where the charge carriers have been depleted by recombination.

Barrier effect: The opposition to the diffusion of majority carriers across a *p-n* junction caused by the inherent emf.

Forward bias: An external emf applied to a *p-n* junction to neutralize the effects of the inherent emf.

Reverse bias: An external emf applied to a *p-n* junction to increase the effects of the inherent emf.

REVIEW QUESTIONS

1-1 Define matter and specify the forms in which it may exist.

1-2 Discuss the Bohr atom and draw a two-dimensional diagram of the silicon atom to illustrate your answer.

1-3 What is the significance of the atomic number and atomic weight of an atom of an element? What is meant by an isotope?

1-4 What are the rules that govern the maximum number of electrons permitted in each shell of an atom?

1-5 The electrical properties of an element are determined by the number of electrons in the valence shell. What are these properties?

1-6 What do you understand by energy levels and energy bands?

1-7 What do you understand by the Pauli Exclusion Principle?

1-8 Under what conditions may an electron change its energy level?

1-9 Explain what you understand by the terms valence band, conduction band and forbidden gap.

1-10 What is meant by an intrinsic semiconductor?

1-11 What is meant by an extrinsic semiconductor, and how can an intrinsic material be made extrinsic?

1-12 What is meant by thermal generation of electron-hole pairs?

1-13 What two factors affect conductivity in an intrinsic semiconductor material?

1-14 Discuss the differences between conductors, semiconductors, and insulators in terms of their energy band diagrams.

1-15 Why are intrinsic semiconductor materials doped? Discuss with the aid of diagrams the two types of doping.

1-16 What is meant by the terms majority carriers and minority carriers? Which are the majority carriers and why in (a) *n*-type materials and (b) *p*-type materials?

1-17 Explain the Fermi level concept.

1-18 Explain what is meant by the space charge effect.

1-19 With the aid of a sketch explain how the depletion region at a *p*-*n* junction is produced.

1-20 When a *p*-*n* junction is forward biased, what are the effects of forward bias upon the depletion region width, the minority carriers, the majority carriers?

1-21 When a *p*-*n* junction is reverse biased, what are the effects of reverse bias upon the depletion region width, the minority carriers, the majority carriers?

1-22 Explain what is meant by carrier injection efficiency.

1-23 What is meant by junction capacitance in a *p*-*n* junction and what factors determine its magnitude?

2

Semiconductor Devices

2-1 INTRODUCTION

Chapter 1 presented the basic principles of semiconductor devices. In this chapter we will review the principles and operating characteristics of the semiconductor devices most commonly encountered in the industrial and power electronic fields.

2-2 THE SEMICONDUCTOR DIODE

The **semiconductor diode** is a two-terminal *p-n* junction device. The semiconductor diode is in effect a high-speed electronic switch with a low impedance when forward biased and a high impedance when reverse biased. Most semiconductor diodes are **diffused junction** silicon devices. The diffused junction is produced by exposing a masked area of *n*-type material to a trivalent vapor, such as gallium, under elevated pressure and temperature. The acceptor atoms are absorbed into a thin layer of the exposed *n*-type material to form the *p-n* junction. The diffused junction silicon diode, which accounts for 90% of the rectifying devices in use, is preferred to germanium because of the higher temperature and current handling capability and the wider energy gap.

Silicon semiconductor diodes are available in two types: (1) The **general purpose diode** with current ratings ranging from 1 to 2200 A with voltage ratings from 600 to 4000 V is used where the reverse recovery time is not critical. Typical industrial applications are ac motor controls, air conditioning, cranes, hoists, conveyors, electric discharge machining, furnaces, heating and melting, lift trucks, battery chargers, electroplating, and dielectric heating. (2) The **fast-recovery**

diode is available in current ratings from 6 to 1400 A with voltage ratings from 600 to 3200 V with reverse recovery times ranging from 200 ns to 5 μs. Typical applications of the fast-recovery diode are uninterruptible power supplies (UPS), inverters, choppers, and bypass diodes. The range of devices produced by Westinghouse Electric Corporation, Semiconductor Division, is illustrated in Figure 2-1.

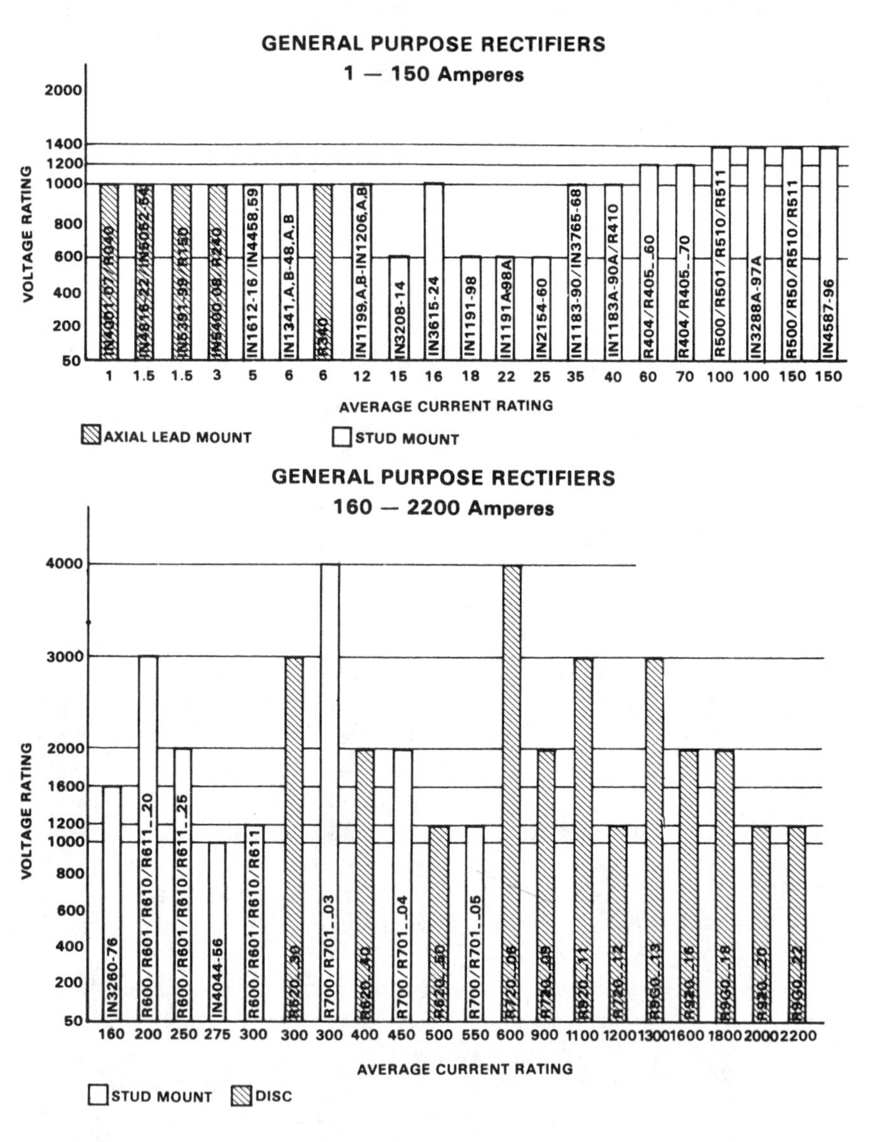

Figure 2-1. Rectifier capability graphs. (Courtesy Westinghouse Electric Corp., Semiconductor Division)

FAST RECOVERY RECTIFIERS
6 — 1400 Amperes

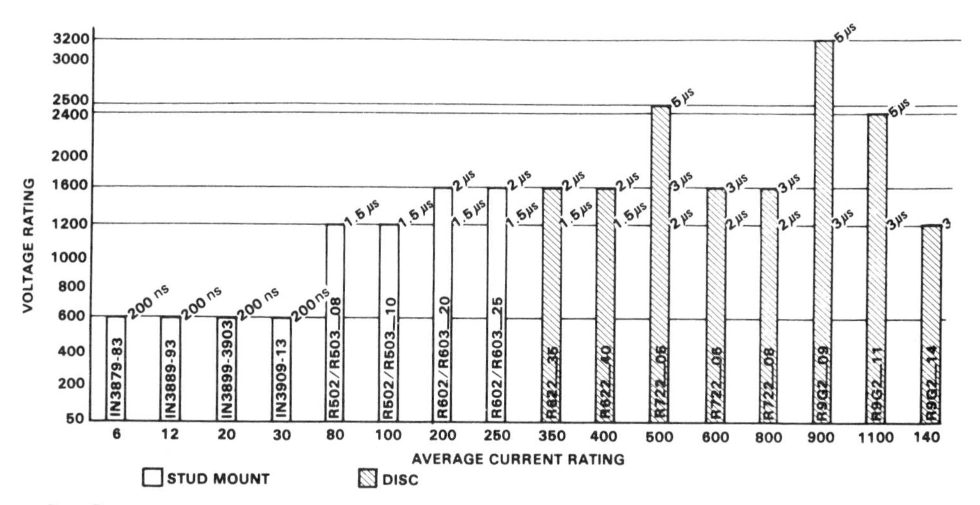

Note: Reverse recovery times shown represent fastest currently available at given voltage rating.

Figure 2-1 continued.

The general purpose diode is available in axial lead mount packages, and the general purpose and fast-recovery diodes are supplied in stud mount (both standard and reverse polarity) and disc mount as shown in Figure 2-2. In addition, diodes are available in modular bridges, air and liquid cooled disc assemblies, high voltage stacks, and custom designed assemblies in single- and three-phase arrangements.

2-3 DIODE CHARACTERISTICS AND PARAMETERS

Under conditions of forward bias, that is, the anode potential is positive with respect to the cathode, forward current will flow and there will be a relatively small forward voltage drop across the device (the magnitude of the forward voltage drop being dependent upon the amount of doping and the device temperature). Under reverse-bias conditions the reverse current is usually in the micro or milliampere range and slowly increases in magnitude until the avalanche of zener voltage is reached. These conditions are illustrated in the steady-state *V-I* characteristic shown in Figure 2-3(b).

There are three power losses that occur in semiconductor diodes:

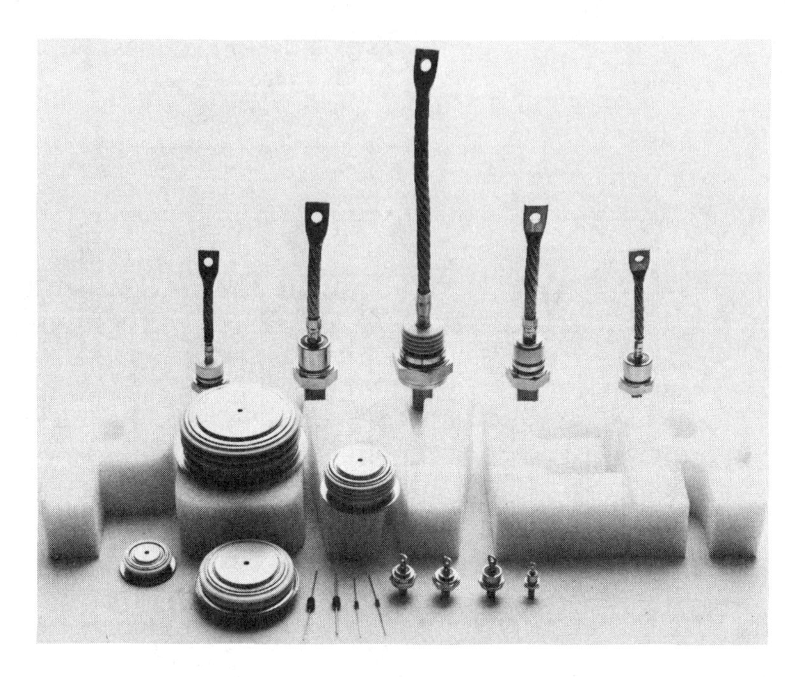

Figure 2-2. Typical silicon diffused junction diodes. (Courtesy Westinghouse Electric Corp., Semiconductor Division)

1. The **forward power loss** is the product of the forward voltage drop and forward current. Since the forward voltage drop is nearly constant, the forward power loss is approximately proportional to the forward current.

2. The **reverse power loss** is the product of the reverse voltage and the reverse leakage current. It should be noted that the reverse leakage current will increase rapidly with increased *p-n* junction temperatures.

3. **High frequency switching loss.** While this loss may be ignored at power line frequencies it becomes appreciable at high frequencies. During the initial period of turn-on, time is required for the minority hole carriers to move across the *p-n* junction. As a result there is an initial voltage spike during this interval, which produces the high frequency switching loss (the product of the voltage spike and the forward current).

These power losses are the source of **junction heating;** for most diffused junction silicon power diodes maximum temperatures of 200°C are permissible. Therefore, the maximum permissible junction temperature defines the current rating of the diode.

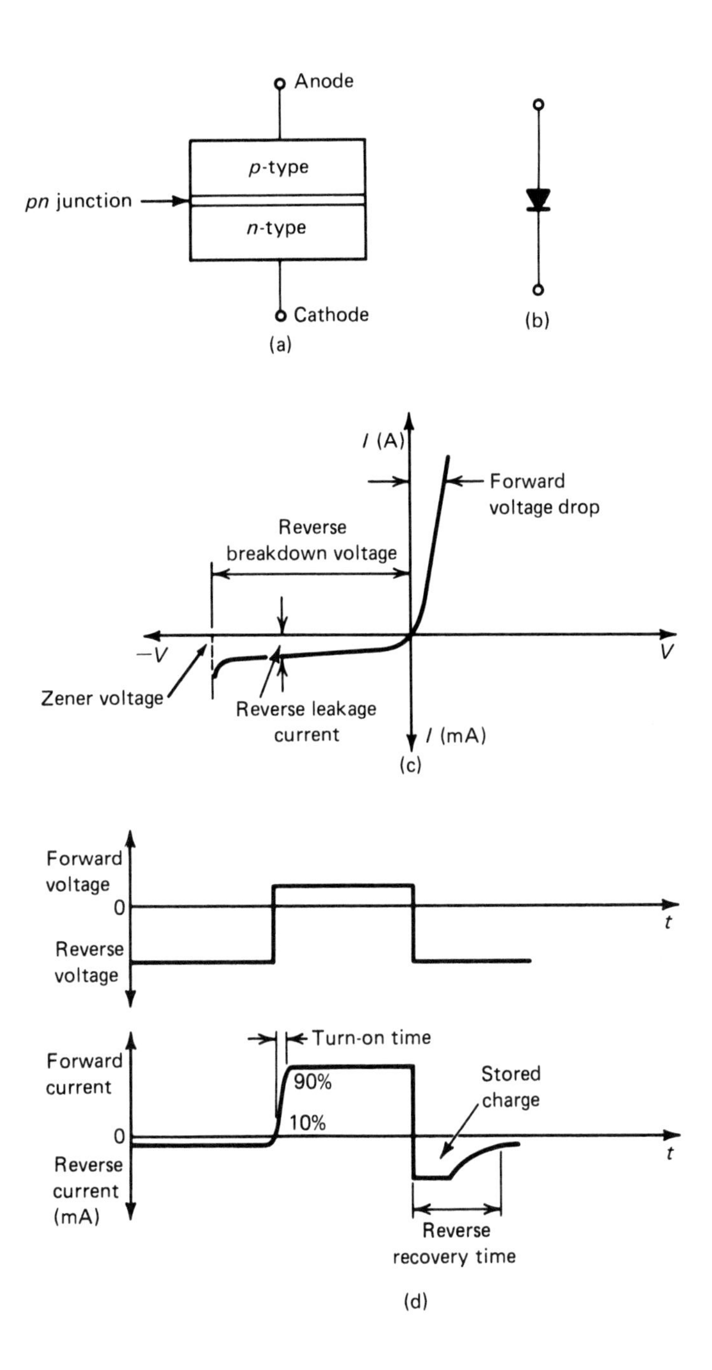

Figure 2-3. A semiconductor diode: (a) sectional view, (b) diode symbol, (c) steady-state *V-I* characteristic, (d) transient *V-I* characteristic.

The turn-on time of a diode is relatively small, usually in the nanosecond range; however, during turn-off, an appreciable time is required for the applied reverse voltage to sweep the stored charges from the junction region before the reverse blocking state is achieved. (See Figure 2-3(d).) The amount of the stored charge is dependent upon the diode design and the amplitude of the forward current. The time interval is called the **reverse recovery time** t_{rr} and ranges from 200 ns to 5 μs for fast-recovery diodes to 2 to 25 μs for general purpose rectifier diodes.

Junction temperatures can be controlled within limits by the use of heat sinks, heat pipes, forced air, or water cooling as is appropriate for the application (see Figure 2-4).

Figure 2-4. Rectifier assemblies with heat sinks. (Courtesy Westinghouse Electric Corp., Semiconductor Division)

2-4 THE DIODE DATA SHEET

The main aid available to determine the correct semiconductor diode to meet a given requirement is the **technical data sheet.** Therefore it is obvious that one must understand the significance of the major parameters used in the rating of semiconductor devices. The following parameters apply to all semiconductor devices.

2-4-1 Maximum Allowable Junction Temperature

$T_{J(max)}$ defines the maximum junction temperature that the device can withstand without failing due to thermal runaway. This is a very critical parameter upon which all device ratings are based. Typical maximum operating junction temperatures for diodes are in the range of 175°C to 200°C; for most transistors the range is 150°C to 175°C; and for most silicon-controlled rectifiers (SCR) the maximum allowable junction temperature is 125°C. In general, operation at lower temperatures will result in an improved performance. However, there is also a mimimum junction temperature, and if the device is operated below this value, the crystal structure of the silicon wafer may be fractured. Normally this is not a problem since the mimimum allowable junction temperatures are in the range of −60°C to −40°C.

2-4-2 Maximum Allowable Case Temperature

Since it is impossible to measure the junction temperature, the case temperature $T_{C(max)}$ provides the only practical means of determining the thermal performance of the device. Graphs are included in the data sheets defining the current-carrying capability as a function of the case temperature.

2-4-3 Thermal Impedance, Junction To Case

$R_{\theta JC}$ is the effective thermal resistance between the junction and the outer case of the device. It is the measure of the ability of the materials and mechanical construction of the device to transfer heat from the junction to the case, and it is usually specified in terms of °C/W. Obviously the lower the value of $R_{\theta JC}$ the more effectively is heat removed from the junction.

The remainder of the parameters are specifically applicable to the diode.

2-4-4 Maximum Full-Cycle Average Forward Current

$I_{F(av)}$ is the maximum allowable value of the average full-cycle forward current at a specified case temperature; that is, it is the maximum current that can be rectified by the diode. Data sheets usually include thermal derating curves showing the relationship between the average forward current and the case temperature. These curves usually provide the relationships for single-, three-, and six-phase configurations (see Figure 2-5).

General Purpose RECTIFIER R610/R611 And R600/R601

200 — 300 A Avg. Up to 3000 Volts

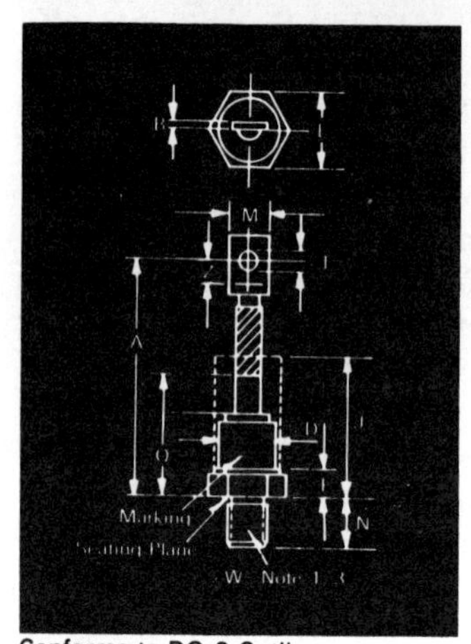

Conforms to DO-9 Outline

Symbol	Inches Min.	Inches Max.	Millimeters Min.	Millimeters Max.
A	5.32	6.00	135.13	152.40
B	.063	.172	1.60	4.37
ϕD	.980	1.065	24.89	27.05
E	1.212	1.250	30.78	31.75
F	.250	.630	6.35	16.00
J	3.250		82.55	
M	.530	.755	13.46	19.18
N	.660	.749	16.76	19.02
Q		2.250		57.15
ϕT	.330	.350	8.38	8.89
Z	.440		11.18	
ϕW	¾-16 UNF-2A			

Creep & Strike Distance:
R600,601—.49 in. min. (12.52 mm).
R610,611—.13 in. min. (3.43 mm).
(In accordance with NEMA standards.)

Finish—Nickel Plate.

Approx. Weight—8 oz. (226g)
R600—Standard Polarity—White Ceramic
R601—Reverse Polarity—Pink Ceramic
R610—Standard Polarity—Gray Glass
R611—Reverse Polarity—Yellow Glass

1. Complete threads to extend to within 2½ threads of seating plane.
2. Angular orientation of terminal is undefined.
3. Pitch diameter of ¾-16 UNF-2A (coated) threads (ASA B1.1-1960).
4. Dimension "J" denotes seated height with lead bent at right angle.

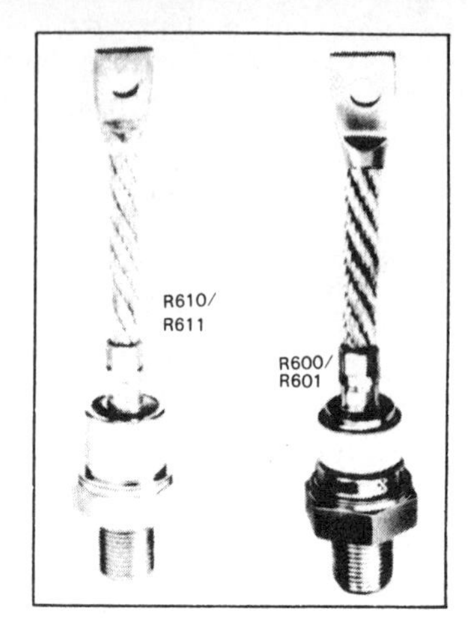

Applications

- Welders
- Battery Chargers
- Electrochemical Refining
- Metal Reduction
- General Industrial High Current Rectification

Features

- Standard and Reverse Polarities
- Flag Lead and Stud Top Terminals Available
- High Surge Current Ratings
- High Rated Blocking Voltages
- Special Electrical Selection for Parallel and Series Operation
- Glazed Ceramic Seal Gives High Voltage Creepage and Strike Paths
- Compression Bonded Encapsulation
- JAN Types Available
- Lifetime Guarantee

Ordering Information

Type	Voltage		Current		Recovery Time		Recovery Time Circuit		Leads	
Code	V_{RRM} (V)	Code	$I_{F(av)}$ (A)	Code	t_{rr} μsec	Code	Circuit	Code	Case	Code
R610 (Standard Polarity)	100	01	200	20	13	X	JEDEC	X	DO-9	YA
	200	02								
	400	04	250	25	11	X				
R611 (Reverse Polarity)	600	06								
	800	08	300	30	9 (typical)	X				
	1000	10								
R600 (Standard Polarity)	1200	12								
	1400	14								
	1600	16								
R601 (Reverse Polarity)	1800	18								
	2000	20								
	2200	22								
	2400	24								
	2600	26								
	2800	28								
	3000	30								

Example

Obtain optimum device performance for your application by selecting proper Order Code.

Type R610 rated at 250A average with $V_{RRM} = 300V$, and standard flexible lead — order as:

Type				Voltage		Current		Time	Circuit	Leads	
R	6	1	0	0	3	2	5	X	X	Y	A

Example

Obtain optimum device performance for your application by selecting proper Order Code.

Type R600 rated at 300A average with $V_{RRM} = 1200V$, and standard flexible lead — order as:

Type				Voltage		Current		trr	Circuit	Leads	
R	6	0	0	1	2	3	0	X	X	Y	A

Figure 2-5. General purpose diode data sheet. (Courtesy Westinghouse Electric Corp., Semiconductor Division)

200 — 300 A Avg. Up to 3000 Volts

General Purpose RECTIFIER R610/R611 And R600/R601

Voltage

<table>
<tr><th>Blocking State Maximums ①</th><th>Symbol</th><th colspan="16"></th></tr>
<tr><td>Repetitive peak reverse voltage , V</td><td>V_{RRM}</td><td>100</td><td>200</td><td>400</td><td>600</td><td>800</td><td>1000</td><td>1200</td><td>1400</td><td>1600</td><td>1800</td><td>2000</td><td>2200</td><td>2400</td><td>2600</td><td>2800</td><td>3000</td></tr>
<tr><td>Non-repetitive transient peak reverse voltage, V ≤ 5.0 m sec</td><td>V_{RSM}</td><td>200</td><td>300</td><td>500</td><td>700</td><td>1000</td><td>1200</td><td>1400</td><td>1600</td><td>1800</td><td>2000</td><td>2200</td><td>2400</td><td>2600</td><td>2800</td><td>3000</td><td>3200</td></tr>
<tr><td></td><td></td><td colspan="7">R600__20</td><td colspan="4">R600__20</td><td colspan="5">R600__20</td></tr>
<tr><td></td><td></td><td colspan="7">R600__25</td><td colspan="4">R600__25</td><td colspan="5"></td></tr>
<tr><td></td><td></td><td colspan="7">R600__30</td><td colspan="9"></td></tr>
<tr><td></td><td></td><td colspan="6">R610 (All types)</td><td colspan="10"></td></tr>
<tr><td>Min., Max. oper. junction temp., °C</td><td>T_J</td><td colspan="6">−65 to 190</td><td colspan="5">−65 to 175</td><td colspan="5">−65 to 150</td></tr>
<tr><td>Min., Max. storage temp., °C</td><td>T_{stg}</td><td colspan="6">−65 to 190</td><td colspan="5">−65 to 190</td><td colspan="5">−65 to 190</td></tr>
<tr><td>Typical Reverse Recovery Time I_{FM} = 785A, tp = 100μs diR/dt = 25A/μs, T_C = 25°C, μs</td><td>trr</td><td colspan="6">9</td><td colspan="5">11</td><td colspan="5">13</td></tr>
<tr><td>Reverse leakage current, mA peak</td><td>I_{RRM}</td><td colspan="16">50</td></tr>
</table>

Current

Conducting State Maximums	Symbol	R600__20 R610__20 R601__20 R611__20	R600__25 R610__25 R601__25 R611__25	R600__30 R610__30 R601__30 R611__30
RMS forward current, A	$I_{F(rms)}$	315	400	470
Ave. forward current, A	$I_{F(av)}$	200	250	300
One-half cycle surge current②, A	I_{FSM}	5500	6000	6500
3 cycle surge current②, A	I_{FSM}	4300	4700	5050
10 cycle surge current②, A	I_{FSM}	3300	3600	3900
I^2t for fusing (for times 8.3 ms) A^2 sec.	I^2t	125,000	150,000	175,000
Forward voltage drop at I_{FM} = 800 A and T_J = 25°C, V	V_{FM}	1.7	1.5	1.4
Forward voltage drop at rated single phase average current and case temperature, V	V_{FM}	1.45	1.45	1.45

Thermal and Mechanical

	Symbol	
Max. mounting torque, in lb.③		360
Thermal resistance ③ Case to sink, lubricated, °C/Watt	$R_{\theta CS}$	.10

① At maximum TJ
② Per JEDEC RS-282, 4.01 F.3.
③ Consult recommended mounting procedures.

RECTIFIER

Figure 2-5 continued.

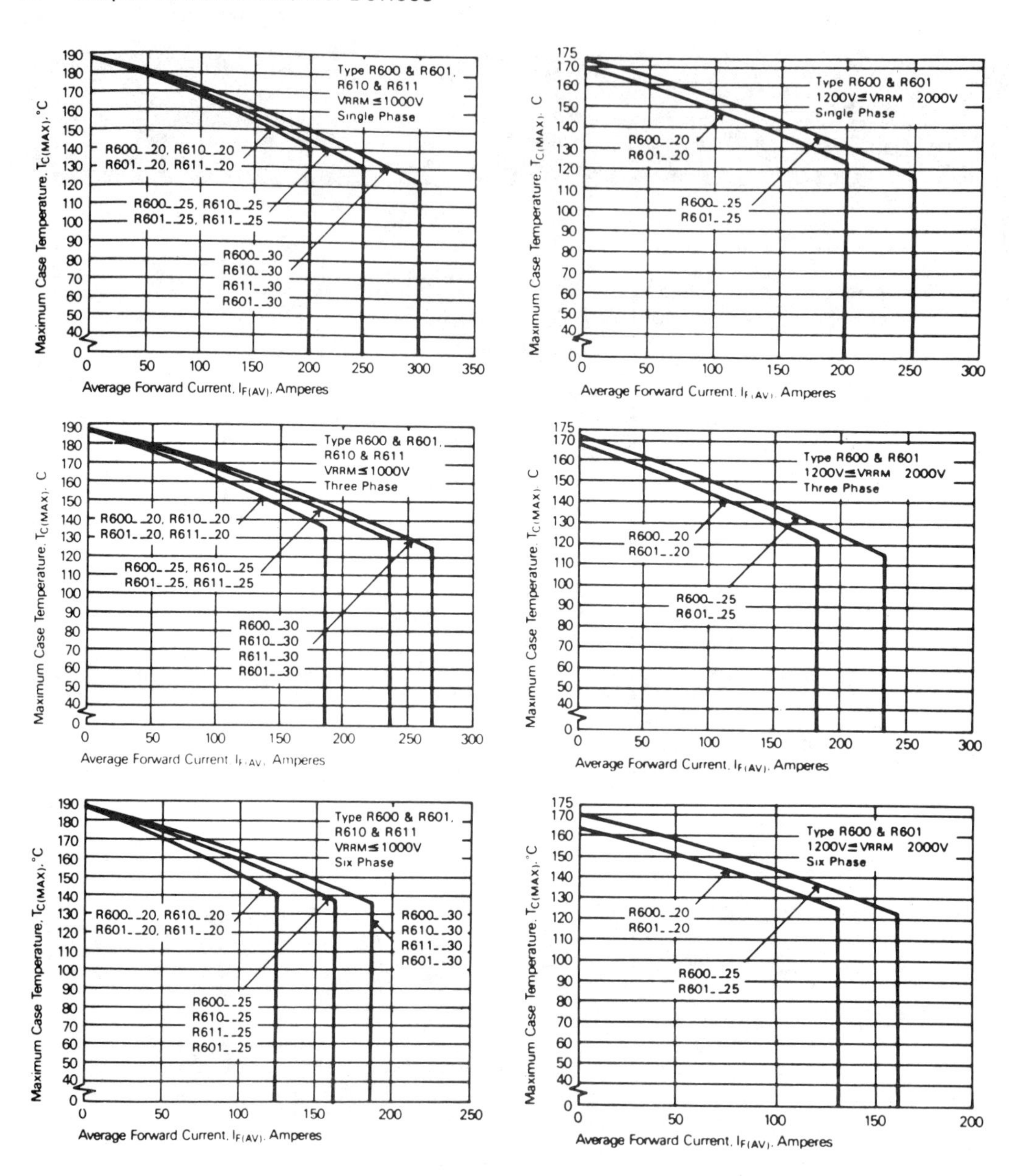

Figure 2-5 continued.

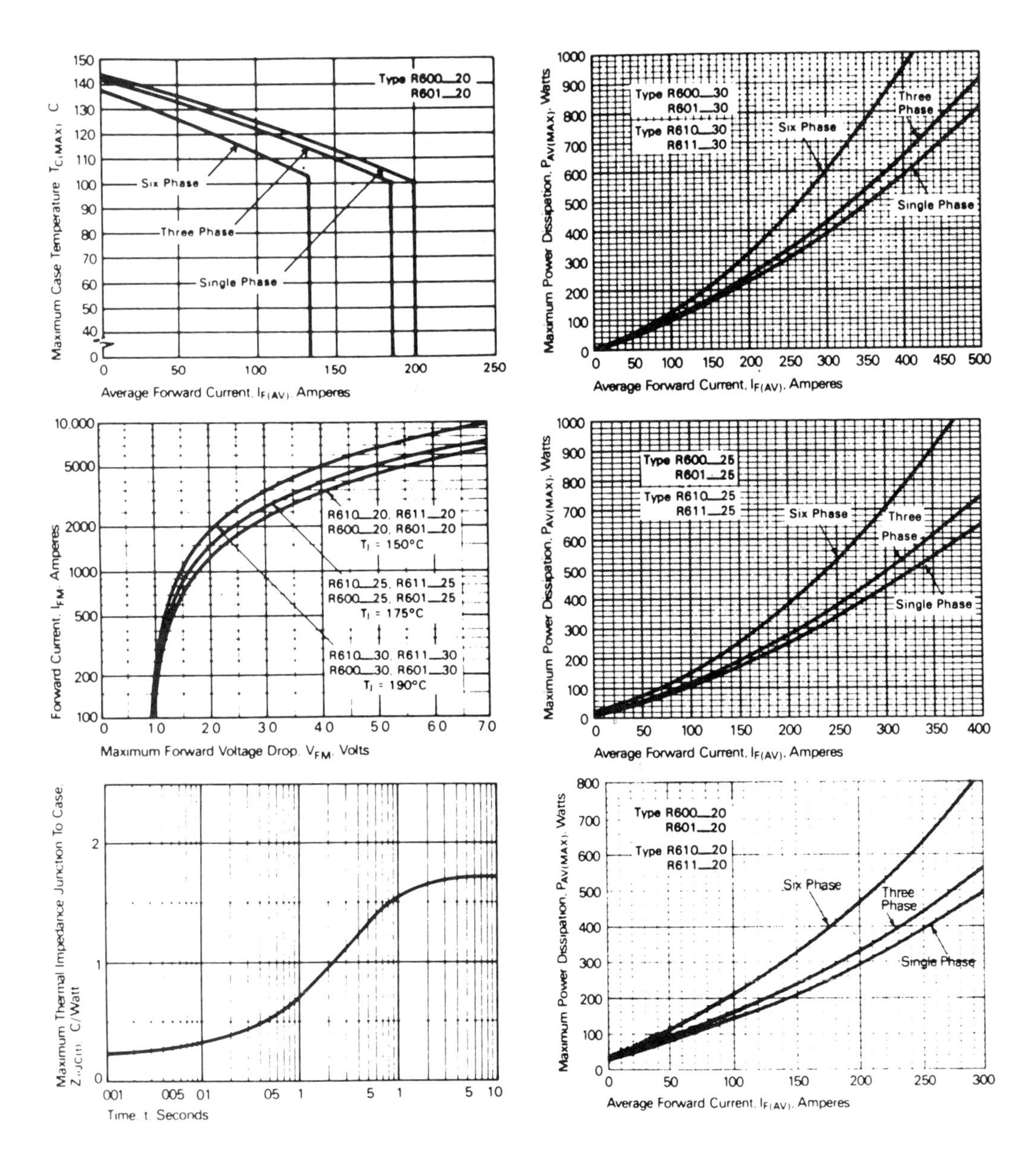

Figure 2-5 continued.

2-4-5 rms Forward Current

All silicon diodes under operating conditions have a small resistance, but under load conditions the I^2R dissipation can cause a very significant amount of heating. The current responsible for the heating effect is the rms forward current $I_{F(\mathrm{rms})}$, and as a result the rms current rating of the diode is the limiting power dissipation factor. This current is independent of the case temperature. The ohmic heating caused by the rms current is independent of the junction heating effects caused by the average forward current; under normal cooling arrangements, for example, heat sinks, etc., the rms current limit is not reached. The average forward current permitted is based on the rms current rating. In the case of the Westinghouse R610/R611, 200 A general purpose diode, $I_{F(\mathrm{av})} = 200$ A and $I_{F(\mathrm{rms})} = 315$ A.

2-4-6 Maximum Repetitive Peak Reverse Voltage

The V_{RRM} parameter defines the maximum allowable instantaneous value of the applied reverse voltage that the diode can block. This rating is determined by the *p-n* junction and the case design.

2-4-7 Nonrepetitive Transient Peak Reverse Voltage

The V_{RSM} parameter specifies the ability of the diode to withstand nonrepetitive transient peak reverse voltages less than 5ms in duration. The amplitude of these peaks ranges from zero to 25% above the values specified for V_{RRM}.

2-4-8 Maximum Forward Voltage Drop

V_{FM} is the maximum forward voltage drop at a specified forward current I_{FM} and junction temperature T_J. In the case of the R610/R611 diodes $V_{FM} = 1.7$ V when $I_{FM} = 800$ A and $T_J = 25°$C.

It should be appreciated that the largest proportion of the diode losses are a direct result of the forward voltage drop, that is,

$$P = V_{FM} I_{F(\mathrm{av})},$$

and as a result, especially in high-current diodes, it is very desirable that V_{FM} be as small as possible.

2-4-9 Maximum Reverse Leakage Current

I_{RRM} is the maximum reverse current that occurs when V_{RRM} is applied; for the R610/R611 diode I_{RRM} is 50 mA. This current is another source of junction heating since $P = V_{RRM} I_{RRM}$, which immediately shows the desirability of achieving low values of the reverse leakage current.

2-4-10 Maximum Nonrepetitive Surge Current

The I_{FSM} parameter defines the maximum one-half cycle (60 Hz) peak load surge currents in excess of the rated operating current that can be carried and still leave the diode capable of blocking V_{RRM} after the surge. The diode is designed to withstand up to 100 such surges during its lifetime. Usually the data sheet will provide maximum surge current values for one-half, three-, and ten-cycle surge durations.

2-4-11 I^2t Rating

The I^2t rating is used to define the thermal capacity of fuses, and in the protection of diodes the I^2t rating of the fuse is selected to be less than the I^2t rating of the diode. These ratings are based on the fuse clearing a fault in less than half a cycle. Prior to clearing, the diode will be subjected to excessive temperatures because the device is acting as a resistance. The specified I^2t rating is that value necessary to enable the correct fuse to be selected so that junction overheating will not occur.

2-4-12 Maximum Reverse Recovery Time

t_{rr} is the maximum time required for the reverse current spike to dissipate after the diode ceases conduction when subjected to a reverse-bias voltage. The reverse current spike is due to the reverse-bias voltage forcing the stored charge carriers across the p-n junction. It will continue until the stored minority carriers in the n-type material either recombine with electrons or are forced across the junction. The time required to complete this action and to restore the blocking capability of the diode is the reverse recovery time t_{rr}.

In high-frequency rectifier applications, that is, frequencies greater than 1 kHz, fast-recovery diodes are used to reduce the effects of diffusion capacitance. Diffusion capacitance is directly proportional to the hole lifetime in n-type material and inversely proportional to the forward conducting resistance. The product of the diffusion capacitance and the forward conducting resistance, that is, the time constant, places an upper limit on the permissible switching frequency. Fast-recovery diodes are specially treated by an irradiation process to reduce the diffusion capacitance. As a result, fast-recovery diodes have reverse recovery times ranging from 200 ns for a 6 A diode to 3 μs for a 1400 A diode.

The adverse effects of high reverse recovery times are diode heating and the introduction of undesirable ripple currents in the rectifier output, both of which reduce the rectification efficiency.

2-4-13 Minimum And Maximum Storage Junction Temperatures

The data sheets also specify the minimum and maximum storage junction temperatures T_{stg} so that the diode is protected against damage during storage, since excessively low or high temperatures may damage the diode. Typical storage temperature ranges are −65°C to 190°C.

2-4-14 Maximum Mounting Torque

While not an electrical characteristic, the maximum mounting torque specifies the torque which may be applied to a stud mount diode without damaging the *p-n* junction or the case, and at the same time achieving the minimum thermal impedance between the case and heat sink. It is usually expressed in in. lbs.

2-5 Schottky Barrier Diodes

The **Schottky barrier diode** is a silicon-to-metal semiconductor junction diode. The barrier metal used to be chrome, but now it may be gold, molybendum, titanium, aluminum, or platinum. Motorola is currently using platinum. The barrier metal is processed to form an intimate nonohmic contact usually with *n*-type silicon. *N*-type silicon is preferred since the mobility of the electrons is greater than the mobility of the holes, and as a result permits a greater high-frequency operating range. In the Schottky diode the minority carriers (holes) have been eliminated. This has the effect of eliminating the transition capacitance with a resultant decrease in the reverse recovery time, thus permitting the use of these diodes in applications up to as high as 40 GHz.

The elimination of minority carriers means that the current flow is entirely by means of majority carriers produced by the barrier metal emitting electrons under the influence of a strong electric field, thus effectively simulating the effect of a *p-n* junction diode. A major disadvantage of the Schottky diode was the relatively low reverse breakdown voltage. However, this problem has been reduced by the use of a *p*-type guard ring around the periphery of the metal-to-*n*-type silicon, and repetitive peak reverse working voltages of the order of 30 to 45 V are now common.

The major advantages of the Schottky diode are: (1) current flow is entirely due to the majority carrier (electrons); (2) it is relatively unaffected by reverse recovery transients caused by the transition capacitance and minority carrier injection; (3) the reverse recovery time is very small; (4) the forward voltage drop is small, e.g., 0.69 V at 75 A, which results in a low forward conduction power loss; and (5) relatively low reverse leakage current.

In the power electronic field the major applications of these devices are in low-voltage high-current dc power supplies and in high-frequency power supply applications. Other applications of the Schottky principle are in transistor-transistor logic (TTL), photodiodes, and power thyristors. The current state of the art is represented by International Rectifier's 85HQ series designed to operate at a maximum junction temperature of 175°C, with a reverse leakage current of 45 mA at 125°C, a repetitive peak reverse working voltage of 30 to 45 V, and a maximum peak one-cycle nonrepetitive surge current at 60 Hz of 1150 A.

2-6 THE ZENER DIODE

Under reverse-bias conditions the normal silicon junction diode will break down, and a relatively large reverse current will flow if the reverse-bias voltage is great enough. This increase of the reverse current is caused by the forcible removal of valence electrons from their parent atoms. At this point the breakdown voltage remains substantially constant even though the current increases rapidly. If the reverse current is limited by a series resistor the diode will not be damaged as long as the power dissipation does not exceed the device rating. Diodes that operate under breakdown conditions are called **zener diodes** and are extensively used as voltage references or voltage regulators.

2-6-1 Avalanche And Zener Breakdown

There are two breakdown mechanisms that can occur in a reverse-biased *p-n* junction: the **avalanche mechanism** or the **zener mechanism.** Either of these two mechanisms may occur together or separately.

Under conditions of increasing reverse bias, the leakage current minority carrier electrons gain sufficient kinetic energy from the applied electric field to dislodge electrons from the covalent bonds in the conduction band. If the reverse-bias voltage is great enough there will also be sufficient energy to dislodge electrons from covalent bonds in the valence band. The result is a rapid increase or avalanche of charge carriers occurring at the avalanche voltage. By careful control of the degree of doping and the junction widths, the voltage at which avalanche occurs can be controlled. An unfortunate side effect of this property is the production of electrical noise caused by the rapid random motion of the electrons. This noise effect covers a wide frequency spectrum and is especially pronounced at avalanche breakdown. It can be minimized by connecting a small capacitor (0.1 μF) across the diode. The minimum voltages at which avalanche occurs are in the region of 6 to 8 V.

The zener diode differs from the avalanche diode in that the *p-n* junction region is made very narrow, and as a result the voltage gradient (V/cm) occurring across the depletion region is sufficiently great that electrons will be forcibly pulled from their covalent bonds, thus creating pairs of electron-hole carriers. This process is termed the **zener breakdown.** Effectively, ionization caused by the electric field has turned the previously high-resistance depletion region into a conductor, thus permitting a relatively large reverse current to flow.

2-6-2 Zener Diode V-I Characteristics

The *V-I* characteristics of a zener diode are shown in Figure 2-6. As can be seen the general characteristics of the *V-I* curves are very similar to the normal *p-n* junction diode. The only difference is that the normal *p-n* junction diode is not designed to operate in the breakdown region. Also the reverse leakage current I_R remains relatively small until the reverse-bias voltage has been increased sufficiently to cause breakdown to occur. At this point there is a rapid increase in the reverse current as the reverse-bias voltage is increased beyond breakdown. After breakdown has occurred the diode is operating in the zener or zener breakdown region. The current flowing through the device when operating in this region is termed the **zener current** I_Z and varies between a minimum I_{ZK} and a maximum I_{ZM}.

The more important points of the reverse characteristic are defined below:

V_Z = zener breakdown voltage

I_{ZT} = zener test current at which V_Z is measured

I_{ZK} = the minimum zener current

I_{ZM} = the maximum zener current

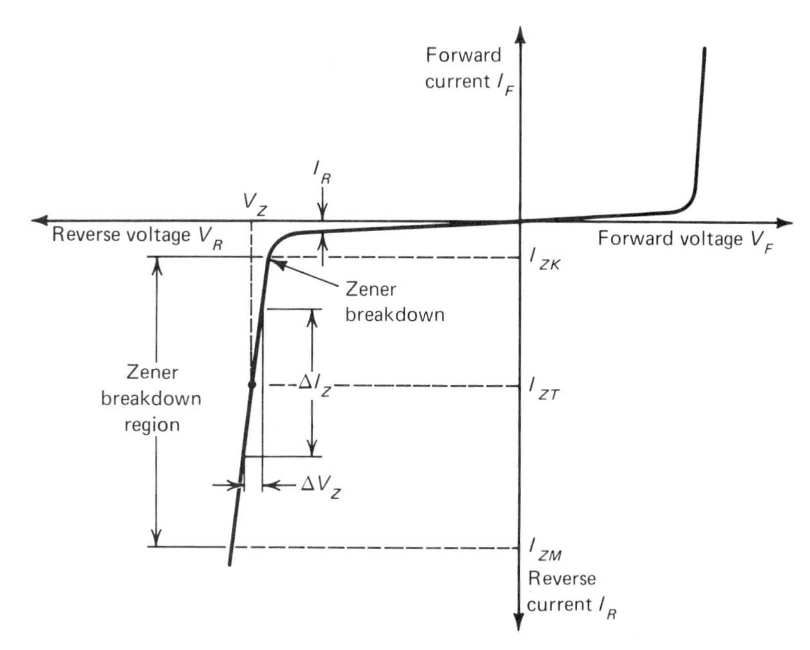

Figure 2-6. Zener diode *V-I* characteristic.

The magnitude of the breakdown voltage V_Z of a zener diode is controlled by the doping techniques during manufacture. Zener diodes are commercially available in a wide range of voltage ratings such as 2·4, 2·7, 3·0, 3·3, 3·6, 3·9, etc., and up to several hundred volts. Because of discrepancies that occur in the manufacturing process, the actual zener breakdown voltage does not always occur at the specified value. The manufacturer will specify the tolerances, usually ± 20%, ± 10%, ± 5%, and on special orders ± 1%. This means that a 3·6V ± 10% zener diode will have a zener breakdown voltage V_Z between 3·24 and 3·96 V. Another important parameter is the maximum power that can be dissipated by the device. These power ratings which are specified in the data sheets vary from as little as 400 mW to as much as 50 W. The power dissipation is specified in terms of a specific ambient temperature, for example 50°C, and also includes a derating factor for operation at temperatures above the specified ambient. The derating factor is usually expressed in terms of mW/°C for each °C above the specified ambient temperature.

Example 2-1

A 1N3997 zener diode has a nominal zener voltage of 5·6 V ± 5% at I_{ZT} = 445 mA and can dissipate 10 W at an ambient temperature of

55°C, derated at 83·3 mW/°C above 55°C. What is the recommended power dissipation when the zener is operated at an ambient temperature of 70°C?

Solution:

$$\Delta T = (70 - 55)°\mathrm{C} = 15°\mathrm{C}.$$

The zener diode must be derated by 15 × 83.3 mW = 1.25 W.

Therefore, the maximum power that should be dissipated is 10 − 1.25 = 8.75 W.

In addition, for zener diodes with axial leads, the power rating is often specified for specific lead lengths. The power dissipation ability of the diode increases as the leads are shortened. See Figure 2-7(a).

2-6-3 Zener Diode Impedance Z_{ZT}

Another important zener diode characteristic is the diode's impedance Z_{ZT}. This is the dynamic impedance and is derived from the V-I characteristic by varying the zener current I_Z above and below the zener test current I_{ZT} and comparing the change of zener current ΔI_Z to the corresponding change of zener voltage ΔV_Z. Then

$$Z_{ZT} = \frac{\Delta V_Z}{\Delta I_Z} \tag{2-1}$$

Usually the manufacturer will also define the zener impedance at the knee of the curve Z_{ZK}. This impedance defines the slope at the knee of the curve.

2-6-4 Temperature Coefficients and Temperature Compensation

When the temperature changes, V_Z also changes. The change of voltage that occurs is usually expressed as a percentage of V_Z for each degree celsius change in temperature and is known as the **zener voltage temperature coefficient** (α_Z). Zener diodes with $V_Z < 5$ V usually have a negative zener voltage temperature coefficient, or expressed in another way, for a constant zener current I_Z the zener voltage V_Z decreases slightly with increasing temperatures. This effect is caused by the increasing temperature raising the valence electrons in the depletion region to a higher energy level, thus increasing the number of charge carriers. For zener breakdown voltages, $V_Z > 7$ V, the avalanche breakdown mechanism predominates and for a constant zener current

I_Z the zener voltage V_Z increases; that is, the device has a positive zener voltage temperature coefficient. Zener diodes with zener breakdown voltages in the range of 5 to 7 V show very little or no change in V_Z for a constant zener current; that is, they have a practically zero zener voltage temperature coefficient.

In applications where a very precise reference voltage is required, a forward-biased silicon diode which possesses a negative temperature coefficient is connected in series with a zener diode that has a positive temperature coefficient in such a way that the silicon diode is forward biased when the zener is reverse biased, and vice versa. This combination is packaged and is available as a temperature compensated zener diode (see Figure 2–7).

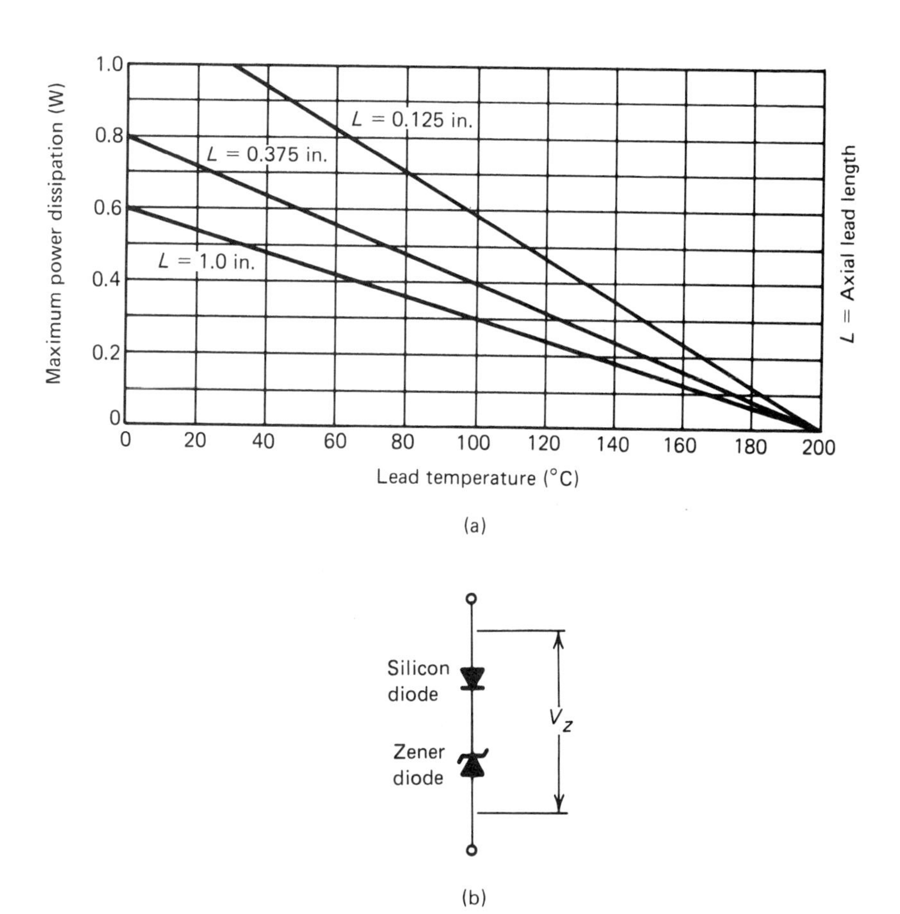

Figure 2–7. Zener diode: (a) power derating curves, (b) temperature compensated zener diode.

2-6-5 Voltage-Regulating Circuits

The most common application of the zener diode is found in voltage stabilization or regulation circuits.

The basic regulator circuit is shown in Figure 2–8, where V_S is an unregulated dc voltage source which may fluctuate above and below the specified value. As a result, the current flowing through the zener and the current limit resistor R_S will fluctuate, but the voltage across the zener diode will remain almost constant. Since the output voltage V_O of the regulator is equal to V_Z, the output voltage is regulated. The circuit shown in Figure 2–8 is very commonly used to provide regulated voltages. However, since the current demands of the load also may vary significantly because of changes in load impedance, it is necessary to design the voltage regulator circuit to accommodate a range of currents as well as a regulated output voltage.

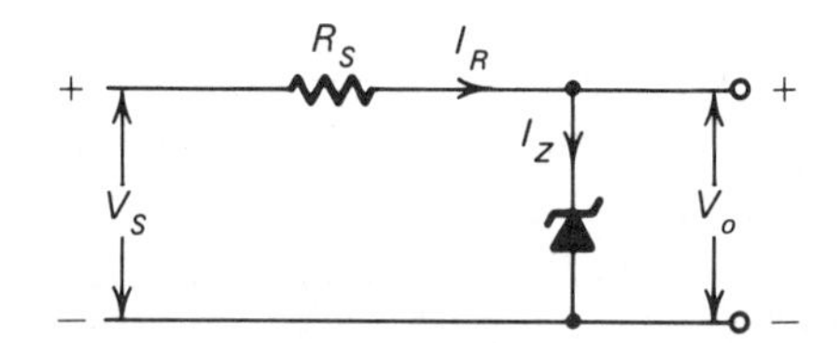

Figure 2–8. The basic zener voltage regulator circuit.

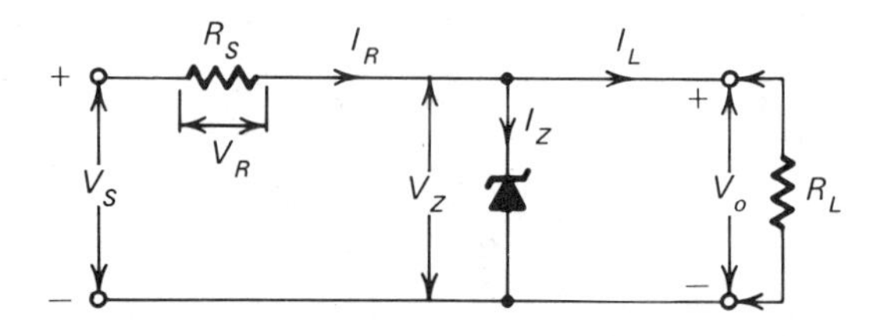

Figure 2–9. A simple zener diode voltage regulator.

Figure 2–9 illustrates the simplest zener diode voltage regulator circuit consisting of a zener diode in series with a current-limiting resistor R_S. The load R_L is connected across the zener diode so that

$$V_O = V_Z,$$
$$V_S = V_R + V_Z,$$

and

$$V_R = V_S - V_Z.$$

Therefore,

$$I_R = \frac{V_S - V_Z}{R_S} = I_Z + I_L,$$

and by Kirchhoff's Current Law,

$$I_Z = I_R - I_L.$$

When R_L is infinite, that is, open-circuited, $I_L = 0$, and

$$I_Z = I_R = I_{Z(\max)}.$$

When R_L is small, I_L will increase in value, and since

$$I_Z = I_R - I_L,$$

I_Z will become very small if we do not make sure that I_Z does not fall below a certain value $I_{Z(\min)}$, where

$$I_{Z(\min)} = I_R - I_{L(\max)}.$$

In actual practice $I_{Z(\max)} = I_{ZM}$. However, recalling that Z_{ZK} is very large at I_{ZK}, it is good practice to ensure that

$$I_{Z(\min)} > I_{ZK}.$$

Example 2-2

Design a zener voltage regulator to produce a regulated output voltage of 5V from a 10V unregulated source. Calculate the minimum value of load resistance that may be connected to the output terminals. Assume that the ambient temperature is 25 °C.

Solution:

From the manufacturer's data sheet, a 1N751 zener diode has a nominal zener voltage of 5.1 V.

Therefore, from Figure 2-9,

$$V_R = V_S - V_O = 10 - 5.1 = 4.9\text{ V}.$$

From the data sheet,

$$I_{ZM} = 70\text{ mA} = I_R.$$

Therefore, $$R_S = \frac{V_R}{I_R} = \frac{4.9\text{ V}}{70\text{ mA}} = 70\,\Omega.$$

The power to be dissipated by R_S is $V_R I_R = 4.9\text{ V} \times 70\text{ mA} = 0.343$ W. Therefore, a ½ W resistor should be selected.

Since I_{ZK} is not specified in the data sheet, we will assume $I_{ZK} = 2$ mA. Therefore, $I_{Z(\min)}$ must be greater than I_{ZK}; so assuming $I_{Z(\min)} = 10$ mA,

$$\begin{aligned} I_{L(\max)} &= I_R - I_{Z(\min)} \\ &= 70\text{mA} - 10\text{mA} = 60\text{ mA}. \end{aligned}$$

Therefore, the minimum acceptable value of R_L is

$$\begin{aligned} R_{L(\min)} &= \frac{V_Z}{I_{L(\max)}} = \frac{5.1\text{ V}}{60\text{ mA}} \\ &= 85\Omega \end{aligned}$$

2-7 THE BIPOLAR OR BIJUNCTION TRANSISTOR

So far we have only discussed *p-n* junction devices. One of the most common semiconductor devices, the transistor, is formed by adding a second *n* or *p* region to a junction diode. The result is a transistor having two *n* regions and one *p* region, known as an *n-p-n* transistor. See Figure 2–10(a). If the transistor has two *p* regions with an *n* region sandwiched between them it is known as a *p-n-p* transistor [see Figure 2–10(b)].

The three regions of a **bipolar junction transistor** (BJT) are called the collector, base, and emitter. The term bipolar refers to the fact that there are two types of charge carriers involved in the current flow process in the transistor. The first charge carrier mechanism, known as **drift**, is the movement of the majority carrier under the influence of the applied electric field. The second mechanism, called **diffusion**, is the result of an accumulation of like charge carriers which diffuse from a region of high concentration to a region of low concentration because of the mutual repulsion between the like charges. These two charge carrier mechanisms are completely independent of each other.

A diagrammatic representation of an *n-p-n* transistor in the common-emitter connection is shown in Figure 2–11(a). Normally the emitter region is more heavily doped than the base region, ensuring that the resistivity of the emitter region is lower than the base region; the concentration of charge carriers is very much greater than in the base region, ensuring that there will be a high injection efficiency. The base region width is made very small compared to the diffusion path length of the minority carriers, thus promoting a high transport efficiency.

However, the concentration of charge carriers in the base region must be greater than the concentration in the collector region to en-

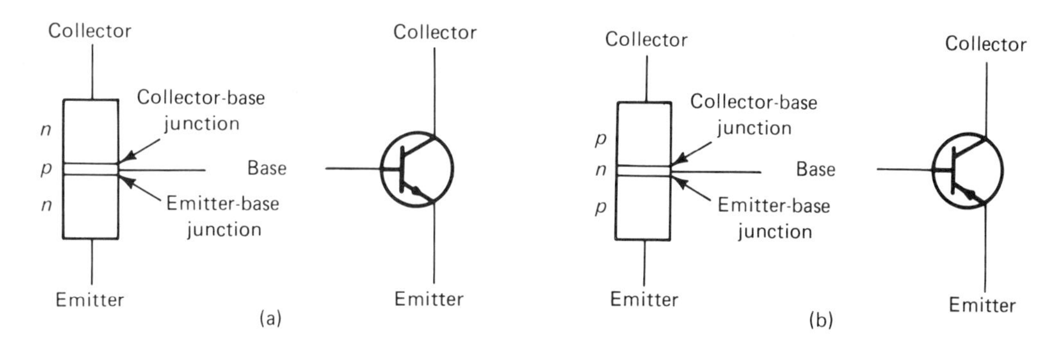

Figure 2-10. Bipolar junction transistors: (a) *n-p-n* transistor junction diagram and symbol, (b) *p-n-p* transistor junction diagram and symbol.

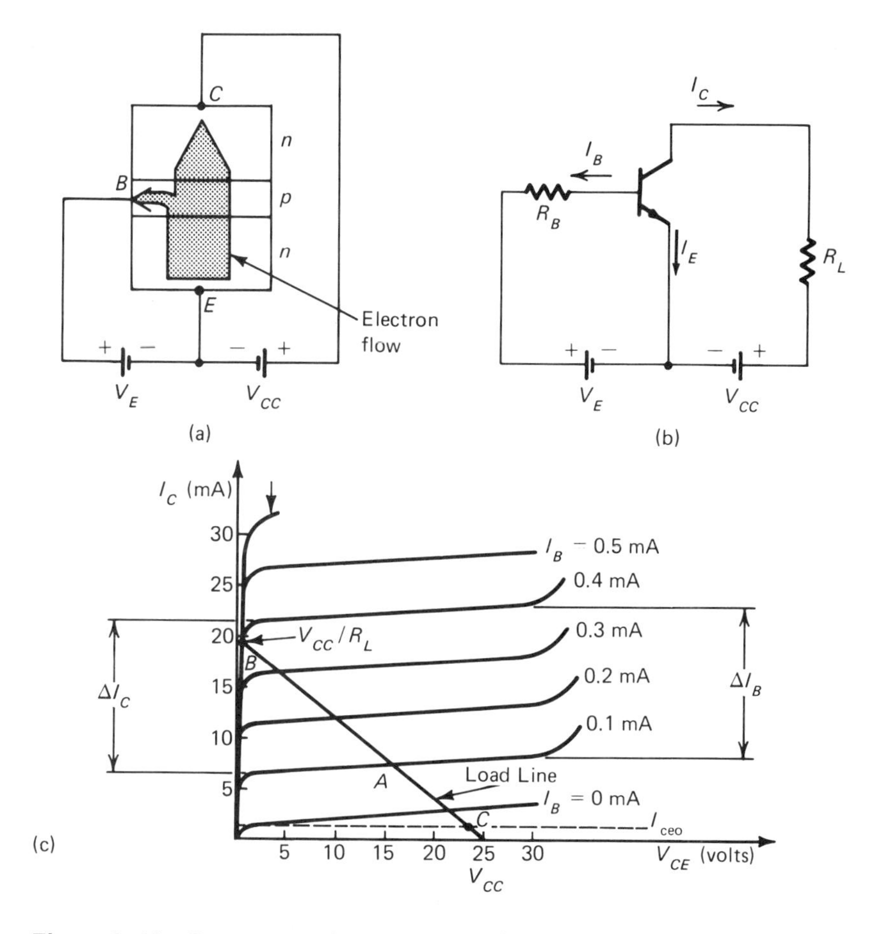

Figure 2-11. Common-emitter *n-p-n* configuration: (a) block diagram, (b) circuit schematic, (c) collector characteristic curves.

sure that the space charge area extends into the collector region. In other words, the resistivity of the base region must be less than the collector region. In turn, the concentration of charge carriers in the collector region must be low to ensure a high resistivity and a high avalanche voltage breakdown. As can be seen from Figure 2–11, the emitter-base junction is forward biased and the collector-base junction is reverse biased, with the collector supply voltage being much greater than the base supply voltage. Under forward-bias conditions the n-region emitter injects electrons into the base region.

At this point the electrons become minority carriers and diffuse across the reverse-biased collector-base junction where they are attracted by the electric field produced by the positive collector terminal voltage. At the same time a limited number of holes flow from the lightly doped base region into the emitter region. In passing through the base region a small number of electrons also recombine with holes. The result of these two recombinations is to form the base current I_B. The current flowing out at the collector is called the **collector current** I_C and the current flowing into the emitter terminal is known as the **emitter current** I_E. The equation relating these currents is

$$I_E = I_B + I_C \tag{2–2}$$

The base current is effectively the input current and the collector current is the output current. The ratio of collector current to base current is known as the current gain factor β, where

$$\beta = \frac{I_C}{I_B} = h_{FE} \tag{2–3}$$

This relationship shows that the collector current I_C is controlled by I_B not by the collector-emitter voltage V_{CE}.

A family of collector current versus collector-emitter voltage curves are shown in Figure 2–11(c). These curves are termed the collector characteristic curves and can be obtained for a particular transistor by using an oscilloscope with a curve tracer adaptor.

From these curves

$$\beta = \frac{\Delta I_C}{\Delta I_B} = \frac{21.5\text{ mA} - 6.5\text{ mA}}{0.4\text{ mA} - 0.1\text{ mA}}$$

$$= \frac{15\text{ mA}}{0.3\text{ mA}} = 50$$

This means that the collector current I_C is 50 times the base current

I_B and is relatively independent of the collector-emitter voltage V_{CE}. In actual practice, β is not constant but varies slightly with both I_C and V_{CE}.

Assuming that $R_L = 800\ \Omega$, we can then draw the load line assuming that V_{CC} is 25 V. The load line has a slope of $1/R_L$. The transistor may be operated as an amplifier, in which case a suitably designed bias network will establish the dc current and voltage so that the transistor operates in the region of A. When operating the transistor as a switch, the input or base signal is increased so that the transistor is on or saturated at point B. The transistor is off or cut off when the base current is removed, that is, at point C. Since the transistor is not an ideal switch, neither the saturation voltage $V_{CE(sat)}$ nor the cutoff current I_{CEO} is zero, the actual values being obtained from the transistor data sheets. Typical ranges of $V_{CE(sat)}$ are from 0.1 V for collector currents in the low milliampere range to as much as 1.5 V when $I_C = 20$ A.

2-7-1 Transistor Data Sheets

Figure 2–12 shows the data sheet for a Westinghouse 2N3429-33 *n-p-n* power transistor, 7.5 A, 50–250 V.

It can be seen that the dc current gain h_{FE} varies from a minimum of 10 to a maximum of 35 at a temperature of 25°C. However, it can be seen from the graph of h_{FE} versus I_C that there are significant β variations when operating at case temperatures varying from $T_C = -65$°C to $T_C = 150$°C with $V_{CE} = 4$V. As a result, there are significant changes in β for collector currents approaching the maximum of 7.5 A.

When the transistor is used as a switch and is unaffected by I_B, that is, $I_C = V_{CC}/R_L$, the transistor is fully on or saturated and $V_{CE(sat)}$ is 1.0 V. This figure is very important in switching applications since most of the power dissipated by the transistor occurs at this point.

In the off state, the voltage of interest is the avalanche voltage that can be applied to the collector base with the emitter connection open, V_{CBO}. If it is not directly specified, the value V_{CEV} may be used. In the on state the commonly used voltage limit is $V_{CEO(SUS)}$, the maximum collector-emitter sustaining voltage with the base open. This is the maximum permissible voltage that may be applied since above this value the collector current increases very rapidly. It should be noted that $V_{CEO(SUS)}$ is usually about 90% of V_{CBO}.

When the transistor is used in switching applications, assuming that the base is grounded or reverse biased when turned off, it should be able to withstand a collector supply voltage V_{CEV}, which for the 2N3429-33 series ranges from 50 to 250 V. The nature of the load also has a significant bearing upon the voltage ratings. For example, if the

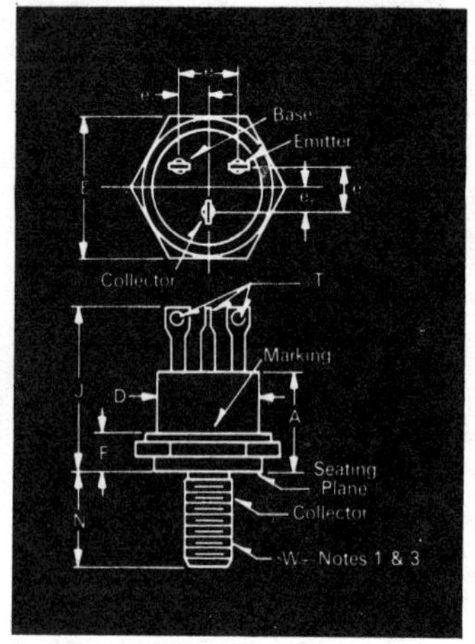

Conforms to MT-52 Outline

Symbol	Inches Min.	Inches Max.	Millimeters Min.	Millimeters Max.
A	.340	.406	8.64	10.31
ϕD	.400	.440	10.16	11.18
e	.160	.190	4.06	4.83
e_1	.080	.095	2.03	2.41
E	.544	.562	13.82	14.27
F	.100	.140	2.54	3.56
J		.710		18.03
N	.422	.453	10.72	11.51
ϕT	.040	.055	1.02	1.40
ϕW	¼-28 UNF-2A			

Finish—Nickel Plate.
Approx. Weight—.25 oz. (7 g).

1. Complete threads to extend to within 2½ threads of seating plane.
2. Contour and angular orientation of terminals is undefined.
3. Pitch dia. of ¼-28 UNF-2A (coated) threads (ASA B1.1-1960).

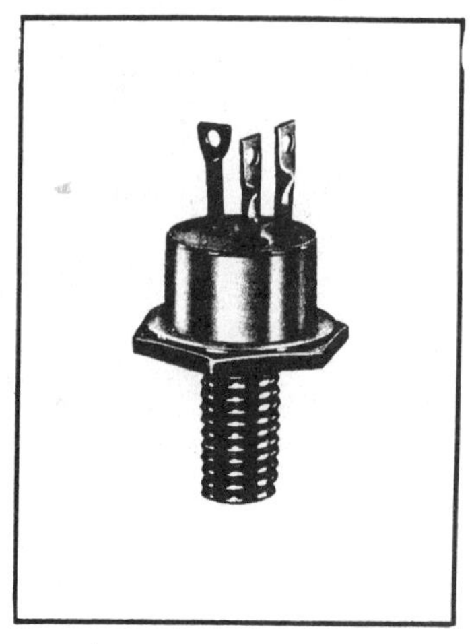

Features:

- Gold Alloy Process
- No forward bias secondary breakdown to 150 volts
- High reverse bias S.O.A. for inductive loads
- Low thermal resistance with copper base
- 150 watt dissipation
- Protection from thermal fatique with hard solder and molybdenum construction
- 25 volt B_{EBO}
- Low $V_{CE(sat)}$
- Lifetime Guarantee

Applications:

- High Power Switching
- Amplifiers
- Servo Systems
- Regulators
- Modulators

Test	Symbol	Test Conditions	2N3429	2N3430	2N3431	2N3432	2N3433
*Collector-Emitter Sustaining Voltage	$V_{CEO(sus)}$	Base Open L=1H I_C=200mA	50	100	150	200	250
*Collector Cutoff Voltage	V_{CEV}	I_{CEV}=2mA V_{EB}=1.5V					
*Collector Cutoff Voltage	V_{CEV}	I_{CEV}=10mA V_{EB}=1.5V T_C=175°C					

Maximum Ratings and Characteristics*
T_c=25°C unless specified

	Symbol	All Types
*Operating and storage temperature		—65°C to 175°C
*Collector-emitter sustaining voltage	$V_{CEO(sus)}$	50 volts to 250 volts
*Emitter-base voltage	V_{EBO}	25 volts
*Collector-emitter voltage, V_{EB}=1.5V, T_C=175°C	V_{CEV}	50 volts to 250 volts
*Continuous collector current	I_C	7.5 amperes
*Continuous base current	I_B	3 amperes
*Linear power derating factor from T_C=60°C		1.33 W/°C
*Thermal resistance	$R_{\theta JC}$	0.75°C/W
*Power dissipation, T_C=60°C	P_T	150 watts

*JEDEC registered parameters

Figure 2-12. Data sheets for Westinghouse *n-p-n* 2N3429-33 power transistors. (Courtesy Westinghouse Electric Corp., Semiconductor Division)

Test	Symbol	Test Conditions	All Types Min.	All Types Max.	Units
*D.C. Current Gain	h_{FE}	V_{CE}=2V, I_C=5A	10	35	
*Collector-Emitter Saturation Voltage	$V_{CE(sat)}$	I_C=5A, I_B=750mA		1.0	V
*Base-Emitter Saturation Voltage	$V_{BE(sat)}$	I_C=5A, I_B=750mA		2.0	V
*Emitter Cutoff Current	I_{EBO}	V_{EB}=25V, T_C=175°C		10	mA
*Turn-On Time	t_{on}	I_C=5A,		5	μsec
*Storage Time	t_s	V_{CC}=12V,		4	μsec
*Fall Time	t_f	I_B(on)=I_B(off)=1.5A		8	μsec
Output Capacitance	C_{ob}	V_{CB}=10V, f=1MHz		750	pF
Gain-Bandwidth	f_T	V_{CE}=10V, I_C=0.5A	250		KHz
*Beta Cutoff Frequency	f_{hfe}	V_{CE}=12V, I_C=5A	20		KHz
Second Breakdown forward Biased, collector current	I_{SB}	V_{CE}=30V, I_C=5A t=1 second, T_C = 25°C		1.5	A
Second Breakdown reversed biased	E_{SB}	L=1mH, V_{BB}=−2V, R_B= 20Ω, I_C=5.6A	15.5		mJ

*JEDEC registered data.

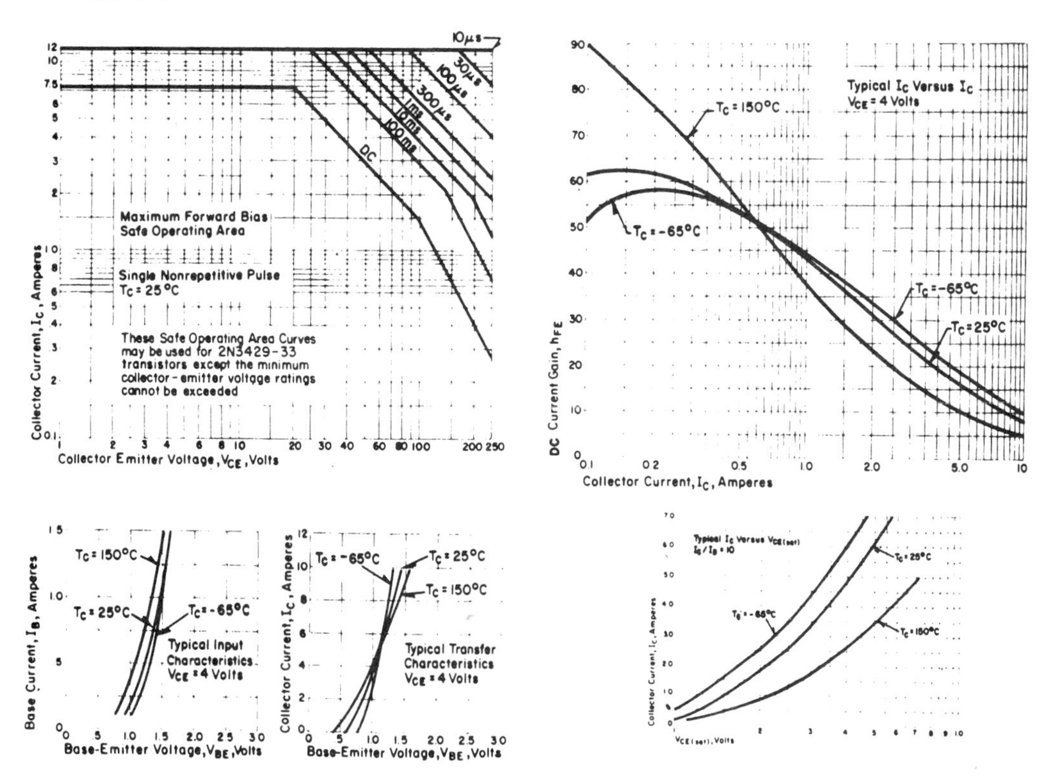

Figure 2-12 continued.

load is purely resistive then the collector supply voltage limit is $V_{CEO(SUS)}$. If the load contains inductance, as it would in motor control applications, then a safe rule of thumb is to ensure that the collector voltage rating is twice the collector supply voltage because of the large induced emfs.

When using power transistors in switching applications, the base drive signal must be great enough to force the transistor into the collector saturation region in order to minimize $V_{CE(sat)}$ and thus minimize power dissipation during the on period. Care must be exercised to ensure that the base-emitter losses do not contribute to heating because of excessive base drive.

There are two breakdown mechanisms that can occur in a transistor; first, the avalanche breakdown or first breakdown that results if the collector-base voltage limits are exceeded and second, especially in power transistors used in switching applications, the so-called second breakdown. Second breakdown is caused by localized heating at the base-emitter junction area, especially noticeable when repeatedly switching highly inductive circuits. It is attributed to localized hot spots caused by current concentration at the center of the junction area because the perimeter of the junction area turns off more rapidly.

The safe operating area (SOA) curves shown in Figure 2–12 provide the designer with a means of ensuring that the transistor will be operated safely, providing the load line lies inside the SOA for the desired operating conditions. Since these curves are specified for $T_C = 25\,°C$, it is necessary to linearly derate the dissipation limit curves to meet the actual case temperature conditions.

The power to be dissipated is approximately equal to $V_{CE}I_C$, neglecting the base power and assuming that the device is mounted on a heat sink maintained at an ambient temperature of $T_A = 25\,°C$, the derating factor is 1.33 W/°C for temperatures above $T_A = 25\,°C$.

In high-speed switching applications, the speed of response of the transistor to the switching signal is very important. The parameters of interest are the **turn-on time** t_{ON} which is affected by the transistor and circuit capacitances as well as the drive circuit resistance. To achieve the shortest turn-on time, the capacitances and base circuit resistance must be kept as low as possible. The second factor is **storage time** t_S, the time interval between the removal of the base signal and collector turn-off commencing. Storage time can be reduced by keeping the base drive resistance small and shunting the resistor by a speed-up capacitor. The third factor, the **fall time** t_f, is the time for the collector current to decrease from 90% to 10% of its original value and is reduced by applying the considerations that pertained to t_{ON} and t_S.

2-8 THE FIELD-EFFECT OR UNIPOLAR TRANSISTOR

The **field-effect transistor** (FET) is a three-terminal device that can provide amplification and is characterized by a high input impedance in contrast to the low input impedance of the bipolar junction transistor. The FET is voltage controlled and is a majority charge carrier device (electrons for the *n*-channel and holes for *p*-channel devices).

As can be seen from Figure 2–13, there are two major groups of devices forming the FET family, the **junction field-effect transistor** (JFET) and the **insulated-gate field-effect transistor** (IGFET). The IGFET is also termed the metal-oxide-silicon field-effect transistor (MOSFET). Both families of FETs are available in complementary forms, that is, in *p*-type and *n*-type materials. Although the JFET only operates in the depletion mode, the MOSFET operates both in the depletion and enhancement modes.

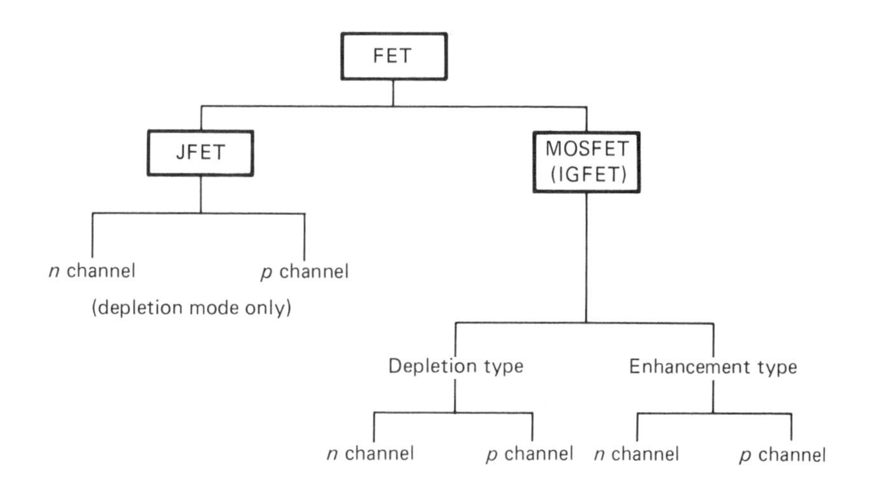

Figure 2–13. The FET family tree.

2-8-1 Junction Field-Effect Transistors

The *n*-channel JFET consists of a lightly doped *n*-type material sandwiched between two heavily doped *p*-type regions that have been formed by diffusion. The connections to the *p*-type regions are usually made internally and a single gate lead is brought out as the gate terminal (G). Ohmic connections at the ends of the channel are brought to the source (S) and drain (D) terminals. See Figure 2–14(a). A similar

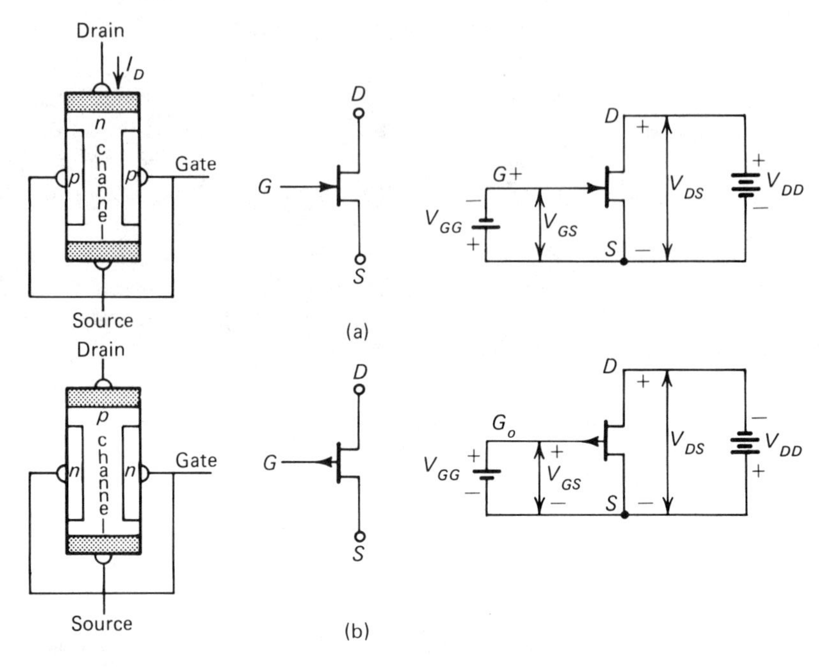

Figure 2-14. JFETs, construction, symbol, and biasing: (a) *n*-channel, (b) *p*-channel.

construction is used to form the *p*-channel JFET. See Figure 2-14(b). Figure 2-14 also shows the circuit symbols and the biasing voltages.

The basic principle of operation of the JFET is relatively easy to understand. Let us in particular consider the *n*-channel JFET. The resistance of the lightly doped *n*-channel is of the order of a few hundred ohms. When a negative voltage is applied to the gate, the gate-to-source *p-n* junction is reverse biased and the area in the *n*-channel immediately adjacent to the junction becomes depleted, that is, electrons have been removed and fixed positive ions remain. The width of the *n*-channel has been decreased as a result. Further increases in the negative bias result in further decreases in the width of the *n*-channel. As a result, electron flow from the source to the drain is restricted, that is, the current is being controlled. If the reverse bias is increased sufficiently, a gate-to-source voltage V_{GS} will be attained in which the conduction region of the *n*-channel has been pinched off. The value of gate voltage at which pinch-off occurs is called the **pinch-off voltage** V_p and is usually found in the manufacturer's data sheet as $V_{GS(\text{off})}$.

The depletion region is wider at the drain end of the channel than at the source end. This effect is caused by the voltage drop along the channel produced by drain current and makes the gate diode at the drain end of the channel more reverse biased than at the source end.

p-channel devices operate in exactly the same manner; the supply voltages are reversed. That is, they operate with a negative drain voltage and a positive gate voltage.

Since the input or gate terminal in either type of JFET is a reverse-biased diode, the input impedance will be very high, typically 100 MΩ. As a result, the input current is actually a leakage current and is typically of the order of 1 nA at 25 °C and approximately 1 μA at 150 °C.

Typical common source, I_D versus V_{DS}, characteristic curves for an n-channel JFET are shown in Figure 2–15. As has been seen, the JFET is a device that controls current flow because of the creation of an electric field extending the depletion region into the conducting channel. JFETs are classified as depletion mode or type A devices and in turn are subdivided into n-channel and p-channel.

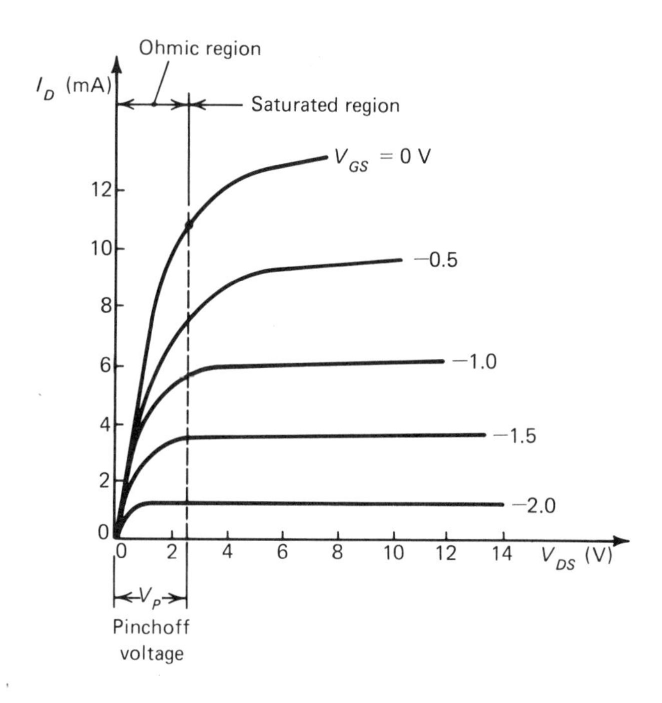

Figure 2–15. An n-channel JFET characteristic curve.

2-8-2 Insulated-Gate Field-Effect Transistor

The IGFET, or MOSFET as it is more commonly known, differs from the depletion mode JFET in the sense that a capacitor consisting of a metal gate contact, a silicon dioxide dielectric, and the conducting channel are used as the input, instead of the reverse p-n junction used with the JFET. The result of using a capacitor input is to greatly increase the input impedance to the order of 10^{10} to 10^{15} Ω and at the same time reduce the leakage current by a factor of 10^2 to 10^3.

When the gate is made positive in an *n*-channel MOSFET a negative charge is induced in the channel which enhances current flow. That is, the device is operating in the enhancement mode and the source-to-drain current is controlled by the gate voltage. See Figure 2–16(a). It should be emphasized that the enhancement mode MOSFET does not have a conducting path between the source and drain until a minimum positive voltage known as the **threshold voltage** $V_{GS(\mathrm{th})}$ is applied to the gate. The enhancement mode MOSFET is also called a type *C* MOSFET.

The depletion mode or type *B* MOSFET performs in a similar manner to the JFET. When a negative potential is applied at the gate it will result in the channel being depleted of majority carriers; for an *n*-channel MOSFET the majority carrier is the electron. As a result of the depletion of majority charge carriers, the conductivity of the channel is reduced and drain current decreases. The depletion mode *n*-channel MOSFET will have maximum drain current flowing when $V_{GS} = 0$. The type *B* MOSFET can also be operated with positive gate potentials since there is not a reverse-biased *p-n* junction as in the JFET. When operated in this manner, the MOSFET is being operated in the enhancement mode; however, the type *B* or enhancement mode MOSFET cannot be operated as a depletion mode device.

Because the silicon dioxide dielectric material is very thin it is easily punctured by static voltage discharges as little as 100 V in amplitude. As a result, MOSFETs are shipped with a shorting ring shorting out all terminals. This shorting ring should be left in place even during soldering, and in fact the soldering iron tip should also be grounded and only the case should be handled.

2-8-3 Power MOSFETs

MOSFETs have been in common use in LSI (large scale integration) and VLSI (very large scale integration) circuit applications for a number of years, though limited in their application because of their limited current-carrying capability and the 10–15 V drain-source breakdown voltage. Now a new type of MOSFET, the **power MOSFET**, has been developed with drain-to-source voltages V_{DS} ranging up to 500 V and continuous drain currents I_D up to 28 A and in pulse applications up to 70 A. The major advantages of these devices as compared to the bipolar power transistors are: (1) significantly reduced input signal power since only the input capacitance has to be

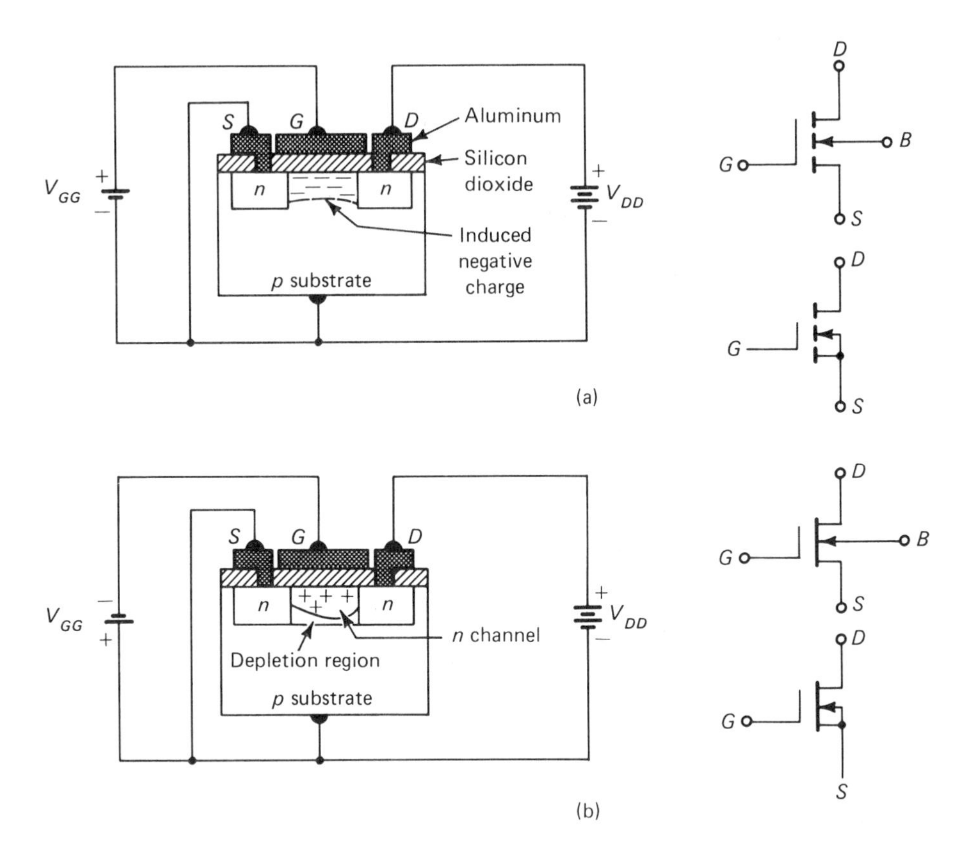

Figure 2-16. MOSFETs with biasing and symbols: (a) *n*-channel enhancement mode, (b) *n*-channel depletion mode.

charged, (2) switching times as little as 100 ns, and (3) the absence of second breakdown.

Power MOSFETs employ a new technology called vertical metal-oxide semiconductor or VMOS which uses a diffused channel and vertical current flow to achieve high voltage and current ratings that are comparable to the power transistor and power Darlington. A cross-section of a VMOS channel is illustrated in Figure 2-17. The function of the n^+ substrate is to provide the drain connection and a low resistance path. The n^- epitaxial layer is provided to increase the drain-to-source voltage breakdown capability by absorbing the depletion region from the reverse-biased drain body junction. An additional benefit of the n epitaxial layer is that it reduces the feedback capacitance. The p^- body and n^+ source are diffused into the epitaxial layer. At this point the V groove is etched into the epitaxial layer, the silicon dioxide layer is formed, and finally the aluminum source and gate connections are made.

When a positive gate voltage is applied, the surface of the p base region is inverted and a channel is formed under the gate electrode surface, which in turn permits current to flow from the drain to source.

To give some idea of the potential of the power MOSFET, International Rectifier is marketing a planar, non-V-groove device called the HEXFET which is available in n- and p-channel configurations. One particular HEXFET, the IRF 150, is rated at 100 V, 28 A continuous and 70 A pulsed, with an on-state resistance of 0.055 Ω maximum.

Another HEXFET, the IRF 350, is rated at 400 V, 11 A continuous and 25 A pulsed; that is, it has the capability of switching 4.4 kW continuously. In addition, the concept of field control is being extended to the thyristor. Research is currently being carried out in the development of a field-controlled thryistor (FCT), of which it is certain that we will hear more about in the near future. Another extremely important advantage is the ability to interface these devices with DTL and TTL.

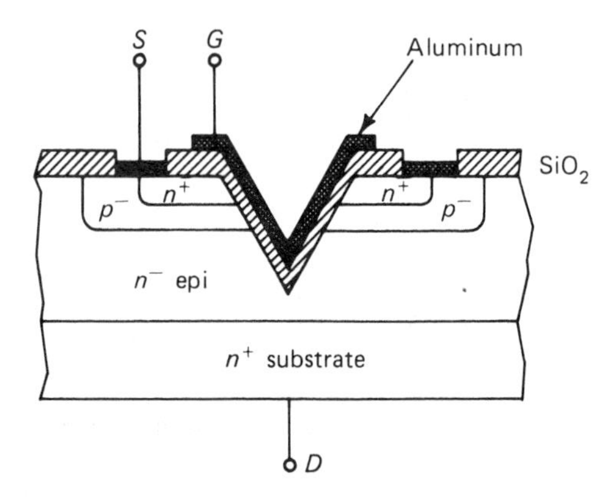

Figure 2-17. A cross-sectional view of a VMOS channel.

2-9 THYRISTORS

The name **thyristor** is applied to a family of devices which can be switched from the nonconducting state to the conducting state, that is, bistable operation, by p-n-p-n regenerative feedback. Thyristors are available as two-, three-, or four-terminal devices in both unidirectional and bidirectional forms.

The best-known member of the family is the SCR, a p-n-p-n, three-terminal, (anode, cathode, and gate) unidirectional device. It is extensively used in thyristor phase-controlled converters in the phase-controlled SCR grade form for dc machine control and in the inverter grade form for inverters used in variable frequency ac motor control applications, and in dc-dc or chopper control for electric traction, etc. Another commonly used bidirectional form is the bidirectional triode thyristor or TRIAC which is extensively used for ac power control. Other members of the unidirectional thyristor family are the light-activated silicon-controlled rectifier (LASCR), the gate turn-off thyristor (GTO) and the programmable unijunction transistor (PUT), to name a few devices.

2-9-1 The Silicon-Controlled Rectifier

The SCR or reverse-blocking triode thyristor is a device that controls power flow by the use of a gate signal. The application of a positive gate signal, usually as a pulse, causes the SCR to change from the high impedance or blocking state to the low impedance or on state.

Silicon-controlled rectifiers are available in a number of different configurations (see Figure 2-18) ranging from the conventional stud mount arrangement, which was later placed in an integral heat sink package to achieve a 40% increase in current rating, to the double-side cooled disc form which achieved an 80% increase in current rating over the same element used in the stud mount arrangement. An added benefit of the disc mount arrangement is that reverse polarity devices are not required since the disc only needs to be turned over.

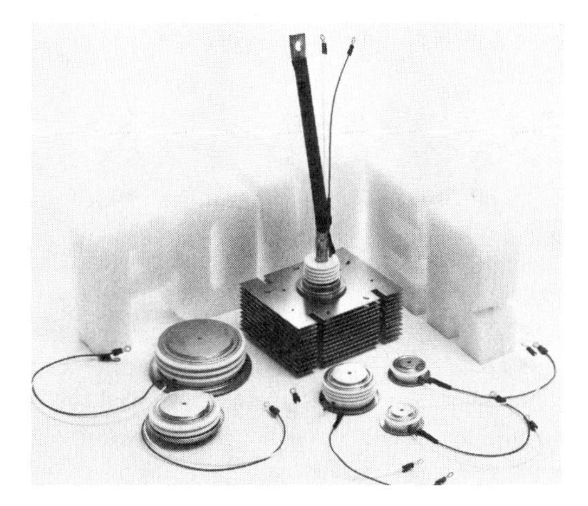

Figure 2-18. Typical SCR configurations. (Courtesy Westinghouse Electric Corp., Semiconductor Division)

The basic construction of the SCR (shown in Figure 2–19) is known as the compression-bonded encapsulation (CBE). This technique completely eliminates hard-soldered connections between the element and the package, the metallic bond being provided by a constant pressure spring washer system which applies a constant load force to the element. Basically the same approach is used for the disc package except that the spring clamp system is supplied externally by the user.

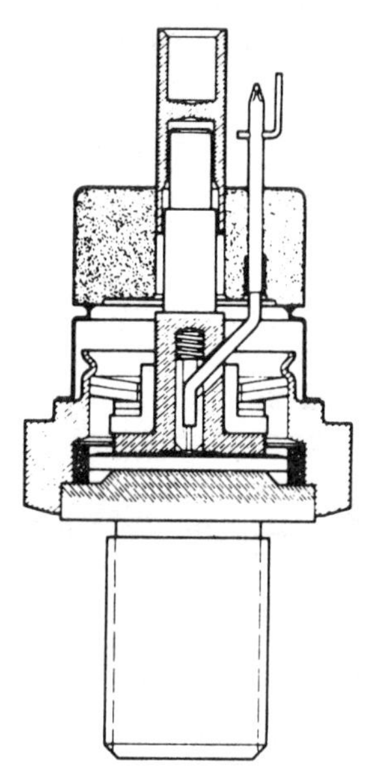

Figure 2–19. Compression-bonded encapsulation (CBE). (Courtesy Westinghouse Electric Corp., Semiconductor Division)

2-9-1-1 The Two-Transistor Analogy of the SCR

The most common method of explaining the operation of the SCR is in terms of the two-transistor analogy as shown in Figure 2–20. The basic construction, shown in Figure 2–20(a), is a four-layer diode. The center section can be divided as shown in Figure 2–20(b) and (c). The SCR can then be considered as two separate complementary transistors, one a *p-n-p* transistor and the other an *n-p-n* transistor, Q_1 and Q_2, respectively. If the gate signal is zero and the anode is positive or negative with respect to the cathode, one of the *p-n* junctions in each transistor is reverse biased. When the SCR is forward biased, junctions J_1 and J_3 are forward biased and J_2 is reverse biased; conversely, when the thyristor is reverse biased, junctions J_1 and J_3 are reverse biased and J_2 is forward biased, and only a small leakage current flows.

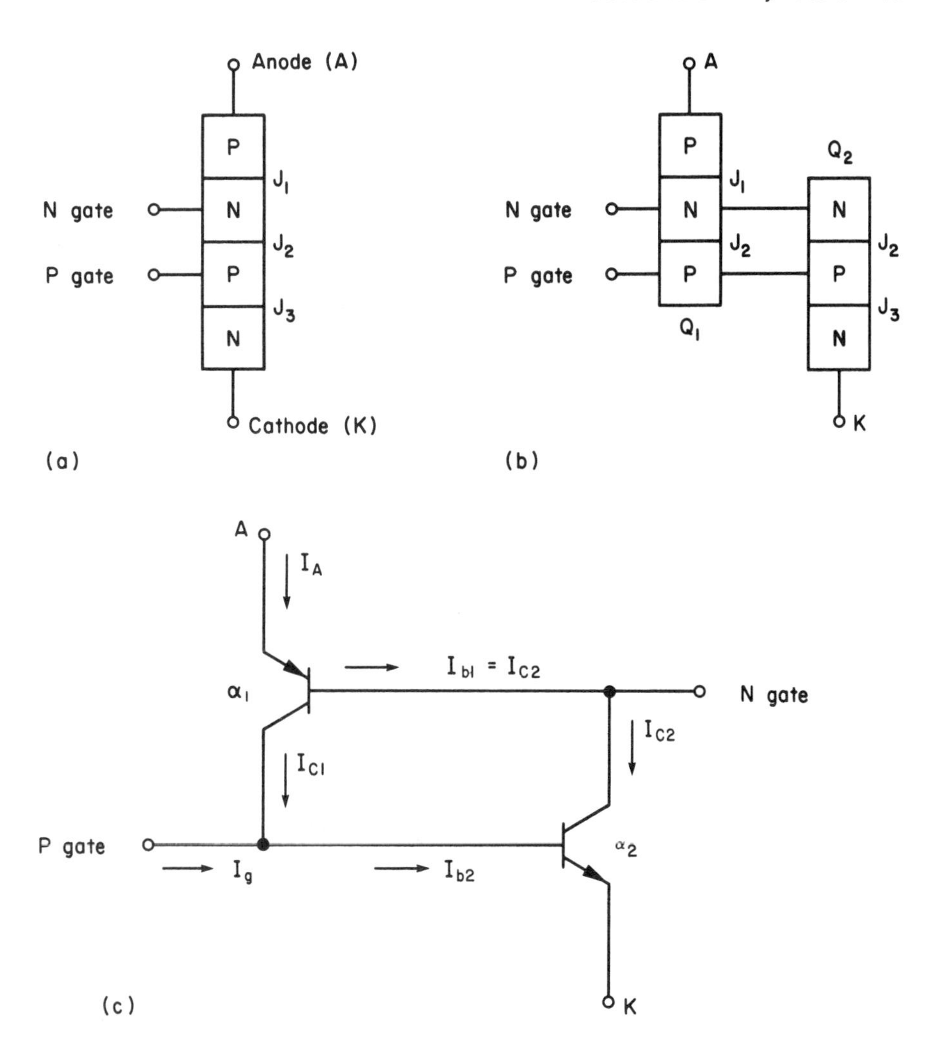

Figure 2-20. The two-transistor equivalent circuit of the thyristor: (a) basic structure, (b) as two complementary transistors, (c) two two-transistor equivalent circuit.

The application of a positive gate signal when the thyristor is reverse biased causes a reverse-anode leakage current to flow approximately equal to the positive gate current, and junction overheating can result in a thermal runaway.

Using the two-transistor analogy, as shown in Figure 2-20(b) and (c), J_1 and J_3 are slightly forward biased, where J_1 is the emitter-base junction of the *n-p-n* transistor. J_2 is the collector-base junction of both Q_1

and Q_2 and is reverse biased. Since the emitter-base junctions are only slightly forward biased, there will be little current flow.

Consider the *p-n-p* transistor Q_1; its emitter current is I_A, the anode current and the base current is

$$I_{b1} = (1 - \alpha_1)I_A - I_{CBO1}, \tag{2-4}$$

where α_1 is the current gain and I_{CBO1} is the leakage current for Q_1.

Now consider the *n-p-n* transistor Q_2:

$$I_{C2} = \alpha_2 I_K + I_{CBO2}, \tag{2-5}$$

where α_2 and I_{CBO2} are current gain and leakage current, respectively, for Q_2. I_K, the cathode current, is equal to the emitter current I_E of Q_2.

Since the base current of Q_1, I_{b1}, and the collector current I_{C2} of Q_2 are the same, then

$$(1 - \alpha_1)I_A - I_{CBO1} = \alpha_2 I_K + I_{CBO2}, \tag{2-6}$$

and since $I_A = I_K$,

$$I_A = I_K = \frac{I_{CBO1} + I_{CBO2}}{1 - (\alpha_1 + \alpha_2)} \tag{2-7}$$

Equation (2-7) forms the basis of explanation for all *p-n-p* devices. When both Q_1 and Q_2 have a very small forward bias of the emitter-base junction, the value of α is $<< 1$, $\alpha_1 + \alpha_2$ is small, and I_A will be small. The sum of $\alpha_1 + \alpha_2$ can be made to momentarily approach 1 by injecting a short duration positive current I_g at the *P* gate, which is the base of Q_2. This causes current to flow in Q_2, and because the collector is positive, collector current will flow in Q_2; this is also the base current of Q_1, and as a result Q_1 will be switched on. At this point, each transistor supplies the base current for the other transistor and the action is regenerative.

Removal of the gate signal will not result in the thyristor turning off as long as there is sufficient forward anode-to-cathode voltage to maintain a holding current I_H. From the above it can be seen that a small-amplitude positive pulse of a few microseconds duration applied when the anode-cathode is forward biased will ensure turn-on of the thyristor. However, once conduction has been initiated, the gate signal serves no useful purpose and may be removed.

2-9-1-2 Initiation of Thyristor Turn-On

Basically a thyristor is turned on by causing an increase in the emitter current. This action can result from any of the following methods:

1. Gate current I_g: The application of a positive pulse of sufficient amplitude and duration to the *P* gate of Figure 2-20.
2. Overvoltage: An increase in the forward anode-cathode voltage above the forward breakover voltage, V_{FBO}, will cause a sufficient increase in the leakage current to initiate regenerative turn-on.
3. *dv/dt*: The *p-n* junctions are effectively capacitive because of the depletion layer during blocking. As a result, whenever there is a rapid rate of change of the anode-cathode voltage (dv/dt), the charging current $i = Cdv/dt$ may be of sufficient magnitude when added to the leakage current to initiate turn-on.
4. Thermal: There is an increase in the number of electron-hole pairs as the temperature of the device increases, which causes an increase in $\alpha_1 + \alpha_2$ resulting in turn-on being initiated at lower forward anode-cathode potentials.
5. Light or radiation: Photons, gamma rays, neutrons, protons, electrons, and hard and soft X-rays when permitted to strike an unshielded thyristor will cause an increase in the electron-hole pairs, thus initiating turn-on. In some devices, such as the LASCR, a window is provided in the shielding to permit the thyristor to be turned on by allowing light to strike the silicon wafer.

Methods 2 and 3 above are not recommended on a repetitive basis since local hot spots may develop in the structure of the SCR which may permanently damage the device.

Normally the SCR will block voltages in both directions, and as a result there will be no power flow. To initiate conduction, a gate signal should only be applied when the SCR is in the forward blocking state (V_{DRM}). As can be seen from Figure 2-21 increasing the gate current I_G from 0 to I_{GT} decreases the SCR's ability to block forward voltage. Also the reverse blocking voltage characteristic V_{RRM} is very similar to that of a reverse-biased diode. It should be noted that it is not good practice to apply a gate signal while the SCR is reverse biased since the thyristor could fail because of increased leakage current.

2-9-1-3 Gate Parameters And Characteristics

The gate parameters can be classified in terms of current and voltage. I_{GT} is the dc current which causes the SCR to latch into conduction and remain on; it is known as the **gate trigger current.** The gate trigger voltage V_{GT} is the dc voltage necessary to produce I_{GT}. I_{GNT}

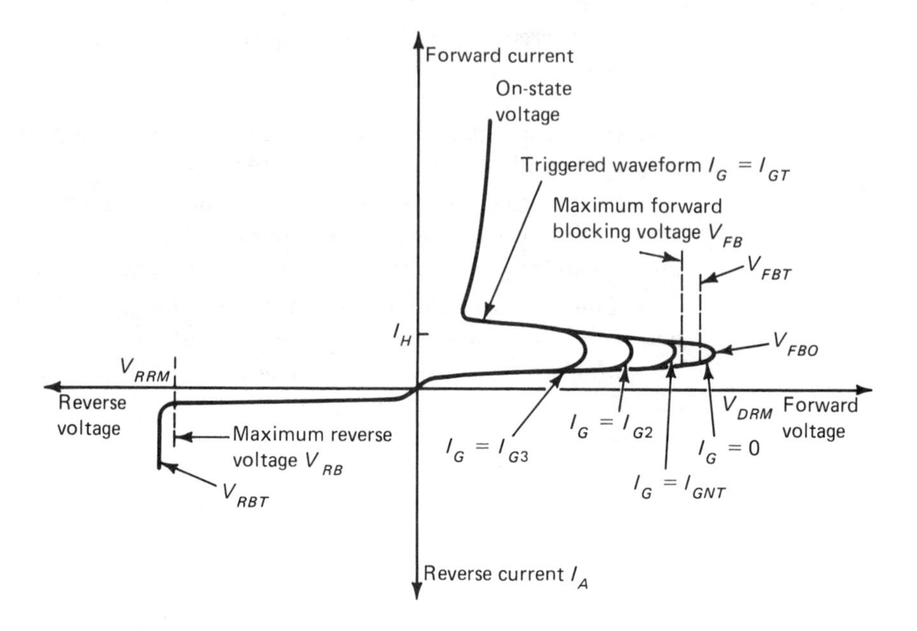

Figure 2-21. SCR static voltage-current (*V-I*) characteristic.

is the dc gate current which when applied will still permit the SCR to block rated V_{DRM}; it is also termed the **nontrigger gate current.** V_{GNT} is the maximum dc gate voltage which when applied to the gate-cathode junction will still permit the SCR to block V_{DRM}. Finally V_{GRM} is the maximum value of negative dc voltage that may be applied without damaging the gate-cathode junction.

In addition, the power dissipation capabilities of the gate-cathode junction must be defined to prevent overheating. The applicable gate power parameters are the peak gate power P_{GM}, the maximum instantaneous product of the gate voltage and current that may be permitted to exist during forward-bias conditions. The average gate power $P_{G(AV)}$ is the maximum gate power dissipation that is permitted at the gate junction over a full cycle. Finally the average reverse gate power $P_{GR(AVE)}$ is the maximum allowable reverse gate power that can be safely dissipated over a complete cycle.

These maximum gate triggering characteristics for a Westinghouse di/namic gate SCR are shown in Figure 2-22. The di/namic or amplifying gate SCR consists of a pilot and main SCR built into the same element. The basic purpose of this structure is to increase the rate of spread of the conducting area to minimize the possibility of local hot spots because of high current density during turn-on.

As can be seen from Figure 2-23, as the junction temperature T_J

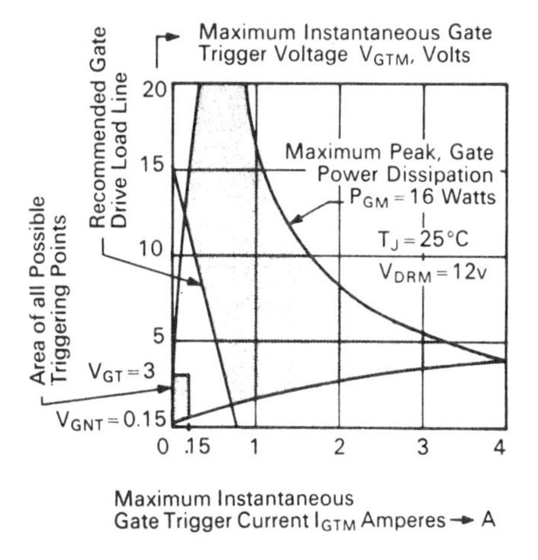

Figure 2-22. Maximum gate triggering characteristics for a dynamic gate SCR. (Courtesy Westinghouse Electric Corp., Semiconductor Division)

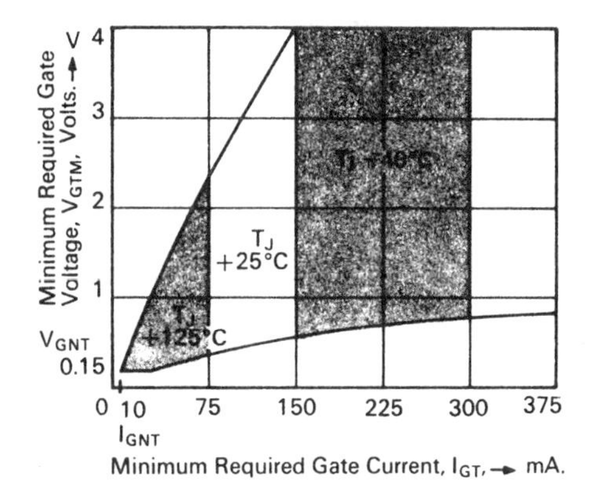

Figure 2-23. Typical gate-triggering range for various junction temperatures. (Courtesy Westinghouse Electric Corp., Semiconductor Division)

increases, the required gate voltage to trigger the SCR decreases, and conversely as the temperature decreases.

In order to minimize gate power dissipation a pulse signal, typically 5 to 10 kHz, is applied instead of a continuous dc gate signal. In order to achieve turn-on it is necessary that the gate drive amplitude be increased when the pulse width is less than 20 μs. This is due to the charge turn-on concept ($q = \int I_{GT} dt_p$). The minimum gate trigger requirements versus pulse widths are shown with respect to junction temperatures in Figure 2-24.

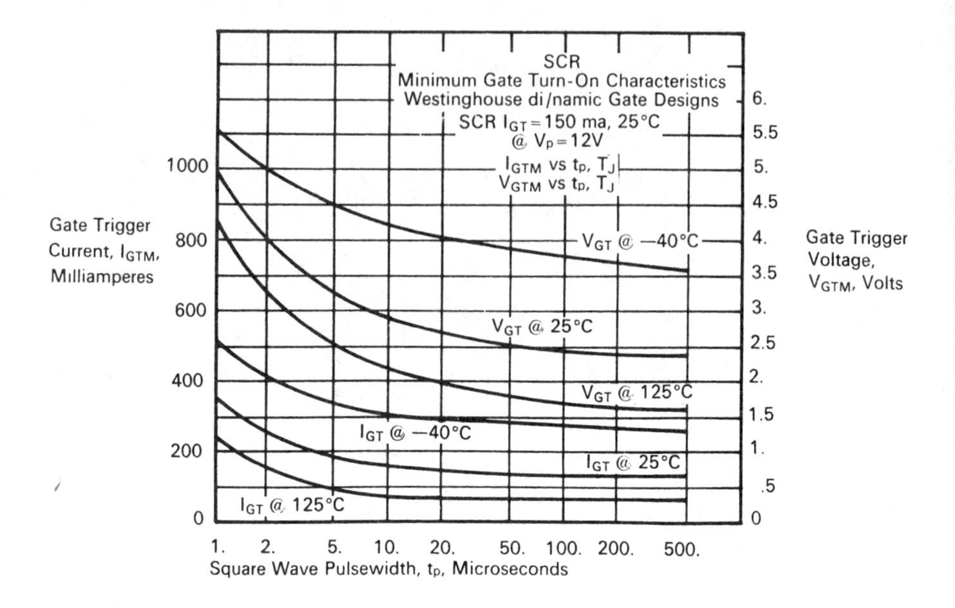

Figure 2-24. Minimum-pulsed gate-trigger requirements. (Courtesy Westinghouse Electric Corp., Semiconductor Division)

2-9-1-4 Switching Time Terminology

The application of a gate pulse to the gate of an SCR does not result in the immediate flow of anode current. Initially, there is no significant increase in anode current. This interval is known as the **delay time** t_d and is defined as the time interval between when the gate current reaches 90% of its final value and when the anode current has risen to 10% of its final value. Similarly, the **rise time** t_r is the time interval it takes the anode current to rise from 10% to 90% of its final value. The turn-on time t_{on} is equal to $t_d + t_r$ and is usually specified in the device data sheets.

Normally the rise time t_r is sufficiently small at normal power frequencies, that is, 60 and 400 Hz, that the SCR is in full conduction before the peak value of the applied anode-cathode voltage is attained. In the case of high *di/dt* applications, conduction will initially be limited to a small area of the silicon wafer in the immediate vicinity of the gate connection, and as a result localized heating may occur with the possibility that the device will fail.

There is a relationship between the t_d and t_r of the SCR and the rise time and amplitude of the initiating gate pulse. In high *di/dt* applications the device delay and rise times can be decreased by using a hard gate drive even with di/namic gate and interdigitated gate geometry SCRs. Typical applications requiring a hard gate drive are: inverters and chopper circuits with capacitive loads where a high repetitive *di/dt* occurs; heavy industrial applications involving thyristor phase-controlled converters and applications where there is a large amount of electrical noise (switching transients, etc.,) requiring gate suppression to prevent inadvertent operation of the SCRs. Figure 2-25(a) shows the requirements that must be met by a hard gate drive, namely a gate pulse with a fast-rise time between 0.1 and 1 μs, with a peak amplitude of 3.5 to 5 times the minimum gate current I_{GT} required to achieve turn-on and to maintain a current pulse no less than I_{GT} for a minimum of 20 μs. As can be seen from Figure 2-25(b), the requirements for a soft gate drive are not as stringent.

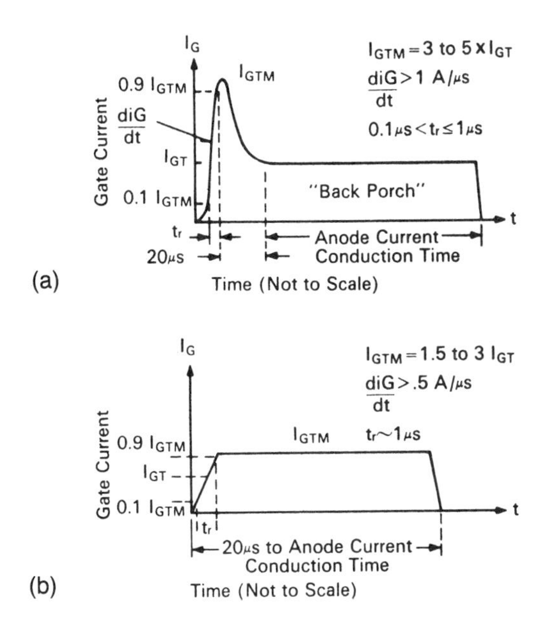

Figure 2-25. Gate drive requirements: (a) hard gate drive, (b) soft gate drive. (Courtesy Westinghouse Electric Corp., Semiconductor Division)

2-10 OTHER MEMBERS OF THE THYRISTOR FAMILY

Since the introduction of the SCR in 1957 there has been a number of other four-layer *p-n-p-n* devices known as thyristors introduced to the commercial market. Some of the more prominent members of this family are briefly described below. However, with the exception of the TRIAC, none of these devices has the power-handling capability of the SCR.

2-10-1 Amplifying Gate SCR

The amplifying gate SCR was developed specifically to meet the requirement for fast turn-on thyristors in high *di/dt* applications such as inverters and choppers. The device consists of an auxiliary SCR built into the thyristor structure as shown in schematic form in Figure 2-26. The auxiliary SCR is gated on by the gate signal and in turn the amplified output is applied as the gate signal to the main SCR. This arrangement provides a hard drive capability with a relatively low power gate input.

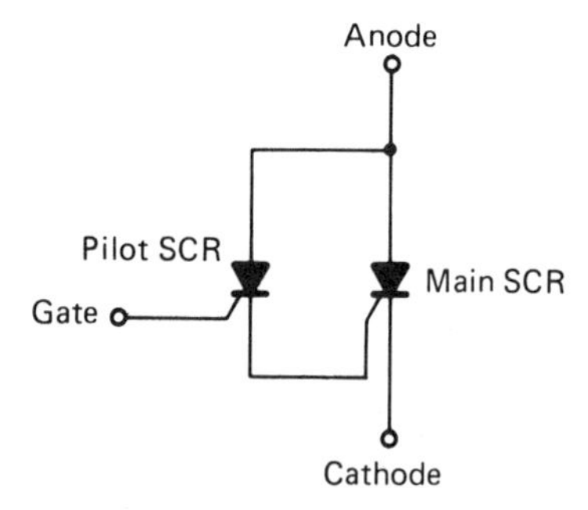

Figure 2-26. The amplifying gate thyristor.

2-10-2 The Gate Turn-Off Thyristor

The GTO effectively combines the characteristics of both the thyristor and transistor and, like the conventional SCR, is turned on by the application of a positive gate signal. However, just as with the transistor, it is turned off by a negative gate signal. The Philips BTW58 GTO, with a V_{DRM} of 1500 V, can be turned on with an approximately 100 mA gate signal and when latched the gate signal can be removed. In turn the BTW58 has a turn-off time of less than 0.5μs when a negative gate voltage of approximately 5 V is applied. GTOs are available up to 1500 V at 50 A. Another important feature of the GTO is its ability to withstand very high rates of reapplied off-state voltages, as much as 10 kV/μs, *dv/dt*. In summary, the GTO has the high blocking voltage capability of the thyristor combined with the ease of turn-on, as well as the switching characteristics of the power transistor and power Darlington.

2-10-3 The Field-Controlled Thyristor (FCT)

The FCT is another newcomer to power electronics. The basic structure of the FCT is very similar to the JFET with the exception that an injecting contact has been placed at the bottom of the wafer. As a result, minority carriers are injected from the anode which control the conductivity during forward conduction. The FCT conducts in the absense of a gate signal, and is normally turned off by applying a reverse bias to the gate.

General Electric has developed a prototype FCT with a surface gate structure that has a breakdown voltage of approximately 1000 V and a current-carrying capability of 10 A with switching speeds less than 1 μs. Hitachi is developing a "static induction thyristor" with a buried grid structure that has breakdown voltages up to 2500 V and current-carrying capabilities up to 500 A but with switching speeds of approximately 6 μs. While at this point in their development it is hard to forecast the future applications, it would appear that a very strong contender is emerging for the ac variable frequency drive field and for dc-dc use in choppers.

2-10-4 The Silicon Unilateral Switch

The **silicon unilateral switch** (SUS) is a three-terminal four-layer *p-n-p-n* device formed from *p-n-p* and *n-p-n* transistors, a resistor, and a zener diode all built into an integrated structure. The SUS will turn on when the anode-cathode voltage exceeds the zener voltage. The gate lead is brought out to permit the device to be turned on when the anode-cathode voltage is less than the zener voltage. The major application of the SUS is as a trigger device for gating on an SCR.

2-10-5 The Silicon Bilateral Switch

The **silicon bilateral switch** (SBS) is a three-terminal, four-layer *p-n-p-n* device which is basically two SUSs connected back-to-back and will turn on and conduct in either direction when the zener voltage is exceeded. However, it can also be gated on at voltages less than the zener voltage by applying a negative voltage to the gate with respect to whichever is the positive anode.

2-10-6 The Shockley Diode

The **Shockley diode** is a two-terminal, four-layer *p-n-p-n* device which acts as a low-power SCR without gate control. The device is turned on by increasing the forward anode-to-cathode voltage. One of the major applications of the Shockley diode is as a trigger device.

2-10-7 The Bidirectional Trigger Diode

The **bidirectional trigger diode** (DIAC) is a two-terminal, three-layer device basically acting as an *n-p-n* transistor without a base connection. The DIAC turns on when the anode-cathode voltage rises above the rated value. Usually DIACs are available in voltage ranges from ± 15 to ± 50 V. Once again a major usage of the DIAC is as an SCR trigger device.

2-10-8 The Bidirectional Triode Thyristor

The TRIAC overcomes one of the major objections to the use of the SCR in ac phase control; it can conduct in both directions and can be controlled by a positive or negative gate signal.

The TRIAC is basically equivalent to two SCRs connected in inverse-parallel on the same silicon wafer. The basic structure is shown in Figure 2-27(c). The inverse-parallel SCRs are gated on by applying a gate signal to SCR1, when MT1 is negative, and a gate signal to SCR2 when MT2 is negative; an arrangement which requires two separate gating sources. See Figure 2-27 (a).

The TRIAC differs from this arrangement in that the two gates are connected together and a single gate connection is brought out. See Figure 2-27(b), (c).

The static switching characteristic is shown in Figure 2-28. The characteristic in quadrant I is identical to that of the SCR, and since effectively the TRIAC is an inverse-parallel arrangement of SCRs, the characteristic in quadrant III is symmetrical to that in quadrant I.

Normally the TRIAC may be turned on by low-power positive or negative gate signals in quadrants I and III, with the gate signal being applied between the gate and the MT1 terminal.

The four operating modes are as shown in Table 2-1.

From these gate current figures, it can be seen that the TRIAC is not equally sensitive to the gate signals in all four modes, being least sensitive to III(+) signals. It can be deliberately made to be inoperative in the III(+) mode, and it will operate with a negative gate in either direction, but with a positive gate signal it will act as an SCR. When operating in this manner the device is called a logic TRIAC.

Since the TRIAC can conduct in both halves of the ac cycle, current ratings are specified in the data sheets for a conduction period of 360°.

Because of its capability of conducting in both halves of the ac cycle,

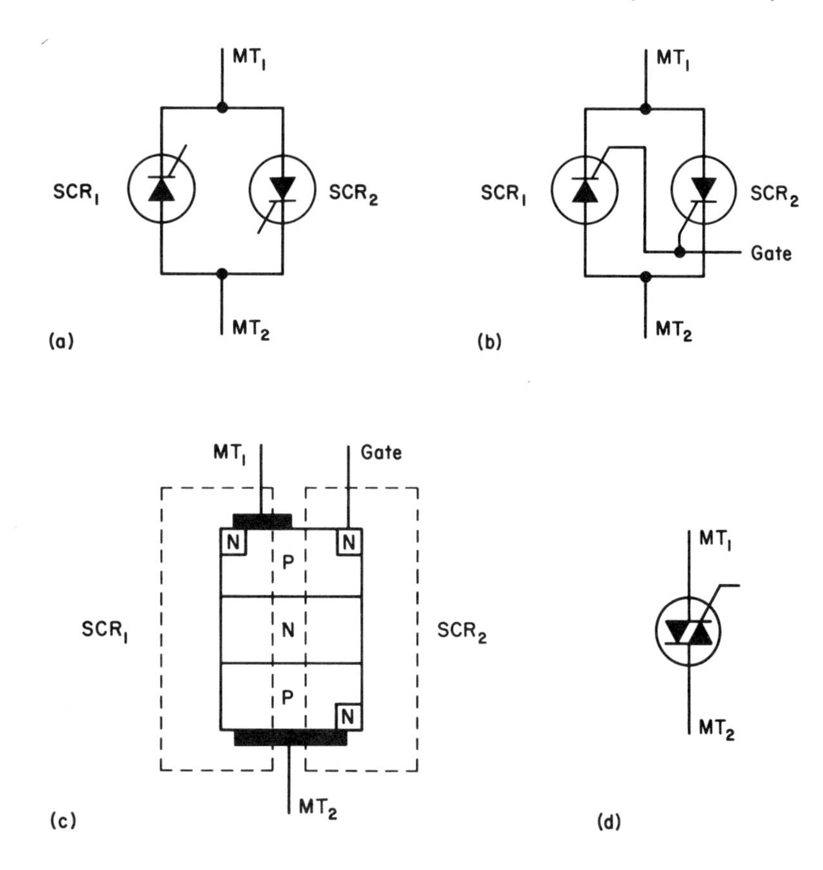

Figure 2-27. Development of the TRIAC: (a) two SCRs in anti-parallel, (b) with a common gate connection, (c) TRIAC pellet structure, (d) circuit symbol.

the TRIAC may have been conducting immediately prior to blocking in the other direction, and it must be able to support the commutating *dv/dt*. In the case of inductive loads, the TRIAC turns off when the load current is zero. However, the voltage across the TRIAC rises very rapidly to the instantaneous ac supply voltage, and as a result the TRIAC is subjected to a high *dv/dt*, which may be in excess of the capability of the device. Under these conditions the device may be subjected to *dv/dt* turn-on, which can be minimized by an *R-C* snubber network across the device. If this is not successful, then the inverse-parallel arrangement of SCRs must be used to control highly inductive loads.

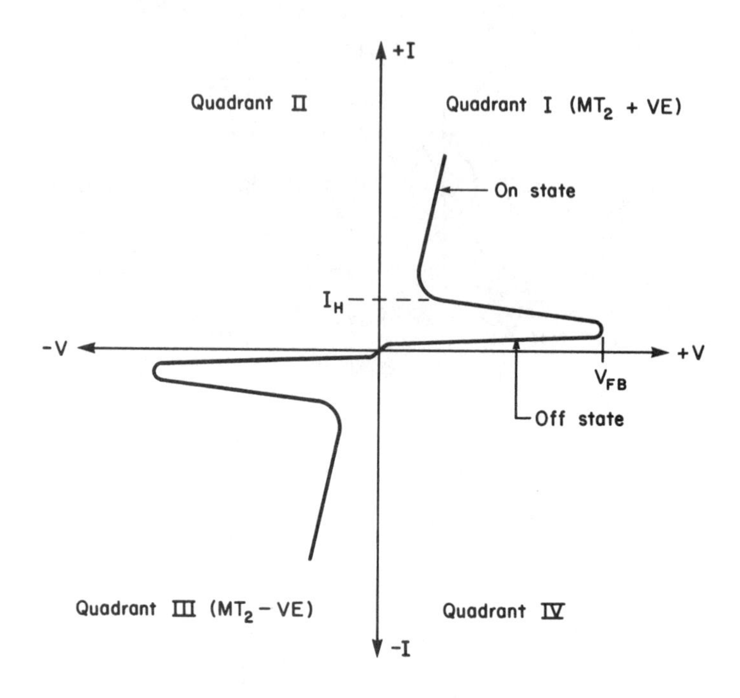

Figure 2-28. Static *V-I* characteristics of a TRIAC.

Table 2-1

Gate to MT1 Voltage	MT2 to MT1 Voltage	Operating Quadrant
Positive	Positive	I(+)(+ I_{GT}, + V_{GT})
Negative	Positive	I(−)(− I_{GT}, − V_{GT})
Positive	Negative	III(+)(+ I_{GT}, + V_{GT})
Negative	Negative	III(−)(− I_{GT}, − V_{GT})

Typical values of gate currents for an RCA 40668, 8 A, 120 V TRIAC are I(+) 10 mA, III(−) 15 mA, I(−) 20 mA, and III(+) 32 mA at a case temperature of 20°C.

2-11 THE UNIJUNCTION TRANSISTOR

The **unijunction transistor** (UJT) is a three-terminal, single-junction device which possesses a negative resistance characteristic. Once again it is a switching device with two stable states, on and off. Figure 2-29(a) shows a cross-sectional view of the UJT. The UJT consists of a

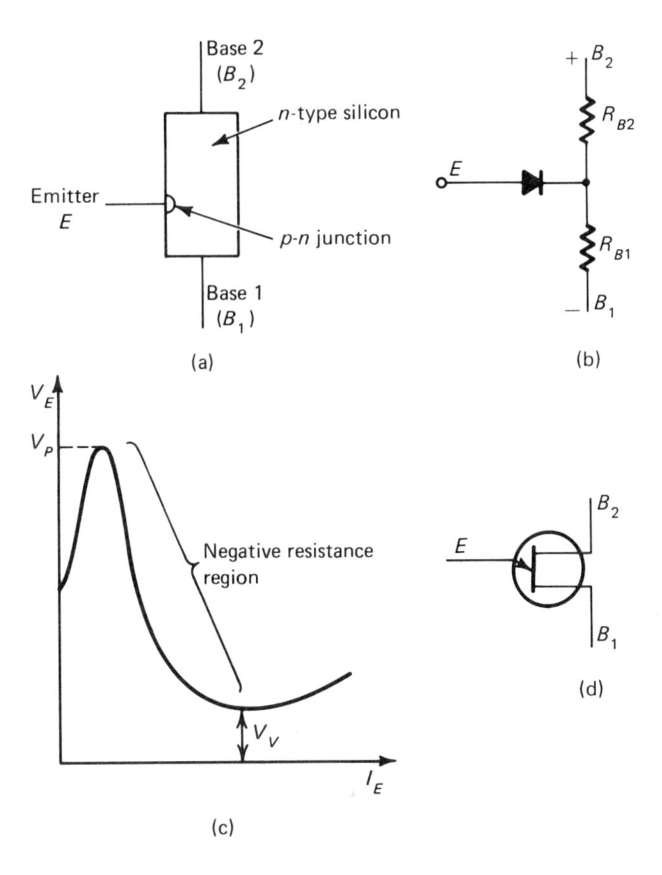

Figure 2-29. The unijunction transistor: (a) schematic, (b) equivalent circuit, (c) emitter characteristic curve, (d) circuit symbol.

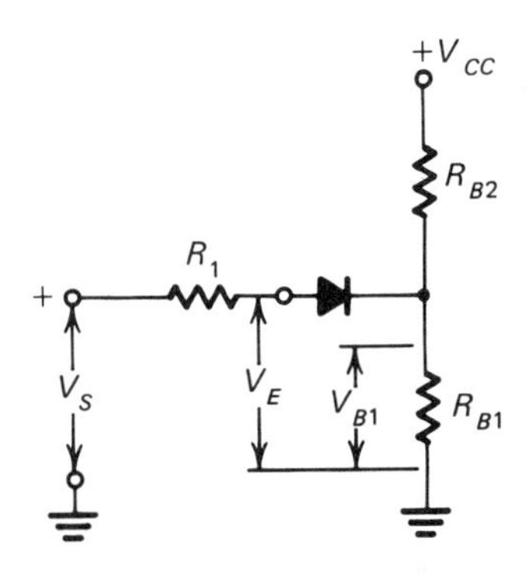

Figure 2-30. A biased UJT.

lightly doped *n*-type high-resistance silicon base bar with an ohmic contact base 1 ($B1$) and base 2 ($B2$) at each end, with an emitter connection E connected to a *p-n* diode alloyed into the base bar. As can be seen from Figure 2-29(b), electrically the UJT consists of two resistors R_{B1} and R_{B2} with a diode connected to the center point between the two base resistors. The operation of the UJT is best described in terms of the biased UJT circuit shown in Figure 2-30. As can be seen the $B2$ connection is positive with respect to $B1$. When the positive source voltage V_S is small, the equivalent diode will be reverse biased and a small current will flow through the device. The amplitude of this current will be $V_{CC}/(R_{B1} + R_{B2})$. The ratio of the potential drop across R_{B1}, V_{B1} to the source voltage is called the intrinsic standoff ratio (η). Then,

$$\text{intrinsic standoff ratio} \quad \eta = \frac{V_{B1}}{V_{CC}} \tag{2-8}$$

and

$$\eta = \frac{R_{B1}}{R_{B1} + R_{B2}} \tag{2-9}$$

Equation (2-9) shows that the intrinsic standoff ratio is determined by the construction of the UJT and is totally independent of V_S or V_{CC}. The value of η for each type of UJT is specified in the data sheets, but it is usually between 0.5 to 0.8. When the intrinsic standoff ratio is known then

$$V_{B1} = \eta V_{CC} \tag{2-10}$$

When V_S is high enough the diode becomes forward biased, which will occur when V_S is greater than $V_{B1} + 0.7$ V. This value of V_E is known as the peak voltage V_p. The value of V_p is determined by the bias voltage V_{CC}; the relationships are expressed by

$$V_p = \eta V_{CC} + V_F \tag{2-11}$$

where V_F = the forward diode drop.

When V_p is applied to the diode, see Figure 2-29(c), the diode conducts and an emitter current I_E flows. The emitter current injects holes into the lightly doped *n*-type base bar and in turn the holes are attracted to the negatively biased B_1 terminal. The effect of the hole charge carriers is to lower the resistance of R_{B1} and as a result more emitter current will flow. This process continues as V_S and V_E decrease as I_E increases. This is the negative resistance region. However, at the bottom of the emitter characteristic V_E stops decreasing and, in fact, starts to increase with increasing emitter current I_E. At the same time there is a small decrease in R_{B2} since some of the injected holes will migrate into that area of the base bar. The point on the emitter characteristic where V_E is at a minimum is known as the valley point and the voltage at this point is termed the valley voltage V_v.

A typical application of the UJT is shown in Figure 2-31; this circuit is known as a **relaxation oscillator.** Capacitor C charges toward V_{CC}, the rate of charging being controlled by the time constant R_1 C. When the capacitor voltage rises to V_p, the emitter junction becomes forward biased and the capacitor discharges through the now low resistance path between the emitter and the $B1$ terminal, producing a very sharp pulse across $R1$. The rate of discharge of the capacitor is very rapid because of the low resistance of $R1$. The result is the waveform shown in Figure 2-31. At the same time, when the UJT is triggered on, the interbase resistance decreases and a negative-going pulse is obtained at the $B2$ terminal. When the capacitor is nearly discharged, the emitter current falls below the holding current and the UJT blocks and C recharges to begin the cycle over again. It should be noted that the output voltage at $B1$ and $B2$ never completely decreases to zero since there is always a small leakage current flowing through the UJT.

The frequency of the output pulses can be changed by varying the timing resistor $R1$. Resistor $R3$ can also take the form of the primary of a pulse transformer, and as a result the secondary winding can be used to provide gate pulse signals to an SCR.

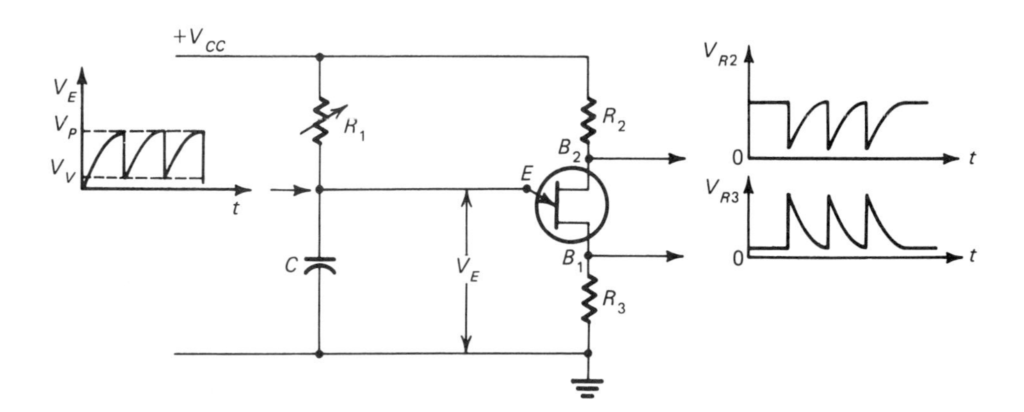

Figure 2-31. A UJT relaxation oscillator.

2-12 THE PROGRAMMABLE UJT (PUT)

The programmable UJT is a three-terminal, four-layer *p-n-p-n* device which basically functions in the same manner as the UJT. The operation of the PUT is best described by considering the relaxation oscillator shown in Figure 2-32, which as can be seen produces the same basic output waveforms as the UJT.

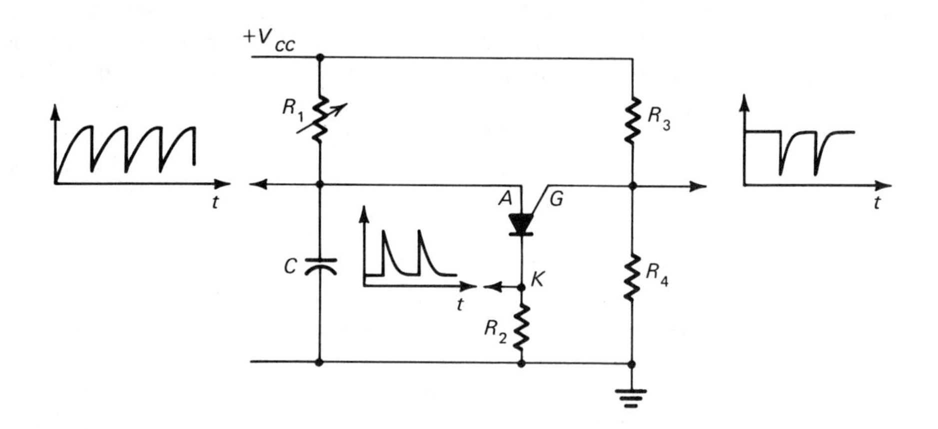

Figure 2-32. A PUT relaxation oscillator.

The PUT will turn on when the anode voltage V_A is greater than the gate voltage V_G by an amount termed the offset voltage V_T, which typically ranges between 0.2 and 1.6 V. This means that the point at which the PUT turns on is determined by the amplitude of V_G which in turn is determined by the voltage divider network $R3 - R4$. With a fixed value of V_G the PUT will turn on when the anode voltage V_A derived from the capacitor charging circuit is equal to $V_A + V_T$. As a result, the capacitor will discharge through the PUT and $R2$ and produce a positive-going pulse at the cathode terminal K and a negative-going pulse at the gate terminal G. As before, the frequency of the output pulses is controlled by varying the timing resistor $R1$.

The PUT relaxation oscillator has an upper frequency limit of approximately 50 kHz as against the 500 kHz for the UJT. Major applications of the PUT are as triggers for high-current SCRs, long duration timers, etc.

GLOSSARY OF IMPORTANT TERMS

Semiconductor diode: A two-terminal *p-n* junction device.

Diffused junction silicon diode: An *n*-type silicon slice that has been exposed to a *p*-type impurity, such as gallium vapor, at increased pressure and temperature.

Reverse recovery time: The time required to sweep the stored charges from the junction region of a reverse-biased diode before reverse blocking is achieved.

Diffusion capacitance: Junction capacitance caused by forward current.

Fast recovery diode: A diode specially treated by irradiation to reduce the effects of diffusion capacitance.

Schottky barrier or hot-carrier diode: A silicon-to-metal semiconductor diode with a forward drop approximately one-half that of a conventional silicon junction diode and an almost zero reverse recovery time.

Zener breakdown: A reverse-biased *p-n* junction breakdown produced by a high voltage gradient across the depletion region forcing electrons from their covalent bonds.

Avalanche breakdown: A reverse-biased *p-n* junction breakdown produced by a high voltage gradient across the depletion region accelerating leakage current carriers so that they gain sufficient energy to dislodge electrons from covalent bonds in the valence band.

Avalanche noise: A wide frequency spectrum electrical noise caused by the collision of rapidly moving electrons at the threshold of avalanche breakdown.

Bipolar or bijunction transistor: An *n-p-n* or *p-n-p* transistor. Bipolar refers to the fact that there are two types of charge carriers involved in the current flow process.

Drift: The movement of the majority charge carrier under the influence of the applied electric field.

Diffusion: Charge carrier movement resulting from an accumulation of like charge carriers.

Common emitter: A transistor connection in which the emitter is common to the base and collector supply voltages.

h_{FE}: The common emitter dc forward current transfer ratio; alternatively it is known as the current gain factor β.

First breakdown or avalanche breakdown: A transistor breakdown mechanism that occurs when the collector-base voltage limit is exceeded.

Second breakdown: A transistor breakdown mechanism that occurs when localized heating takes place at the base-emitter junction region.

Safe operating area (SOA) curves: Provide a means of ensuring that a transistor is operated within the designed limits.

Turn-on time t_{ON}: The time required for the collector current to increase from 10% to 90% of its final value.

Storage time t_S: The time interval between the removal of the base signal and collector turn-off commencing.

Fall time t_f: The time required for the collector current to decrease from 90% to 10% of its original value.

n-channel JFET: A field effect transistor consisting of *n*-type channel and *p*-type gate regions in which electrons are the majority charge carrier.

p-channel JFET: A field effect transistor consisting of *p*-type channel and *n*-type gate regions in which holes are the majority charge carrier.

Pinch-off voltage $V_p(V_{GS(off)})$: The value of drain voltage at which the two depletion regions are almost in contact, and current flow almost ceases.

Pinch-off region: The region in which the drain current remains almost constant for a constant gate source voltage V_{GS}.

Insulated-gate FET (IGFET): A FET with a very thin layer of silicon dioxide between the metal gate electrode and the channel.

MOSFET: Metal oxide semiconductor field effect transistor. The more common term for the IGFET.

Type A or depletion mode devices: A Type A or depletion-mode-only FET is designed not to be operated with forward gate bias. The JFET is classified as a Type A device and in turn is further subdivided into *n*- or *p*-channel.

Type B or depletion-enhancement mode devices: A Type B MOSFET is constructed with both an insulated gate and a conducting channel between the source and drain, that is, they are partially conductive and will conduct with forward or reverse gate bias.

Type C or enhancement mode devices: Are MOSFETs that are normally nonconductive between source and drain unless forward biased.

VMOS or vertical MOS: MOSFETs with a V-shaped gate and diffused channel with vertical current flow.

Power MOSFET: A higher power VMOS.

Thyristor: The generic name for *p-n-p-n* devices that can be switched from the nonconducting to conducting state by regenerative feedback.

Silicon controlled rectifier (SCR): A three-terminal *p-n-p-n* unidirectional device. The most common member of the thyristor family. Sometimes known as a reverse blocking triode thyristor.

Amplifying gate SCR: A fast turn-on thyristor with an auxiliary SCR built into the thyristor structure which provides a hard drive capability in high *di/dt* applications with a minimal gate power input.

Gate turn-off thyristor (GTO): A fast switching thyristor which is turned on by a low power positive gate signal and turned off by a negative voltage.

Field controlled thyristor (FCT): A fast switching high power device similar to a JFET which conducts in the absence of a gate signal.

Silicon unilateral switch (SUS): A three-terminal, four-layer *p-n-p-n* device with a zener diode connected between the gate and cathode which turns on when the anode-cathode voltage exceeds the zener voltage.

Silicon bilateral switch (SBS): A three-terminal, four-layer *p-n-p-n* device consisting of two SUSs in antiparallel and that will turn on in either direction when the zener voltage is exceeded. It may also be turned on with voltages less than the zener voltage by applying a negative voltage from the gate to the currently positive anode.

Shockley diode: A two-terminal, four-layer *p-n-p-n* device which is turned on by increasing the forward anode-to-cathode voltage. It is basically similar to the SUS but switches on in about 0.1 μs

Bidirectional trigger diode (DIAC): A two-terminal, three-layer device that acts as an *n-p-n* transistor without a base connection.

Bidirectional triode thyristor (TRIAC): A three-terminal, five-layer device which is basically two SCRs connected in antiparallel. It can be turned on by either positive or negative gate signals.

Unijunction transistor (UJT): A three-terminal, single-junction bistable switching device.

Programmable UJT (PUT): A three-terminal, four-layer *p-n-p-n* device which basically functions in a similar manner to the UJT, except that it turns on when the anode voltage V_A is greater than the gate voltage V_G by an amount termed the offset voltage V_T. V_T ranges typically between 0.2 and 1.6V.

REVIEW QUESTIONS

2-1 Explain how a diffused junction silicon diode is produced.

2-2 Why are diffused junction silicon diodes preferred to germanium diodes?

2-3 Discuss the two types of silicon diodes and list typical applications for each type.

2-4 Discuss the three types of power loss that occur in semiconductor diodes.

2-5 What is the significance of the I^2t rating of a semiconductor diode?

2-6 What factor determines the reverse recovery time of a semiconductor diode? How may this time be reduced and what are the adverse effects of high reverse recovery times?

2-7 Explain the principle of the Schottky barrier diode. Discuss their advantage and list typical applications.

2-8 Explain the difference between zener and avalanche breakdown.

2-9 What is meant by avalanche noise and how may it be minimized?

2-10 Sketch the V-I characteristics of a zener diode. Define the following quantities V_Z, I_{ZT}, I_{ZK}, I_{ZM} and show how they may be determined from the V-I characteristics.

2-11 Explain what is meant by a temperature compensated zener diode.

2-12 What are the two charge carrier mechanisms in a bipolar transistor?

2-13 Explain how a transistor may be operated as an amplifier, a switch, or cutoff.

2-14 What are the two breakdown mechanisms that can occur in a transistor?

2-15 What is the function of SOA curves?

2-16 In high speed switching applications the speed of response of the transistor is very important. What factors affect the speed of response and how can their effects be minimized?

2-17 Explain the construction and the principle of operation of an n-channel JFET.

2-18 Explain the construction and principle of operation of a MOSFET.

2-19 Explain the construction and principle of operation of a power MOSFET.

2-20 Explain the two-transistor analogy of the SCR.

2-21 Discuss the various ways that thyristor turn-on may be initiated.

2-22 Discuss the various methods used to reduce gate power dissipation during thyristor turn-on.

2-23 Explain the principle of operation and list typical applications of an amplifying gate SCR.

2-24 Explain the principle of operation and the features of the GTO.

2-25 Explain the basic principle of operation of both types of FCT.

2-26 Discuss the construction and principle of operation of the SUS. What is the principle application of an SUS?

2-27 Explain the principle of operation and the application of an SBS.

2-28 Explain the principle of operation and the application of a Shockley diode.

2-29 Explain the principle of operation and the application of a DIAC.

2-30 Explain the construction and principle of operation of a TRIAC. What is a major limiting factor in its application and how may this effect be reduced?

2-31 Explain the construction and principle of operation of a UJT.

2-32 With the aid of a sketch explain the operation of a UJT relaxation oscillator.

2-33 With the aid of a sketch explain the operation of a PUT relaxation oscillator.

PROBLEMS

2-1 A 1N4000 zener diode has $V_Z = 7.5$ V ± 10% at I_{ZT}=335mA and can dissipate 10W at an ambient temperature of 55°C, derated at 83.3mW/°C above 55°C. What is the recommended power dissipation when the zener is operated at 75°C?

2-2 Design a simple zener voltage regulator to provide 5V from a 12V source. What is the minimum value of load resistance that can be connected across the output terminals? Assume the regulator is operating at an ambient temperature of 25°C.

3

Rectifier Circuits

3-1 INTRODUCTION

The function of a **rectifier** is to convert ac power to dc power. In addition to the basic rectifier auxiliary equipment such as transformers, protective devices, cooling equipment as required, filters to remove unwanted ripple from the dc output, and switching equipment are usually included in this requirement.

3-1-1 Single-Phase Half-Wave Rectifier With Resistive Load

The **single-phase rectifier** is the simplest half-wave configuration and normally it is not used in industrial configurations; however, it is useful in that the basic principles of rectification can be easily demonstrated.

The single-phase rectifier circuit with a resistive load is shown in Figure 3-1(a), the transformer secondary voltage V_s is shown in Figure 3-1(b). Since the voltage and current are in phase with a resistive load, the load voltage and load current waveforms are as illustrated in Figure 3-1(c).

In the following analysis of a half-wave rectifier it is assumed that the forward resistance of the diode is negligible. Then the instantaneous open circuit voltage of the transformer secondary is derived:

$$v_s = V_m \sin \omega t, \; 0 \leq \omega t \leq \pi \tag{3-1}$$

and the resulting peak diode and load current will be

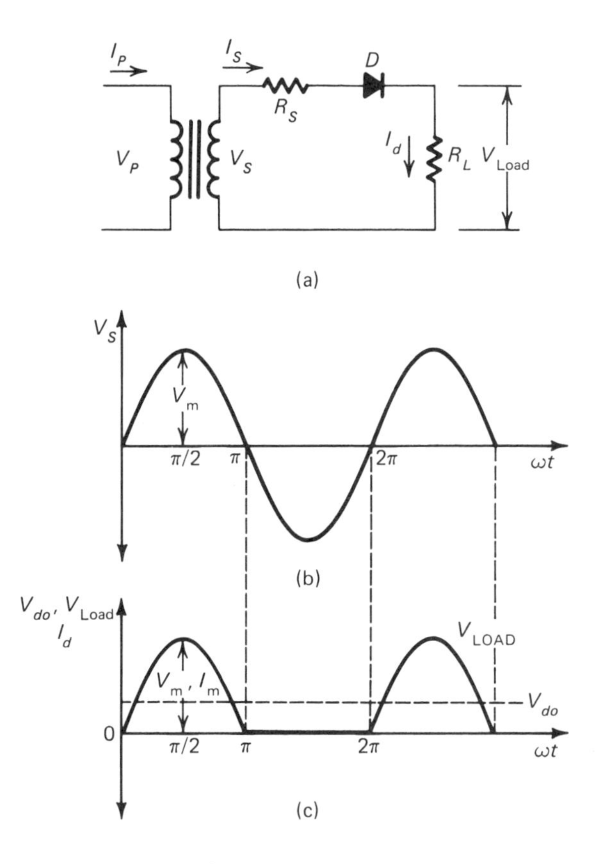

Figure 3-1. Single-phase half-wave rectifier with resistive load: (a) schematic, (b) secondary voltage waveform, (c) load voltage and load current waveforms.

$$I_m = \frac{V_m}{R_S + R_L} \tag{3-2}$$

where R_S = the resistance of the transformer secondary winding, and R_L = the load resistance.

The average value of the output dc current can be obtained by

$$I_d = \frac{I_m}{\pi}, \tag{3-3}$$

and the effective or root mean square (rms) value of the total load current I_{rms} is expressed as

$$I_{rms} = \frac{I_m}{2}. \tag{3-4}$$

The output dc voltage developed across R_L will be

$$V_{do} = I_d R_L \tag{3-5}$$

$$= \frac{I_m}{\pi} R_L \tag{3-6}$$

$$= \frac{\frac{V_m}{\pi} R_L}{R_S + R_L} = \frac{V_m}{\pi}\left(\frac{R_L}{R_S + R_L}\right), \tag{3-7}$$

and if $R_L >> R_S$, then

$$V_{do} \simeq \frac{V_m}{\pi} \tag{3-8}$$

or alternatively,

$$I_d = \frac{I_m}{\pi} = \frac{\frac{V_m}{\pi}}{R_S + R_L},$$

$$I_d(R_S + R_L) = \frac{V_m}{\pi},$$

$$I_d R_L = \frac{V_m}{\pi} - I_d R_S,$$

$$V_{do} = \frac{V_m}{\pi} - I_d R_S. \tag{3-9}$$

Equation (3-9) illustrates quite clearly that the output dc voltage V_{do} is dependent upon the load current I_d and also upon the transformer and diode resistances.

From Figure 3-1(b), it can be seen that the diode must be able to withstand a **peak inverse voltage** (PIV); that is,

$$\text{PIV} = V_m = \sqrt{2}\, V_s = 3.14\, V_{do} \tag{3-10}$$

The **average power** P_{av} is the product of the average dc output voltage and current and is calculated as

$$P_{av} = V_{do}I_d = \left[\frac{V_m}{\pi}\left(\frac{R_L}{R_S + R_L}\right)\right]\left[\frac{V_m}{(R_S + R_L)}\right]$$

$$= \frac{V_m{}^2R_L}{\pi^2(R_s + R_L)^2} \tag{3-11}$$

The rectifier efficiency (η), which is a figure of merit, permits the effectiveness of rectifier configurations to be compared, and

$$\eta = \frac{P_{av}}{P_{ac}} = \frac{V_{do}I_d}{V_{rms}I_{rms}}$$

$$= \left[\frac{V_m{}^2R_L}{\pi^2(R_S + R_L)^2}\right]\left[\frac{4(R_S + R_L)}{V_m{}^2}\right]$$

$$= \frac{4}{\pi^2(R_S + R_L)} \tag{3-12}$$

$$= 0.4053 \text{ or } 40.53\%.$$

The load voltage waveform consists of two components, a dc component with an ac component or **ripple component** superimposed on it. The ratio of the effective (rms) value of the ac ripple to the average value of the dc component is called the **ripple factor** γ.

$$\gamma = \frac{\text{effective value of the ac ripple component}}{\text{average value of the dc component}}.$$

The relationship between the combined ac and dc components V_{rms} is expressed as

$$V_{rms} = \sqrt{V_{do}{}^2 + V_{ac}{}^2} \tag{3-13}$$

where V_{rms} = the effective or rms value of the combined ac and dc components,

V_{do} = the average value of the dc component, and

V_{ac} = the effective or rms value of the ac ripple component.

Then,

$$V_{ac} = \sqrt{V_{rms}{}^2 - V_{do}{}^2}. \tag{3-14}$$

Therefore,

$$\gamma = \frac{\sqrt{V_{rms}{}^2 - V_{do}{}^2}}{V_{do}} \tag{3-15}$$

$$= \sqrt{\left(\frac{V_{rms}}{V_{do}}\right)^2 - 1} \tag{3-16}$$

which in the case of the half-wave rectifier is

$$\gamma = \left[\left(\frac{V_m}{2} \cdot \frac{\pi}{V_m}\right)^2 - 1\right]^{1/2}$$

$$\gamma = \left[\left(\frac{\pi}{2}\right)^2 - 1\right]^{1/2}$$

$$= 1.21 \text{ or } 121\%$$

A similar relationship can also be derived in terms of current and is expressed as

$$\gamma = \sqrt{\left(\frac{I_{rms}}{I_d}\right)^2 - 1} \tag{3-17}$$

where $I_{rms} = \frac{I_m}{2}$ since current only flows for one half-cycle.

The form factor F is the ratio of the effective value to the half-cycle mean value; thus

$$F = \frac{V_{rms}}{V_{do}} \tag{3-18}$$

and in turn

$$\gamma = \sqrt{F^2 - 1} \cdot \tag{3-19}$$

Equation (3-13) shows that the ripple factor can be determined by connecting a dc voltmeter (ammeter) and an ac rms reading voltmeter (ammeter) to the output terminals of the rectifier. From these readings the ripple factor can be determined.

Example 3-1

By measurement the output load voltage of a single-phase half-wave rectifier is found to consist of an average dc voltage component of 100 V and an ac rms component of 15 V. Calculate: (1) V_{rms}, (2) the ripple factor, and (3) the form factor of the waveform.

Solution:

(1) From Equation (3-13)

$$V_{rms} = \sqrt{V_{do}^2 + V_{ac}^2}$$

$$= \sqrt{100^2 + 15^2}$$

$$= 101.12 \text{ V}$$

(2) From Equation (3-16)

$$\gamma = \sqrt{\left(\frac{V_{rms}}{V_{do}}\right)^2 - 1}$$

$$= \sqrt{\left(\frac{101.12}{100}\right)^2 - 1}$$

$$= 0.225 \text{ or } 2.25\%$$

(3) From Equation (3-18)

$$F = \frac{V_{rms}}{V_{do}}$$

$$= \frac{101.12}{100} = 1.011$$

In any rectifier or power supply design it is also necessary to determine the transformer characteristics in order to obtain the transformer ratings. The transformer characteristics are as follows.

The rms secondary voltage V_s is calculated as

$$V_s = \frac{V_m}{\sqrt{2}} = \frac{V_{do}}{2 \sin \pi/2} \tag{3-20}$$

$$= 2.22V_{do},$$

the rms secondary current as

$$I_s = \frac{\pi I_d}{2} = 1.57I_d \tag{3-21}$$

the maximum value of the secondary current as

$$I_{sm} = \pi I_d = 3.14 I_d \tag{3-22}$$

and in turn the rms primary current as

$$\begin{aligned} I_{d(\text{rms})} &= \sqrt{I_s^2 - I_d^2} \\ &= \sqrt{1.57 I_d^2 - I_d^2} \\ &= 1.21 I_d \end{aligned}$$

Therefore,

$$I_p = 1.21 I_d (N_s/N_p) \tag{3-23}$$

The secondary VA rating is derived by the expression

$$\begin{aligned} V_s I_s &= (2.22 V_{do})(1.57 I_d) \\ &= 3.49\ V_{do} I_d \end{aligned} \tag{3-24}$$

and the transformer primary VA rating by

$$\begin{aligned} V_p I_p &= V_s (N_p/N_s)(N_s/N_p)\, 1.21 I_d \\ &= (2.22 V_{do})(1.21 I_d) \\ &= 2.7 V_{do} I_d \end{aligned} \tag{3-25}$$

From the above analysis of the single-phase half-wave rectifier with a resistance load, it can be seen that the maximum rectification efficiency is 40.53%; that is, assuming no diode loss only 40.53% of the ac input power is converted to dc power as input to the load. The balance is present in the load in the form of ac power. The ripple factor of 121% indicates that filtering is required to smooth out the dc output voltage.

From our study of the single-phase half-wave rectifier it can be seen that the current flowing in the secondary of the input transformer has a duration of one half-cycle and is not sinusoidal; however, the current flows in the primary for the complete cycle. As a result, the I^2R heating effect is different in both the primary and secondary windings because of the difference in the effective currents. In addition, the primary and secondary voltage and current waveforms are also different. Normally transformers are rated in terms of voltage and current, that is, VA or kVA ratings, rather than against the true power supplied to the load. The reason for rating transformers in this manner is because the heating effects are proportional to the current squared.

The **transformer utilization factor** (TUF) is defined as follows:

$$\text{TUF} = \frac{\text{rectified dc power supplied to the load}}{\text{VA rating of the transformer secondary}}. \tag{3-26}$$

A similar relationship, the **primary utilization factor**, can also be derived as the ratio of the rectified dc output power to the VA rating of the primary winding. From Figure 3-1(c) it can be seen that the duration of the secondary current is one half-cycle.

Assuming that $R_L \gg R_s$, then

$$P_{dc} = I_d^2 R_L = \frac{I_m^2}{\pi^2} R_L$$

$$P_{ac} = \left[\frac{V_m}{\sqrt{2}}\right]\left[\frac{I_m}{2}\right]$$

but $$V_m = I_m R_L$$

Therefore, $$P_{ac} = \frac{I_m^2 R_L}{2\sqrt{2}}$$

and $$\text{TUF} = \left[\frac{I_m^2 R_L}{\pi^2}\right]\left[\frac{2\sqrt{2}}{I_m^2 R_L}\right] \tag{3-27}$$

$$= \frac{2\sqrt{2}}{\pi^2} = 0.287 \tag{3-28}$$

since $$I_d = \frac{I_m}{\pi} \text{ and}$$

$$V_{do} = \frac{V_m}{\pi},$$

$$P_{dc} = \frac{V_m I_m}{\pi^2} = 0.101 V_m I_m \tag{3-29}$$

The transformer secondary VA is also calculated as

$$\text{VA} = \left[\frac{V_m}{\sqrt{2}}\right]\left[\frac{I_m}{2}\right]$$
$$= 0.354 V_m I_m \tag{3-30}$$

Therefore,

$$\text{TUF} = \frac{0.101 V_m I_m}{0.354 V_m I_m} \tag{3-31}$$
$$= 0.286$$

The implication of Equation (3-31) is that the transformer secondary winding in a single-phase half-wave rectifier configuration must be 1/TUF; that is, 3.50 times larger than when the transformer is being used to deliver ac power.

A further factor that must be considered is that the transformer secondary winding is carrying both dc and ac current components. The effect of the dc current component is to cause saturation of the transformer core, and as a result the primary excitation current will be greater than when the transformer is used to deliver ac power only. Therefore, the primary winding utilization factor will be less than 1, and in fact is of the order of 0.38.

The significance of the transformer utilization factor is that the windings of rectifier transformers must be wound with larger cross-section wire than would be the case for a normal ac power transformer. As will be seen in Section 3-5-1, improvements in the TUF are obtained with filtering since this results in the current flow interval being greater than one half-cycle.

3-1-2 Single-Phase Half-Wave Rectifier With An RL Load

When a voltage V_S is applied, the rate of change of current in an inductive load is at its maximum and the induced voltage load (180° out-of-phase with the applied voltage) delays the build-up of current. As the current builds up, energy is stored in the magnetic field surrounding the inductor coil. When the applied voltage v_s is constant, the induced voltage v_L is zero and the current is at a maximum. When the applied voltage decreases, the current starts decreasing and the collapsing magnetic field links with the inductor winding and induces a voltage which opposes the decrease in the applied voltage; that is, the induced voltage opposes the change producing it at all times (Lenz's Law). In the case of a pure inductance in an ac circuit, the induced voltage lags the applied voltage by 180° and the current lags the applied voltage by 90°.

In the case of a **resistive-inductive** (*RL*) load, which is the practical case, the current lags behind the applied voltage by the phase angle ∅, where ∅ is less than 90°.

Consider the single-phase rectifier supplying an *RL* load. See Figure 3-2(a), (b), and (c). During the time interval from 0 to $\pi/2$, the applied voltage v_s is increasing from zero to its positive maximum, while the induced voltage v_L across the inductor is opposing the change of current through the load. In the interval $\pi/2$ to π, the applied voltage is decreasing from its positive maximum to zero. At the same time the polarity of the induced voltage has reversed and is opposing the decrease in current; that is, it is maintaining forward diode current.

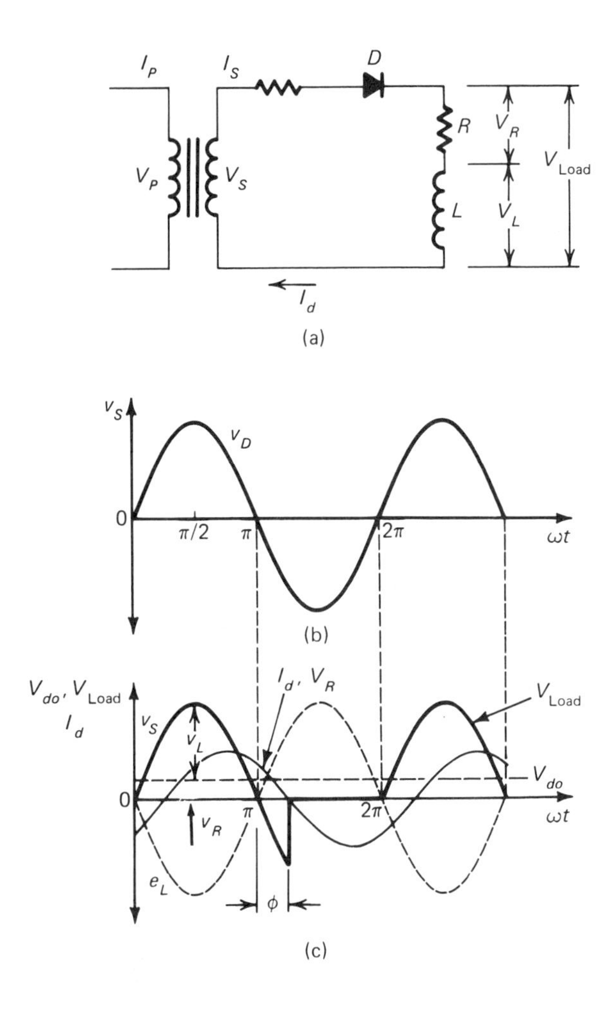

Figure 3-2. Single-phase half-wave rectifier with an *RL* load: (a) schematic, (b) secondary voltage waveform, (c) load voltage and load current waveforms.

At π the applied voltage reverses and commences to build up to a positive maximum; however, the induced voltage across the inductance is still positive and will maintain forward conduction of the diode until the induced voltage decreases to zero at which point the diode blocks. That is, there is forward current flowing through the diode until an angle of $\pi + \varnothing$ has been reached, even though the diode is reverse biased. This situation is the result of the energy stored in the magnetic field being returned to the source. The angle $\varnothing$ is dependent upon the values of L and R. The average dc voltage V_{do} is given by

$$V_{do} = \frac{V_m}{2\pi} \int_0^{\pi + \varnothing} \sin \omega t \, (d\omega t) \tag{3-32}$$

$$= \frac{V_m}{2\pi} \left[-\cos\omega \, t \right]_0^{\pi + \varnothing}$$

$$= \frac{V_m}{2\pi} \left[1 - \cos(\pi + \varnothing) \right] \tag{3-33}$$

from which it can be seen that as $\varnothing$ approaches 180°, V_{do} approaches zero; or in other words, the effect of inductance is to reduce the average dc voltage applied to the load.

It can be shown that the instantaneous value of the current I_d is calculated as

$$i_d = \frac{V_m}{Z} \left[\sin(\omega t - \varnothing) + e^{(-Rt/L)} \sin \varnothing \right] \tag{3-34}$$

where $\phi = \tan^{-1} \left[\frac{\omega L}{R} \right]$

and $Z = \sqrt{R^2 + (\omega L)^2}$

The conduction angle $\pi + \varnothing$ is dependent upon $\omega L/R$, and as a result the average value of the current I_d will decrease as the inductance of the circuit increases,

$$I_d = \frac{V_m}{2\pi R} \left[1 - \cos(\pi + \varnothing) \right] \tag{3-35}$$

The average current I_d can be increased by the use of a freewheel diode $D2$. See Figure 3-3(a). The effect of the freewheel diode is to prevent negative ac voltage appearing at the load terminals, and as a result it increases the average output voltage V_{do} appearing at the load terminals, as well as the average current I_d. As long as V_s is greater than V_L, $D1$ conducts; when v_L is greater than v_s, $D1$ becomes reverse biased and commutates off and $D2$ being forward biased conducts. That is, the stored magnetic energy is dissipated in the path $D2$ and the load.

In summary, the single-phase half-wave rectifier is not very satisfactory due to

1. the excessive ripple (121%),
2. the low rectification ratio (40.5%),

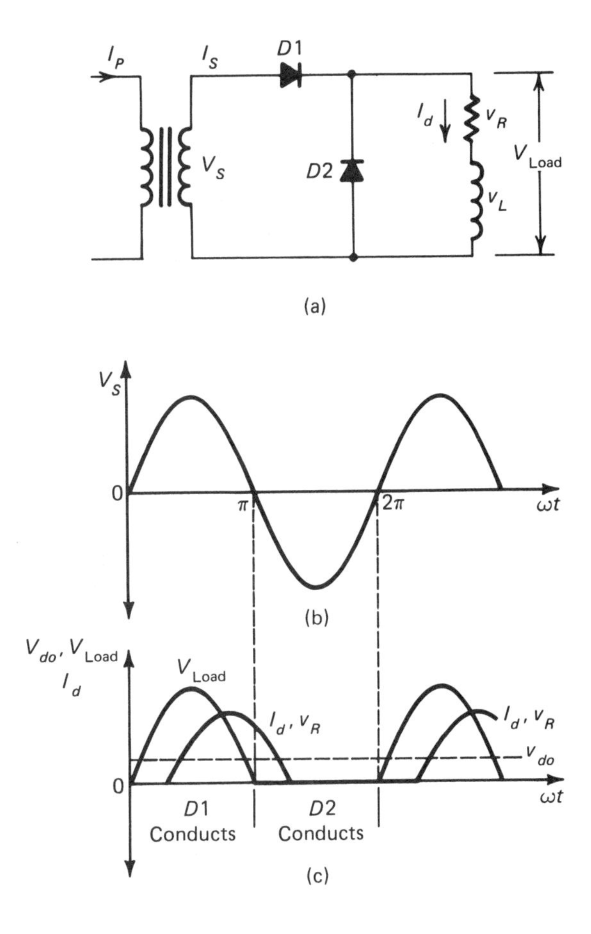

Figure 3-3. Single-phase rectifier with an *RL* load and freewheel diode: (a) schematic, (b) secondary voltage waveform, (c) load voltage and current waveforms.

3. the poor transformer utilization factor (0.286), and
4. the dc saturation of the transformer core.

3-1-3 Single-Phase Full-Wave Rectifiers

Most of the disadvantages of the half-wave rectifier can be overcome by single-phase full-wave rectification. There are two basic configurations of **full-wave rectifiers**; the single-phase midpoint rectifier and the single-phase bridge rectifier. Both rectifier configurations have essentially the same characteristics, but the peak inverse voltage and transformer ratings are different. Full-wave rectifiers, because of their higher average voltages and currents and reduced ripple factor, are more commonly used than are single-phase half-wave rectifiers.

3-1-3-1 Single-Phase Midpoint Full-Wave Rectifier

The **single-phase midpoint** or **push-pull rectifier** is shown in Figure 3-4(a). It is basically the same as the half-wave rectifier with an additional diode and transformer winding. From Figure 3-4(a), it can be seen that the transformer secondary is center-tapped and that the voltage between A and B is equal to the voltage between B and C, and the voltage from A to C is twice the voltage of the half-wave circuits. When point A is positive with respect to B, diode $D1$ conducts and $D2$ is reverse biased. When C becomes positive with respect to B, diode $D2$ conducts and $D1$ commutates off. The resulting output voltage waveform has a higher average value, a ripple frequency of $2f$, where f is the ac supply frequency, and a lower ripple factor.

The maximum peak diode current is

$$I_m = \frac{V_m}{R_S + R_L} \tag{3-36}$$

which is the same as for the half-wave rectifier.

The average load current is

$$I_d = \frac{2V_m}{\pi(R_S + R_L)} \tag{3-37}$$

$$= \frac{2I_m}{\pi} \tag{3-38}$$

twice the current in the half-wave rectifier.

The average output voltage is

$$V_{do} = I_d R_L = \frac{2V_m R_L}{\pi(R_S + R_L)} \simeq \frac{2V_m}{\pi} \tag{3-39}$$

and the average power delivered to the load is

$$P_{av} = V_{do} I_d = \left[\frac{2V_m R_L}{\pi(R_S + R_L)}\right] \left[\frac{2V_m}{\pi(R_S + R_L)}\right]$$

$$= \left(\frac{4}{\pi^2}\right) \left[\frac{V_m^2 R_L}{(R_S + R_L)^2}\right] \tag{3-40}$$

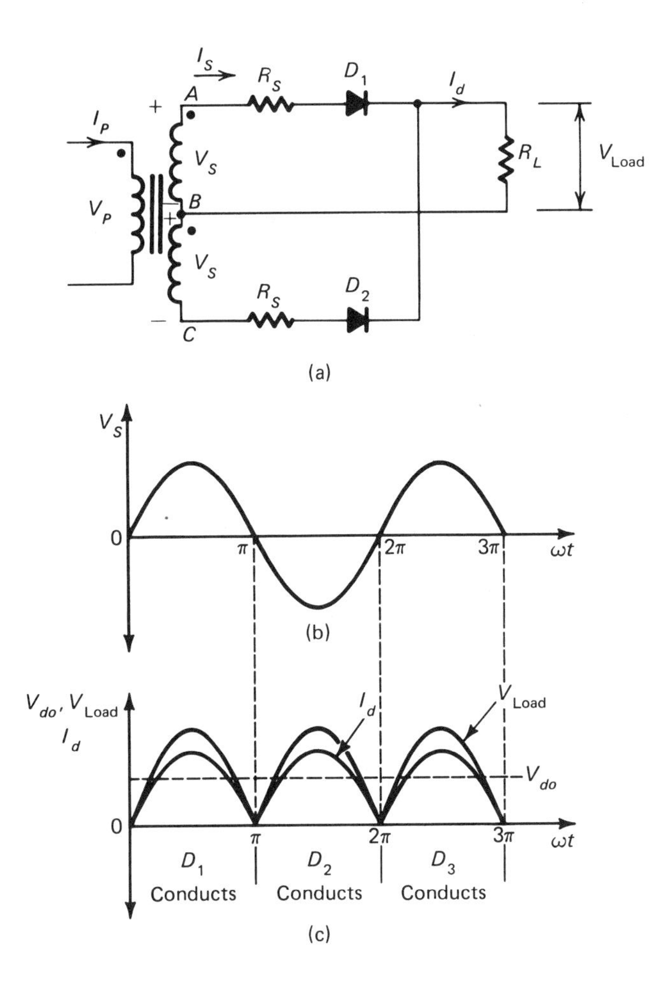

Figure 3-4. Single-phase full-wave midpoint rectifier: (a) schematic, (b) secondary voltage waveform, (c) load voltage and current waveforms resistive load.

The rectifier efficiency is

$$\eta = \frac{P_{av}}{P_{ac}} = \frac{\left[\frac{4}{\pi^2}\right]\left[\frac{V_m{}^2 R_L}{(R_S + R_L)^2}\right]}{\left[\frac{V_m}{\sqrt{2}}\right]\left[\frac{V_m}{\sqrt{2}(R_S + R_L)}\right]}$$

$$= \frac{8}{\pi^2}\left[\frac{1}{1 - R_S/R_L)}\right] \tag{3-41}$$

and the ripple factor is

$$\gamma = \left[\left(\frac{I_{rms}}{I_d}\right)^2 - 1\right]^{1/2}$$

$$= \left[\left(\frac{I_m}{\sqrt{2}}\right)\left(\frac{\pi}{2I_m}\right)^2 - 1\right]^{1/2}$$

$$= \left[\left(\frac{\pi}{2\sqrt{2}}\right)^2 - 1\right]^{1/2} = 0.4834 \text{ or } 48.34\% \qquad (3\text{-}42)$$

which is a significant improvement over the 121% of the half-wave rectifier.

The PIV rating of the diodes is twice the peak voltage V_{AB}. Therefore,

$$\text{PIV} = 2V_m \qquad (3\text{-}43)$$

since only one-half of the secondary voltage is applied across the load, but the full secondary voltage $2V_s$ is applied to each diode for one half-cycle.

For the rms secondary voltage V_s, the transformer characteristics for the single-phase midpoint rectifier are

$$V_s = \frac{V_m}{\sqrt{2}} = \frac{\pi V_{do}}{2\sqrt{2}} = 1.11V_{do} \qquad (3\text{-}44)$$

The rms secondary current I_s is

$$I_s = \frac{I_m}{2} = \frac{\pi I_d}{4} = 0.785I_d \qquad (3\text{-}45)$$

and the rms primary current I_p is

$$I_p = \sqrt{2}\left[0.785I_d\right]\left[N_s/N_p\right]$$
$$= 1.11\left[N_s/N_p\right] I_d \qquad (3\text{-}46)$$

As a result, the VA rating of the transformer secondary is

$$2V_sI_s = 2\left[1.11V_{do}\right]\left[0.785I_d\right]$$
$$= 1.74V_{do}I_d \qquad (3\text{-}47)$$

and the primary winding VA rating is

$$V_pI_p = [N_p/N_s][V_s][1.11][N_s/N_p]\,I_d$$
$$= 1.23V_{do}I_d \qquad (3\text{-}48)$$

The transformer utilization factor is found to be 0.636 as compared to 0.286 for the half-wave rectifier. Additionally, since current flows in the center-tapped secondary in opposite directions there is no dc saturation of the transformer core.

Example 3-2

Assume that in Figure 3-4 the voltage V_{AC} is 230V and R_L is 100Ω. Calculate: (1) V_{do}, (2) V_{rms} of the output voltage, (3) the ripple factor, (4) the maximum diode current, (5) the average diode current, and (6) the diode PIV.

Solution:

(1) From Equation (3-39)

$$V_{do} = \frac{2V_m}{\pi}$$

where V_m is one-half of the maximum voltage of V_{AC}. Therefore,

$$V_m = \frac{\sqrt{2}\cdot 230}{2},$$

$$V_{do} = \left[\frac{2}{\pi}\right]\left[\frac{\sqrt{2}\cdot 230}{2}\right]$$
$$= 103.54\text{ V}$$

(2) $$V_{rms} = \frac{230}{2} = 115\text{V}$$

(3) $$E_{ac} = \sqrt{V_{rms}^{\,2} - V_{do}^{\,2}}$$
$$= \sqrt{115^2 - 103.54^2}$$
$$= 50.05\text{ V}.$$

Therefore, $$\gamma = \frac{50.05}{115} = 0.4352$$

(4) The maximum current per diode is:

$$I_m = \frac{V_m}{R_L} = \frac{162.63}{100} = 1.63 \text{ A}$$

(5) The diode only conducts for one half-cycle so that the average current passing through each diode is $I_d/2$. Therefore,

$$I_{av} = I_d/2 = \frac{V_{do}}{2R_L} = \frac{103.54}{2 \times 100} = 0.52 \text{ A}$$

(6) The diode PIV is:

$$\text{PIV} = 2V_m = 230\sqrt{2} = 325.27 \text{ V}$$

3-1-3-2 Single-Phase Full-Wave Bridge Rectifier

The single-phase **diode bridge rectifier** is illustrated in Figure 3-5(a). The major differences of the bridge rectifier as compared to the midpoint rectifier are as follows:

1. There are always two diodes in conduction at the same time, resulting in a device voltage drop which is double that of the midpoint rectifier.
2. The peak inverse voltage applied to the rectifier is the peak-applied voltage, and therefore the peak inverse voltage applied to each diode is V_m or $1.57V_{do}$.

Equations (3-36) through (3-42) apply to the bridge rectifier; however, the transformer characteristics differ as follows:
The secondary rms voltage is

$$V_s = \frac{V_m}{\sqrt{2}} = \frac{\pi V_{do}}{2\sqrt{2}} = 1.11V_{do} \tag{3-49}$$

while the secondary rms current is

$$I_s = \frac{I_m}{2} = \frac{\pi I_d}{2\sqrt{2}} = 1.11I_d \tag{3-50}$$

The peak or maximum value of the secondary current is

$$I_{sm} = \frac{\pi I_d}{2} = 1.57I_d \tag{3-51}$$

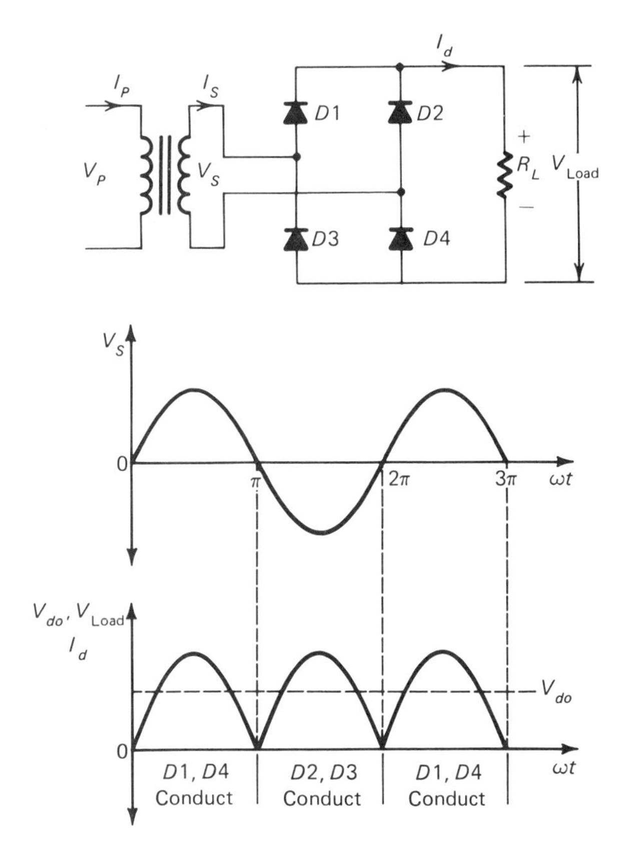

Figure 3-5. Single-phase full-wave bridge rectifier: (a) schematic, (b) input voltage waveform, (c) load voltage and load current waveforms resistive load.

The primary rms current is

$$\begin{aligned} I_p &= I_s \left[N_s/N_p\right] \\ &= 1.11 I_d \left[N_s/N_p\right] \end{aligned} \tag{3-52}$$

and the VA rating of the secondary winding is

$$V_s I_s = \left[1.11 V_{do}\right]\left[1.11 I_d\right] = 1.23 V_{do} I_d \tag{3-53}$$

and the primary winding VA is

$$\begin{aligned} V_p I_p &= \left[1.11 V_{do}\right]\left[N_p/N_s\right]\left[1.11 I_d\right]\left[N_s/N_p\right] \\ &= 1.23 V_{do} I_d. \end{aligned} \tag{3-54}$$

Summarizing, midpoint rectifiers require a transformer with a center-tapped secondary, each half of which conducts for half the time. Also the PIV across the rectifier is high. The bridge rectifier requires two extra diodes which increase cost and losses, but if the bridge rectifier is transformer fed it does not require a specially rated transformer. The bridge rectifier can also be directly supplied from the ac source without the use of a transformer.

Both full-wave rectifier configurations are significantly more efficient than half-wave rectifiers, and the ripple factor is reduced from 121% to 48% without external filters, although in many cases even the reduced ripple factor is not acceptable and filtering is incorporated. In most industrial applications single-phase full-wave rectifiers are not used for loads greater than 2 kW.

3-1-3-3 Single-Phase Full-Wave Midpoint Rectifier With An RC Load

Referring to Figure 3-6(a) and (b) and assuming that the voltage at point A is zero and increasing positively, then diode $D1$ is forward biased and conducts, and the instantaneous voltage v_s is applied across the parallel combination of R_L and C from 0 to $\pi/2$. At $\pi/2$ the capacitor C is fully charged to the peak value of v_s. After $\pi/2$ the instantaneous voltage v_s starts decreasing, and the voltage across the capacitor also starts decreasing. However, the rate of decrease of the capacitor voltage is determined by the time constant R_LC of the load. If the time constant is sufficiently large, the instantaneous voltage v_s will decrease faster than the capacitor voltage so that, at some time t_1, the capacitor voltage will be greater than the input voltage. At this point the diode $D1$ becomes reverse biased and ceases conduction. However, the capacitor discharge current supplies current to the load until time t_2. At π the instantaneous voltage at point C is becoming positive and at time t_2 the instantaneous voltage v_s is greater than the capacitor voltage. $D2$ becomes forward biased and remains in conduction until time t_3, at which point the capacitor voltage is greater than the input voltage v_s and $D2$ is reverse biased and ceases conduction.

This sequence of the diodes alternately conducting continues until power is removed. The current carried by the diodes supplies not only load current I_d but capacitor-charging current I_c as well. The magnitude of I_c is proportional to capacitance C. As a result, as capacitance C is increased the current through the diode increases but the conduction interval decreases. The average current drawn from the ac source equals the average load current I_d, and since the conduction time of the diodes is relatively small, the peak diode current must be substantially greater than the peak load current. It should also be noted that unlike the rectifier with a resistive or, as will be seen, with an RL load the di-

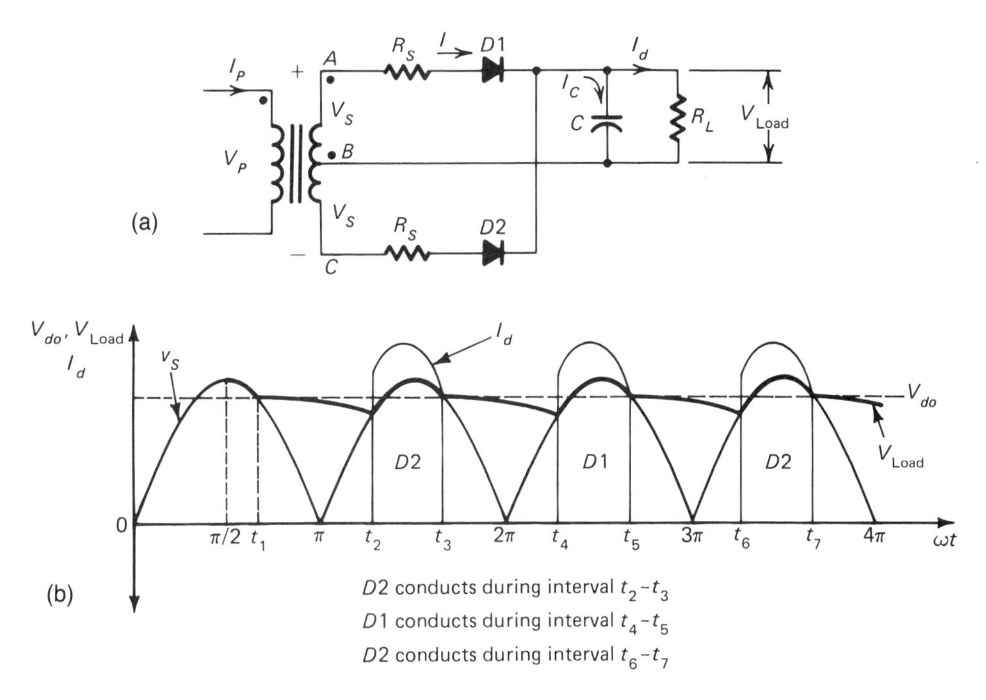

Figure 3-6. Single-phase full-wave midpoint rectifier with an *RC* load: (a) schematic, (b) load voltage and current waveforms.

odes commutate off prior to the voltage zero crossings.

It can be seen from Figure 3-6(b) that the capacitor is charged twice per cycle of the input source voltage, and intuitively the average load voltage V_{do} will be greater than with a resistive load. Mathematically V_{do} can be obtained based on the following assumptions: (1) the discharge time constant is greater than the periodic time of the output waveform ($1/f$ for a half-wave rectifier and $1/2f$ for a full-wave rectifier), i.e., $R_LC \gg T$; (2) the capacitor charges instantly to the peak secondary voltage V_{sm}; and (3) the capacitor discharges linearly, that is, in effect the capacitor is supplying the load current and as a result the output voltage waveform is triangular (see Figure 3-7).

From Figure 3-7 the average value of the dc load current I_d will be the average value of the capacitor discharge current over time interval T. As a result, the charge lost by the capacitor is:

$$Q_{\text{discharge}} = I_d T \tag{3-55}$$

and by assumption 2 this charge is replaced instantly. The voltage variation across the capacitor δV_{do} is the peak-to-peak ripple voltage.

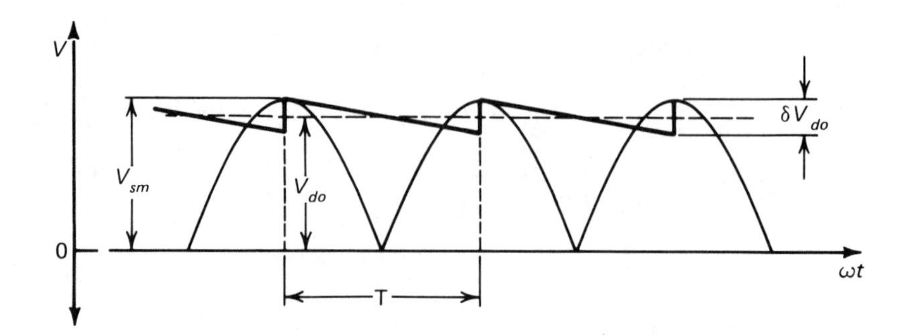

Figure 3-7. Determination of the average output voltage from a full-wave rectifier.

Then from $Q = CV$, the charge supplied to the capacitor is equal to the discharge $Q_{\text{discharge}}$; that is,

$$Q_{\text{charge}} = \delta V_{do} C \tag{3-56}$$

therefore,

$$\delta V_{do} C = I_d T$$

which yields

$$\delta V_{do} = \frac{I_d T}{C} \tag{3-57}$$

Since $T = \frac{1}{2}f$ for a full-wave rectifier, then

$$\delta V_{do} = \frac{I_d}{2fC}, \tag{3-58}$$

and

$$V_{\text{rms (ripple)}} = \frac{\delta V_{do}}{2\sqrt{3}} = V_{\text{ac}} \tag{3-59}$$

but

$$I_d = \frac{V_{do}}{R_L}$$

Therefore,

$$V_{\text{ac}} = V_{\text{rms (ripple)}} = \frac{I_d}{4\sqrt{3}fC}$$

$$= \frac{V_{do}}{4\sqrt{3}fR_L C} \tag{3-60}$$

By definition,

$$\text{ripple factor} \quad \gamma = \frac{V_{\text{rms (ripple)}}}{V_{do}} = \frac{V_{ac}}{V_{do}}$$

Therefore,

$$\gamma = \frac{1}{4\sqrt{3}fR_L C} \tag{3-61}$$

The average value of the output voltage is

$$V_{do} = V_{sm} - \frac{\delta V_{do}}{2}$$

but

$$\delta V_{do} = \frac{I_d}{2fC} \quad \text{and} \quad I_d = \frac{V_{do}}{R_L}$$

Therefore,

$$\begin{aligned} V_{do} &= V_{sm} - \frac{I_d}{4fC} \\ &= V_{sm} - \frac{V_{do}}{4fR_L C} \\ &= V_{sm}\left[\frac{4fR_L C}{4fR_L C + 1}\right] \end{aligned} \tag{3-62}$$

In the case of the half-wave rectifier with a capacitor input filter the equivalent relationship for the output voltage is

$$V_{do} = V_{sm}\left[\frac{2fR_L C}{2fR_L C + 1}\right] \tag{3-63}$$

and

$$\gamma = \frac{1}{2\sqrt{3}fR_L C} \tag{3-64}$$

From Equations (3-62) and (3-63) it can be seen that the dc output voltage of a full-wave rectifier with an RC load is significantly greater than that of a half-wave rectifier with an RC load. In addition, from Equations (3-61) and (3-64), the amplitude of the ripple component of voltage is half that of the half-wave rectifier; the ripple frequency is $2f$, or twice that of the half-wave rectifier.

Example 3-3

For the full-wave rectifier shown in Figure 3-6, $R_L = 500\Omega$, $C =$

100F, $V_{sm} = 163$ V, and $f = 60$ Hz. Calculate: (1) the average dc voltage supplied to the load; (2) the ripple factor; and (3) the power delivered to the load.

Solution:

(1) From Equation (3-62)

$$V_{do} = V_{sm}\left[\frac{4fR_LC}{4fR_LC + 1}\right]$$

$$= 163\left[\frac{4 \times 60 \times 500 \times 100 \times 10^{-6}}{4 \times 60 \times 500 \times 100 \times 10^{-6} + 1}\right]$$

$$= 163 \times 0.9231$$

$$= 150.46 \text{ V}$$

(2) From Equation (3-61)

$$\gamma = \frac{1}{4\sqrt{3}fR_LC}$$

$$= 0.0481 \text{ or } 4.81\%$$

(3) From Equations (3-13) and (3-60)

$$V_{rms} = \sqrt{V_{do}^2 + V_{ac}^2}$$

$$= \sqrt{150.46^2 + 7.24^2}$$

$$= 150.63 \text{ V}$$

$$P_{dc} = \frac{(V_{rms})^2}{R_L}$$

$$= \frac{150.63^2}{500} = 45.38 \text{ W}$$

3-1-3-4 Single-Phase Full-Wave Midpoint Rectifier with an RL Load

In Section 3-1-2 the effects of an RL load on a half-wave rectifier were studied, and it was noted that the load current continued to flow for a period of time after the diode was reverse biased, with a resultant reduction in the amplitude of the output dc voltage. From Figure 3-8 it can be seen that when $\omega t = 0$, the applied voltage v_s is increasing in the positive sense until $\omega t = \pi/2$. During this time interval the load reactance is opposing the flow of current and energy is being stored in the magnetic field surrounding the inductance. The voltage across the resistance portion of the load v_R is the difference voltage between the applied voltage v_s and the induced voltage v_L across the inductance. As the applied voltage decreases, then by Lenz's Law the induced voltage opposes any decrease in load current; that is, between $\omega t = \pi/2$ and π the applied voltage and the induced voltage act together to oppose the reduction in current.

As a result it can be seen in Figure 3-8(b) that the load current is a maximum when $v_s = 0$, and a minimum when $V_s = V_{sm}$, and then increases to a maximum when $v_s = 0$, the process being repeated for every half-cycle of the rectified sine wave. It is to be noted that the load current never decreases to zero because the energy stored in the magnetic field acts to maintain current flow.

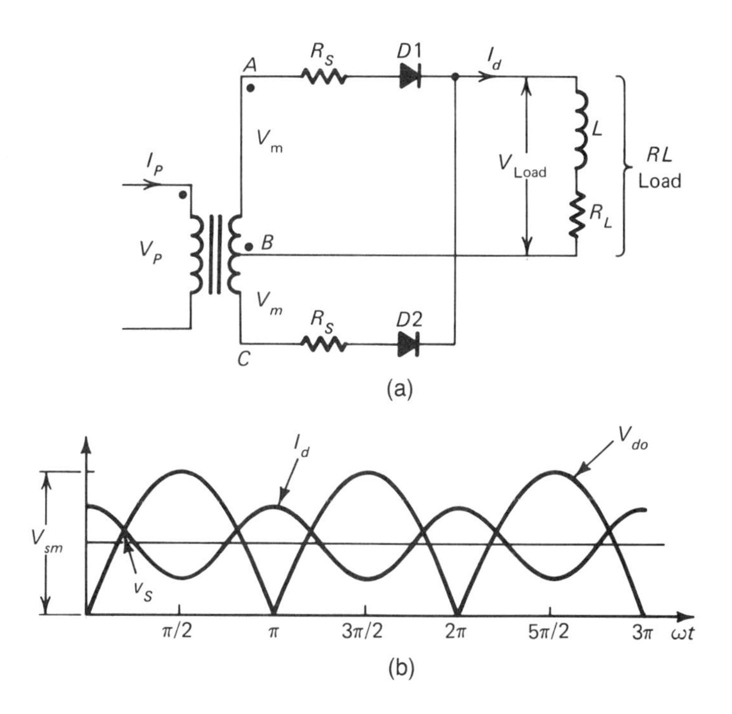

Figure 3-8. Single-phase full-wave midpoint rectifier with an *RL* load: (a) schematic, (b) load voltage and load current waveforms.

The output dc voltage V_{do} is

$$V_{do} = \frac{2V_{sm}}{\pi} \tag{3-65}$$

and the average load current I_d will be

$$I_d = \frac{2V_{sm}}{\pi R_L} \tag{3-66}$$

Assuming that L is sufficiently large, I_d will remain nearly constant, and the power P_{dc} applied to the load will be

$$P_{dc} = V_{do} I_d = \frac{4V^2_{sm}}{\pi^2 R_L} \tag{3-67}$$

The dc rectifier voltage has a nonsymmetrical ac component, which by Fourier analysis can be shown to contain only even harmonics, the most important of which has an amplitude of $4V_{sm}/3\pi$. It should be noted that the fundamental is not present in the output wave.

The ac current will then be

$$I_{ac} = \frac{\dfrac{4V_{sm}}{3\pi} \dfrac{1}{\sqrt{2}}}{\sqrt{R_L^2 + (2\omega L)^2}} \tag{3-68}$$

$$= \frac{0.3001}{\sqrt{R_L^2 + (2\omega L)^2}}$$

By definition,

ripple factor $$\gamma = \frac{I_{ac}}{I_d}$$

$$= \frac{0.3001}{\sqrt{R_L^2 + (2\omega L)^2}} \left(\frac{\pi R_L}{2V_{sm}}\right)$$

$$= \frac{0.4714\, R_L}{\sqrt{R_L^2 + (2\omega L)^2}}$$

$$= \frac{0.4714}{\sqrt{1 + \left(\dfrac{2\omega L}{R_L}\right)^2}} \tag{3-70}$$

Example 3-4

For the full-wave rectifier shown in Figure 3-8, $R_L = 500\ \Omega$, $V_{sm} = 163$ V, and $f = 60$ Hz. Calculate the value of inductance to produce a ripple factor of 5%.

Solution:

From Equation (3-70)

$$\gamma = \frac{0.4714}{\sqrt{1 + \left(\frac{2\omega L}{R_L}\right)^2}}$$

Therefore,

$$0.05 = \frac{0.4714}{\sqrt{1 + \left(\frac{120\pi L}{500}\right)^2}}$$

then squaring both sides

$$0.0025 = \frac{0.2222}{1 + \left(\frac{120\pi L}{500}\right)^2}$$

Therefore, $$0.2222 = 0.0025 + 0.0025\left(\frac{120\pi L}{500}\right)^2$$

Then $$\frac{120\pi L}{500} = \sqrt{\frac{0.2222 - 0.0025}{0.0025}}$$

$$0.754\,L = 9.37$$

$$L = \frac{9.37}{0.754} = 12.43\text{ H}$$

3-2 HARMONIC PRODUCTION

From Figures 3-1, 3-2, 3-3, 3-4, 3-5, 3-6, and 3-8 it can be seen that the output waveform consists of a dc voltage with a superimposed ac ripple voltage. In most cases the ripple voltage is not sinusoidal and in fact consists of a number of **harmonics** of different frequencies and amplitudes which are dependent upon both the frequency of the supply voltage and the type of rectifier connection. The number and amplitude of the harmonics can be calculated from a Fourier analysis of the individual output waveforms. In the case of a single-phase full-wave rectifier supplied from a 60 Hz source, the predominant harmonic is the second (120 Hz) with the fourth (240 Hz), the sixth (360 Hz), the eighth (480 Hz), etc., harmonics being present. However, as the order of the harmonics increases, the amplitude of the harmonic component decreases. The harmonics present in the output waveform may be removed or greatly reduced by the use of filter circuits.

3-2-1 Filter Circuits

From our review of RC and RL loads it has been seen that an inductance L smooths the current but has little effect on the load voltage, while the capacitance smooths the voltage but introduces short duration current spikes.

To obtain improved rejection of the harmonics generated by the rectification process, it is necessary to use more sophisticated filter circuits. In the **choke input** or **L-section filter** shown in Figure 3-9(a), the inductor L presents a high impedance ($2\pi fL$) to the harmonic currents while offering little opposition to dc currents. At the same time the capacitor C bypasses the remaining ripple current away from the load. Under light load conditions, the inductor has minimal effect, the circuit acts as a **capacitor input filter**, and the diode currents will be discontinuous during the capacitor discharge.

During the capacitor discharge interval, the output voltage will drop off rapidly, as shown in Region 1 of Figure 3-9(b), until the average current reaches some value I_k. Region 1 represents the discontinuous conduction region of the rectifier. Region 2 represents the continuous conduction mode of the rectifier resulting from the fact that the increased current flow causes more energy to be stored in the magnetic field of the inductor. This in turn results in an increased diode conduction angle so that diode current only ceases when the diode is reverse biased. The problem of discontinuous conduction with its rapid drop-off of output voltage can be minimized by shunting a bleeder resistance across the load. The bleeder resistance is selected to ensure that a current equal or greater than I_k flows through the bleeder resistance. With the bleeder resistance the rate of decrease of output voltage is decreased, and most of the drop-off in the output voltage is attributable to the leakage reactance and resistance of the choke and input transformer.

A further improvement in the performance of the choke input filter can be obtained by placing another capacitor shunted across the input

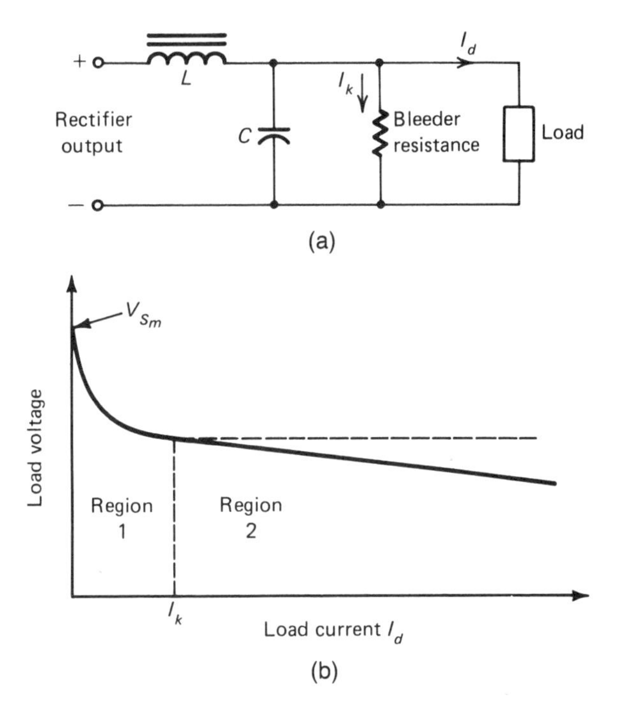

Figure 3-9. Choke input filter: (a) schematic, (b) load characteristic.

terminals. This type of filter is called a π **filter** and is illustrated in Figure 3-10(a). In the π filter a significant proportion of thediode current flows into capacitor $C1$, the ripple voltage output across $C1$ is then smoothed by L and $C2$. The resulting output voltage is greater than for the choke input filter, with the additional benefit of a further reduction in the ripple content and additionally discontinuous conduction is eliminated. The penalty introduced by the added capacitor is that the peak diode currents have increased.

In applications where the dc current is small, the inductor in Figure 3-10(a) may be replaced by a resistor to reduce costs; however, this is achieved at the expense of a greater IR drop and increased power loss.

The filters considered to date are all mounted on the dc side of the rectifier and their function is to reduce harmonics resulting from the rectification process. It must be appreciated that harmonic currents are also produced on the ac side as a result of the rectifier action; the magnitude and order of these harmonics are dependent upon the type of rectifier connection. Usually the order of the harmonics is identified, and resonant LC filters are connected to the ac side for each of the major harmonics.

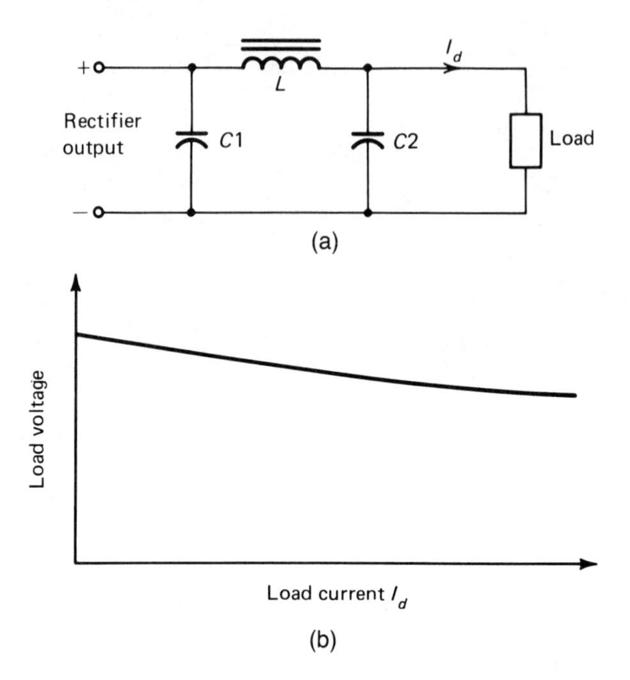

Figure 3-10. The π filter: (a) schematic, (b) load characteristic.

3-3 VOLTAGE MULTIPLIERS

There are a number of requirements for dc power supplies to provide voltages greater than the ac source voltage, but with a low load current demand. For example, voltage levels in the kV region are required in an oscilloscope.

This requirement can be met by **voltage-multiplying circuits**. These circuits all basically operate on the same principle, namely, capacitors are charged on alternate half-cycles of the ac source voltage and arranged so that their voltages are added together to form the output. By suitably arranging the diodes and capacitors, output voltages approaching two, three, and four times the peak amplitude of the ac source voltage may be obtained.

3-3-1 Voltage Doubling Circuits

As illustrated in Figure 3-11, there are several methods of doubling the output voltage. Figure 3-11(a) shows what is termed the conventional or **full-wave doubler**. When diode $D1$ is forward biased, capacitor $C1$ charges to V_m. During the next half-cycle $D2$ is forward biased and capacitor $C2$ charges to V_m. The output is taken across $C1$ and $C2$ in series and is approximately equal to $2V_m$. The resistance R is connected in parallel with the capacitors to provide a discharge path for the capacitors after the ac source voltage is removed, thus reducing the risk of injury to users. The resistance R is typically in the range of 10 to 20 MΩ. After $C1$ and $C2$ have been initially charged, current will flow through the diodes when the voltage across the diode is greater than the voltage across the capacitor which is in series with it. This occurs only for a short time interval during each half-cycle, and as a result the diode peak currents will be quite large. The doubler is limited to applications where the load current is in the milliampere range. If the load current demand is too great the capacitors will discharge excessively, the output voltage will drop off rapidly, and the peak diode currents will increase substantially.

The ripple frequency of the output dc voltage is twice the source frequency and since the circuit is similar to that of the full-wave rectifier, it is called a full-wave doubler. Capacitors $C1$ and $C2$ must have volt-

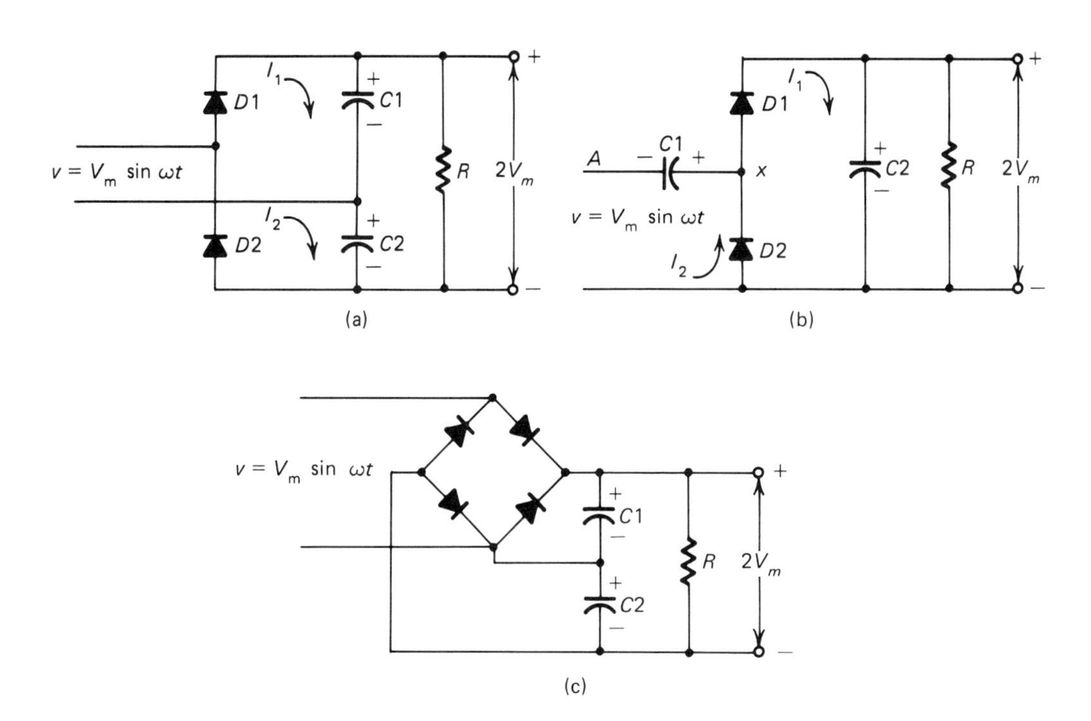

Figure 3-11. Voltage doubling circuits: (a) conventional doubler, (b) cascade doubler, (c) bridge doubler.

age ratings in excess of V_m, and the voltage and peak inverse ratings of diodes $D1$ and $D2$ must be greater than $2V_m$.

Another type of voltage doubler known as the **cascade doubler** is shown in Figure 3-11(b). During the negative half-cycle, capacitor $C1$ is charged to V_m via $D2$; thus point X is at V_m less the voltage drop across diode $D2$. During the next half-cycle, as the potential at point A becomes more positive, the potential at point X always remains positive with respect to A by the voltage across $C1$; when point A reaches V_m the potential at X is $2V_m$. During the positive half-cycle, diode $D2$ is reverse biased and diode $D1$ conducts, allowing capacitor $C1$ to discharge and charge capacitor $C2$ to $2V_m$. During the next negative half-cycle, the charge on $C1$ is replaced via diode $D2$, and $C2$ supplies current to the load. As before the load current must be limited to the milliampere range to prevent excessive discharging of the capacitors when charging currents are not flowing with a resulting decrease in the output voltage.

The cascade doubler is also called the **half-wave doubler** because diode $D1$ acts as a half-wave rectifier to the voltage whose peak value is doubled.

The output dc voltage has a ripple component of the same frequency as the ac source, which results in the ripple component having a greater amplitude than is the case with the full-wave doubler. In this circuit, capacitor $C1$ must have a voltage rating greater than V_m, and $C2$ must have a voltage rating greater than $2V_m$. Additionally the diodes must have voltage ratings greater than $2V_m$.

The major advantage of the cascade doubler is that the common connection between the input and output permits cascaded stages to be added to give higher voltage multiplications.

In the case of the bridge doubler circuit, shown in Figure 3-11(c), capacitor $C2$ is charged up to V_m during the negative half-cycle and current is supplied to the load. During the positive half-cycle the ac source voltage and the charge voltage of capacitor $C2$ are supplied to the output terminals to give a potential difference of approximately $2V_m$ across the output terminals, with a ripple frequency double that of the ac source. The voltage ratings of the diodes and capacitors must not be less than V_m. The bridge doubler has an improved voltage regulation as compared to both the full-wave and half-wave doublers. However, the improved regulation is achieved at the expense of increased diode losses because of the extra diodes as well as the increased cost resulting from the increased number of components.

The major disadvantage of all the doubler circuits is that the amplitude of the ripple component increases with increased current loading, thus limiting their use to low current applications.

3-3-2 Voltage Tripler Circuits

Two examples of **voltage tripler circuits** are illustrated in Figure 3-12. The **full-wave tripler** is shown in Figure 3-12(a). This circuit is essentially a cascade doubler combined with a half-wave rectifier. Diodes $D1$ and $D2$ together with capacitors $C1$ and $C2$ form the cascade doubler to give a voltage level of $2V_m$ across $C2$. $D3$ acts as the half-wave rectifier and charges capacitor $C3$ up to V_m. The outputs of the cascade doubler and the half-wave rectifier combine to produce an output voltage of $3V_m$ at the terminals. Capacitors $C1$ and $C3$ must have voltage ratings greater than V_m and the rating of $C2$ must be greater than $2V_m$. Diodes $D1$, $D2$, and $D3$ must have voltage ratings in excess of $2V_m$. The ripple frequency of the output voltage is twice that of the source frequency.

As was previously mentioned, the cascade doubler can be used to produce higher multiples of the source voltage by the addition of extra cascaded sections. A **cascade tripler** circuit is shown in Figure 3-12(b). During the positive half-cycle, $D1$ is forward biased and $C1$ is charged to V_m. During the negative half-cycle, $D1$ and $D3$ are reverse biased, $D2$ is forward biased, and capacitor $C3$ is charged to $2V_m$. The output voltage $3V_m$ is the sum of the voltages on capacitors $C1$ and $C3$.

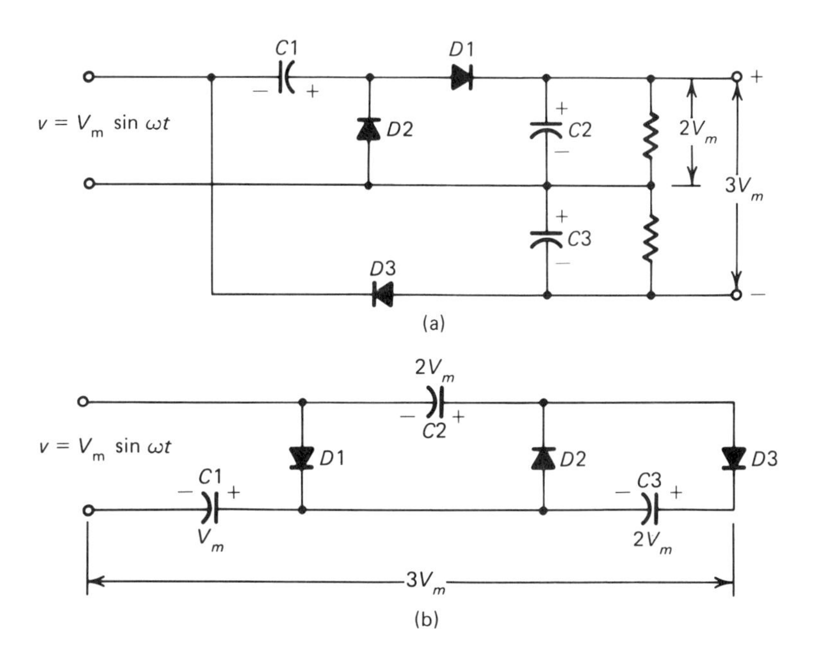

Figure 3-12. Voltage tripler circuits: (a) full-wave tripler, (b) cascade tripler.

The voltage ratings of capacitors $C2$ and $C3$ must be greater than $2V_m$ and the voltage rating of $C1$ must be greater than V_m. The output voltage has a ripple frequency the same as the ac source frequency, in common with all half-wave multiplier circuits. In addition the amplitude of the ripple content of the cascade tripler is greater than that of the full-wave tripler, while its regulation is poorer.

3-3-3 Voltage Quadrupling Circuits

The **full-wave quadrupler** illustrated in Figure 3-13(a) is effectively two cascade doublers in series. During the negative half-cycle, capacitors $C1$ and $C2$ are charged to V_m, and during the positive half-cycle they discharge in series with the source voltage to charge capacitors $C3$ and $C4$ to $2V_m$ to produce a potential of $4V_m$ across the output terminals. All the capacitors should have a voltage rating of at least $2V_m$, and the diodes must be able to withstand a peak inverse voltage greater than $2V_m$. Since this is a full-wave circuit it will have a reduced ripple content with a ripple frequency of twice the ac source frequency, and in common with all voltage multipliers the current drain must be limited to the milliampere range to prevent a considerable reduction in output voltage.

The **half-wave cascade quadrupler** shown in Figure 3-13(b) is basically the cascade tripler with an extra stage added. Capacitor $C1$ should have a voltage rating of at least V_m and capacitors $C2$, $C3$, and $C4$ should be rated for at least $2V_m$. The peak inverse voltage rating of all diodes must be greater than $2V_m$. This circuit will have a greater amplitude ripple voltage with a frequency equal to the source frequency and a poorer voltage regulation than the full-wave quadrupler.

3-4 MULTIPHASE RECTIFIERS

Even though the majority of rectifier circuits are single-phase, the practical upper limit of single-phase rectifiers is 2 kW, mainly because the amplitude of the ripple content is not acceptable. In nearly all industrial applications where high-power dc outputs with a low ripple content are required, **polyphase rectifiers** are used. The major reasons are that the ripple content of the output voltage is significantly reduced combined with a higher mean output dc voltage, and that in most industrial applications the incoming power to the plant is usually three-phase. Another important factor which determines the choice of rectifier configurations is the frequency of the ripple content; the high-

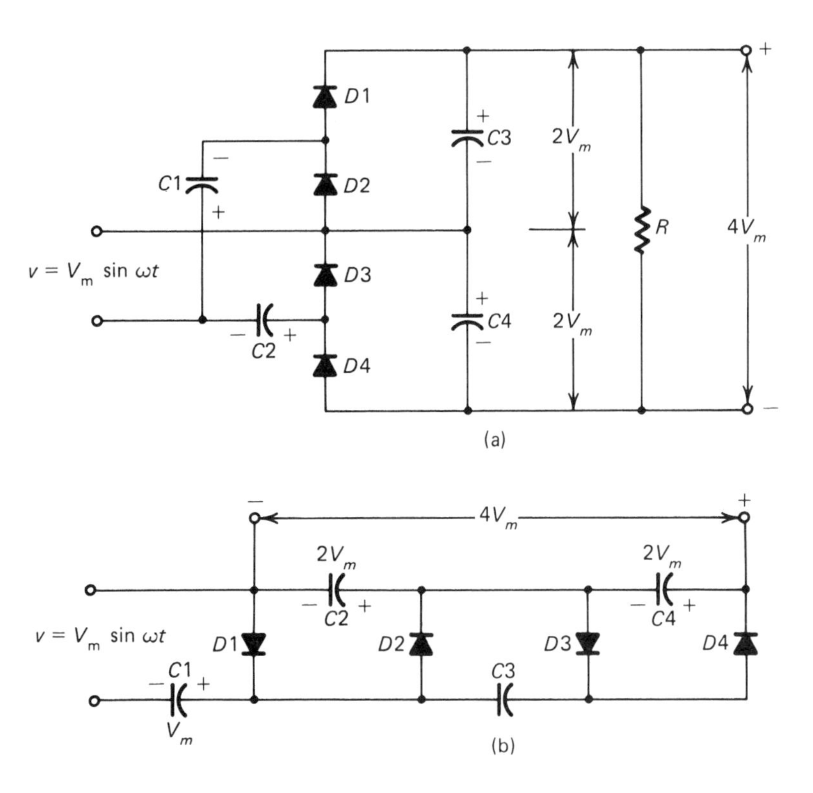

Figure 3-13. Voltage quadrupler circuits: (a) full-wave cascade, (b) half-wave cascade.

er the frequency the cheaper become any required filtering arrangements.

Rectifiers are also classified in terms of their **pulse number**, where the pulse number is the ratio of the fundamental ripple frequency present in the dc output to the frequency of the ac source voltage.

$$\text{Pulse Number } m = \frac{\text{fundamental dc output ripple frequency}}{\text{ac source frequency}} \qquad (3\text{-}71)$$

In terms of the pulse number, full-wave single-phase rectifiers are called two-pulse rectifiers, while three-phase rectifiers are either three-pulse or six-pulse rectifiers.

Prior to a study of three-phase rectifier circuits our analysis will be made easier by the use of the following simplifying assumptions:

1. Diodes do not have a voltage drop or permit leakage current to flow.
2. Turn-on and turn-off times of the diodes are almost instantaneous.
3. ac source voltage is sinusoidal.
4. dc load current is constant over each cycle; that is, the rectifier is connected to an ideal filter (infinite inductance).

3-4-1 Three-Phase Star Rectifier

The simplest three-phase rectifier configuration is the **three-phase star circuit**, which is a three-pulse rectifier. Shown in Figure 3-14(a), this configuration is a three-phase half-wave rectifier. The rectifier is supplied by a delta-star transformer arrangement. When the instantaneous voltage of phase a is greater than that of phases b and c, diode $D1$ is forward biased and conducts. As shown in Figure 3-14(b), this occurs between the points marked 30° and 150°. At 150° the instantaneous voltage of phase b becomes greater than that of phase a and diode $D1$ becomes reverse biased and commutates off as diode $D2$ becomes forward biased and turns on. At 270° the instantaneous voltage of phase c becomes greater than that of phase b, and diode $D2$ becomes reverse biased and commutates off. At the same time diode $D3$ becomes forward biased and turns on. At 390° or 30° diode $D3$ commutates off and $D1$ goes into conduction to repeat the cycle. It should be noted that the diodes assuming a pure resistive load conduct for 120°. The voltage V_{LOAD} seen by the load follows the top of the individual phase voltages and consists of a dc component V_{do} and a ripple component with a frequency of $3f$, where f is the frequency of the ac source voltage. The load current I_d remains relatively constant.

General equations have been developed as follows:

The output dc voltage V_{do} is

$$V_{do} = V_m \left(\frac{m}{\pi} \sin \frac{\pi}{m}\right) \tag{3-72}$$

and in the case of the three-pulse rectifier,

$$V_{do} = V_m \left(\frac{3}{\pi} \sin \frac{\pi}{3}\right) = 0.827\ V_m, \tag{3-73}$$

where V_m = the maximum value of the peak-to-neutral voltage, and m = the pulse number or number of diodes.

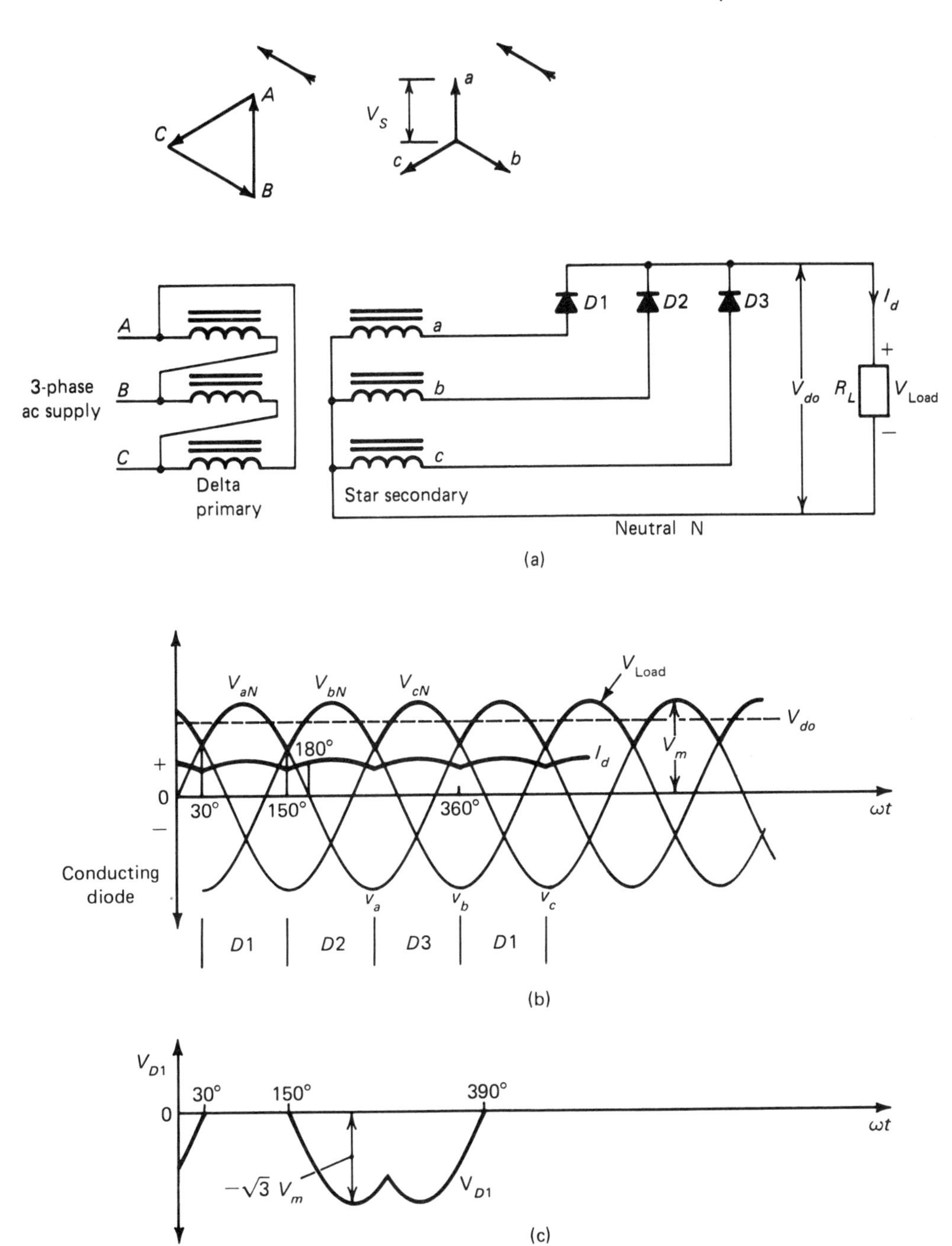

Figure 3-14. Three-phase star rectifier: (a) basic circuit, (b) phase and load voltage, and load current waveforms, (c) voltage waveform across $D1$.

And in turn $$V_{do} = 0.827\sqrt{2}V = 1.17V \tag{3-74}$$

where V = rms value of the peak-to-neutral voltage,

and $$V = V_{do}/1.17 = 0.855V_{do} \tag{3-75}$$

The rms value of the transformer secondary current I_s is

$$I_s = I_m \left[\frac{1}{2\pi}\left(\frac{\pi}{m} + \frac{1}{2}\sin\frac{2\pi}{m}\right)\right]^{1/2} \tag{3-76}$$

and $$I_m = \frac{\pi I_d}{m \sin(\pi/m)} \tag{3-77}$$

where I_m = peak diode current, and I_d = average value of dc current.

Therefore, from Equation (3-77)

$$I_d = \frac{mI_m}{\pi}\cdot\sin\frac{\pi}{m} = \frac{3I_m}{\pi}\sin\frac{\pi}{3} \tag{3-78}$$

$$= 0.827I_m$$

And substituting for I_m in Equation (3-76),

$$I_s = \frac{I_d\pi\left[(\frac{1}{2}\pi)(\pi/m + \frac{1}{2}\sin 2\pi/m)\right]^{1/2}}{m\sin(\pi/m)} \tag{3-79}$$

$$= I_m\left[\tfrac{1}{2}\pi\ (\pi/3 + \tfrac{1}{2}\sin 2\pi/3)\right]^{1/2}$$

$$= 0.486\, I_m \tag{3-80}$$

When $m \geq 3$, Equation (3-79) is approximately equal to

$$I_s = \frac{I_d}{\sqrt{m}} = \frac{I_d}{\sqrt{3}} = 0.586\, I_d \tag{3-81}$$

From Equation (3-78)

$$I_m = 1.21\, I_d \tag{3-82}$$

and the PIV applied to the diodes is

$$\text{PIV} = \sqrt{3}V_m = \sqrt{3}V_{do}/0.827$$
$$= 2.094V_{do} \tag{3-83}$$

By Fourier analysis it can be shown that the fundamental ripple factor for a multiphase rectifier is

$$\gamma = \frac{V_{ac}}{\sqrt{2}V_{do}} = \frac{\sqrt{2}}{[m^2 - 1]} \tag{3-84}$$

where V_{ac} = peak amplitude of the ripple voltage

$$\gamma = \frac{\sqrt{2}}{3^2 - 1} = \frac{\sqrt{2}}{8} = 0.177 \tag{3-85}$$

The efficiency of rectification η is the ratio of the ac power input to the dc power output. Therefore, the rms value of the total load current is

$$I_{rms} = \sqrt{3}I_s = \sqrt{3}\,(0.586)I_d$$
$$= 1.015I_d. \tag{3-86}$$

Therefore,

$$\eta = \frac{P_{av}}{P_{ac}}(100)$$
$$= \frac{I_d^2 R_L}{(1.015I_d)^2 R_L} \times 100$$
$$= 97.07\% \tag{3-87}$$

The rms value of the ac component of the transformer secondary current is

$$\sqrt{I_s^2 + I_{av}^2} = \sqrt{(0.586I_d)^2 - (I_d/3)^2}$$

$$= 0.484I_d, \tag{3-88}$$

and the resulting primary phase current I_p is

$$I_p = 0.484I_d\,\left[N_s/N_p\right] \tag{3-89}$$

From the above the secondary VA rating of the transformer is

$$\begin{aligned} 3V_sI_s &= (3)(0.855)V_{do}(0.586)I_d \\ &= 1.50V_{do}I_d \end{aligned} \tag{3-90}$$

and the VA rating of the transformer primary is

$$\begin{aligned} 3V_pI_p &= (3)(N_p/N_s)V_s(0.484)(N_s/N_p)I_d \\ &= (3)(0.850)V_{do}(0.484)I_d \\ &= 1.24V_{do}I_d \end{aligned} \tag{3-91}$$

As can be seen above, for the three-phase star rectifier the ripple factor has been reduced and the rectification efficiency improved as compared to the single-phase rectifier circuits. In addition, the mean dc voltage V_{do} has increased. Also the diodes are subjected to a peak inverse voltage of $\sqrt{3}V_m$ as against $2V_m$ for the single-phase rectifier. The major problem with this circuit is that only one phase at a time of the transformer secondary is carrying load current, and as a result the primary windings must be delta connected.

When supplying an *RL* load, the effect of the series inductance is to produce a more constant load current and reduce the amplitude of the ac ripple voltage component. The reduction is defined by the smoothing factor δ, where

$$\delta = \sqrt{1 + (X_L/R)^2} = \sqrt{1 + Q^2} \tag{3-92}$$

Therefore, from Equation (3-85),

$$\gamma = \frac{0.177}{\sqrt{1 + Q^2}},$$

from which it can be seen that if X_L is large, the load current I_d is approximately constant.

It can be shown that the secondary transformer utilization factor of a half-wave rectifier is

$$\text{TUF}_s = \frac{\frac{2m}{\pi}\left(\sin^2\frac{\pi}{m}\right)}{\sqrt{\pi\left(\frac{\pi}{m} + \sin\frac{\pi}{m}\cos\frac{\pi}{m}\right)}} \tag{3-93}$$

which for the 3-pulse rectifier becomes

$$TUF_s = \frac{\frac{2 \times 3}{\pi}\left(\sin^2 \frac{\pi}{3}\right)}{\sqrt{\pi\left(\frac{\pi}{3} + \sin \frac{\pi}{3} \cos \frac{\pi}{3}\right)}}$$

$$= \frac{1.4324}{2.1564} = 0.6643 \tag{3-94}$$

Example 3-5

A three-phase star rectifier delivers 25 A to a resistive load at 125V. Determine:

1. the average diode current;
2. the PIV per diode;
3. the VA rating of the transformer secondary;
4. the effective secondary voltage of the transformer secondary;
5. the effective phase current of the transformer secondary; and
6. the amplitude of the ripple voltage.

Solution:

(1) The average diode current is

$$\frac{I_d}{3} = \frac{25}{3} = 8.33 \text{ A}$$

(2) The peak inverse voltage per diode is $\sqrt{3}V_m$; thus it may be found from Equation (3-73) where

$$V_{do} = 0.827V_m$$

Therefore,

$$V_m = \frac{125}{0.827} = 151.15 \text{ V}$$

and

$$\text{PIV} = \sqrt{3} \times 151.15 = 261.8 \text{ V}$$

(3) The VA rating of the transformer secondary is

$$3 \times (V_m/\sqrt{2})(I_s)$$
$$= \frac{3V_m}{\sqrt{2}} \times (0.586 I_d)$$
$$= \frac{3 \times 151.15}{\sqrt{2}} \times 0.586 \times 25$$
$$= 4.70\,\text{kVA}$$

(4) The effective (rms) voltage rating of the transformer secondary is

$$V_s = V_m/\sqrt{2}$$
$$= 151.15/\sqrt{2}$$
$$= 106.88\text{ V}$$

(5) From Equations (3-80) and (3-78), the rms secondary current per phase is

$$I_s = 0.486 I_m$$
$$I_d = 0.827 I_m$$
$$I_s = \frac{0.486}{0.827} I_d$$
$$= \frac{0.486}{0.827} \times 25 = 14.7\text{A}$$

(6) From Equation (3-84), the amplitude of the ripple voltage is

$$\gamma = \frac{V_{ac}}{\sqrt{2} V_{do}}$$

Therefore,

$$0.177 = \frac{V_{ac}}{\sqrt{2} \times 125}$$

and

$$V_{ac} = 0.177 \times \sqrt{2} \times 125$$
$$= 31.29\text{V}$$

3-4-2 Three-Phase Interstar Rectifier (Three-Phase Zigzag Rectifier)

The **three-phase interstar** circuit is a modification of the three-phase star circuit and is designed to prevent dc saturation of the transformer core that is inherent in the star circuit. The transformer may be delta or star-zigzag connected. The zigzag connection requires two separate

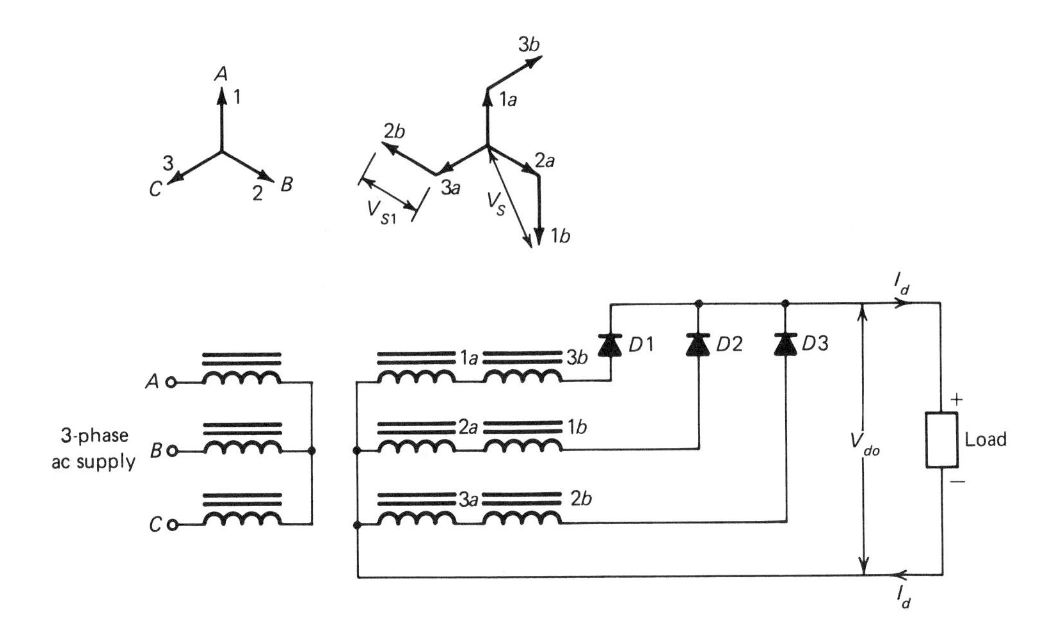

Figure 3-15. The three-phase interstar or zigzag connected rectifier.

identical secondary windings for each phase, and for the same output voltage requires more copper than a single secondary winding. It prevents dc saturation of the transformer core by permitting equal and opposite currents to flow in each half of the secondary winding. The basic rectifier circuit is shown in Figure 3-15.

It can be seen that each diode is turned on when it becomes forward biased; that is, 30° after the phase voltage crosses the zero axis. As a result, the ac input voltage with the greatest instantaneous value is applied to the load terminals. The conduction period of each diode is 120°, and it blocks reverse voltage for 240°, the maximum blocking voltage being equal to the line-to-line voltage of the transformer secondary. The frequency of the ac ripple component is $3f$, where f is the frequency of the ac source voltage.

The relationships developed for the three-phase star circuit all apply to the interstar rectifier except for those applying to the transformer ratings. The rating of the primary transformer winding is reduced because each primary phase winding carries current twice in each cycle; however, the secondary rating of the zigzag connected windings is increased because the line-to-neutral voltage V_s is $\sqrt{3}/2$ greater than that of the star-connected winding. As a result, the secondary rating is $2/\sqrt{3}$ times greater than the star-connected circuit.

The ratings are obtained as follows:

$$V_s = \sqrt{3}V_{s1}$$
$$V_{s1} = V_s/\sqrt{3} = V_{do}/2$$

The primary phase winding is carrying current twice per cycle and is

$$I_p = \sqrt{2}\left[N_s/N_p\right] I_s$$
$$= 0.825\left[N_{s1}/N_p\right] I_d \tag{3-95}$$

and the secondary VA rating is

$$6V_{s1}I_s = 6\left[V_{do}/2\right]\left[0.586I_d\right]$$
$$= 1.75V_{do}I_d \tag{3-96}$$

while the primary VA rating is

$$3V_pI_p = 3\left[N_p/N_{s1}\right] V_s\left[0.825\right]\left[N_{s1}/N_p\right] I_d$$
$$= 3\left[V_{do}/2\right]\left[0.825I_d\right]$$
$$= 1.24\, V_{do}I_d \tag{3-97}$$

The ripple factor and transformer utilization factor remain unchanged.

3-4-3 Three-Phase Full-Wave Bridge Circuit

Even though the ripple factor, and thus the amplitude of the ac ripple voltage, was considerably improved by the three-phase half-wave or three-pulse rectifiers, there are many applications where the amplitude of the ripple voltage must be still further reduced. This is accompanied by using the **three-phase full-wave bridge rectifier,** see Figure 3-16(a), or six-pulse rectifier. The three-phase bridge rectifier may be supplied directly from a three-phase source or it can be supplied by a delta-star, star-delta, or delta-delta transformer arrangement.

The operation of a three-phase bridge rectifier is probably easier to understand if it is considered to be two three-phase half-wave star rectifiers each with a neutral. Diodes $D1$, $D3$, and $D5$ form the positive group, and each diode conducts for 120° and blocks for 240°. The voltage at the cathodes of the diodes is positive with respect to the neutral and forms the positive group. Similarly the bottom group of diodes $D4$, $D6$, and $D2$ form the negative group, with each diode conducting for 120° and blocking for 240°; the voltage at the anodes of the diodes being negative with respect to the neutral. The two sets are shown in Figure 3-16(b).

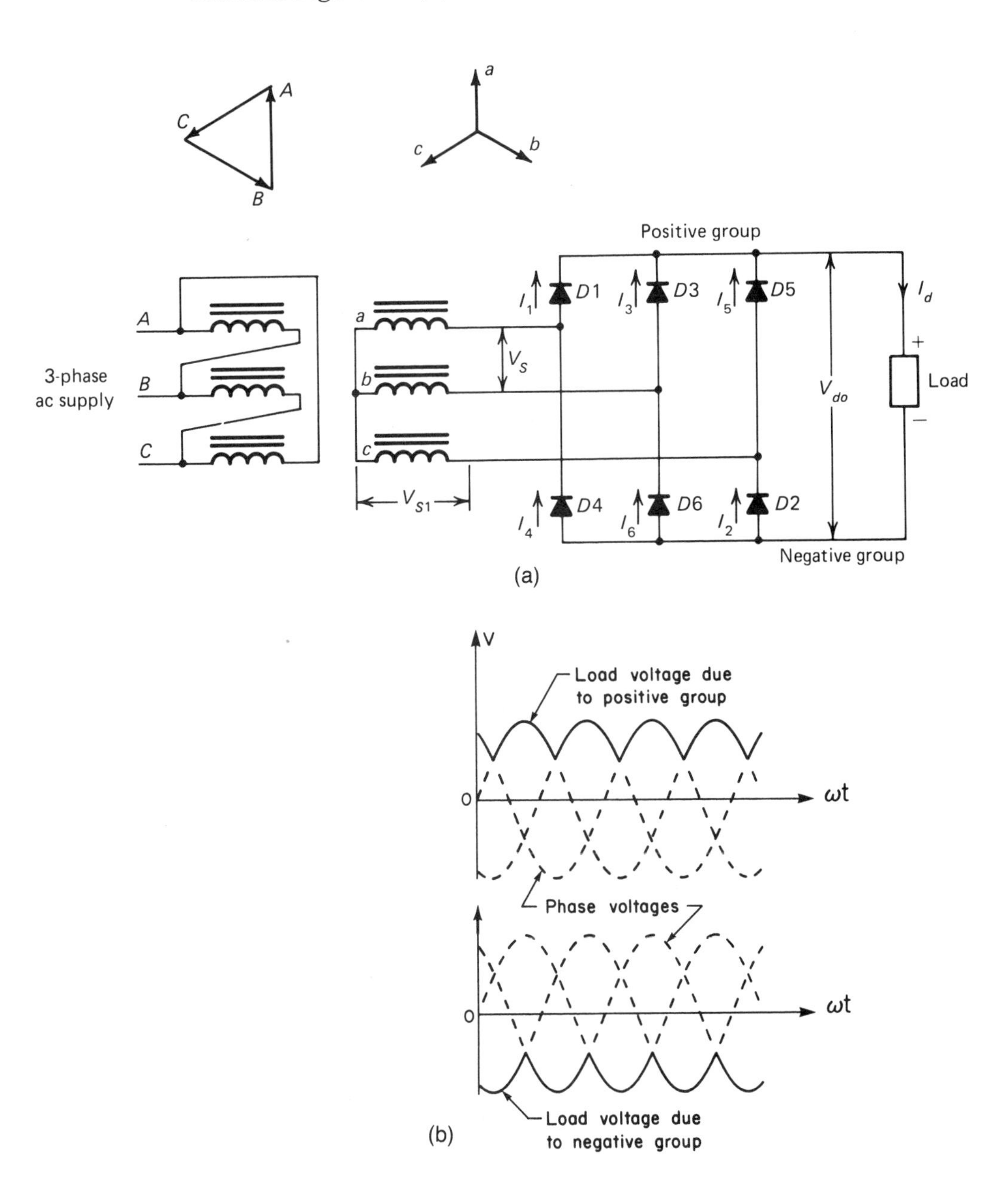

Figure 3-16. Three-phase full-wave bridge rectifier: (a) basic circuit, (b) voltage due to positive and negative groups.

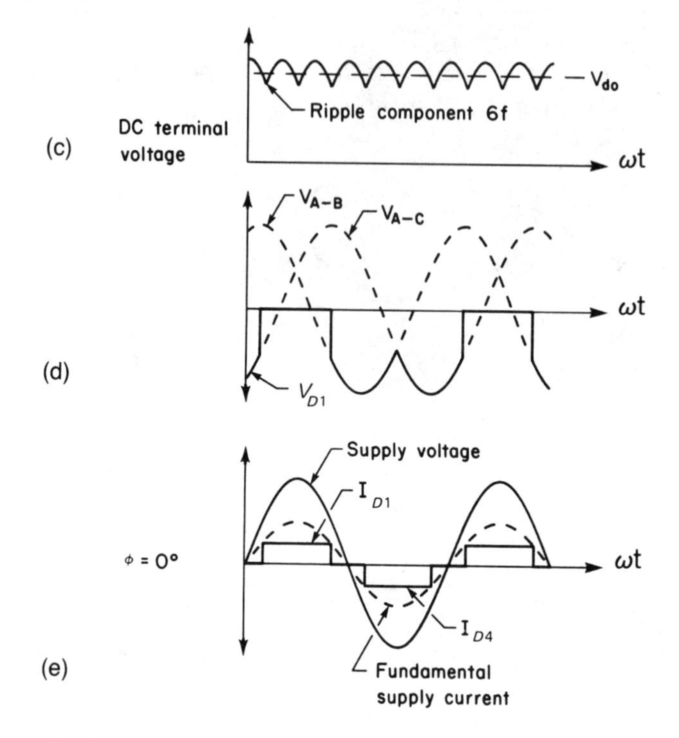

Figure 3-16 continued. (c) load voltage, (d) diode voltage waveform, (e) secondary current waveform.

When the load is connected between the positive and negative groups as in Figure 3-16(a), the average dc output voltage doubles and the diodes commutate alternately to form a six-pulse output; that is, the ripple frequency is $6f$. See Figure 3-16(c). Note that the diodes are numbered in the sequence they conduct. Figure 3-16(e) shows the current in each secondary phase of the transformer, the current being a square wave of amplitude I_d, and consists of a 120° positive pulse and zero current flow for 60°, followed by a 120° negative pulse and another 60° of zero current before the cycle is repeated. If the transformer primary is star connected, the current waveform will be repeated on the primary side. However, if the primary is delta connected, the line current waveform will consist of $+I_d/2$ for 60°, $+I_d$ for 60°, $+I_d/2$ for 60°, $-I_d/2$ for 60°, $-I_d$ for 60°, and $-I_d/2$ for 60°. This is also called a six-step waveform.

In our analysis of the three-phase bridge rectifier circuit, if the pulse number $m = 6$ and V_s is the line-to-line voltage applied to the rectifier, then

$$V_{do} = V_m \left[\frac{6}{\pi} \sin \frac{\pi}{6} \right] = 0.956 V_m$$

$$= 0.956\sqrt{2}V_s = 1.35V_s \qquad (3\text{-}98)$$

$$V_s = \sqrt{3}V_{s1} \qquad (3\text{-}99)$$

$$V_{do} = 1.35\sqrt{3}V_{s1} = 2.34V_{s1} \qquad (3\text{-}100)$$

$$I_d = 0.956I_m \qquad (3\text{-}101)$$

and

$$I_m = 1.05I_d \qquad (3\text{-}102)$$

Each diode carries current twice per cycle, and each transformer secondary phase carries current four times per cycle; therefore, the rms value of the diode current is

$$I_{rms} \quad \text{per diode} = 0.58I_d \qquad (3\text{-}103)$$

and the secondary phase current is

$$I_s = \sqrt{2}I_{rms} \quad \text{per diode} = 0.82I_d \qquad (3\text{-}104)$$

The peak inverse voltage experienced by each diode is

$$\text{PIV} = 3V_{s1m} = \left[\frac{\sqrt{3} \times \sqrt{2}}{2.34}\right] V_{do}$$
$$= 1.04V_{do}. \qquad (3\text{-}105)$$

The primary phase current I_p is

$$I_p = [N_s/N_p]\, I_s$$
$$= 0.82I_d\,[N_s/N_p] \qquad (3\text{-}106)$$

and the secondary VA rating for a star-connected secondary is

$$3\,[V_s/\sqrt{3}]\,I_s = 3\,[0.740V_{do}/\sqrt{3}]\,[0.82I_d]$$
$$= 1.05V_{do}I_d \qquad (3\text{-}107)$$

This is also equal to the primary VA rating.

The ripple factor γ is

$$\gamma = \frac{\sqrt{2}}{[m^2 - 1]} = 0.0404$$
$$= 4.04\% \qquad (3\text{-}108)$$

and the transformer secondary utilization factor is

$$\text{TUF}_s = 0.955 \tag{3-109}$$

With a 4.04% ripple factor and a high transformer utilization factor, the three-phase bridge rectifier can be seen to be one of the most effective rectifier arrangements.

Example 3-6

A three-phase bridge rectifier supplies a 100 A load at 240 V. Determine: (1) the average diode current; (2) the diode PIV; (3) the VA rating of the transformer secondary; (4) the secondary voltage rating; (5) the rms current in each phase of the transformer secondary; and (6) the rms value of the ac ripple voltage.

Solution:

(1) The average diode current is

$$I_{\text{av}} = \frac{100}{3} = 33.33 \text{ A}$$

(2) The PIV per diode is from Equation (3-104)

$$\begin{aligned} \text{PIV} &= 1.04 V_{do} \\ &= 1.04 \times 240 = 249.6 \text{ V} \end{aligned}$$

(3) From Equation (3-107) the transformer secondary VA rating is

$$\begin{aligned} VA_s &= 1.05 V_{do} I_d \\ &= 1.05 \times 240 \times 100 \\ &= 25.2 \text{ kVA} \end{aligned}$$

(4) From Equation (3-100) the secondary phase voltage rating is

$$\begin{aligned} V_{s1} &= \frac{V_{do}}{2.34} = \frac{240}{2.34} \\ &= 102.56 \text{ V} \end{aligned}$$

(5) From Equation (3-104) the rms secondary phase current is

$$I_s = 0.82 I_d$$

$$= 0.82 \times 100 = 82 \text{ A}$$

(6) From Equation (3-108) the amplitude of the ac ripple voltage is

$$V_{ac} = 0.0404\ V_{do}$$

Therefore,

$$V_{ac} = 0.0404 \times 240$$
$$= 9.70 \text{ V}$$

3-4-4 Three-Phase Double-Star Rectifier With Interphase Reactor

The **three-phase double-star** rectifier configuration illustrated in Figure 3-17 combines the outputs of two three-phase star rectifiers operating in parallel through an interphase reactor. Each rectifier operates independently of the other, with the load current being shared equally by the rectifiers and the reactor absorbing the difference in the instantaneous voltage. The frequency of the ac ripple content of the individual converters is $3f$; however, because of the phase difference of the ac ripple voltage between the two rectifiers, the mean output dc voltage at the center tap of the reactor has a ripple frequency of $6f$. As a result, for each cycle of the primary supply there are six segments of dc voltage in the mean output dc voltage, that is, it is a six-pulse rectifier.

Under reasonable loads the diodes conduct for 120°, thus giving a high transformer utilization factor. However, the performance is dependent upon the interphase reactor carrying sufficient dc current to maintain a satisfactory core flux level. Under conditions of light load the rectifier arrangement will operate as if the reactor was not present and the diodes will conduct for 60°, producing a discontinuous output with a reduced dc output voltage and a low transformer utilization factor. This problem can be reduced by shunting a bleeder resistance across the output terminals, thus ensuring that there will be sufficient dc current through the interphase reactor.

Many of the characteristics of the three-phase star rectifier arrangement are also the same for the double-star rectifier. The mean dc output voltage is

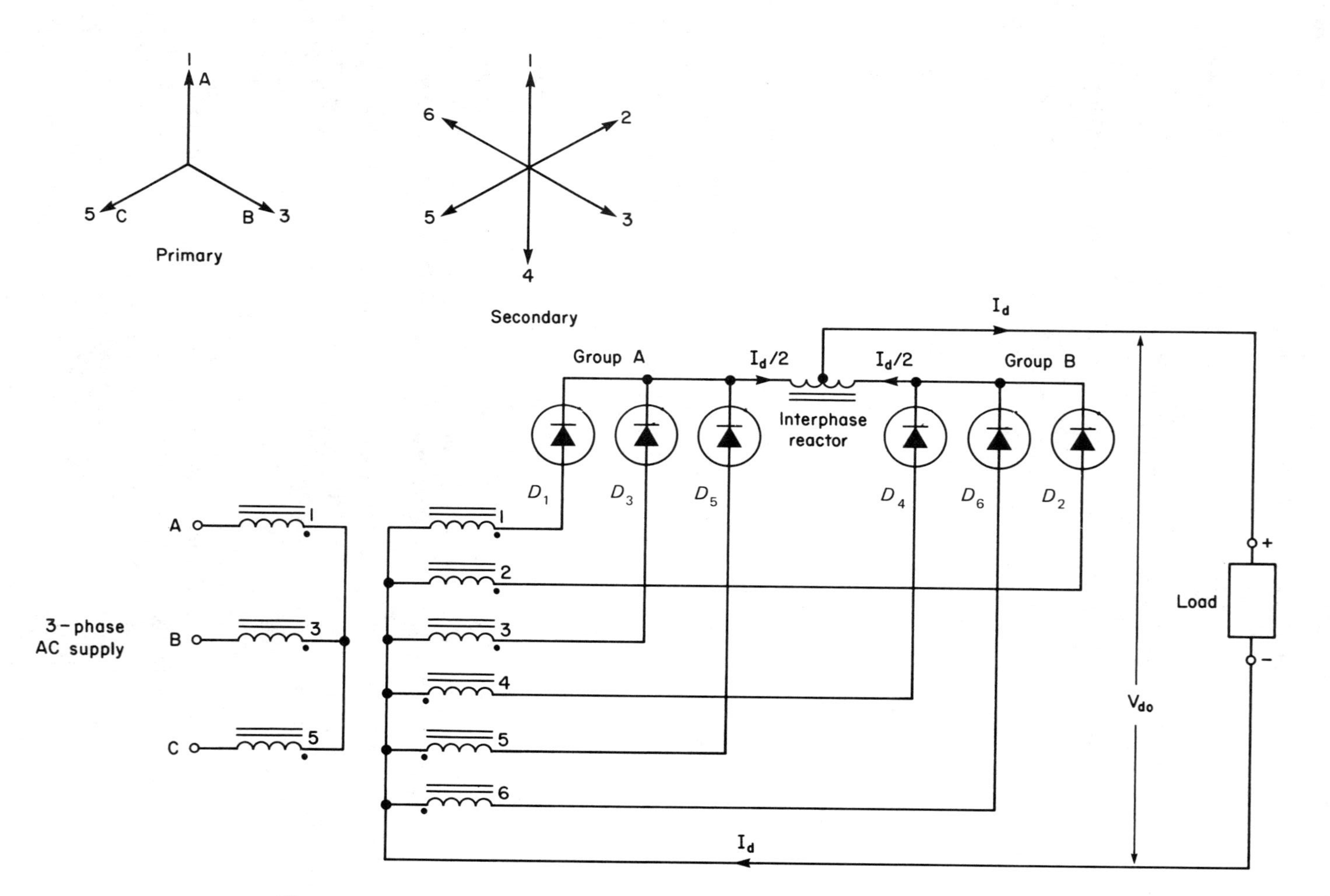

Figure 3-17. Three-phase double-star rectifier with interphase reactor.

$$V_{do} = 1.17V_s \quad (3\text{-}110)$$

and

$$V_s = 0.855V_{do} \quad (3\text{-}111)$$

since each star group supplies half of the load current. Then the secondary phase currents are

$$I_s = 0.293I_d \quad (3\text{-}112)$$

and the transformer primary current is

$$\begin{aligned} I_p &= \sqrt{2}\,[N_s/N_p]\,I_s \\ &= 0.414I_d\,[N_s/N_p] \end{aligned} \quad (3\text{-}113)$$

The secondary VA rating is

$$\begin{aligned} 6V_sI_s &= 6\,[0.855V_{do}]\,[0.293I_d] \\ &= 1.50V_{do}I_d \end{aligned} \quad (3\text{-}114)$$

and the primary VA rating is

$$\begin{aligned} 3V_pI_p &= 3\,[0.855V_{do}]\,[N_p/N_s]\,[0.414I_d]\,[N_s/N_p] \\ &= 1.06V_{do}I_d \end{aligned} \quad (3\text{-}115)$$

The interphase reactor increases the PIV experienced by the diodes and

$$\text{PIV} = 2.42V_{do} \quad (3\text{-}116)$$

As was previously mentioned a minimum current must flow between the two rectifiers to ensure that the interphase reactor core remains magnetized. This current has a pronounced third harmonic. The load current carried by each half of the reactor must be greater than the third harmonic current, that is,

$$I_d/2 \geq I_{3m} = \frac{V_{3m}}{3\,\omega\,\text{L}} \quad (3\text{-}117)$$

where

$$V_{3m} = \frac{3\sqrt{3}V_{sm}}{4\pi} \quad (3\text{-}118)$$

and the minimum value of the inductance L is

$$L_{\min} \geq \frac{\sqrt{3}V_{sm}}{2\pi\omega I_d} = \frac{0.390V_s}{\omega I_d} \tag{3-119}$$

3-4-5 Six-Phase Star Rectifier

The **six-phase rectifier,** shown in Figure 3-18, is transformer supplied and consists of a three-phase delta-connected primary with a three-phase star-connected secondary whose secondaries are center tapped and commoned to form a six-phase output.

The major application for this arrangement is when a common anode or cathode connection combined with a low ripple factor is required. Another feature is the cancellation of the dc currents in the transformer secondary, thus removing any tendency for core saturation to occur. This circuit, however, is not too commonly used nowadays.

Summary

Diode rectifiers are most commonly used in applications where the control of the dc output voltage is not required or where the output voltage may be controlled external to the rectifier.

Typical applications of diode rectifiers are low power dc supplies in electronic equipment, battery chargers, dc welders, dc traction drives, brushless excitation systems for large alternating current generators, and for dc supplies to electrochemical processes.

3-5 SERIES OR HIGH-VOLTAGE OPERATION

A number of **high-voltage applications** exist where the reverse voltage capability of a single silicon diode cannot meet the requirements of the circuit. In this situation, it is common practice to connect diodes in series so that the total reverse voltage is divided equally between the seriesed diodes.

There are a number of factors that cause unequal voltage sharing, the most important of which are: (1) unequal reverse resistances and thus unequal leakage currents; (2) unequal storage times between the diodes in the string; and (3) the variations in the diode capacitances with respect to ground.

Voltage sharing can be assisted by placing a resistor in parallel with the diode. The purpose of these resistances is to equalize the reverse resistances of the diodes. A rule-of-thumb method of determining the value of the shunting resistance is to select a resistor that has an

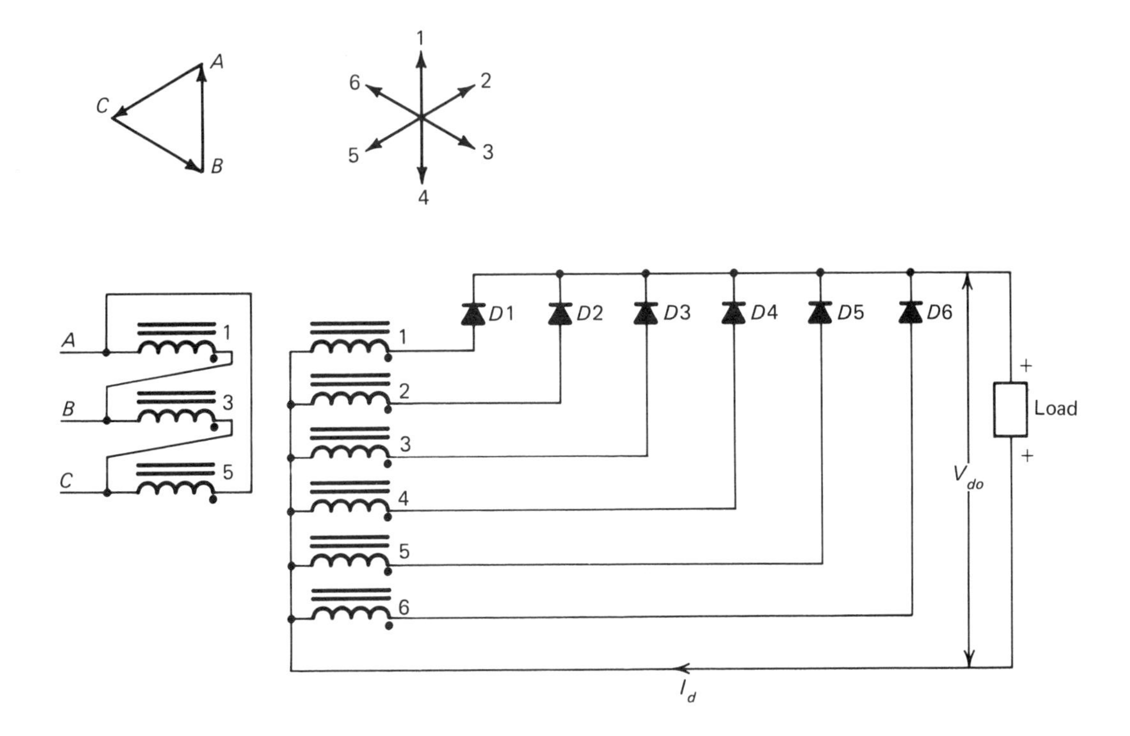

Figure 3-18. Six-phase star rectifier.

ohmic value of one-half of the minimum reverse resistance; typically this value will lie between 0.5 to 2 MΩ.

Voltage sharing can also be attempted by matching the reverse currents at a particular voltage. However, this method is not very reliable since voltage fluctuations and transients will change the operating voltage and the reverse currents may not be equal at that point.

The junction capacitance affects the storage time, so that when a reverse voltage is applied to a seriesed diode string the diode with the smallest junction capacitance will recover first and, for a short period of time, will have the total applied voltage across it. This is a dynamic problem and can be minimized by connecting capacitors in parallel with the diodes to equalize the different values of junction capacitance. Since the value of the shunting capacitance is directly proportional to the recovery time of the diode, it may be found that the appropriate high-voltage high-capacitance capacitors may be expensive, and a smaller value of capacitance may be used in conjunction with fast recovery diodes. See Figure 3-19.

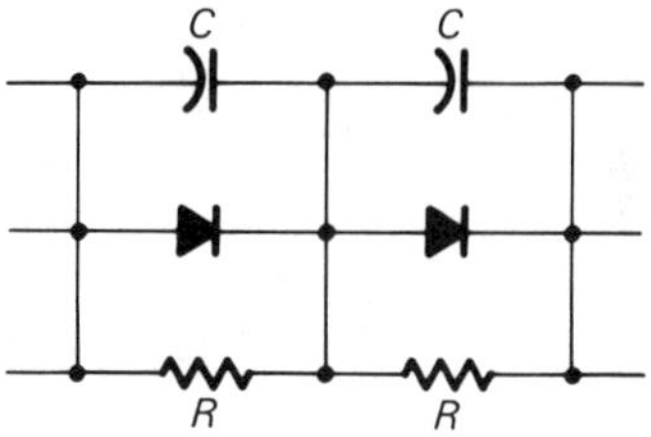

Figure 3-19. Voltage sharing of seriesed diodes in high-voltage applications.

3-6 PARALLEL OR HIGH-CURRENT OPERATION

In the industrial field there are many applications of silicon diodes where the current requirements are in excess of the current-carrying capability of an individual diode. In applications of this nature it is common practice to parallel diodes and share the load current between them. The problem experienced when paralleling diodes is that the forward resistances of the diodes are not equal, resulting in unequal current sharing between the paralleled diodes and further complicated by the negative resistance temperature coefficient characteristic of silicon diodes. The diode carrying the greatest current will experience an increase in junction temperature which will result in a decrease of forward resistance because of the negative temperature characteristic of silicon. In turn, an increase in diode current and eventual destruction by thermal runaway is caused. Additionally, under conditions of reverse bias each paralleled diode must be able to withstand the reverse voltage; however, if one diode fails, its failure does not produce a failure among the other diodes.

Among the numerous methods of promoting current sharing are the use of series resistors, matching forward characteristics and current balancing reactors.

Series or ballast resistors may be placed in series with each diode; however, while they promote current sharing these resistors increase the power losses and reduce the overall efficiency of the rectifier.

Matching the forward resistance of diodes in itself requires careful measurement and identification of the diodes. However, in the installation of the diodes, unequal resistances may be created by the mechanical mounting and connection of the paralleled diodes. Additionally, although temperature coefficient differences between diodes can cause unequal current sharing, it can be minimized by using a common heat sink and derating the diodes.

The most effective method of forcing equal current sharing is by the use of current-balancing reactors, which in most cases proves to be the most economical and reliable method. The principle of the current-balancing reactor is shown in Figure 3-20. The current-balancing reactor is basically a center-tapped reactor in which the diode carrying the greater current will cause an unbalance in the ampere-turns in its half of the reactor. This in turn creates an aiding voltage that will cause an increase of current through the diode which originally carried the lower current, and as a result promotes current sharing between the paralleled diodes.

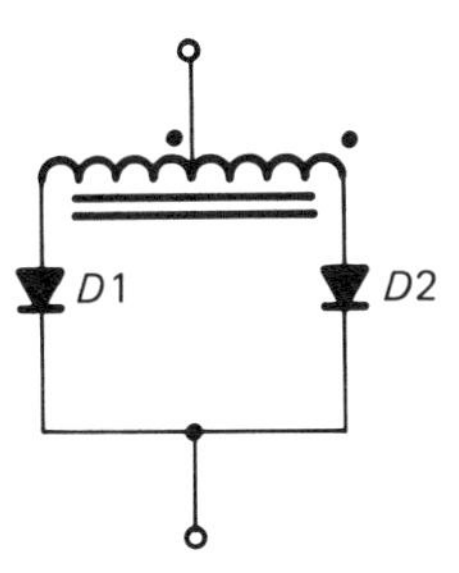

Figure 3-20. Forced current sharing of paralleled diodes using a current-balancing reactor.

Table 3-1 Summary of Rectifier Circuits and Parameters

Type of Rectifier	V_{do}	No. of Diodes	PIV	Current Per Diode	Ripple Freq.	Ripple Factor	VA Secondary	VA Primary	TUFs
Single-Phase Half-Wave	$0.318V_m$	1	$3.14V_{do}$	I_d	f	121%	$3.49V_{do}I_d$	$2.7V_{do}I_d$	0.286
Single-Phase Midpoint	$0.636V_m$	2	$3.14V_{do}$	$I_d/2$	$2f$	48%	$1.74V_{do}I_d$	$1.23V_{do}I_d$	0.636
Single-Phase Bridge	$0.636V_m$	4	$1.57V_{do}$	$I_d/2$	$2f$	48%	$1.23V_{do}I_d$	$1.23V_{do}I_d$	0.90
Three-Phase Half-Wave	$0.827V_m$	3	$2.094V_{do}$	$I_d/3$	$3f$	17.7%	$1.50V_{do}I_d$	$1.24V_{do}I_d$	0.664
Three-Phase Zigzag Half-Wave	$0.827V_m$	3	$2.094V_{do}$	$I_d/3$	$3f$	17.7%	$1.75V_{do}I_d$	$1.24V_{do}I_d$	0.664
Three-Phase Bridge	$0.953V_m$	6	$1.04V_{do}$	$I_d/3$	$6f$	4.04%	$1.05V_{do}I_d$	$1.05V_{do}I_d$	0.955
Three-Phase Double Star With Interphase Reactor	$0.827V_m$	6	$2.42V_{do}$	$I_d/6$	$6f$	Very Small	$1.50V_{do}I_d$	$1.06V_{do}I_d$	0.67
Six-Phase Star	$0.96V_m$	6	$2.094V_{do}$	$I_d/6$	$6f$	4.04%	$1.82V_{do}I_d$	$1.28V_{do}I_d$	0.951

GLOSSARY OF IMPORTANT TERMS

Peak inverse voltage (PIV): The maximum reverse voltage capability of the semiconductor diode.

Rectifier efficiency (η): Sometimes called the ratio of rectification. It is the measure of comparing the effectiveness of rectifier configurations and is the ratio of dc power supplied to the load to the ac power input to the rectifier.

Ripple factor (γ): The measure of purity of the rectifier output. It is the ratio of the rms value of the ac ripple component to the average value of the dc component.

Transformer utilization factor (TUF): The factor that describes the utilization of a transformer winding. It is the ratio of the dc power output of the rectifier to the volt-ampere rating of the transformer winding.

Midpoint rectifiers: Single-phase and multiphase rectifier circuits in which one terminal of the dc circuit is connected to the center-tap or neutral point of the transformer winding.

Bridge rectifiers: Single-phase or multiphase rectifier circuits in which there is no direct connection between the dc circuit and the ac supply. They may be supplied with ac directly from the ac lines or via transformer windings.

Voltage multipliers: These are circuits in which capacitors are charged on alternate half-cycles of the ac supply voltage and by suitable arrangements of diodes and capacitors produce dc output voltages two, three, and/or four times the peak ac source voltage.

Multiphase rectifiers: Midpoint and bridge rectifiers supplied usually from three-phase ac sources.

Pulse number: The ratio of the ripple voltage frequency to the ac source voltage frequency.

REVIEW QUESTIONS

3-1 Explain the operation of a half-wave rectifier.

3-2 Explain what is meant by ripple and ripple factor.

3-3 What is the significance of the form factor?

3-4 Single-phase half-wave rectifiers are rarely used in industrial applications. Why?

3-5 What is the significance of the transformer utilization factor?

3-6 Explain with the aid of waveforms the operation of a half-wave rectifier supplying an RL load.

3-7 What are the advantages of using a single-phase full-wave rectifier as against the use of a single-phase half-wave rectifier?

3-8 Explain with the aid of waveforms the operation of a single-phase full-wave midpoint converter supplying an RC load.

3-9 Repeat Question 3-8 for an RL load.

3-10 Explain the operation of a choke input or L-section filter.

3-11 What are the advantages of using a π filter?

3-12 Explain with the aid of sketches (a) a full-wave doubler, and (b) a cascade doubler. What are the advantages and disadvantages of each type, and the limitations of the circuit elements?

3-13 Repeat Question 3-12 for a full-wave and cascade tripler.

3-14 Repeat Question 3-12 for a full-wave and cascade quadrupler.

3-15 In what applications would voltage multipliers be used and what are their limitations?

3-16 What are the advantages of using three-phase rectifier configurations?

3-17 What is the disadvantage of using the three-phase star rectifier connection? How may this problem be overcome?

3-18 Explain the operation of a three-phase bridge rectifier.

3-19 What are the advantages of using the three-phase double-star rectifier configuration? What precaution must be taken under light load conditions?

3-20 What are the advantages of using a six-phase star rectifier configuration?

3-21 In high-voltage applications a number of diodes are connected in series. What precautions must be taken to ensure voltage sharing across the diode string?

3-22 In high-current applications what precautions must be taken to ensure equal current sharing between paralleled diodes?

PROBLEMS

3-1 The output voltage of a half-wave rectifier consists of an average dc component of 125V, and an ac rms component of 10V. Calculate (1) V_{rms}, (2) the ripple factor, and (3) the form factor of the waveform.

3-2 A single-phase midpoint rectifier is shown in Figure 3-4. The secondary voltage V_{AC} is 220V and R_L is 150Ω. Calculate (1) V_{do}, (2) V_{rms}, (3) the ripple factor, (4) the maximum diode current, (5) the average diode current, and (6) the diode PIV.

3-3 A single-phase midpoint rectifier (see Figure 3-6) has a secondary voltage V_{AC}=208V, R_L=200Ω, C=50μF and F=60Hz. Calculate: (1) V_{do}, (2) the ripple factor, and (3) the power delivered to the load.

3-4 A single-phase midpoint rectifier as shown in Figure 3-8 has a secondary voltage V_{AC} of 208V, R_L=1000Ω, L=10H and f=60Hz. Calculate: (1) V_{do}, (2) the average load current I_d, (3) the power supplied to the load P_{dc}, and (4) the ripple factor.

3-5 A single-phase full-wave center-tapped rectifier (see Figure 3-8) supplies a maximum of 220V across the load when R_L=1000Ω and f=60Hz. Calculate the value of inductance required to produce a ripple factor of 5%.

3-6 A full-wave rectifier supplies 20W to a load consisting of a 500Ω resistor in parallel with a capacitor. If the ripple factor of the load voltage is 15%, calculate (1) V_{do}, and (2) the peak-to-peak value of the ac ripple voltage.

3-7 A three-phase midpoint rectifier supplies 100A to a resistive load at 600V. Calculate (1) the average diode current, (2) the PIV per diode, (3) the secondary winding VA rating, (4) the primary winding VA rating, (5) the rms secondary current, (6) the secondary transformer utilization factor, and (7) the amplitude of the ripple voltage.

3-8 A six-pulse or three-phase bridge rectifier is supplied via a transformer from a three-phase 550V, 60Hz source. If the load on the rectifier is 300kW at 250V dc and the transformer utilization factor is 0.955 calculate (1) the kVA and voltage ratings of the transformer, (2) the rms diode current, (3) the diode PIV, and (4) the frequency and amplitude of the ripple component.

4

Thyristor Phase-Controlled Converters

4-1 INTRODUCTION

Thyristor phase-controlled converters are natural or line commutated ac to dc converters that produce a variable dc output voltage whose amplitude is varied by phase control techniques, that is, the duration of the conduction period is controlled by varying the time at which a gate pulse signal is applied to the thyristor. In the past, the principle has been applied to the control of thyratrons, mercury arc rectifiers, ignitrons, and excitrons, and now with refinements has been applied to the thyristor.

In most converters the power flow is from the ac source to the dc load, which is the rectification process. However, in some converter configurations when the dc load is, for example, a motor operating under regenerative conditions, the power flow can be from the dc load to the ac source. This process is called **synchronous inversion**, as distinct from the inversion of dc to variable frequency ac by means of a dc link converter which utilizes forced commutation techniques. A major advantage of power conversion by the phase-controlled converter is the high efficiency of the process (usually in excess of 95% since the only losses are the small amounts dissipated by the thyristors). In addition, because the gate power requirements of the thyristors are relatively small, extensive use is made of low-power devices such as signal diodes, transistors, and digital and linear integrated circuits for the control and firing circuits.

Phase-controlled converters are classified into two major groups, single- and three-phase converters. These groups in turn are further subdivided into one-quadrant, two-quadrant, and four-quadrant or dual converters (see Figure 4-1).

One-quadrant, or semiconverters or half-controlled, converters operate only as rectifiers; that is, they have only one voltage polarity and a unidirectional current flow and are not capable of synchronous inversion. They operate as rectifiers in quadrant I or quadrant III. Two-quadrant or full converters are capable of operating with voltage polarity reversal and unidirectional current flow, that is, they operate in the rectifier mode in quadrant I or in the synchronous inversion mode in quadrant IV. Four-quadrant or dual converters operate as rectifiers in quadrants I and III and as synchronous inverters in quadrants II and IV. The dual converter is the power electronic equivalent of the Ward-Leonard system.

Thyristor phase-controlled converters are available in a wide range of output ratings suitable for the control of fractional horsepower (kilowatt) dc or universal motors, ranging up to high horsepower (kilowatt)

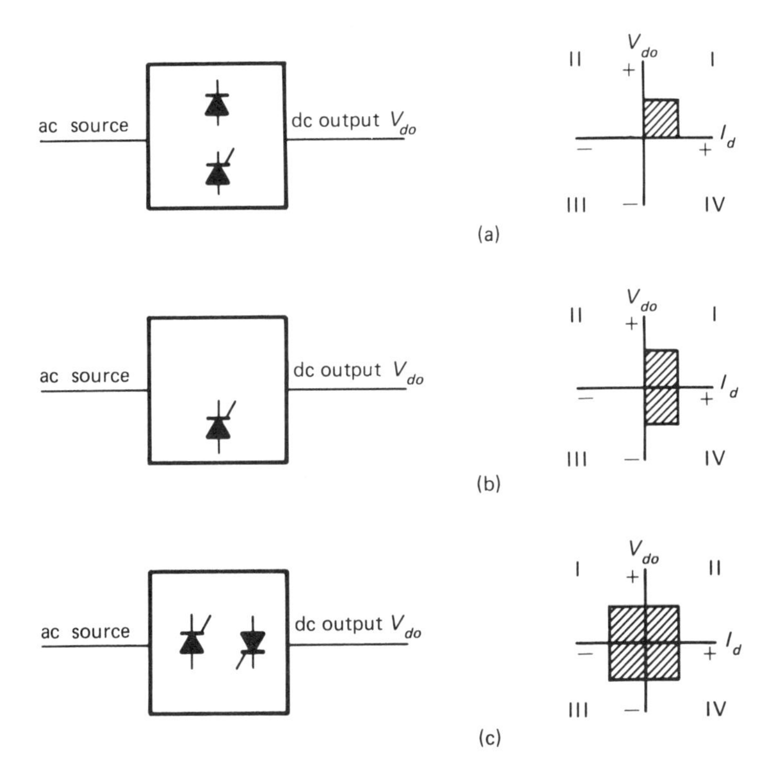

Figure 4-1. Thyristor phase-controlled converter configurations: (a) one-quadrant or semiconverter, (b) two-quadrant or full converter, (c) four-quadrant or dual converter.

reversing adjustable-speed dc motor drives used in a variety of applications in the metal fabricating and paper industries. Other applications include constant voltage power supplies, battery chargers, and high-voltage dc transmission. By far, the most common application is variable speed control of dc machines, where in conjunction with closed-loop control techniques it can meet torque, acceleration, and positioning requirements. Another use of the thyristor converter is as the source of variable voltage dc power for variable-frequency inverters for polyphase ac motor speed control.

4-2 THE BASIC PRINCIPLE OF PHASE CONTROL

The concepts and terminology used in **phase control** can best be described as a result of considering a half-wave phase-controlled thyristor with resistive and inductive loading as illustrated in Figure 4-2.

During the negative half-cycle of the ac supply voltage, the thyristor blocks the flow of load current, and no voltage is applied to the load. During the positive half-cycle, the thyristor is forward biased and will conduct if gated on. Assuming that the thyristor is turned on at time t_1, load current will flow, and the voltage applied to the load will be the supply voltage minus the voltage drop across the thyristor of approximately 1 V. See Figure 4-2 (b), (c). With a pure resistive load when the thyristor becomes reverse biased at time t_2, the flow of load current I_d will cease, and the load voltage will be zero until the thyristor is again forward biased and gated on at time t_4. The amplitude of the mean dc output voltage $V_{do\alpha}$ is controlled by the firing delay angle α.

With an inductive load (see Figure 4-2 (d), (e)), if the thyristor is turned on at time t_1, load current will flow and voltage will be applied to the load. However, at time t_2, when the thyristor is reverse biased, the stored magnetic energy in the load inductance is returned and will maintain a decaying forward current through the thyristor (Lenz's Law). At time t_3, the forward current has decayed to zero, and the thy-

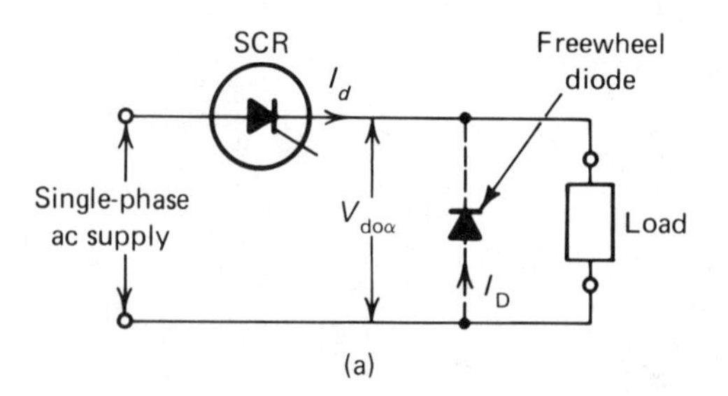

Figure 4-2. Single-phase, half-wave phase control: (a) circuit;

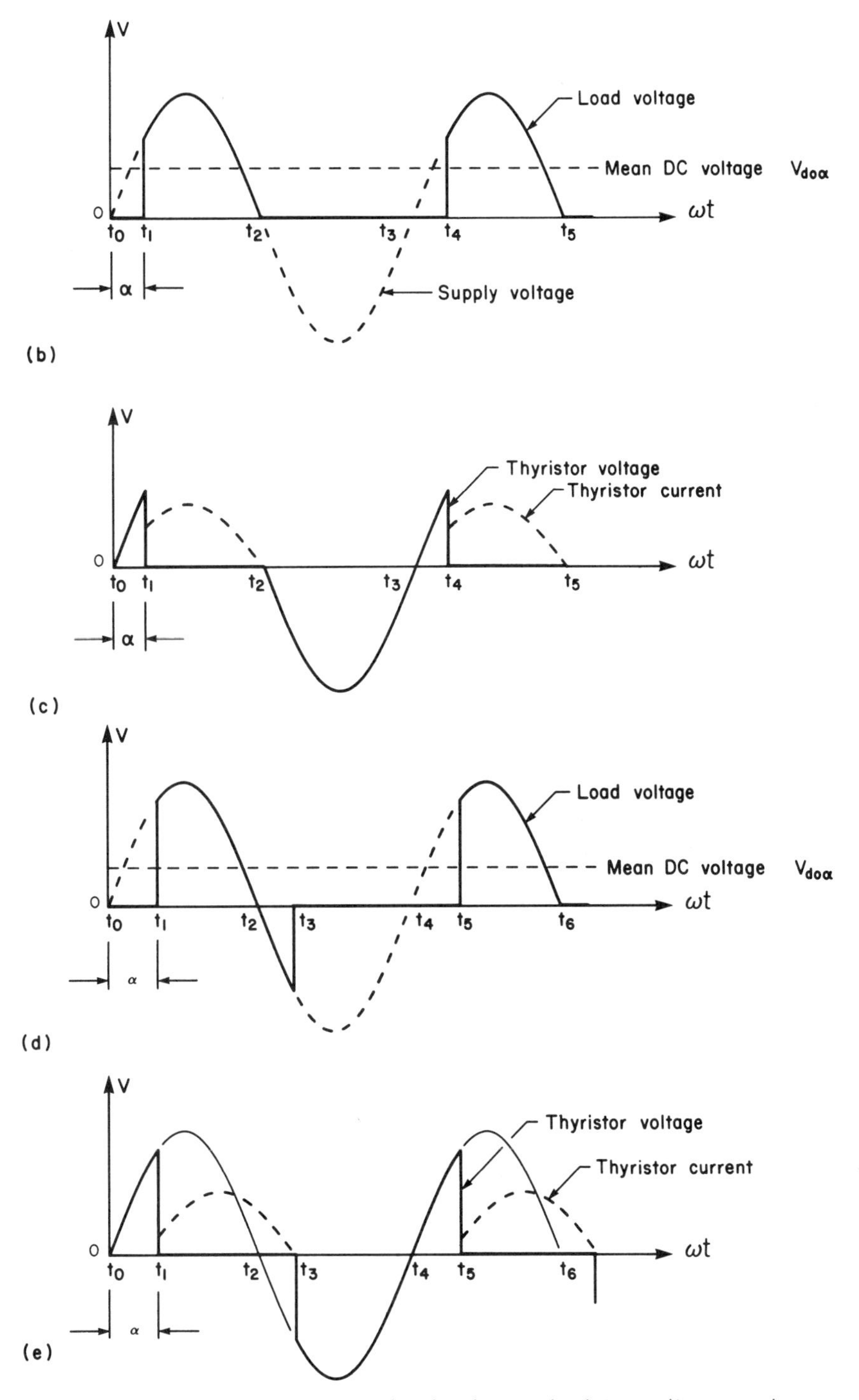

Figure 4-2. continued. (b) and (c) load voltage, thyristor voltage, and current waveforms, resistive loading; (d) and (e) load voltage, thyristor and current waveforms, inductive loading.

ristor turns off and resumes the blocking state. Between times t_3 and t_5 there is no voltage or current applied to the load.

As can be seen from the waveforms, the load voltage contains a large amplitude ac ripple component of the same frequency as the ac source; the load current I_d is also discontinuous. The peak forward and reverse voltages applied to the thyristor are equal to the maximum value of the ac source voltage, i.e., $V_m = \sqrt{2}V$, where V is the rms value of the ac source voltage.

The mean output dc voltage $V_{do\alpha}$ is

$$V_{do\alpha} = \frac{V_m}{2\pi}(1 + \cos\alpha)$$

$$= 0.23V(1 + \cos\alpha), \quad 0° < \alpha < 180° \qquad (4\text{-}1)$$

with an average thyristor current I_d of

$$I_d = \frac{V_m}{\pi R} = 0.45\frac{V}{R} \qquad (4\text{-}2)$$

Equations (4-1) and (4-2) apply for a pure resistive load, or an RL load. The rms thyristor current is

$$I_{rms} = 0.707\, I_d \qquad (4\text{-}3)$$

Sometimes with an inductive load or a motor load a freewheel diode is connected across the load terminals, so that when the thyristor becomes reverse biased at time t_2, the freewheel diode will dissipate the stored energy in the load. The load voltage will be zero during the time that the freewheel diode is conducting. With a freewheel diode in circuit, the average load current becomes

$$I_d = \frac{V_m}{2\pi R} = 0.23\frac{V}{R} \qquad (4\text{-}4)$$

the current carried by the diode I_D is

$$I_D = 0.54\frac{V_m}{\pi R} = 0.24\frac{V}{R} \qquad (4\text{-}5)$$

Example 4-1

A single-phase half-wave converter supplied from a 115V, 60Hz source is used to charge a 24Vdc battery bank whose resistance is

0.25Ω. Calculate (1) the maximum delay angle α if the charging current is not to exceed 15A, and (2) the thyristor PIV.

Solution:

(1)
$$I = \frac{V_{do\alpha} - E}{R}$$

then

$$V_{do\alpha} = E + IR = 24 + 15 \times 0.25$$
$$= 27.75\text{V}$$

From Equation (4-1)

$$V_{do\alpha} = 0.23\text{V}(1 + \cos\alpha)$$

then

$$\cos\alpha = \frac{V_{do\alpha}}{0.23\text{V}} - 1$$
$$= \frac{27.75}{0.23 \times 115} - 1$$
$$= 0.0491$$

and

$$\alpha = \cos^{-1} 0.0491 = 87.18^{0}$$

(2)
$$\text{PIV} = V_m = \sqrt{2}V$$
$$= \sqrt{2} \times 115 = 162.63\text{V}$$

4-3 TWO-QUADRANT CONVERTERS

For **two-quadrant** operation, that is, unidirectional current flow and both polarities of dc voltage, it is essential that there are thyristors in all positions.

To simplify our discussions, the following assumptions are made:

1. The voltage drop across a conducting thyristor or diode is negligible.
2. There is no leakage current when the device is in the blocking state.
3. Turn-on and turn-off occur instantly.
4. The output dc terminals are connected to an infinite inductance

(i.e., an ideal filter), thus producing a ripple free constant amplitude current.

In general, two-quadrant converters can be classified as:

1. Midpoint converters which are supplied from tapped transformers.
2. Bridge converters which may be supplied directly from the ac source.

4-3-1 Two-Pulse Midpoint Converter

Figure 4-3 shows the basic circuit arrangement of a single-phase **two-pulse midpoint converter** together with the associated current and voltage waveforms. Consider Figure 4-3(b), where the firing delay angle $\alpha = 0°$.

During the positive half-cycle of the ac supply voltage, SCR1 is forward biased and the load current is carried by the transformer secondary S_1, and SCR2 is reverse biased, 180°. Later SCR2 is forward biased and the load current is carried by the transformer secondary S_2 with SCR1 being reverse biased. During the period that each thyristor conducts, the load voltage is equal to the instantaneous ac voltage across the appropriate half of the secondary winding of the transformer. When the firing delay angle $\alpha = 0°$ the converter acts as a two-pulse midpoint diode rectifier. The load voltage waveform has a dc component V_{do} and a superimposed ac ripple voltage with a frequency double that of the ac source frequency, accounting for the name two-pulse converter, where

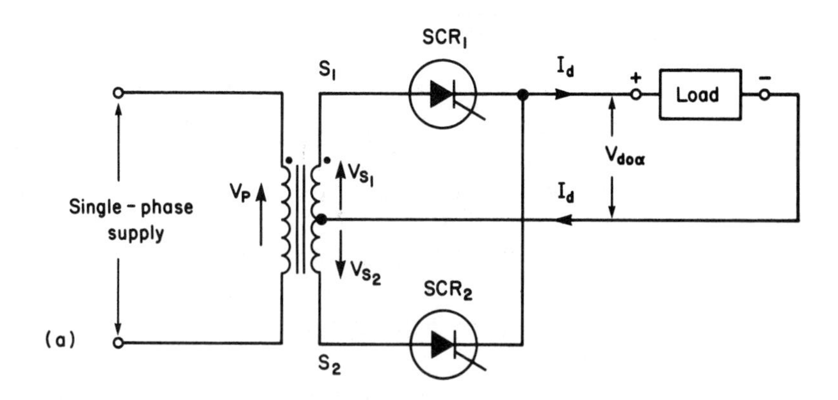

Figure 4-3. Two-pulse midpoint converter; (a) basic circuit.

$$\text{pulse number } m = \frac{\text{ripple voltage frequency}}{\text{source frequency}}. \tag{4-6}$$

As a general rule, most single-phase converters are two-pulse, and most three-phase converters are three- or six-pulse, although in some applications they may be modified to form twelve or even twenty-four pulse converters. In general the smaller the amplitude of the ac ripple voltage the higher its frequency.

The load current I_d is supplied in turn by SCR1 and SCR2 and, by assumption 4, it is constant in value and is in phase with the secondary voltage. As a result, the converter is being supplied from the ac source at unity power factor.

In Figure 4-3(c), where $\alpha = 45°$, the effect of increasing the firing delay angle is to reduce the mean output dc voltage $V_{do\alpha}$. Since it has been assumed that the load is inductive, SCR2 remains in conduction

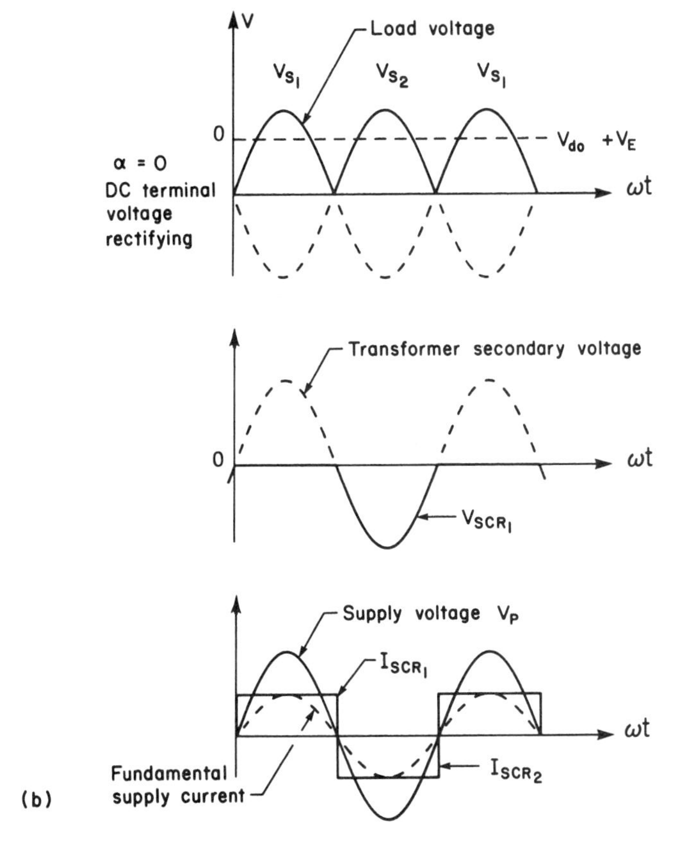

Figure 4-3 continued. (b) load voltage, V_{SCR_1}, I_{SCR_1}, and I_{SCR_2} and fundamental current waveforms for $\alpha = 0°$.

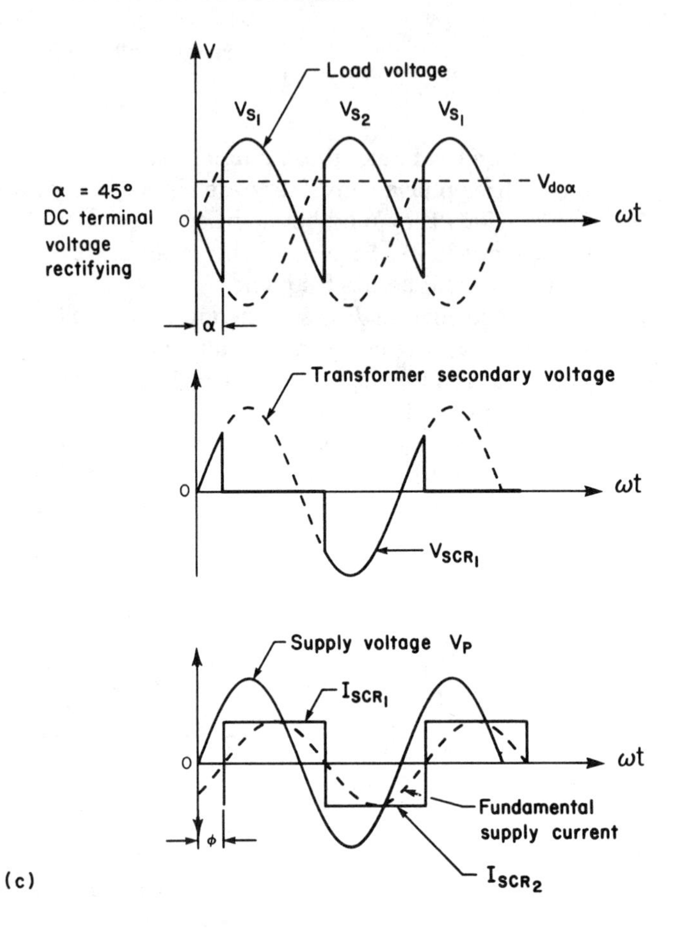

Figure 4-3 continued. (c) load voltage, V_{SCR_1}, I_{SCR_1}, and I_{SCR_2} and fundamental current waveforms for $\alpha = 45°$.

up to the point where SCR1 is turned on at $\alpha = 45°$, even though SCR2 was reverse biased, a condition which may be observed by connecting an isolated oscilloscope across SCR2. This is the regeneration condition and the power flow is from the load to the source. Up to the point of conduction, the voltage across SCR1 is positive and equal to the sum of the voltage across both halves of the transformer or twice the load voltage. When SCR1 conducts, SCR2 is commutated off and has a voltage equal to twice the load voltage applied across it.

Each SCR in turn conducts and supplies the load current I_d for 180°. The load currents through the SCRs, and thus the supply current, is lagging the ac supply voltage by 45°, and the ripple content of the load voltage has increased. If the load is resistive, then as each SCR is reverse biased it will commutate off and load current will cease until the

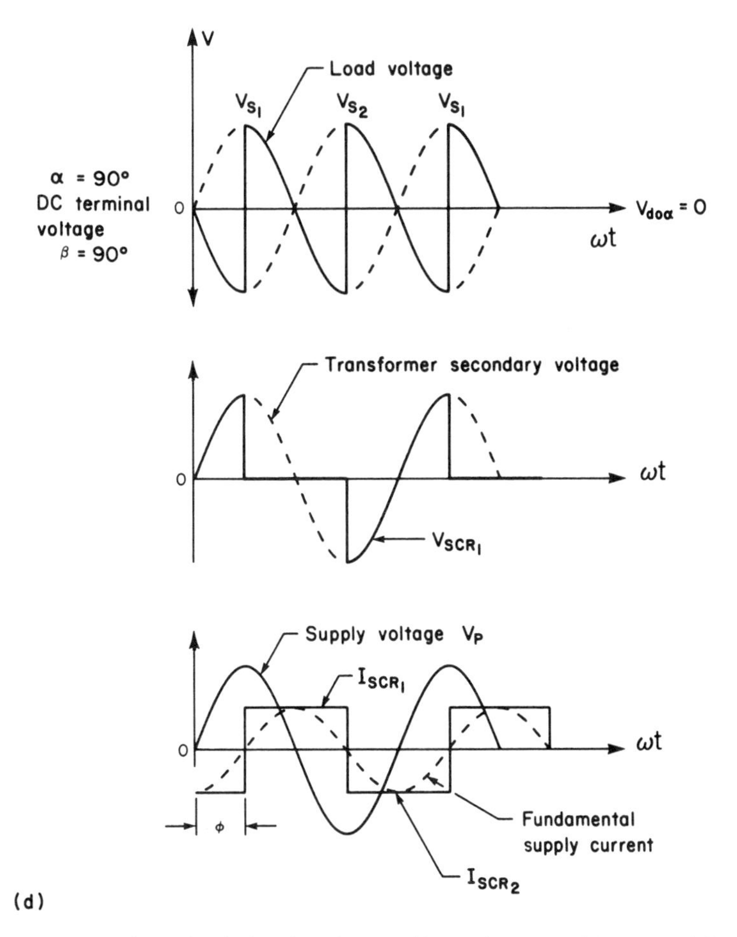

Figure 4-3 continued. (d) load voltage, V_{SCR_1}, I_{SCR_1}, and I_{SCR_2} and fundamental current waveforms for $\alpha = 90°$.

next SCR is turned on, that is, the current supplied to the load will be discontinuous.

When $\alpha = 90°$ [see Figure 4-3(d)], the average dc voltage $V_{do\alpha}$ is zero, and the load voltage consists entirely of the ac ripple component. With an inductive load the SCRs remain in conduction for 180° and the respective SCR currents lag the anode-cathode voltage by 90°. This in turn is reflected to the transformer primary and the supply current lags the ac source voltage by 90°, and there is no power transfer to the load ($P = VI \cos \phi = VI \cos 90° = 0$).

In Figure 4-3(e), when the firing delay angle is progressively increased beyond $\alpha = 90°$, there will be a flow of load current only if the load itself presents a negative voltage, which will occur in case of a dc motor under overhauling load conditions. At $\alpha = 135°$, the mean dc voltage $V_{do\alpha}$ is negative, and load current flows in each SCR for 180° in its original direction. However, since the voltage has reversed polarity, the power flow is from the dc load to the ac source and the converter is operating as a line commutated inverter. This process is known as synchronous inversion. Since the SCR currents lag by 135°, the supply current also lags the ac source voltage by 135°.

In Figure 4-3(f), with $\alpha = 180°$, the mean dc voltage $V_{do\alpha}$ is at its negative maximum and the ac ripple content has decreased from a maximum at $\alpha = 90°$ to a minimum at $\alpha = 180°$, and the supply current lags the source voltage by 180°.

In actual practice, this situation cannot be achieved, since the period

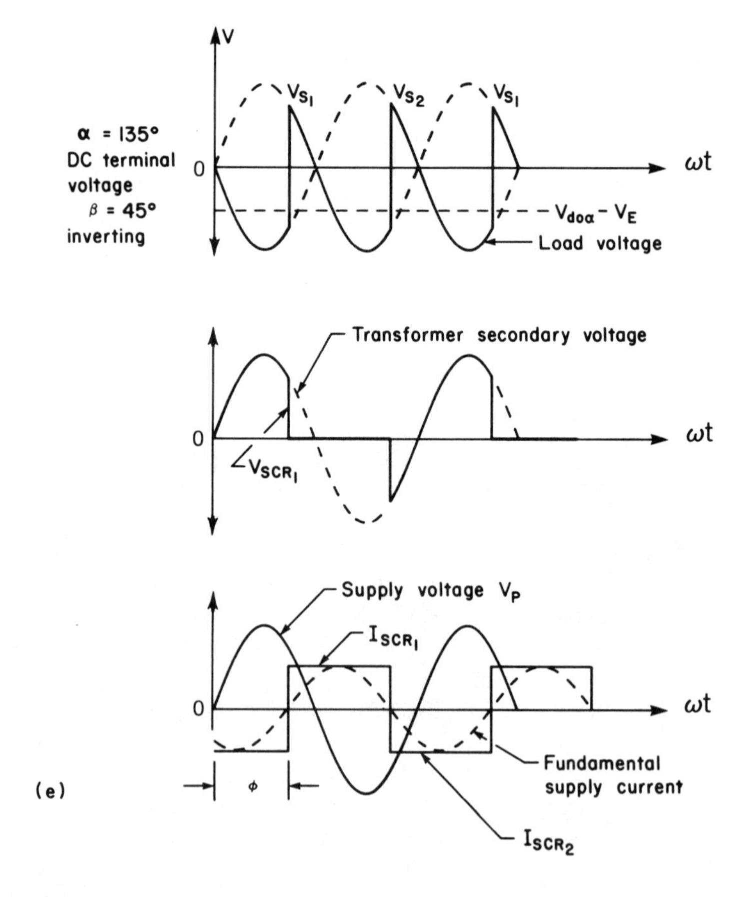

Figure 4-3 continued. (e) load voltage, V_{SCR_1}, I_{SCR_1}, and I_{SCR_2} and fundamental current waveforms for $\alpha = 135°$.

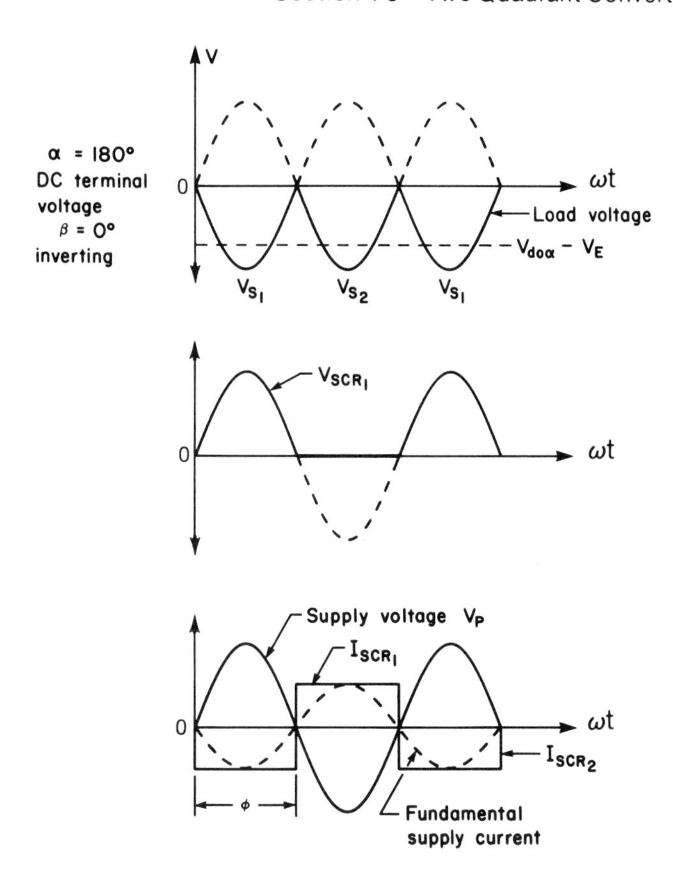

Figure 4-3 continued. (f) load voltage, V_{SCR_1}, I_{SCR_1}, and I_{SCR_2} and fundamental current waveforms for $\alpha = 180°$.

of reverse bias is continually decreasing as α approaches 180°, and sufficient time must be provided for the thyristors to turn off and achieve a blocking state before they are subjected to the reapplied forward voltage. In a 60 Hz system the practical limit to the retardation of the firing pulse is 160° to 175°.

When operating in the inversion mode, the inversion region is specified in terms of the inverter advance angle β, where $\beta = 180° - \alpha$.

In summary, the following points should be noted with respect to the operation and performance of a two-pulse midpoint converter.

1. For firing delay angles $0° < \alpha < 90°$, the converter is operating in the rectifying mode; for $90° < \alpha < 180°$ the converter is operating in the inverting mode.

2. In the rectifying mode the net transfer of power is from the ac source to the dc load. In the inversion mode, provided there is a source of negative voltage at the dc load, the net transfer of power is from the dc load to the ac source.

3. As the firing delay angle is increased, there is a corresponding lag of the supply current with respect to the ac supply voltage.

4. The mean dc voltage $V_{do\alpha}$ decreases from a positive maximum to zero when $\alpha = 90°$, to a negative maximum at $\alpha = 180°$. At the same time the ac ripple content increases from a minimum at $\alpha = 0°$ to a maximum at $\alpha = 90°$ to a minimum when $\alpha = 180°$.

5. With a sufficiently inductive load the thyristors will conduct for 180°, and the load current I_d will be continuous. In the case of a resistive load, conduction ceases when the conducting thyristor becomes reverse biased and the load current will be discontinuous.

The peak forward and reverse voltages applied to the thyristors are $2V_m = 2.8$ V.

The mean dc voltage $V_{do\alpha}$ is

$$V_{do\alpha} = \frac{2V_m}{\pi} \cos\alpha = 0.90 \text{ V} \cos\alpha \tag{4-7}$$

and the average thyristor current I_d is

$$I_d = \frac{V_m}{\pi R} = 0.45\frac{V}{R} \tag{4-8}$$

and the rms thyristor current is

$$I_{rms} = 0.707\, I_d \tag{4-9}$$

The frequency of the ripple component is $2f$, where f is the ac source frequency. Since each half of the secondary winding is idle for 50% of the time, the VA rating of the transformer secondary is high and in fact is

$$VA_{\text{sec}} = 1.57P_{do} \tag{4-10}$$

where

$$P_{do} = V_{do\alpha} I_d \tag{4-11}$$

since the secondary windings are carrying currents with a dc component.

The primary winding is not carrying the dc current component, and as a result the VA rating is

$$VA_{pri} = 1.11P_{do} \tag{4-12}$$

Example 4-2

A two-pulse midpoint converter with a center tap to line effective voltage of 115V, 60Hz is used to charge a 24V dc battery whose resistance is 0.25Ω. Calculate (1) the maximum delay angle if the charging current is not to exceed 15A, (2) the thyristor PIV, and (3) the average thyristor current.

Solution:

(1) From Example 4-1 $V_{do\alpha} = 27.75\text{V}$

From Equation (4-7)

$$V_{do\alpha} = 0.9V\cos\alpha$$

therefore

$$\cos\alpha = \frac{V_{do\alpha}}{0.9V} = \frac{27.75}{0.9 \times 115} = 0.27$$

then

$$\alpha = \cos^{-1}0.27 = 74.45^{\circ}$$

(2)

$$\text{PIV} = 2.8V = 2.8 \times 115$$
$$= 322\text{V}$$

(3) From Equation (4-8)

$$I_d = 0.45V/R$$
$$= 0.45 \times 115/0.25$$
$$= 207A$$

Midpoint converters are used in applications where electrical isolation from the ac source is required and it is desirable to reduce the amplitude of the ac ripple component by increasing the number of pulses. The single-phase midpoint converter has been used to illustrate the concept, but in actual practice it is rarely used since its power capability is limited by the single-phase source. When used in industrial applications, it is normal practice to use three-phase midpoint converters.

Midpoint converters, while they use half the number of thyristors required by a bridge converter and have a less complicated and costly control section, have the savings reduced because of the increased cost of the higher voltage thyristors.

4-3-2 Two-Pulse Bridge Converter

A single-phase **two-pulse bridge converter** arrangement is shown in Figure 4-4. This circuit eliminates the requirement for an input transformer. In this circuit, diagonally opposite pairs of thyristors are turned on together and commutate off together. Except for the fact that two thyristors are gated on at the same time by the firing control, the performance of the two-pulse bridge converter is identical to that of the two-pulse midpoint converter.

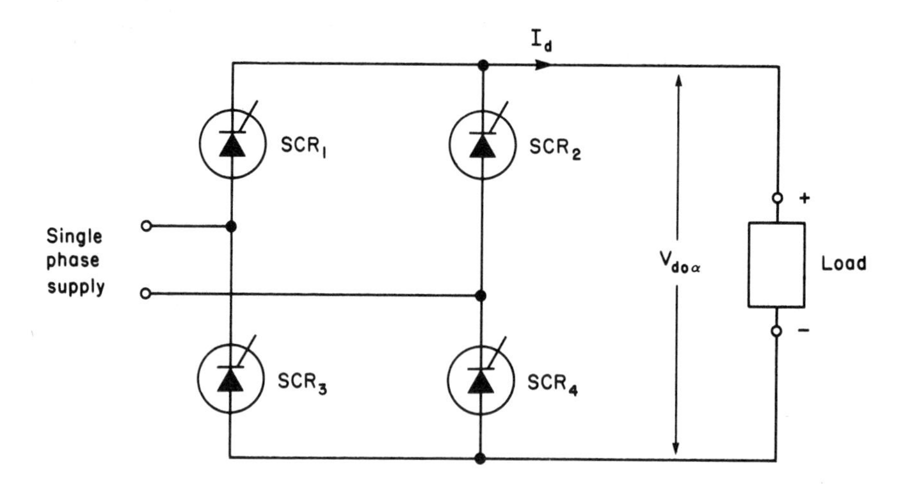

Figure 4-4. Two-pulse bridge converter.

The major differences of the bridge converter as compared to the midpoint converter are:

1. There are always two thyristors in conduction at the same time, which results in a device voltage drop double that of the midpoint converter.
2. The control circuitry is of necessity slightly more complex.

The peak forward and reverse voltages applied to the thyristors are equal to the maximum value of the ac source voltage.

The mean dc voltage $V_{do\alpha}$ is

$$V_{do\alpha} = \frac{2V_m}{\pi} \cos \alpha = 0.90 \text{ V} \cos \alpha \tag{4-13}$$

and the average thyristor current I_d is

$$I_d = \frac{V_m}{\pi R} = 0.45 \frac{V}{R} \tag{4-14}$$

It should be noted that the frequency of the ac ripple voltage is $2f$. The rms thyristor current is

$$I_{\text{rms}} = 0.707 I_d$$

4-3-3 Three-Pulse Midpoint Converter

The major advantage of the single-phase circuits is their relative simplicity; however, they are limited by the power capability of the single-phase source and the ripple content of the output voltage. Since most industrial applications have readily available three-phase power, most thyristor phase-controlled converters are supplied with three-phase power. The major benefits are an increased mean dc output voltage and a reduction in the amplitude of the ac ripple component.

The three-phase ac supply combined with suitable transformer connections produces an increase in the pulse number, or an increase in the number of dc voltage segments produced for each cycle of the ac supply voltage.

The simplest form of the **three-pulse midpoint converter** is shown in Figure 4-5.

The thyristors are supplied from a star-zigzag transformer. The zigzag-connected secondary requires two separate identical secondary windings on each phase, and for the same output voltage requires more copper than a conventional single winding secondary. The zigzag connection is used to prevent dc magnetization of the core by the dc component of the ac supply by permitting equal and opposite currents to flow in each half of the secondary winding.

From Figure 4-5(b) it can be seen that each thyristor is fired at the point where it becomes forward biased, that is, $\alpha = 0°$, or $30°$ after the phase voltage crosses the zero axis. As a result, the ac input phase voltage with the greatest instantaneous value is applied to the load, and the mean output dc voltage V_{do} is at its positive maximum. Each thyristor conducts for $120°$ and blocks reverse voltage for $240°$, the maximum reverse voltage equal to the line-to-line voltage of the transformer secondary, that is, $\sqrt{3}$ times the phase voltage. The ripple frequency is $3f$, where f is the ac supply frequency. The duration of current is $120°$ and the amplitude of the thyristor current is $I_d/3$. When $\alpha = 0°$, the supply voltage and the fundamental component of the sup-

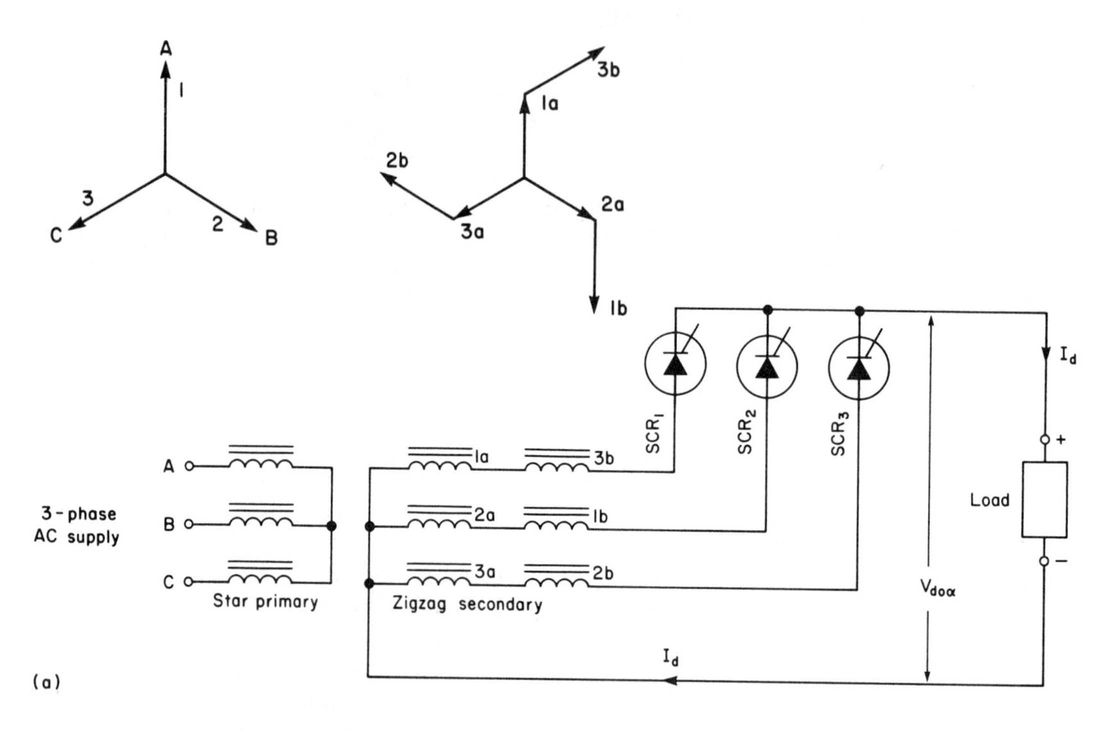

Figure 4-5. Three-pulse midpoint converter: (a) Basic circuit.

ply current are in phase, and the converter presents a unity power factor load to the ac supply.

In Figure 4-5(c), when $\alpha = 45°$, the thyristors are blocking for 45° from the natural commutation point, and the mean output dc voltage $V_{do\alpha}$ has been reduced; the supply current lags the ac supply voltage by 45°.

In Figure 4-5(d), when $\alpha = 90°$, the thyristors are blocking the forward and reverse voltages for equal periods and the mean output dc voltage is zero; the fundamental supply current lags the ac supply voltage by 90°.

To operate with firing delay angles greater than 90°, it is necessary to have a counter voltage source at the load, for example, a dc motor under regenerative conditions. When $\alpha = 135°$, see Figure 4-5(e), the thyristors are blocking voltage in the forward direction, and the mean output dc voltage $V_{do\alpha}$ is progressively becoming more negative. When $\alpha \simeq 175°$ the thyristors block forward voltage for almost the whole 180° allowing only sufficient time for turn-off after a very short conduction period.

For $0° < \alpha < 90°$ the converter is operating in the rectifying mode,

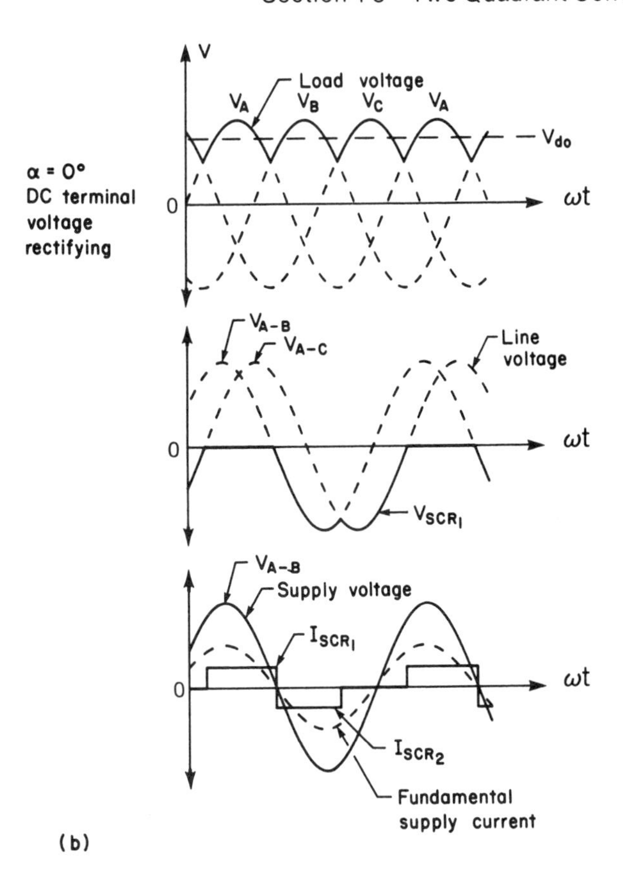

Figure 4-5 continued. (b) load voltage, V_{SCR_1}, I_{SCR_1}, and I_{SCR_2} and fundamental current waveforms for $\alpha = 0°$.

and for $90° < \alpha < 180°$ it is operating in the inversion mode as a synchronous inverter.

The peak forward and reverse voltages applied to the thyristors are $\sqrt{3}$ times the peak transformer secondary phase voltage V_m, or the peak line-to-line voltage.

The mean output dc voltage $V_{do\alpha}$ is

$$V_{do\alpha} = \frac{3\sqrt{3}\ V_m \cos \alpha}{2\pi}$$

$$= 1.17\ V \cos \alpha \tag{4-16}$$

and the average thyristor current is $I_d/3$, i.e.,

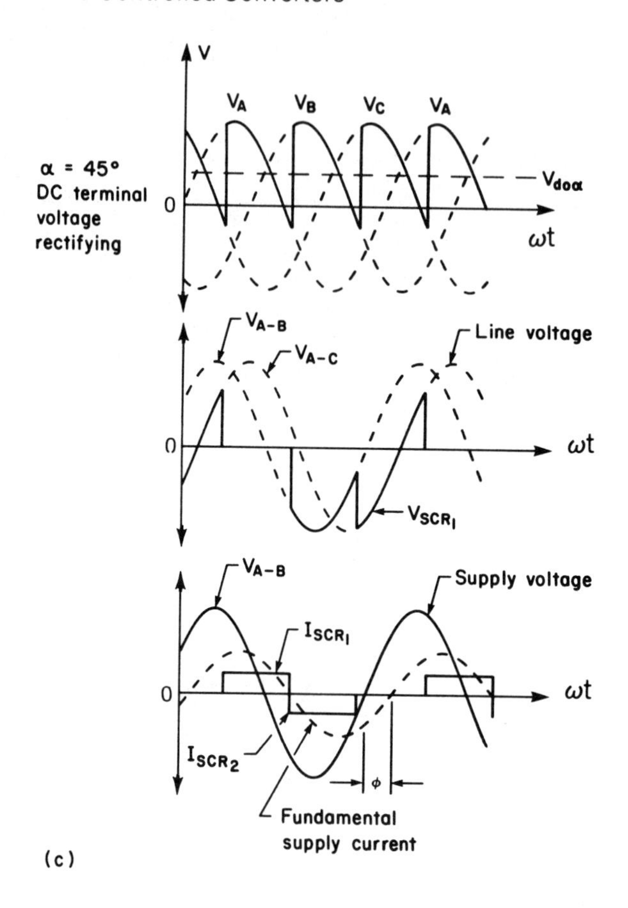

Figure 4-5 continued. (c) load voltage, V_{SCR_1}, I_{SCR_1}, and I_{SCR_2} and fundamental current waveforms for $\alpha = 45°$.

$$I_d/3 = \frac{\sqrt{3}\,V_m}{2\pi R} = 0.39\,\frac{V}{R} \tag{4-17}$$

and the rms thyristor current is

$$I_{\text{rms}} = 0.23\,\frac{V}{R} \tag{4-18}$$

4-3-4 Six-Pulse Midpoint Converters

When smoother load voltage is required the amplitude of the ac ripple component can be reduced by using a **six-pulse midpoint converter,** or a six-pulse midpoint converter with an interphase reactor.

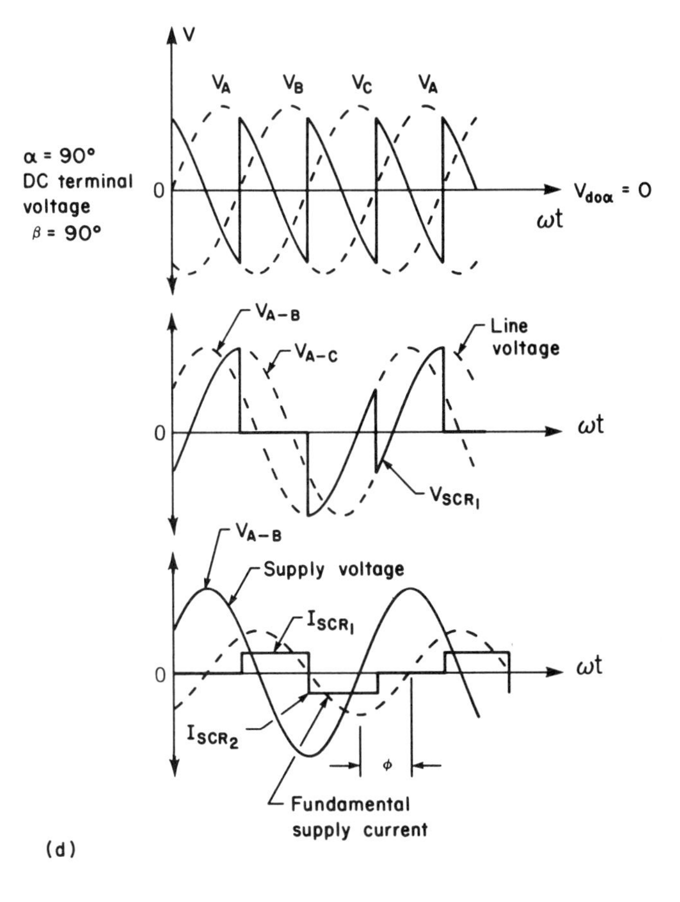

Figure 4-5 continued. (d) load voltage, V_{SCR_1}, I_{SCR_1}, and I_{SCR_2} and fundamental current waveforms for $\alpha = 90°$.

4-3-4-1 Six-Pulse Midpoint Converter

The simplest version of the six-pulse midpoint converter is shown in Figure 4-6. The thyristors are supplied by a three- to six-phase transformer. The major disadvantage of this arrangement is that each thyristor conducts only for 60°, resulting in poor utilization of the thyristors and the transformer secondaries because of the high ratio of rms to average current. As a result this configuration is rarely used.

4-3-4-2 Six-Pulse Midpoint Converter with Interphase Reactor

This converter arrangement consists basically of two three-pulse midpoint converters connected in parallel through an interphase reactor. Each converter operates independently of the other, the load cur-

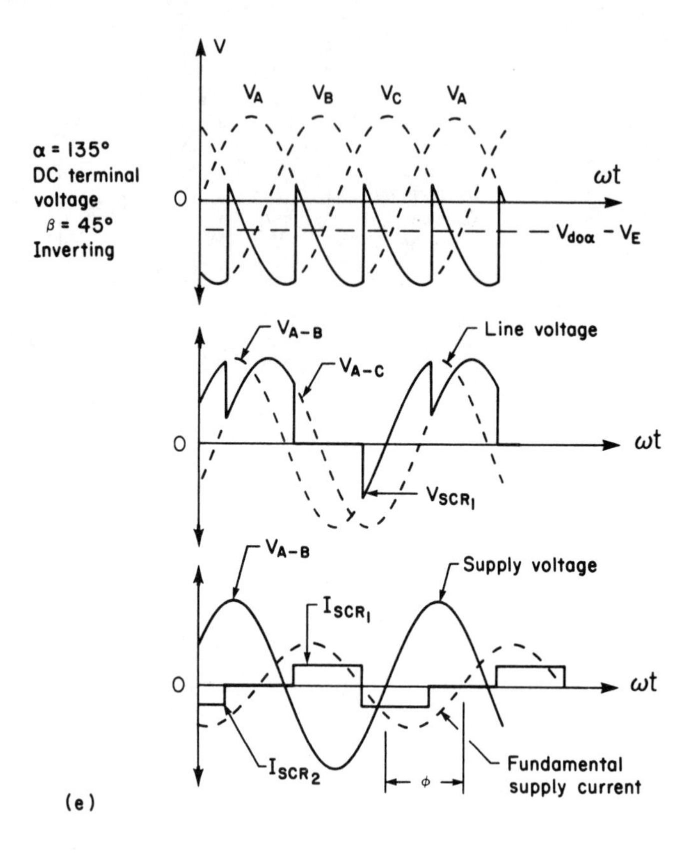

Figure 4-5 continued. (e) load voltage, V_{SCR_1}, I_{SCR_1}, and I_{SCR_2} and fundamental current waveforms for $\alpha = 135°$.

rent being shared equally between the converters, with the interphase reactor absorbing the difference between the out-of-phase voltages of the two converters. The frequency of the ac ripple component of each of the converters is $3f$; however, because of the phase difference between the ac ripple voltages, the output voltage at the center tap of the reactor has a ripple frequency of $6f$. As a result, for each cycle of the ac source voltage there are six segments of dc voltage in the mean dc output voltage. Unlike the previous converter, each thyristor conducts for 120°, resulting in improved device and transformer utilization.

The operation depends upon the interphase reactor carrying sufficient dc current to maintain an adequate flux level in the core. On light loads the core flux level will be low enough that the converter will operate as a normal six-pulse midpoint converter with the thyristors conducting for 60° with a discontinuous output and reduced mean dc output voltage. This problem can be overcome by connecting a bleeder resistance across the output terminals to ensure a minimum dc current through the reactor.

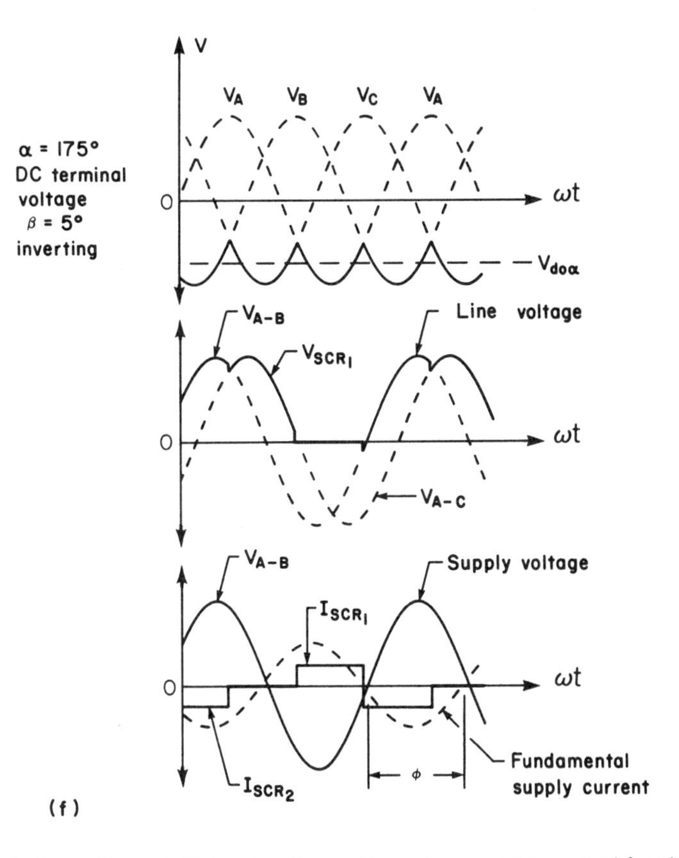

Figure 4-5 continued. (f) load voltage, V_{SCR_1}, I_{SCR_1}, and I_{SCR_2} and fundamental current waveforms for $\alpha = 175°$.

The basic circuit and the associated waveforms are shown in Figure 4-7 for firing delay angles $\alpha = 45°$ and $\alpha = 135°$. It should be noted that the transformer primary current no longer has a dc component present, and as a result more closely approximates a sine wave than has been the case with the previous converter configurations.

4-3-5 Higher Pulse Number Midpoint Converters

The pulse number can be increased to twelve by simply connecting two six-pulse midpoint converters with interphase reactors in parallel with each other through an interphase reactor. Similarly this combination can be extended to a twenty-four–pulse system. The major reason for increasing the pulse number is to reduce the amplitude of the ac ripple component and increase its frequency to permit simplified filtering if necessary.

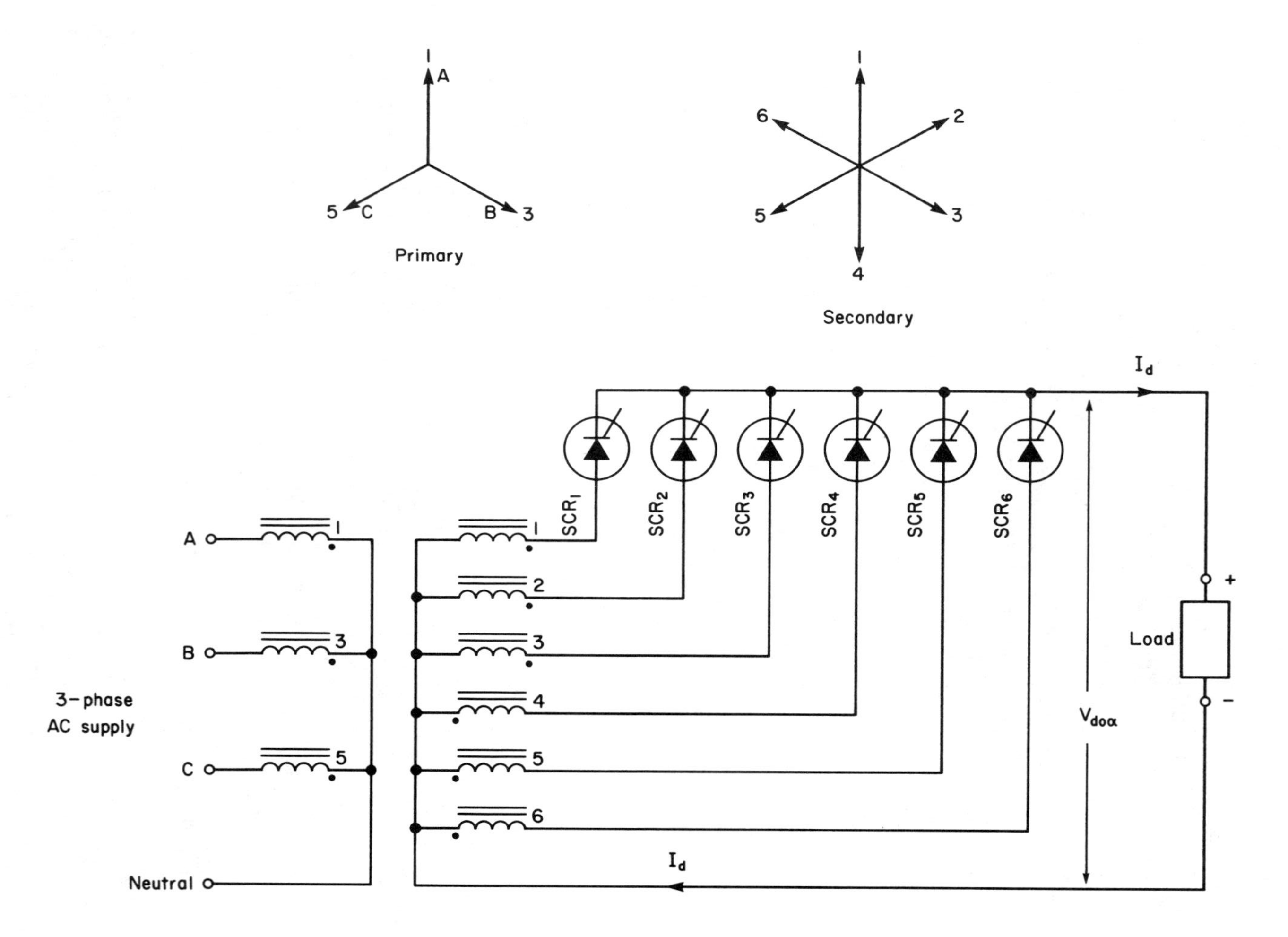

Figure 4-6. Six-pulse midpoint converter, basic circuit.

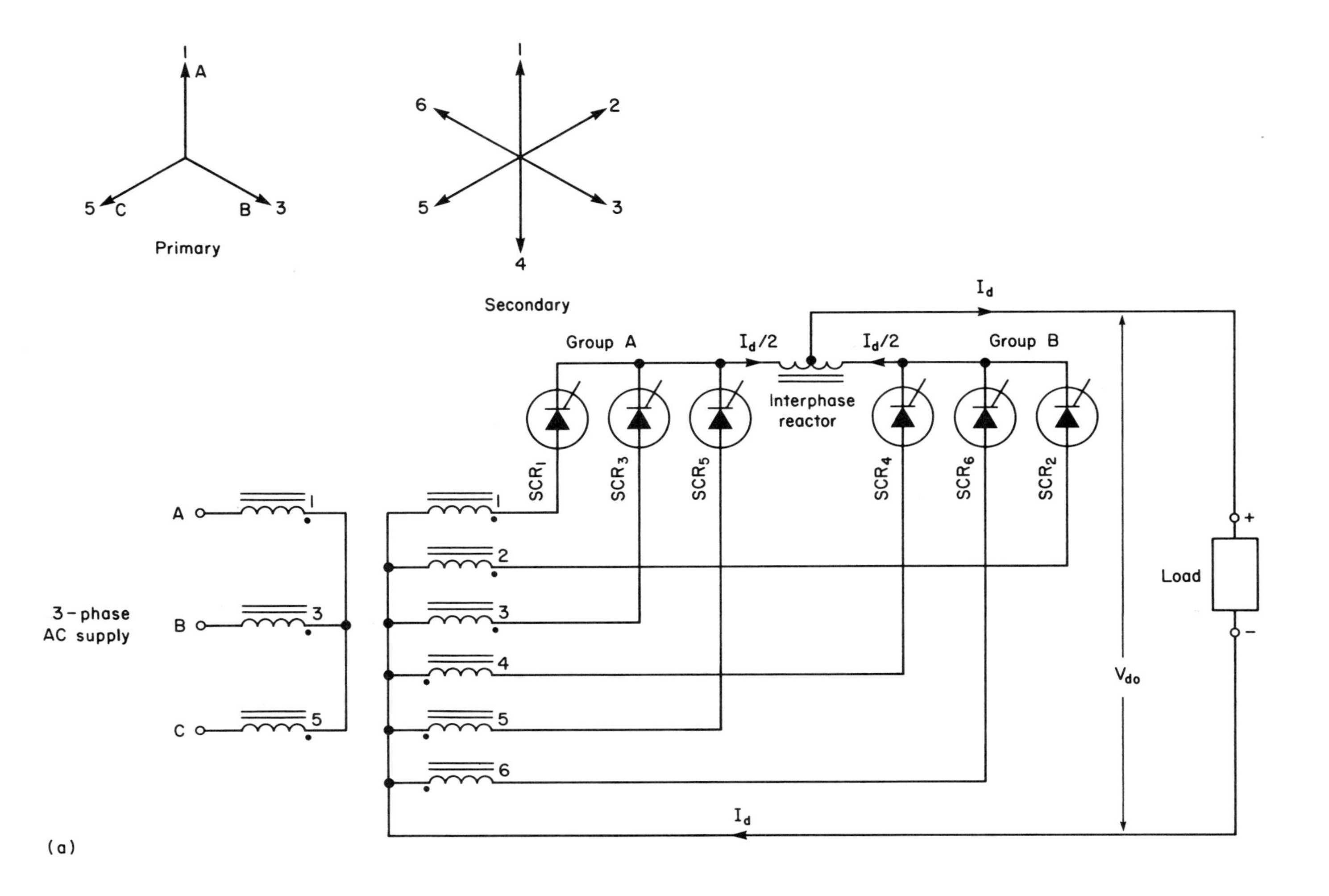

Figure 4-7. Six-pulse midpoint converter: (a) basic circuit.

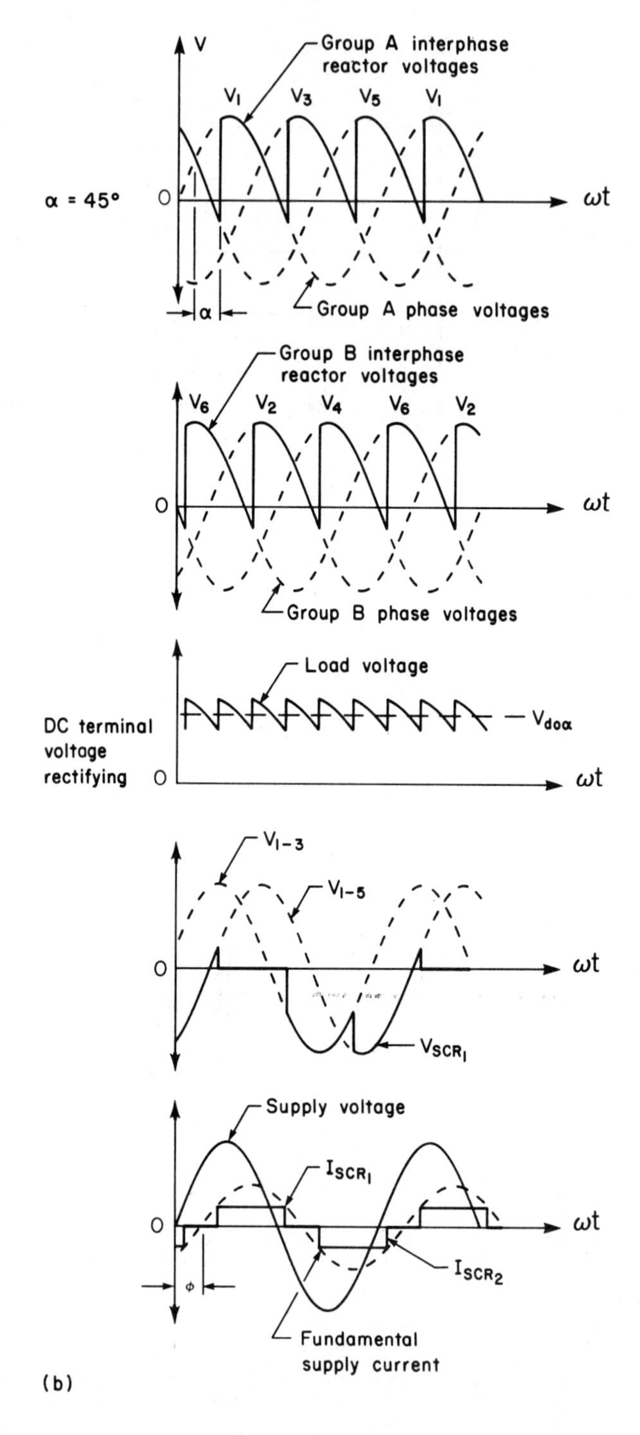

Figure 4-7 continued. (b) interphase reactor voltages, load voltage, V_{SCR_1}, I_{SCR_1}, and I_{SCR_2} and fundamental current waveforms for $\alpha = 45°$.

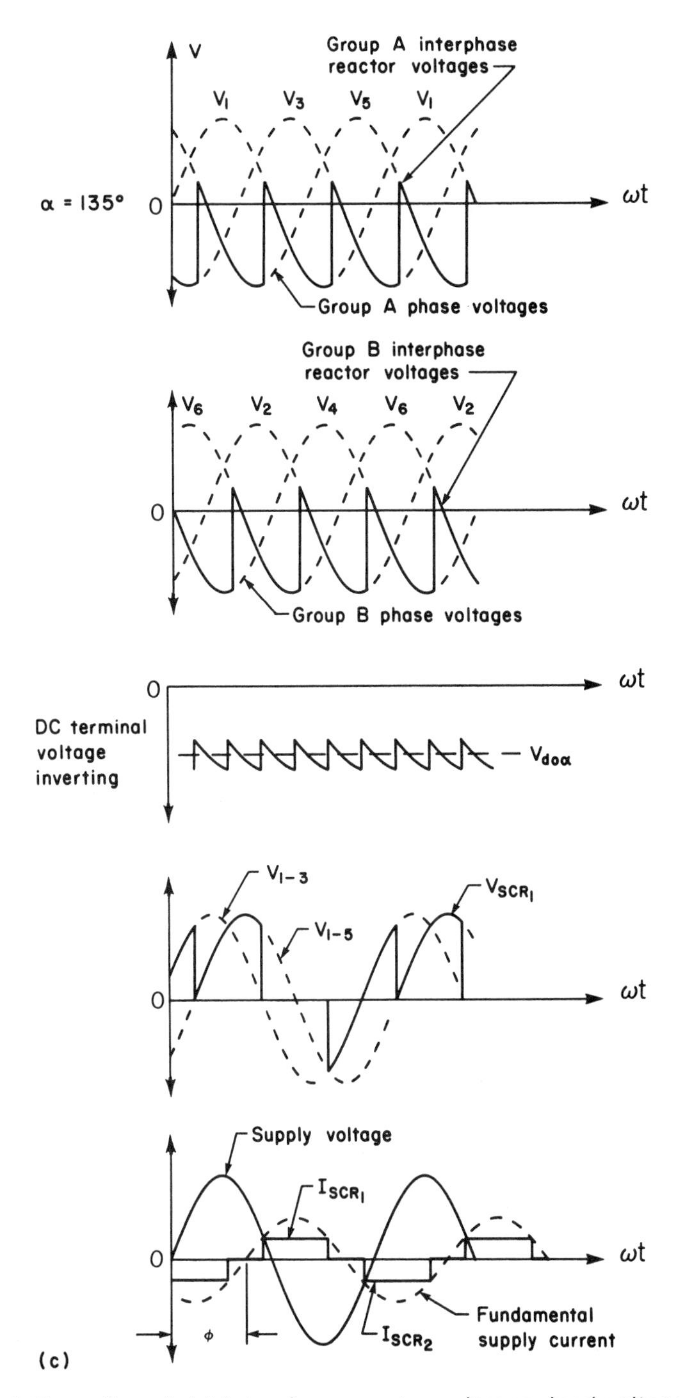

Figure 4-7 continued. (c) interphase reactor voltages, load voltage, V_{SCR_1}, I_{SCR_1}, and I_{SCR_2} and fundamental current waveforms for $\alpha = 135°$.

4-3-6 Six-Pulse Bridge Converters

The **six-pulse bridge converter** is basically two three-pulse midpoint converters connected in series, as shown in Figure 4-8(a). The neutral line has become redundant since the currents that flow in the neutral are equal and opposite. The converter configuration shown in Figure 4-8(b) is the result of eliminating the neutral. The benefits derived from this arrangement as compared to the three-pulse midpoint converter are:

1. The mean dc output voltage has been doubled from the same ac supply.
2. The amplitude of the ac ripple component has been more than halved and its frequency has been doubled to $6f$.
3. The converter can be supplied directly from the ac source thus eliminating the need of an input transformer.

From Figure 4-8(c) when $\alpha = 0°$, it can be seen that the mean output dc voltage V_{do} is the sum voltage of the two series-connected three-pulse midpoint converters. The positive group with the common cathode connection will have the thyristor with the most positive anode conducting. Similarily the thyristor with the most negative cathode will be conducting making the common anode connection negative. The mean output dc voltage V_{do} is the algebraic sum of the instantaneous voltages at the anode and cathode busbars and consists of segments of the three-phase line-to-line voltages. The frequency of the ac ripple component is $6f$. The magnitude of V_{do} is therefore double that of the individual three-pulse midpoint converters. Each thyristor conducts and carries the load current I_d for 120° and blocks for 240° and there is no dc current component in the ac line current. There must always be two thyristors in conduction at the same time, although the conduction periods of the positive and negative groups do not occur simultaneously. Assuming a supply voltage phase sequence ABC, the firing order of the thyristors is SCRs 1 and 2, SCRs 2 and 3, SCRs 3 and 4, SCRs 4 and 5, and SCRs 5 and 6, with commutations occurring alternately in the positive and negative groups every 60°.

When $\alpha = 45°$, the mean dc output voltage $V_{do\alpha}$ is reduced [Figure 4-8(d)] and is still twice that of a three-pulse midpoint converter. Since the current flow in the converter consists of positive and negative blocks of current 120° in duration, there is no dc component of current in the ac supply current.

As the firing delay angle is increased to $\alpha = 90°$ $V_{do\alpha}$ decreases to zero. Assuming that the load contains a source of negative voltage, further increases in α causes the mean output dc voltage $V_{do\alpha}$ to be-

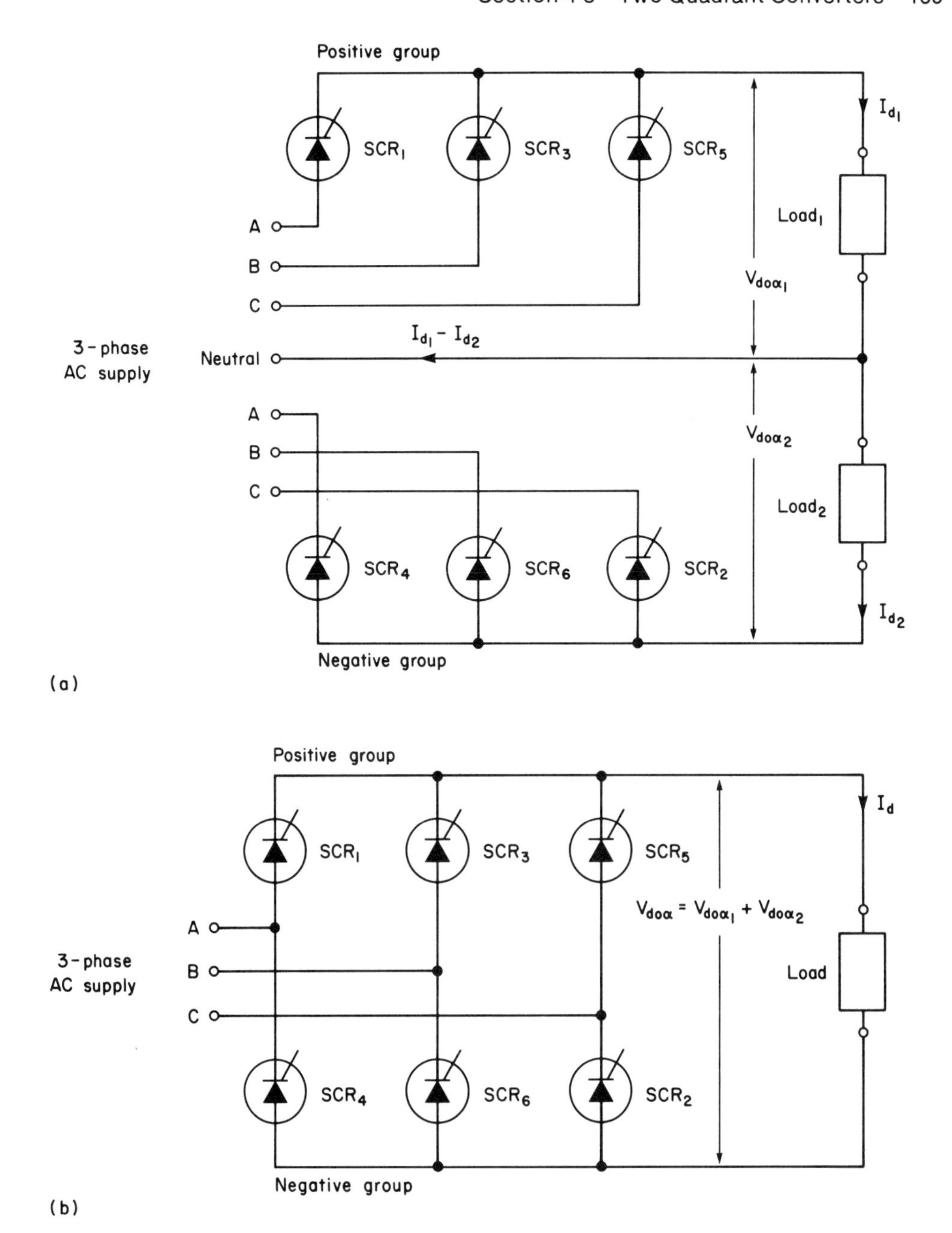

Figure 4-8. Six-pulse bridge converter: (a) basic concept; (b) basic circuit.

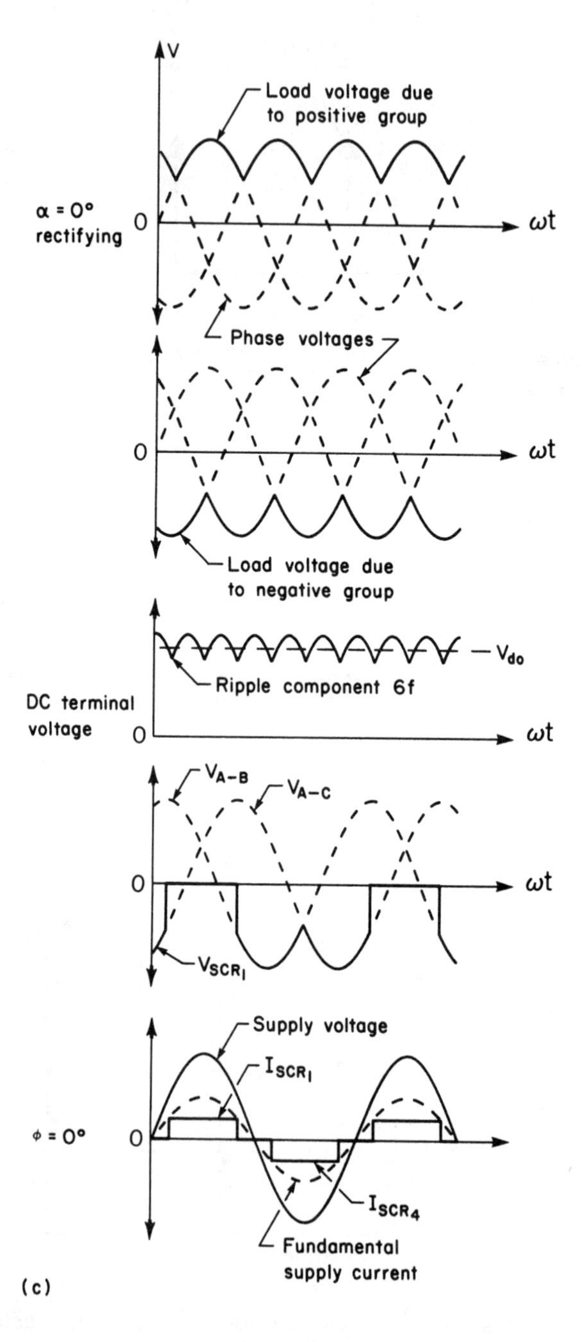

Figure 4-8 continued. (c) load voltages due to positive and negative groups, load voltage, V_{SCR_1}, I_{SCR_1}, and I_{SCR_4}, supply and fundamental current waveforms for $\alpha = 0°$.

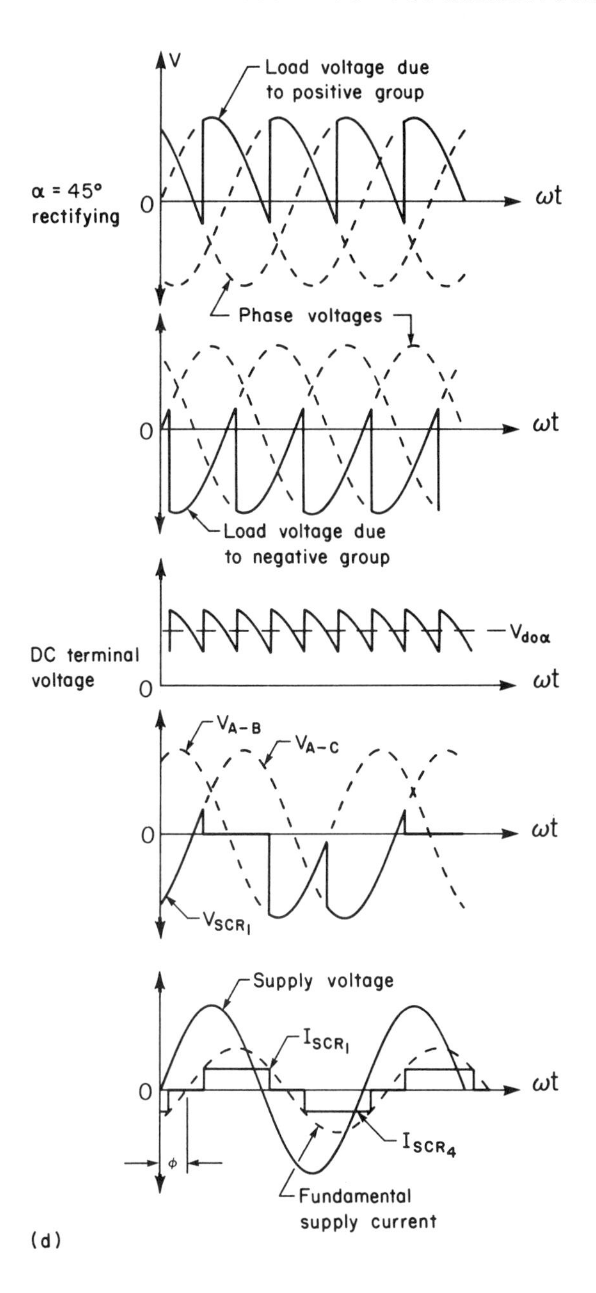

Figure 4-8 continued. (d) load voltages due to positive and negative groups, load voltage, V_{SCR_1}, I_{SCR_1}, and I_{SCR_4}, supply and fundamental current waveforms for $\alpha = 45°$.

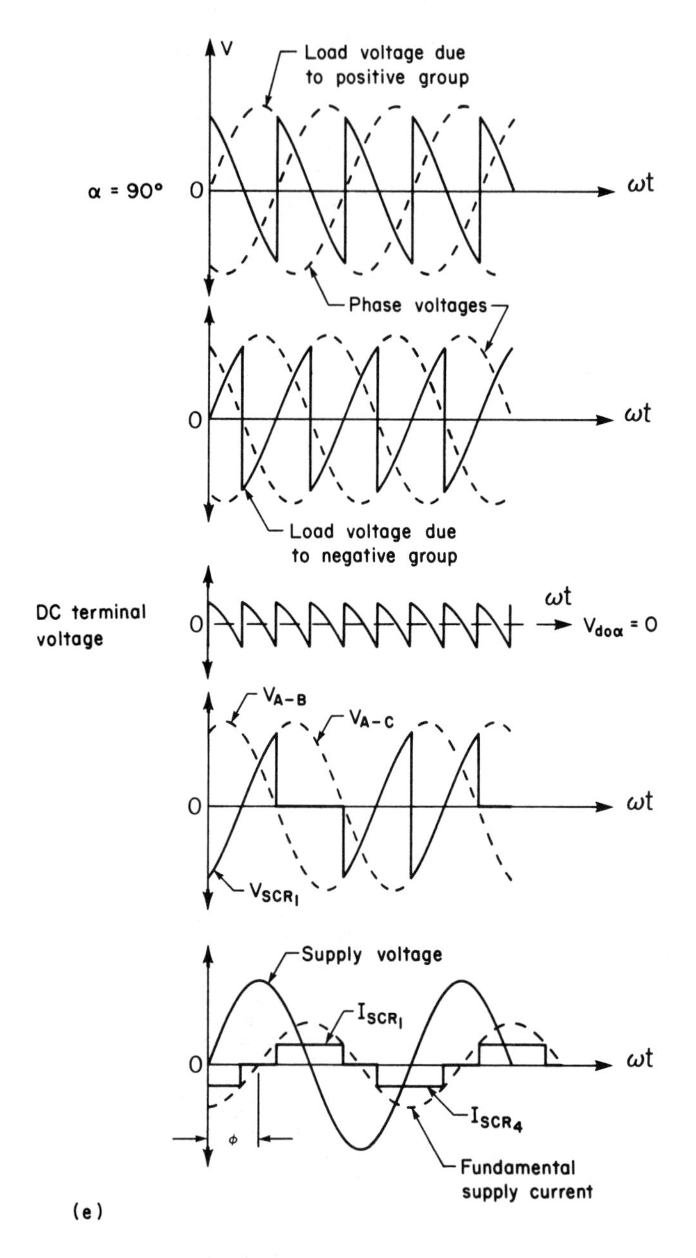

Figure 4-8 continued. (e) load voltages due to positive and negative groups, load voltage, V_{SCR_1}, I_{SCR_1}, and I_{SCR_4}, supply and fundamental current waveforms for $\alpha = 90°$.

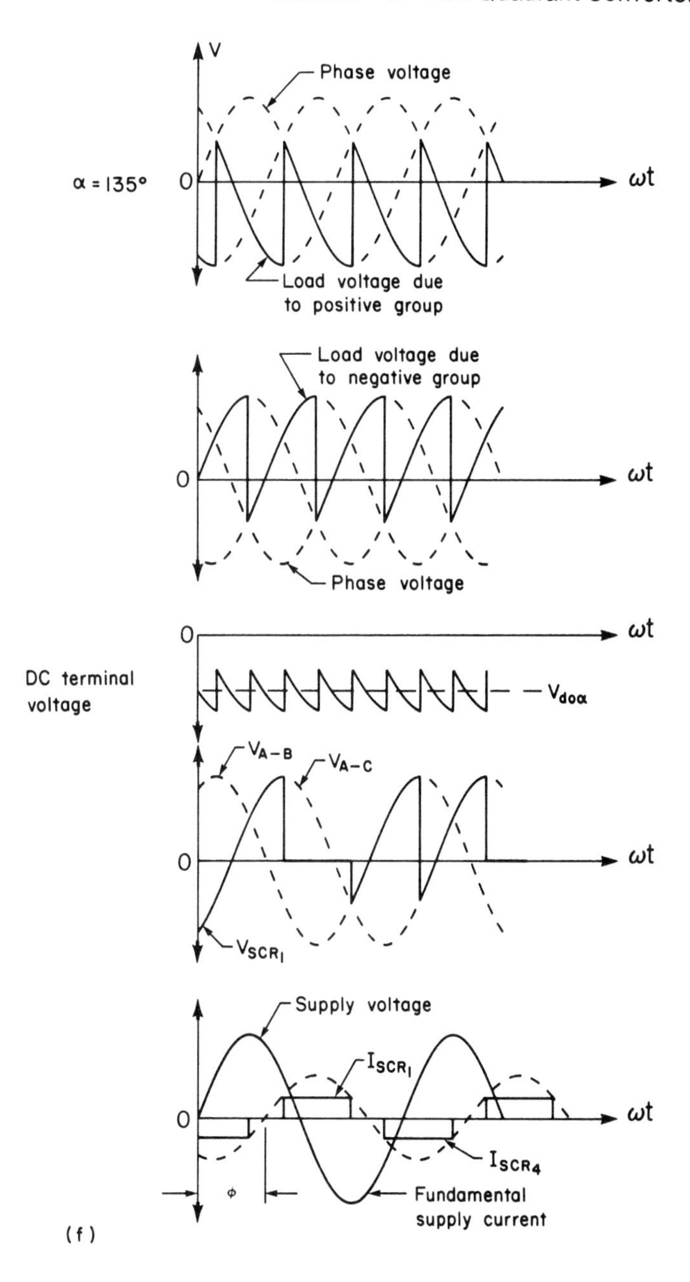

Figure 4-8 continued. (f) Load voltage due to positive and negative groups, load voltage, V_{SCR_1}, I_{SCR_1}, and I_{SCR_4}, supply and fundamental current waveforms for $\alpha = 135°$.

come increasingly more negative and the converter is operating in the synchronous inversion mode. $V_{do\alpha}$ will be at a negative maximum when $\alpha \simeq 180°$.

To summarize:

1. When $0° < \alpha < 90°$, the six-pulse bridge converter operates in the rectifying mode. When a source of negative voltage is present at the load it then operates in the inversion mode for $90° < \alpha < 180°$.

2. The peak inverse voltage applied to the thyristors is half that of the six-pulse midpoint converter, and the load current carried by the thyristors is double that of the midpoint converter.

3. The ripple frequency is $6f$, where f is the source frequency.

4. There is no dc component in the ac supply current, and the phase angle of the ac supply current lags the supply voltage by approximately the same angle as the firing delay angle α.

The peak forward and reverse voltages applied to the thyristors are $\sqrt{2}V$, where V is the rms line-to-line voltage.

The mean dc output voltage $V_{do\alpha}$ is

$$V_{do\alpha} = \frac{3\sqrt{2}V}{\pi} \cos \alpha$$

$$= 1.35\, V \cos \alpha \tag{4-19}$$

and the current carried by each thyristor is

$$\frac{I_d}{3} = 0.45 \frac{V}{R} \tag{4-20}$$

and the rms current carried by each thyristor is

$$I_{rms} = \frac{1}{\sqrt{3}} I_d = 0.78 \frac{V}{R} \tag{4-21}$$

Example 4-3

The three-phase six-pulse converter of Figure 4-8 is connected to a three-phase 460 V, 60 Hz supply. The dc load consists of a 480 V dc source with an internal resistance of 3 Ω. Calculate: (1) the power supplied and thyristor current for firing delay angles of (a) 30° and (b) 60°, and (2) the PIV applied to each thyristor.

Solution:

(1) (a)

$$\alpha = 30°, \text{ from Equation (4-19)}$$

$$V_{do\alpha} = 1.35\ V \cos \alpha$$

$$= 1.35 \times 460 \times \cos 30°$$

$$= 537.80 \text{ V}$$

The voltage drop across the internal resistance of the load is

$$V = V_{do\alpha} - E$$

$$= 537.80 - 480 = 57.80 \text{ V}$$

The dc load current I_d is

$$I_d = \frac{V}{R} = \frac{57.80}{3} = 19.27 \text{ A}$$

The power supplied to the load is

$$P_{do} = V_{do\alpha} \times I_d$$

$$= 537.80 \times 19.27$$

$$= 10.36 \text{ kW}$$

The thyristor current is

$$\frac{I_d}{3} = \frac{19.27}{3} = 6.42 \text{ A}$$

(b)

$$\alpha = 60°$$

$$V_{do\alpha} = 1.35\ V \cos \alpha$$

$$= 1.35 \times 460 \times \cos 60°$$

$$= 310.50 \text{ V}$$

Since the dc output voltage of the converter is less than the load voltage, this implies that the load current would be reversed. Since this is impossible, the load current I_d is zero and the power supplied by the converter is zero.

(2) The PIV applied to the thyristors is

$$\begin{aligned} \text{PIV} &= \sqrt{2}V \\ &= \sqrt{2} \times 460 \\ &= 650.54 \text{ V} \end{aligned}$$

When a six-pulse converter is operating in the inversion mode two requirements must be met: first, the mean output dc voltage $V_{do\alpha}$ must be negative, and second, the load must be active; that is, a dc voltage source of the correct polarity to ensure that load current will flow through the thyristors in the forward direction. These conditions are met in the case of a dc motor load under overhauling conditions (see Figure 4-9). Under overhauling conditions the motor counter emf E_c is opposite in polarity to the converter output voltage $V_{do\alpha}$ and greater in amplitude. As a result the load current I_d will be

$$I_d = (E_c - V_{do\alpha})/R \tag{4-22}$$

and since the load current is being produced by the active load, the power output from the dc motor acting as a generator is $P_{do} = V_{do\alpha} I_d$. Some of the generated power is dissipated as heat in the armature circuit of the motor, but the greater proportion is fed back through the converter to the ac source neglecting the small losses in the thyristors of the converter. The firing delay angle α is in the range $90° < \alpha < 180°$; however, in actual practice, α is not allowed to exceed approximately 165° since there is danger that the thyristors will not have sufficient time to commutate off and a short-circuit will result.

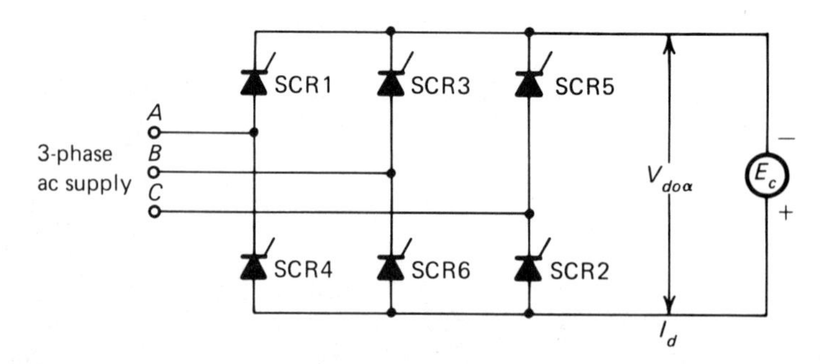

Figure 4-9. Six-pulse bridge converter operating in the synchronous inversion mode.

From our previous analysis of the waveforms, it was noted that the phase angle ϕ of the fundamental component of the ac line current with respect to the ac source voltage lagged by approximately the same angle as the firing delay angle α. Since the dc load absorbs only true power, that is, $P_{do} = V_{do\alpha} I_d$, the converter must absorb reactive power, the reactive power Q is

$$Q = P_{do} \tan \alpha \tag{4-23}$$

where

Q = reactive power, vars (kVARs)

P_{do} = true power, watts (kW)

α = the firing delay angle (deg).

Example 4-4

In Example 4-3 what is the reactive power absorbed by the converter when the firing delay angle is 30°, and what is the input power factor?

Solution:

From Equation (4-23)

$$\begin{aligned} Q &= P_{do} \tan \alpha \\ &= 10.36 \tan 30^\circ \\ &= 5.98 \text{ kVAR} \end{aligned}$$

The input power factor (PF) is

$$\begin{aligned} \text{PF} &= \cos \alpha \\ &= \cos 30^\circ = 0.866 \text{ lag} \end{aligned}$$

Example 4-5

A 240 V dc separately excited motor with an armature resistance of 0.2 Ω has its armature circuit supplied by a six-pulse converter connected to a 208 V, three-phase, 60 Hz source. The motor is operating under an overhauling load condition and the counter emf E_c is 245 V dc while the load current is 50A. Calculate: (1) the converter terminal voltage when in the inversion mode; (2) the required firing angle; (3) the true power returned to the ac source; (4) the reactive power; and (5) the PF of the system.

Solution:

(1)
$$
\begin{aligned}
V_{do\alpha} &= E_c - I_d r_a \\
&= 245 - 50 \times 0.2 \\
&= 235 \text{ V}
\end{aligned}
$$

(2) Since the ac source line-to-line voltage is 208 V and

$$
\begin{aligned}
V_{do\alpha} &= 1.35 \text{ V} \cos \alpha , \\
\cos \alpha &= \frac{V_{do\alpha}}{1.35 V} = \frac{235}{1.35 \times 208} = 0.84, \\
\alpha &= \cos^{-1} 0.84 \\
&= 33.19^\circ
\end{aligned}
$$

This is the firing delay angle required to produce 235 V when operating in the rectifying mode. The firing delay required to produce this voltage when operating as an inverter is

$$\alpha = 180^\circ - 33.19^\circ = 146.81^\circ$$

(3) The power supplied to the converter by the dc motor acting as a generator is

$$
\begin{aligned}
P_{do} &= V_{do\alpha} I_d \\
&= 235 \times 50 \\
&= 11.75 \text{ kW}
\end{aligned}
$$

(4)
$$
\begin{aligned}
Q &= P_{do} \tan \alpha \\
&= 11.75 \text{ kW} \times \tan 146.81^\circ \\
&= -7.69 \text{ kVARs}
\end{aligned}
$$

(5)
$$
\begin{aligned}
\text{PF} &= \cos \alpha \\
&= \cos 146.81^\circ \\
&= -0.84 \\
&= 0.84 \text{ lead}
\end{aligned}
$$

4-4 ONE-QUADRANT CONVERTERS

As has been seen in the previous section, two-quadrant converters operate with positive and negative dc load voltages, and in the rectifying mode convert ac power to dc power, and in the synchronous inversion mode will remove dc power from an active load and return it to the ac source as ac power.

There are a large number of applications that only require ac power to be converted to dc power, that is, the converter is only required to operate in the rectifying mode. Thyristor phase-controlled converters that operate in this manner are called **one-quadrant converters** or alternatively half-controlled or semiconverters. Basically, a one-quadrant converter consists of a bridge converter in which half the thyristors in each leg have been replaced by diodes, with a consequent overall reduction in the cost of the semiconductors and a simplified firing control circuitry.

Alternatively, one-quadrant operation can be obtained in both midpoint and bridge converters by connecting a freewheel diode across the output terminals of the converter. One-quadrant operation also results in a reduction of the ripple content of the mean output dc voltage and a reduction in the lag of the fundamental ac supply current with respect to the supply voltage.

The two-quadrant or full converter and the one-quadrant or semiconverter are the major connection arrangements used in industrial dc drive applications.

4-4-1 Two-Pulse Half-Controlled Bridge Converters

The simplest form of the two-pulse half-controlled bridge converter, shown in Figure 4-10, consists basically of two separate two-pulse midpoint converters connected in series. The positive group with a common cathode connection is a two-pulse midpoint converter with its output voltage being controlled by phase shift control of SCRs 1 and 2. The negative group with a common anode connection is uncontrolled and consists of diodes $D1$ and $D2$.

When the firing delay angle is varied, the mean output dc voltage $V_{do\alpha}$ is varied from a positive maximum to nearly zero as the firing delay angle is varied from $0°$ to nearly $180°$.

When $\alpha = 0°$, SCR1 and $D2$ conduct when point A is positive with respect to point B; when point B is positive with respect to point A, SCR2 and $D1$ are the conducting devices. In either case the potentials at the common anode and cathode connections are equal and opposite; therefore the voltage applied to the load is twice that of either converter, and the bridge converter is acting as a diode bridge. See Figure 4-10(b).

When $\alpha = 45°$, Figure 4-10(c), for the first 45° the voltages at the common anode and cathode connections are equal and have the same polarity; therefore the voltage applied to the load is zero. When SCR1 is turned on at $\alpha = 45°$, the voltage applied to the load is the sum of the voltages across both converters, and load current will flow through

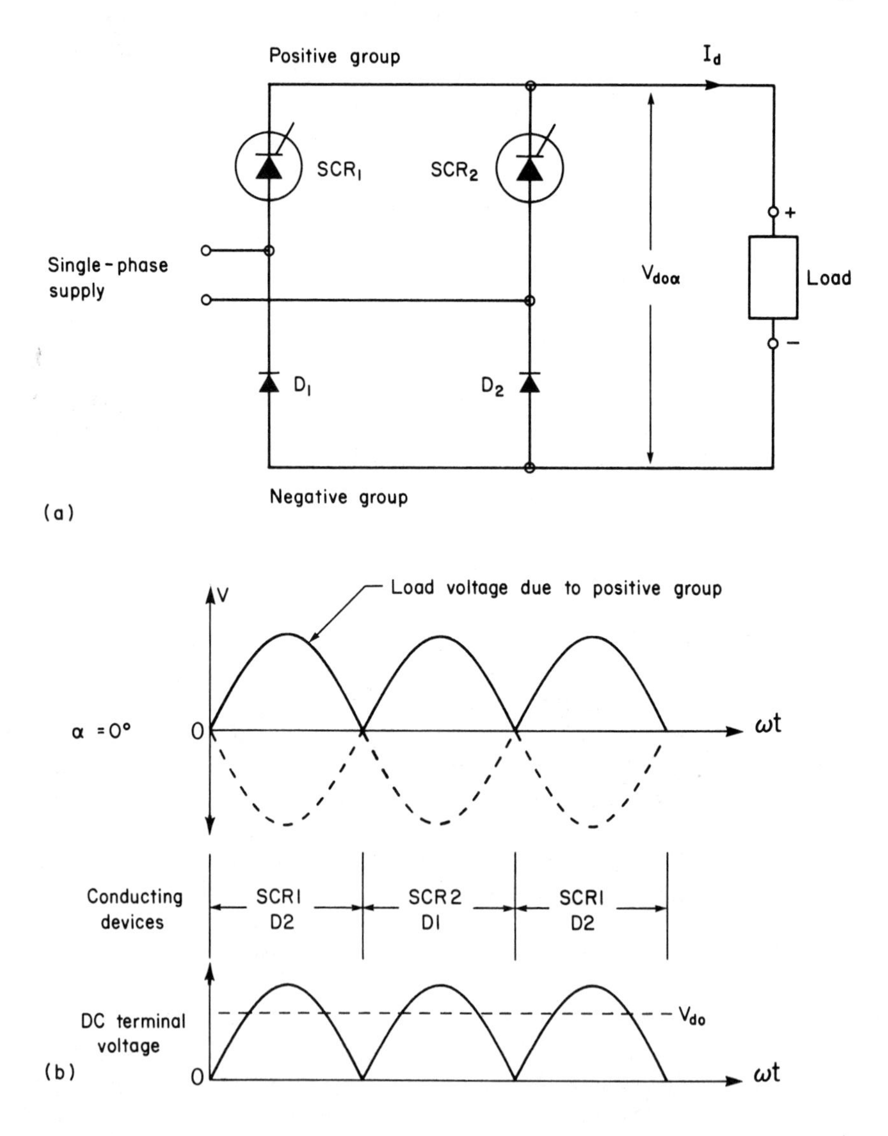

Figure 4-10. Two-pulse half-controlled bridge converter: (a) basic circuit; (b) load voltages due to positive and negative groups and dc terminal voltage for $\alpha = 0°$.

SCR1 and $D2$. When SCR1 becomes reverse biased, if an inductive load is assumed, SCR1 will remain in conduction until SCR2 is gated on; and the load current will freewheel through the path offered by SCR1 and $D1$, SCR1 remaining in conduction for 180°. As soon as SCR2 is turned on, SCR1 commutates off and load current flows

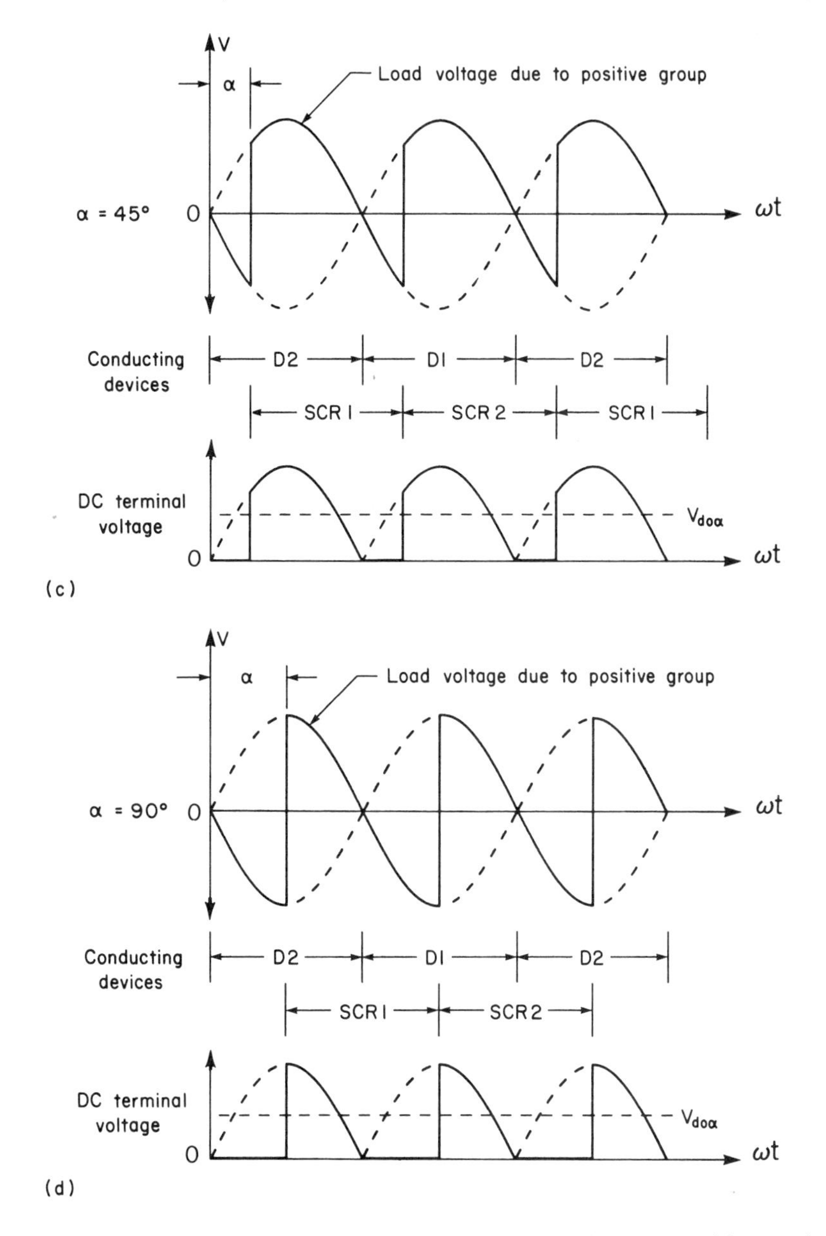

Figure 4-10 continued. (c) and (d) load voltages due to positive and negative groups and dc terminal voltage for $\alpha = 45°$, and 90°.

through SCR2 and $D1$ until they are reverse biased. Load current will freewheel through SCR2 and $D2$ until SCR1 is turned on again. It should be noted that during the intervals when the load current is freewheeling there is no current being drawn from the ac source.

When $\alpha = 90°$, exactly the same conditions prevail—see Figure 4-10(d)—except that there has been a reduction in the mean dc voltage applied to the load.

Continually increasing the firing delay angle results in further reductions in the mean dc voltage applied to the load. However, since the practical limit to increasing the firing delay angle is approximately 160° if commutation failure is not to occur, and since the load voltage is the sum of the voltages across the controlled and uncontrolled midpoint converters, it is impossible to reduce the mean dc output voltage to zero.

With this configuration, at small firing delay angles when supplying a highly inductive load when SCR1 becomes reverse biased, the load current will freewheel through SCR1 and $D1$, assuming that the load current has not decayed to zero during the time interval that SCR1 is reverse biased. Then as SCR1 again becomes forward biased with the gate signal removed, it will immediately conduct as if $\alpha = 0°$ and load current will flow through SCR1 and $D2$ for a complete half-cycle; the converter is said to be "half-waving." The only way the bridge can be turned off under this condition is to reapply the gate signal and increase the firing delay angle to reduce the mean output dc voltage and load current so that the load current will be insufficient to maintain the thyristor in conduction during the reverse-biased period.

Half-waving can be eliminated by rearranging the bridge circuit so that the freewheel path consists entirely of diodes as in Figure 4-11. This results in the conduction periods of the diodes increasing and the conduction periods of the thyristors decreasing. In fact the thyristors will block for 180°. As a result, the mean output dc voltage can be reduced to zero and removal of the gate signals will not result in "half-waving."

Other possible configurations of the two-pulse bridge converter operating as one-quadrant converters are shown in Figure 4-12.

4-4-2 Three-Pulse Half-Controlled Bridge Converter

The basic circuit of a **three-pulse half-controlled converter** is shown in Figure 4-13(a). It consists of a positive controlled three-pulse midpoint converter and a negative uncontrolled three-pulse midpoint converter. It should be noted that the relative positions of the two converters can be reversed without affecting performance. By varying the firing delay angle of the controlled midpoint converter over its full

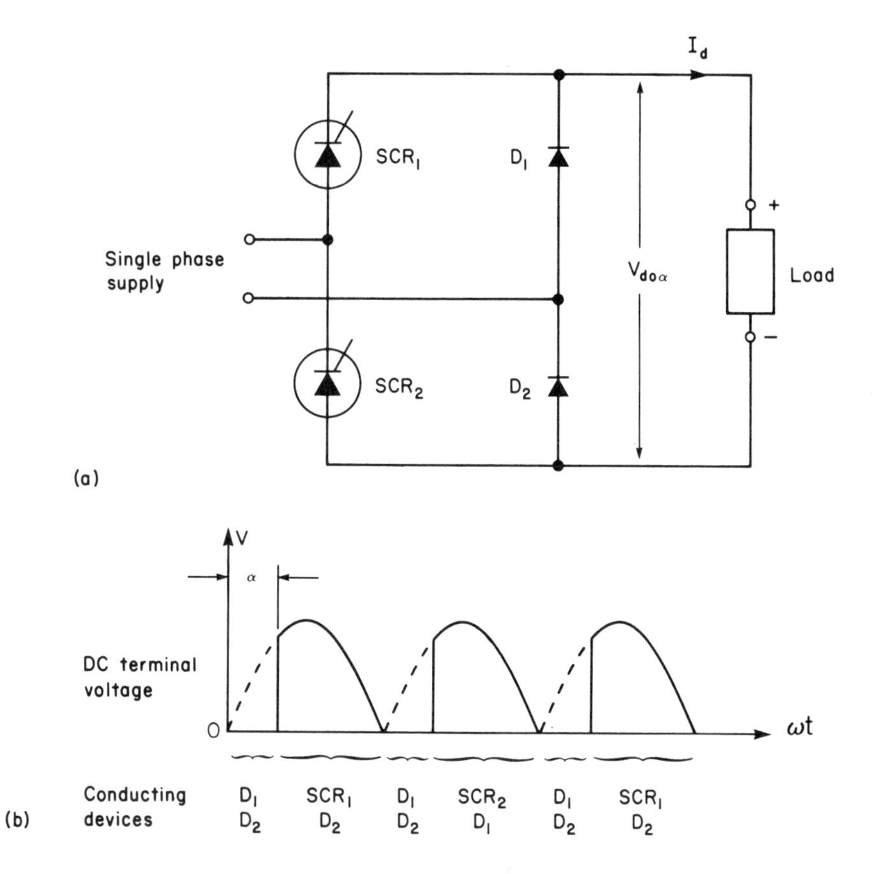

Figure 4-11. Two-pulse half-controlled bridge converter eliminating half-waving: (a) basic circuit; (b) dc terminal voltage.

range, the converter will operate in the rectification and inversion ranges; however, since its voltage is added to that of the uncontrolled converter, the mean output dc voltage $V_{do\alpha}$ can only be varied from a maximum to nearly zero.

When $\alpha = 0°$, the three-phase half-controlled converter acts as a three-phase bridge rectifier and the mean output dc voltage V_{do} is at its maximum and the ripple frequency of the ac component is $6f$. Each device conducts for 120° and the ac supply current consists of alternating components of 120° in duration in phase with the source voltage, see Figure 4-13(b).

Increasing the firing delay angle so that $\alpha = 60°$ results in the mean output dc voltage $V_{do\alpha}$ being the difference voltage between the positive controlled converter and the negative uncontrolled converter. However, the ac input current for the controlled converter has been retarded by 60° and then flows for 120°, while the ac current drawn by

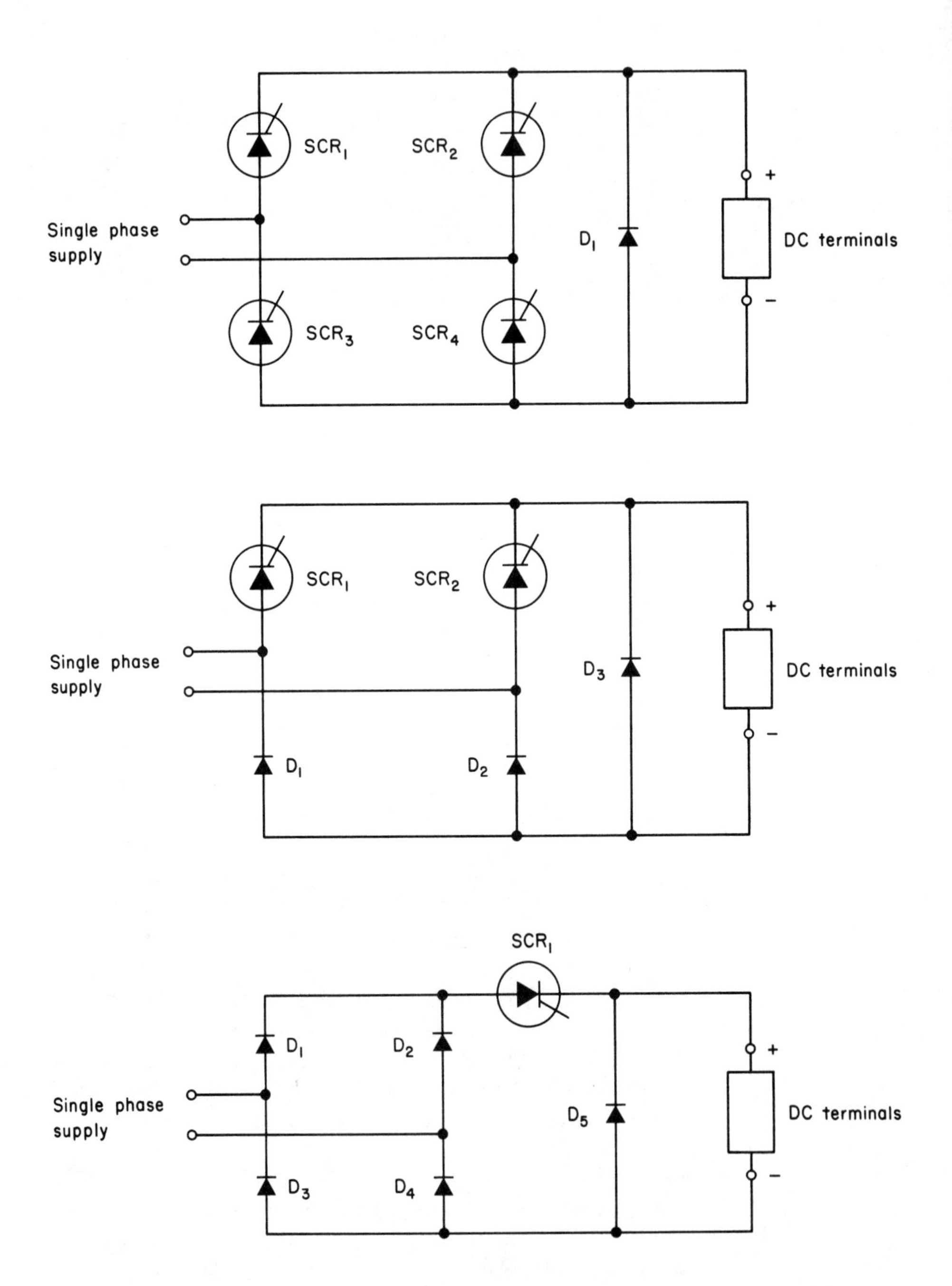

Figure 4-12. Two-pulse bridge converters, alternative arrangements.

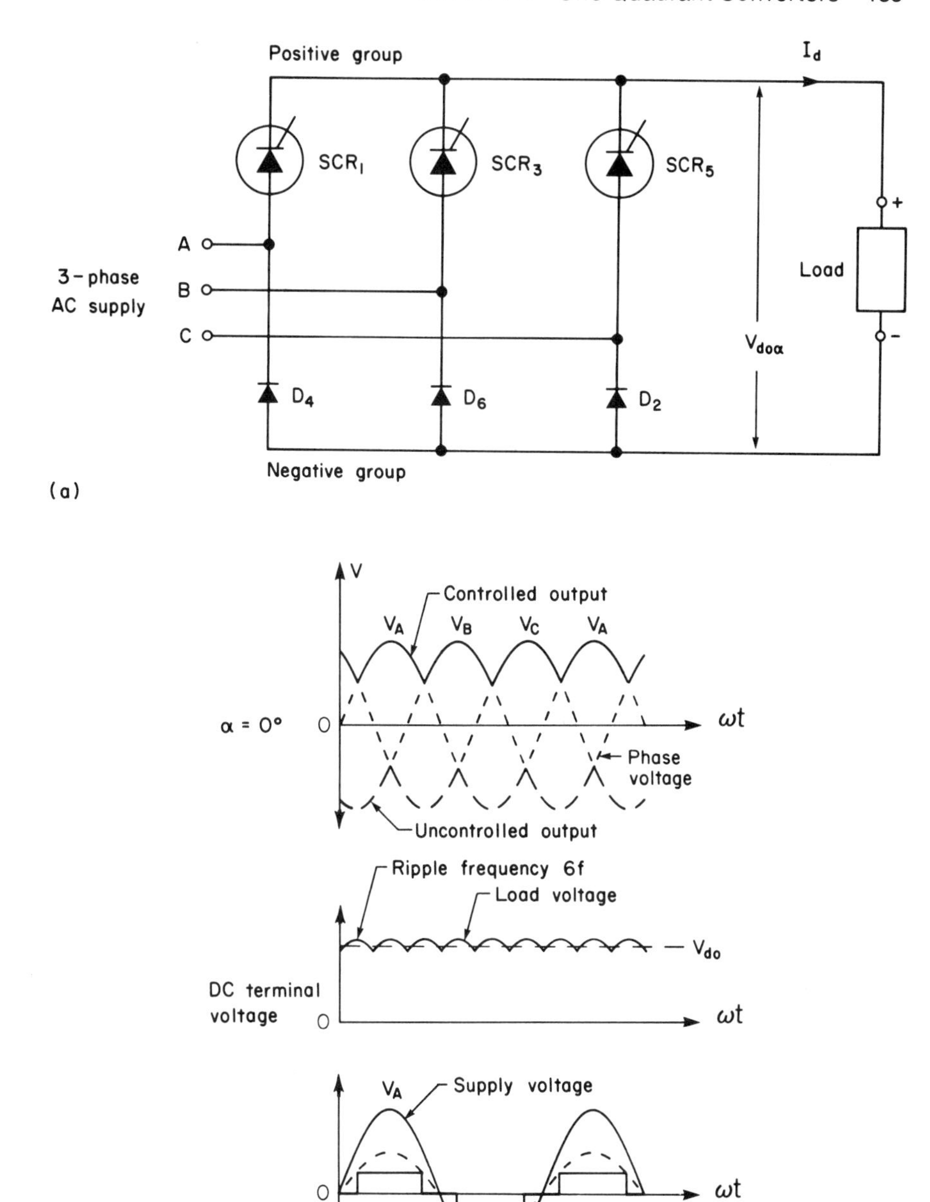

Figure 4-13. Three-pulse half-controlled bridge converter: (a) basic circuit. (b) load voltages due to controlled and uncontrolled groups, dc terminal voltage for $\alpha = 0°$.

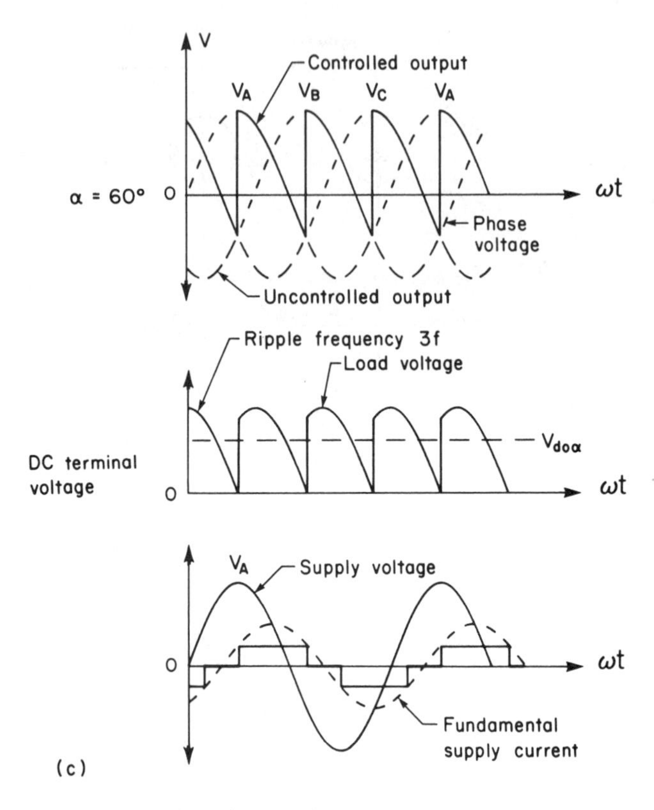

Figure 4-13 continued. (c) load voltage due to controlled and uncontrolled groups, dc terminal voltage for $\alpha = 60°$.

the uncontrolled converter remains in phase with the source voltage and flows for a duration of 120°. The resultant supply current is alternating, but the net phase shift of the supply current with respect to the source voltage is less than would be the case with a full-controlled converter. It should be noted that the ripple frequency of the load voltage decreases to $3f$ when $\alpha \geq 60°$. This is due to the appearance of a third harmonic component shortly after $\alpha = 0°$, and which becomes quite pronounced when $\alpha = 30°$.

When $\alpha = 60°$ the thyristor and diode in the same leg are both in conduction; increases in the firing delay angle cause this condition to continue with the load current freewheeling through a thyristor and diode. During the freewheeling interval the load voltage is zero. As can be seen from the load voltage waveforms, the amplitude of the ac ripple component is less, since there are no negative components, and the ripple frequency is half that of the full-controlled converter.

With a highly inductive load the three-phase half-controlled con-

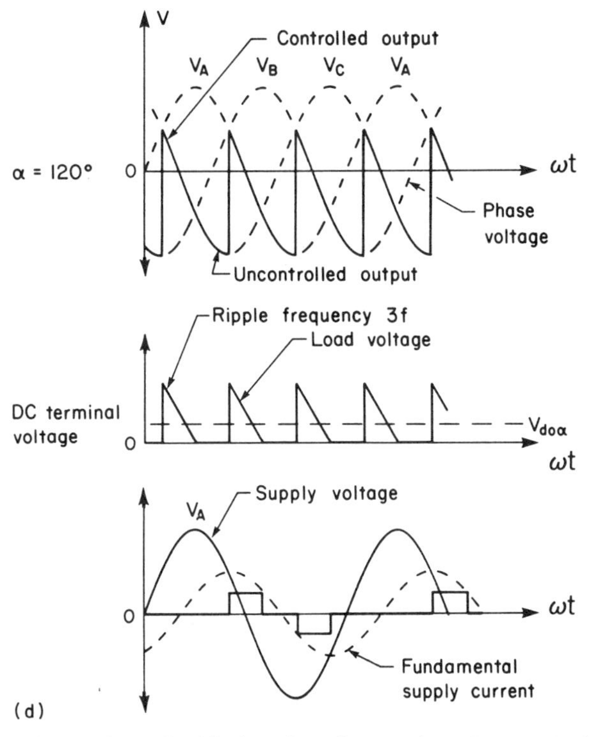

Figure 4-13 continued. (d) Load voltage due to controlled and uncontrolled groups, dc terminal voltage for $\alpha = 120°$.

verter can also "half-wave" at small firing delay angles. As a corrective measure, a diode is connected across the output terminals to prevent half-waving and at the same time it shifts the freewheeling path away from the thyristors.

The peak forward and reverse voltages applied to the thyristors and diodes is $\sqrt{2}V$, where V is the rms line-to-line voltage.

The mean dc output voltage $V_{do\alpha}$ is

$$V_{do\alpha} = \frac{3\sqrt{2}V}{\pi}(1 + \cos\alpha), 0° < \alpha < 180°$$

$$= 1.35\,V(1 + \cos\alpha), 0° < \alpha < 180° \tag{4-24}$$

and the current carried by each thyristor is

$$\frac{I_d}{3} = 0.45\frac{V}{R} \tag{4-25}$$

and the rms current carried by each thyristor and diode is

$$I_{\text{rms}} = \frac{1}{\sqrt{3}} I_d = 0.78 \frac{V}{R} \tag{4-26}$$

The current carried by the freewheel diode is

$$I_{\text{FWD}} = \frac{3\sqrt{2}\,V}{\pi} \frac{0.14}{R}$$

$$= 0.19 \frac{V}{R} \tag{4-27}$$

The major advantages of using half-controlled or one-quadrant converters as compared to full-controlled converters for undirectional applications are:

1. They are cheaper since only half the active devices are thyristors.
2. The firing control circuitry is simpler and therefore less costly.
3. The amplitude of the ac ripple component is less since there are no negative portions to the load voltage waveform.
4. The mean output dc voltage varies from a positive maximum to practically zero for $0° < \alpha < 180°$.
5. The ripple frequencies are half those of a full-controlled converter.

It should also be noted that a full-controlled converter can be made to operate as a one-quadrant converter by connecting a freewheel diode across the output terminals or by replacing half the thyristors with diodes. However, in the case of midpoint converters they can only be made to operate as one-quadrant converters by connecting a freewheel diode across the output terminals.

4-5 FOUR-QUADRANT OR DUAL CONVERTERS

There are a number of dc motor drive applications that require acceleration, speed control, and regenerative braking capability in both directions, that is, **four-quadrant** operation. Typical applications are mine hoists, reversing mill drives, and electric traction applications.

There are several methods by which forward and reverse operation of a dc drive motor may be obtained. First, the output of a two-quadrant converter can be connected to the motor armature by means of a two-pole reversing contactor, adding the complication of a time delay in the operation of the contactor, during which time control of the motor is lost. Second, the output of a two-quadrant converter remains connected to the armature circuit of the dc motor, but the polarity of the separately excited field is reversed. The major disadvantage of this method is that, because of the high time constant of the highly inductive field, the response of the drive will be slow unless the field current reversal is forced by applying a forcing voltage two or three times the normal field voltage. Third, the most effective method is to supply the armature circuit of a separately excited dc motor from two two-quadrant converters connected back to back as illustrated in Fig. 4-14. This system achieves control in both directions as well as regeneration without requiring the armature or field circuits to be reversed.

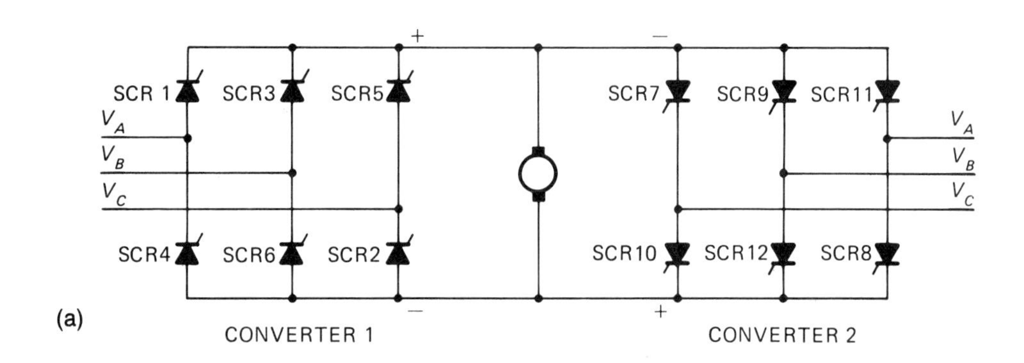

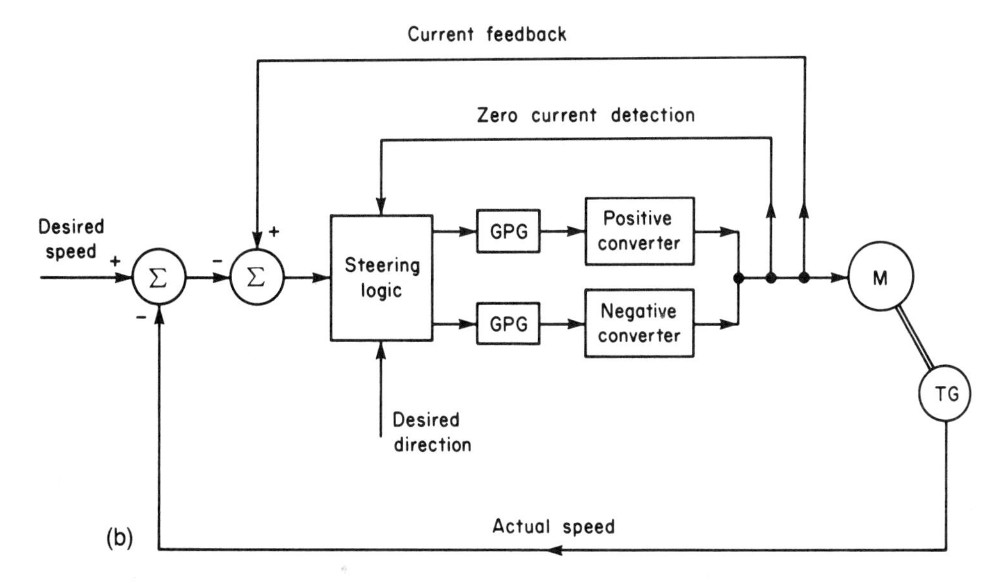

Figure 4-14. Four-quadrant or dual-converter: (a) basic circuit, (b) block diagram of a closed-loop control system.

By simultaneous firing angle control of both converters, very rapid reversals of direction and torque can be achieved. When operated in this manner converters are known as dual converters, and the current through the motor armature can flow in either direction. Ideally the firing delay angles of both converters are adjusted to give the same output voltage at their output terminals and therefore

$$v_{d\mathrm{POS}} + v_{d\mathrm{NEG}} = 0 \tag{4-28}$$

where $v_{d\mathrm{POS}}$ and $v_{d\mathrm{NEG}}$ are the instantaneous voltages of the positive and negative converters, and

$$V_{d\mathrm{POS}} \text{ and } V_{d\mathrm{NEG}} = 0 \tag{4-29}$$

where $V_{d\mathrm{POS}}$ and $V_{d\mathrm{NEG}}$ are the average voltages of the positive and negative converters.

Because the ac ripple components of the output dc voltages do not occur simultaneously, the condition defined by Equation (4-28) cannot be met. As a result, in normal practice, even though the firing control circuits generate gate pulse signals to both converters to meet the condition

$$\alpha_p + \alpha_N = 180° \tag{4-30}$$

the gate pulse signals to the negative converter are blanked out as long as the motor armature current is positive, and the gate pulse signals to the positive converter are blanked out as long as the motor armature current is negative. As a result, only one converter at a time is in operation, the inactive converter presents a high impedance. The transition from motoring to regeneration, that is, from quadrant I to quadrant IV, or quadrant II to quadrant III or vice versa (see Figure 4-1) takes place effectively in three steps. The current in the rectifying converter is reduced to zero by increasing the firing delay angle; then the gating signals are blanked out, and those of the other converter are restored and phased forward to increase the armature current in the opposite direction from zero. The polarity of the motor counter emf E_c remains unchanged. This method of operation is called the "circulating current free" mode.

A closed-loop control scheme for a dual converter operating in the circulating current free mode is illustrated in block diagram form in Figure 4-14(b). A voltage representing the actual speed is derived from a tachogenerator coupled to the drive motor shaft; this speed voltage is compared against another voltage representing the desired speed of the drive motor. The resulting error signal voltage is compared against a voltage representing the actual motor armature circuit current (usually this signal is derived from a shunt in series with the armature circuit). If the drive motor is not operating in current or torque limit, the error signal is supplied to the steering logic. The steering logic determines which converter will be conducting, determined by an applied voltage signal designating the desired direction of rotation of the drive motor armature. In addition a current sensor (shunt) in the armature circuit senses the direction of current flow and provides an inhibit signal which will prevent a changeover from the conducting converter until the current in the drive motor armature circuit is zero. Provided all of these conditions are met, a reference signal is applied to the appropriate gate pulse generator, which in turn will apply correctly sequenced firing pulses to the converter thyristors with the correct firing delay angle.

4-6 FIRING CIRCUITS

To date we have not considered how gating signals are produced and applied to the thyristors at precisely the correct time. Equally important because of the relatively low voltage level of most thyristor gate signals, is the requirement that the gate circuit be free of all unwanted or spurious signals, since a noise signal may be of sufficient amplitude and duration to turn on a thyristor.

4-6-1 Electrical Noise

Electrical noise consists of fast rise-time electrical transients that appear in supply and control wiring. In the industrial environment the principal sources of electrical noise are associated with high rates of change of voltage or current. The sources are switches, relay contacts, commutator noise from dc machines, inductive devices such as coils, solenoids and relays, induction heating, and thyristor switching. Noise can be transmitted by capacitive or inductive coupling, conduction, radiation, or common line injection. Normally the effects of electrical noise can be minimized by shielding, screening, isolating (both electrically and physically), filtering, and separating power wiring from the control wiring.

Some of the more common methods of suppressing noise are:

1. Noise suppression at its source. Usually it is not practical to try to eliminate noise, but minimization is a realistic objective. Some methods are to replace electromagnetic relays with solid-state relays with zero voltage switching capabilities. Capacitors connected across

the relay contacts help to remove transients. Diodes, varactors, zeners, and RC snubbers across inductive coils also help in the reduction of transient spikes.

2. Reducing the sensitivity of the control circuitry. In logic controls the use of high threshold logic (HTL) reduces the circuit sensitivity at the expense of speed capability.

3. Reducing the coupling between control circuits and noise sources. In most situations involving noise suppression, this area yields the most effective results at a minimum cost. Some of the techniques that should be followed in all solid-state converter applications either during manufacture or during installation are:

(a) Provide the greatest possible physical distance between the control circuits and the power-carrying circuits. This minimizes magnetic and capacitive coupling.
(b) Use the shortest distance between connections and avoid grouping wiring runs.
(c) Magnetic coupling can be reduced, for example, by twisting conductors with a minimum of one twist for every two inches.
(d) Capacitive coupling can be minimized by using shielded cable, with the shielding being grounded at one point only.
(e) Never ground to the electrical common but to the control system common.
(f) If the control system is supplied from the same ac source as the solid-state converter, the source should be filtered.
(g) In closed-loop control systems (for example, in current detection circuits) transformers operating at power line frequencies have a significant interwinding capacitance, and the screening between the windings should be grounded.

4-6-2 Firing Circuit Parameters

In general, the following criteria should be met in all firing circuit designs:

1. The applied gate pulse must be of sufficient amplitude and duration so that the thyristor is turned on when required.

2. Provide accurate firing angle control over the required range.

3. Minimize the time delay between a change in the controlled variable (e.g., speed) and the corrective change in the firing delay angle.

4. Achieve a linear relationship between changes in the control signal and the converter output.

5. In polyphase applications accurately maintain the phase relationship between the firing signals and ensure that there is simultaneous firing of all series- and/or parallel-connected thyristors in the same leg or phase.

4-6-3 Firing Delay Angle Control Techniques

Numerous techniques of varying the firing delay angle α have been developed for three-phase systems; however, we intend to concentrate only on the commonly used approaches that form the basis for control in industrial applications. The basic concepts of cosine crossing, integral and phase-lock control will be examined.

4-6-3-1 Cosine Crossing

The **cosine-crossing technique** compares a variable dc reference voltage V_R to a cosine-timing signal so that under continuous conduction conditions there is a linear relationship between the dc reference voltage V_R and the mean dc output voltage $V_{do\alpha}$.

For a two-quadrant converter with phase control the mean dc output voltage is

$$V_{do\alpha} = V_{do} \cos \alpha \tag{4-31}$$

and for a one-quadrant half-controlled converter with phase control the mean dc output voltage is

$$V_{do\alpha} = V_{do} (1 + \cos \alpha) \tag{4-32}$$

The dc voltage ratio for a two-quadrant converter is

$$\frac{V_{do\alpha}}{V_{do}} = \cos \alpha \tag{4-33}$$

and for the one-quadrant converter is

$$\frac{V_{do\alpha}}{V_{do}} = (1 + \cos \alpha) \tag{4-34}$$

Equation (4–31), when plotted as shown in Figure 4–15(a), is a cosine curve.

Assuming that the maximum value of the cosine-timing wave is V_m and that the maximum value of the dc reference voltage V_R is equal to V_m, then we have

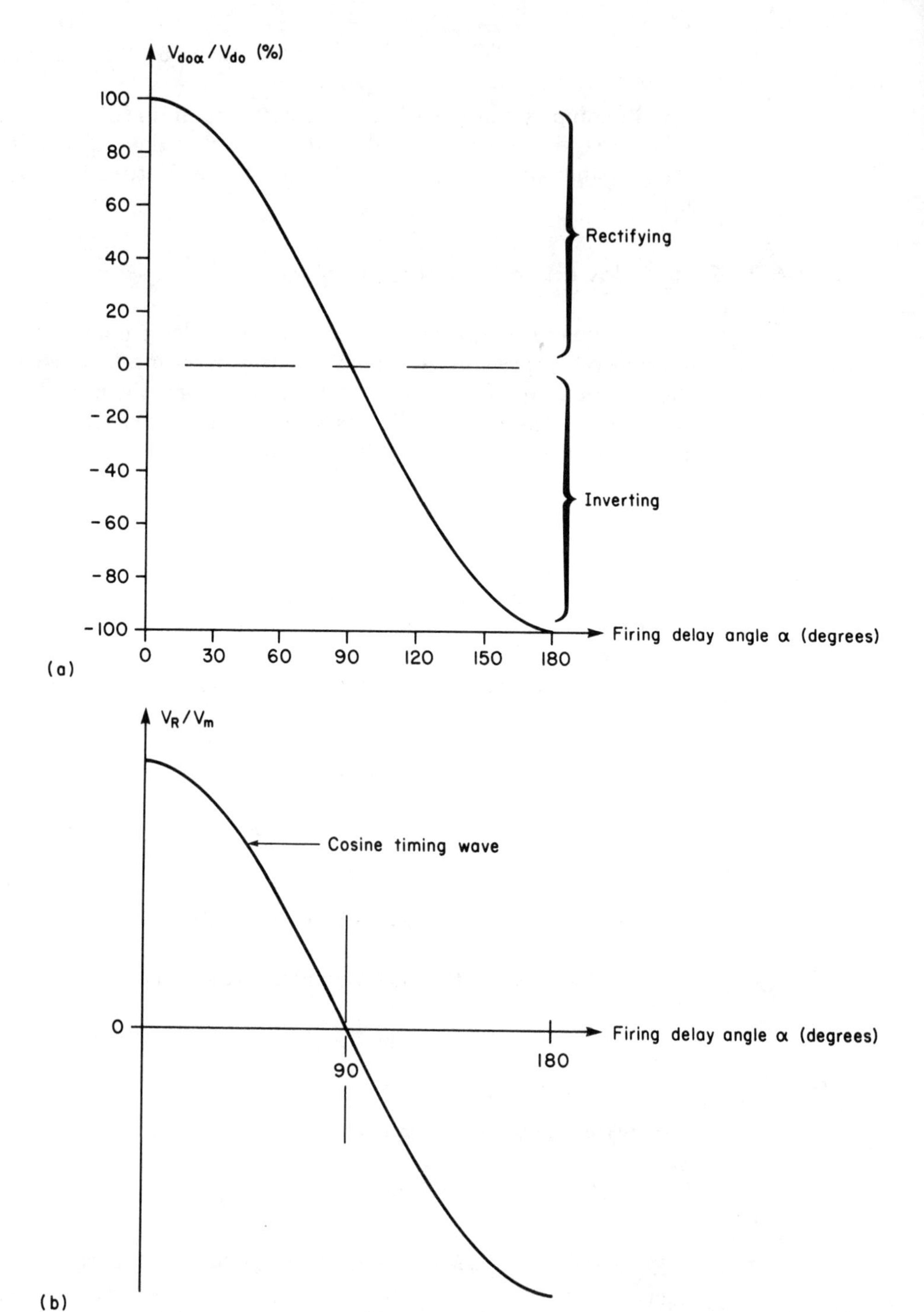

Figure 4-15. Variation of $V_{do\alpha} / V_{do}$ versus firing delay angle α for a two-quadrant converter.

$$V_m \cos\ \alpha = V_R \tag{4-35}$$

and

$$\cos\ \alpha = \frac{V_R}{V_m} \tag{4-36}$$

This relationship is illustrated in Figure 4–15(b), and from Equations (4-33) and (4-36)

$$\frac{V_{do\alpha}}{V_{do}} = \frac{V_R}{V_m} = \cos\ \alpha \tag{4-37}$$

and

$$\frac{V_R}{V_{do}} = \text{a constant} \tag{4-38}$$

The transfer function is linear. A similar relationship can be developed from Equation (4–34) and (4–36) for half-controlled converters by biasing V_R so that $V_R = 0$ when $\alpha = 180°$. Then

$$\frac{V_{do\alpha}}{V_{do}} = \frac{V_R}{V_m} = (1 + \cos\ \alpha)$$

and

$$\frac{V_R}{V_{do}} = \text{a constant} \tag{4-39}$$

The resulting transfer function is linear.

The basic concept of the cosine-crossing method of controlling the firing delay angle is shown in block diagram form in Figure 4–16.

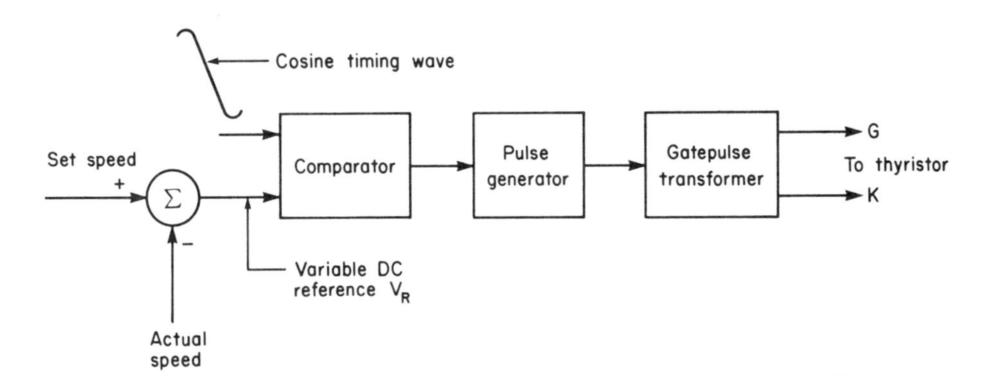

Figure 4–16. Block diagram of cosine-crossing firing control.

The dc reference voltage V_R is the error signal derived by comparing a reference input signal, designating the desired speed, against a negative feedback signal representing the actual motor speed. This signal is obtained from a voltage divider network sensing the motor counter emf or from a tachogenerator coupled to the motor.

The cosine-crossing method, in order to accurately time and position the gating pulses to the SCRs, requires that the following conditions be met:

1. Control must be exerted to ensure that $0° < \alpha < 180°$; this is known as the "limit" signal or "end-stop" control. Failure to maintain the firing delay angle within these limits will either cause the SCRs not to be triggered, or a commutation failure will occur, which will probably cause device failure. A practical way to achieve this control is to ensure that the variation of the dc reference voltage V_R is slightly less than the peak values of the cosine-timing wave. A simple but effective way to ensure that this is within the required limits is to clamp the maximum positive and negative values of V_R by means of a pair of inverse-parallel–connected zener diodes at the comparator input. This method has the added advantage that the operation of the converter, when in the synchronous inversion mode, can be controlled to ensure that there is sufficient time for the SCR to recover its forward blocking capability prior to being forward biased.
2. A reference signal V_R, or "advance-retard" signal should be available to advance or retard the firing delay angle.
3. When $V_R = 0$ the firing delay angle should be retarded so that the mean dc output voltage $V_{do\alpha} = 0$.
4. A biasing signal should be available to ensure that the converter will not operate in the rectifying mode irrespective of the firing delay angle when the converter is operating under current limit conditions.

These signal conditions are interpreted by the "gate pulse generator," which at the same time must provide a linear relationship between the reference signal and the mean output dc voltage $V_{do\alpha}$

In three-phase converters the limit or end-stop signals are derived from the line-to-line ac source voltage, usually $L1$–$L2$ as a reference transformed down to a suitable voltage level, phase-shifted and filtered by an RC network, and then applied to the gate pulse generator, so that it will permit gate pulses to be generated when it is greater than 0 V.

Similarly, at the same time the cosine-timing waves are generated by phase-shifting techniques but leading the limit signals by 90°. The firing delay angle is obtained by comparing the positive or negative dc reference signal against the appropriate cosine-timing wave by means of a comparator. As V_R is increased positively, the firing delay angle is reduced; that is, the firing point is advanced. Similarly, if V_R is increased negatively, the firing point is retarded.

The system in block diagram form is shown in Figure 4-17. The output of the comparator initiates the production of a gate pulse signal, provided that the limit signal is permissive. The gate pulse signal is in turn amplified by a gate pulse amplifier before being applied to the primary of the gate pulse transformer. If optoelectronic couplers are used, the gate pulse amplifier may be eliminated, since these are compatible with TTL logic. Modern practice is to use NAND logic for the gate pulse generator. It is also common practice to provide a bias signal that ensures that the firing delay angle is fully retarded when the reference input signal designating the desired speed is zero.

The major advantage of the cosine-crossing method of gate control is that it automatically responds to variations in the ac source voltage. A decrease in the source voltage will reduce the amplitude of the cosine-timing wave and, since the dc reference signal is independent of this voltage and will remain constant, the firing delay angle will be reduced and the mean output dc voltage will remain constant. Similarly, with an increase in the ac source voltage the firing delay angle will be in-

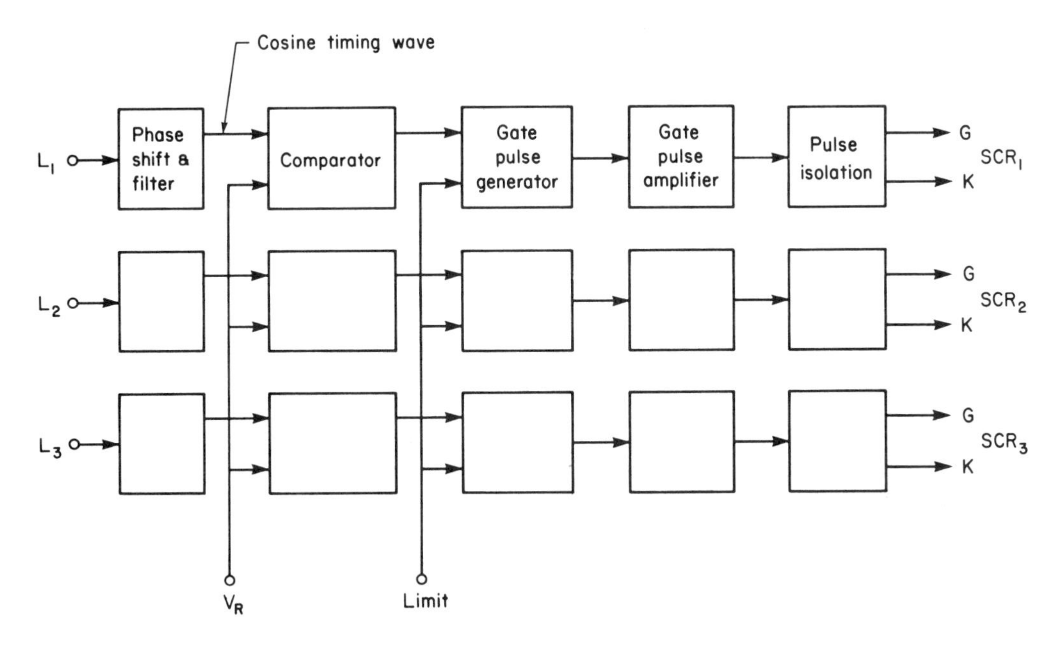

Figure 4-17. Block diagram of cosine-crossing method applied to a three-phase half-controlled converter.

creased, with the result that once again the mean output dc voltage will remain constant.

The major disadvantages of the cosine-crossing method are, first, since the cosine-crossing wave is derived from the ac source voltage, any harmonics caused by system transients will be reproduced in the cosine-timing wave with the possibility of the gate pulses being mistimed, thus producing a nonlinear transfer characteristic and even instability. Second, if there are variations in the ac source frequency the cosine relationship will vary, since the RC phase-shift network is frequency sensitive.

4-6-3-2 Integral Control

The two major disadvantages of the cosine-crossing method of firing control, namely, distortions of the ac source voltage waveform producing a nonlinear transfer characteristic and the phasing of the cosine signals being changed by ac source frequency variations, are overcome by the **integral system** of firing angle control. A block diagram illustrating the application of integral control to a three-pulse half-controlled bridge converter is shown in Figure 4–18(a). The difference voltage between the reference voltage V_R and an attenuated feedback voltage from the converter output terminals is applied to the integrator and integrated, see Figure 4–18(b). The integrator output is zero at the desired firing points and has a frequency three times that of the ac source frequency. In turn, the integrator output is supplied to a comparator, which produces a train of pulses 120° apart at each zero point of the input. The comparator output is supplied in turn to a clock pulse generator, the output of which is applied to a three-stage ring counter and then distributed by means of pulse transformers or optocouplers as gate trigger pulses to the thyristors.

If there is a decrease in the converter output voltage, the feedback voltage will decrease and the positive and negative integrals will also decrease; the zero points of the integrator output will advance. As a result, the trigger pulses to the thyristors will also advance, and the converter output voltage will increase. The system is then operating as a closed-loop control system responsive to variations in load demand.

In addition, because of the integrator, variations in the supply voltage waveform, the feedback voltage waveform, and ac source frequency do not affect the system.

4-6-3-3 Phase-Locked Loop Control

The **phase-locked loop** oscillator control of the production of thyristor gating pulses in converter control produces gate pulses at

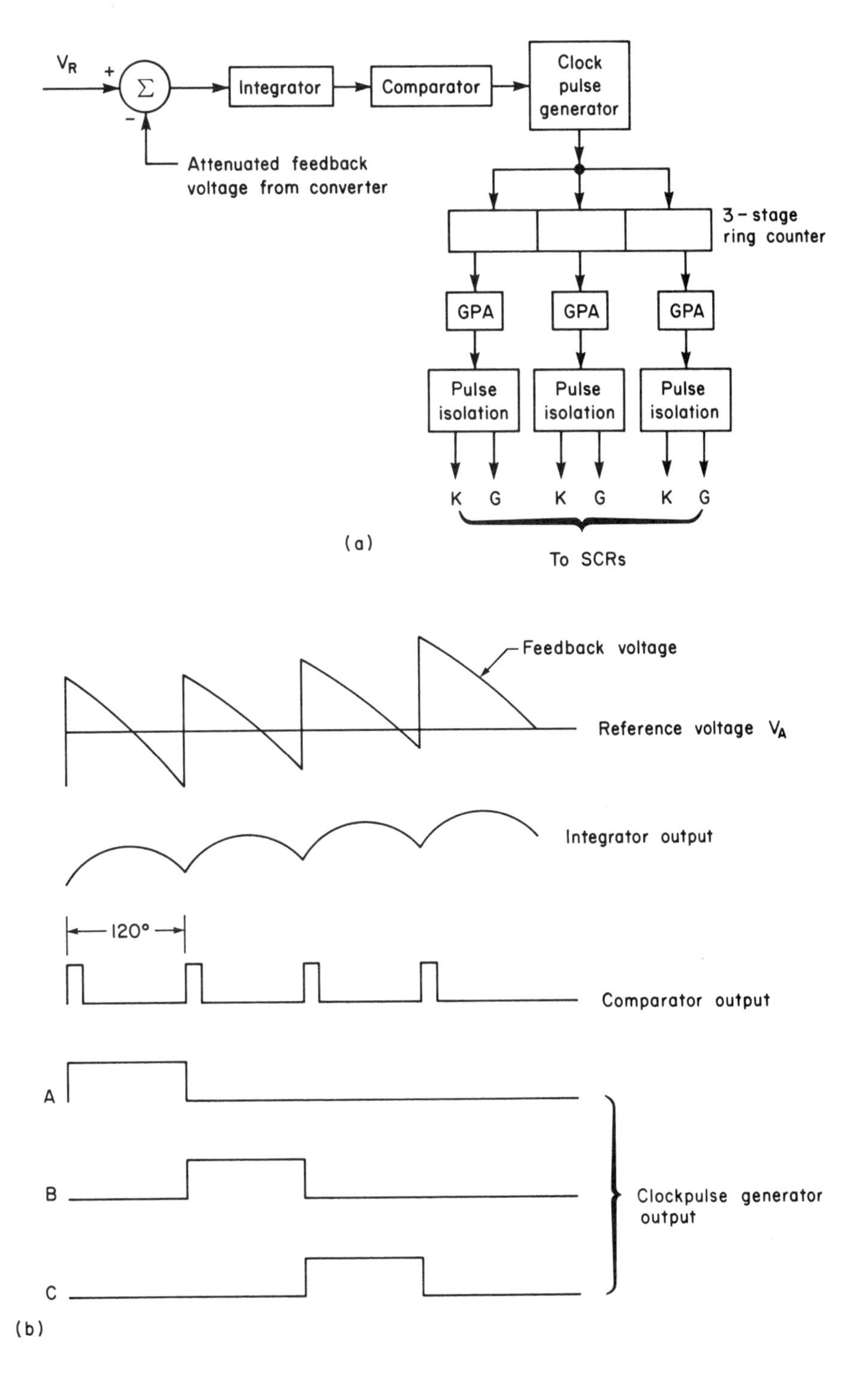

Figure 4-18. Integral firing control: (a) block diagram, (b) waveforms.

very accurate intervals and is not affected by voltage disturbances in the ac input to the converter.

The basic concept of the control system is shown in Figure 4-19 as applied to a three-phase, six-pulse bridge converter. The major components are a voltage-controlled oscillator (VCO), and a six-stage ring counter. The VCO consists of an integrator, a comparator, and a reset or phase-lock. The VCO generates a train of pulses with a pulse repetition frequency f_1, which is directly controlled by a bias voltage E; in this case $f_1 = 6f$ where f is the ac source frequency. The VCO output is supplied to the six-stage ring counter, which in turn distributes the pulses in the correct sequence to the gates of the thyristors, the pulses being spaced 60° apart.

Because of the possibility of frequency drift the system is locked by means of a negative feedback loop in the VCO. End-stop control is also applied to the integrator to limit the amount of retardation of the firing delay angle. See Figure 4–10(b).

Any variations in the ac source frequency will cause a change in the firing delay angle and the mean dc voltage output of the converter. The resulting change in the error signal fed into the VCO will cause the oscillator frequency to change until it is once again $6f_2$, where f_2 is the new source frequency. However, the firing delay angle and the mean output dc voltage will be returned to their original values. As a result, the VCO can accommodate quite wide frequency variations but still maintain an accurate spacing of the firing pulses.

When operating as a closed-loop control system, if there is a decrease in speed, the feedback signal will decrease; the increased error voltage

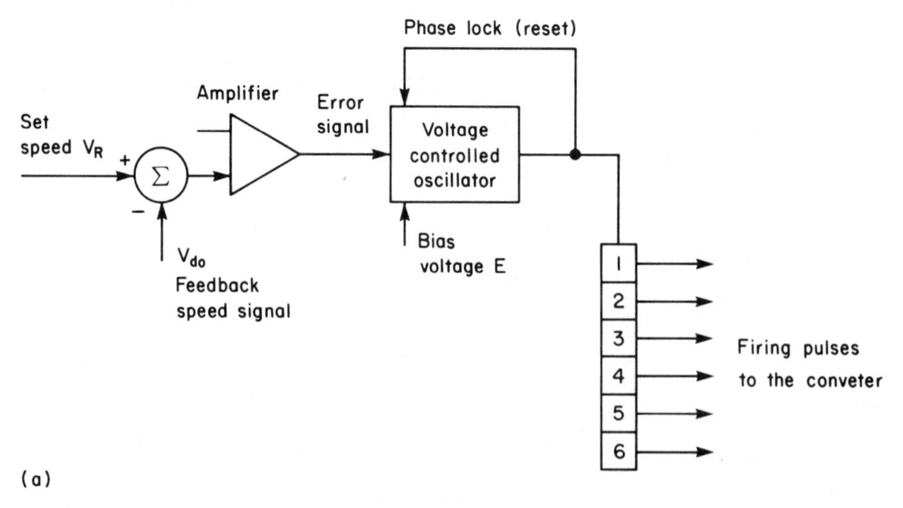

Figure 4-19. The principle of phase-locked loop control: (a) block diagram of system.

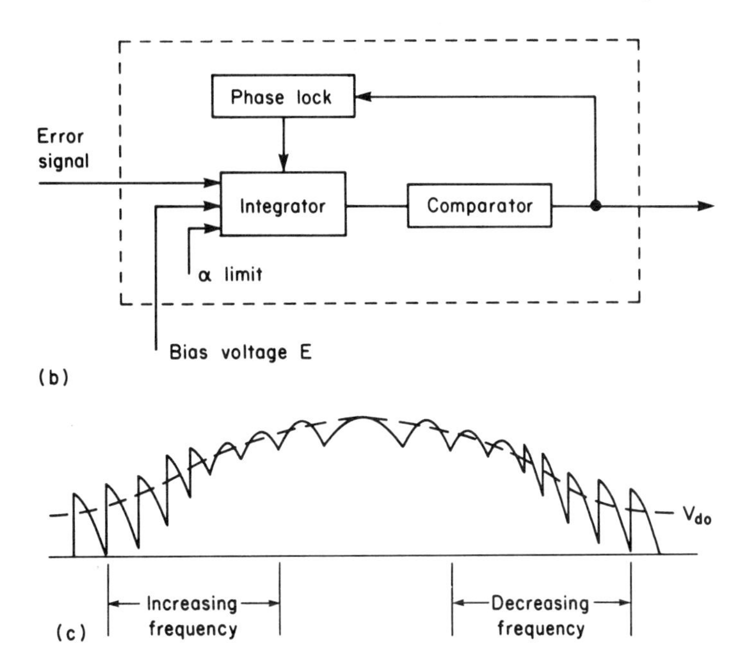

Figure 4-19 continued. (b) Block diagram of voltage-controlled oscillator, (c) the oscillator frequency variations.

applied to the VCO will cause the frequency of the output pulses to increase, resulting in the firing delay angle being advanced. In the case of an increase in the motor speed, the frequency of the VCO output will be decreased and the firing delay angle will be retarded. See Figure 4–19(c).

The phase-locked loop control system is a closed-loop system that provides very accurate response to changes in the controlled variable (motor-speed, voltages, current, etc.).

4-7 FIRING PULSES

The shape, amplitude, and duration of the firing pulse is determined by the gating requirements of the thyristor and the nature of the controlled load. For example, in high *di/dt* applications, V_{GT} can range from 3 to 20 V and I_{GT} from as low as 0.25 to 4 A. At the same time, the rise time of the gate pulse should be less than one μs to ensure rapid spreading of the conducting area in the thyristor, and of sufficient duration to ensure that the latching current has been achieved. The pulse duration usually is as long as 50 μs, although in some applications the duration may be as long as 200 μs.

Pulses meeting these requirements may result in overheating of the gate-cathode junction. This effect can be minimized by a two-stage pulse (see Figure 4–20).

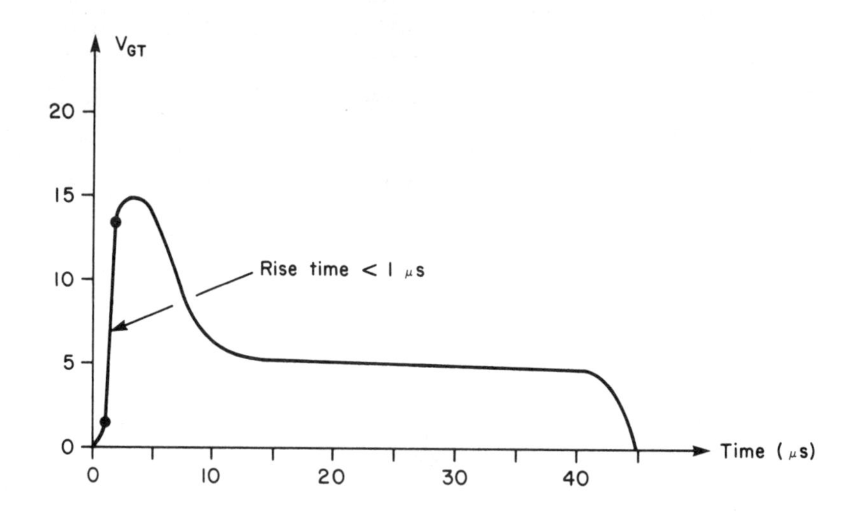

Figure 4–20. Waveform of a two-stage firing pulse for high *di/dt* applications.

4-7-1 Double Pulsing and Long Pulses

In any full-controlled converter, when completion of the circuit depends upon two or more thyristors being in conduction at the same time and as conduction is transferred from one thyristor to the next as its anode voltage becomes positive (or during discontinuous conduction), one short pulse will not guarantee that all devices are triggered. One solution is to provide two pulses per cycle to each thyristor, the spacing of pulses being 2π/pulse number. In the case of a three-phase, full-controlled bridge converter the spacing is 60°. See Figure 4–21(a). An alternative solution is to use long pulses. Long pulses are usually >60°, see Figure 4–21(b), but they still have the disadvantage that inductive loads will delay the current buildup in the thyristor, so that at the end of the pulse a latching current has not been achieved. The only way to ensure that the thyristors are turned on is to supply a gating pulse of 120° duration, i.e., equal to the conduction period of the thyristor. There are two major disadvantages to this scheme: First, it is difficult to produce a pulse of 120° duration; and second, it increases the probability of overheating the gate-cathode junction.

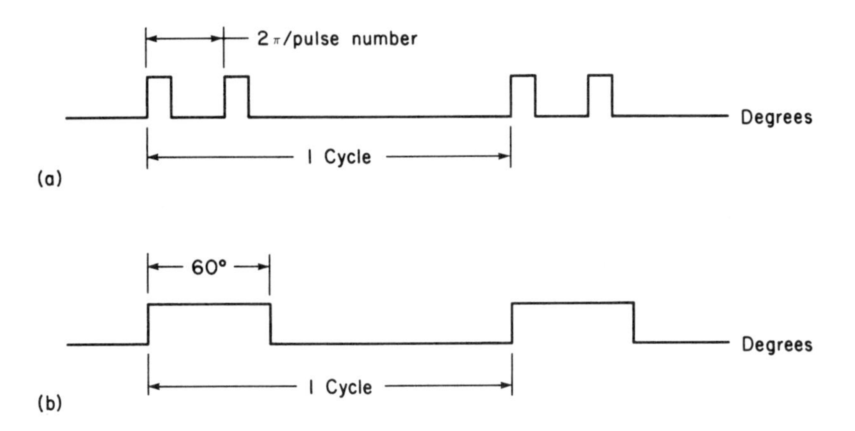

Figure 4-21. Gate pulses for (a) double pulsing and (b) long pulses.

4-8 PULSE ISOLATION

In any power converter, differences in potential between the various gates and between the individual thyristors and the gate pulse generators exist. Protection must be provided by **electrical insulation** or **isolation** between the thyristors and the gate pulse-generating circuits.

4-8-1 Pulse Transformers

A commonly used method of providing electrical insulation is by the use of a **pulse transformer**. Pulse transformers are available in a number of different mounting arrangements suitable for printed circuit board applications or dual-in-line packages (DIP) containing two, three, or four transformers for integrated circuit applications.

Normally there is a primary winding and one or more isolated secondary windings, permitting simultaneous gating signals to be applied to thyristors in series and parallel configurations. These windings are usually tested at 2,500 V to ensure that all windings are electrically isolated from each other.

The electrical requirements of a pulse transformer are, first, that the windings must be tightly coupled to minimize leakage inductance and thereby ensure that the output pulse will have a fast rise time. Second, that the insulation be great enough to provide the isolation required. This latter requirement interferes with the first, and as a result the design must be an acceptable compromise.

4-8-2 Optocouplers

Optocouplers are a combination of optoelectronic devices, normally a light-emitting diode (LED) and a junction-type photoconductor such as a phototransistor assembled into one package, such as a DIP integrated circuit arrangement. This arrangement permits the coupling of a signal from one electronic circuit to another, and at the same time maintains an almost complete electrical isolation between the circuits.

The most commonly used optocouplers consist of an input LED operating in the infrared or near-infrared region and an output silicon photodetector, such as a phototransistor, a photo-Darlington, or a photo-SCR (see Figure 4–22).

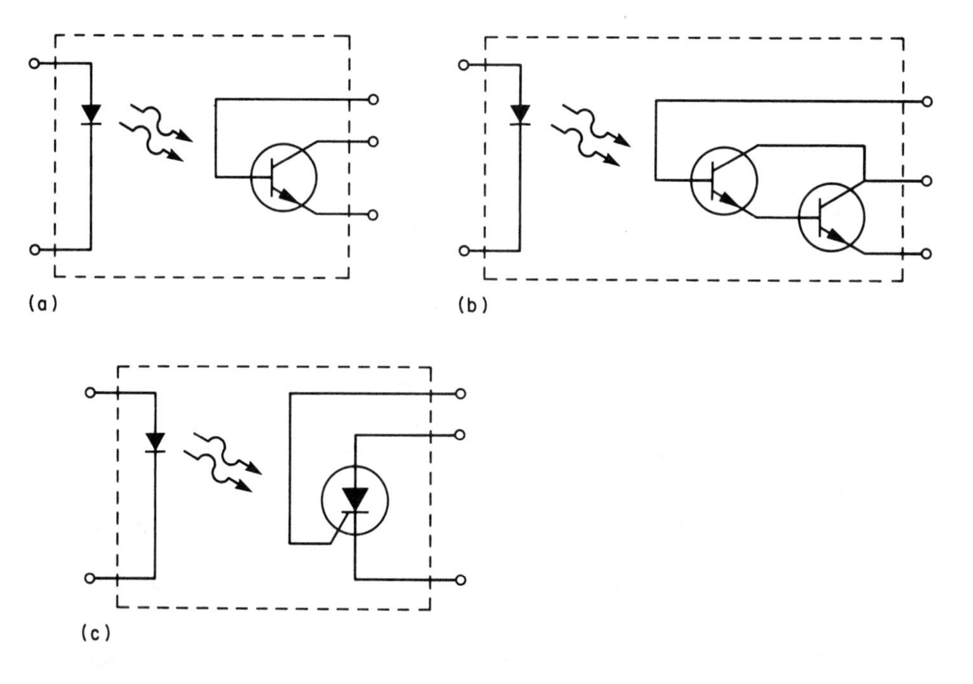

Figure 4–22. Typical optocouplers: (a) phototransistor, (b) photo Darlington, (c) photo SCR.

When using a pulse transformer, gate trigger power is transferred from the trigger source to the thyristor gate and there is no leakage in the gate circuit. In the case of an optocoupler, however, a power source is required on the load side. If this source is derived from the anode supply of the thyristor, the blocking voltage of the SCR cannot exceed that of the optocoupler. This disadvantage can be overcome by one of two means: first, by providing a separate power supply which provides greater flexibility at an added expense (more than one power supply will be required, if the load-carrying SCRs do not have common reference points), and second, by deriving the optocoupler power from a voltage divider across the input to the SCR, which will have the disadvantage of providing a leakage path around the SCR during the nonconducting period.

Typical operating speeds are 100 to 500 kHz for phototransistors, 2.5 to 10 kHz for photo-Darlingtons, and 2 to 20 μs turn-on time for photo-SCRs. The type selected depends upon the power requirements of the gate circuit.

4-8-3 Output Stages

Probably the simplest application of the pulse transformer is shown in Figure 4–23(a), where an *n-p-n* transistor is connected in series with the primary. When a signal is applied to the base of the transistor from the gate pulse generator in the firing control, the transistor saturates and the voltage V is applied across the transformer primary, inducing a voltage pulse at the secondary terminals which is applied to the SCR. When the drive pulse to the base of the transistor is removed, the transistor turns off, and the current caused by the collapsing magnetic field in the transformer flows through the diode D connected across the primary; the voltage across the primary momentarily reverses.

The core flux is unidirectional, and the core-magnetizing current as well as the pulse-forming current both have to be carried by the transistor. As a result, in order to minimize the core size, this type of configuration is usually limited to the production of pulses between 25 to 50 μs, and cannot be used in long pulse applications.

In order to produce two-stage pulses for high *di/dt* applications, the circuit is modified as shown in Figure 4–23(b). The circuit operates as before, except that the charging of the capacitor prolongs the duration of the overall pulse.

Long pulses can be simulated by using a blocking oscillator configuration, as shown in Figure 4–24. This application requires a pulse transformer with two secondaries and will produce a train of pulses, as long as there is a base drive on the transistor, by means of a positive feedback signal to the base of the transistor. The major disadvantage of this circuit in three-phase applications is the fact that it is necessary to provide simultaneous pulses to two or more thyristors, but there is no guarantee that the thyristors will receive these pulses at exactly the same instant. A secondary disadvantage is that the regenerative circuit is very susceptible to noise and may cause inadvertent triggering of the thyristors.

A much more effective way of producing pulse trains for long puls-

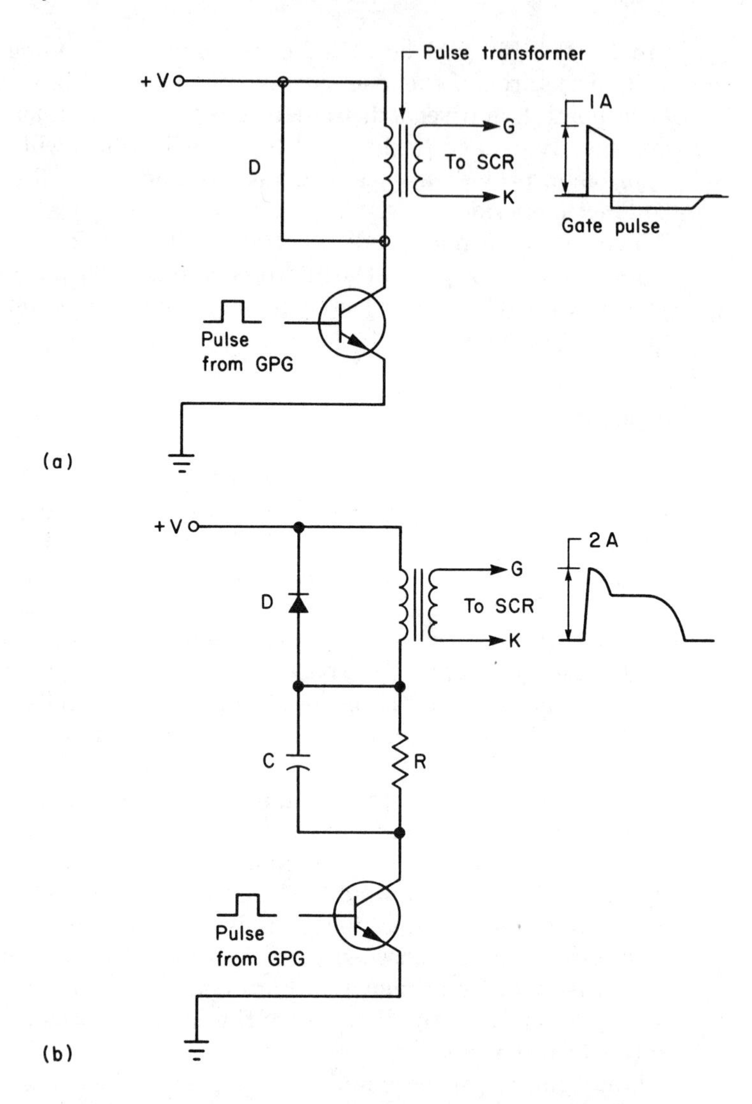

Figure 4-23. Gate pulse isolation: (a) simple pulse, (b) two-stage pulse.

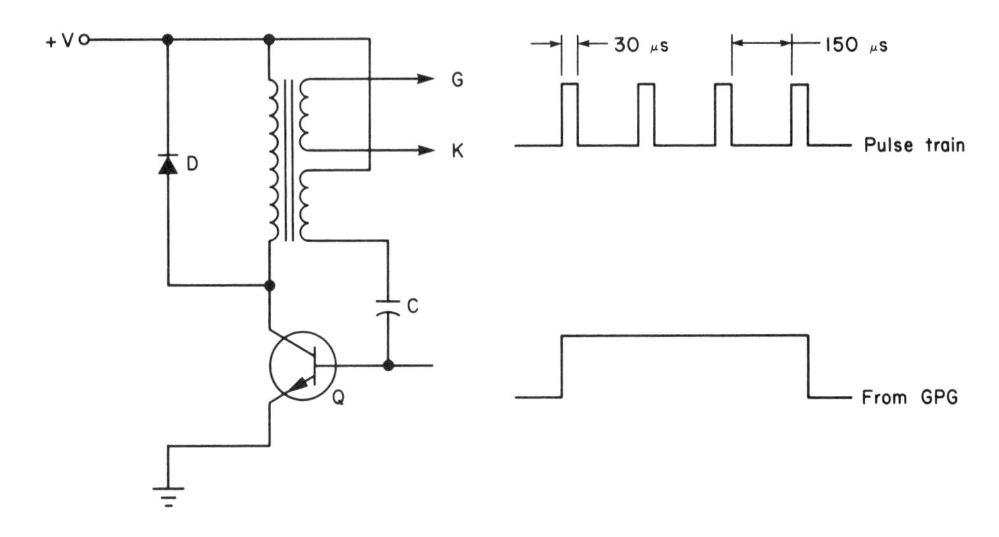

Figure 4–24. Gate pulse isolation using a blocking oscillator to simulate long pulses.

ing is to use a common master oscillator and apply the output to the thyristors as dictated by the firing control circuit. This scheme is illustrated in Figure 4–25 where the master oscillator produces a train of high-frequency pulses, typically between 5 and 20 kHz. The center-tapped pulse transformers are connected in parallel across the master oscillator output; base signals are applied to the gating transistors from the firing control circuit. The resulting pulse trains are applied to the thyristors for 120° conduction, and since the pulses are all derived from the master oscillator, the potential conducting thyristors will receive simultaneous gating signals.

Optocouplers provide an excellent means of coupling TTL logic or a microprocessor output to thyristors, since the output currents may not be sufficient to cause the thyristor to latch. Figure 4–26 illustrates the use of a photo-Darlington optocoupler which because of its higher current gain, can be used for triggering a thyristor.

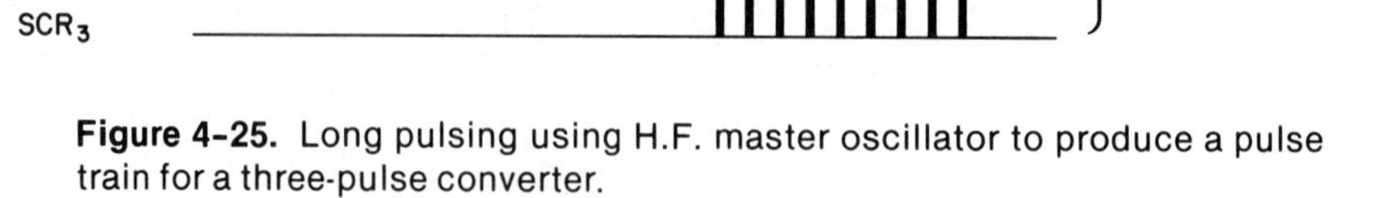

Figure 4-25. Long pulsing using H.F. master oscillator to produce a pulse train for a three-pulse converter.

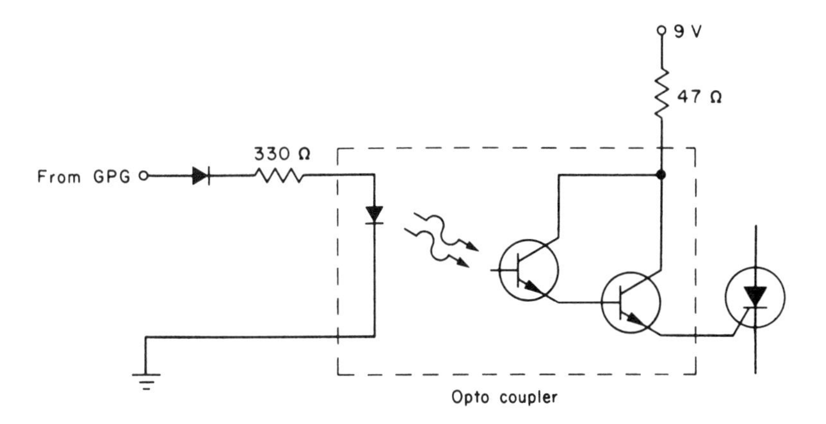

Figure 4-26. Gate pulse isolation using an optocoupler.

GLOSSARY OF IMPORTANT TERMS

Converter: A circuit that converts ac to dc and is usually used to control a dc motor drive.

Synchronous inversion: The return of power from an active dc load to the ac source.

Semiconverter, half-controlled, or one-quadrant converter: Converters that operate only in the rectifying mode; that is, they have only one voltage polarity and a unidirectional current flow.

Full or two-quadrant converters: Converters that may operate with voltage polarity reversal and unidirectional current flow. These converters may operate in the synchronous inversion mode when there is an active dc load.

Dual converter: A converter system that operates in all four quadrants. It is the power electronic equivalent of the Ward-Leonard system.

Pulse number: The ratio of the ripple frequency to the source frequency. Most single-phase converters are two-pulse and most three-phase converters are three- or six-pulse.

Midpoint converter: A transformer supplied converter with the load connected between the common anode or cathode connection and the center tap or neutral of the transformer.

Bridge converter: A converter which may be supplied from a transformer or directly from the ac source with the load connected between the common anode and cathode connections.

Electrical noise: Fast rise-time electrical voltage transients that appear in supply and control wiring.

Cosine-crossing firing delay control: Compares a variable dc reference voltage to a cosine timing wave to produce a linear relationship between the reference voltage and the converter output voltage.

Integral firing control: Compares the actual converter output to a reference voltage and integrates the error and produces a firing signal whenever the integral is zero.

Phase-locked loop control: The basic principle depends upon the fact that for any specific firing delay angle the firing pulses are accurately spaced at a specific frequency. If the frequency is increased the firing pulses will be advanced or vice versa.

REVIEW QUESTIONS

4-1 Explain what is meant by the following terms: a one-quadrant converter, a two-quadrant converter, a dual or four-quadrant converter.

4-2 What is meant by synchronous inversion?

4-3 With the aid of a schematic and waveforms discuss the operation of a single-phase half-wave phase-controlled converter when supplying first, a pure resistive load and second, an RL load.

4-4 Discuss with the aid of a schematic and waveforms the operation of a single-phase two-pulse midpoint converter in the rectifying and inverting modes. List the advantages and disadvantages of this configuration.

4-5 Repeat Question 4-4 for a single-phase two-pulse bridge converter.

4-6 Discuss the advantages and disadvantages of midpoint converters versus bridge converters.

4-7 Explain why three-phase converters are preferred to single-phase converters.

4-8 Explain with the aid of a schematic and waveforms the operation of a three-phase midpoint converter.

4-9 Why is it preferable to supply a three-pulse midpoint converter from a three-phase star-zigzag connected transformer?

4-10 Why is it desirable to use a six-pulse midpoint converter with an interphase reactor?

4-11 What is the advantage of using twelve- and twenty-four-pulse midpoint converters?

4-12 With the aid of a schematic and waveforms explain the operation of a six-pulse bridge converter operating in the rectifying and inverting modes.

4-13 How may a two-quadrant converter be converted to a one-quadrant converter?

4-14 What are the advantages and disadvantages of using one-quadrant converters?

4-15 Explain with the aid of a schematic the possible arrangements of a single-phase half-controlled converter.

4-16 What is meant by "half-waving" and how may it be eliminated?

4-17 What is the function of a freewheel diode?

4-18 Explain with the aid of a schematic and waveforms the operation of a three-pulse half-controlled bridge converter.

4-19 What is meant by a dual converter? Discuss the methods with their advantages and disadvantages by which a reversing dc motor drive can be controlled.

4-20 With the aid of a sketch explain how a dual converter driving a reversing dc motor may be operated in the circulating current free mode.

4-21 What is the function of a firing circuit?

4-22 What is meant by electrical noise? How is electrical noise produced and how may its effects be minimized?

4-23 What are the characteristics that must be met by a firing circuit?

4-24 Explain with the aid of a sketch the principle of the cosine crossing technique of firing circuit control. What are the advantages and disadvantages of this technique?

4-25 Repeat Question 4-24 for integral firing control.

4-26 Repeat Question 4-24 for phase-locked loop control.

4-27 Explain what is meant by double phasing.

4-28 What is meant by long pulsing and what are the disadvantages to this type of gate signal?

4-29 What is meant by pulse isolation?

4-30 Why are optocouplers preferable to pulse transformers as the method of applying gate signals to a thyristor?

4-31 What are the advantages of using a two-stage gate pulse?

4-32 Discuss the interfacing of TTL or a microprocessor to the gate of a thyristor.

PROBLEMS

4-1 An ac source having an effective voltage of 230V, 60Hz is connected to a single-phase half-wave converter supplying a 10Ω load resistance. If the load current is not to exceed 18A calculate (1) the maximum delay angle, (2) the average thyristor current, and (3) the PIV of the thyristor.

4-2 Repeat Problem 4-1 using a two-pulse midpoint converter. The center tap-to-line voltage being 230V, 60Hz.

4-3 A two-pulse full-controlled bridge converter supplied from a 230V, 60Hz single-phase source is used to control a 1hp (0.746kW) 115V, 1750rpm (183.26rads/s) separately excited dc motor. The full load armature current is 9A and the armature resistance is 0.85Ω. Calculate (1) the firing delay angle when the motor is running under rated load conditions, (2) the firing delay angle for the motor to run at 1000rpm (104.72rads/s).

4-4 The speed of a 5hp (3.73kW), 250V, 2250rpm (235.62rads/s) separately excited motor is controlled by a three-phase fully controlled bridge converter. The converter is supplied from a 3-phase, 208V, 60Hz source. The rated armature current of the motor is 19A and the armature resistance is 0.42Ω. Calculate (1) the firing delay angle for the terminal voltage to be 250V, (2) the firing delay angle for the motor to operate at 1125rpm (117.81rads/s), (3) the firing delay angle required to operate the motor at rated torque at 550 rpm (57.60rads/s), (4) the average thyristor current, and (5) the thyristor PIV.

4-5 The motor of Problem 4-4 is controlled by a three-phase half-controlled converter supplied from a three-phase, 208V, 60Hz source. Repeat Problem 4-4.

4-6 A 10hp (7.46kW), 250Vdc, 1200rpm (125.66rads/s) separately excited motor is controlled by a three-phase fully controlled bridge converter. The converter is supplied from a three-phase, 208V, 60Hz source. The armature resistance is 0.29Ω and the rated armature current is 40A. Calculate (1) the firing delay angle for the motor to operate at rated voltage, (2) the firing delay angle if the starting current is to be limited to 50A, and (3) the reactive power absorbed by the converter when operating at rated load.

5

Variable-Frequency Conversion

5-1 INTRODUCTION

Drive motors used in machine tool control, steel and paper drives, variable delivery pumps and fans, mine hoists, and cranes require an adjustable speed capability combined with continuous and accurate speed control, high stability, and good transient performance. These requirements have been traditionally met by the use of closed-loop adjustable speed dc drives using armature voltage and field-control techniques based on the Ward-Leonard system. Unfortunately, the high initial and on-going maintenance costs of the dc motor have caused the dc drive systems to be at an economic disadvantage as compared to the **ac variable-frequency motor drive.** The ac variable-frequency motor drive offers benefits such as accurate speed regulation and improved energy efficiency as well as variable speed control.

Although the principle of variable-frequency static inverters and cycloconverters has been known for a long time, it has only been economically feasible to develop ac drives since the introduction of the thyristor. Rapidly decreasing thyristor costs and the development of low-cost digital and analog integrated circuits have accelerated the development of ac drives. This, combined with lower initial motor costs and drastically reduced maintenance costs, is making the ac variable-frequency drive the logical replacement for the static dc drive.

Other factors that weight the selection of a drive in favor of ac are:

1. Many squirrel cage induction motors (SCIMs) can accept frequencies up to 200 Hz, resulting in shaft speeds of approximately 12,000 rpm (1257 rad/sec) with a two-pole machine.

2. Speed control ranges from 6:1 up to approximately 20:1 are obtainable.

3. Squirrel cage motors possess a relatively low inertia, which permits good dynamic response.

4. Standard motors may be used.

5. Speed regulation is improved, since the variable-frequency output is not load dependent.

6. The ability to operate under open-loop control results in a significant reduction in control-circuit complexity and costs.

7. Multimotor drive systems are easily synchronized.

8. Electronic reversal of the direction of rotation is easily achieved.

9. dc motor braking is possible.

10. Four-quadrant operation is readily achieved.

5-2 BASIC POLYPHASE INDUCTION MOTOR THEORY

Prior to discussing static frequency conversion methods, it is necessary to determine the requirements of the load, in particular the **polyphase induction motor.**

When a balanced three-phase supply is applied to the stator of a three-phase induction motor, a constant amplitude rotating magnetic field is produced. The angular velocity of this field is given by

$$S = 120 \frac{f}{P} \text{ rpm} \tag{5-1E}$$

or

$$\omega = 4\pi \frac{f}{P} \text{ rad/sec} \tag{5-1SI}$$

where S or ω is the synchronous speed of the rotating magnetic field in the appropriate units, f is the frequency (Hz), and P is the number of stator poles/phase, from which it can be seen that the synchronous speed of the rotating magnetic field for a given machine is

$$S \text{ or } \omega \propto f \tag{5-2}$$

Under normal operating conditions the ac source voltage and frequency are constant. When the rotor is stationary, the synchronous rotating magnetic field will induce emfs at the supply frequency in the conductors of the shorted rotor. If the rotor conductors are purely resistive, the rotor emf and rotor current will be in phase, and the torque contribution of each rotor conductor is proportional to the product of the conductor current and the flux density at the conductor; that is, the conductor with the greatest current will also be exposed to the greatest air-gap flux and maximum electromagnetic torque will be produced. If, on the other hand, the rotor circuit was purely inductive, the rotor currents would lag the induced rotor voltages by 90°, and the electromagnetic torque would be zero.

In actual practice, the rotor circuit possesses both resistance and inductance; an electromagnetic torque will be produced which causes the rotor to accelerate from standstill in the direction of rotation of the rotating magnetic field. Since the electromagnetic torque depends upon the interaction between the rotating magnetic field and the rotor field produced by the rotor currents, it is obvious that, if the rotor turned at the same speed as the rotating magnetic field, there would be zero emf induced in the rotor and thus no rotor current. As a result, there would be zero electromagnetic torque produced. This means that the rotor must turn at an angular velocity S_r or ω_r less than that of the synchronous rotating magnetic field. The difference,

$$S - S_r$$

or

$$\omega - \omega_r$$

is called the slip speed of the motor and is usually expressed as a fraction of the synchronous speed to give either the fractional or percentage slip

$$s = \frac{S - S_r}{S} \tag{5-3E}$$

or

$$s = \frac{\omega - \omega_r}{\omega} \tag{5-3SI}$$

where

s = decimal slip, or percentage slip when multiplied by 100,
s or ω = synchronous speed
s_r or ω_r = actual rotor speed

Under operating conditions the rotating magnetic field or the air-gap flux moves past the rotor conductors at slip speed. The frequency of the induced rotor emfs is

$$f_r = \left(\frac{S - S_r}{S}\right) f = sf \tag{5-4E}$$

or

$$f_r = \left(\frac{\omega - \omega_r}{\omega}\right) f = sf \tag{5-4SI}$$

When the induction motor is operating under normal load conditions, the frequency of the rotor emfs and currents is relatively low, 2 to 5 Hz, and the rotor reactance is small. The rotor current is then mainly limited by the rotor resistance. Assuming that the air-gap flux ϕ is constant, the induced rotor emf is proportional to the fractional slip, and the rotor current and torque are also proportional to the slip. As a result, under low slip conditions, the polyphase induction motor has a torque-speed characteristic similar to the dc shunt motor; that is, speed decreases linearly with increasing load torque with a resulting increase in the induced rotor emf, rotor current, and electromagnetic torque until a lower equilibrium speed is reached.

5-2-1 Equivalent Circuit of a Polyphase Induction Motor

When operated under steady-state conditions the performance of a polyphase induction motor can be predicated from an analysis of the equivalent circuit shown in Figure 5-1.

The polyphase induction motor is basically a polyphase transformer with a rotating secondary winding. The stator current produces a mutual flux, which links with the rotor winding, and a stator leakage flux, which links only with the stator winding. The leakage flux induces a stator counter emf. The applied stator emf, V_1, is greater than the counter emf, E_1, by the voltage drop in the stator leakage impedance, which on a per-phase basis is

$$V_1 = E_1 + I_1 (r_1 + jx_1) \tag{5-5}$$

where

V_1 = applied stator voltage

E_1 = counter emf produced by the resultant air-gap flux

I_1 = stator current

r_1 = effective stator resistance

x_1 = stator leakage reactance

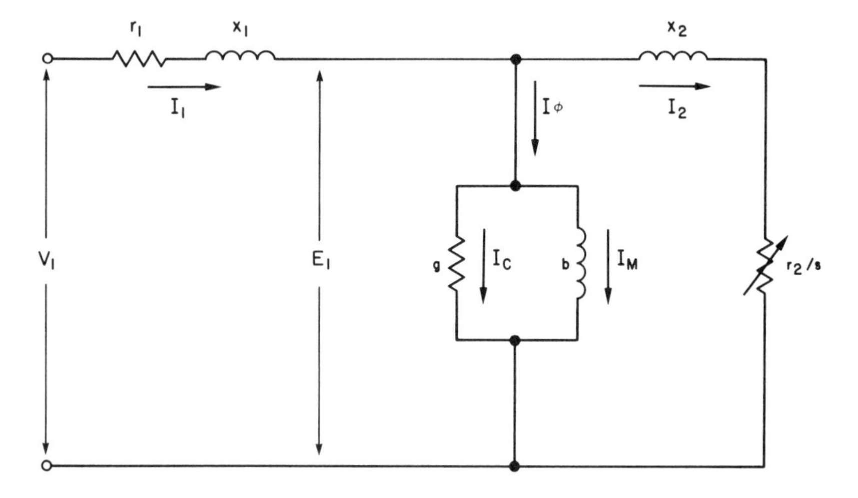

Figure 5-1. Equivalent circuit of a polyphase induction motor on a per-phase basis.

The air-gap flux is the resultant of the fluxes produced by the stator and rotor currents. The stator current can be resolved into two components, the load current I_2, which produces an emf that exactly counterbalances the emf produced by the rotor current, and a no-load current I_ϕ consisting of the magnetizing component I_M and the core-loss component I_c. The rotor leakage reactance is dependent upon the frequency of the rotor currents, i.e., $f_r = sf$; therefore, the rotor leakage reactance is sx_2, where x_2 is the locked rotor leakage reactance. The rotor emf, at standstill in the equivalent circuit, is equal to the stator emf; therefore, at slip s, the rotor emf $E_2 = sE_1$. If r_2 is the rotor resistance per phase in the equivalent circuit and x_2 is the rotor standstill reactance per phase, then the rotor current I_2 is calculated as

$$I_2 = \frac{E_2}{r_2 + jsx_2} = \frac{sE_1}{r_2 + jsx_2} \tag{5-6}$$

in terms of slip frequency; thus

$$I_2 = \frac{E_1}{(r_2/s) + jx_2} \tag{5-7}$$

in terms of the supply frequency.

5-2-2 Power Balance Equations

From the equivalent circuit of Figure 5-1 for an m-phase machine,

$$\text{input power} = mV_1I_1\cos\phi_1$$

$$\text{stator copper loss} = mI_1^2r_1 \tag{5-8}$$

$$\text{rotor power input (RPI)} = mI_2^2 r_2/s \tag{5-9}$$

$$\text{rotor copper loss (RCL)} = mI_2^2r_2 \tag{5-10}$$

$$\text{rotor power developed (RPD)} = (5\text{-}9) - (5\text{-}10) = \text{RPI} - \text{RCL}$$

$$= mI_2^2r_2\left(\frac{1-s}{s}\right) \tag{5-11}$$

and

$$\text{RPD} = \text{RPO} + P_{\text{ROT}}$$

where RPO is the rotor mechanical power output and P_{ROT} the rotational losses, i.e., windage and friction.

If ω_r is the angular velocity of the rotor and T is the electromagnetic torque, then

$$T\omega_r = \text{RPD} = mI_2^2r_2\frac{1-s}{s}$$

and

$$T = \frac{mI_2^2r_2}{\omega_r}\left(\frac{1-s}{s}\right) \tag{5-12}$$

where T is the internal motor torque and is greater than the actual shaft torque by the amount required to overcome the rotational losses.

Since the angular velocity of the rotating magnetic field is

$$\omega = \frac{\omega_r}{1-s} = \frac{4\pi f}{P}$$

then Equation (5-12) can be rewritten as

$$T = \frac{mI_2^2r_2}{\omega_r}\left(\frac{1-s}{s}\right) = \frac{mI_2^2r_2}{\omega_s}$$

$$= \frac{mP}{4\pi f}\cdot I_2^2\cdot\frac{r_2}{s} \tag{5-13}$$

From

$$I_2 = \frac{E_1}{r_2/s + jx_2} = \frac{E_1}{\sqrt{[(r_2/s)^2 + x_2^2]}}$$

and $$s = f_r/f,$$

then
$$T = \frac{mP}{4\pi} \cdot \left[\frac{E_1}{f}\right]^2 \cdot \frac{f_r r_2}{\left[r_2^2 + s^2 x_2^2\right]} \tag{5-14}$$

The air-gap flux must be maintained constant at all frequencies to produce a constant torque. When the ratio E_1/f is constant, a constant air-gap flux is produced. If the stator leakage impedance is small, then $E_1 \simeq V_1$, and the air-gap flux is approximately constant when the V_1/f ratio is constant. This is known as the **constant V/Hz** method of operation and is most commonly used in open-loop control of static inverters and cycloconverters.

However, at low frequencies, since the stator resistance is the predominant component (see Figure 5-2), there will be a decrease in air-gap flux with a consequent reduction in motor torque. If the low-speed performance is not acceptable with a constant V/Hz ratio, then the control must be modified to increase the V/Hz ratio at low frequencies. This action will not cause an undue increase in core losses, even though the stator iron may be saturated.

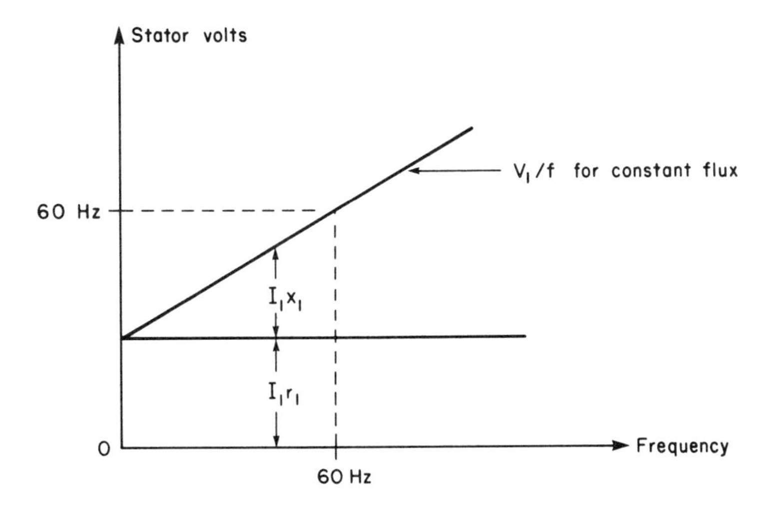

Figure 5-2. Stator voltage versus frequency.

Equation (5-14) may be written in terms of the rotor frequency as

$$T = \frac{mP}{4\pi}\left[\frac{E_1}{f}\right]^2\left[\frac{f_r r_2}{r_2^2 + (2\pi f_r \ell_2)^2}\right] \tag{5-15}$$

and differentiating with respect to rotor frequency f_r and equating to zero gives the relation for the rotor breakdown frequency as

$$f_b = \pm \frac{r_2}{2\pi \ell_2} \tag{5-16}$$

where f_b = the rotor frequency at which maximum torque occurs. Substituting in Equation (5-15), we find that the maximum or breakdown torque is

$$T_b = \pm \frac{mP}{4\pi} \left[\frac{E_1}{f}\right]^2 \cdot \frac{1}{4\pi \ell_2} \tag{5-17}$$

where f_b is the rotor breakdown frequency, T_b the torque at breakdown, and ℓ_2 the rotor leakage inductance (− sign implies the motor is operating as a generator).

Since the air-gap flux is proportional to E_1/f, the breakdown torque is proportional to the air-gap flux squared and inversely proportional to the rotor leakage inductance. It should be noted that the rotor resistance does not affect the magnitude of the breakdown torque, but from Equation (5-16), it will affect the rotor frequency at which breakdown occurs.

It is very often convenient to express torque in terms of the torque ratio t, the ratio of torque T at any rotor frequency f_r to the maximum torque T_b at the rotor breakdown frequency f_b. Substituting Equations (5-16) and (5-17) in Equation (5-15) produces

$$t = \frac{T}{T_b} = \frac{2}{(f_r/f_b) + (f_b/f_r)} \tag{5-18}$$

for constant flux operation. Also for a given machine,

$$T_b = \text{a constant} \tag{5-19}$$

When a polyphase induction motor is operated under constant flux conditions, the torque supplied is greater at all supply frequencies than when supplied at rated voltage and frequency.

As a result, a variable-frequency induction motor drive will have a greater starting torque at low frequencies, and for the same operating slip, the torque will be greater at higher frequencies. For a given load torque, the horsepower output is proportional to frequency, and as the frequency is increased, the efficiency increases, since

$$\eta = \frac{\text{RPO} - P_{\text{ROT}}}{\text{RPI} + \text{stator copper losses} + \text{core losses}} \tag{5-20}$$

$$\simeq \frac{\text{RPO}}{\text{RPI}} \simeq \frac{mI_2^2r_2\left(\frac{1-s}{s}\right)}{mI_2^2r_2/s}$$

$$\simeq (1 - s) \tag{5-21}$$

From Figure 5-3 it can be seen that at a constant load torque as the supply frequency is increased, the slip s decreases, and thus the motor efficiency increases.

In summary, the operation of a polyphase induction motor, either squirrel cage or wound rotor, with a constant V/Hz ratio, results in a higher starting and breakdown torque. With the same full load slip the torque is greater at higher frequencies than at the lower frequencies. In addition, the horsepower (kW) output and efficiency are greater at the higher frequencies. Failure to maintain the constant V/Hz ratio will either affect the constant torque by the square of the air-gap flux density or permit the stator current to increase and overheat the motor.

Because a solid-state static frequency converter can be programmed to provide the optimum voltage and frequency to a polyphase induction motor for any desired operating speed, there is no requirement for reduced voltage starting equipment; therefore, starting inrush currents are eliminated, and the maximum torque per kVA is developed.

5-3 MOTORING AND REGENERATION

When a polyphase rotating magnetic field is set up, the rotor will turn at some speed less than the synchronous speed of the field. If the rotor is connected to an overhauling load, the rotor will accelerate and eventually exceed the speed of the synchronous rotating magnetic field. At this point, the rotor conductors are sweeping past the rotating field and the induced rotor emfs and rotor currents are reversed. The reflected stator currents are also reversed, and the motor acts as an induction generator. When the motor is operating as an induction generator, regeneration is occurring and the overhauling load is being slowed by the energy being fed back to the power supply (see Figure 5-4).

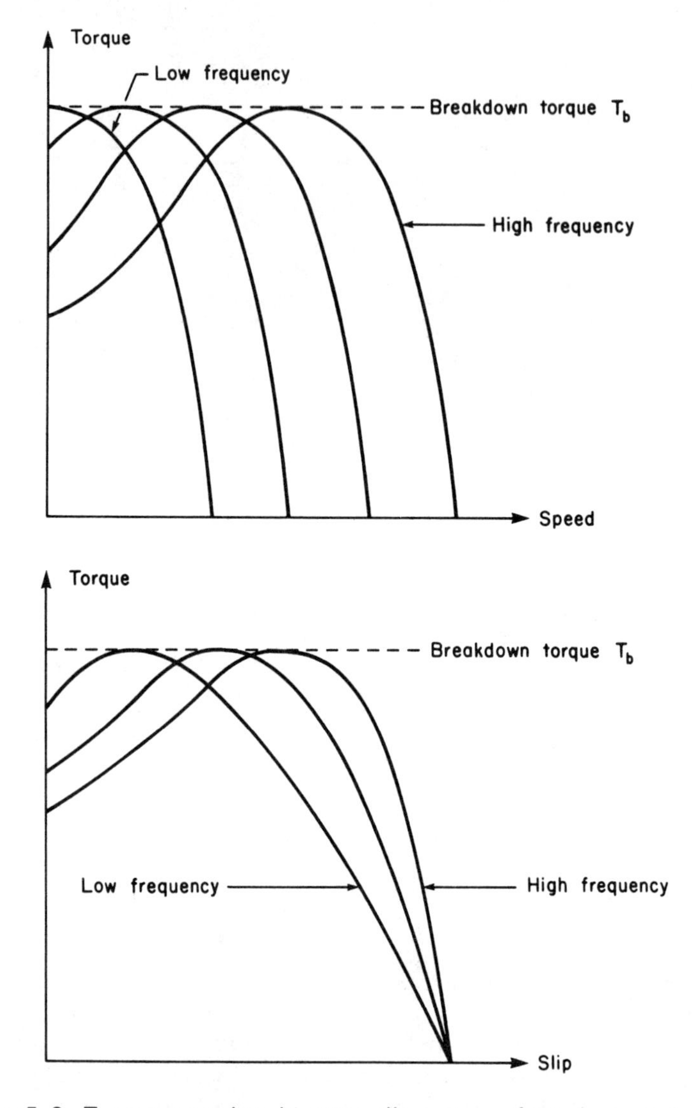

Figure 5-3. Torque-speed and torque-slip curves of a polyphase induction motor with a variable-frequency source.

5-4 FORWARD AND REVERSE ROTATION

In order to reverse a polyphase induction motor, it is necessary to reverse any pair of the stator leads. In addition, provided that the rotor can be accelerated through synchronous speed, the motor will also operate as an induction generator in the reverse direction; that is,

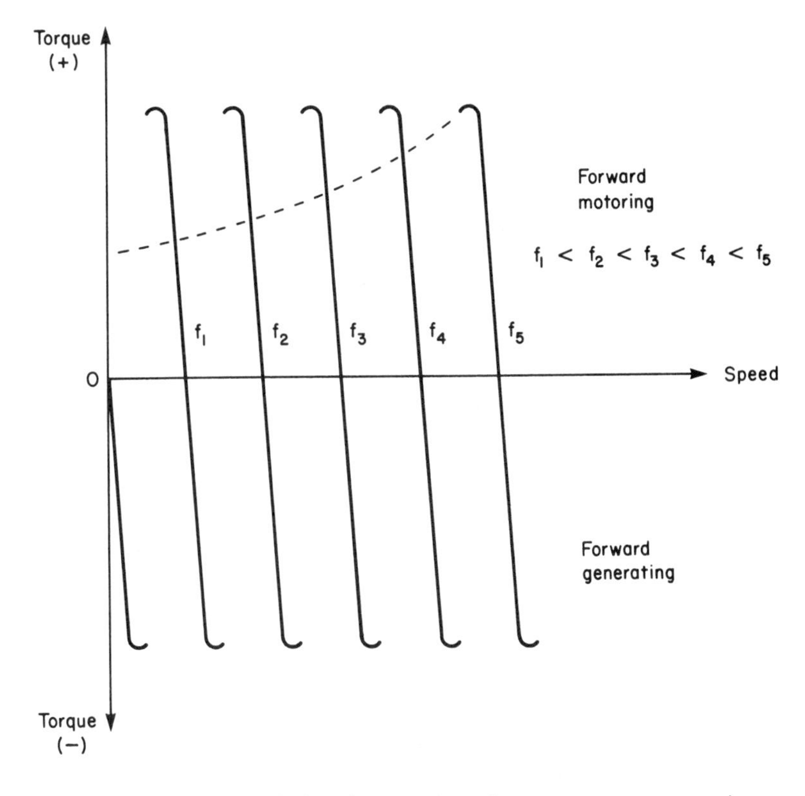

Figure 5-4. Polyphase induction motor, torque versus speed curves, motoring, and regeneration with a variable-frequency supply.

the polyphase induction motor can be operated in all four quadrants with the necessity of changing only the phase sequence of the supply to the stator. The only other drive that has this capability is a dc Ward-Leonard system supplied by a two-quadrant converter with a reversing switch at its output or a dual-converter (four-quadrant).

5-5 THE dc LINK CONVERTER

The **dc link converter,** one of the two major methods of obtaining a variable-frequency supply, consists of a dc rectifier and an inverter. Usually, dc rectification is accomplished by a three-phase phase-controlled thyristor bridge converter or a three-phase diode bridge whose dc output is supplied to a three-phase inverter. Voltage control to obtain a constant V/Hz relationship can be obtained by phase-angle control of the phase-controlled converter or by voltage control internally in the inverter.

5-5-1 Single-Phase Bridge Inverter

The principle of inverter operation is best introduced by considering the **single-phase inverter** shown in Figure 5-5. It can be seen immediately that only one thyristor in each leg can be on at any one time; that is, SCR1 or SCR2 may be on, but not both simultaneously, or a short circuit will be applied across the dc source. The thyristors are connected in a bridge configuration with the load connected between points A and B. If the load is reactive, that is, either inductive or capacitive, a return path for the reactive energy must be provided to the dc source. The function of the feedback diodes connected in inverse-parallel with the thyristors is to provide a path for the reactive energy.

The operation of the inverter is as follows: All SCRs are gated on and turned off in the appropriate sequence for a period of time corresponding to 180° of the output ac cycle. Then if SCR1 is turned on, point A will be positive with respect to the negative bus. When SCR1 is turned off and SCR2 is turned on, point A will be at the potential of the negative bus. Similarly, when SCR3 is on, point B is at the positive bus potential. Turning off SCR3 and gating on SCR4 places point B at the negative bus potential. Thus cycling SCRs 1 and 2 on and off alternately produces a series of positive pulses of voltage V_{AN}; cycling SCRs 3 and 4 produces a series of positive pulses of voltage V_{BN} displaced 180° from those of V_{AN}.

If SCRs 1 and 4 are now gated on for 180° and then turned off, and SCRs 2 and 3 are gated on for 180° and then turned off, the voltage applied to the load V_{AB} is produced, which is an alternating square wave of peak amplitude V_d.

In the case of a pure resistive load, the load current I_L will have exactly the same waveform and be in phase with the load voltage. If the load is inductive, the load current will lag the inverter output voltage. The function of the feedback diodes is to return the reactive energy from the load back to the dc source. For example, if SCRs 1 and 4 are turned off, the reactive energy will return via diodes $D2$ and $D3$; when SCRs 2 and 3 are turned off, the reactive energy is returned via diodes $D1$ and $D4$.

Another function of the diodes is to prevent the peak amplitude of the inverter output voltage exceeding that of the dc source, and as a result the output voltage will always have a constant amplitude.

The gate-firing circuits and commutation circuitry for the SCRs have been deliberately omitted at this point to simplify the circuit explanation.

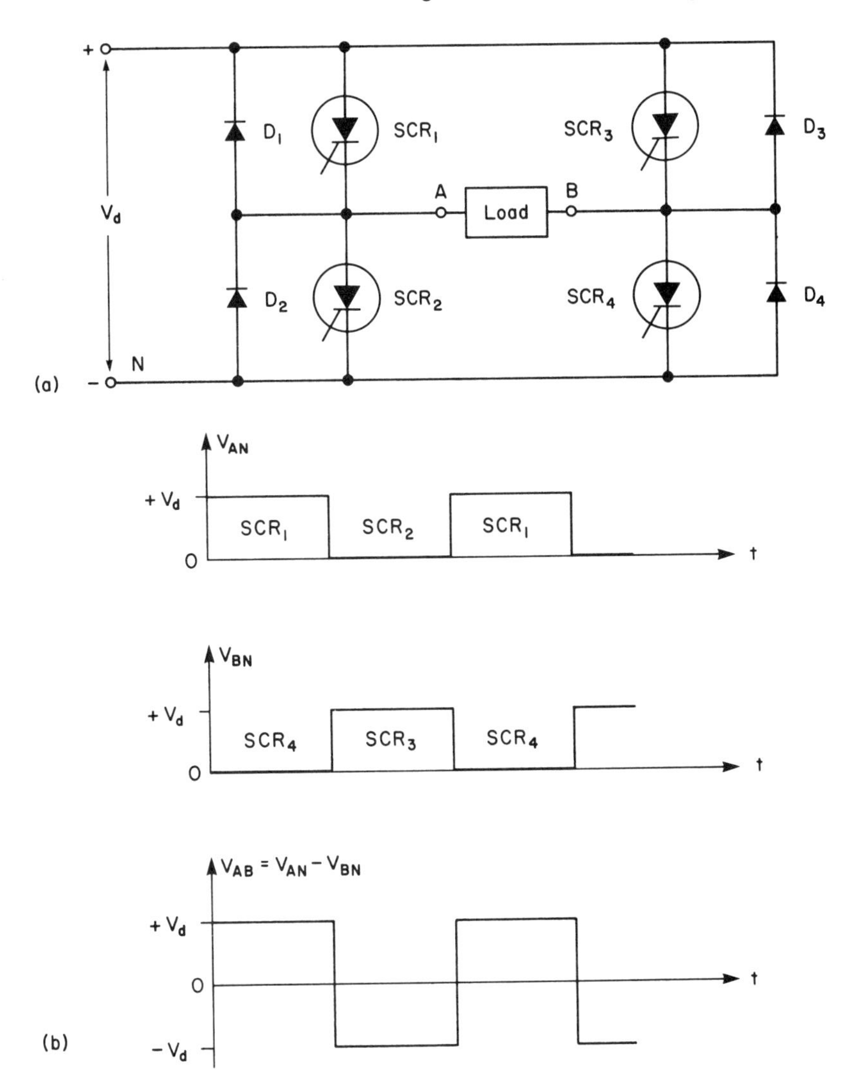

Figure 5-5. Single-phase bridge inverter: (a) basic circuit, (b) voltage waveforms.

5-6 SINGLE-PHASE INVERTER VOLTAGE CONTROL

The basic concepts of inverter output voltage control can be most easily understood by considering the single-phase inverter. Most inverter applications require a means of controlling the output voltage. This may be achieved by controlling either the dc voltage supplied to the inverter or the voltage within the inverter.

5-6-1 Variable-Input Voltage Control

The application of dc link converters to variable-frequency ac motor drive systems requires a constant V/Hz ratio in order that a constant torque output can be obtained.

One method of controlling the V/Hz ratio is by variation of the dc input voltage to the inverter input terminals.

If the source voltage is ac, then the dc input voltage to the inverter may be controlled by a number of methods, namely:

1. A phase-controlled converter.
2. An uncontrolled rectifier with a variable dc output voltage being obtained by
   (a) Variation of the input ac voltage by means of an induction regulator or a variable autotransformer. Either of these methods requires mechanical adjustment of the device and results in a poor dynamic response for the overall inverter system and increased control complexity.
   (b) Variation of the output dc voltage by means of a dc-dc controller or chopper.

If the source voltage is dc, then the major method of varying the dc input voltage to the inverter is by means of a chopper.

As can be seen, there are a number of methods of obtaining a variable dc voltage at the inverter input terminals; however, the system in most common use is the thyristor phase-controlled converter.

The major advantage of controlling the dc input voltage to the inverter input terminals is that the inverter output waveform and its harmonic content do not vary greatly as the dc voltage level is changed. However, the disadvantages outweigh the advantages, namely:

1. The current commutating capability of the inverter varies with the dc voltage level because a reduction in voltage without a corresponding reduction in the current reduces the time available for turn-off. This problem can be offset to some extent by increasing the commutating capacitance or by providing a fixed dc commutating supply with a resulting increase in circuit complexity.

2. The control of the dc voltage adds to the complexity of the overall dc link converter.

3. The dc voltage should be smooth and requires filtering of the dc output voltage, which in turn increases the response time of the system.

5-6-2 Voltage Control Within The Inverter

Because of the disadvantages of the variable-voltage input control techniques, the most efficient method of voltage control is by varying the ratio between the dc voltage at the inverter input terminals and the ac output voltage of the inverter. In general, the methods in common usage differ only in the effect that they have on the harmonic content of the output ac voltage waveform, with the harmonic content increasing as the output voltage is decreased. The most common methods of internal voltage control all use pulse width modulation (PWM) techniques and are:

1. Pulse width control.
2. Pulse width modulation (PWM).

5-6-2-1 Pulse Width Control

In the pulse width control technique the inverter output voltage is varied by controlling the width of the output pulse. In the circuit of Figure 5-5, if SCRs 1 and 4 are turned on for one half-cycle, and SCRs 2 and 3 are turned on for the next half-cycle, the output voltage waveform shown in Figure 5-6(a) will result. Voltage control can be obtained by varying the phase relationship between the firing of SCRs 3 and 4 with respect to SCRs 1 and 2. See Figure 5-6(b). It can also be seen that varying the retardation angle γ from $0°$ to $180°$ will vary the inverter output voltage from a maximum to zero, with only one commutation of each SCR for each cycle of the inverter output voltage, thus reducing the commutation losses to a minimum.

The major disadvantage of pulse width control is that as the output pulse width δ decreases, the mean output voltage decreases and the harmonic content of the output voltage waveform increases, until at the point where the fundamental is 20% of its maximum and the third, fifth, and seventh harmonics are nearly equal in value to the fundamental. The problem of the high harmonic content at reduced output levels can be greatly reduced by using PWM.

5-6-2-2 Pulse Width Modulation

Instead of reducing the pulse width to obtain voltage variations, as is done in pulse width control, the output of the inverter is switched on and off rapidly a number of times during each half-cycle, producing a train of constant amplitude pulses.

The amplitude of the inverter output voltage is controlled by varying the ratio of the total on-time of the pulses to total off-time. There

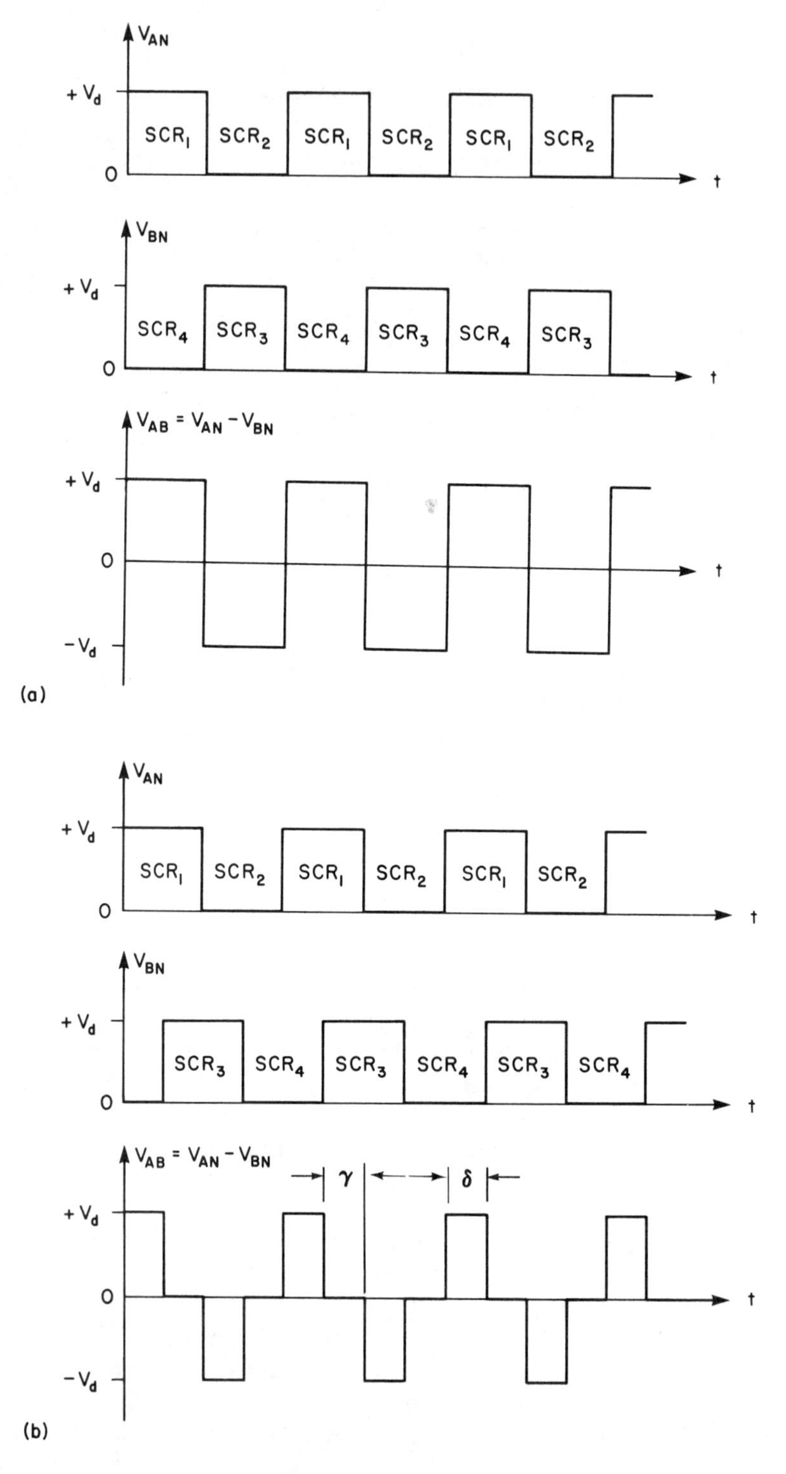

Figure 5-6. Pulse width voltage control: (a) 0° retard, (b) 90° retard.

are two basic approaches: maintaining a constant pulse width and varying the number of pulses per half-cycle or varying the pulse width for a constant number of pulses per half-cycle. The repetition rate of the pulses is known as the carrier frequency. The carrier frequency can be synchronized to the fundamental inverter frequency, or it can be independent of the fundamental inverter frequency.

The production of constant width pulses is accomplished by comparing a variable dc voltage level against a synchronized triangular carrier signal. See Figure 5-7(a). However, the inverter output voltage still contains a significant harmonic content.

A reduction in harmonic content is obtained by varying the pulse widths in a cyclic manner. This is accomplished by comparing a sine wave of the same frequency as the inverter output against a synchronized triangular carrier. See Figure 5-7(b).

The control techniques to obtain constant width pulses are simpler

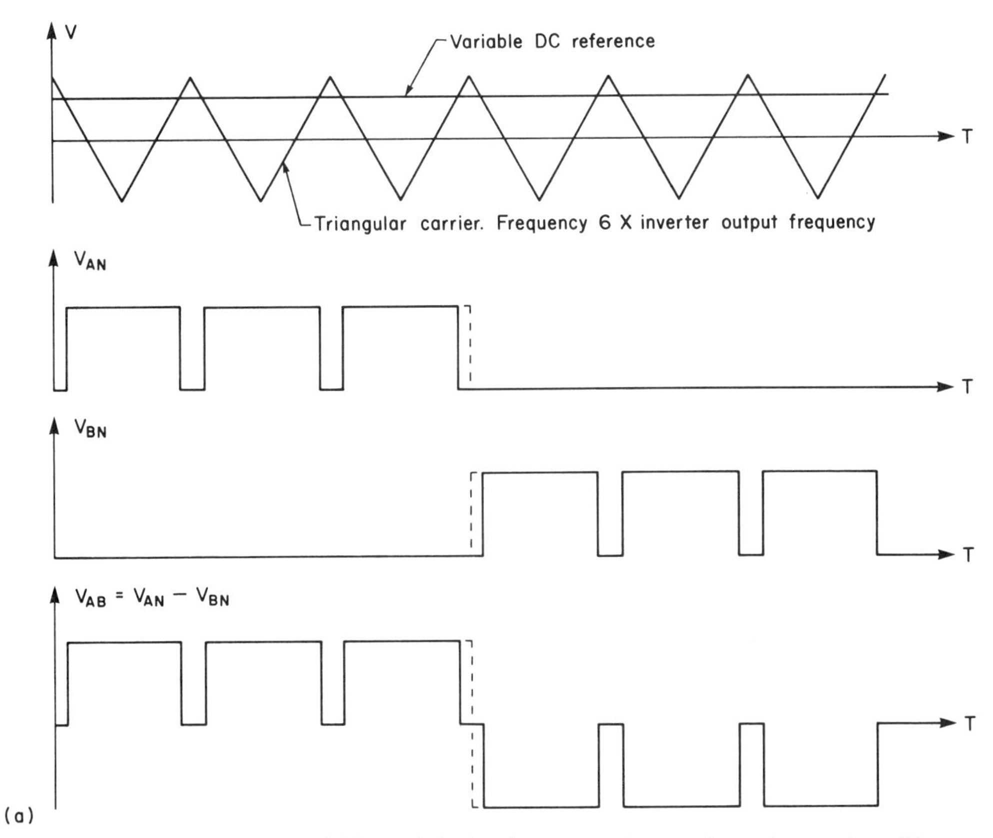

Figure 5-7. Pulse width modulation by comparing a triangular carrier with (a) a variable dc reference.

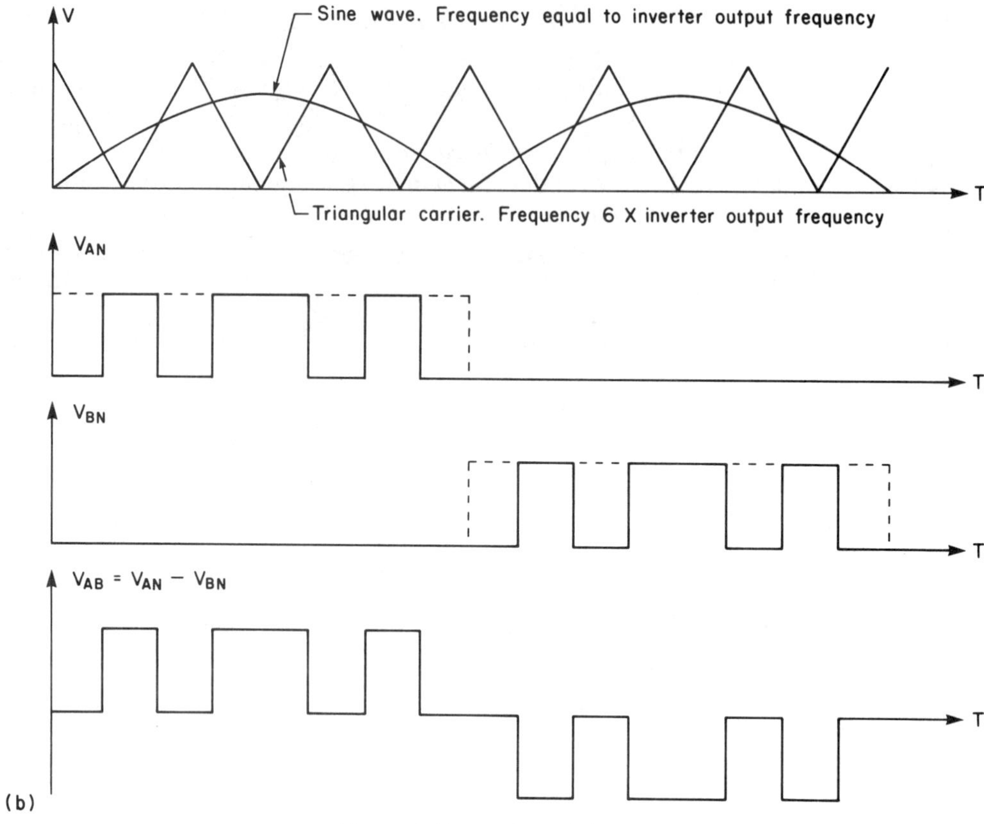

Figure 5-7 continued (b) a sine wave.

than those required for varying pulse widths. The ratio of the carrier frequency to the fundamental frequency of the inverter output should be maintained as high as possible. However, the turn-on and turn-off times of the SCRs, as well as the resetting time for the commutation components, place a practical limit on the number of pulses per half-cycle. The commutation losses in the SCRs and commutating circuits also rise as the pulse rate increases, with a consequent reduction in the inverter efficiency and kVA output.

5-7 FORCED-COMMUTATION TECHNIQUES

Once a thyristor is turned on, the gate loses control. The process of turning off a thyristor is known as **commutation**. Turn-off may be accomplished by interrupting the current flow by mechanical means, or by applying a reverse anode-to-cathode voltage across the thyristor.

The application of a reverse anode-to-cathode voltage reduces the forward current to zero and establishes a reverse current, which sweeps the current carriers from the J_1 and J_3 end junctions. At this point the forward current has decreased to zero, and the reverse-biased end junctions block the reverse voltage. Forward blocking capability is not reestablished until the current carriers have recombined at the J_2 junction.

Circuits using thyristors are classified in terms of the method of obtaining commutation. Natural commutation occurs in ac-supplied circuits. In dc-supplied circuits, the forward current must be reduced to zero by auxiliary components. This process is called **forced commutation**, and the circuits and components achieving forced commutation are called the **commutation circuits**.

Commutation in variable-frequency dc link converters is usually achieved by impulse techniques. Impulse commutation requires the production of an impulse that reverse biases the anode-cathode of an SCR to achieve turn-off. The amplitude of the pulse must be great enough to decrease the forward current to zero and of sufficient duration to restore forward blocking capability. The pulse is usually produced by an oscillatory inductance-capacitance (*LC*) network, whose natural period of oscillation is related to the SCR turn-off time, the dc source voltage, and the maximum value of load current to be commutated.

In our considerations of impulse commutation methods the following assumptions are made to simplify the explanations:

1. The load is inductive, which is reasonable since the text is devoted to motor control, and the load current remains substantially constant during the commutation period. This is valid, since typical turn-off times vary from 10 to 80 μ s.

2. The thyristor turn-on time is very small and can be neglected, which is reasonable since inverter-grade thyristors are being used with turn-on times typically between 4 to 8 μ s, and *di/dt* is limited by the circuit components.

3. The forward voltage drop of the thyristor is negligible.

Two methods of impulse commutation will be considered as applied to inverters.

5-7-1 Auxiliary Impulse-Commutated Inverter

The auxiliary impulse-commutated inverter employs a method of forced commutation in which the commutating pulse is produced by circuitry separate from the main power source. The major benefit of this method is that a loss of the gating signals will not cause commutation failure, provided that the auxiliary commutating circuit turns off the last conducting SCR. The method that will be described is also called the McMurray circuit.

With reference to Figure 5-8, SCRs 1 and 2 are the main load-carrying thyristors; SCRs 1A and 2A are called the auxiliary thyristors and are used to commutate SCRs 1 and 2 by switching the high-Q LC pulse-generating network in parallel with the conducting SCR.

The operation of the circuit is as follows: It is assumed that SCR1 is in conduction and that the right-hand plate of the capacitor is positive. SCR1 is turned off by gating SCR1A on; the discharge current pulse

(a)

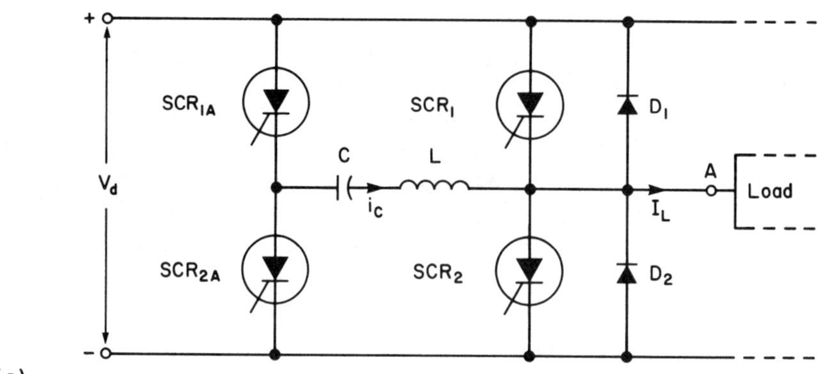

(b)

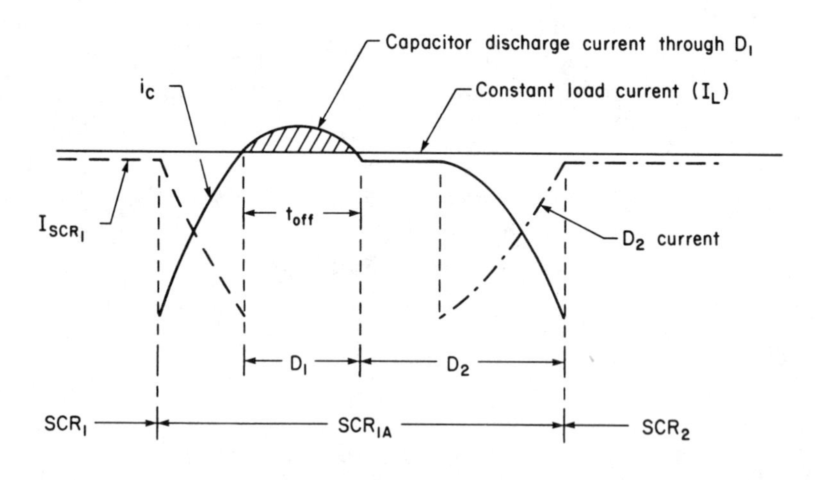

Figure 5-8. Auxiliary impulse commutated (McMurray) inverter: (a) basic circuit (single phase), (b) current waveforms with an inductive load.

flows through SCR1A, C, and L and opposes the load current I_L flowing through SCR1, initiating turn-off. As the discharge current i_c builds up to exceed I_L, the excess current flows through the feedback diode $D1$, thus applying a reverse-bias voltage across SCR1 equal to the drop across $D1$. The discharge current i_c peaks when the capacitor voltage has decreased to zero, and then decreases as the capacitor begins to recharge with reversed polarity. At the same time, when i_c becomes less than I_L, current flow through $D1$ ceases and the reverse bias is removed from SCR1.

After SCR1 is turned off, load current is transferred to $D2$ if the load is inductive. SCR2 may be turned on any time after SCR1 achieves its forward-blocking capability. At the same time, if it is assumed that SCR1A is still conducting, a small charging current flows through SCR1A, C, L, and SCR2 to complete the charging of capacitor C with its left-hand plate positive. At this point, SCR1A turns off, and SCR2 is ready to be commutated when SCR2A is fired.

The current relationships during the commutation cycle are shown in Figure 5-8(b).

In the McMurray circuit the no-load commutating losses are small, and the capacitor voltage increases with load current, thus ensuring an improved commutating capability. This commutation method is capable of being used over a range of frequencies up to approximately 5 kHz, thus ensuring its suitability with PWM voltage control techniques.

5-7-2 Complementary Voltage Commutation, McMurray-Bedford Circuit

In the McMurray-Bedford circuit shown in Figure 5-9(a) the turning on of SCR2 turns off SCR1, and vice versa. This technique is known as **complementary commutation** and, compared to the McMurray circuit, has the advantage of eliminating the auxiliary thyristors but at the expense of additional inductance and capacitance.

In Figure 5-9(a) the commutating capacitors $C1$ and $C2$ are equal, and the two halves of the commutating inductance $L1$ and $L2$ are equal and tightly coupled by being wound on the same core. For porposes of analysis, one half-cycle of operation is divided into five intervals, A, B, C, D, and E. The sequence of events taking place during the commutation process is shown in Figure 5-9(b) for one phase of a three-phase bridge inverter supplying an inductive load.

Interval A

It is assumed that SCR1 is conducting. Since the load is inductive, the rate of change of load current I_L will be small, and as a result the in-

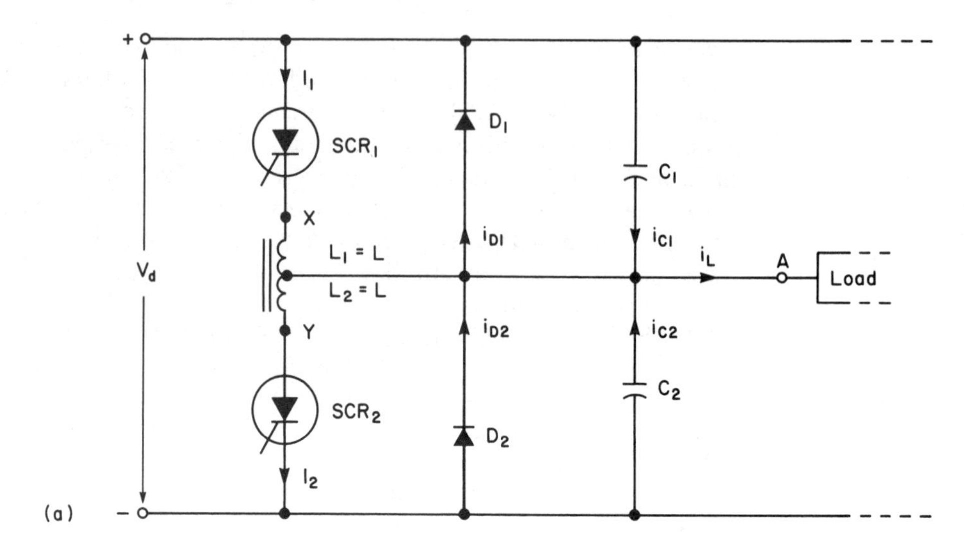

Figure 5-9. One phase of a McMurray-Bedford commutated inverter: (a) basic circuit.

duced voltage across $L1$ will be small. If the device voltage drop is neglected, the voltage at point A will be equal to the positive busbar voltage. At the same time, the voltage across capacitor $C1$ will be zero, and the voltage across capacitor $C2$ is equal to the supply voltage V_d.

Interval B

When SCR2 is turned on, the voltage across $C2$ is applied across inductor $L2$ and will in turn appear across $L1$. In other words, point Y is at the negative busbar potential, and point X is at a potential of $2V_d$ with respect to the negative busbar. SCR1 is reverse-biased, and turn-off is initiated. Since the emf in the core of the inductors $L1$ and $L2$ cannot change instantly, the load current I_L previously carried by $L1$ will be transferred immediately to $L2$. The load current i_L in the load is maintained by capacitor currents i_{c1} and i_{c2}. During this period, capacitor $C2$ is discharging and thus supports the voltage across $L2$ and the induced voltage $L1$; capacitor $C1$ is also charging to the dc source potential V_d. In the course of the charging cycle it will exceed the reverse bias across SCR1 produced by the voltage across $L1$, and SCR1 will be in a forward-blocking state. During this same interval of time, the induced voltage across $L2$ will be carrying the oscillatory current in the $L2$-$C2$ circuit and the current through SCR2, I_{SCR2}, will increase to I_m at the point when $C2$ is fully discharged.

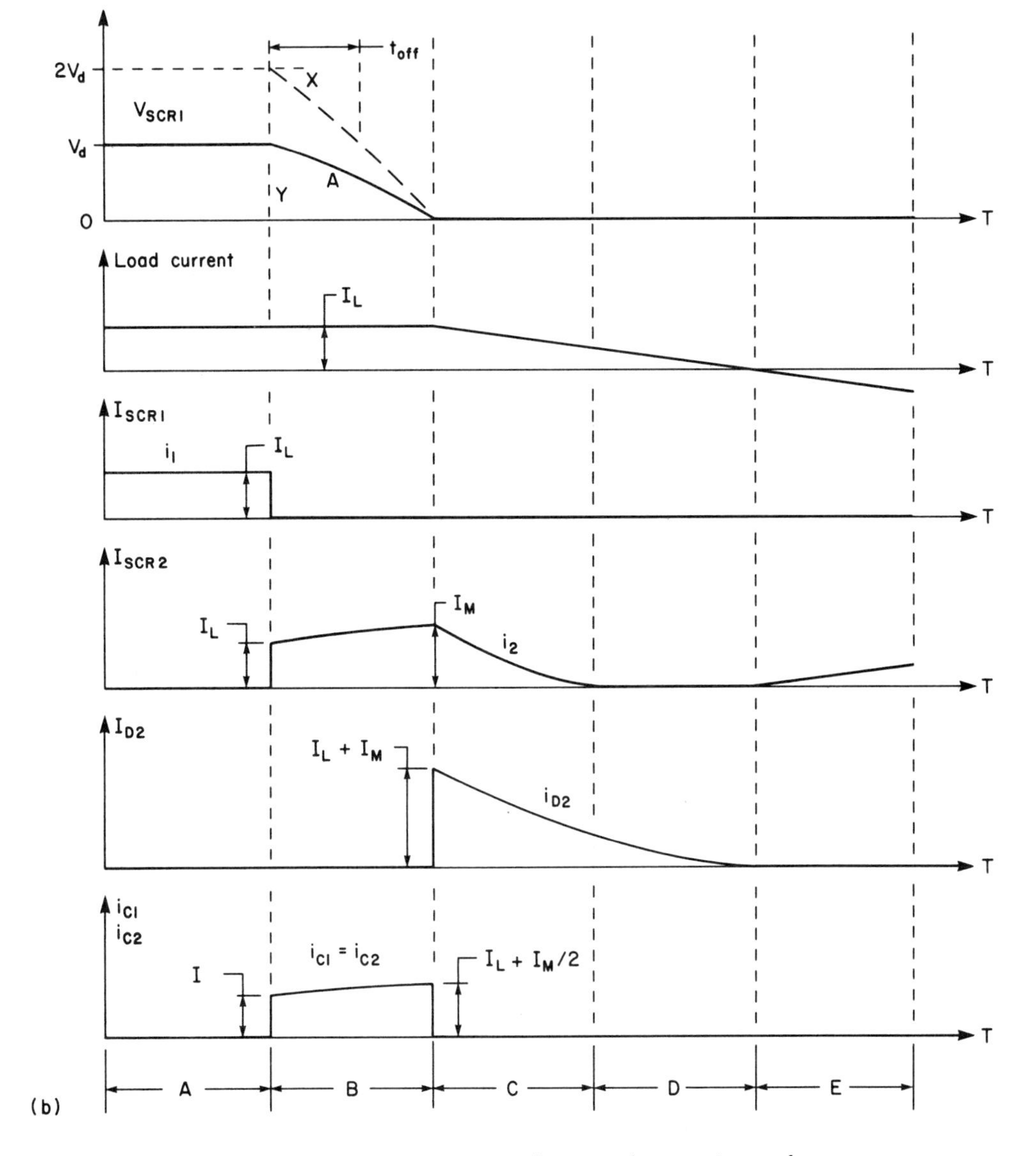

Figure 5-9 Continued (b) voltage and current waveforms.

Interval C

Capacitor $C1$ is now fully charged. The inductive load current i_L flows through diode $D2$, and point A is clamped to the negative busbar. Any stored energy remaining in $L2$ is dissipated as an I^2R loss through $L2$, SCR2, and $D2$.

Interval D

During this interval the current through $L2$ and SCR2 has decayed to zero, but because of the inductive nature of the load current, load current will still flow through $D2$.

Interval E

The load current flow through $D2$ decreases to zero and $D2$ becomes reverse biased; the load current is transferred to SCR2, if it is assumed that it is gated on. At this point the transfer of load current from SCR1 to SCR2 is completed. The next half-cycle would then be initiated by gating SCR1, and the process would be repeated.

This commutation technique is known as the McMurray-Bedford circuit and has the disadvantage that, under conditions of light load, the commutating losses are high; therefore, the inverter efficiency is low under light load conditions.

5-8 FREQUENCY CONTROL

Variable-speed ac motor control by the use of static inverters requires precise speed regulation over a wide range of frequencies. The output frequency of a static inverter is controlled by the rate at which the thyristors are gated on and off. The desired output frequency is controlled by a low-power master oscillator that generates a train of timing pulses, which, by means of logic control, can be directed to control the gating and turn-off circuits. Normally, with ac motor control the frequency control system is operated as an open-loop system and will provide satisfactory results even if there are variations in the ac source voltage and frequency.

The most common type of master oscillator is the UJT relaxation oscillator which can provide accuracies of $\pm 0.05\%$ for any set frequency under normal operating conditions as met in industrial applications.

5-9 THE SIX-STEP THREE-PHASE INVERTER

The simplest form of three-phase inverter is the three-phase bridge configuration shown in Figure 5-10, in which the firing and commutating circuits have been omitted to simplify the explanation. Some commutation techniques, explained in Section 5-7 for single-phase circuits, can be readily extended to three-phase applications.

The thyristors in Figure 5-10 are numbered in the sequence in which

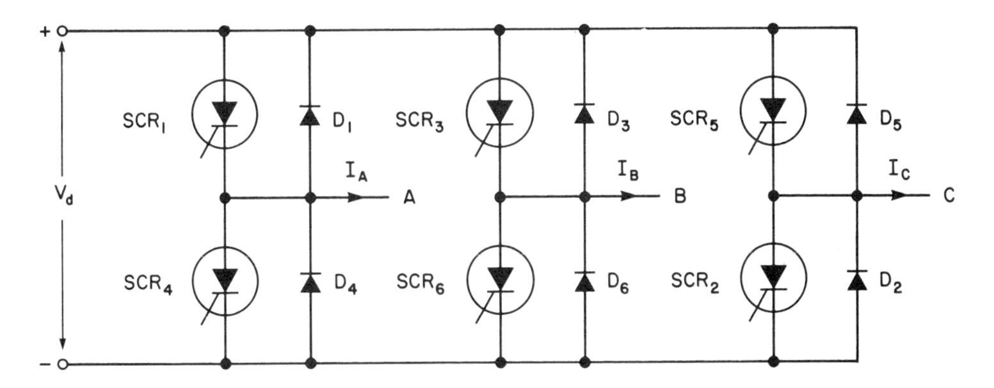

Figure 5-10. Basic six-step, three-phase inverter.

they are fired to produce positive-phase sequence voltages V_{AB}, V_{BC}, and V_{CA} at the output terminals A, B, and C. The output of the inverter can be obtained by either of the following gating firing sequences:

1. three thyristors in conduction at the same time or
2. two thyristors in conduction at the same time.

In either case, gating signals are applied and removed every 60° of the output waveform, and as a result there are six distinct steps in every cycle.

5-9-1 Three Thyristors in Conduction at the Same Time

Gating the thyristors of Figure 5-10 in the sequence SCR1, SCR2, SCR3, SCR4, SCR5, and SCR6 every cycle and leaving each in conduction for 180° of the output cycle will produce voltages, with respect to the negative busbar, V_{AN}, V_{BN}, and V_{CN} at the output terminals A, B, and C. The line-to-line voltages V_{AB}, V_{BC}, and V_{CA} are obtained by subtraction as follows: $V_{AB} = V_{AN} - V_{BN}$, $V_{BC} = V_{BN} - V_{CN}$, and $V_{CA} = V_{CN} - V_{AN}$, and are displaced 120° from each other (see Figure 5-11).

The line-to-line voltages can be shown by Fourier analysis to be

$$V_{AB} = \sum_{n=1,3,5}^{\infty} \frac{4V_d}{n\pi} \cos\frac{n\pi}{6} \sin n\left(\omega t + \frac{\pi}{6}\right) \text{V} \tag{5-22}$$

$$V_{BC} = \sum_{n=1,3,5}^{\infty} \frac{4V_d}{n\pi} \cos\frac{n\pi}{6} \sin n\left(\omega t - \frac{\pi}{2}\right) \text{V} \tag{5-23}$$

$$V_{CA} = \sum_{n=1,3,5}^{\infty} \frac{4V_d}{n\pi} \cos\frac{n\pi}{6} \sin n\left(\omega t + \frac{5\pi}{6}\right) \text{V} \tag{5-24}$$

Considering V_{AB}, we can simplify Equation (5-22) to

$$V_{AB} = \frac{2\sqrt{3}}{\pi} V_d \left[\sin \omega t - \frac{1}{5} \sin 5\omega t - \frac{1}{7} \sin 7\omega t + \frac{1}{11} \sin 11\omega t + \cdots \right] \tag{5-25}$$

and similarly for V_{BC} and V_{CA}. Since the line-to-line voltages are the differences of two square waves, the triplen harmonics ($n = 3,6,9, \ldots,$ etc.) in Equations (5-22), (5-23), and (5-24) are all zero.

The rms value of V_{AB}, V_{BC}, and V_{CA} is $\sqrt{(2/3)}V_d$, or $0.816V_d$, and the rms value of the fundamental is $\sqrt{6}V_d/\pi$, or $0.78V_d$.

If it is assumed that the load connected to the output terminals A, B, and C of Figure 5-10 is delta-connected, then the load-phase currents can be calculated from Equations (5–22), (5–23), and (5–24).

If the connected load is star- or wye-connected, the line-to-neutral voltages can be developed as shown in Figures 5-12 and 5-13. The load phase voltage is defined as

$$V_{AO} = \frac{3}{\pi} V_d \left[\sin \omega t + \frac{1}{5} \sin 5\omega t + \frac{1}{7} \sin 7\omega t + \frac{1}{11} \sin 11\omega t + \cdots \right] \tag{5-26}$$

and similarly for V_{BO} and V_{CO} and thus the load-phase currents can be calculated.

For either a delta- or star-connected load, a six-step waveform is produced as a result of applying six evenly spaced gating and commutating signals.

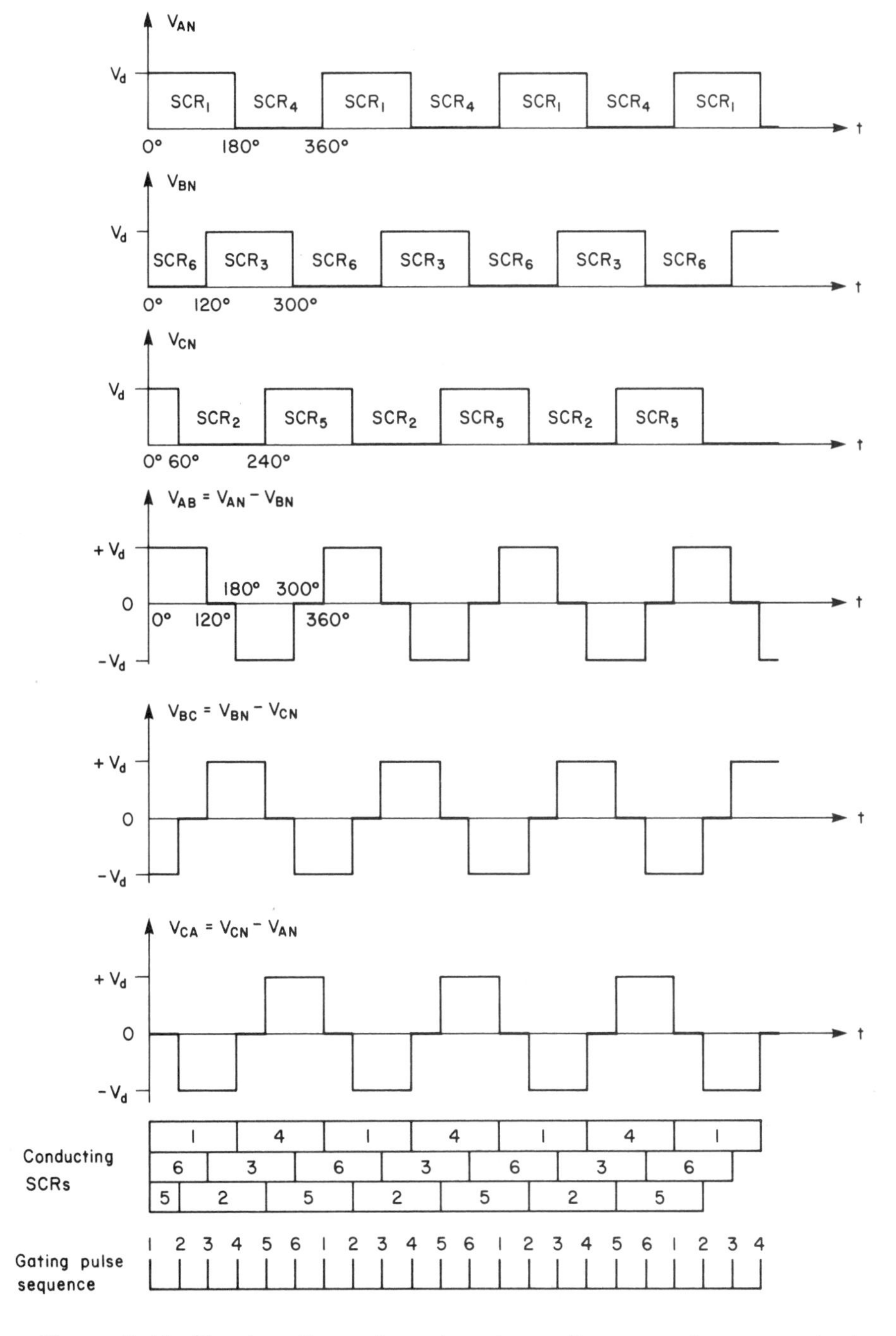

Figure 5-11. Six-step, three-phase inverter, voltage waveforms, conducting thyristors, and gating sequence.

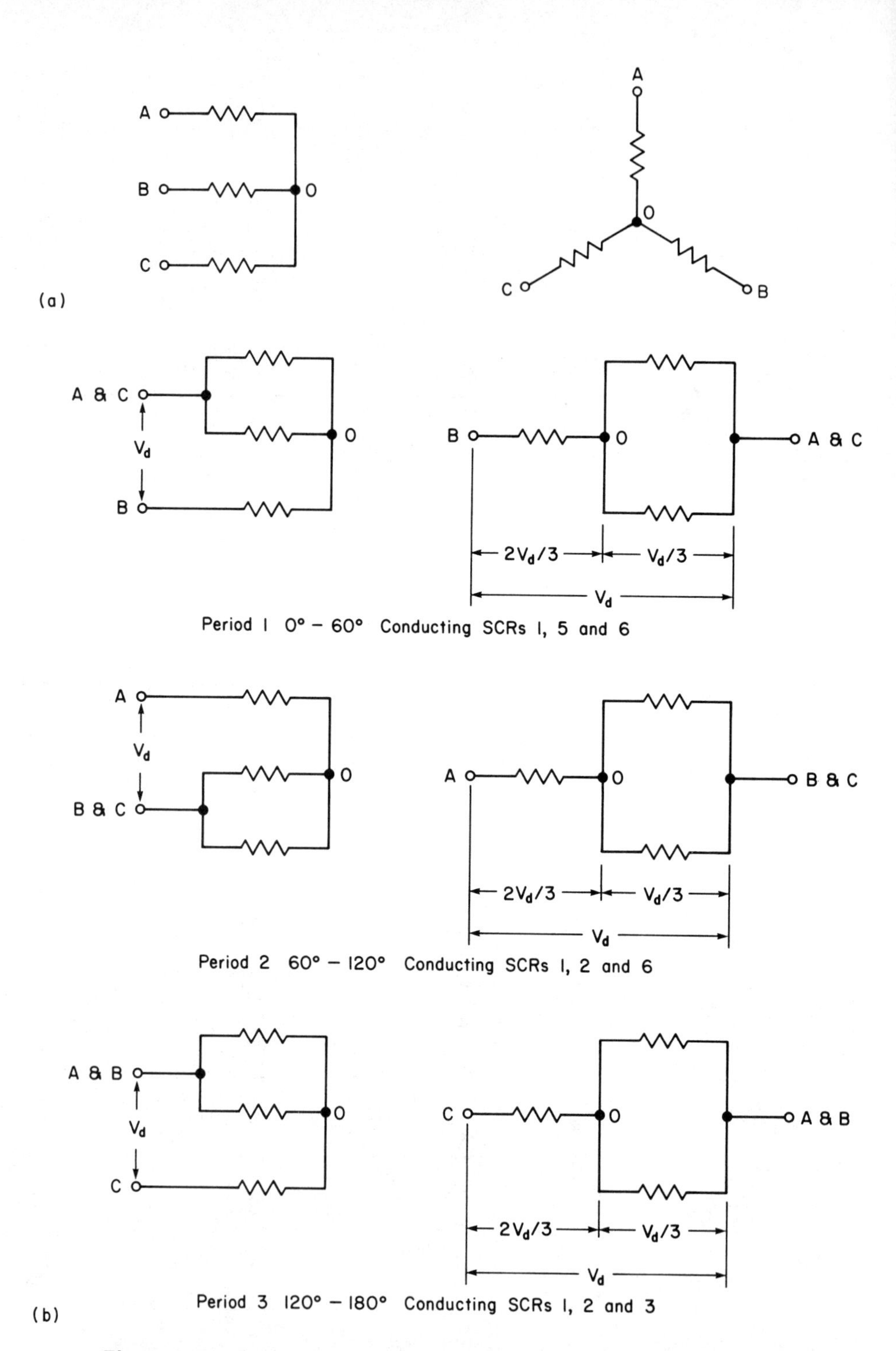

Figure 5-12. Determination of line-to-neutral voltages for a star-connected load: (a) equivalent circuit, (b) phase voltages every 60° interval.

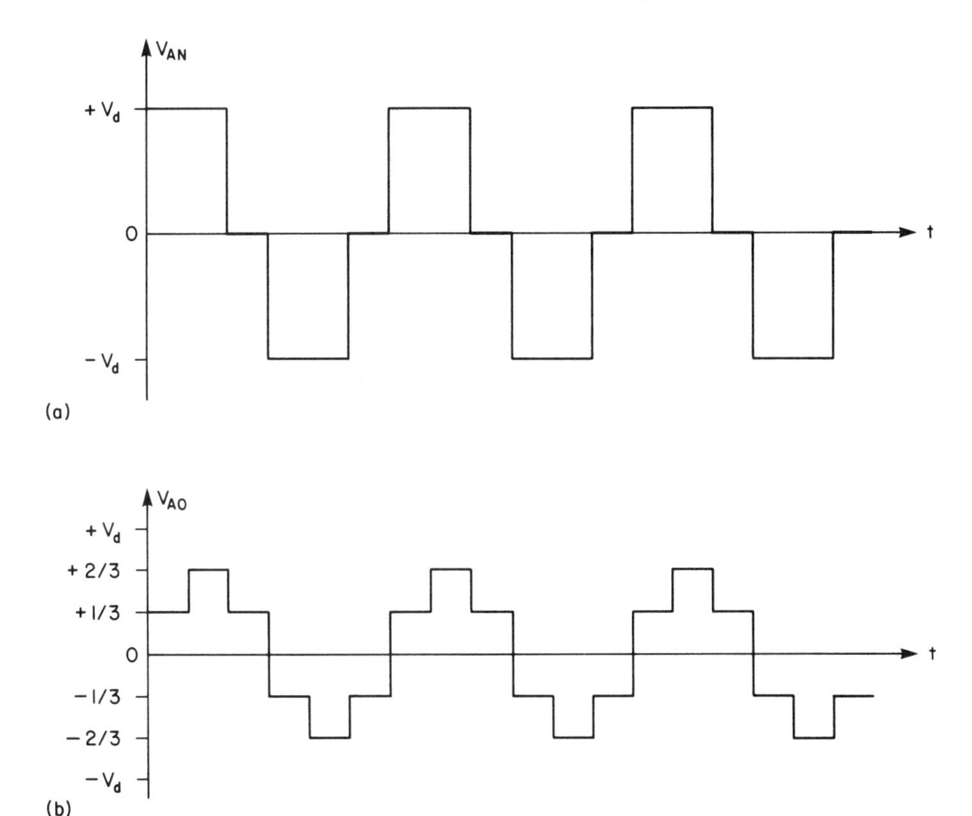

Figure 5-13. Phase voltages: (a) delta, (b) star-connected load.

5-9-2 Two Thyristors in Conduction at the Same Time

From Figure 5-11, it can be seen that SCR1 conducts from 0° to 180°, and SCR4 conducts from 180° to 360°. This condition assumes that SCR1 is commutated instantly prior to SCR4 being turned on; if there is any delay in the turn-off of SCR1, a short circuit or "shoot through" will exist between the positive and negative busbars through SCRs 1 and 4. It can be seen that this arrangement of gating the SCRs is a possible cause of failure. The possibility of a "shoot through" condition can be reduced by arranging for a 60° delay prior to turning on SCR4 and after SCR1 has had commutation initiated. Six commutations are still required per cycle of inverter output voltage. The associated voltage waveforms and gating sequences are shown in Figure 5-14.

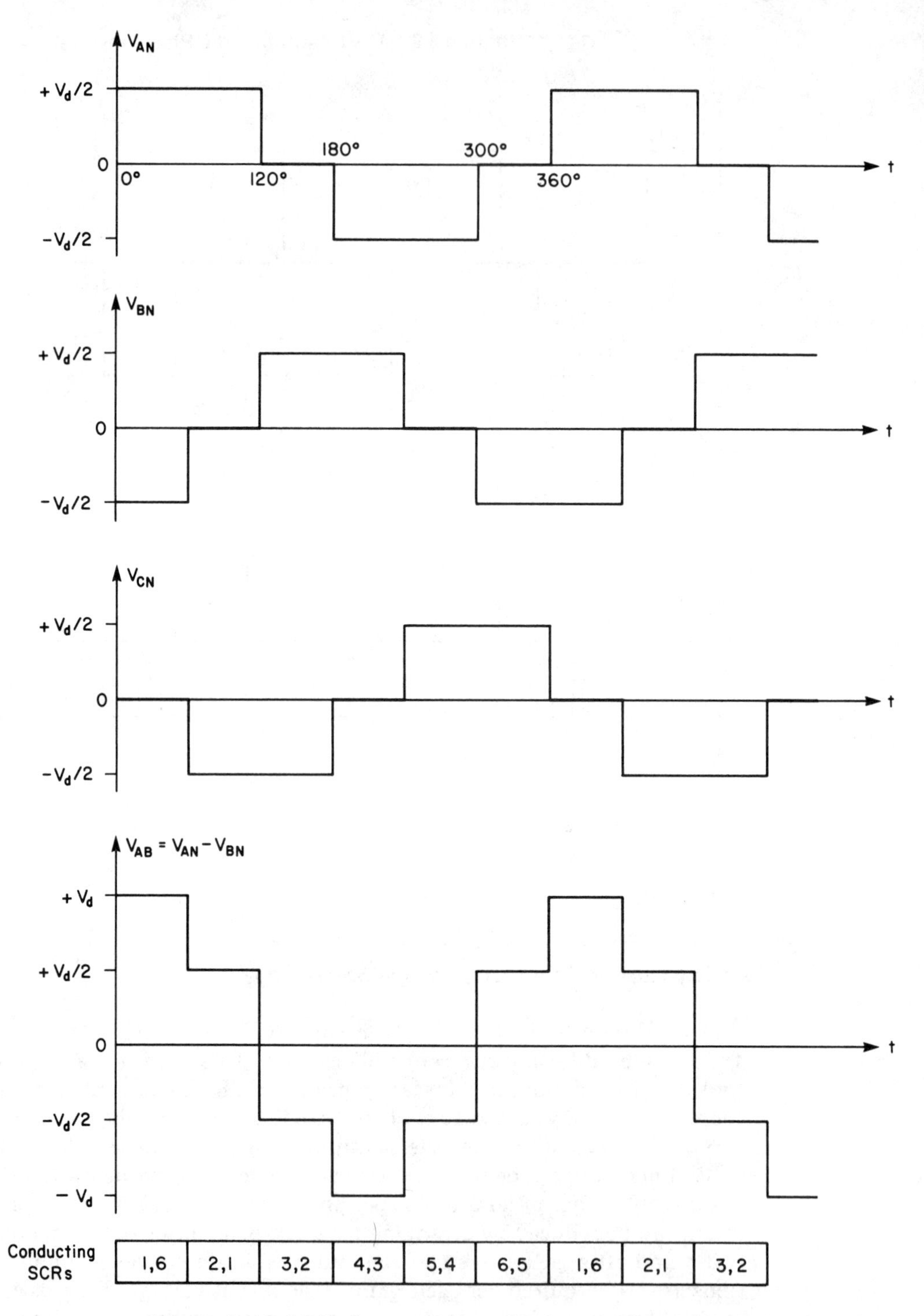

Figure 5-14. Voltage waveforms with two thyristors in conduction at the same time.

5-9-3 Current Waveforms

To complete the analysis of a six-step three-phase inverter, the relationship and nature of the phase and line currents supplying a typical balanced delta-connected inductive load are considered.

Since the voltage V_{AB} supplied to each phase is a square wave with voltage levels of $+V_d$, zero, and $-V_d$ (refer to Figure 5-15(a)) the step-changes of voltage will result in the load-phase current rising and decaying in an exponential manner. The load-phase currents i_A and i_B will be identical but displaced 120° from each other. The line current drawn from the inverter is $I_A = i_C - i_A$. Similar load-phase and line-current waveforms can be developed for a star-connected load.

The currents in the SCRs and diodes can be developed from Figures 5-15(e) and 5-10. At the instant that SCR1 is turned on and SCR4 is commutated off, I_A is negative; thus the line current must flow through $D1$ and in turn reduce the current drawn from the dc source. As long as $(i_C - i_A)$ is negative during the period of commutation, i_A will flow through $D1$ and SCR5 until i_C increases, making $(i_C - i_A)$ positive. At this point SCR1 conducts, provided there is a gating signal. Normally the gate signal is applied for 180° for inductive loads to ensure thyristor turn-on.

After 180° SCR1 is turned off, but since the line current is positive and lags the voltage, the line current is carried by $D4$ (see Figure 5-16). The greater the lag of the current with respect to the voltage, the greater will be the magnitude of the current being carried by the thyristor at the instant of commutation.

In Figure 5-15, the load-phase currents consist of two components, the charging component and the discharge component. The charging component flows through the dc source, and the discharge component circulates through a conducting SCR and diode. The dc current drawn from the dc source is the sum of the charging current components of the phase currents. When $(i_C - i_A)$ is negative during and just after the commutation period, current flows through $D1$ and reduces the current drawn from the dc source. During the interval 60° to 120°, SCR1 is the only thyristor connected to the positive dc busbar, and if $(i_C - i_A)$ is still negative, which would be the case for a low lagging power factor load, the dc supply current reverses, indicating a return of stored energy through the feedback diodes. Solid-state rectifiers will not permit the reversal of the dc current, and as a result a large capacitor is usually connected in parallel across the inverter input to prevent a rise in the dc voltage supplied to the inverter.

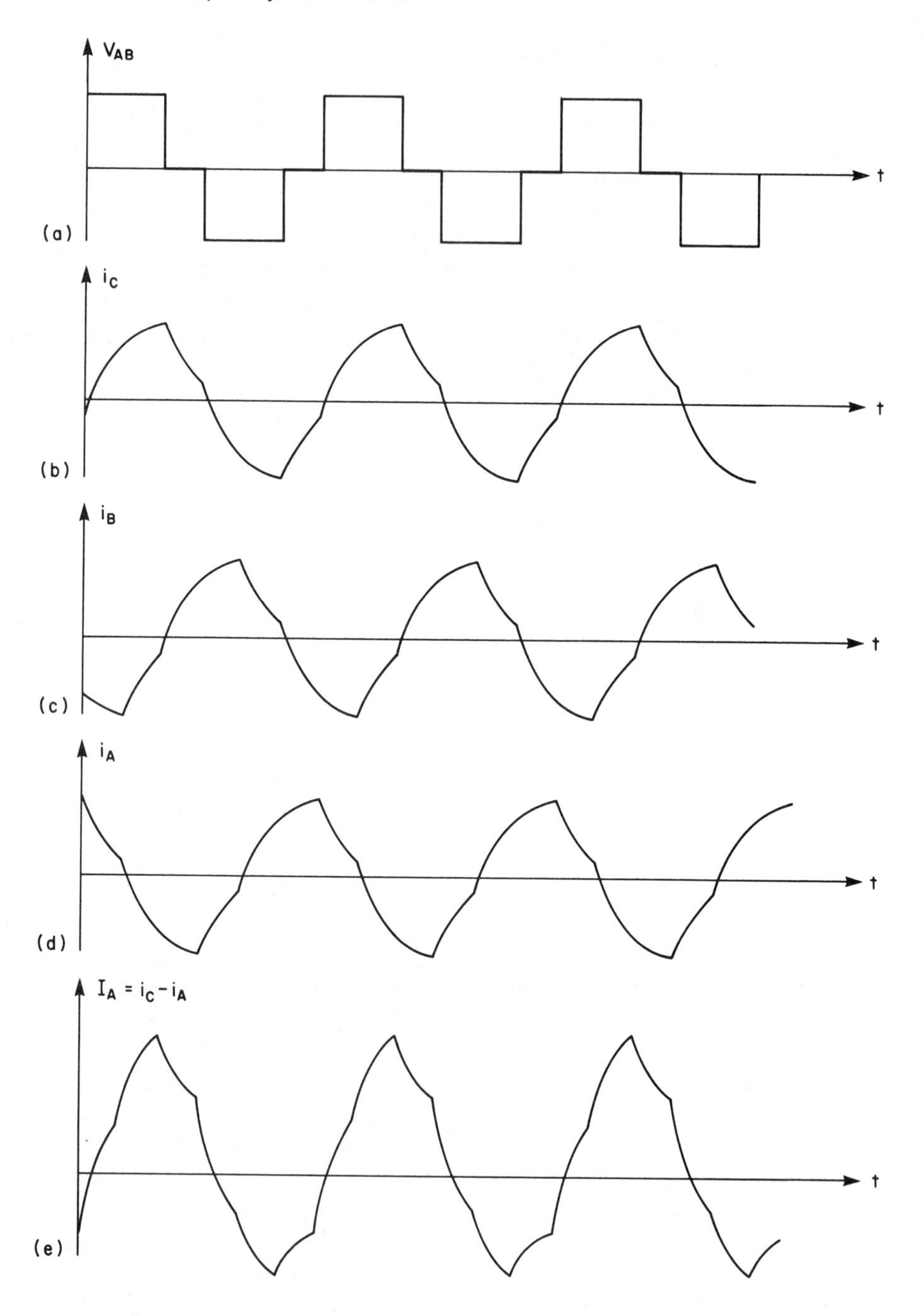

Figure 5-15. Voltage and current waveforms for a six-step, three-phase inverter supplying a delta-connected inductive load: (a) line voltage; (b), (c), and (d) load phase currents; (e) line current.

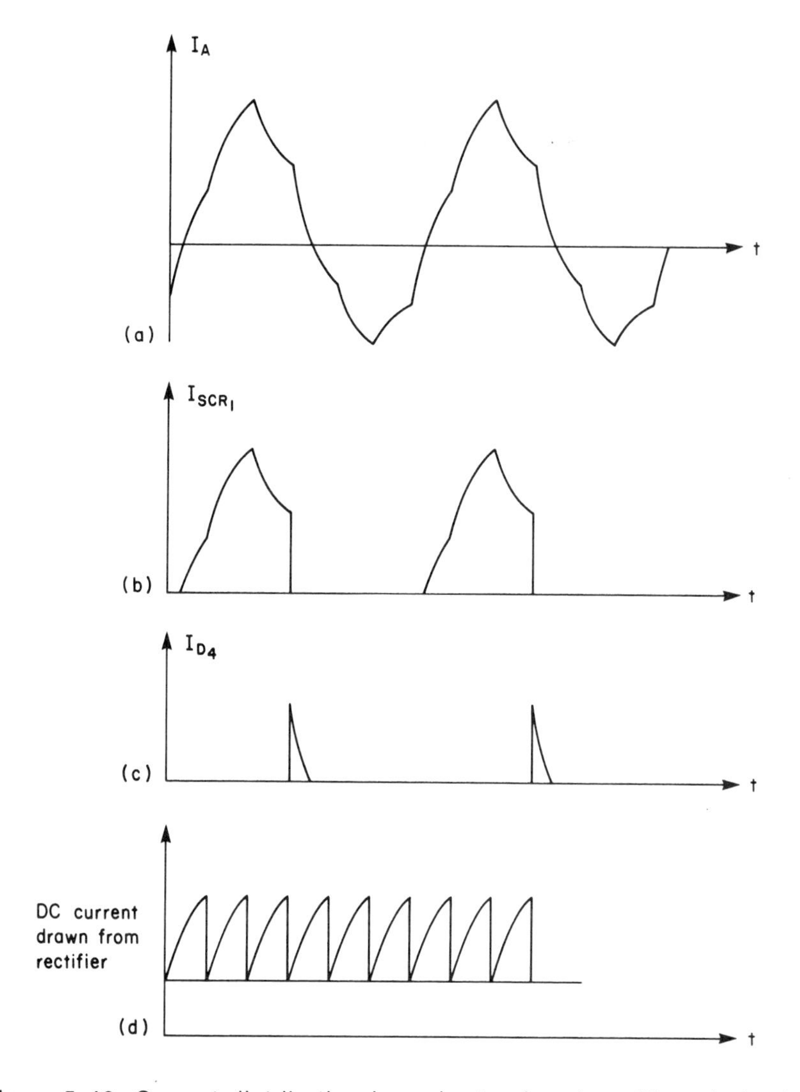

Figure 5-16. Current distribution in a six-step inverter with an inductive load: (a) line current waveform, (b) SCR1 current, (c) diode *D*4 current, (d) current drawn from rectifier.

5-10 HARMONIC NEUTRALIZATION

The output of a dc link converter has a high harmonic content, since the converter is designed to operate over a wide frequency range, typically 10 to 200 Hz; hence it is not practical to attempt to design and apply filtering externally to the converter. There are several ways to minimize the output harmonic content. One way is to use PWM techniques within the inverter. An alternative method is to combine a number of square-wave inverters, each of which is phase shifted and fired at the desired output frequency. This method is known as **harmonic neutralization.**

Figure 5-17 shows a three-phase inverter which consists of three single-phase bridge inverters. The transformer primaries that constitute the load of each converter have their secondaries connected in star and supply the phase load. This arrangement eliminates the third and triplen harmonics in the output waveform, although the higher harmonics remain. As can be seen from Figure 5-18, a commutation occurs every 60°, resulting in six commutations per cycle, producing a six-step waveform.

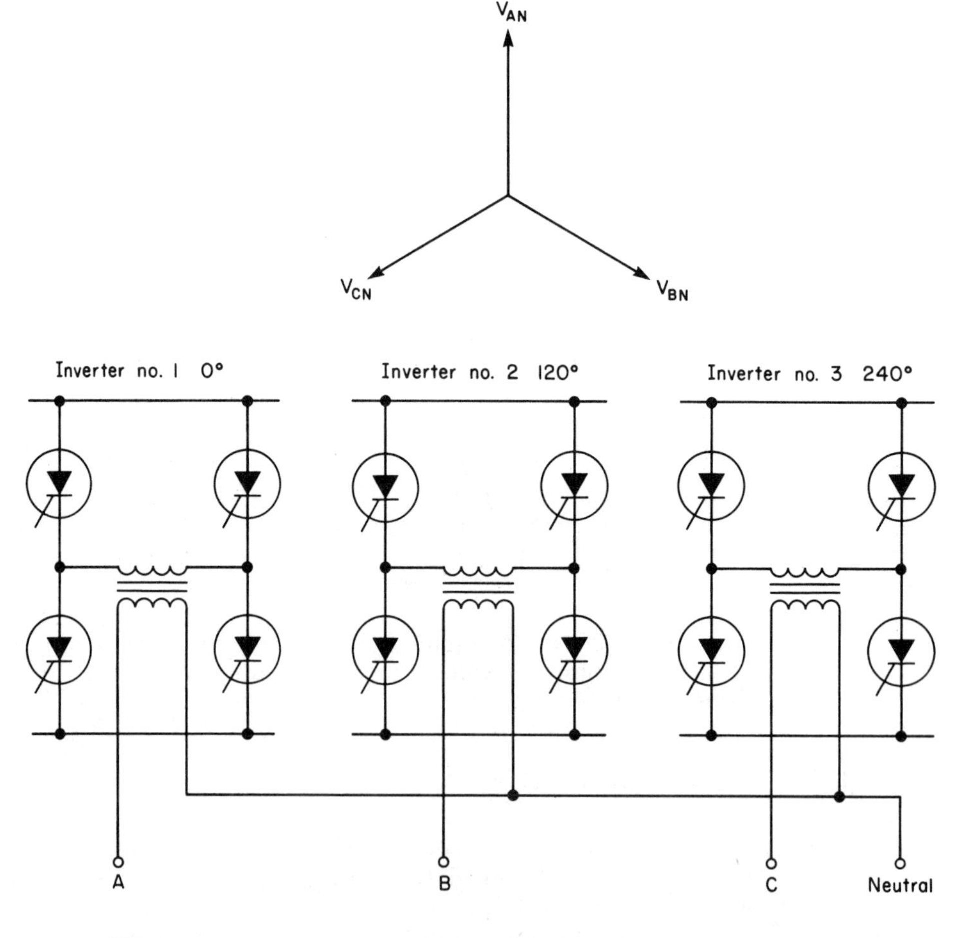

Figure 5-17. A six-step inverter formed from three single-phase bridge inverters.

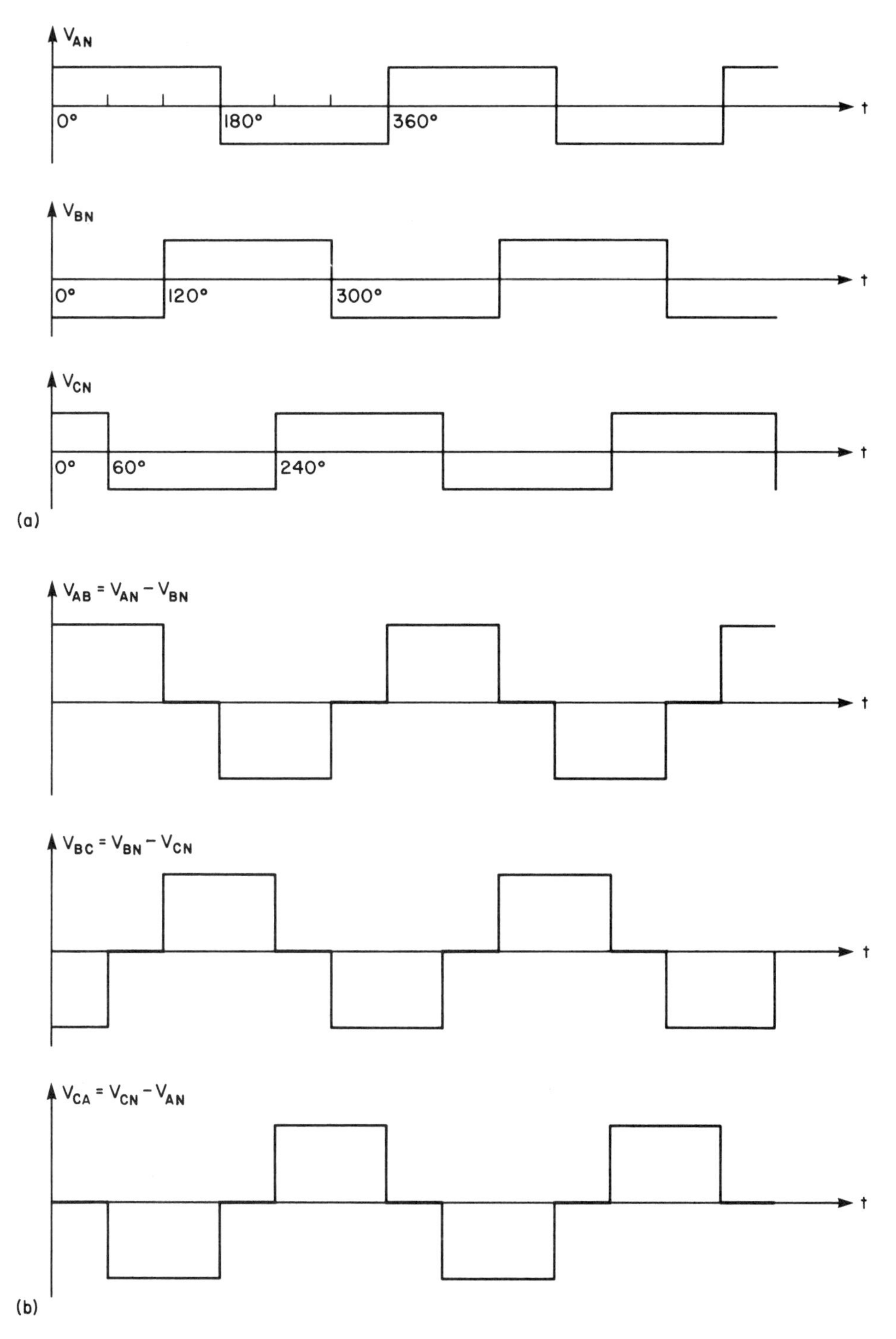

Figure 5-18. Six-step inverter waveforms with 180° pulse width, 0° retard: (a) line-to-neutral, (b) line-to-line.

It can be proved that there is a relationship between the number of commutations N per cycle of the output waveform and the lowest harmonic present. The relationship is that the lowest harmonics present are the $(N \pm 1)$ harmonics; for example, in the six-step output waveform, the lowest harmonics present are the fifth and seventh.

If the output voltage amplitude is varied by pulse width control, then the voltage waveforms applied to the primaries of the transformers can vary in width from 180° to 0°. Examples of the output waveforms for varying pulse widths are shown in Figure 5-19.

A twelve-step output can be obtained by using two six-step inverters, with the gating signals of the second inverter being displaced 30° with respect to the first; that is, at 30°, 150°, and 270°. This combination results in twelve commutations per cycle of the output, and the lowest harmonics present will be the eleventh and thirteenth. By careful choice of the primary and secondary transformer winding ratios, the amplitudes of the steps of the output waveform can be made to closely approximate the average value of the stepped intervals of a sine wave; thus the output waveform will be a very good approximation of a sine wave, even under minimum pulse width conditions.

If it is required to increase the kVA rating of an inverter, it is obvious that more benefit is gained by adding together simple inverters and offsetting them to create more steps to improve the output waveform, than is to be gained by paralleling the inverters. It must be appreciated, however, that the greater the number of steps, the greater the cost and complexity of the inverter, which results in the higher-step inverters being primarily used in high kVA applications.

CONVERTERS

The **cycloconverter** is a direct frequency changer that converts ac power at one frequency to ac power at another frequency by ac-to-ac conversion, without an intermediate dc link converter.

The concept of the cycloconverter is not new; it was originally developed in the 1930s and utilized grid-controlled mercury-arc rectifiers to reduce a standard three-phase 50 Hz source to single-phase power at 16⅔ Hz for electric traction purposes by the German State Railways.

However, it was not until the introduction of the thyristor with its compact size, lower voltage drop, ability to be operated in any position, mechanical ruggedness, and high-power and high-frequency switching capabilities, combined with logic and microprocessor control systems, that the cycloconverter reemerged as a viable frequency changer.

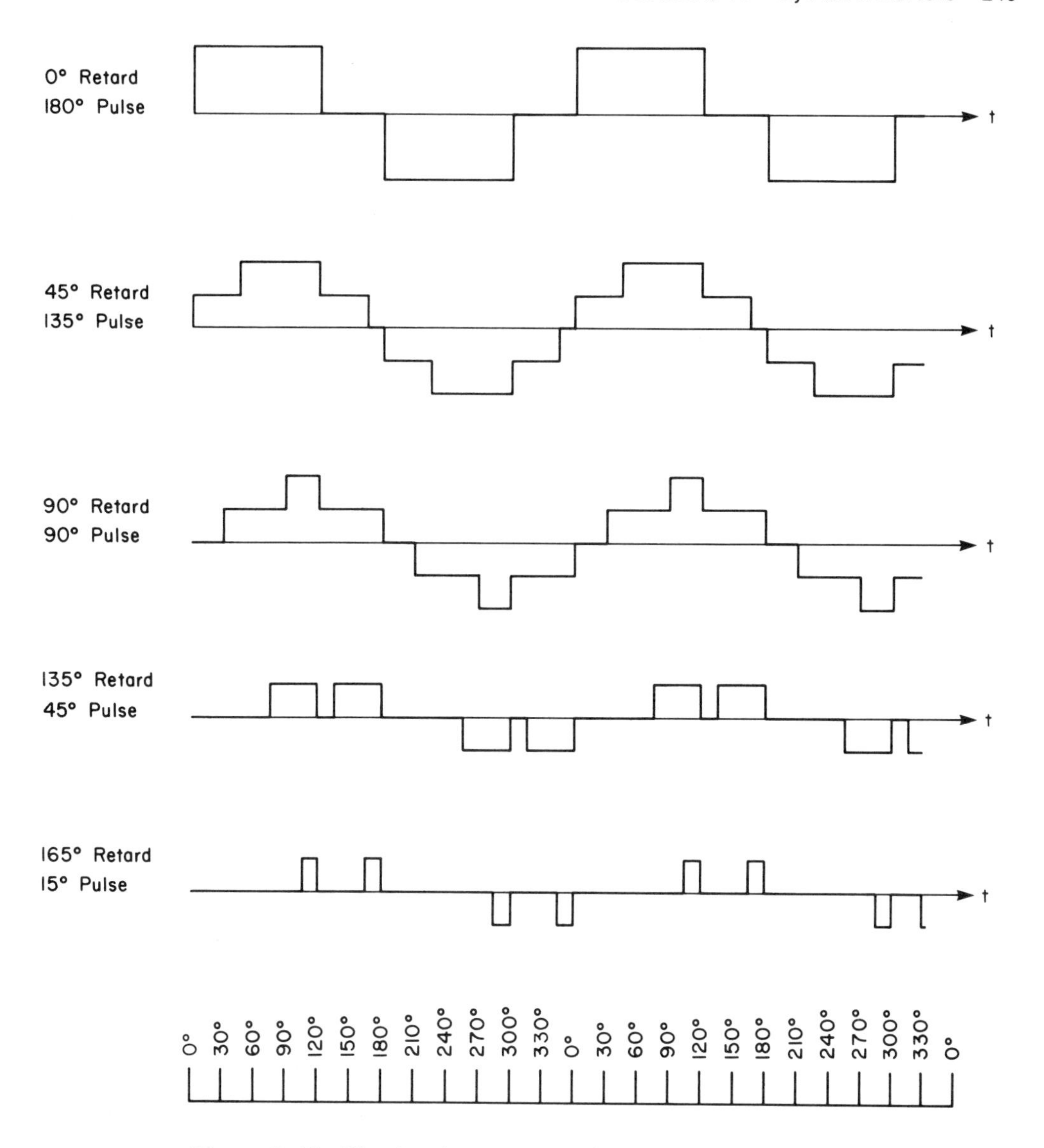

Figure 5-19. Six-step inverter waveforms with pulse widths varying from 15° to 180°.

The major applications of cycloconverters are low-speed multimotor drives in the range up to 300 hp (224 kW) and large motor drives in the range of 2,000 to 20,000 hp (1,500 to 15,000 kW) with supply frequencies from 0 to 20 Hz. The system can provide reverse operation and regeneration, and the motor drive is started by reducing the stator voltage and frequency.

Another application used extensively in the aircraft industry is the variable-speed, constant-frequency (VSCF) system, which provides, by a closed-loop control system, regulated output voltage at constant frequency irrespective of speed changes in the engine-driven generator system.

Normally cycloconverters are operated in the frequency range of zero to one-third of the supply frequency in order to keep the harmonic content of the output voltage waveform within acceptable limits. It may be operated as a frequency multiplier, but the harmonic content is greatly increased and the converter efficiency greatly reduced, although by the use of complicated rectifier circuits the output waveform may be improved.

Cycloconverters can produce a variable-frequency output by the use of phase-controlled converters or a fixed frequency output by envelope control.

The major advantages of using a cycloconverter are as follows:

1. Elmination of the dc link converter improves overall converter efficiency.

2. Voltage control is achieved within the converter.

3. Line commutation is obtained, with the total elimination of forced commutation components and circuitry.

5-11-1 The Single-Phase to Single-Phase Cycloconverter

The simplest way to understand the principle of operation of a cycloconverter is by studying the seldom used single-phase to single-phase cycloconverter.

The basic circuit shown in Figure 5-20(a) consists of two two-pulse midpoint phase-controlled converters, one forming the positive group and the other the negative group, a dual-converter configuration. The output current from each group flows only in one direction; therefore, in order to produce an alternating current in the load, the positive and negative groups must be connected in inverse-parallel. The positive group permits load current to flow during the positive half-cycle, and the negative group permits current flow during the negative half-cycle.

With reference to Figure 5-20(b), it is assumed that the connected load is inductive, and that the secondary voltages V_{AN} and V_{BN} are antiphase as shown. At time t_0, line A is positive with respect to line B, and the load current is negative; SCR4 is conducting and remains in conduction because of the return of the stored energy in the reactive

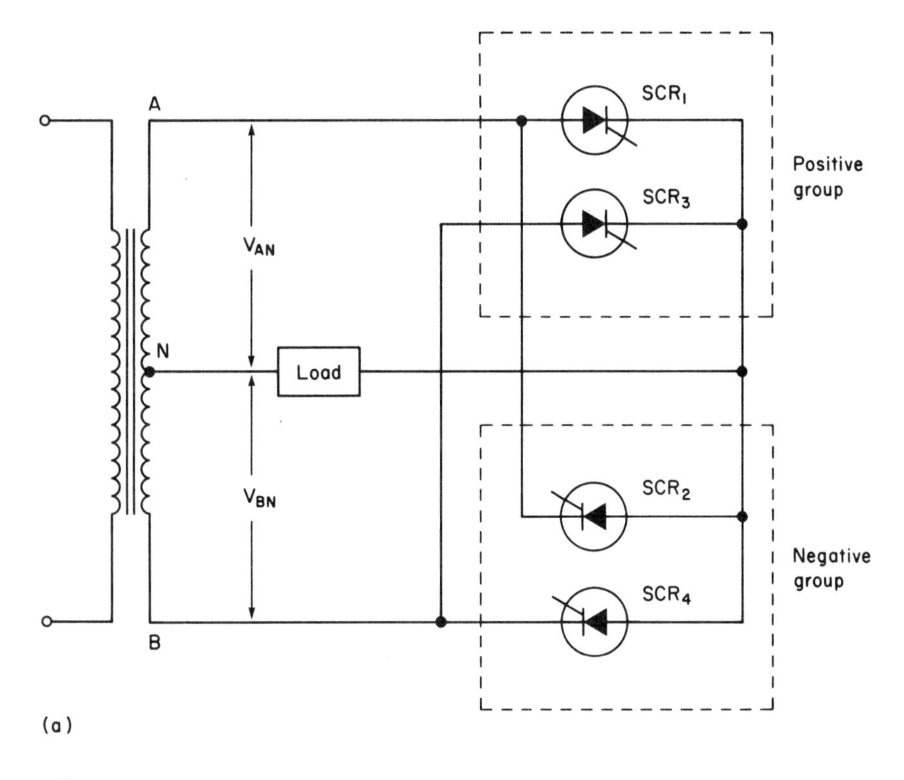

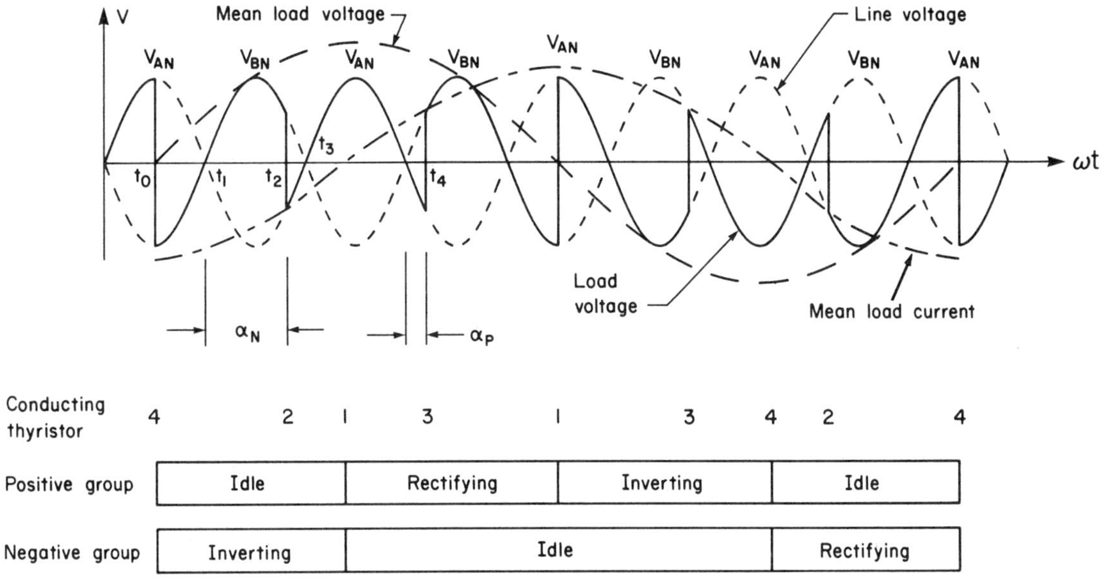

Figure 5-20. Single-phase to single-phase two-pulse cycloconverter: (a) basic circuit, (b) load voltage waveform, and thyristor conducting sequence.

load, even though the anode-cathode became reverse biased at time t_1. At time t_2, SCR2 is turned on and SCR4 turns off. When the load current becomes positive at t_3, SCR2 turns off and SCR1 is gated on and remains in conduction until time t_4, at which point SCR3 is grated on and SCR1 commutates off.

By varying the firing points of the thyristors, the amplitude $V_{do\alpha}$ of the mean output load voltage can be varied. Obviously, at time t_0, when the firing delay angle is 90°, the output voltage is zero; it is a maximum when the firing delay is zero, and similarly for the negative half-cycle output voltage. Since the mean output voltage is

$$V_{do\alpha} = V_{do} \cos \alpha,$$

the mean output voltage can be made to vary sinusoidally by suitably varying the firing delay angles α_P and α_N of the positive and negative groups so that the voltages will have the same amplitudes and frequency at their output terminals. As a result, load current can flow in either direction irrespective of the polarity of the mean load voltage. From this, it can be seen that the direction of mean power flow can be into the load (the rectifying mode) or from the load to the ac source (the inverting mode). Regardless of the direction of power flow, the direction of current flow is such that positive load current is carried by the positive converter, and negative load current is carried by the negative converter.

As a result, each two-quadrant converter for any load other than unity power factor will be operating as a rectifier, or as an inverter, or will be idling and will contribute to the positive and negative half-cycles of the mean output load voltage. See Figure 5-20(b). Consequently, the cycloconverter is capable of handling loads of any power factor from zero leading to zero lagging.

Since a firing delay of 90° results in a zero-load voltage output, varying the firing delay angles about 90° in a sinusoidal manner at the desired output frequency will result in a mean output load voltage that is sinusoidal at the desired frequency. Voltage control is obtained by controlling the amount of variation of the sinusoidally modulated firing delay angle. The mean load current has been assumed to be sinusoidal, but in fact unless the load is highly inductive or filtering is applied, this will not be the case.

Normally the firing delay angles α_P and α_N of the positive and negative converters, respectively, are controlled simultaneously. In order to increase the mean output voltage of the positive converter, α_P is reduced and α_N is increased. Similarly, to increase the mean output voltage of the negative group, α_N is reduced and α_P is increased. The firing delay angles are advanced and retarded with respect to the zero output voltage position, that is, the 90° point, and

$$\alpha_P = 180° - \alpha_N$$

resulting in the output voltages of both converters being equal. However, the sum of the instantaneous voltages is not equal, and as a result a ripple voltage is produced which will drive current through both converters. The magnitude of this current must be limited by the use of a current-limiting reactor, the "circulating current mode of operation," or eliminated by blocking the firing signals to the converter that is not carrying load current, the "circulating current-free mode of operation."

5-11-2 Three-Phase Cycloconverter Arrangements

The output of the single-phase two-pulse midpoint converter contains a high-amplitude ripple content. Increasing the pulse number will cause a smaller amplitude ripple content in the load voltage waveform, as was the case with phase-controlled rectifiers.

The pulse number can be increased in a number of ways; for example, by using six three-pulse midpoint converters connected to form three dual-converters, as shown in Figure 5-21. This configuration permits only unidirectional operation and requires 18 thyristors. A further increase in the pulse number can be obtained by using six six-pulse midpoint converters with interphase reactors (Figure 5-22), requiring a total of 36 thyristors; once again, only unidirectional operation is permitted.

The midpoint converter configurations all have a common output point to which the three-phase loads are connected. The bridge converter arrangement, on the other hand, requires the load to be connected across the common output terminals, thus requiring either the individual output phases or the three-phase inputs to the bridge converters to be isolated. The usual configuration is for the individual loads to be isolated (Figure 5-23). Once again, this configuration requires 36 thyristors; however, four-quadrant operation is now obtainable.

5-12 CIRCULATING CURRENTS

As has been previously discussed, the output frequency is controlled by the rate of variation of the firing delay angles about 90°, and the amplitude of the mean output voltage is varied by the amount of variation of the firing delay angles about 90°, continuously maintaining the relationship $\alpha_N + \alpha_P = 180°$.

As a result, the mean output voltages of the positive and negative

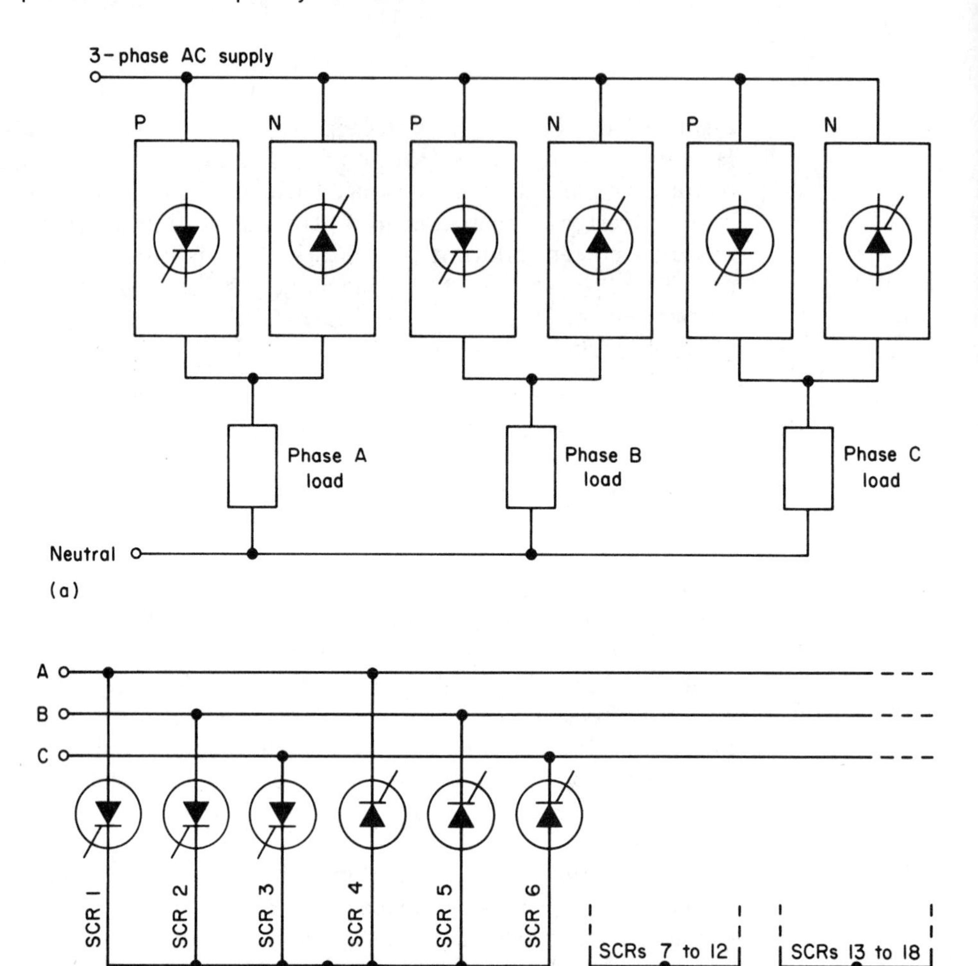

Figure 5-21. Three-phase to three-phase three-pulse cycloconverter: (a) schematic, (b) basic circuit.

groups are equal. However, the instantaneous values of these two voltages are not equal (Figure 5-24). As a result, it is possible for large harmonic currents to circulate which will increase the losses in the load circuit and increase the thyristor loading. There are two techniques by which the circulating current effects may be reduced: by an intergroup reactor and by gate-signal blanking.

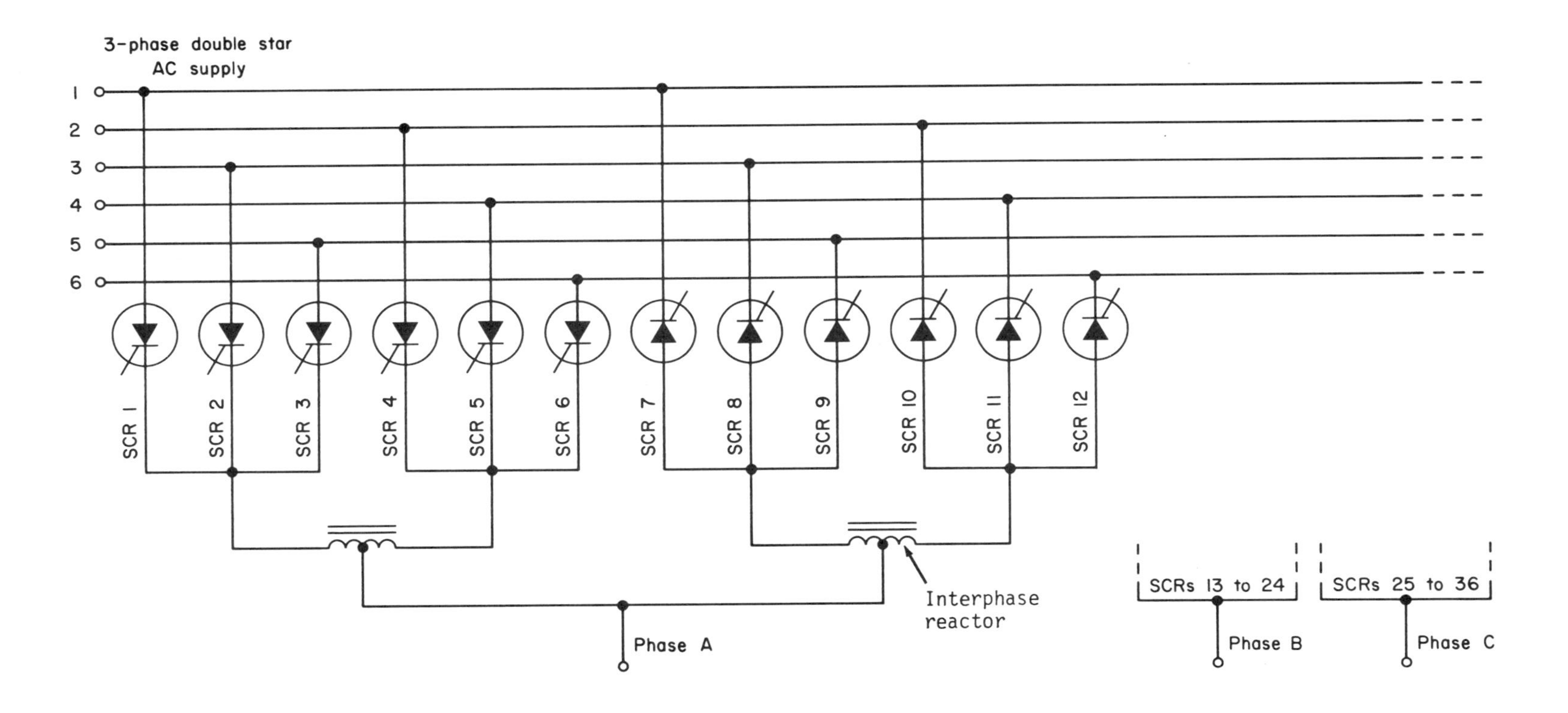

Figure 5-22. Three-phase to three-phase six-pulse cycloconverter.

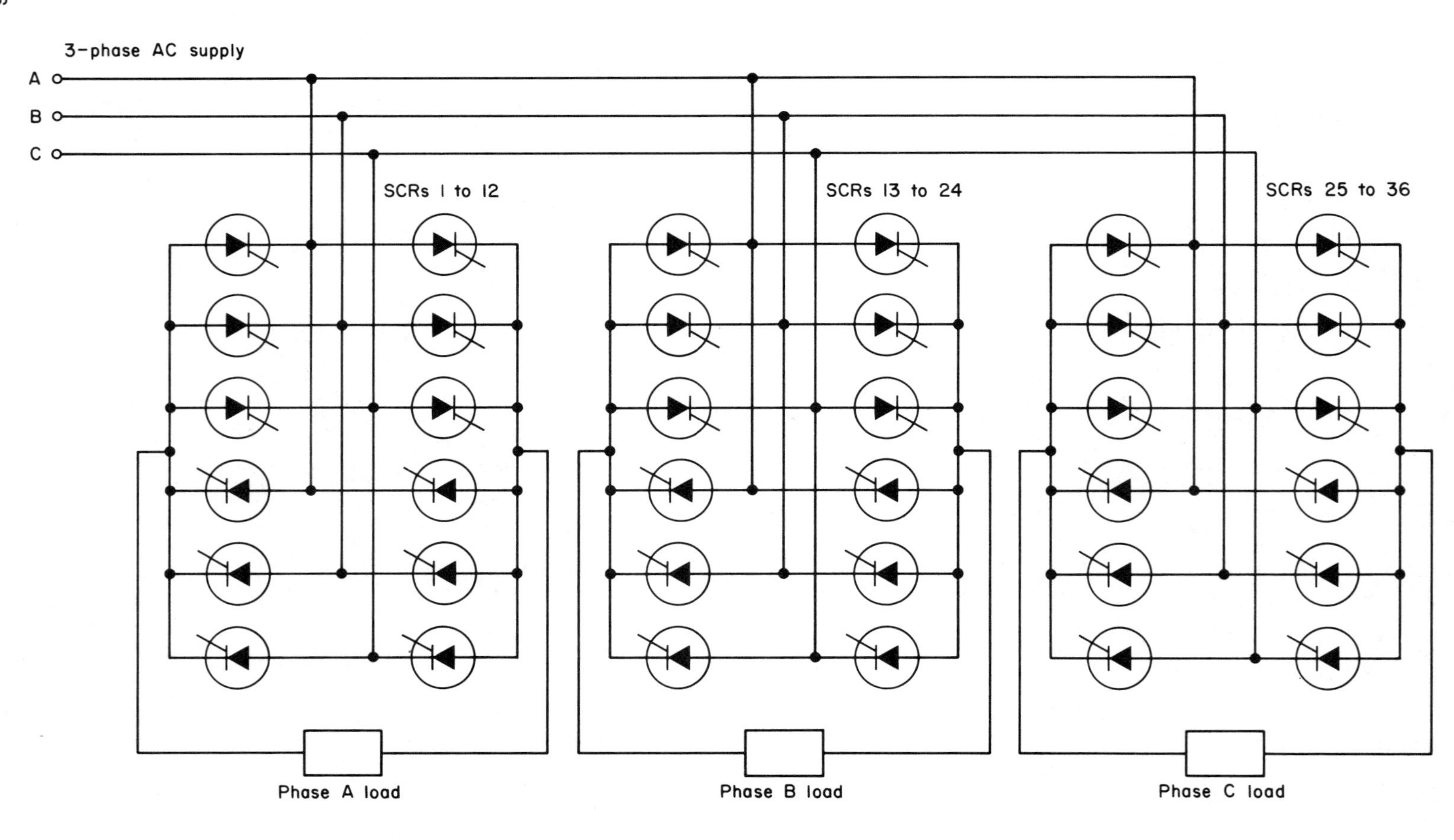

Figure 5-23. Three-phase to three-phase six-pulse cycloconverter with isolated load.

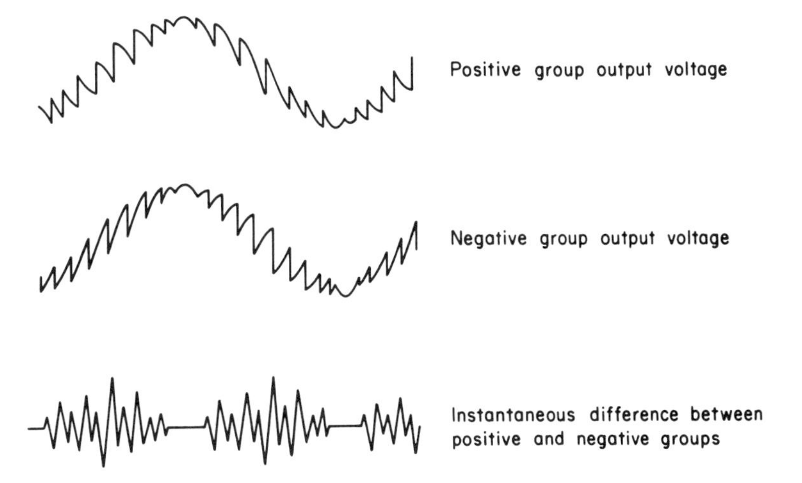

Figure 5-24. Cycloconverter positive and negative converter voltage waveforms and instantaneous voltage difference waveform.

5-12-1 Circulating Current Reduction by Intergroup Reactor

A current-limiting reactor is connected between the positive and negative groups and the load is connected to its center tap (Figure 5-25). As a result, the reactor presents its full reactance to the passage of the circulating currents and a quarter of its reactance to the passage of the load currents.

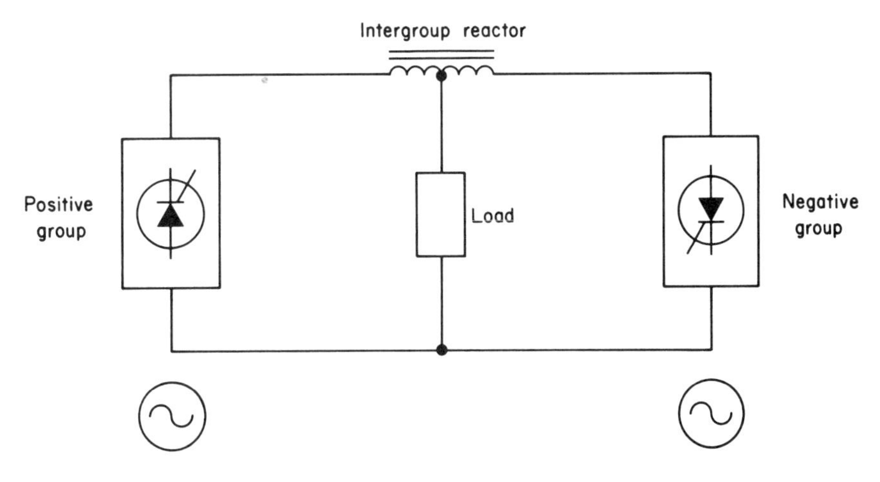

Figure 5-25. Cycloconverter with intergroup reactor to reduce circulating currents.

5-12-2 Circulating Current Elimination by Gate-Signal Blanking

Gate-signal blanking totally eliminates the circulating current by blanking the gating pulses to the inactive converter. This is accomplished by using current detectors in each phase.

This is the preferred method, but it is achieved by adding further complexity to the firing control circuitry. Under light load or unity power factor loads the mean output voltage distortion may be greater because of discontinuous conduction.

There are a number of other factors that affect the harmonic content of the output waveform. These are:

1. The pulse number of the converter; the higher the pulse number, the less the harmonic content.

2. The ratio of the source frequency to the output frequency; the lower the output frequency, the more closely will the output waveform approach the desired sinusoidal waveform, since the output waveform will be composed of more and more segments of the source waveform.

3. A high load reactance; the load current will be continuous and result in a reduction in the harmonic content.

4. The source reactance; the source reactance causes commutation overlap, which in turn increases the harmonic content.

5-13 ENVELOPE CYCLOCONVERTERS

The discussion so far has presented the application of cycloconverters for continuously variable-frequency applications. There are a number of applications where the output frequency is a fixed percentage of the source frequency or a limited number of fixed but lower frequencies. In these applications a cheaper and less complex configuration is obtained by the use of an **envelope cycloconverter.** There are two basic approaches used in envelope cycloconverters.

5-13-1 The Synchronous Envelope Cycloconverter

The single-phase configuration of Figure 5-20 can, with a suitable logic control of the SCR gate signals, be made to produce output waveforms of 2:1, 3:1, 4:1, etc., reductions of the source frequency, as shown in Figure 5-26. The disadvantage of the single-phase configuration is that it has a high harmonic content. By the use of three-phase configurations the harmonic content can be reduced. See Figure 5-26(c). When a variable-ratio input transformer is used, the amplitude of the output waveform can be varied as desired; however, the output waveform is still not a good approximation to a sine wave.

A better output waveform can be obtained when operating at one fixed output frequency by the use of a six-pulse configuration fed by a star-double-star transformer with four different secondary output voltages, which will then produce an output waveform that is a good approximation to a sine wave (Figure 5-27).

With either of the above envelope cycloconverter configurations, the function of the thyristors is to ensure that the conducting group is turned off prior to the other group being turned on, thus preventing a short-circuit being produced across the ac source. In fact, these configurations operate satisfactorily with a resistive load, but with an inductive load they are incapable of regeneration. As a result, if they are to be used with an inductive load, a current detection sensor must be used and changes to the firing control circuitry must be made to

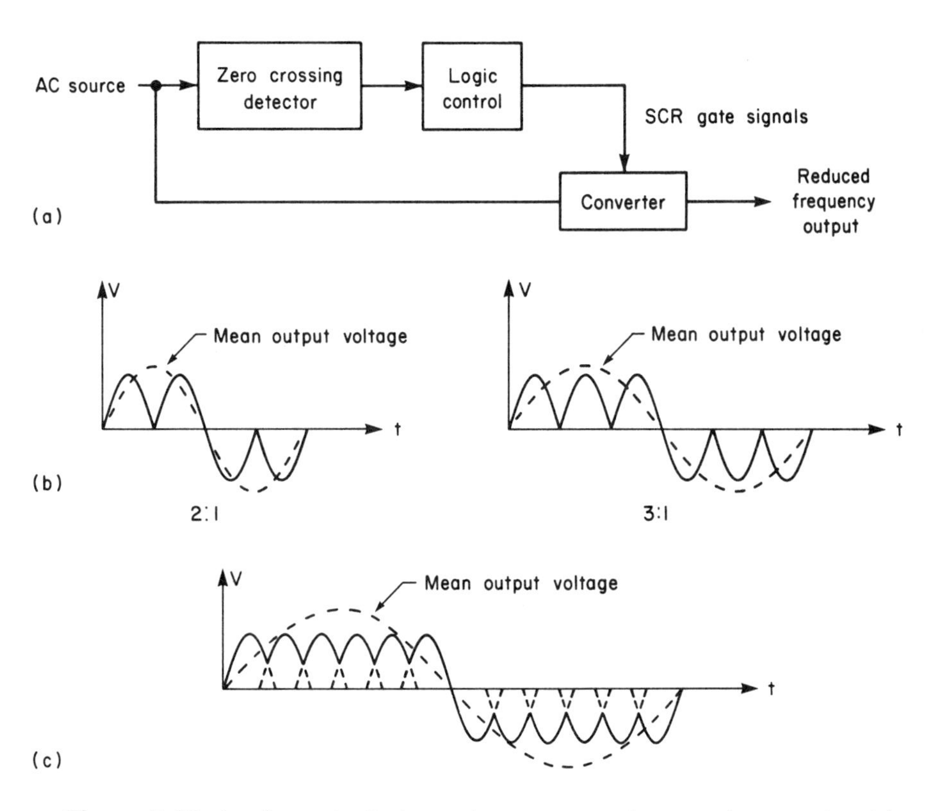

Figure 5-26. Logic-controlled synchronous envelope cycloconverter: (a) block diagram, (b) single-phase outputs, (c) three-phase output.

ensure that firing signals to the incoming converter are blanked out until the load current goes to zero.

The basic disadvantage of the envelope cycloconverter is that it can operate only at a fixed frequency or at a number of discrete frequencies when controlled by a variable modulus counter. The major advantage, however, is a reduction in the complexity of the firing control circuitry.

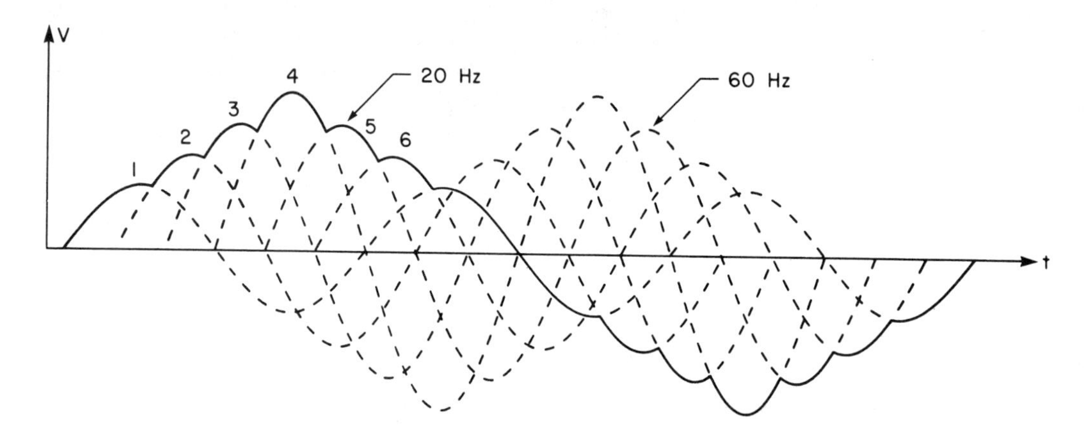

Figure 5-27. Synchronous cycloconverter, mean output voltage synthesized from the outputs of a six-pulse midpoint converter with varying amplitude voltages.

5-14 CYCLOCONVERTER FIRING CONTROL

The mean output voltage of a cycloconverter must vary from zero to a positive maximum to zero to a negative maximum and back to zero for each cycle of the output frequency. This is accomplished by biasing the firing delay angle to 90° and then oscillating the firing delay angle ±90° with respect to this point. As a result, for the positive half-cycle the positive converter firing-delay angle is advanced, and the negative converter delay angle is retarded by the same amount. Similarly, during the negative half-cycle, the firing-delay angle of the negative converter is advanced and that of the positive converter is retarded by the same amount, so that $\alpha_P = 180° - \alpha_N$ at all times. Consequently, the mean output voltages of both converters are equal and opposite in phase.

A commonly used method of generating the desired gate-firing signals for the thyristors is known as the cosine-crossing method. For a phase-controlled converter, this method consists of comparing a cosine wave generated from and synchronized to the ac source with a variable dc reference signal by means of a comparator, as shown in Figure 5-28. It can be seen that as the amplitude of the dc reference signal is varied, the point of intersection with the cosine-timing curve will vary, and a firing signal can be generated corresponding to the point of intersection. As a result, the firing-delay angle can be varied as desired. However, unless the dc reference signal is varying, the mean output voltage $V_{do\alpha}$ will remain constant.

In the case of a cycloconverter, the cosine-crossing method requires a minor modification in order to produce a sinusoidal mean output voltage. Instead of having a dc reference signal, a sinusoidal reference signal of the desired output frequency of the cycloconverter is compared to the cosine-timing curves; consequently, the mean output waveform is modulated by the sinusoidal reference voltage. Voltage control is achieved by varying the amplitude of the reference voltage. If the cycloconverter is being operated in the circulating-current free mode, then current detectors are required to sense the current and

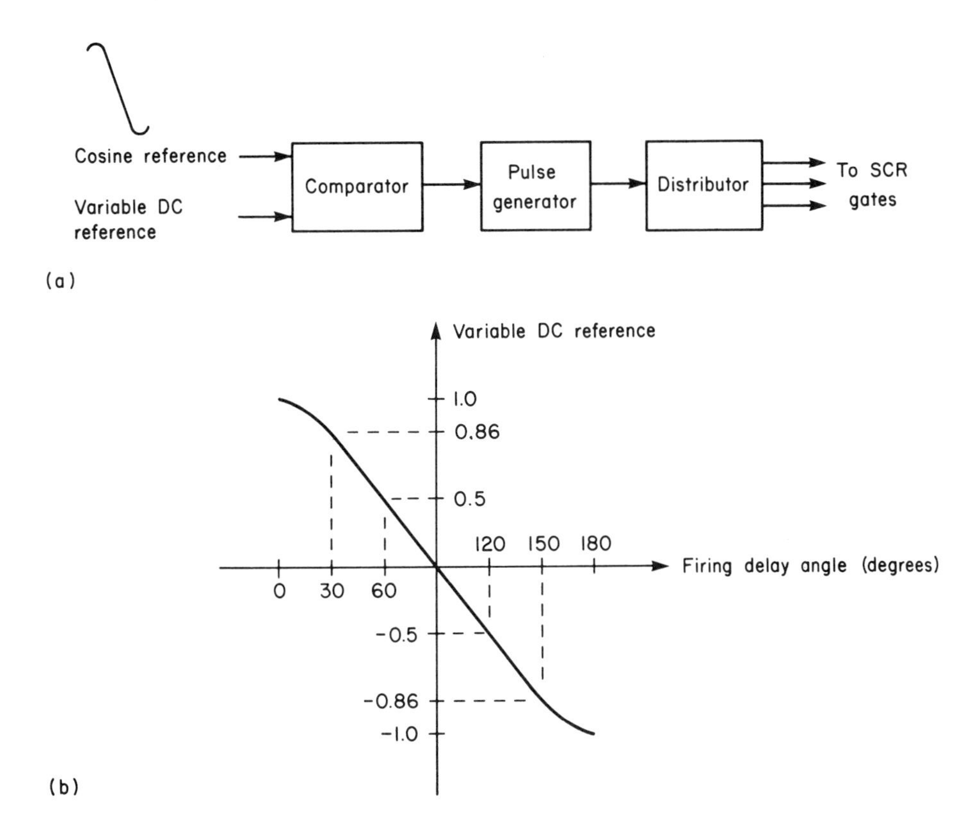

Figure 5-28. Concept of cosine-crossing firing control: (a) block diagram, (b) cosine reference curve.

produce blanking signals for the inactive converter. A block diagram of this control method is shown in Figure 5-29.

There are a number of firing-control techniques that may be applied to both phase-controlled converters and cycloconverters, such as the cosine-crossing method (sometimes called the bias-cosine or ac-dc comparison method), the ramp crossing (also known as ramp and bias or ramp and pedestal method), integral control, and phase-locked loop controls. In addition, considerable progress has been made in applying the microprocessor to the control of these devices.

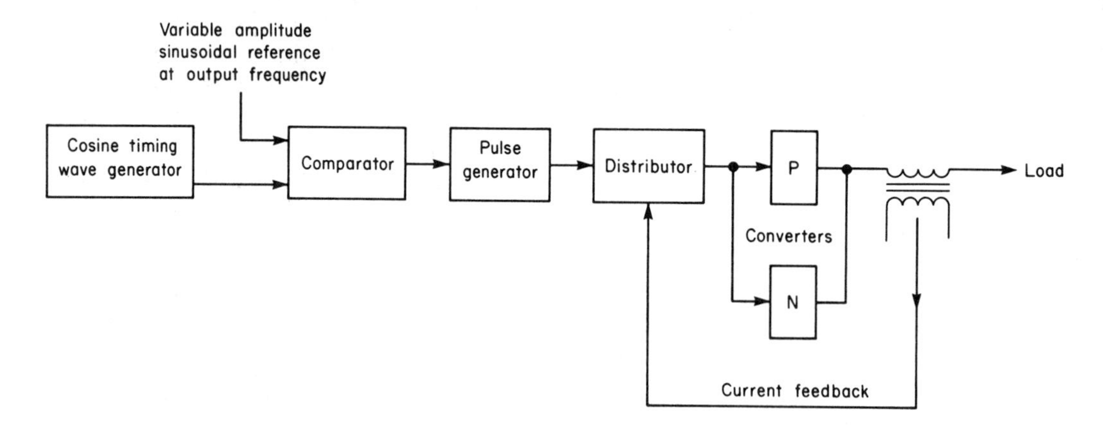

Figure 5-29. Block diagram of a cycloconverter control system.

5-15 SUMMARY

At this point it would appear to be advantageous to summarize the relative merits and capabilities of the dc link converter and the cycloconverter.

In both cases the frequency of the output ac voltage is controlled by a reference oscillator system, operated as an open-loop control, and thus maintains a stable output frequency irrespective of load fluctuations. Voltage control, and thus torque control, is obtained in the dc link converter by PWM techniques and in the cycloconverter by controlling the amplitude of the reference signal. Also, a change of direction of rotation is accomplished by changing the firing sequence of the thyristors.

The differences between the dc link converter and the cycloconverter are as follows:

1. The cycloconverter is an ac-to-ac frequency changer that is line commutated and used to produce a low frequency in one stage. The dc link converter, however, has two power conversions and requires forced commutation; as a result, the cycloconverter is more efficient.

2. The cycloconverter output frequency is usually limited to from zero to one-third of the source frequency in order to maintain the harmonic content of the output within acceptable limits. The reason that the harmonics are reduced at the lower frequencies is that the output waveform is composed of more segments of the source waveform than would be the case of higher frequencies. In the dc link converter pulse width or PWM techniques are used to reduce the amplitude of the harmonics, but the range of frequency variation is usually from 10 to 200 Hz with the variation being continuously variable in either case.

3. A bidirectional cycloconverter is capable of four-quadrant operation over its whole frequency range.

4. If it is assumed that a six-pulse phase-controlled converter is used for the dc source, the maximum number of load-carrying thyristors in a three-phase dc link converter is 12. The cycloconverter, on the other hand, requires a minimum of 18 thyristors for unidirectional operation and 36 thyristors for bidirectional operation.

5. The reduced number of thyristors in a dc link converter, in spite of the increased cost of inverter-grade thyristors, combined with the less-complicated control circuitry, gives it a competitive advantage compared to the cycloconverter, even though the lower duty cycle of the cycloconverter thyristors would reduce their cost.

6. A dc link converter supplied by a diode bridge presents a high input power factor, whereas the phase-controlled cycloconverter presents a low input power factor to the source, especially when operated at low output voltages.

In general, the cycloconverter is used in low-frequency, low-speed, high-horsepower applications, whereas the dc link converter may be used in all applications where wide frequency variations are required. With the recent introduction of high-current switching transistors, it is anticipated that the range of transistorized inverters will be extended into far higher horsepower (kW) ranges than are currently available.

GLOSSARY OF IMPORTANT TERMS

Inverter: An arrangement of solid-state switching devices that converts a dc source voltage into an ac output voltage. It may be operated to produce a fixed or variable-frequency output.

dc-link converter: A major method of producing a variable-frequency ac supply; it consists of a dc rectifier and an inverter. Usually the dc rectification is accomplished by a three-phase, phase-controlled thyristor converter, or a three-phase diode bridge, in either case the dc output is supplied to a three-phase inverter. The amplitude of the output ac voltage is controlled either by phase angle control of the phase-controlled converter or internally in the inverter.

V/Hz ratio: At reduced frequencies the stator impedance of a polyphase induction motor decreases and since the iron of the magnetic circuit is operated at near saturation, it is necessary to reduce the applied stator voltage proportionally with the reduction in frequency to prevent excessive iron losses and magnetizing current.

Pulse width control: Control of the amplitude of output ac voltage of an inverter by reducing the width of the output pulse. This is achieved at the expense of an increased harmonic content in the output voltage waveform.

Pulse width modulation (PWM): The rapid switching on and off of the dc source voltage a large number of times during each half-cycle of the ac output voltage. As a result PWM directly controls the amplitude of the output ac voltage by the ratio of the on-time to the off-time and significantly reduces the harmonic content of the output waveform.

Commutation: The process of turning off a thyristor.

Natural commutation: Occurs in ac supplied circuits since the anode-to-cathode voltage is reversed once every cycle.

Forced commutation: The process of artificially achieving a reverse anode-to-cathode voltage to achieve turn-off in dc supplied circuits.

Impulse commutation: A reverse-bias pulse produced by an LC oscillatory circuit is used to achieve thyristor turn-off.

Auxiliary impulse commutation: The commutating pulse is produced by circuitry separate to the main power source. This method is also called the McMurray circuit.

Complementary voltage commutation: Forced commutation is initiated by turning on the incoming thyristor. This method is known as the McMurray-Bedford circuit.

Harmonic neutralization: One method of minimizing the harmonic content of the ac output waveform. It is achieved by combining a number of square-wave inverters each of which is phase-shifted and fired at the desired frequency.

Cycloconverter: A direct frequency changer that converts power at one frequency to ac power at another frequency by ac-to-ac conversion without an intermediate dc-link converter.

Envelope cycloconverter: A fixed frequency cycloconverter.

REVIEW QUESTIONS

5-1 Discuss the advantages of selecting a variable-frequency ac drive as compared to selecting a variable-speed dc drive for a given application.

5-2 What requirements must be met to satisfactorily operate a polyphase induction motor from a variable-frequency source?

5-3 With the aid of torque-speed and torque-slip curves, discuss the advantages of constant torque control on starting torque, slip, and efficiency of a polyphase induction motor when operated under variable-frequency control.

5-4 What are the advantages of using a static variable-frequency inverter as against using a variable-frequency motor-alternator set?

5-5 Explain how four-quadrant operation of a polyphase induction motor can be obtained when using a variable-frequency inverter.

5-6 With the aid of a schematic and waveforms show how a six-step ac output waveform may be obtained from a three-phase inverter.

5-7 What are the advantages and disadvantages of controlling the amplitude of the output ac voltage of an inverter by varying the input dc voltage?

5-8 Discuss the control of the amplitude of the output ac voltage of inverter by (1) pulse width control and (2) pulse width modulation control.

5-9 With the aid of a schematic and waveforms explain the McMurray-Bedford forced commutation technique. What are its disadvantages as compared to the McMurray circuit?

5-10 With the aid of a schematic and waveforms explain the McMurray forced commutation technique. What are its advantages as compared to the McMurray-Bedford circuit?

5-11 What is meant by harmonic neutralization and how may it be achieved?

5-12 With the aid of waveforms explain how the voltage and frequency of a phase-controlled cycloconverter are controlled.

5-13 What factors affect the harmonic content of the output waveform of a cycloconverter?

5-14 Discuss two methods of reducing circulating currents in cycloconverters.

5-15 What is an envelope cycloconverter? What are its advantages and disadvantages, and how may the output waveform be improved?

5-16 With the aid of a block diagram discuss the firing control of a cycloconverter supplying an RL load.

5-17 Discuss the advantages and disadvantages of cycloconverters versus dc-link converters.

6

Logic Control

6-1 INTRODUCTION

During the past twenty years there has been a major move to the use of digital techniques in every area of industry. The major reasons for this shift are improved accuracy and resolution, larger dynamic range, improved stability, and greater convenience in reading displays. Processes may be fully automated, and the design and application of circuits using digital integrated circuits is relatively easy.

Digital systems are in common use in every major engineering, scientific, commercial, and domestic field, and as a result it is necessary for all designers and maintainers to have a sound understanding of digital techniques.

6-2 THE BINARY NUMBER SYSTEM

Digital systems depend upon two possible states or conditions; for example, a diode is conducting or not conducting, a switch is closed or open, a voltage is high or low. In each of these examples the two states may be specified in terms of 1 or 0. For example, when the diode is conducting this may be defined as the binary 1, when not conducting as the binary 0.

The feature that defines a number system is the base or radix. The base specifies the number of digits used to represent a number. The decimal number system which we use for all our everyday needs has a **radix** or **base of 10**, that is, quantities are represented by the ten digits 0 through 9; for example 1983 is represented by

$$(1 \times 10^3) + (9 \times 10^2) + (8 \times 10^1) + (3 \times 10^0)$$

$$= 1{,}000 + 900 + 80 + 3$$

$$= 1983$$

The value of the number is determined by multiplying each digit by the weight of its position and adding.

In binary arithmetic, that is, base 2, only two numbers are required, namely, 0 and 1. Just as in the base 10 number system, the position weights are the base raised to a power. The weights reading from right to left are $2^0 = 1$, $2^1 = 2$, $2^2 = 4$, $2^3 + 8$, $2^4 = 16$, $2^5 = 32$, $2^6 = 64$, etc. Therefore, 1101101_2 is

$$(1 \times 2^6) + (1 \times 2^5) + (0 \times 2^4) + (1 \times 2^3) + (1 \times 2^2) + (0 \times 2^1) + (1 \times 2^0)$$

$$= 64 + 32 + 8 + 4 + 1$$

$$= 109_{10}$$

The **most significant bit (MSB)** is on the left, and the **least significant bit (LSB)** is on the right, exactly the same as in the base 10 system.

The number systems are extended so that fractional numbers may be defined. For example, the decimal number 273.15 can be written as

$$(2 \times 10^2) + (7 \times 10^1) + (3 \times 10^0) + (1 \times 10^{-1}) + (5 \times 10^{-2})$$

$$= 200 + 70 + 3 + 0.1 + 0.05$$

$$= 273.15_{10}$$

In the case of fractional binary numbers the weights of the fractional positions are determined by the negative powers of 2; for example, $2^{-1} = 0.5$, $2^{-2} = 0.25$, $2^{-3} = 0.125$, $2^{-4} = 0.0625$, etc. A binary point is used to separate the whole and fractional parts of the binary number.

Therefore 1001.11_2 is

$$(1 \times 2^3) + (0 \times 2^2) + (0 \times 2^1) + (1 \times 2^0) + (1 \times 2^{-1}) + (1 \times 2^{-2})$$

$$= 8 + 0 + 0 + 1 + 0.5 + 0.25$$

$$= 9.75_{10}$$

6-2-1 Converting Decimal to Binary

Decimal or base 10 numbers are converted to binary by repeatedly dividing the decimal number by 2. As an example convert 255_{10} into its binary equivalent.

	Remainder	
$255 \div 2 = 127$	1	(LSB)
$127 \div 2 = 63$	1	
$63 \div 2 = 31$	1	
$31 \div 2 = 15$	1	
$15 \div 2 = 7$	1	
$7 \div 2 = 3$	1	
$3 \div 2 = 1$	1	
$1 \div 2 = 0$	1	(MSB)

Therefore, $225_{10} = 11111111_2$.

In order to convert a decimal fraction to binary, the fraction is multiplied successively by 2 and any resulting integers are recorded as an overflow. For example, to convert 0.7854 to its binary equivalent, it is multiplied by 2 repeatedly as follows:

$0.7854 \times 2 = 1.5708 = 0.5708$ with overflow 1 (MSB)

$0.5708 \times 2 = 1.1416 = 0.1416$ with overflow 1

$0.1416 \times 2 = 0.2832 = 0.2832$ with overflow 0

$0.2832 \times 2 = 0.5664 = 0.5664$ with overflow 0

$0.5664 \times 2 = 1.1328 = 0.1328$ with overflow 1 (LSB)

This process is repeated until the required precision is obtained or the product is 0. The overflows are rearranged and written with the MSB starting at the binary point and ending with the LSB. Therefore, $0.7854_{10} = 0.11001_2$.

In the case of a decimal number consisting of both integer and fractional parts, the integer and fractional parts are separated at the decimal point, and the appropriate method of conversion is applied to each part.

Example 6-1

Convert 12.566_{10} to its binary equivalent.

Solution:

$$12.566_{10} = 12_{10} + 0.566_{10}$$

$12 \div 2 = 6$ with remainder 0 (LSB)

$6 \div 2 = 3$ with remainder 0

$3 \div 2 = 1$	with remainder 1	
$1 \div 2 = 0$	with remainder 1	(MSB)

$$12_{10} = 1100_2$$

$0.566 \times 2 = 1.132 = 0.133$	with overflow 1	(MSB)
$0.132 \times 2 = 0.264 = 0.264$	with overflow 0	
$0.264 \times 2 = 0.528 = 0.528$	with overflow 0	
$0.528 \times 2 = 1.056 = 0.056$	with overflow 1	(LSB)

$$0.566_{10} = 0.1001_2$$

Therefore,
$$\begin{aligned} 12.566_{10} &= 12_{10} + 0.566_{10} \\ &= 1100_2 + 0.1001_2 \\ &= 1100.1001_2. \end{aligned}$$

6-3 WORD LENGTH

Binary numbers or, as they are sometimes called, binary words are made up of bits. The binary number 1110 is a 4-bit word, with the left-hand bit being the MSB and the right-hand bit being the LSB. Binary words are classified in terms of the number of bits required to form the word, for example, 4-, 8-, 16-, and 32-bit words. The number of bits in a word determines the resolution and the maximum magnitude of the decimal number that can be represented. The largest number that can be represented by n bits is

$$N = 2^n - 1 \tag{6-1}$$

For example, the largest number that can be represented by a 4-bit word is

$$N = 2^4 - 1 = 15$$

This is shown in Table 6-1.

Resolution defines the ability and accuracy of transferring information from digital encoders and instrumentation. For example, a shaft encoder using 4-bits has an angular resolution of 2^4 or 1 in 16 or 22.5°, 8-bit words have a resolution of 1 in 256 or 1.41°, and 16-bit words have a resolution of 1 in 65536 or 0.0055°.

Table 6-1 Short List of Decimals and Binary Equivalents

Decimal	Binary
0	0000
1	0001
2	0010
3	0011
4	0100
5	0101
6	0110
7	0111
8	1000
9	1001
10	1010
11	1011
12	1100
13	1101
14	1110
15	1111

Decimal and binary equivalents of a 4-bit word.

6-4 BINARY ARITHMETIC

The binary system uses the same concepts of position and value of the binary digits as does the decimal system, which would lead us to expect that the binary system has the ability to add, subtract, multiply, and divide.

6-4-1 BINARY ADDITION

The process of addition in the binary system is similar to decimal addition. It can be simplified by remembering four rules:

		Sum	Carry
1.	$0 + 0$	0	0
2.	$0 + 1$	1	0
3.	$1 + 0$	1	0
4.	$1 + 1$	0	1

Binary addition is best illustrated by an example.

Example 6-2

Add 1100_2 (12_{10}) and 1110_2 (14_{10})

Solution:

CARRY	11000	
ADDEND	1100	(12_{10})
AUGEND	+ 1110	+ (14_{10})
SUM	11010_2	(26_{10})

Similarly fractional binary numbers may also be added.

Example 6-3

Add 0.0011_2 ($3/16_{10}$) and 0.0110_2 ($3/8_{10}$)

Solution:

CARRY	1100	
ADDEND	0.0011	($3/16_{10}$)
AUGEND	+0.0110	+($3/8_{10}$)
SUM	0.1001_2	($9/16_{10}$) $= 0.5625_{10}$

And finally mixed numbers may also be added.

Example 6-4

Add 101101.11_2 (45.75_{10}) and 100.01_2 (4.25_{10})

Solution:

CARRY	1001.1	
ADDEND	101101.11	(45.75_{10})
AUGEND	100.01	(4.25_{10})
SUM	110010.00_2	(50.00_{10})

As can be seen, binary addition is basically no more complicated than normal decimal addition.

6-4-2 Binary Subtraction

In binary subtraction, we subtract column by column and borrow as necessary from a higher position column. As before the process can be simplified by remembering four rules:

		Remainder	Borrow
1.	0 − 0	0	0
2.	1 − 0	1	0
3.	0 − 1	1	1
4.	1 − 1	0	0

Example 6-5

Solve 101_2 (5_{10}) − 10_2 (2_{10}).

Solution:

BORROW	1	
MINUEND	101	(5_{10})
SUBTRAHEND	− 10	$-(2_{10})$
DIFFERENCE	11_2	(3_{10})

As may be suspected binary subtraction is more complicated than binary addition, and more circuitry is required in a digital computer as compared to that necessary for addition. As a result methods have been developed where the binary numbers are converted to complements, and the operation becomes one of addition.

6-4-3 Subtraction by Complementing

One method of subtracting is by **complementing** the larger number and adding the smaller number to it, and then complementing the sum. A binary number is complemented by changing all the zeros to ones and all the ones to zeros. The complement of a number A is written as A.

Example 6-6

Find 10110101_2 -1110011_2 by complementing the larger number.

Solution:

A 10110101_2	$\overline{A}$	1001010_2	MINUEND
$-B$ 1110011_2	$+B$	1110011_2	SUBTRAHEND
	SUM $=$	10111101_2	DIFFERENCE
	$\overline{\text{SUM}}$ $=$	1000010_2	

6-4-4 Subtraction By Ones Complement

In Example 6-6 the larger number was complemented and added, and in turn the sum was complemented. However, we determined the larger number. In computer applications, if the computer is required to determine the larger number, the operation may take longer than the subtraction. The **ones-complement method** was devised to eliminate this problem; the subtrahend is complemented irrespective of the size of the number and then the numbers are added. However, if the minuend is greater than the subtrahend there is a carry involved which is greater than the highest significant figure involved in the subtraction. Instead of forming the MSB, the carry is added to the LSB. This process is known as the end-around carry (EAC). The EAC will only occur if the minuend is greater than the subtrahend.

Example 6-7

Using the ones-complement method find $1110_2 - 1010_2$.

Solution:

A	$1110_2\,(14_{10})$	A	1110	MINUEND
$-B$	$1010_2\,(10_{10})$	$+\overline{B}$	0101	SUBTRAHEND
		S	(1)0011	DIFFERENCE
			1	(EAC)
			$0100 = 4_{10}$	

If there is no EAC, this means that $A<B$, and the subtrahend is complemented and added.

Example 6-8

Using the ones-complement method find $1000_2 - 1100_2$.

Solution:

A	1000_2 (8_{10})	A	1000	MINUEND
$-B$	1100_2 (12_{10})	$+\overline{B}$	0011	SUBTRAHEND
R		S	1011	DIFFERENCE

The remainder is $R = \overline{S} = -100_2 = -4_{10}$.

6-4-5 Subtraction By Twos Complement

The **twos complement** of a number is obtained by adding 1 to the ones complement of the number, that is, the twos complement of A is $\overline{A} + 1$. The procedure varies slightly depending upon whether the difference is positive or negative. The techniques are best illustrated by an example.

Example 6-9

Using the twos-complement method, find (1) $1101_2 - 111_2$ and (2) $111_2 - 1101_2$.

Solution:

(1) A>B (positive difference)

A	(13_{10})	1101_2	A	1101_2	MINUEND
$-B$	(-7_{10})	$-\ 111_2$	$+(\overline{B}+1)$	1001_2	SUBTRAHEND
R	6_{10}		SUM	$(1)0110_2 = 6_{10}$	

The twos complement of the subtrahend is added to the minuend. The carry indicates that the answer is positive and is discarded.

(2) A<B (negative difference)

A	(7_{10})	111_2	A	0111_2	MINUEND
$-B$	(-13_{10})	-1101_2	$+(\overline{B}+1)$	0011_2	SUBTRAHEND
R	-6_{10}		SUM	1010_2	

The absence of a carry indicates a negative difference, and to find the difference we take the twos-complement of the sum. Therefore, the difference R is $\overline{\text{SUM}} + 1 = 101_2 + 1_2 = -110_2 = -6_{10}$.

The subtraction operation can be performed equally well by ones or twos-complement methods, however, the twos-complement method is preferred since it permits signed arithmetic operations to be performed.

6-4-6 Binary Multiplication

In binary multiplication since there are only two multipliers 1 and 0, binary multiplication is easier to perform than its decimal equivalent. The following four rules are required to perform the multiplication operation.

1. $0 \times 0 = 0$
2. $0 \times 1 = 0$
3. $1 \times 0 = 0$
4. $1 \times 1 = 1$

Example 6-10

Multiply 11001_2 by 1101_2.

Solution:

MULTIPLICAND	11001_2	(25_{10})
MULTIPLIER	$\times\ 1101_2$	$\times(13_{10})$
	11001	
	00000	Shift and multiply by 0
	11001	Shift and multiply by 1
	11001	Shift and multiply by 1
	101000101_2	(325_{10})

In our digital circuits we will see that the shift operation is performed by a shift register.

Example 6-11

Multiply 11.101_2 by 1.11_2.

Solution:

```
    11.101
   × 1.11
   -------
     11101
    11101
   11101
  ---------
  110.01011₂
```

6-4-7 Binary Division

Binary division is also simpler than decimal division and is the reverse of multiplication.

Example 6-12

Divide 1111101_2 by 101_2.

Solution:

```
                         11001        QUOTIENT
        DIVISOR     101 ) 1111101     DIVIDEND
                          101
                           101        REMAINDER
                           101
                             101      REMAINDER
                             101
                               0      REMAINDER
```

Therefore, $\frac{125_{10}}{5_{10}} = \frac{1111101_2}{101_2} = 11001_2 = 25_{10}$

6-5 THE OCTAL NUMBER SYSTEM

In the octal number system the radix is 8, and each digit in an octal number is assigned a positional weight; for example, 327.24_8 is represented in positional form as $(3 \times 8^2) + (2 \times 8^1) + (7 \times 8^0) + (2 \times 8^{-1}) + (4 \times 8^{-2})$. The decimal equivalent is found by multiplying each digit by the appropriate positional weight and adding the result. Therefore, 327.24_8 is

$$(3 \times 64) + (2 \times 8) + (7 \times 1) + (2 \times 0.125) + (4 \times 0.015625)$$

$$= 215.3125_{10}$$

An abbreviated listing of the powers of 8 is shown in Table 6-2.

Table 6-3 gives the first 20 decimal numbers and their octal and binary equivalents.

6-5-1 Decimal To Octal Conversion

A decimal number is converted to octal in a similar manner to the conversion to binary, except the base is now 8 instead of 2.

Table 6-2 A Short Listing of the Powers of 8

$$8^4 = 4096_{10}$$
$$8^3 = 512_{10}$$
$$8^2 = 64_{10}$$
$$8^1 = 8_{10}$$
$$8^0 = 1_{10}$$
$$8^{-1} = \frac{1}{8} = 0.125_{10}$$
$$8^{-2} = \frac{1}{16} = 0.015625_{10}$$
$$8^{-3} = \frac{1}{512} = 0.001953_{10}$$
$$8^{-4} = \frac{1}{4096} = 0.000244_{10}$$

Table 6-3 Decimal-Octal-Binary Equivalents

Decimal	Octal	Binary
0	0	0
1	1	1
2	2	10
3	3	11
4	4	100
5	5	101
6	6	110
7	7	111
8	10	1000
9	11	1001
10	12	1010
11	13	1011
12	14	1100
13	15	1101
14	16	1110
15	17	1111
16	20	10000
17	21	10001
18	22	10010
19	23	10011
20	24	10100

Example 6-13

Convert 226_{10} to octal.

Solution:

$$226 \div 8 = 28 \text{ with remainder } 2 \qquad \text{(LSD)}$$
$$28 \div 8 = 3 \text{ with remainder } 4$$
$$3 \div 8 = 0 \text{ with remainder } 3 \qquad \text{(MSD)}$$

Therefore, $226_{10} = 342_8$.

A decimal fraction is converted to octal by multiplying the fraction successively by 8.

Example 6-14

Convert 0.53125_{10} into the octal equivalent.

Solution:

$$0.53125 \times 8 = 4.25 = 0.25 \text{ with overflow } 4 \qquad \text{(MSD)}$$
$$0.25 \times 8 = 2.00 = 0 \text{ with overflow } 2 \qquad \text{(LSD)}$$

Therefore, $0.53125_{10} = 0.42_8$.

As was shown in binary conversion a decimal number consisting of an integer and a fraction requires that the integer and fraction be separated and the appropriate octal conversion be carried out on each part.

Example 6-15

Convert 126.46875_{10} into the octal equivalent.

Solution:

$$126.46875_{10} = 126_{10} + 0.46875_{10}$$
$$126 \div 8 = 15 \text{ with remainder } 6 \qquad \text{(LSD)}$$
$$15 \div 8 = 1 \text{ with remainder } 7$$
$$1 \div 8 = 0 \text{ with remainder } 1 \qquad \text{(MSD)}$$
$$0.46875 \times 8 = 3.75 = 0.75 \text{ with overflow } 3 \qquad \text{(MSD)}$$
$$0.75 \times 8 = 6.00 = 0 \text{ with overflow } 6 \qquad \text{(LSD)}$$

Therefore, $126.46875_{10} = 176.36_8$.

6-5-2 Octal To Decimal Conversion

The conversion from octal to decimal simply involves multiplying each digit by the appropriate positional weight and adding the result.

Example 6-16

Convert 242.36_8 to decimal.

Solution:

$$(2 \times 8^2) + (4 \times 8^1) + (2 \times 8^0) + (3 \times 8^{-1}) + (6 \times 8^{-2})$$

$$(2 \times 64) + (4 \times 8) + (2 \times 1) + (3 \times 0.125) + (6 \times 0.015625)$$

$$= 162.46875_{10}.$$

6-5-3 Binary To Octal Conversion

Binary to octal conversion simply consists of splitting the binary number into groups of three starting at the decimal point and moving to the left and right as appropriate, using zeros as necessary to complete the groups.

Example 6-17

Convert 111010.011101_2 to octal.

Solution:

111010_2	111(MSB)	010 (LSB)
	7	2
	$= 72_8$	
0.011101_2	0.011 (MSB)	101 (LSB)
	3	5

Therefore, $111010.011101_2 = 72.35_8$.

The above example illustrates the advantage of the octal system. The octal expression requires four characters, but the equivalent binary expression requires twelve characters. In addition the conversion of a decimal number requires fewer divisions by 8 to convert to octal as compared to the number of divisions by 2 required to convert the same number to binary.

6-6 THE HEXADECIMAL NUMBER SYSTEM

The hexadecimal number system is similar to the octal number system except the radix or base is 16. It is used extensively by microprocessors and permits easy conversions to the decimal and binary number systems as well as simplifying data entry. The hexadecimal system uses the numbers 0 to 9 inclusive and the letters A to F inclusive, as illustrated in Table 6-4.

Table 6-4 Decimal-Hexadecimal-Binary Equivalents

Decimal	Hexadecimal	Binary
0	0	0000
1	1	0001
2	2	0010
3	3	0011
4	4	0100
5	5	0101
6	6	0110
7	7	0111
8	8	1000
9	9	1001
10	A	1010
11	B	1011
12	C	1100
13	D	1101
14	E	1110
15	F	1111
16	10	10000
17	11	10001
18	12	10010
19	13	10011
20	14	10100
21	15	10101
22	16	10110
23	17	10111
24	18	11000
25	19	11001
26	1A	11010
27	1B	11011
28	1C	11100
29	1D	11101
30	1E	11110
31	1F	11111
32	20	100000

As is the case with all our numbering systems the position of each digit carries a positional weight which determines the magnitude of the number. By way of illustration consider the hexadecimal number 4C5.A3 which may be written as

$$(4 \times 16^2) + (C \times 16^1) + (5 \times 16^0) + (A \times 16^{-1}) + (3 \times 16^{-2})$$

$$(4 \times 256) + (12 \times 16) + (5 \times 1) + (10 \times 1/16) + (3 \times 1/256)$$

$$1024 + 192 + 5 + 0.625 + 0.011719$$

$$= 1221.636719_{10}$$

Therefore, $4C5.A3_{16} = 1221.63719_{10}$

An abbreviated listing of the powers of 16 is shown in Table 6-5

Table 6-5 A Short Listing of the Powers of 16

$16^4 =$	65536_{10}
$16^3 =$	4096_{10}
$16^2 =$	256_{10}
$16^1 =$	16_{10}
$16^0 =$	1_{10}
$16^{-1} =$	0.062500_{10}
$16^{-2} =$	0.003906_{10}
$16^{-3} =$	0.000244_{10}
$16^{-4} =$	0.000015_{10}

6-6-1 Decimal To Hexadecimal Conversions

In decimal to hexadecimal conversions the procedure is similar to that used in converting decimal to binary and octal.

Example 6-18

Convert 11562.136_{10} into hexadecimal.

Solution:

The conversion is carried out in two parts, namely, the integer part and the fractional part.

$$11562.136_{10} = 11562_{10} + 0.136_{10}$$

$$11562 \div 16 = 722 \text{ with remainder } 10 = A \quad \text{(LSD)}$$
$$722 \div 16 = 45 \text{ with remainder } 2 = 2$$
$$45 \div 16 = 2 \text{ with remainder } 13 = D$$
$$2 \div 16 = 0 \text{ with remainder } 2 = 2 \quad \text{(MSD)}$$
$$0.136 \times 16 = 2.176 = 0.176 \text{ with overflow } 2 = 2 \quad \text{(MSD)}$$
$$0.176 \times 16 = 2.816 = 0.816 \text{ with overflow } 2 = 2$$
$$0.816 \times 16 = 13.056 = 0.056 \text{ with overflow } 13 = D$$
$$0.056 \times 16 = 0.896 = 0.896 \text{ with overflow } 0 = 0$$
$$0.896 \times 16 = 14.336 = 0.336 \text{ with overflow } 14 = E \quad \text{(LSD)}$$

Therefore $11562.136_{10} = 2D2A.22DOE_{16}$.

6-6-2 Conversions Between Binary and Hexadecimal Number Systems

Conversions between the two number systems requires a 4-bit binary number to represent 16_{10}, which means that a 4-bit binary number is replaced by a one-digit number in the hexadecimal system.

Example 6-19

Convert (1) 10111101100.010111_2 to hexadecimal and
(2) $6F.32_{16}$ to binary.

Solution:

(1) Commencing at the binary point divide the binary number into groups of four and then convert each group into its hexadecimal equivalent. Groups are completed by the addition of zeros.

0101	1110	1100	.	0101	1100
5	E	C	.	5	C

Therefore, $10111101100.010111_2 = 5EC.5C_{16}$.

(2)

6	F	.	3	2
0110	1111	.	0011	0010

Therefore, $6F.32_{16} = 1101111.0011001_2$.

6-7 BINARY CODED DECIMAL (BCD)

In our normal everyday life we have become very familiar with the decimal system. On the other hand while it is relatively easy to convert from the decimal system to the binary, it is not so easy to look at a binary number and quickly convert it to the decimal equivalent. This is a disadvantage, so as a result because so many systems and devices accept decimal inputs and provide decimal outputs, the binary coded decimal (BCD) code was developed, even though it introduced some hardware disadvantages.

The BCD code represents the decimal digits 0 to 9 inclusive by a 4-bit binary code. It uses the standard weighting system of the pure binary number system, namely the 8421 positional weights. The 4-bit binary equivalents of the decimal numbers 10 to 15 inclusive are not utilized in the BCD code, which results in a system that is inefficient in the use of circuits. Table 6-6 shows a comparison between the decimal, 8421 BCD, Gray, and binary codes. As can be seen from the table it is very easy to mentally convert a BCD number to its decimal equivalent or vice versa.

Example 6-20

Convert (1) 6.28_{10} to 8421 BCD and
(2) 0101 0100 0111.0010 0101 to its decimal equivalent.

Solution:

(1) 6.28_{10} = 0110.0010 1000

(2)	0101	0100	0111	.	0010	0101
	5	4	7	.	2	5_{10}

6-8 GRAY CODE

The most commonly encountered codes are the pure binary and the 8421 BCD. However, another code, the Gray coding, is used quite extensively in angle measurement using shaft encoders. The major advantages of the Gray code is the fact that only one bit changes at a time (see Table 6-6) as compared to the BCD and binary codes. The Gray code minimizes errors since it requires an appreciable time for circuit components to change states which limits the capability of the devices to respond to high speed changes. Since with the Gray code only one bit changes at a time the speed of operation can be increased

Table 6-6 Decimal-BCD-Gray-Binary Equivalents

Decimal	BCD	Gray	Binary
0	0000	0000	0000
1	0001	0001	0001
2	0010	0011	0010
3	0011	0010	0011
4	0100	0110	0100
5	0101	0111	0101
6	0110	0101	0110
7	0111	0100	0111
8	1000	1100	1000
9	1001	1101	1001
10	0001 0000	1111	1010
11	0001 0001	1110	1011
12	0001 0010	1010	1100
13	0001 0011	1011	1101
14	0001 0100	1001	1110
15	0001 0101	1000	1111

and the introduction of error is reduced. A major disadvantage of the Gray code is the difficulty in performing mathematical operations. If it is necessary to perform calculations the Gray code representation is usually converted to pure binary.

6-9 DIGITAL INTEGRATED CIRCUITS

Digital integrated circuits form the building blocks of digital control systems. There are two main types of semiconductor devices used as switching elements in integrated circuits, namely the bipolar transistor and the metal-oxide-semiconductor field-effect transistor (MOSFET). Each type of switching element has its own distinctive parameters and applications.

To be able to understand and compare the different integrated-circuit (IC) logic families it is necessary to understand the following logic circuit characteristics: logic levels, propagation delay, power dissipation, noise immunity, and fan-out.

6-9-1 LOGIC LEVELS

A **logic level** is a voltage level that is in the form of a pulse or square wave signal that represents binary 1 or 0, or logical true or false. With most digital ICs the binary states are represented by 0 V for a binary 0 and by a positive voltage for a binary 1. This is termed positive logic which is also extended to embrace all combinations where the binary 1 is represented by a voltage that is more positive than the voltage representing the binary 0. On the other hand, if the voltage representing the binary 1 is more negative than that representing the binary 0, this is termed negative logic. Table 6-7 summarizes the logic types.

Table 6-7 Logic Types

Binary	Logic Type			
Digit	Positive	Positive	Positive	Negative
1	+5 V	0 V	+6V	−5 V
0	0 V	−12 V	−6 V	0 V

The types of signals usually encountered in digital applications are shown in Figure 6-1. They are defined in terms of amplitude or level and duration. Control signals maintain a voltage level for comparatively long time intervals before changing level. See Figure 6-1(a). Pulses on the other hand are relatively short duration changes in voltage level and are classified as positive and negative polarity pulses as well as being positive or negative going. See Figure 6-1(b), (c).

6-9-2 Propagation Delay

An extremely important characteristic of a digital system is the speed of operation; the greater the speed the greater the number of logic operations that can be performed. Speed is measured in terms of switching time. There are two switching-time parameters, namely, propagation delay time t_{PHL} from a binary 1 (logical 1) to a binary 0 (logical 0) at the output, and the propagation delay time t_{PLH} from a logical 0 to a logical 1 at the output. For TI 54/74 TTL logic gates typical values are $t_{PHL} \simeq 7$ ns and $t_{PLH} \simeq 11$ ns. As can be seen from Figure 6-2 the propagation delays are measured from the 50% amplitude points to the leading and trailing edges of the input pulse. Another important characteristic is the rise time t_r, the time it takes for the output pulse to rise from 10% to 90% of its peak amplitude, and the fall time t_f, the time it takes for the pulse to fall from 90% to 10% of its original value. The propagation delays also take into account storage times t_s and t_d. It should be noted that the propagation delays are additive when devices are cascaded.

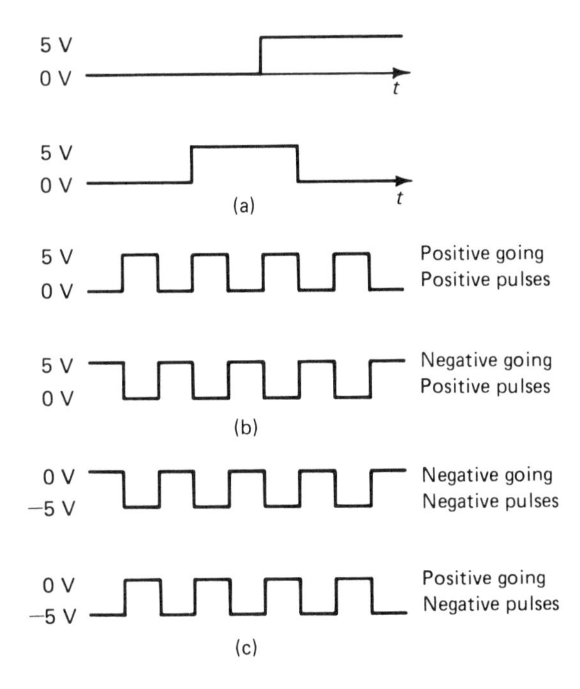

Figure 6-1. Digital signal waveforms: (a) control signals, (b) positive pulses, (c) negative pulses.

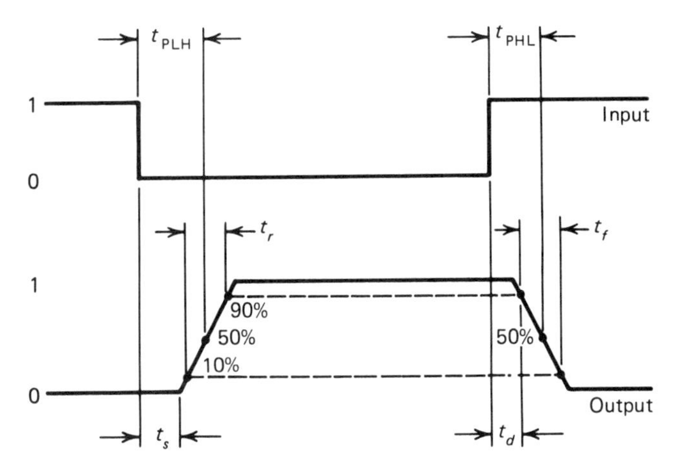

Figure 6-2. Propagation delay.

6-9-3 Power Dissipation

The power consumed by a logic gate determines the design and cost of the power supply. Additionally the power dissipation per logic gate, which is an average figure of the power dissipation for the logical 1 and logial 0 output conditions, determines the number of logic gates that may be incorporated into one package. The power dissipation of the normal IC is usually about 500 mW, and as a result the power dissipation per logic gate determines the total number of logic gates per IC if safe operating temperatures are not to be exceeded. Generally it can be stated that high-speed logic such as transistor-transistor-logic (TTL) will consume more power per gate, typically 20 mW per gate, as compared to low-speed logic such as complementary-metal-oxide-semiconductor (CMOS) which in the inactive state (quiescent) consumes hardly any power at all, although its power consumption rises linearly with an increase in operating frequency.

Normally logic families are compared on the basis of a figure of merit which is the product of their switching speed and the power consumption per gate. In general it will be found that high-speed logic gates utilize nonsaturating bipolar transistors, which as a result have high emitter-collector voltage drops and this combined with low circuit resistances produces high power consumptions per logic gate. Typical TTL logic is high speed (25 to 50 MHz) with propagation delays typically 7 ns per gate and power dissipations of 20 mW per gate. On the other hand the high impedance of the CMOS reduces power consumption to less than 1 mW per gate with propagation delays of the order of 250 ns per gate and speeds up to 2 MHz.

6-9-4 Noise Immunity

Noise in digital circuits can be best described as any electrical signal that masks a desired signal. Industrial systems are prime sources of electrical noise ranging from voltage transients caused by switching to noise inductively or capacitively coupled to a circuit via power lines and ground paths. This type of noise can be minimized by the use of *r-f* line filters and transient noise suppressors and by ensuring that control circuits are kept as far as possible from power circuits. These noise voltages may produce a voltage spike of sufficient amplitude to cause a logic gate to switch states.

A measure of the ability of a logic gate to not respond to noise spikes is the dc noise margin. The dc noise margin is the potential difference between the input and output switching voltage levels. For example, for a typical TTL logic gate a 2.0 V logical 1 input with a guaranteed

400-mV noise margin means that the logical 1 output will not be less than 2.4 V. However, typical noise margins for series 54/74 gates are usually in excess of 1.0 V.

6-9-5 Fan-Out

Very often the output of a logic gate is connected to the inputs of logic gates of the same logic family. The term **fan-out** defines the number of inputs that can be driven by a logic gate while maintaining its correct logical 1 and 0 levels. A typical fan-out capability for TI series 54/74 TTL logic is 10. The fan-out capability of a specific logic gate is shown in manufacturers' data sheets and have been normalized to show the device's ability to supply a number of selected unit loads.

6-10 LOGIC GATES AND FUNCTIONS

In 1847 George Boole developed an algebra which is used to deal with variables that exist in only two forms, 0 and 1, or false and true, or low and high. This algebra, termed **Boolean algebra**, enables us to express all logical functions in a convenient mathematical form. The use of Boolean algebra provides a method of understanding and designing logic circuits. In this section we will study the basic logic functions and gates, and how to minimize the number of devices by the use of Boolean algebra and Karnaugh mapping.

Boolean algebra is most commonly used to combine the basic logic operations of AND, OR, and NOT into the design of more complex applications.

6-10-1 The AND Gate

The AND gate by definition has an output only when all inputs are present. A common example of an AND is shown in Figure 6-3(a), where the lamp L can only be turned on if switches A and B are closed. Other examples of the AND occur in many different applications and involve a number of devices that can be in either of two states such as, transistors, SCRs, relays, switches, etc. The standard logic symbol for a 2-input AND is shown in Figure 6-3(b). Figure 6-3(c) shows two versions of a 2-input AND truth table. In the first version each row defines the circuit condition in terms of the switches as open or closed and the lamp on or off. The first version of the truth table is cumbersome and is replaced in the second version by the generally accepted form, where 1 represents the switch in the closed position and 0 represents the switch in the open position. From Figure 6-3(c) it should be

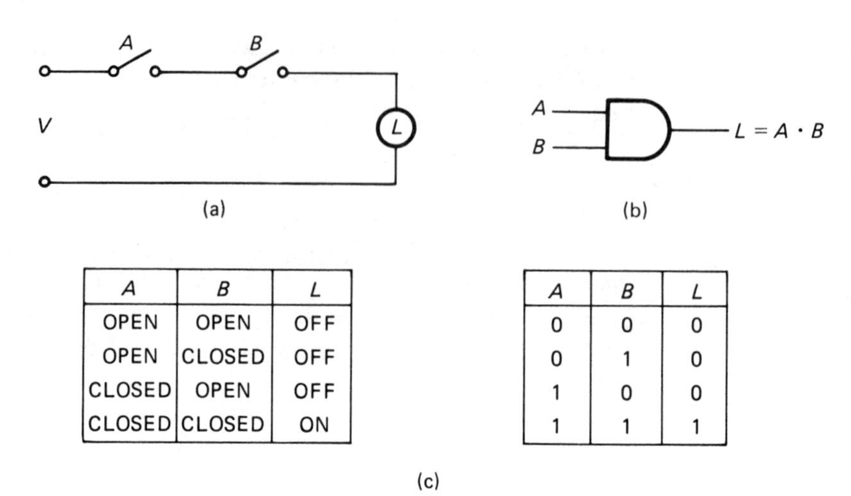

A	B	L
OPEN	OPEN	OFF
OPEN	CLOSED	OFF
CLOSED	OPEN	OFF
CLOSED	CLOSED	ON

A	B	L
0	0	0
0	1	0
1	0	0
1	1	1

Figure 6-3. The AND function: (a) 2-switch AND circuit, (b) 2-input AND logic symbol, (c) truth tables.

noted that the number of rows or combinations is equal to 2^n, where n is the number of inputs to the AND; for example, a 3-input AND would have $2^3 = 8$ combinations.

Usually the AND operations illustrated in Figure 6-3 are written in an algebraic form as follows:

$$A \text{ AND } B = L, \qquad (6\text{-}2a)$$

or

$$A \times B = L, \qquad (6\text{-}2b)$$

or

$$A \cdot B = L, \qquad (6\text{-}2c)$$

or

$$AB = L. \qquad (6\text{-}2d)$$

If we refer back to Section 6-4-6 and compare the rules for binary multiplication to the truth table in Figure 6-3(c), it immediately becomes evident that the Boolean AND is binary multiplication.

6-10-2 The OR Gate

The **OR gate** by definition produces an output whenever an input is present. Going back to our switches and lamp analogy, a two input OR circuit is represented by the two switches A and B in parallel, and as can be seen if either switch or both switches are closed, the lamp L will be turned on [see Figure 6-4(a)]. The standard logic symbol for a 2-

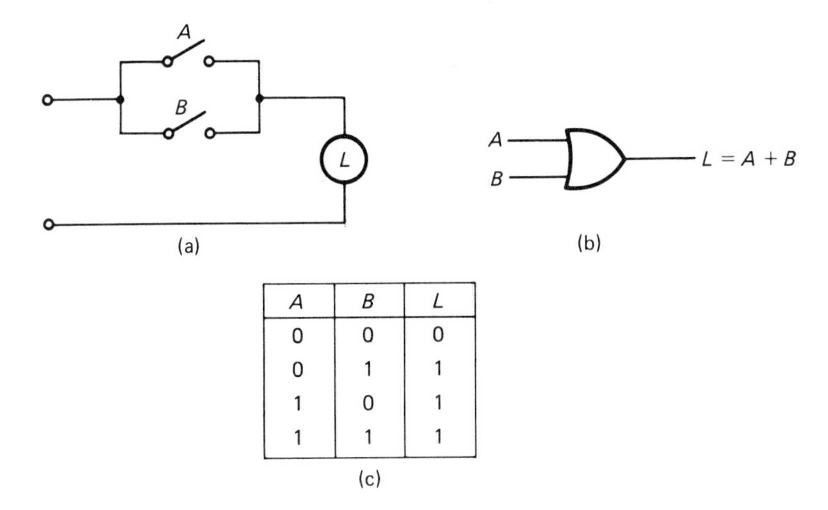

A	B	L
0	0	0
0	1	1
1	0	1
1	1	1

(c)

Figure 6-4. The OR function: (a) 2-switch OR, (b) 2-input OR logic symbol, (c) truth table.

input OR is shown in Figure 6-4(b), while the truth table for a 2-input OR is shown in Figure 6-4(c).

The algebraic form of the OR is written as

$$A \text{ OR } B = L \tag{6-3a}$$

or the preferred form

$$A + B = L \tag{6-3b}$$

If the truth table of Figure 6-4(c) is compared to the rules of binary addition in section 6-4-1, there is agreement for the first three rows of the truth table; however, in the case of the fourth row there is no agreement. It will be seen when we study the rules of Boolean algebra that there are other instances where there is a deviation from the rules of normal aglebra. The OR gate is not restricted to a minimum of two inputs and as was the case with the AND the number of combinations is equal to 2^n, where n = the number of inputs.

6-10-3 The NOT Or Inverter

The **NOT** operation or inverter completes the three basic operations of Boolean algebra. The NOT operation may also be termed the complement operation, that is, the output is the complement of the input.

The logic symbol and truth table of the NOT are shown in Figure 6-5. The negation or complement may be obtained by the use of a common emitter amplifier using an *n-p-n* transistor for positive logic or *p-n-p* transistor for negative logic. It should be emphasized that the NOT is not a logic gate, but strictly inverts or complements the input. It will be used later to also produce the NOT AND or NAND and the NOT OR or NOR logic gates. A major advantage of using the common emitter amplifier is that it may be used to amplify the output and increase the drive capability of the NOT.

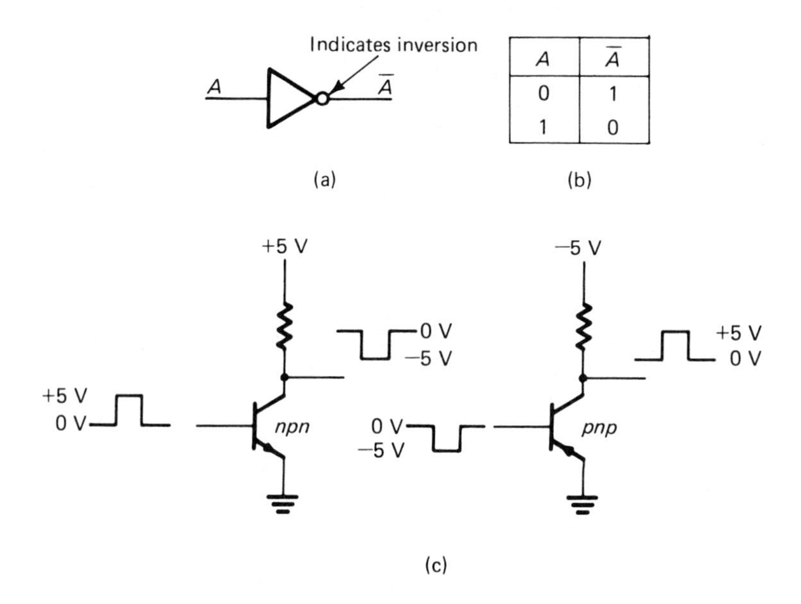

A	$\overline{A}$
0	1
1	0

Figure 6-5. The NOT function: (a) logic symbol, (b) truth table, (c) common-emitter transistor amplifier NOT circuits.

6-10-4 The NAND Gate

The **NAND gate** is functionally an AND followed by a NOT, that is, when all the inputs are logical 1 then the output is logical 0. The NAND logic symbol is basically the AND symbol with a circle at the output indicating the NOT function, Figure 6-6(a). The truth table for a 2-input NAND is shown in Figure 6-6(b).

6-10-5 The NOR Gate

The **NOR gate** is functionally an OR gate whose output is inverted by a NOT, that is, when all inputs are at logical 0, the output is at logical 1. The NOR gate symbol is an OR symbol with a circle at the output indicating the NOT function, Figure 6-7(a). The truth table for a 2-input NOR is shown in Figure 6-7(b).

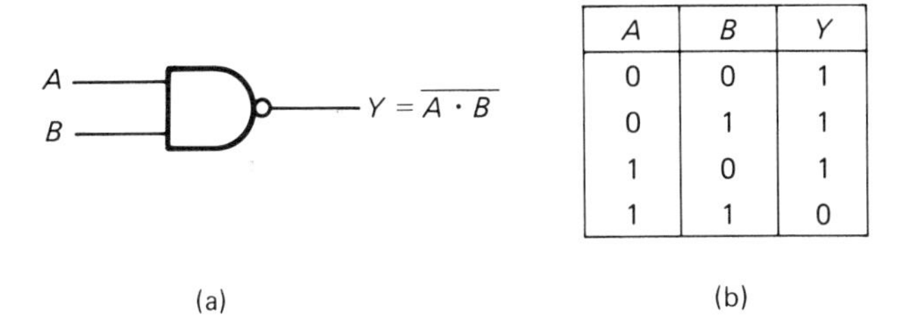

A	B	Y
0	0	1
0	1	1
1	0	1
1	1	0

Figure 6-6. The NAND function: (a) the logic symbol, (b) the truth table for a 2-input NAND.

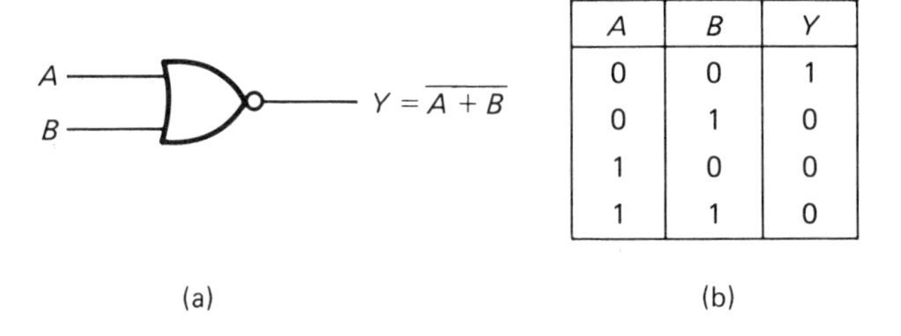

A	B	Y
0	0	1
0	1	0
1	0	0
1	1	0

Figure 6-7. The NOR function: (a) the logic symbol, (b) the truth table for a 2-input NOR.

6-10-6 The Exclusive-OR

The **exclusive-OR** function has a logical 1 output if any one input is a logical 1, but has a logical 0 output if both inputs are at logical 1 or are logical 0. The exclusive-OR is used in binary addition, and as can be seen from the truth table in Figure 6-8(b), the truth table would be obtained by adding two binary digits. The exclusive-OR is also called a half-adder since there is no provision for the carry bit.

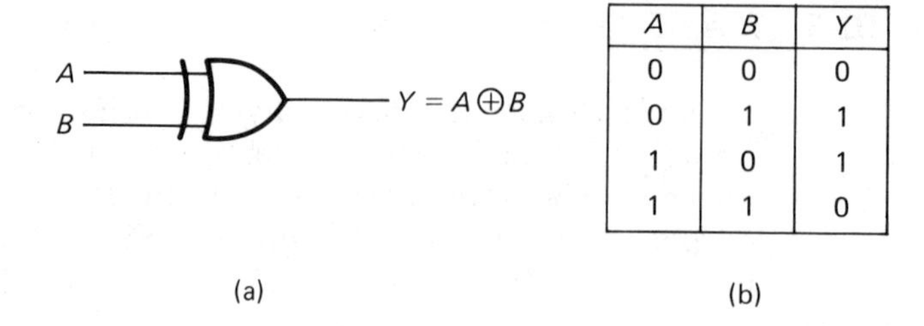

A	B	Y
0	0	0
0	1	1
1	0	1
1	1	0

Figure 6-8. The exclusive-OR function: (a) the logic symbol, (b) the truth table for a 2-input exclusive-OR.

6-10-7 The Exclusive-NOR (Comparator)

The **exclusive-nor** has a logical 1 output when the two inputs are equal, that is, $A = 1$, $B = 1$, or $A = 0$, $B = 0$. The exclusive - NOR is also called a comparator or coincidence function (see Figure 6-9).

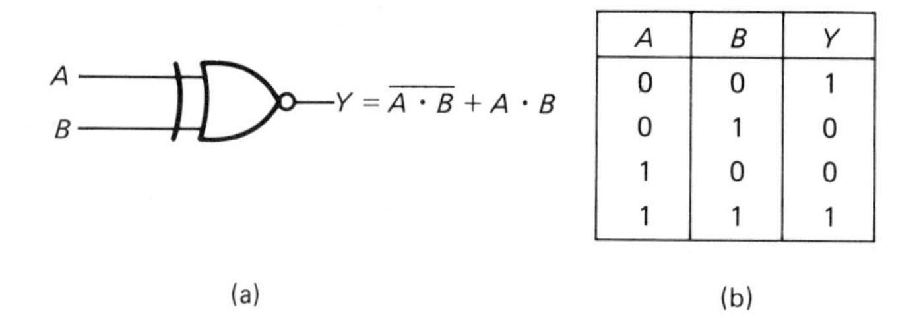

A	B	Y
0	0	1
0	1	0
1	0	0
1	1	1

Figure 6-9. The exclusive-NOR function: (a) the logic symbol, (b) the truth table for a 2-input exclusive-NOR.

6-11 BOOLEAN ALGEBRA

So far in our studies we have shown the basic logic functions and their truth tables. To be able to solve a specific problem we must be able to use the postulates and theorems of Boolean algebra to determine the required logic functions. In the postulates and theorems listed in Table 6-8, the letters A, B, C, etc. represent two-valued variables; for example, when $A = 1$ a high voltage level or a pulse is present, and when $A = 0$ there is a low voltage level or a pulse is absent.

Table 6-8 Boolean Algebra Postulates and Theorems

	No.	Rule	
Postulates			
	1.	$A = 1 \text{ if } A \neq 0$	
	2.	$A = 0 \text{ if } A \neq 1$	
*	3.	$0 \cdot 0 = 0$	
	4.	$1 + 1 = 1$	
*	5.	$1 \cdot 1 = 1$	
*	6.	$0 + 0 = 0$	
*	7.	$1 \cdot 0 = 0$	
*	8.	$1 + 0 = 1$	
	9.	$\overline{0} = 1$	
	10.	$\overline{1} = 0$	
Theorems			
*	1.	$A + 0 = A$	
*	2.	$A \cdot 0 = 0$	
	3.	$A + 1 = 1$	
*	4.	$A \cdot 1 = A$	
	5.	$A + A = A$	
	6.	$A \cdot A = A$	
	7.	$A + \overline{A} = 1$	
	8.	$A \cdot \overline{A} = 0$	
*	9.	$A + B = B + A$	Commutative Property
*	10.	$A \cdot B = B \cdot A$	
*	11.	$A + (B + C) = (A + B) + C$	Associative Property
*	12.	$A \cdot (B \cdot C) = (A \cdot B) \cdot C$	
*	13.	$A \cdot (B + C) = A \cdot B + A \cdot C$	Distributive Property
	14.	$A + (B \cdot C) = (A + B) \cdot (A + C)$	
	15.	$\overline{A + B + C} = \overline{A} \cdot \overline{B} \cdot \overline{C}$	De Morgans Theorems
	16.	$\overline{A \cdot B \cdot C} = \overline{A} + \overline{B} + \overline{C}$	

The rules marked with an * are the properties of ordinary algebra that are equally valid for Boolean algebra.

At this point it may be useful to illustrate the application of logic gates to prove some of the theorems of Boolean aglebra.

Example 6-21

Develop logic circuits to prove the validity of the following:

(1) $A + 0 = A$

(2) $A \cdot 0 = 0$

(3) $A + 1 = 1$

(4) $A \cdot 1 = A$

(5) $A + A = A$

(6) $A \cdot A = A$

(7) $A + AB = A$

(8) $A(A + B) = A$

(9) $\overline{\overline{A}} = A$

(10) $\overline{A \cdot B} = \overline{A} + \overline{B}$

(11) $\overline{A + B} = \overline{A} \cdot \overline{B}$

Solutions:

(1) $A + 0 = A$ see Figure 6-10(a)

(2) $A \cdot 0 = 0$ see Figure 6-10(b)

(3) $A + 1 = 1$ see Figure 6-10(c)

(4) $A \cdot 1 = A$ see Figure 6-10(d)

(5) $A + A = A$ see Figure 6-10(e)

(6) $A \cdot A = A$ see Figure 6-10(f)

(7) $A + AB = A$ see Figure 6-10(g)

(8) $A(A + B) = A$ see Figure 6-10(h)

(9) $\overline{\overline{A}} = A$ see Figure 6-10(i)

(10) $\overline{A \cdot B} = \overline{A} + \overline{B}$ see Figure 6-10(j)

(11) $\overline{A + B} = \overline{A} \cdot \overline{B}$ see Figure 6-10(k)

It should be noted that (10) and (11) represent De Morgan's first and second theorems. From the logic diagrams and truth tables, it can be seen that for (10) the output of a 2-input NAND with inputs A and B is identical to the output of a 2-input OR with inputs A and B, and for

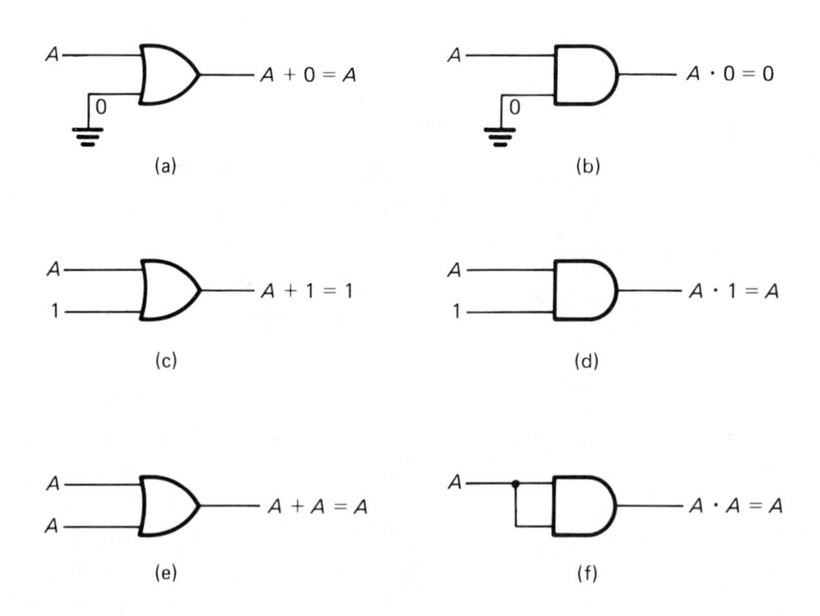

Figure 6-10. Logic diagrams for Example 6-21.

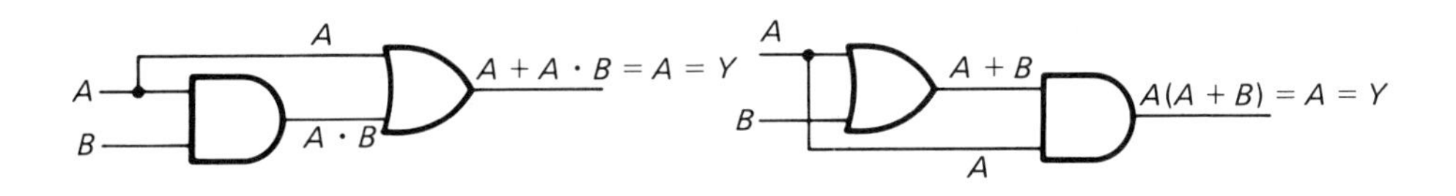

A	B	Y
0	0	0
0	1	0
1	0	1
1	1	1

$A = Y$
i.e., $A + A \cdot B = A$

(g)

A	B	Y
0	0	0
0	1	0
1	0	1
1	1	1

$A = Y$
i.e., $A(A + B) = A$

(h)

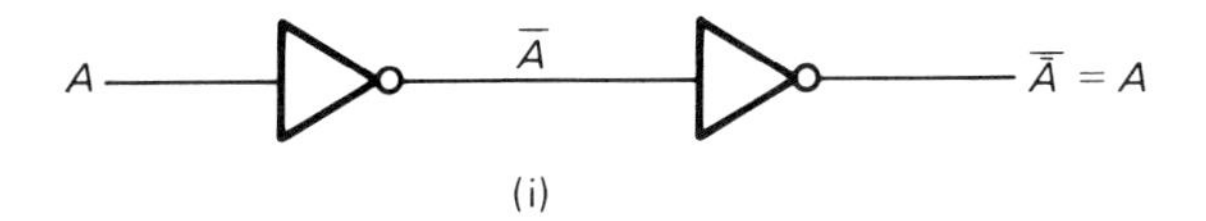

(i)

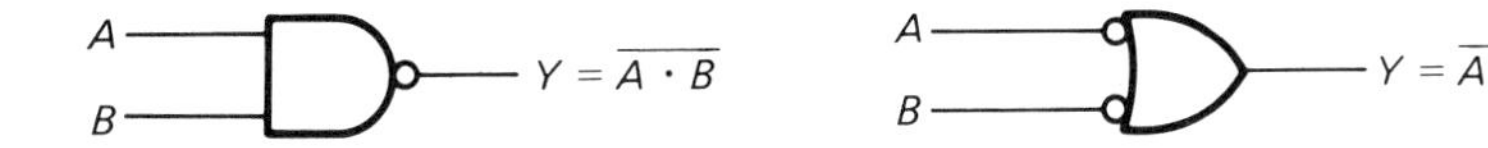

A	B	Y
0	0	1
0	1	1
1	0	1
1	1	0

$\overline{A \cdot B} = \overline{A} + \overline{B}$

A	B	Y
0	0	1
0	1	1
1	0	1
1	1	0

(j)

A
B
$Y = \overline{A + B}$

A	B	Y
0	0	1
0	1	0
1	0	0
1	1	0

$\overline{A + B} = \overline{A} \cdot \overline{B}$

A	B	Y
0	0	1
0	1	0
1	0	0
1	1	0

(k)

Figure 6-10 continued.

(11) the output of a 2-input NOR with inputs A and B is identical to the output of a 2-input AND with inputs A and B.

Example 6-22

Draw the logic circuit whose output is $\overline{A}\overline{B}C + A\overline{B} + AB\overline{C}$. Simplify if possible.

Solution [see Figure 6-11(a)]

$$\begin{aligned}
\overline{A}\overline{B}C + A\overline{B} + AB\overline{C} &= \overline{A}\overline{B}C + A(\overline{B} + B\overline{C}) \\
&= A(\overline{B} + \overline{C}) + \overline{A}\overline{B}C \\
&= \overline{B}(A + A\overline{C}) + A\overline{C} \\
&= \overline{B}(A + C) + A\overline{C} \\
&= A\overline{B} + \overline{B}C + A\overline{C} \\
&= A\overline{B}(C + \overline{C}) + \overline{B}C + A\overline{C} \\
&= A\overline{B}C + A\overline{B}\overline{C} + \overline{B}C + A\overline{C} \\
&= A\overline{C}(1 + \overline{B}) + \overline{B}C(1 + A) \\
&= A\overline{C} + \overline{B}C.
\end{aligned}$$

This is represented by the logic circuit shown in Figure 6-11(b). As can be seen the solution represented in Figure 6-11 (b) uses one less NAND than the solution to the Boolean expression shown in Figure 6-11(a).

6-11-1 Simplification and Minimization

Logic equations describe the solutions to logic problems, that is, the functions and operations required to be performed by a digital system. The physical complexity of the digital system is related to the complexity of the logic equations representing the solution. The logic equations may be simplified by applying the rules and postulates listed in Table 6-8, and the resulting simplification in turn reduces the complexity and the number of components required by the digital system. However, simplifying complex logic equations by applying the theorems and postulates of Boolean algebra can be very tedious, as demonstrated in Example 6-22. As a result special graphical techniques have been developed to describe and simplify Boolean logic expressions. The method that will be used is termed a Karnaugh map. The **Karnaugh map** consists of a rectangle divided into as many subdivisions as there are variables to be represented.

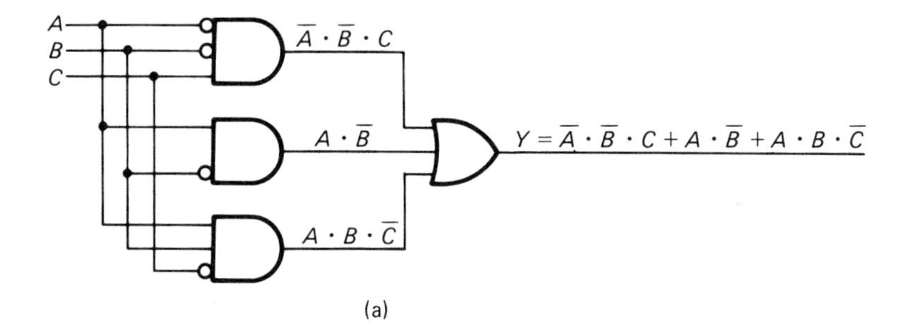

(a)

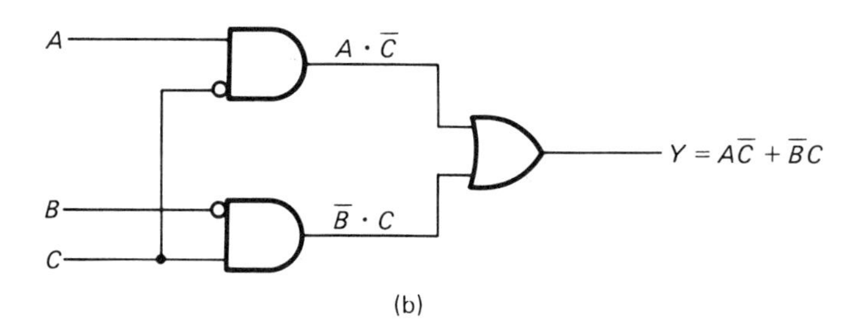

(b)

Figure 6–11. Logic diagrams for Example 6–22.

6-11-1-1 Karnaugh Map For A Single Variable

A single variable exists in either of two forms A or $\overline{A}$. Referring to Figure 6-12(a) and (b) the presence of a variable is indicated by placing a 1 in the appropriate square; implicitly an empty box indicates a 0 or the absence of the variable. If both squares of Figure 6-12(c) were filled then the two adjacent squares would be ORed, that is, $A + \overline{A} = 1$. In applying Karnaugh mapping to functions with more than one variable it is necessary to provide 2^n subdivisions or boxes where n is the number of variables; that is, four boxes for two variables, eight boxes for three variables, etc.

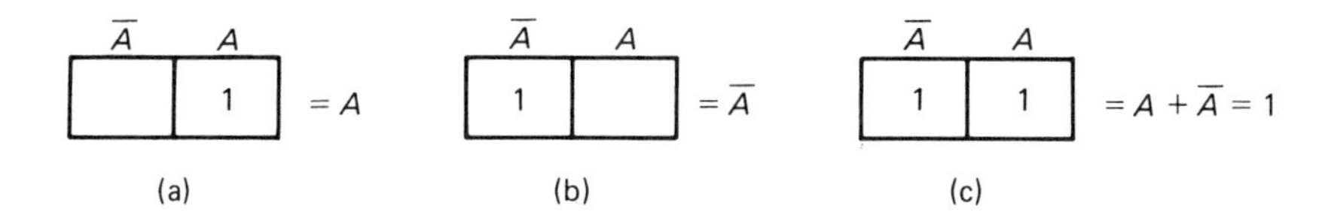

Figure 6–12. A single-variable Karnaugh map.

6-11-1-2 Two-Variable Karnaugh Maps

In Karnaugh maps the Boolean expression must be expressed in the form of an OR function, although it is permissible for the individual terms to be AND functions, which eliminates terms in parentheses. For example, $A\bar{B}C$ is permissible but $A(\overline{BC})$ is not.

The general form of a two-variable Karnaugh map is shown in Figure 6-13(a). The two axes are labelled with 0's and 1's which permits direct transfer of an expression from a truth table to the map to permit possible simplification. Figure 6-13(b) shows the truth table for $Y = A\bar{B}$ and this in turn is represented by the mapping shown in Figure 6-13(c) where the minterms have been identified. If we have an expression such as $\bar{A}\bar{B} + \bar{A}B$, it would be mapped as shown in Figure 6-13(d) which, since the 1's are in adjacent horizontal boxes, indicates that the simplified expression is $\bar{A}$, which follows from $\bar{A}\bar{B} + \bar{A}B = \bar{A}(\bar{B} + B) = \bar{A}$. Similarly, $\bar{A}\bar{B} + A\bar{B}$ would result in two adjacent vertical boxes showing that $\bar{A}\bar{B} + A\bar{B}$ simplifies to $\bar{B}$. However, $\bar{A}\bar{B} + AB$, or $A\bar{B} + \bar{A}B$ would place 1's in diagonally adjacent boxes and cannot be simplified. Even though the examples of two-variable expressions are hardly challenging, two basic rules have been established.

1. Expressions resulting in 1's in horizontally or vertically adjacent boxes may be simplified and expressed as one variable.
2. Expressions resulting in 1's in diagonally adjacent boxes cannot be simplified.

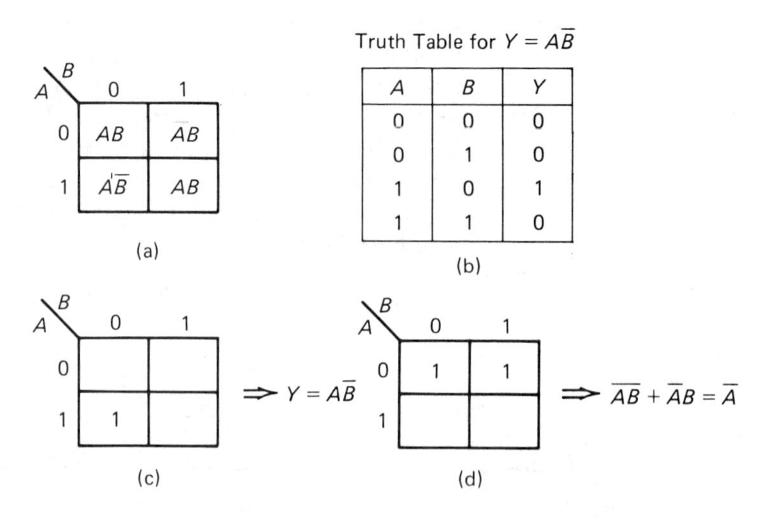

A	B	Y
0	0	0
0	1	0
1	0	1
1	1	0

Figure 6-13. Two-variable Karnaugh maps: (a) a two-variable map, (b) truth table for $Y = A\bar{B}$, (c) Karnaugh map for $Y = A\bar{B}$, (d) mapping for $\bar{A}\bar{B} + \bar{A}B$.

6-11-1-3 Three-Variable Karnaugh Maps

Three-variable expressions require a map with eight boxes as shown in Figure 6-14 (a). Figure 6-14(b) is a mapping of $A\overline{B}C + A\overline{B}\overline{C}$ and, as can be seen, there are two adjacent vertical boxes, and as a result the expression may be simplified to $A\overline{B}$. The expression $A\overline{B}\overline{C} + \overline{A}B\overline{C} + \overline{A}\overline{B}C + \overline{A}BC$ is mapped in Figure 6-14(c). The 1's in the diagonally adjacent boxes show that $A\overline{B}\overline{C} + \overline{A}\overline{B}C$ cannot be simplified; however, the 1's in the vertically adjacent boxes show that $\overline{A}B\overline{C} + \overline{A}BC = \overline{A}B$. In addition, there is adjacency between the two end boxes in the bottom horizontal row since the map is effectively in the form of a cylinder. As a result $\overline{A}\overline{B}C + \overline{A}BC$ simplifies to $\overline{A}C$. The whole expression $A\overline{B}\overline{C} + \overline{A}B\overline{C} + \overline{A}\overline{B}C + \overline{A}BC$ simplifies to $\overline{A}B + \overline{A}C + A\overline{B}\overline{C}$. It should be noted that the term $\overline{A}BC$ was linked twice, which is permitted provided that each loop links with another unlinked term. Figure 6-14(d) shows that $\overline{A}\overline{B}\overline{C} + A\overline{B}\overline{C} + AB\overline{C} + \overline{A}B\overline{C} = \overline{C}$, that is, four adjacent boxes simplify down to a single variable. The four boxes do not have to be in a single row but could equally as well form a square.

6-11-1-4 Four-Variable Karnaugh Maps

Four-variable expressions require sixteen boxes as shown in Figure 6-15(a). Figure 6-15(b) shows a mapping of $\overline{A}\overline{B}\overline{C}\overline{D} + \overline{A}\overline{B}\overline{C}D + A\overline{B}CD + \overline{A}\overline{B}C\overline{D}$. A study of Figure 6-15(b) shows that $A\overline{B}CD$ will not simplify because of the diagonal relationship with $\overline{A}\overline{B}\overline{C}D$. Similarly, terms $\overline{A}\overline{B}\overline{C}\overline{D}$ and $\overline{A}\overline{B}C\overline{D}$ can be looped eliminating C leaving $\overline{A}\overline{B}\overline{D}$. Similarly, terms $\overline{A}\overline{B}\overline{C}\overline{D}$ and $\overline{A}\overline{B}\overline{C}D$ may be looped in turn eliminating D leaving $\overline{A}\overline{B}\overline{C}$. Thus $\overline{A}\overline{B}\overline{C}\overline{D} + \overline{A}\overline{B}\overline{C}D + A\overline{B}CD + \overline{A}\overline{B}C\overline{D}$ simplifies to $\overline{A}\overline{B}\overline{D} + \overline{A}\overline{B}\overline{C} + A\overline{B}CD$. Similarly, Figure 6-15(c) shows the simplification of $\overline{A}\overline{B}\overline{C}\overline{D} + A\overline{B}\overline{C}\overline{D} + \overline{A}\overline{B}C\overline{D} + A\overline{B}C\overline{D} + \overline{A}\overline{B}CD + A\overline{B}CD + \overline{A}\overline{B}\overline{C}D + A\overline{B}\overline{C}D + \overline{A}BCD + \overline{A}B\overline{C}D$. The loop of eight eliminates three variables and leaves $\overline{B}$. The loop of four in the bottom two rows eliminates two variables and leaves $\overline{A}D$. As a result our expression has been simplified to $\overline{B} + \overline{A}D$. In general, the following should be noted:

1. Diagonally adjacent or isolated terms cannot be simplified.

2. Horizontally and vertically adjacent terms can be looped and simplified. If two terms are looped, one variable is eliminated; if four terms are looped, two variables are eliminated; and if eight terms are looped three variables are eliminated.

3. Looped turns may be linked with unlooped terms.

4. A four-variable mapping is continous in both the horizontal and vertical directions.

Karnaugh mapping may be extended to the simplification of expressions involving five variables. However, we have developed sufficient capability to realize that Karnaugh maps provide a very powerful tool to simplify logic expressions and as a result simplify the corresponding logic circuits.

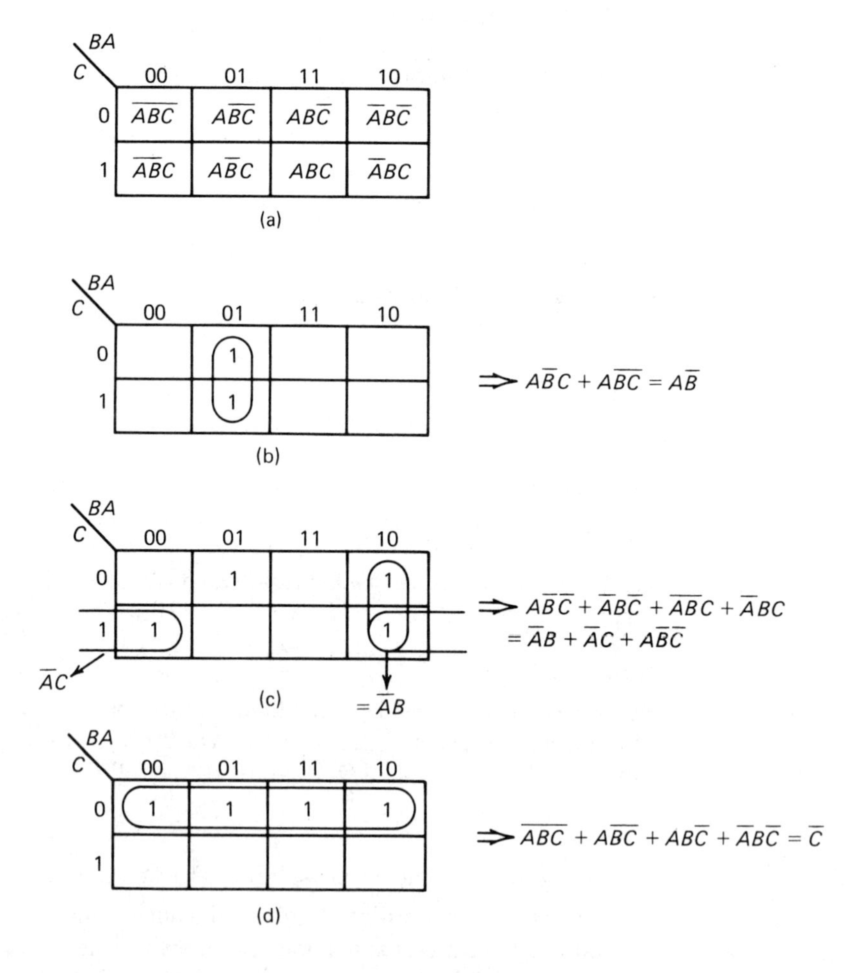

Figure 6-14. Three-variable Karnaugh maps: (a) three-variable map, (b) the mapping for $A\overline{B}C + A\overline{B}\,\overline{C}$, (c) the mapping for $A\overline{B}\,\overline{C} + \overline{A}B\overline{C} + \overline{A}\,\overline{B}C + \overline{A}BC$, (d) the mapping for $\overline{A}\,\overline{B}\,\overline{C} + A\overline{B}\,\overline{C} + AB\overline{C} + \overline{A}B\overline{C}$.

(a)

DC \ BA	00	01	11	10
00	$\overline{ABCD}$	$A\overline{BCD}$	$AB\overline{CD}$	$\overline{A}B\overline{CD}$
01	$\overline{AB}C\overline{D}$	$A\overline{B}C\overline{D}$	$ABC\overline{D}$	$\overline{A}BC\overline{D}$
11	$\overline{AB}CD$	$A\overline{B}CD$	$ABCD$	$\overline{A}BCD$
10	$\overline{ABC}D$	$A\overline{BC}D$	$AB\overline{C}D$	$\overline{A}B\overline{C}D$

(b)

DC \ BA	00	01	11	10
00	1			
01	1			
11		1		
10	1			

$$\Rightarrow \overline{ABCD} + \overline{ABC}D + A\overline{B}CD + \overline{AB}CD + \overline{ABC}\,\overline{D}$$
$$= \overline{ABD} + \overline{ABC} + A\overline{B}CD$$

(c)

DC \ BA	00	01	11	10
00	1	1		
01	1	1		
11	1	1		1
10	1	1		1

$$\Rightarrow \overline{ABCD} + A\overline{BCD} + \overline{AB}C\overline{D} + A\overline{B}C\overline{D}$$
$$+ \overline{AB}CD + A\overline{B}CD + \overline{ABC}D + A\overline{BC}D$$
$$+ \overline{A}BCD + \overline{A}B\overline{C}D = \overline{B} + \overline{A}D$$

Figure 6-15. Four- variable Karnaugh maps: (a) a four-variable map, (b) the mapping of $\bar{A}\bar{B}\bar{C}\bar{D} + \bar{A}\bar{B}\bar{C}D + A\bar{B}CD + \bar{A}\bar{B}C\bar{D}$, (c) the mapping of $\bar{A}\bar{B}\bar{C}\bar{D} + A\bar{B}\bar{C}\bar{D} + \bar{A}\bar{B}C\bar{D} + A\bar{B}C\bar{D} + \bar{A}\bar{B}CD + A\bar{B}CD + \bar{A}\bar{B}\bar{C}D + A\bar{B}\bar{C}D + \bar{A}BCD + \bar{A}B\bar{C}D$.

6-12 INTEGRATED CIRCUIT LOGIC FAMILIES

Initially the basic logic functions were performed by discrete components; then these concepts were incorporated into the earlier and more simple ICs. As the manufacturing processes improved the ICs rapidly became more complex, until now we classify ICs in terms of the number of transistors involved in their design, as follows:

1. SSI: small-scale integration; less than 10 transistors.
2. MSI: medium-scale integration; between 10 and 100 transistors.
3. LSI: large-scale integration; in excess of 100 transistors.

The major IC logic families are:

1. RTL: resistor-transistor logic.
2. DTL: diode-transistor logic.
3. TTL (T^2L): transistor-transistor logic.
4. ECL: emitter-coupled logic.
5. PMOS and NMOS: *p*-and *n*-channel metal-oxide-semiconductors.
6. CMOS: complementary metal-oxide-semiconductor.

Currently RTL and DTL are practically obsolete resulting from the research and development to produce faster, lower energy consuming and smaller devices. The workhorse is the TTL family, but there is rapid development occurring with the remaining members of the major logic families.

6-12-1 Resistor-Transistor Logic (RTL)

Basic RTL circuits are shown in Figure 6-16, and as can be seen it consists of resistors and transistors. In the case of the RTL NOR gate shown in Figure 6-16(a), with an input signal at either *A* or *B* or both, *Q*1 or *Q*2 or both will be on, and the output at *Y* will be low. If both *A* and *B* inputs are low, then the output *Y* will be high. These conditions are shown in the associated truth table, which will be seen to be that of a NOR gate. While we have shown the device as a 2-input NOR it can have more inputs, but the deciding factor is that increased leakage current will increase the voltage drop across R_c and lower the value of the output voltage.

An RTL NAND gate configuration is shown in Figure 6-16(b). As can be seen there will only be a current flow through *Q*1 and *Q*2 if inputs *A* and *B* are both high, and as a result the output will be low. If only one of the inputs is high then the nonconducting transistor will block current flow through R_c and the output will be high. These conditions are reflected in a NAND truth table. Once again the number of inputs may be increased; however, the output voltage level is actually the sum of the $V_{CE(\text{sat})}$'s of the individual transistors, and the effect of increasing the number of inputs is to decrease the output voltage level, and thus reduce the fan-out capability, usually to no more than 4.

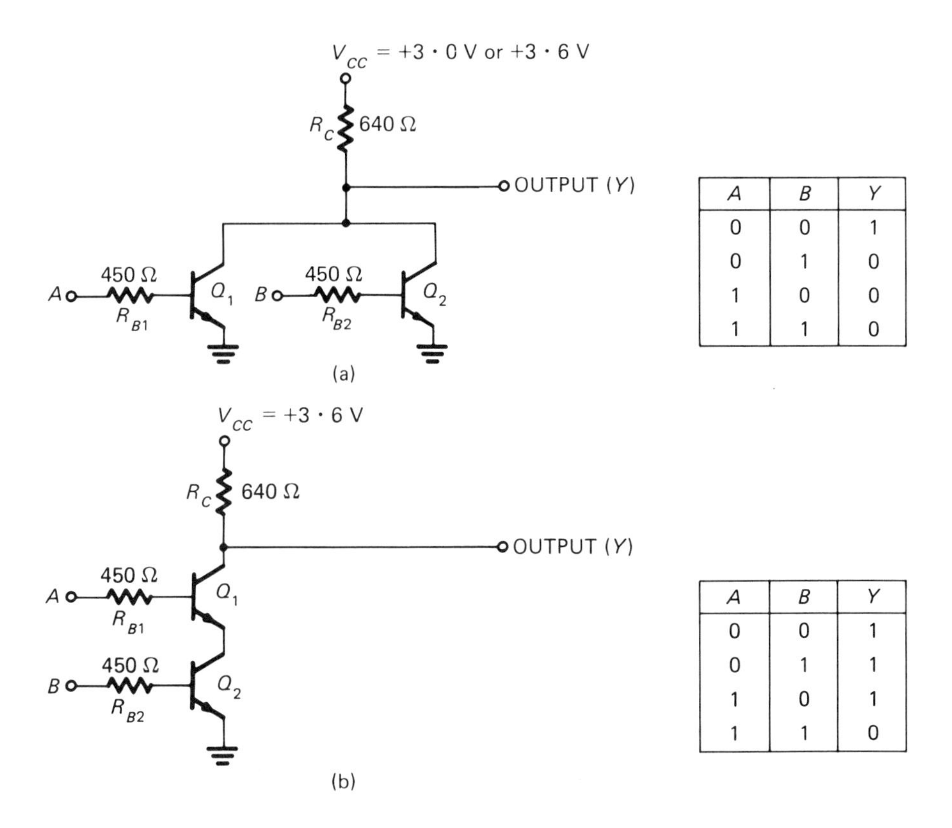

A	B	Y
0	0	1
0	1	0
1	0	0
1	1	0

A	B	Y
0	0	1
0	1	1
1	0	1
1	1	0

Figure 6-16. Resistor-transistor logic: (a) basic circuit of a 2-input NOR and the truth table, (b) basic circuit of a 2-input NAND and the truth table.

The major disadvantage of RTL ICs is the relatively high propagation time, typically of the order of 50 ns. A reduction in propagation time can be achieved by connecting a speed-up capacitor in parallel with the input resistors. The effect of the speed-up capacitor is to achieve the reduction in propagation time by bypassing the steep-fronted leading and trailing edges of the input pulse around the base resistance. This approach, which is relatively cheap and easy to achieve in discrete logic circuits, becomes relatively expensive as well as occupying a relatively large area when attempted in an IC. This type of logic was termed resistor-capacitor-transistor logic (RCTL). Apart from the relatively large propagation delay, RTL logic cannot be interfaced with other types of logic because of its voltage, namely 3.6 V. Additionally, RTL logic has an inherently low noise margin, typically 0.2 V. Both the RTL and RCTL logic families are no longer in production but may be encountered in some older equipments.

6-12-2 Diode-Transistor Logic

Diode-transistor logic (DTL) is the oldest and was the most widely used form of bipolar logic. It was originally conceived to reduce costs because switching transistors were originally quite expensive. The overall cost was reduced by using diodes for each gate input as against the use of transistors in RTL devices. Figure 6-17(a) shows a typical DTL NAND gate, and as can be seen diodes $D1$ and $D2$ form an AND gate, and transistor $Q1$ acts as an inverter to form the NOT. When A or B or both are low, the appropriate diode(s) will be forward biased and conduct, and the forward voltage drop of approximately 0.7 V will place the potential at point N at approximately 0.7 V. Diodes $D3$ and $D4$ each require about 0.7 V to conduct, and before Q1 can be turned on, V_{BE} must be at least 0.5 V; this means that the potential at N must be at least 1.9 V, which provides a noise margin of $1.9 - 0.7 = 1.2$ V. With only one diode, $D3$, the noise margin would be $1.2 - 0.7 = 0.5$ V.

Increasing the number of diodes in series or substituting a zener diode will result in an increase in the noise margin. When inputs A and B are both high, diodes $D1$ and $D2$ are reverse biased, the potential at point N will rise sufficiently to drive transistor $Q1$ into saturation, and the output will be low. In order to increase the fan-out capability, currently limited by diodes $D1$ and $D2$, the configuration shown in Figure 6-17(b) is used. The circuit shown is a modification of the original circuit shown in Figure 6-17(a); $Q1$ has been introduced to increase the base drive to $Q2$, thus permitting a higher collector current and an increased fan-out, usually up to 10.

As compared to RTL, the advantages of DTL are improved propagation time, typically 25-30 ns per gate, but with an increased power consumption approximately 12 mW per gate. They also possess a good noise margin, increased fan-out, and are cheaper and can be interfaced with TTL. A disadvantage is that the fan-in capability is limited, since effectively a reverse-biased input diode ($D1$ or $D2$) acts as a capacitor with the result that increasing the number of inputs increases the capacitance and greatly increases the charging time, thus reducing the switching speed. When the logic has to operate in an electrically noisy environment, the noise margin can be increased by substituting a reverse-biased zener diode for $D3$ in Figure 6-17(b). This zener is designed not to allow base current to flow to $Q2$ until the input voltage exceeds 7.5 V. This type of logic which operates from a 15 V supply is known as high-threshold logic (HTL).

6-12-3 Transistor-Transistor Logic

Transistor-transistor logic (TTL or T^2L) occupies by far the largest proportion of the market for IC logic families. Its major appeal stems

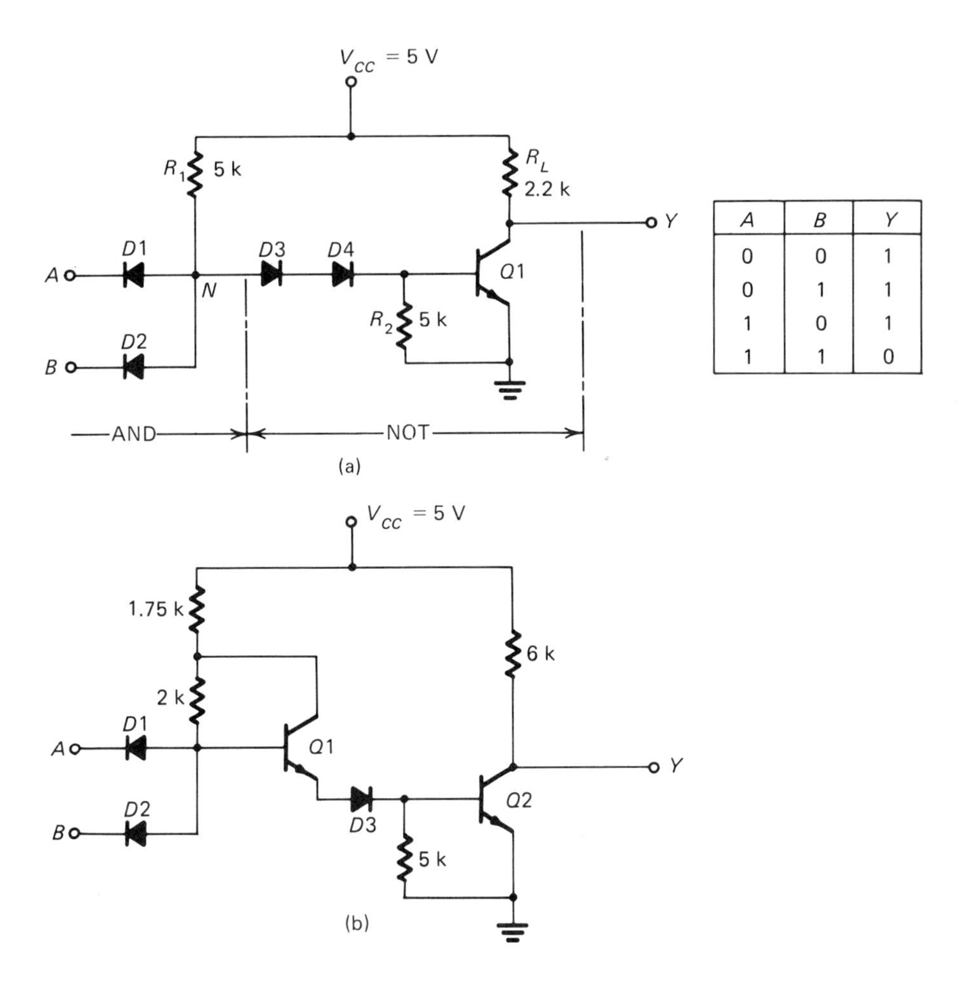

A	B	Y
0	0	1
0	1	1
1	0	1
1	1	0

Figure 6-17. A DTL 2-input NAND gate: (a) schematic and truth table, (b) configuration for increased fan-out capability.

from the relatively low cost and the wide range of standard circuits available in SSI and MSI, combined with high performance and good interfacing capability at switching speeds up to 30 MHz.

The circuit shown in Figure 6-18 operates with $V_{CC} = +5$ V, the multiple emitter transistor $Q1$ is used to improve the operating speed and functions basically in the same manner as the input to a DTL NAND gate. When either, or both, inputs A and B are low the base-emitter junction of $Q1$ is forward biased. $Q1$ collector current is supplied from V_{CC} through $R2$ and the reverse-biased collector-base junction of $Q2$. The collector current is of the order of a few micro-amperes when $Q1$ is saturated. The forward-biased collector-base junction forms a low impedance path for the removal of the base charge from $Q2$, with a consequent reduction in the turn-off time. The multiple emitter design removes the clutter of resistors, diodes, and transistors that were present in DTL and RTL, and as a result reduces the space requirement and parasitic capacitance effects as well as achieving a significant increase in the switching speed.

The operation of the TTL NAND is such that when either or both of the inputs are low, $Q1$ is saturated and $Q2$ and $Q4$ are OFF because of the low base voltage. $Q3$ saturates because of the high base voltage received from the collector of $Q2$ and the output Y is high; $Q3$ is acting as an emitter-follower providing a low impedance driving source.

When inputs A and B are both high, $Q1$ is OFF since the base-emitter junction is reversed biased. At the same time the base-emitter junction of $Q2$ is forward biased with base current being supplied through $R1$ and the base-collector junction of $Q1$. The potential at the emitter of $Q2$ is sufficient to turn on $Q4$ which saturates. The emitter-collector saturation voltage of $Q4$ is approximately 0.4 V or less and therefore the output will be low. The collector potential of $Q2$ is the emitter potential plus the emitter-collector saturation voltage, that is, approximately 0.8 V. If diode D was not in the emitter circuit of $Q3$, the potential at the base of $Q3$ would be sufficient to turn it on.

The output circuit consisting of transistors $Q3$ and $Q4$ is known as a totem-pole or active pull-up output and is designed to ensure that the output will have basically similar driving characteristics whether it is in the high or low state. This is achieved because $Q4$ acts as a shunt transistor switch and $Q3$ and D act as an active pull-up resistor with an effective resistance of approximately 120 to 130 Ω when $Q4$ is OFF, thus reducing the charging time of a capacitive loading, and as a result ensures that the switching time gains made at $Q1$ are not lost at the output.

Under no-load conditions the output high voltage V_{OH} is approximately 3.8 V. When load current is drawn from the output, the output voltage decreases to the order of 2.4 V. The current drawn is known as the sourcing current, $I_{IH} \leq 40\ \mu$ A, and with a typical fan-out of 10 this means that the output can supply 400 μ A. With a logic 0 output $V_{OL} \simeq 0.4$ V and $I_{OL} = I_{SINK} = 16$ mA or 1.6 mA per unit load with a fan-out of 10. The sink load is effectively a resistor between the output and ground. See Figure 6-18(b). The effect of a sink load is to reduce the logic 1 output voltage level and improve the logic 0 output voltage.

One problem that occurs under fast switching conditions is that when switching from a high to a low, $Q4$ will turn on faster than $Q3$ turns off, so as a result there is a short interval of time during which there is a low impedance from V_{CC} through $Q3$, D, and $Q4$ to ground.

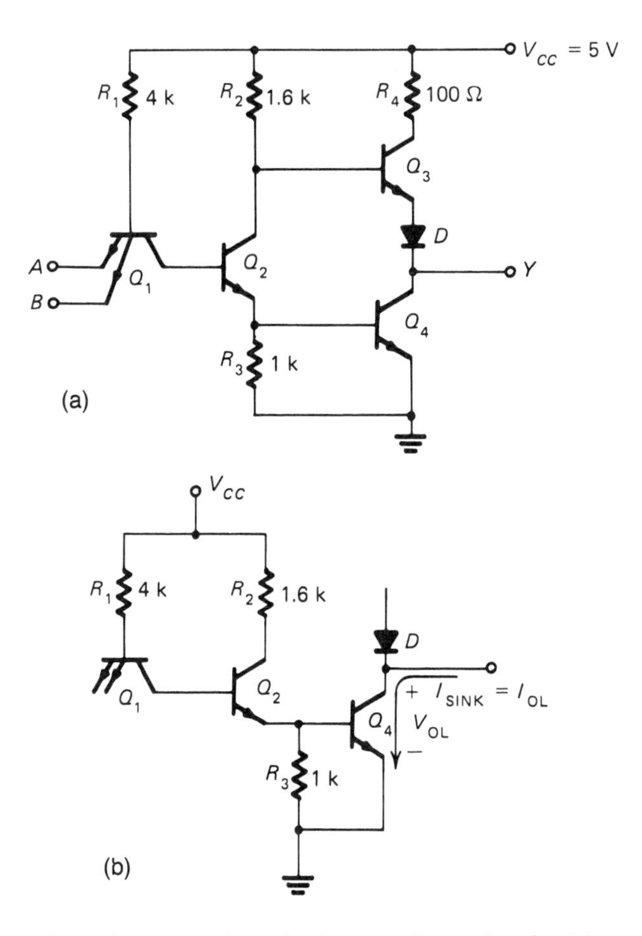

Figure 6-18. Transistor-transistor logic: (a) schematic of a 2-input NAND, (b) current-sinking.

During this condition there will be a short-duration high-current spike supplied by the power supply, which will generate electrical noise during the switching. The effect of these current spikes is minimized by connecting a 0.1 μF or 0.01 μF capacitor in parallel across the power supply leads at the rate of one for every 20 gates.

The major benefits derived from the totem-pole arrangement of the output transistors are a low output impedance in both output logic states, which in turn gives the device the capability of driving highly capacitive loads with a reasonable noise margin, and high speed-switching capability. Unfortunately this capability also gives us problems when it is necessary to AND the output of several gates in what is called the wired-AND connection. See Figure 6-19(a). In the wire-

AND, the NAND gates must be of the open collector type.

An open collector gate is one in which the collector resistance $R4$, $Q3$, and D have been omitted. See Figure 6-18(a). The collector of $Q4$ forms the output terminal. When the outputs are connected as shown in Figure 6-19(a), the common point acts as an AND gate. The output Y will be high only when all the inputs to the NAND gates are low. This configuration eliminates a NAND and a NOT, see Figure 6-19(b), but at the expense of a reduction in the speed of response. If the wire-AND connection is made using NAND gates with a totem-pole output, the output Y would have to sink current from the active pull-up transistors ($Q3$) in all the other gates, with the possibility of damaging the output transistors under overcurrent conditions.

There is another form of TTL known as three-state or tristate TTL; this is basically the standard TTL NAND with a modification which permits both the output transistors $Q3$ and $Q4$ to be turned off. With both $Q3$ and $Q4$ off, the output has a high impedance and is effectively an open collector device and as a result can be wire-ANDed. The three output states are low, high, and high impedance or OFF. This TTL configuration is often found in conjunction with data bus systems.

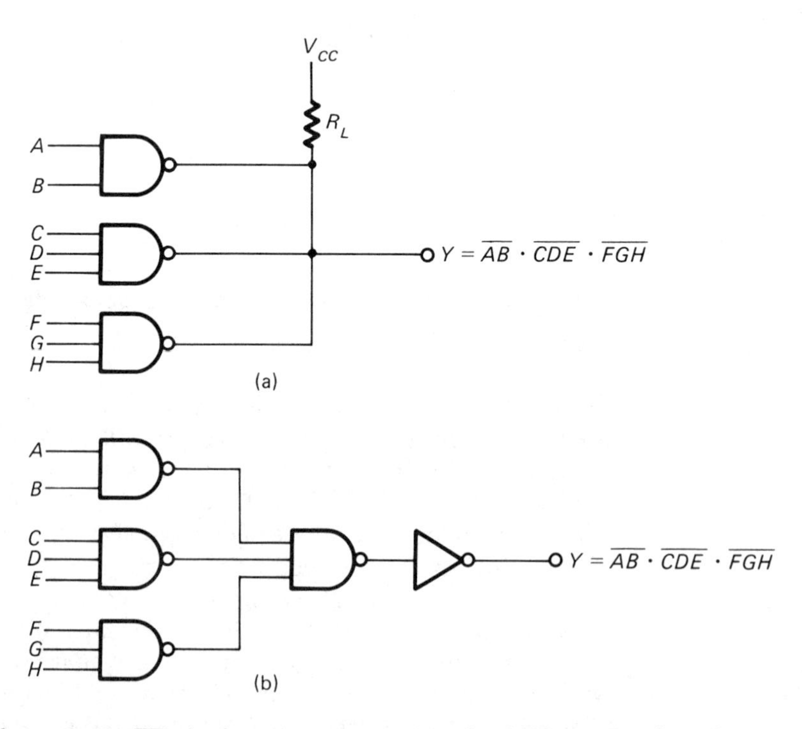

Figure 6-19. TTL logic gate outputs: (a) wire-AND connection, (b) totem-pole output connections.

The TTL family of logic circuits is by far the most commonly used because of its high performance to cost ratio, and the range of devices is constantly being expanded. The major manufacturers are Texas Instruments with the 5400/7400 series, Signetics and its 8000 series, and Fairchild with the 9300 series. Whether TTL ICs are designed for SSI or MSI, combinational or sequential logic, the basic gate is that shown in Figure 6-18. A number of variations of the basic TTL gate have been developed to achieve reductions in power consumption, increased operating speeds, as well as meeting the industrial requirements for special logic functions.

6-12-3-1 Low-Power TTL

The circuit configuration of low-power TTL is similar to the basic TTL logic gate, except that all the resistances are ten times greater, and as a result the power consumption has been reduced to approximately one-tenth of that of the standard TTL logic gate. Texas Instruments Series 54/74L devices have typical power dissipations per gate of 1 mW and propagation delays of the order of 33 ns. The principal application of the Series 54/74L is when power consumption and heat dissipation are critical.

6-12-3-2 High-Speed TTL

The high-speed TTL gate shown in Figure 6-20 is basically the standard gate with modifications. First, all the resistances are lower; second, clamping diodes have been added to each input to reduce transmission effects that become apparent with the decreased rise and fall times of the input pulses; and third, the output section now includes a Darlington pair $Q3$ and $Q4$. The result of these modifications in the Series 54/74H is to reduce the propagation delay to the order of 6 ns because of the transistor action and the lower output impedance. The penalty for increased speed is increased power consumption, typically 22 mW per gate. The high-speed TTL gate is being replaced by the Schottky clamped TTL which gives even higher operating speeds with reduced power consumption.

6-12-3-3 Schottky Clamped TTL

The Schottky clamped TTL gate is the highest-speed TTL gate. The basic Schottky clamped TTL gate is shown in Figure 6-21. The Schottky barrier diodes (SBDs) are used as a clamp between the base and collector of transistors $Q1$, $Q2$, $Q3$, and $Q5$. It should be recalled from Chapter 2 that the SBD has no minority carriers and as a result

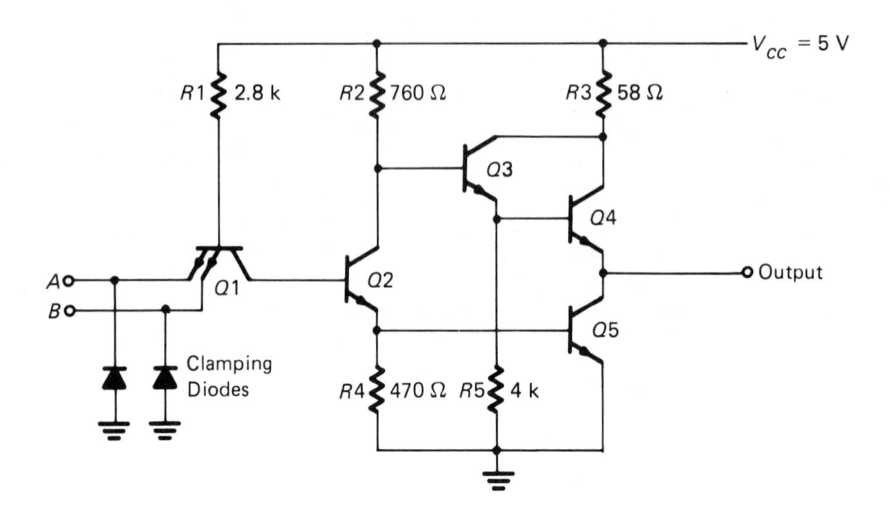

Figure 6-20. High-speed TTL gate.

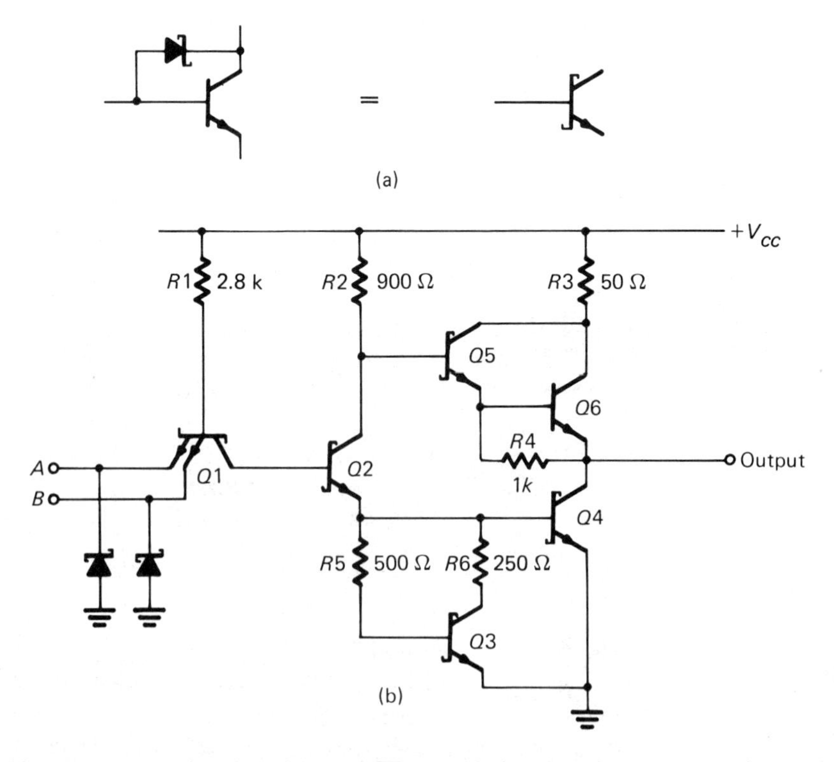

Figure 6-21. Schottky clamped TTL: (a) Schottky clamped transistor, (b) schematic.

there is no stored charge; in addition, it has a lower forward voltage drop than the conventional silicon *p-n* junction. The SBD clamped transistor will have any excess base current diverted away from it, and the transistor will not operate in saturation. As a result of eliminating the stored charges, storage time has been greatly reduced with a consequent improvement in the transistor switching time. The Series 54/74S has very low propagation delays, 3 ns is typical, with power consumptions of the order of 20 mW per gate, slightly less than that for high-speed TTL.

6-13 EMITTER-COUPLED LOGIC (ECL)

The major problem in saturated bipolar TTL is the stored charges present in the base region when the base-emitter and base-collector junctions are forward biased to drive the transistor into saturation. The time necessary to remove this stored charge is a major component of the propagation delay time, and if it could be reduced or eliminated there would be considerable improvement in the switching speed. ECL gates achieve the improvement in switching speed by not permitting the switching transistors to go into saturation. However, the reduction in the propagation delay is achieved at the expense of an increased power consumption. ECL is sometimes also referred to as current-mode logic (CML). The basic circuit of an ECL OR/NOR logic gate is shown in Figure 6–22.

The input transistors $Q1$ and $Q2$ have their emitters and collectors connected in parallel, and together with $Q3$ form a differential amplifier. When the inputs are at binary 0, that is $A = B = -1.55$ V, $Q1$ and $Q2$ are off, however $Q3$ is biased on but kept out of saturation. The resistor $R2$ is selected so that the voltage drop across it is approximately 0.8 V, and therefore the base of $Q5$ is approximately -0.8 V with respect to ground. The emitter of $Q5$ will be at approximately -1.55 V with respect to ground which is binary 0, or the output expected at the OR output terminal. At the same time, since both inputs are low, there will be no current flow through $R1$ and the collector and base of $Q6$ will be at the same potential; as a result $Q6$ will act as a forward-biased diode with the emitter potential being approximately -0.75 V, or binary 1, which is the output expected at the NOR terminal. Resistors $R3$, $R5$, diodes $D1$ and $D2$, and transistor $Q4$ form a temperature-stabilized voltage source that maintains a voltage of approximately -1.3 V on the base of $Q3$. The emitter of $Q3$ and the emitters of $Q1$ and $Q2$ will be approximately 0.75 V more negative, that is, the emitter potentials will be about -2.05 V.

If the inputs A or B or both are high, that is approximately -0.75 V, the potential on the common emitter connection of $Q1$, $Q2$, and $Q3$ will

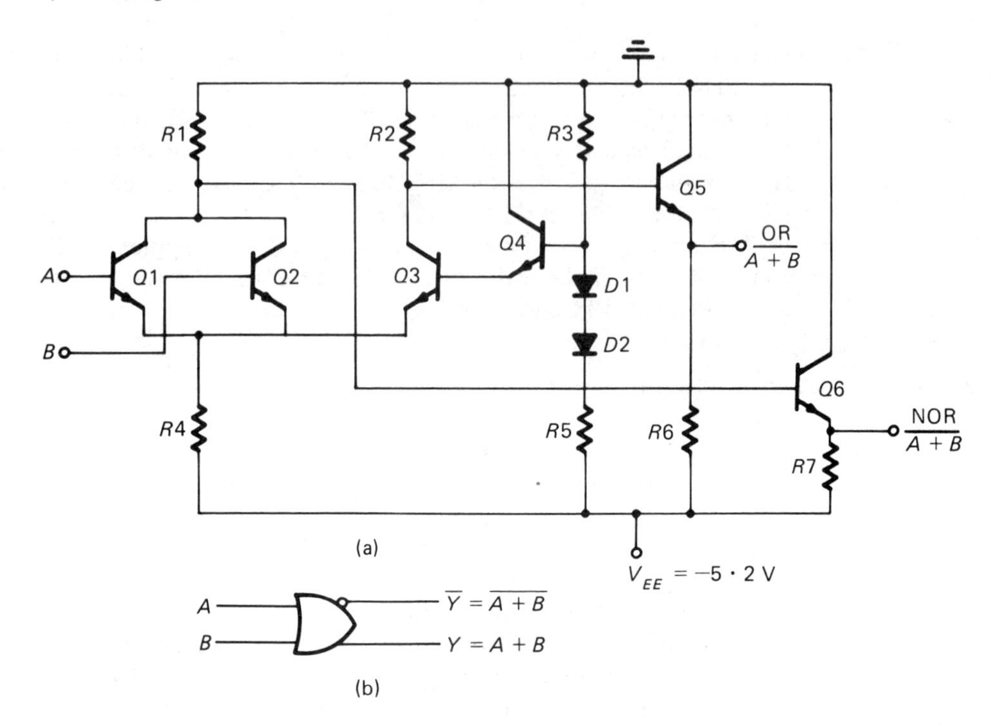

Figure 6-22. Emitter coupled OR/NOR logic gate: (a) schematic, (b) logic symbol.

rise from −2.05 V to about −1.7 V, which will result in $Q3$ being biased off and current flow transferred from $R2$ to $R1$. The collectors of $Q1$ and $Q2$ will be low and the collector of $Q3$ will be high. As a result emitter followers $Q5$ and $Q6$ will switch states, that is, the emitter of $Q5$ will be high and that of $Q6$ will be low.

The major characteristics of ECL circuits are very low propagation delays, usually 1 to 3 ns; a high fan-out, usually 10 to 25; a high noise margin; and the availability of both the OR and NOR outputs. The major disadvantages are high power consumption, about 25 mW per gate, and a −5.2 V power supply which eliminates the possibility of interfacing with other IC logic families.

6-14 METAL-OXIDE-SEMICONDUCTOR INTEGRATED CIRCUITS

There are many advantages to be gained by using MOSFETs. The principal advantages being:

1. The high degree of electrical isolation that exists between MOS

devices because all p-n junctions are reverse or zero biased. As a result, the space requirement is substantially reduced to about 15% of that required by a bipolar transistor.

2. Enhancement mode devices are normally off, which makes them especially suitable for logic elements.

3. An input impedance of the order of $10^{14}\ \Omega$, which means the device draws practically no current to maintain it in the on or off state.

4. A high fan-out capability, typically 20.

5. The reduced number of production steps combined with the extremely small device size makes them most suitable for medium and large scale integrated circuits.

6. The relatively low power consumption, typically 10 mW per gate.

The major disadvantage of the MOSFET is the high propagation time which is of the order of 300 ns.

There are two types of MOSFETs, the p-channel (PMOS) and the n-channel (NMOS). Both forms are used in the enhancement mode, in which the device is normally off until a voltage is applied between the source and gate, at which point the device switches on, that is, it presents an extremely high impedance in the off state and an extremely low impedance in the on state.

In the original application of MOSFETs to digital ICs, the enhancement mode PMOS was most commonly used since they were simpler to manufacture. However, NMOS technology has developed enhancement mode devices in which the electron mobility is approximately three times as fast as the hole mobility of the PMOS. As a result, the propagation delay has been decreased and this, combined with a lower threshold voltage, has made the devices compatible with bipolar TTL. The circuit symbol for an NMOS is shown in Figure 6–23 in the two possible configurations, first where the substrate to source connection

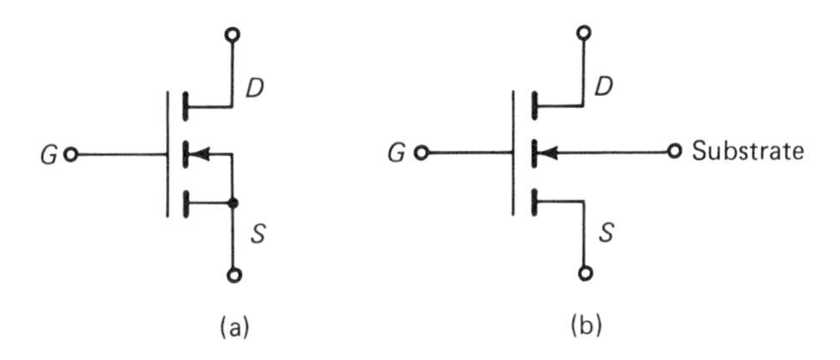

Figure 6-23. Enhancement mode MOSFET: (a) circuit symbol for n-channel source and substate internally connected, (b) externally connected.

is made internally, and second, externally. The circuit symbol for a PMOS device is identical except that the arrowhead direction is reversed.

Figure 6–24(a) shows the basic circuit of a PMOS logic inverter using a resistive load R. The load resistor can be replaced by a PMOS transistor $Q1$ so that when a sufficiently negative voltage V_{GG} is applied to the gate of $Q1$ it will turn on and act as a low-value load resistance. When the logic value of the input signal A is less negative than the gate-source threshold voltage of -5 V, Q2 is off and is in the high impedance state and the output voltage will be negative. If the input signal is more negative than the threshold voltage of $Q2$, Q2 conducts and because of its low on-state impedance will cause the output voltage to be close to ground potential. The equivalent circuit for an NMOS is shown in Figure 6–24(c). The basic circuit operation is similar to that of the PMOS inverter, except that the polarity of V_{DD} is positive and the gate and drain of $Q1$ have been connected together. The advantage of using a MOSFET as the load resistance stems from the fact that it occupies less space than an IC resistor and reduces circuit complexity.

The application of NMOS to NOR and NAND gates is shown in Figure 6–25. In Figure 6–25(a) two NMOS transistors $Q1$ and $Q2$ are connected in parallel and share the common load transistor $Q3$. When either or both $Q1$ and $Q2$ are turned on, the output signal is low; if both input signals to $Q1$ and $Q2$ are less than the threshold voltage, $Q1$ and $Q2$ are in the high impedance state and the output voltage will be approximately equal to V_{DD}. This circuit is acting as a NOR gate with positive logic. In Figure 6–25(b) $Q1$ and $Q2$ are connected in series to ground. If neither one of the input signals A and B are low, the outpout will be high; however, if both A and B are high, the output will be low. This circuit acts as a NAND for positive logic.

Another form of MOSFET logic circuits is known as the complementary MOS or CMOS and combines both PMOS and NMOS enhancement mode transistors on the same chip. The major advantages of this configuration are that only one power supply is required and one transistor is always in the on state; this results in a significantly reduced power consumption (as little as 1 μW per gate in the quiescent state) and a significantly reduced propagation delay, typically 70 ns. As a result the CMOS configuration produces the best speed-power product of all MOS/LSI, with fan-outs as great as 100 and full TTL compatibility, as well as noise margins of approximately 40% of the supply voltage. The disadvantages are greater complexity in fabrication and reduced circuit density.

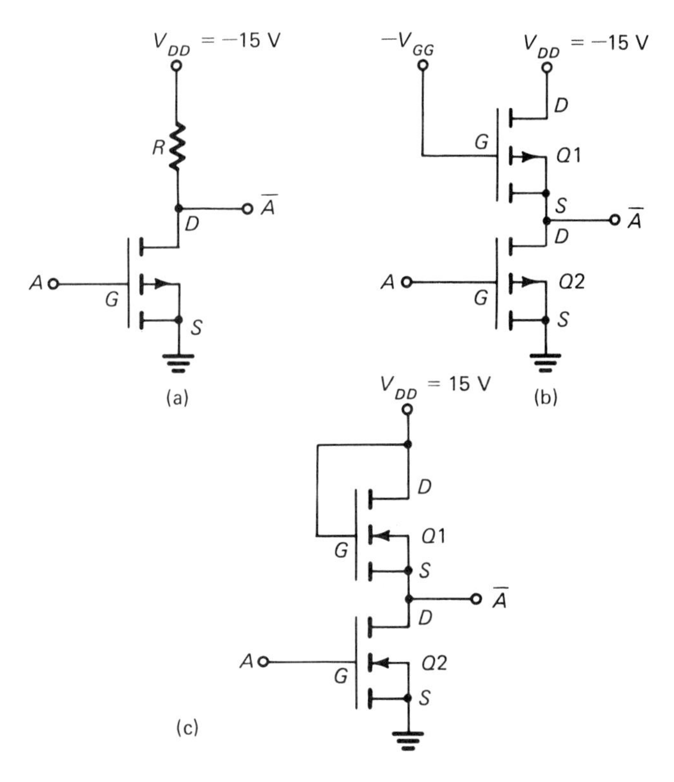

Figure 6-24. MOSFET inverters: (a) PMOS inverter with passive load resistance, (b) PMOS inverter with active load resistance, (c) NMOS inverter with active load resistance.

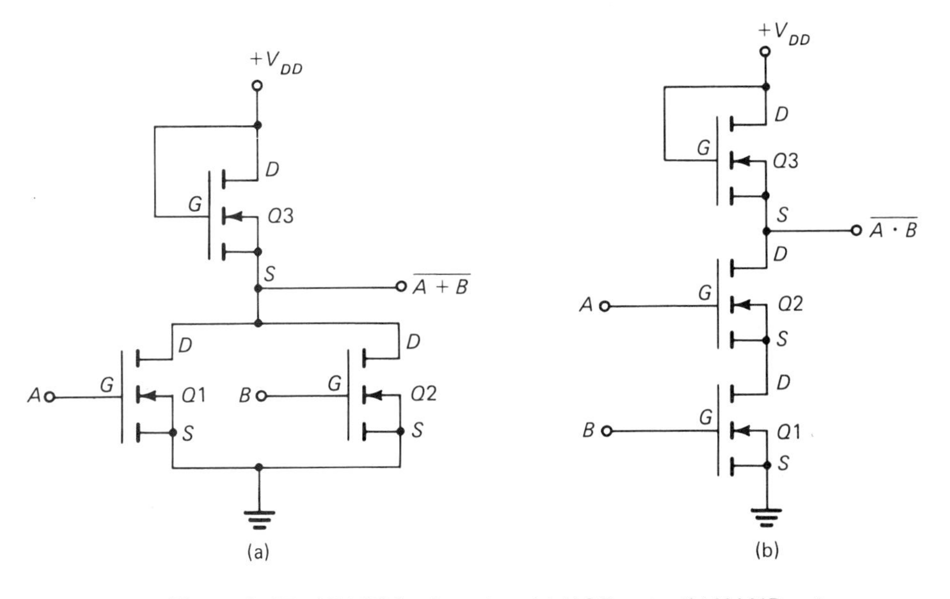

Figure 6-25. NMOS logic gates: (a) NOR gate, (b) NAND gate.

Figure 6–26(a) shows a CMOS inverter, where $Q1$ is a PMOS and $Q2$ is an NMOS device. The input signal A is applied simultaneously to the gates of $Q1$ and $Q2$. When the input signal is low, that is, at zero volts, the gate-source voltage is below the threshold voltage of $Q2$, and Q2 will be off. At the same time, however, the threshold voltage of $Q1$ is exceeded and will be turned on, thus applying V_{DD} to the output. When the input signal A is high, usually $+V_{DD}$, the threshold voltage of $Q2$ is exceeded and will turn on. At the same time the threshold voltage of $Q1$ is not exceeded and will be in the high impedance or off state. As a result the output voltage will be approximately zero, that is, the device is operating as inverter.

A typical 2-input NOR CMOS logic gate is illustrated in Figure 6–26(b). The gate consists of two PMOS transistors $Q1$ and $Q2$ in series with two NMOS transistors $Q3$ and $Q4$ in parallel. When either or both of the inputs A and B are high, that is $+V_{DD}$, the appropriate NMOS transistors $Q3$ and/or $Q4$ will be turned on and the output voltage level will be approximately zero or binary 0. At the same time the appropriate PMOS transistors $Q1$ and/or $Q2$ will be cut off. When both inputs A and B are low, $Q3$ and $Q4$ will not conduct since the gate-source voltage is below the threshold voltage; however, the gate-source voltages of $Q1$ and $Q2$ will be approximately equal to V_{DD}, and as a re-

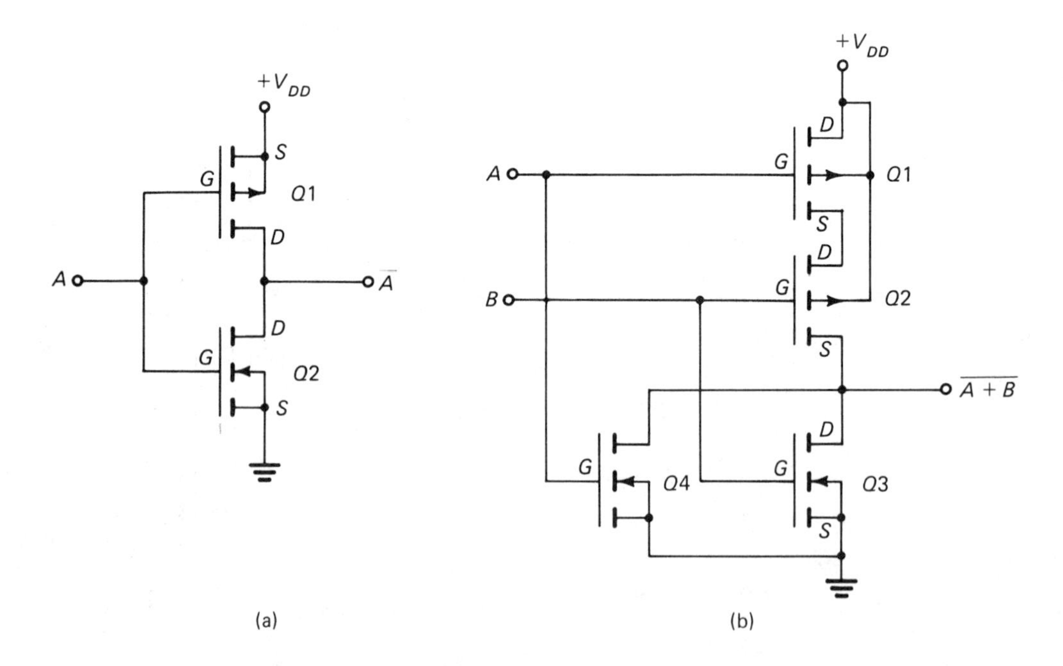

Figure 6–26. Complementary or CMOS: (a) logic inverter, (b) 2-input NOR.

sult $Q1$ and $Q2$ will conduct and the output will be approximately V_{DD} or binary 1.

To summarize, CMOS logic circuits possess characteristics that make them the most versatile and suitable of all the available types of logic. They have a low power consumption, typically 1 mW per logic gate; large fan-out capability, 50 or more; an acceptable operating speed; a high noise margin, usually about 0.45 V_{DD}; a wide range of power supply voltages, usually between +3 to +15 V; and are relatively easy to interface with other logic families such as TTL and DTL.

6-15 BISTABLE MULTIVIBRATORS (FLIP-FLOPS)

Bistable multivibrators or as they are more commonly called flip-flop (FFs) have two stable states. The FF will remain in one of the two stable states until an input has been applied. The removal of the input signal will not cause the FF to change state, that is, the FF is a one-bit memory storage. Flip-flops are commonly used in digital circuits where they are used to form memories, counters, shift registers, and frequency dividers. There are many different forms of FFs such as the reset-set (*R-S*), the clocked *R-S*, the clocked *J-K*, *J-K* master-slave, *D* types, and *T* types. In turn the characteristics of the FF also depend upon the type of logic family, for example, TTL or CMOS. Nearly all FFs are clocked, that is, the change of state of the device occurs at a definite rate.

6-15-1 The R-S Flip-Flop

The *R-S* FF is the simplest of the FFs and may consist of two NAND gates as shown in Figure 6–27. Figure 6–27(a) shows the NAND configuration and the associated truth table. The *S* or set input sets the FF, that is, the *Q* output is high and the FF is storing a binary 1. The *R* or reset input resets the FF and stores a binary 0. The operation of the *R-S* FF can be understood from the truth table of Figure 6–27(b) as follows:

Line 1 $S = 1, R = 0; R = 0$, then $\overline{Q} = 1$; $\overline{Q} = 1$ and $S = 1$, then $Q = 0$; $Q = 0$ and $R = 0$, then $\overline{Q} = 1$. This means that $R = 0$ is the reset condition that causes the FF state to be $Q = 0$ and $\overline{Q} = 1$.

Line 2 $S = 0$, $R = 1$; $S = 0$, then $Q = 1$; $Q = 1$ and $R = 1$, then $\overline{Q} = 0$. This means that $S = 0$ is the set condition that causes the FF state to be $Q = 1$ and $Q = 0$.

Line 3 $S = 1, R = 1; S = 1$, then $Q = 0$; $Q = 0$ and $R = 1$, then $\overline{Q} = 1$; $\overline{Q} = 1$ and $S = 1$ then $Q = 0$. This means that the outputs of the FF are unchanged.

Line 4 $S = 0, R = 0; S = 0$, then $Q = 1$; $Q = 1$ and $R = 0$, then $\overline{Q} = 1$. This condition is not allowed.

These conditions are best illustrated by a timing diagram as shown in Figure 6–27(c). An *R-S* FF may also be constructed from two interconnected NOR gates.

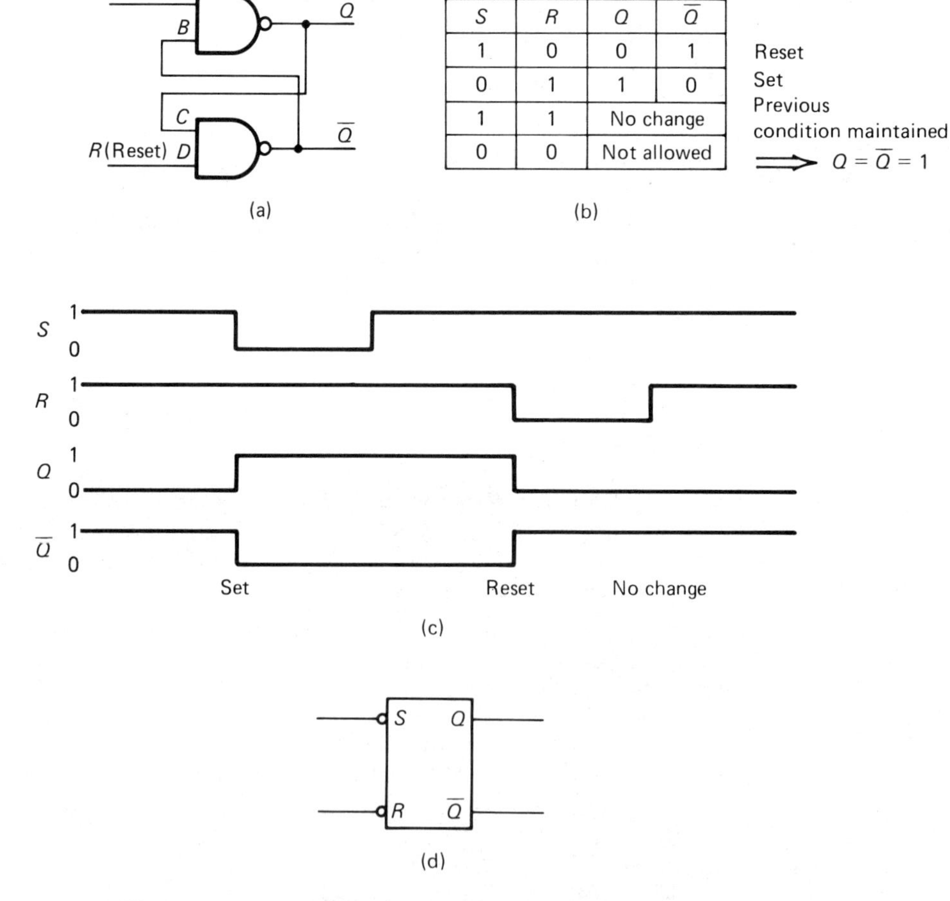

S	R	Q	$\overline{Q}$
1	0	0	1
0	1	1	0
1	1	No change	
0	0	Not allowed	

Figure 6–27. Basic *R-S* flip-flop: (a) NAND gate FF, (b) truth table, (c) timing diagram, (d) logic symbol.

6-15-2 Clock Pulses

Most sequential logic circuits depend upon timing or clock pulses to achieve the desired cycling or sequencing of the logic system. In our study of FFs clock pulses will be used to transfer data from the input to the output of an FF. Clock pulses are square voltage pulses that may be generated by an astable multivibrator, a unijunction transistor, or a crystal oscillator if higher frequencies are desired. The most commonly used clock pulse generator is the astable multivibrator which will produce accurate clock pulses over a frequency range of 1000 Hz to 1 MHz.

6-15-2-1 The Astable Multivibrator

The astable or free-running multivibrator is a modification of an FF. It has two unstable states and switches from one state to the other at a rate determined by the time constants of the RC networks. The astable multivibrator is shown in Figure 6–28. The basic circuit consists of two transistor inverters $Q1$ and $Q2$ with their outputs ac coupled by means of capacitors $C1$ and $C2$ to the base of the other. Resistors $RB1$ and $RB2$ are used to bias the transistors into saturation.

Assume that $Q1$ is in saturation and conducting and $Q2$ is cut off. Capacitor $C2$ will charge up to approximately V_{CC} with its right-hand plate negative, thus applying a negative potential to the base of $Q2$, and as a result $Q2$ is cut off. Since $Q1$ is saturated its collector-emitter voltage is approximately zero and the output $\overline{C_p} = 0$. At the same time the collector-emitter voltage of $Q2$ is approximately V_{CC} and the output $C_p = V_{CC}$. At this point capacitor $C2$ has its left-hand plate at approximately zero volts, and the right-hand plate will charge toward V_{CC} through resistor $RB2$, the rate of charge being determined by the time constant $RB2 \bullet C2$. As the right-hand plate of $C2$ becomes positive the base of $Q2$ becomes positive and $Q2$ is turned on. The collector-emitter voltage is (then decreasing in a negative direction to) approximately zero, and the drop in the collector voltage of $Q2$ is coupled to the base of $Q1$ via $C1$.

As a result, $Q1$ is cut off and its collector voltage rises toward V_{CC}, and in turn this positive-going voltage is coupled to the base of $Q2$ via $C1$ to cause $Q2$ to snap into saturation. At this point $C2$ will charge through the emitter-base junction of $Q2$ and $RC1$ to $+V_{CC}$. Capacitor $C1$ discharges through $RB1$ and maintains $Q1$ in the cut-off state. When the voltage across $C1$ becomes zero it starts to recharge with its polarity reversed, and when the voltage is approximately equal to 0.7 V, $Q1$ is turned on. As it turns on it places the left-hand side of C_2 at approximately zero and applies a negative voltage to the base of $Q2$

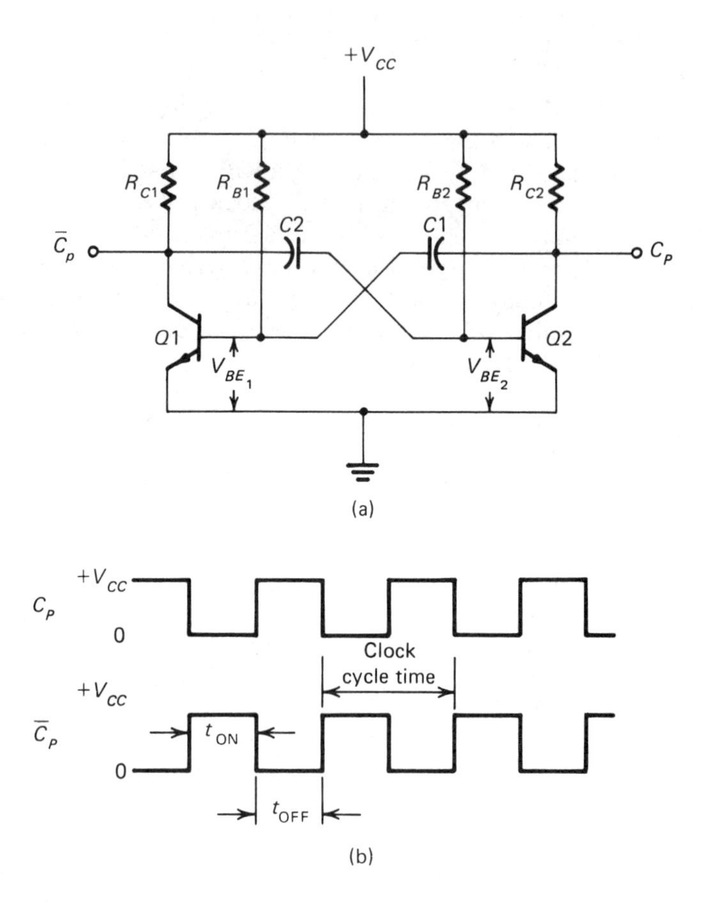

Figure 6-28. A transistor astable multivibrator: (a) basic schematic, (b) associated clock pulse waveforms.

causing it to turn off. $C2$ then discharges through $RB2$. Simultaneously $C1$ recharges through the emitter-base junction of $Q1$ and $RC2$. When the potential on $C2$ reaches zero it recharges to V_{CC}, and the cycle begins again. The frequency of oscillation (f) is determined by the values of $RB1$, $RB2$, $C1$, and $C2$. For $RB1 = RB2 = RB$, and $C1 = C2 = \mathrm{C}$; then

$$\mathrm{f} = 1/1.4\,RB.C. \tag{6-4}$$

If $C1 = C2$ a 50% duty cycle results. The duty cycle, that is, the ratio of pulse-on to pulse-off time, may be varied by ensuring that $C1 \neq C2$. The frequency of oscillation is also called the pulse repetition rate (PRR) and if $RB1 \neq RB2$ and $C1 \neq C2$, then

$$f = \frac{1}{0.7(RB1C1 + RB2C2)} \qquad (6\text{-}5)$$

Example 6-23

In Figure 6–28 $RB1 = 15\ \text{k}\Omega$, $RB2 = 15\ \text{k}\Omega$: (1) find the PRR when $C1 = C2 = 0.01\ \mu\text{F}$ and (2) the PRR when $C1 = 0.01\ \mu\text{F}$ and $C2 = 0.1\ \mu\text{F}$.

Solution:

(1) From Equation (6-4)

$$f = \frac{1}{1.4 \times (15 \times 10^{-3} \times 0.01 \times 10^{-6})}$$
$$= 4762 \text{ pulses per second.}$$

(2) From Equation (6-5)

$$f = \frac{1}{0.7(15 \times 10^{3} \times 0.01 \times 10^{-6} + 15 \times 10^{3} \times 0.1 \times 10^{-6})}$$
$$= 865.8 \text{ pulses per second}$$

While the astable multivibrator will produce a good-quality square wave the wave does not have to be symmetrical, that is, $t_{ON} = t_{OFF}$ [see Figure 6–28(b)]. The major requirement is that, where $T = t_{ON} + t_{OFF}$, the clock-cycle time which is equal to the periodic time T, be consistently the same. The clock-cycle time is extremely important since the action of all logic gates, flip-flops, sequential logic, etc., must be completed during one clock cycle.

Example 6-24

(1) What is the clock-cycle time of a 1 MHz clock? (2) A device has a propagation delay time of 200 ns. What is the maximum permissible clock frequency?

Solution:

(1) The clock-cycle time is equal to the periodic time of the 1 MHz clock. Therefore,

$$T = \frac{1}{f} = \frac{1}{1\text{ MHz}} = 1 \times 10^{-6}\text{ sec}$$
$$= 1\,\mu\text{s}$$

(2) The maximum permissible clock frequency for a device with a 200 ns propagation delay time is

$$f = \frac{1}{T} = \frac{1}{200 \times 10^{-9}} = 5\ \text{MHz}$$

6-15-2-2 Integrated Circuit Astable Multivibrator

An astable multivibrator can also be constructed using TTL inverters as shown in Figure 6–29. The principle of operation is as follows. When pin 4 of the output of inverter *M* is low, capacitor *C* couples this low to the input pin 1 of inverter *L*. As a result the output, pin 2 of the inverter goes high and the output of inverter *M* is low. At the same time the capacitor charges through *R* from the high at the output of inverter *L*. When the capacitor is charged to approximately 1.5 V, pin 2 of inverter *L* goes low and the output, pin 4 of inverter *M* goes high. This high output is applied to *C* and insures that the input of inverter *L* is maintained high and, as a result, its output pin 2 is low. The capacitor *C* will now discharge through *R*, the input to inverter *L* will go low, and its output pin 2 will be high. This cycle of operations will be continually repeated. The pulse repetition rate will be

$$f = \frac{1}{3RC}, \qquad (6\text{-}6)$$

and the clock-cycle time *T* will be

$$T = 3RC \qquad (6\text{-}7)$$

The function of inverter *N* is to provide a good square wave output as well as isolating the multivibrator from the remainder of the circuit.

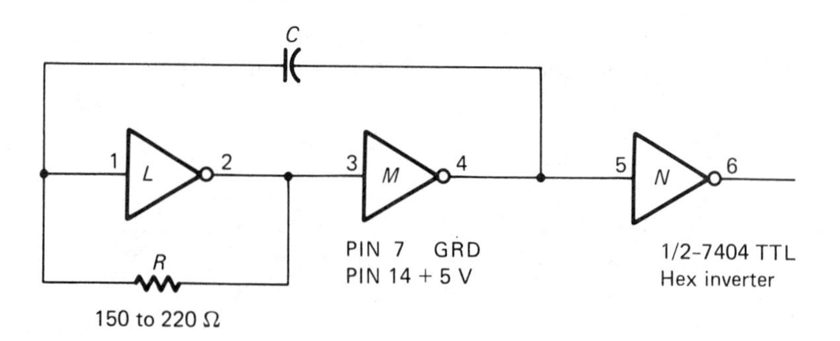

Figure 6–29. Integrated circuit multivibrator.

6-15-3 The Clocked R-S Flip-Flop

The clocked *R-S* FF is basically the *R-S* FF with the addition of two more gates which permit the use of a clock-pulse input to determine when the changes in the complementary outputs will occur. The NAND gate form of the clocked *R-S* FF is shown in Figure 6–30 and is used to form the basic memory unit in a semiconductor memory. The block diagram of the clocked NAND *R-S* FF is shown in Figure 6–30(a). It consists of four 2-input NAND gates. NAND gates $G1$ and $G2$ are steering gates, the set (S) and reset (R) inputs determine the output state of the FF, and the clock pulse (C_p) determines when the change of output will occur. As long as the clock pulse is low there will be no change in the output. The *R-S* FF will only change the output state when C_p is high. Effectively, the NAND gates $G3$ and $G4$ forming the *R-S* FF act as a memory and will hold a 1-bit word either 0 or 1 at the Q output as long as C_p remains low. The operation is best understood by referring to the timing diagram [Figure 6–30(c)]. Sometimes the clocked *R-S* FF is referred to as a latch. Also, since there are two input data lines S and R, the data on the two data lines is referred to as double-rail data. The operation of the clocked *R-S* FF described is referred to as **synchronous operation**. Referring to Figure 6–30(b) it should be noted that there are two extra inputs, direct set (S_D) and direct clear (C_D), and depending upon the manufacturer one or both of these inputs may be available. The function of the direct-clear and direct-set inputs is to permit the FF to be cleared or set without a clock pulse. When operated in this manner the FF is said to be operated asynchronously. Care should be exercised to ensure that S and R are not high together since this represents an ambiguous output condition, that is, both Q and $\bar{Q}$ are high.

6-15-4 D-Type FFs

To overcome the ambiguous condition in the clocked *R-S* FF, that is, $R = 1$ and $S = 1$, an inverter is connected between the R and S input lines so that the S and R inputs can never be the same at the same time. As a result only one data line is required and it is referred to as single-rail data and is called the D input. This arrangement is shown in block diagram form in Figure 6–31(a), the arrangement being known as the D or data FF. The timing diagram, Figure 6–31(c), shows that the data on the D input is transferred to the Q output by the leading edge of the next clock pulse. In other words there is a delay in the change of state of Q which gives rise to the alternative term of the delay FF or latch. As can be seen from the truth table, the no change and indeterminate conditions of the clocked *R-S* FF have been eliminated, which results from the use of inverter $G1$. An example of the use of a D-type FF would be the storage of a BCD digit. Since a BCD digit is represented by four bits, a D-type FF would be needed for each bit. This could be accomplished using a quad IC such as a TTL SN54/7475 IC.

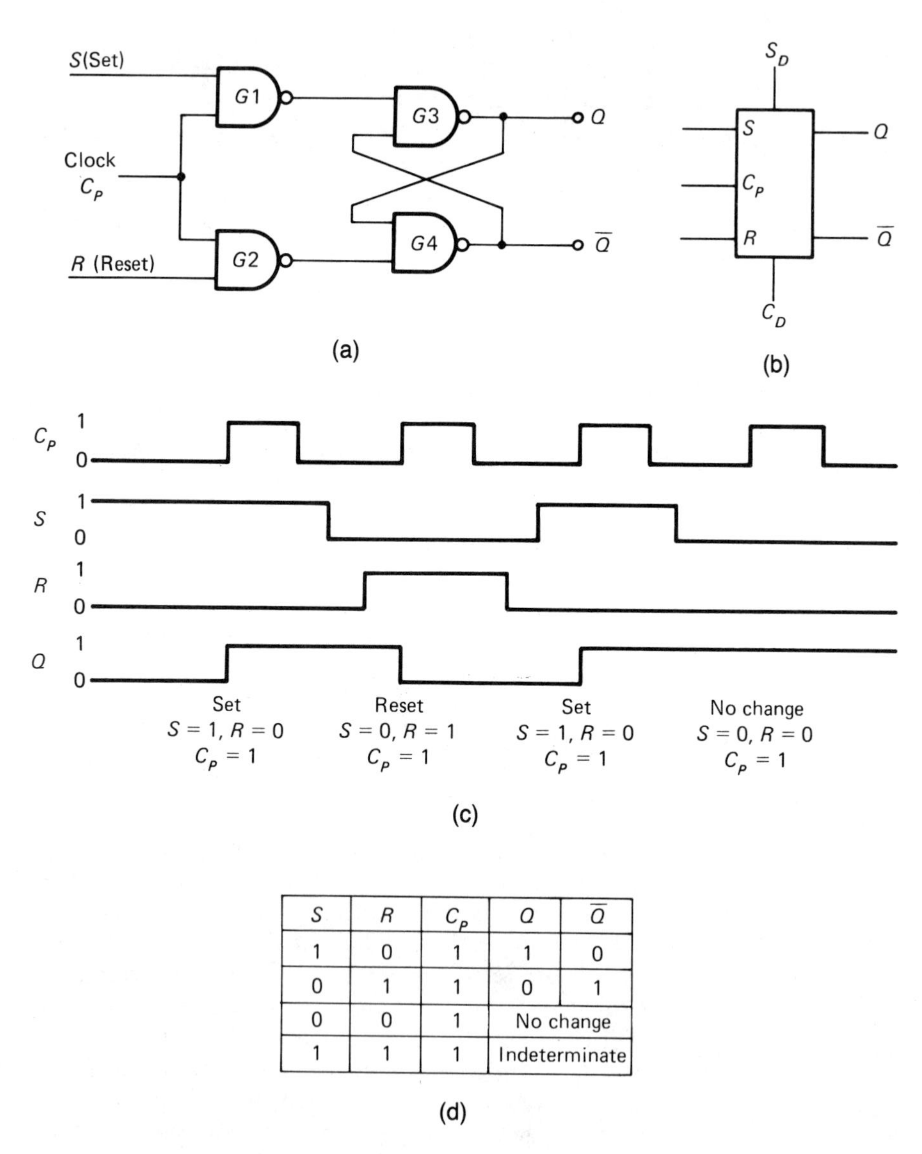

S	R	C_P	Q	$\overline{Q}$
1	0	1	1	0
0	1	1	0	1
0	0	1	No change	
1	1	1	Indeterminate	

(d)

Figure 6-30. Clocked NAND R-S flip-flop: (a) block diagram, (b) logic symbol, (c) timing diagram, (d) truth table.

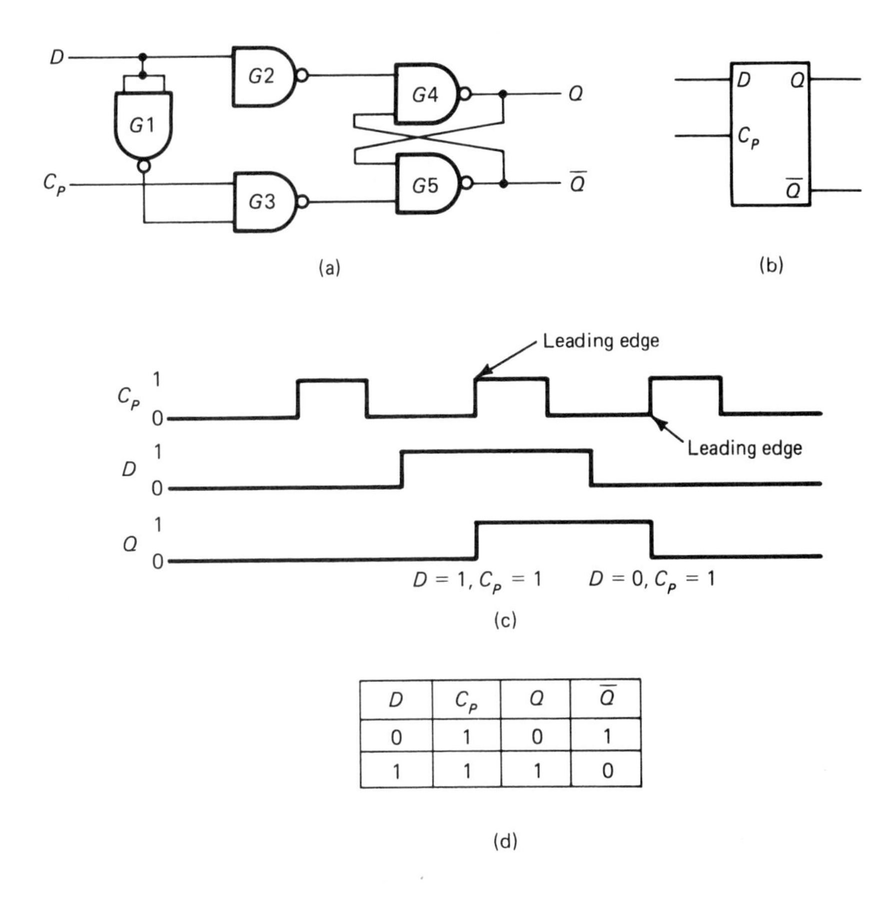

D	C_P	Q	$\overline{Q}$
0	1	0	1
1	1	1	0

Figure 6-31. The D or data type flip-flop: (a) block diagram, (b) logic symbol, (c) timing diagram, (d) truth table.

6-15-5 The Master-Slave R-S FF

One of the major problems associated with the R-S FF is the fact that there is insufficient isolation between the inputs and outputs. In the case of the clocked R-S FF a race problem develops when several FFs are connected in cascade to form a shift register. The data will race through the various stages when they are simultaneously clocked with a positive clock pulse, instead of advancing one stage at a time for each clock pulse. The first solution to this problem that comes to mind is to use a short-duration pulse which would permit data to move through the register one stage at a time. However, while this is a solution to the problem, the width of the clock pulse is very critical and becomes unstable with temperature variations.

These problems associated with the R-S and clocked R-S FFs may be overcome by using the master-slave R-S FF. It consists basically of two R-S FFs in cascade with clock pulses arranged to be complementary. This is accomplished by using an inverter. See Figure 6–32(a). When the clock pulse is low, the clock signal to the master R-S FF is high and data will be accepted at the R and S inputs. At the same time the clock signal to the slave R-S FF is low and data will not be accepted. When the clock pulse goes high the data from the master R-S FF is transferred to the slave R-S FF. A number of master-slave R-S FFs may be cascaded and the race problem is eliminated; that is, data is shifted one stage at a time for each clock pulse.

From the timing diagram, Figure 6–32(c), it can be seen that the Q output is responding to the trailing edge of the clock pulse C_p. The reason is that when C_p goes high the master R-S FF is enabled, and the complement of C_p is applied to the slave R-S FF which disables it. However, because of the inverter, as the clock pulse to the master goes low the clock pulse to the slave goes high; that is, the Q output of the slave is responding to the trailing edge of the clock pulse. Comparing the truth table of Figure 6–32(d) to that of Figure 6–30(d), it can be seen that they are identical, that is, the master-slave R-S FF and the clocked R-S FF are identical in their outputs. The major gain is that the race problem has been solved, and data will only move one stage per clock pulse.

6-15-6 The T or Toggle FF

The T or toggle FF can be very easily produced from the master-slave R-S FF by respectively cross-connecting the outputs Q and $\overline{Q}$ of the slave to the R and S inputs of the master. See Figure 6-33(a). As a result, when the outputs of the slave R-S FF are $Q = 1$ and $\overline{Q} = 0$, the inputs of the master R-S FF are $S = 0$ and $R = 1$. When the clock pulse goes high, the slave R-S FF outputs will change to $Q = 0$ and $\overline{Q} = 1$ on the trailing edge of C_p. As can be seen from the timing diagram, the outputs Q and $\overline{Q}$ change state for each clock pulse and are said to toggle. It should be noted that the output frequency of the toggle or T FF is half that of the clock frequency, which immediately suggests that the T FF may be used to divide by 2. Because of the ease with which a T can be constructed from the master-slave R-S FF, or as will be seen later from the D or J-K FFs, the T FF is not normally available as a TTL IC.

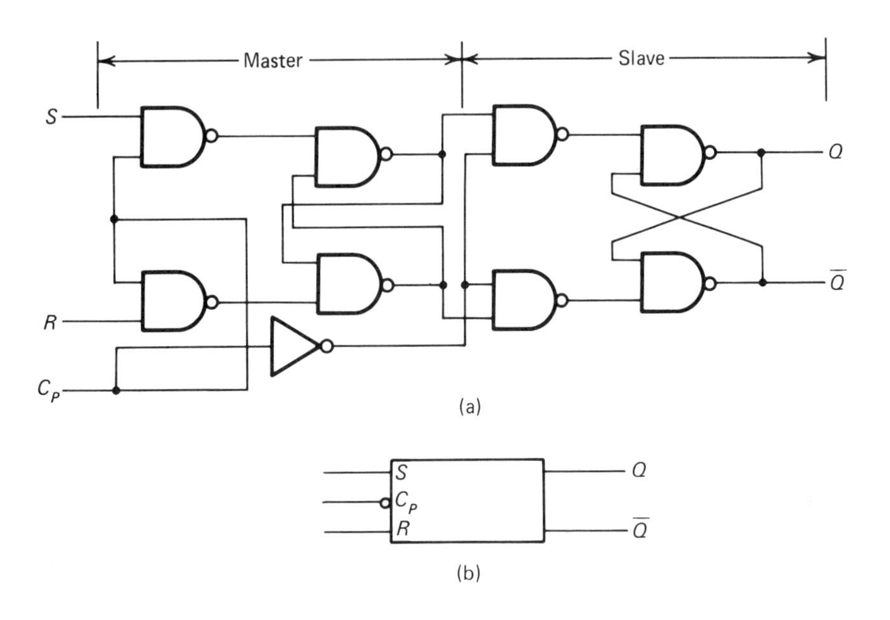

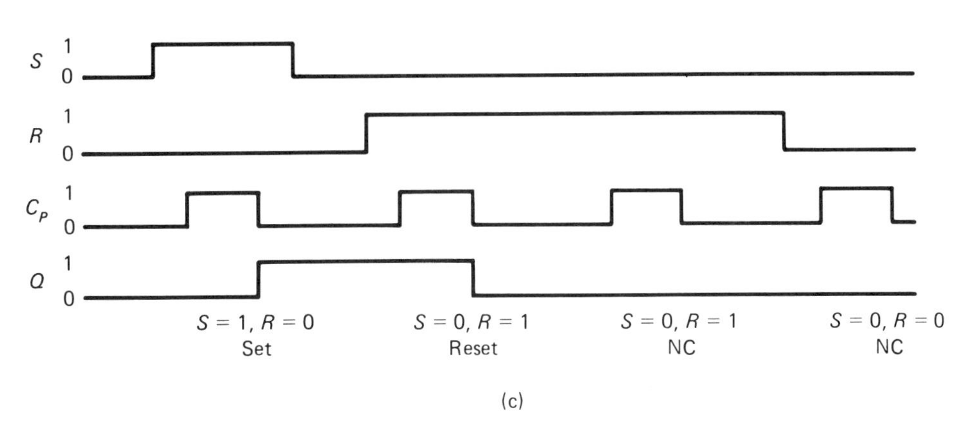

(c)

S	R	C_P	Q	$\overline{Q}$
1	0	1	1	0
0	1	1	0	1
0	0	1	No change	
1	1	1	Indeterminate	

(d)

Figure 6-32. Master-slave *R-S* flip-flop: (a) block diagram, (b) logic symbol, (c) timing diagram, (d) truth table.

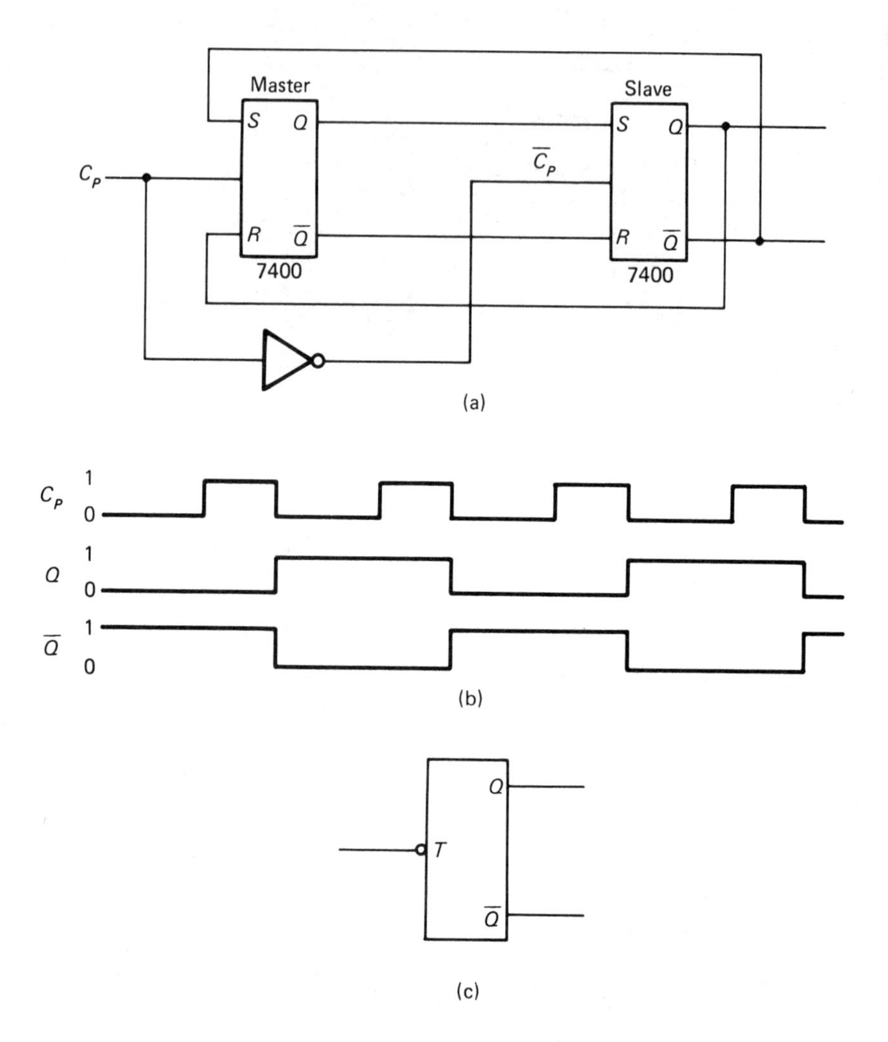

Figure 6-33. The toggle or *T* flip-flop: (a) *R-S* flip-flop connections, (b) timing diagram, (c) logic symbol.

6-15-7 The J-K Master-Slave FF

When the *R-S* FF and *T* FF are combined, the result is called the *J-K* master-slave FF (see Figure 6-34). The major advantage gained by this configuration is that the indeterminate condition of the *R-S* FF has been eliminated and the FF toggles when the inputs *J* and *K* are high or equal to 1, as can be seen from the truth table in Figure 6-34(c). It should be noted that the truth table has been modified slightly so that

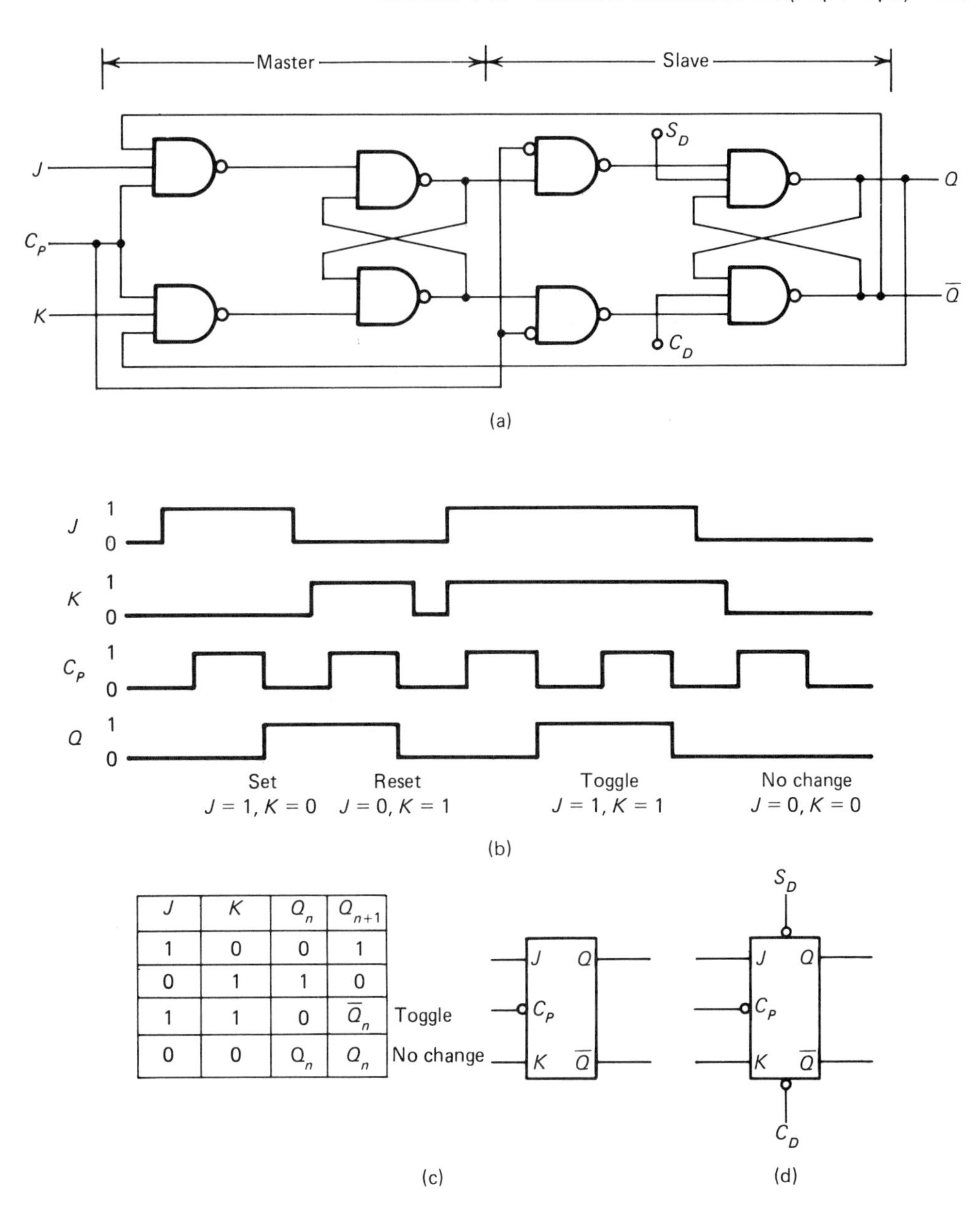

J	K	Q_n	Q_{n+1}	
1	0	0	1	
0	1	1	0	
1	1	0	$\overline{Q}_n$	Toggle
0	0	Q_n	Q_n	No change

Figure 6-34. The *J-K* master-slave flip-flop: (a) block diagram, (b) timing diagram, (c) truth table, (d) logic symbol, (e) logic symbol of *J-K* master-slave flip-flop with direct set and direct clear.

Q_n represents the state of the output prior to the clock pulse and Q_{n+1} represents the output after the clock pulse. It should be noted that in the logic symbol [Figure 6-34(d)] the circle present at the clock pulse input indicates that the FF outputs will change on the trailing edge of the clock pulse. If the circle is absent the outputs will change on the leading edge of the clock pulse.

The *J-K* FF still requires further modification to be of practical use, namely, the provision of direct inputs in order that the FF may be set initially in the desired state. These direct inputs are direct set (S_D) and direct reset (C_D) and are connected into the slave NAND gates as shown in Figure 6-34(a).

The direct inputs modify the operation of the *J-K* master-slave FF as follows:

1. If $S_D = C_D = 1$, the performance of the FF remains as shown in the truth table.

2. If $S_D = 0$ and $C_D = 1$, the Q output is high, that is $Q = 1$. The complementary output $\overline{Q}$ remains unchanged.

3. If $C_D = 0$ and $S_D = 1$, $Q = 0$ and $\overline{Q} = 1$.

4. If $S_D = C_D = 0$, both Q and $\overline{Q}$ are high and the outputs are no longer complementary. This condition should not be allowed to happen.

As can be seen the FF may be set to the desired initial state by the use of the appropriate direct input. In the normal operating state both direct inputs are maintained high.

The most commonly used *J-K* master-slave FFs are the 7473, 7476, and the 74107; however, the 7473 and 74107 do not have the direct-set capability. All of these *J-K* master-slave FFs can be operated with clock frequencies ranging from very low to a maximum of about 20 MHz. The clock pulse widths should be a minimum of about 30 ns wide since the time required to move data is about 25 ns. The rise and fall times of the clock pulse should be 5 μs or less to minimize noise.

The 7473, 7476, and 74107 are level clocked, which means that the *J* and *K* inputs can only be changed once immediately after clocking. As a result, their principle applications are restricted to situations where the inputs remain constant or only change after clocking. Typical applications are binary counters and dividers.

It should be noted that the *J-K* master-slave FF is only one form of the *J-K* FF. Another type that is more commonly used is the edge-triggered *J-K* FF which has the advantage that the *J* and *K* inputs can be changed at any time but the FF outputs will not change until the clock pulse changes from a 0 to 1 positive edge or from a 1 to 0 negative edge clocking.

6-16 MONOSTABLE MULTIVIBRATORS

The multivibrators that have been considered so far have two stable output states. The monostable multivibrator has one stable and one unstable state. The monostable remains in the stable state until an input pulse is applied, driving the output into the unstable state. It will remain in the unstable state for a period of time determined by the charging time of an externally connected RC combination. The monostable is used to delay, increase, or decrease pulse widths and shape pulses.

6-16-1 Integrated Circuit Monostable

It is possible to build monostables from gates such as the 7400 or 7402 combined with RC networks; however, circuits are not normally built up from discrete components. Instead specially designed monostables such as the 74121, 74122, or 74123 are used almost exclusively.

The 74121 monostable, for example, has a dual negative transition-triggered input (A_1 and A_2), single positive transition-triggered input (B), and complementary outputs Q and $\bar{Q}$ [see Figure 6-35(a)]. It should be noted that input B is a Schmitt trigger which provides a hysteresis effect to accommodate slowly rising or noisy input signals. The A_1 and A_2 inputs are ORed, and the output is ANDed with the B input. The output pulse length of the 74121 can be varied from 40 ns to 28 sec by suitable choices of R and C, which can be determined from the manufacturer's data sheets. As can be seen from the function table, Figure

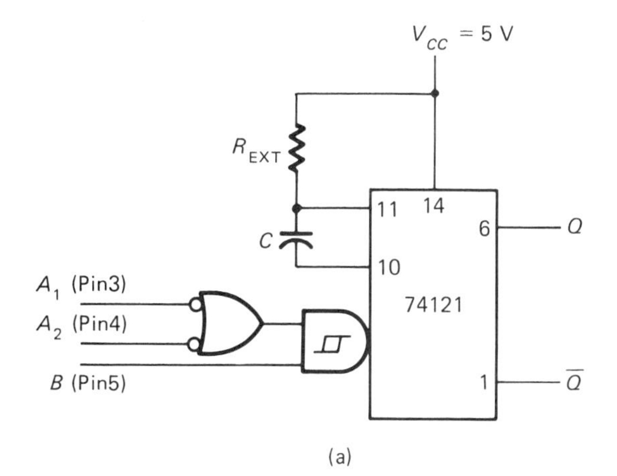

Function Table				
Inputs			Outputs	
A_1	A_2	B	Q	$\bar{Q}$
L	X	↑	⎍	⊔
X	L	↑	⎍	⊔
↓	H	H	⎍	⊔
H	↓	H	⎍	⊔
L	X	↑	⎍	⊔
X	L	↑	⎍	⊔

(b)

Figure 6-35. 74121 monostable multivibrator: (a) block diagram, (b) function table.

6-35(b), the monostable will be triggered by a negative-going pulse transition, or either of the A inputs with the other A input and the B input high, or by a positive-going transition at the B input. Once the 74121 has been triggered the output pulse width is solely determined by the externally connected RC combination and is completely unresponsive to transitions of the A and B inputs. An internal resistance together with an externally connected capacitor will give an output pulse with a pulse width between 30 and 35 ns.

The 74122 and 74123 monostable multivibrators are retriggerable, that is, a new timing cycle will be initiated if there are suitable input signals during the output pulse. The output pulse can be extended as long as desired, providing the monostable is triggered during the duration of the output pulse. However, if the input trigger signals are removed, the output pulse will be terminated one pulse width after the last input signal transition. In addition, the 74122 and 74123 monostables have an overriding clear capability which permits the output pulse to be immediately terminated. The major applications of retriggerable or negative recovery monostables is the generation of very long pulses or as a missing pulse detector.

6-16-2 The 555 Monostable Multivibrator

The 555 timer was specifically designed for applications requiring precise timing, but has been adapted for a number of applications including the monostable multivibrator. Normally, the trigger input (pin 2) is held at +5 V, the monostable action is initiated by applying a carefully shaped negative transition pulse to pin 2. The trigger pulse switches from an OFF state greater than 2/3 V_{CC} to an ON state at less than 1/3 V_{CC}. The actual triggering occurs when the voltage at pin 2 drops through 1/3 V_{CC}. The input pulse width must be greater than 100 ns and less than the desired output pulse width. The width of the positive pulse at pin 3 is determined by the external resistance and capacitance connected, as shown in Figure 6-36. Initially C is short-circuited by an internal clamping transistor; however, when the output (pin 3) goes positive after the 555 is triggered, the short is removed from C and it commences to charge toward V_{CC} via R. When C has charged to 2/3 V_{CC} the internal threshold comparator (pin 6) senses this voltage, changes state, and switches the internal R-S FF to its original state, which in turn switches the internal clamping transistor ON and shorts the external capacitor to ground. Concurrently the output at pin 3 goes low. The 555 monostable multivibrator is not retriggerable, but the output pulse may be terminated at any time by applying a negative-going pulse to the reset (pin 4); however, when not in use the reset input should remain at V_{CC}.

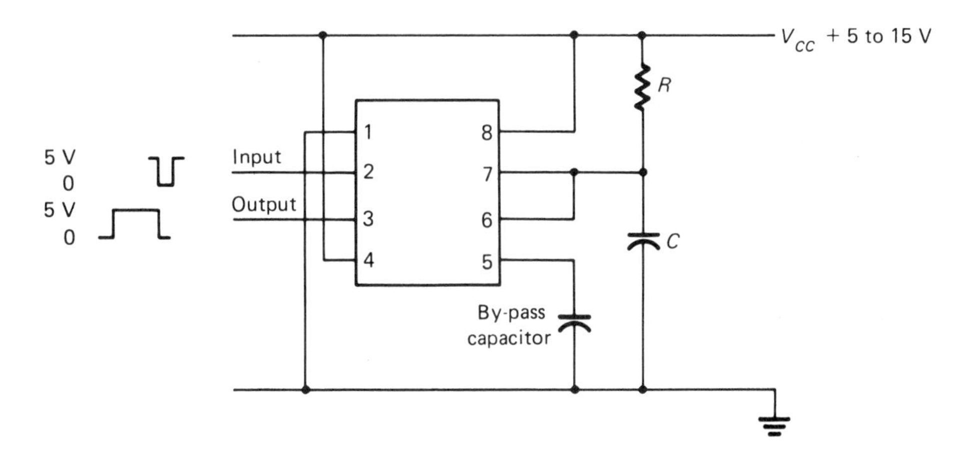

Figure 6-36. The 555 monostable multivibrator.

The pulse width of the output pulse can be calculated from

$$t = 1.1\,RC$$

where t = milliseconds (ms)

R = external resistance (typical range, 10k to 10M) and

C = external capacitor (typical range, 100 pF to 1000 μF)

It should be noted that C must be a low leakage capacitor. Normally the duty cycle should not exceed 80% with large values of R or 50% with small values of R. It is also good practice to ensure that the input pulse width does not exceed 25% of the output pulse width. Normally pin 5, the control voltage input, is bypassed to ground by the bypass capacitor to reduce the possibility of instability in the output pulse.

6-16-3 The Schmitt Trigger

One of the benefits of the 555 monostable multivibrator is the fast rise and fall times, typically 100 ns, of the output waveform. Another circuit which provides fast rise and fall times is the Schmitt trigger, primarily used to generate square wave outputs from slowly changing input signals. TTL gates, one shots, and FF require input signal waveforms with fast rise and fall times. The Schmitt trigger is a regenerative bistable multivibrator with one input and one output and is used to convert slowly changing input signal waveforms to a square wave output with fast rise and fall times.

The principle of operation is best considered by studying a discrete component form of the Schmitt trigger circuit. See Figure 6-37(a). The circuit consists of an emitter-coupled amplifier with a common emitter resistance R_E and a regenerative feedback network $R1$ and $R2$ connected between the transistors. Under operating conditions only one of the transistors will be in conduction at any one time. Assuming that the potential of the input signal V_{in} applied to the base of $Q1$ is less than the potential of the common emitter V_E, the base-emitter junction of $Q1$ is reverse biased and $Q1$ is cut off. The collector potential V_{C1} will be at a maximum. The base voltage V_{B2} for $Q2$ is supplied from V_{CC} via the voltage divider $RL1$, $R1$, and $R2$, and as a result $Q2$ will be driven into saturation. The resulting collector current I_{C2} will produce a voltage drop across $RL2$, so that the output voltage is $V_{\text{out}} = (V_{CC} - I_{C2}RL2)$.

As the potential of the input signal to the base of $Q1$ increases, $Q1$ will turn on as the base-emitter junction becomes forward biased, $V_E = V_{\text{in}} - V_{BE}$. The resulting collector current I_{C1} produces a voltage drop across $RL1$, and as a result the base voltage V_{B2} of $Q2$ becomes less than V_E, $Q2$ is cut off, and $I_{C2} = 0$. The output voltage V_{out} is approximately equal to V_{CC} since there is no voltage drop across $RL2$.

The lowest value of the input voltage $V_{\text{in}} = V_E + V_{BE}$ that causes $Q1$ to turn on is called the upper trigger point (UTP). As $Q1$ starts to turn on the flow of collector current I_{C1} causes V_{C1} to decrease and, as a result, V_{B2} decreases. As $Q1$ turns on the base-emitter voltage of $Q2$, V_{BE2} decreases rapidly, and $Q2$ is rapidly cut off as $Q1$ saturates. This process is known as regeneration.

The emitter current of $Q1$, I_{E1}, is $I_{E1} = (V_{\text{in}} - V_{BE1})/R_E$. From this relationship it can be seen that a reduction of V_{in} causes a corresponding decrease of I_{E1} and since $I_{C1} \simeq I_{E1}$, I_{C1} will also decrease. Since $V_{C1} = (V_{CC} - I_{C1}RL1)$, as I_{C1} decreases, V_{C1} increases and V_{B2} applied to $Q2$ increases. Transistor $Q2$ will turn on again when $V_{\text{in}} = V_{B2}$. The input voltage at which this occurs is known as the lower trigger point (LTP). As $Q2$ turns on, the common emitter potential V_E rises and assists the turn-off of $Q1$, which in turn causes $Q2$ to turn on rapidly because of regeneration. Capacitor C, which is initially discharged before V_{in} is applied, will charge up to $V_{\text{in}} - V_{BE1}$ as V_{in} increases. The capacitor-charging current is applied to the base of $Q2$ and assists in the rapid turn-on of $Q2$; as a result the rise time of V_{out} is minimized. When $Q2$ is turning off the capacitor discharges and a reverse base current flows, causing Q2 to snap off, which in turn causes the V_{out} pulse to have a rapid fall time. Capacitor C is known as the speed-up capacitor.

The difference in potential between V_{in} at UTP and LTP is known as the hysteresis voltage V_H; that is, $V_H = \text{UTP} - \text{LTP}$ volts. The value of V_H can be varied by making $RL1 \neq RL2$ or by changing the ratio of

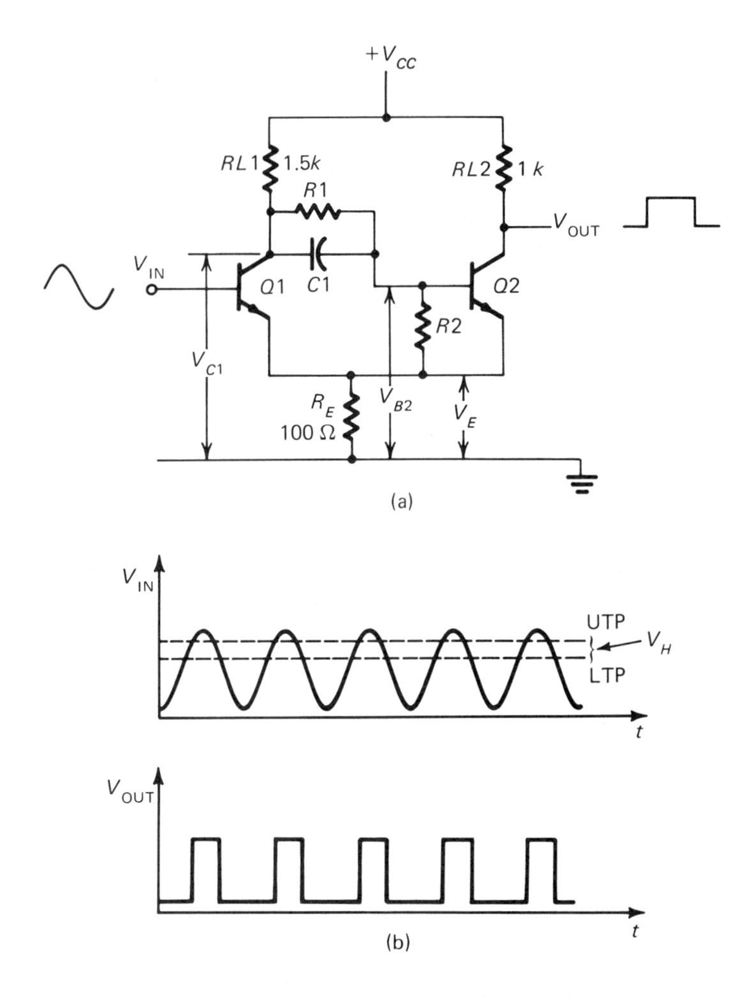

Figure 6-37. Schmitt trigger: (a) schematic, (b) trigger waveforms.

$R1$ to $R2$. Hysteresis is desirable since, if the input signal is derived from a high resistance source when Q1 turns on, the flow of base current into $Q1$ could decrease V_{in} to a potential below UTP and $Q1$ would turn off. Hysteresis prevents this from happening.

The Schmitt trigger acts as a level detection circuit and turns on at UTP and off at LTP producing a fast-rise, fast-fall output pulse, and is independent of the frequency of the input signal.

6-17 SEQUENTIAL LOGIC CIRCUITS

Sequential logic circuits are used to time, sequence, and store data. The most commonly used components in sequential logic circuits are

FFs, in which binary data are stored or changed by logic input signals, the sequence of operations controlled by clock signals.

The two most common applications of sequential logic are counters and shift registers. These devices are readily available from most manufacturers in TTL, MSI and CMOS integrated circuits.

6-17-1 Counters

In the industrial scene there are many applications in computers, instrumentation, and digital control circuits where there is a need to count time, count the number of events in a given interval of time, count pulses, etc. Most counters are constructed from *J-K*, toggle, or *D*-type FFs. There are a number of different types of counters, the two main types being the binary and 8421 BCD counters.

It should be recalled that flip-flops have two possible states, 1 (set) and 0 (reset). The maximum possible number of combinations of 1 and 0 states that can be obtained by a counter is called the modulus. The maximum modulus of a counter formed from N FFs is 2^N; however, in actual practice we must allow for all FFs to be reset so the actual count would be $2^N - 1$. For example, a counter consisting of four FFs will have a maximum count of $2^4 - 1 = 15$, assuming that the 0 count is not included.

6-17-1-1 Binary Counters

A four-bit binary counter can be constructed by cascading four *J-K* FFs as shown in Figure 6-38(a). In this application two 7476 dual *J-K* FFs with preset and clear are used to form the counter. The normal output of each FF is connected to the *T* or toggle input of the next FF. The *J-K* inputs are left open, that is, they are high. Recall that a *J-K* FF, when *J* and *K* are high, will toggle on each negative-going transition of the input pulse, that is, when the input signal goes from the 1 to 0 state.

Assume that initially all outputs are reset, that is the *A*, *B*, *C*, and *D* outputs are 0. When the first pulse is applied to FF/*A* it will be set, that is, its output *A* will be 1. Since this is the first binary number stored FF/*A* represents the LSB, and the outputs are read from right to left, that is, *DCBA*. After the first clock pulse the counter reads 0001 [see the count sequence table in Figure 6-38(c)]. When clock pulse 2 is applied FF/*A* toggles and resets and also causes FF/*B* to set; therefore after two clock pulses the outputs are 0010. When clock pulse 3 is applied FF/*A* sets and $Q_A = 1$. Since this is a positive-going transition it is ignored by FF/*B*. The count is now 0011. When clock pulse 4 is applied FF/*A* resets and Q_A changes from 1 to 0 and toggles FF/*B*

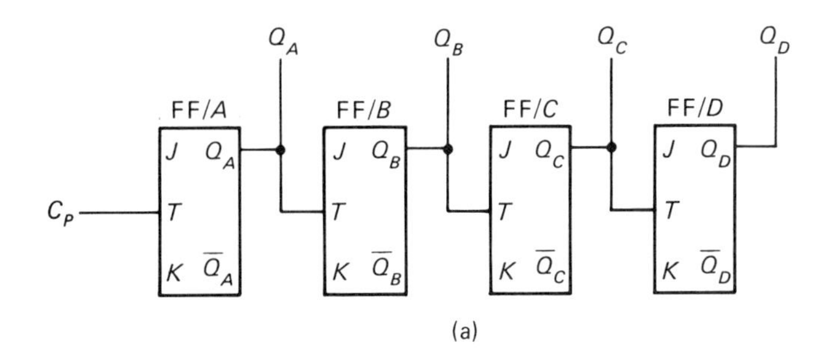

(a)

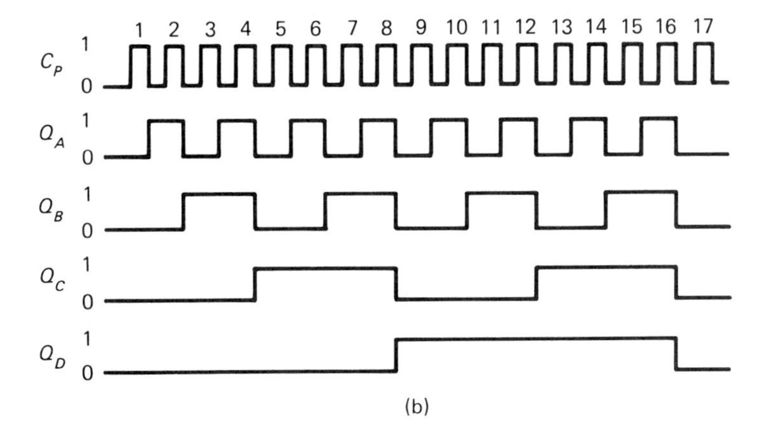

(b)

Clock Pulse	Flip-Flop Outputs			
	Q_D	Q_C	Q_B	Q_A
0	0	0	0	0
1	0	0	0	1
2	0	0	1	0
3	0	0	1	1
4	0	1	0	0
5	0	1	0	1
6	0	1	1	0
7	0	1	1	1
8	1	0	0	0
9	1	0	0	1
10	1	0	1	0
11	1	0	1	1
12	1	1	0	0
13	1	1	0	1
14	1	1	1	0
15	1	1	1	1
16	0	0	0	0

(c)

Figure 6-38. A 4-bit binary counter: (a) block diagram, (b) clock input and output waveforms, (c) count sequence table.

which resets; $Q_B$changes from 1 to 0 which sets FF/C. The count is now 0100.

Referring to Figure 6-38(c), it can be seen that the counting sequence will continue until the 15th clock pulse has been applied; at this point the count is 1111, which is the maximum four-bit number and the maximum count capacity of the counter. When clock pulse 16 is applied, all FFs change state and the counter recycles back to a count of 0000.

It should be noted from the waveforms that the frequency of FF/A's frequency output is $C_p/2$, FF/B's frequency output is $C_p/4$, FF/C's frequency output is $C_p/8$ and FF/D's output frequency is $C_p/16$. Also it is not necessary for the input signal to be a constant frequency or for the clock pulse to be periodic, that is equally spaced.

The 4-bit counter just described is known as an up-counter since the binary output count increases by one for each clock pulse. When the binary output of the counter decreases by one for each clock pulse, the counter is known as a down-counter. A down-counter is easily arranged by using the complementary outputs instead of the normal outputs. A down-counter table with associated waveforms and count sequence is shown in Figure 6-39, assuming that the counter has been reset and the count is 0000. When clock pulse 1 is applied then FF/A which was reset with $\overline{Q}_A = 1$ will set and $\overline{Q}_A = 0$. This negative transition will toggle FF/B, its complementary output $\overline{Q}_B$ changes from 1 to 0 and in turn toggles FF/C, and its complementary output $\overline{Q}_C$ changes from 1 to 0 which in turn toggles FF/D so that its complementary output $\overline{Q}_D = 0$. The first clock pulse has therefore recycled the counter from 0000 to 1111. The second clock pulse causes FF/A to reset and $\overline{Q}_A$ switches from 0 to 1, FF/B ignores the positive transition and the count is 1110, or in other words the counter has decremented by one. By observing the count-sequence table it can be seen that the count decreases by one for each clock pulse until after the 15th pulse when the count is 0000.

The up-counter and down-counter can be combined as shown in Figure 6-40. With a little modification the up- or down-counters can be made to perform as an up-down counter. In Figure 6-40 the normal outputs from FF/A, FF/B, and FF/C are applied to the 2-input ANDs, the other input being connected to the count control. If the arrangement is to operate as an up-counter, with the count control line high as the normal outputs go high, the output of the AND is high and will result in the OR output being high, and the next FF will be toggled; the arrangement is then operating as an up-counter. When the count control line is low, the inverted input to the 2-input ANDs connected to the complementary outputs will be high. Therefore, as the complementary outputs go high, the next FF will be toggled and the counter is acting as a down-counter.

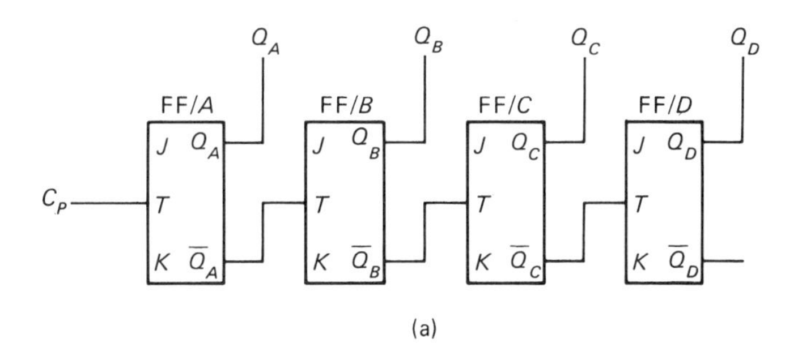

(a)

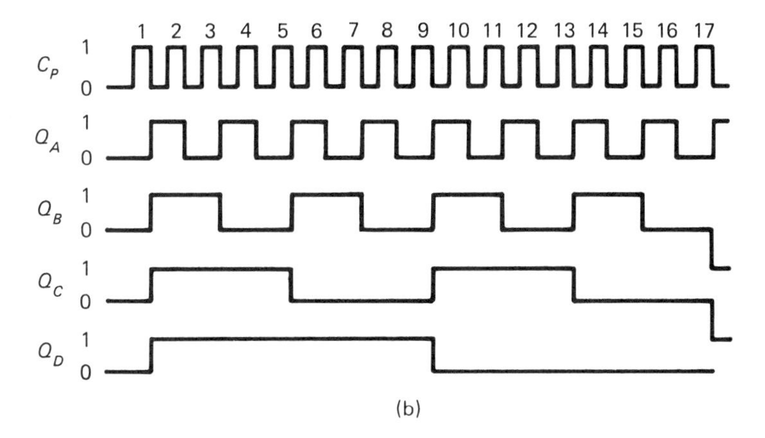

(b)

Clock Pulse	Flip-Flop Outputs			
	Q_D	Q_C	Q_B	Q_A
0	1	1	1	1
1	1	1	1	0
2	1	1	0	1
3	1	1	0	0
4	1	0	1	1
5	1	0	1	0
6	1	0	0	1
7	1	0	0	0
8	0	1	1	1
9	0	1	1	0
10	0	1	0	1
11	0	1	0	0
12	0	0	1	1
13	0	0	1	0
14	0	0	0	1
15	0	0	0	0
16	1	1	1	1

(c)

Figure 6-39. A 4-bit binary down-counter: (a) block diagram, (b) clock input and output waveforms, (c) count sequence table.

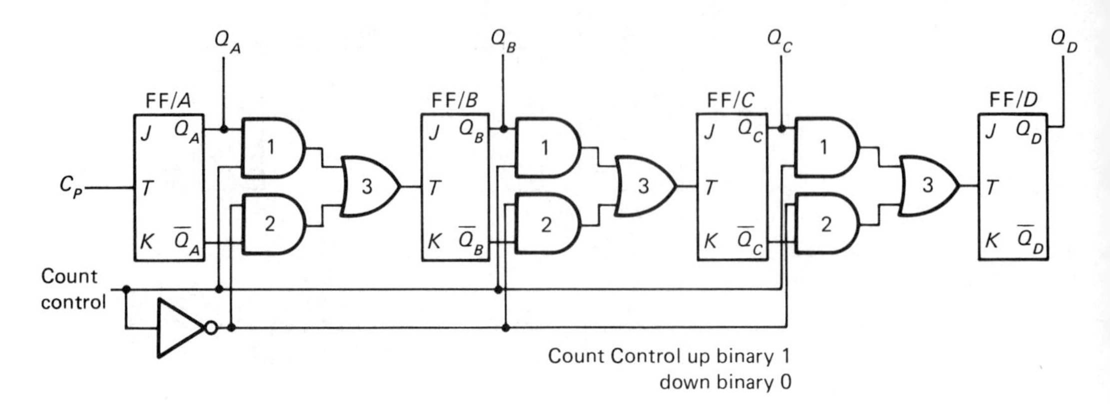

Figure 6-40. Up-down binary counter.

The counters that have been described in this section are called **ripple** or **asynchronous counters.** In an asynchronous counter the FF associated with the LSB produces the clock input for the next stage, then the next stage produces the clock input for the next stage, and so on. The operation just described easily explains the more common name of ripple counter associated with this method of operation. The major advantage of a ripple counter is its simplicity. The main limitation is the maximum counting speed. Since each FF in turn is triggered from the previous FF, the time taken for a pulse to travel through the chain of cascaded FFs will be equal to the sum of the propagation delays for each FF. For example, if the propagation delay of each FF of a 4-bit binary counter is 20 ns, then the minimum time required for the counter to change state is 4 × 20 ns = 80 ns. In turn, the maximum frequency of the ripple counter is

$$f = \frac{1}{Nt \times 10^{-9}}. \qquad (6\text{-}9)$$

Example 6-25

What is the maximum counting speed of a 4-bit binary counter which is composed of FFs with a propagation delay of 25 ns?

Solution:

$$f = \frac{1}{4 \times 25 \times 10^{-9}} = 10 \text{ MHz}$$

The implication of this figure is that if counting speeds higher than this are attempted, the counter will lag behind and counting errors will result. Another source of counting errors results from intermittent input pulses.

There are counters available that can achieve counting speeds of 500 MHz. However, to achieve the necessarily short propagation delays means resorting to ECL, which immediately increases the cost and the power consumption.

A relatively simple approach to reducing the effects of propagation delay is the **synchronous counter.** In the synchronous counter all FFs are triggered simultaneously by the input signal and the total propagation delay is that of one FF. The synchronous counter is more complicated than the ripple or asynchronous counter but can be readily achieved in MSI integrated circuits. As a result the synchronous counter is the most commonly used type.

A typical synchronous binary counter is illustrated in Figure 6-41. The clock input C_p is common to all FFs. The J and K inputs to FF/A are high, that is, $J = 1$ and $K = 1$. This enables the FF, and as a result it will be toggled each time a negative transition of the clock input occurs. The J and K inputs to FF/B are supplied from the normal output of FF/A, and FF/B will only change state when $Q_A = 1$. The J and K inputs to FF/C are controlled by the normal outputs of FF/A and FF/B; therefore for FF/C to toggle both Q_A and $Q_B = 1$, so that FF/C will toggle on the trailing edge of the first clock pulse after $Q_A = Q_B = 1$ occurs. The J and K inputs to FF/D are high when $Q_A = Q_B = Q_C = 1$. The relationship of the input and output wave-

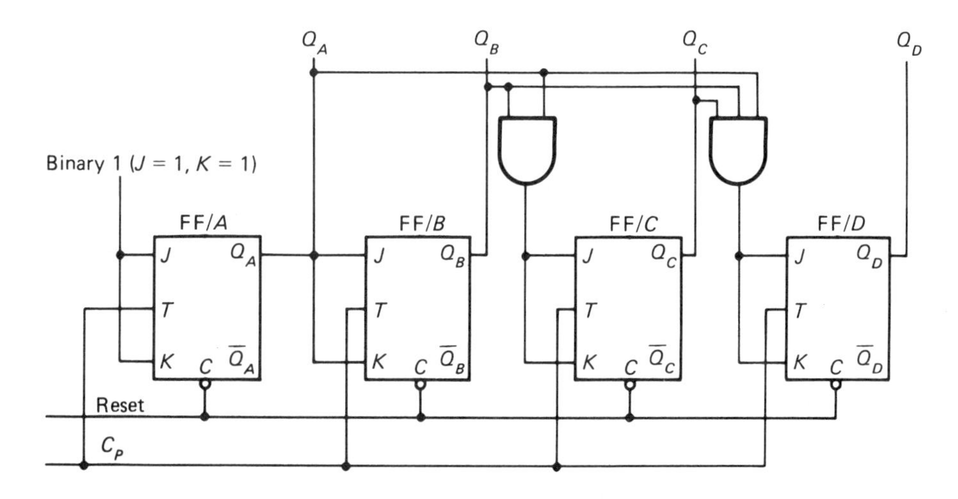

Figure 6-41. A synchronous 4-bit binary counter.

forms and the count sequence are identical to those of the 4-bit binary ripple counter of Figure 6-38(b) and (c). The only difference in the counters is that the FFs of the synchronous counter all change together, the only delay being that of the longest propagation delay of the FFs.

To summarize, the synchronous counter is preferred because it has a significantly higher counting speed as compared to the asynchronous counter, and the possibility of errors caused by accumulative propagation delays has been eliminated. Synchronous up- and/or down-counters can be expanded as required to meet the bit requirements. The only rule that must be followed is that the additional FFs are controlled through AND gates by all the preceding FFs.

Normally counters are not constructed from *J-K* FF integrated circuits but implemented by using readily available MSI ICs in TTL or CMOS.

6-17-1-2 Binary Coded Decimal Counters

The binary coded decimal (BCD) counter is the most commonly used mod-*N* counter, since it is a mod-10 or decade counter; that is, it has 10 count states (0 to 9) and divides by 10. As can be seen from Figure 6-42, the BCD counter is similar to the 4-bit binary ripple counter, and in fact operates in exactly the same manner up to 9 (0000 to 1001). At this point the BCD counter recycles to 0000 and repeats the count. As can be seen six counts have been omitted, namely 10 to 15 inclusive. The BCD counter is a weighted counter and is frequently referred to as an 8421 counter.

The 7490A BCD or decade counter shown in Figure 6-42(a) is an asynchronous or ripple counter constructed from *J-K* FFs. The Q_A output is not internally connected to the input of the *B* FF. To obtain a BCD count, Q_A is connected externally to the *B* input and $\overline{Q}_D$ is connected to the *J* input of the *B* FF. Also internally Q_B and Q_C are connected by means of a 2-input AND to the set terminal of the *D* FF. Gate 1 is used to preset the counter to 9 (1001) when the two inputs are high, a feature which is particularly useful when the BCD counter is being used for 9's complement arithmetic. Similarly gate 2 will preset the counter to 0000 when the two inputs are high. The maximum count frequency of the 7490A is 42 MHz and it may be used as a decade counter or scaler (divide by 10).

The *J* and *K* inputs to FFs *A* and *C* and the *K* input to the B FF are permanently high. With the counter reset (0000), $\overline{Q}_D$ and J_B are high which will permit FF/*B* to toggle when Q_A changes state. When pulse 1 is applied, FF/*A* toggles and Q_A is high. The count is 0001. When pulse 2 is applied FF/*A* toggles and FF/*B* toggles as Q_A goes low; then the count is 0010. When pulse 3 is applied, only FF/*A* toggles and the

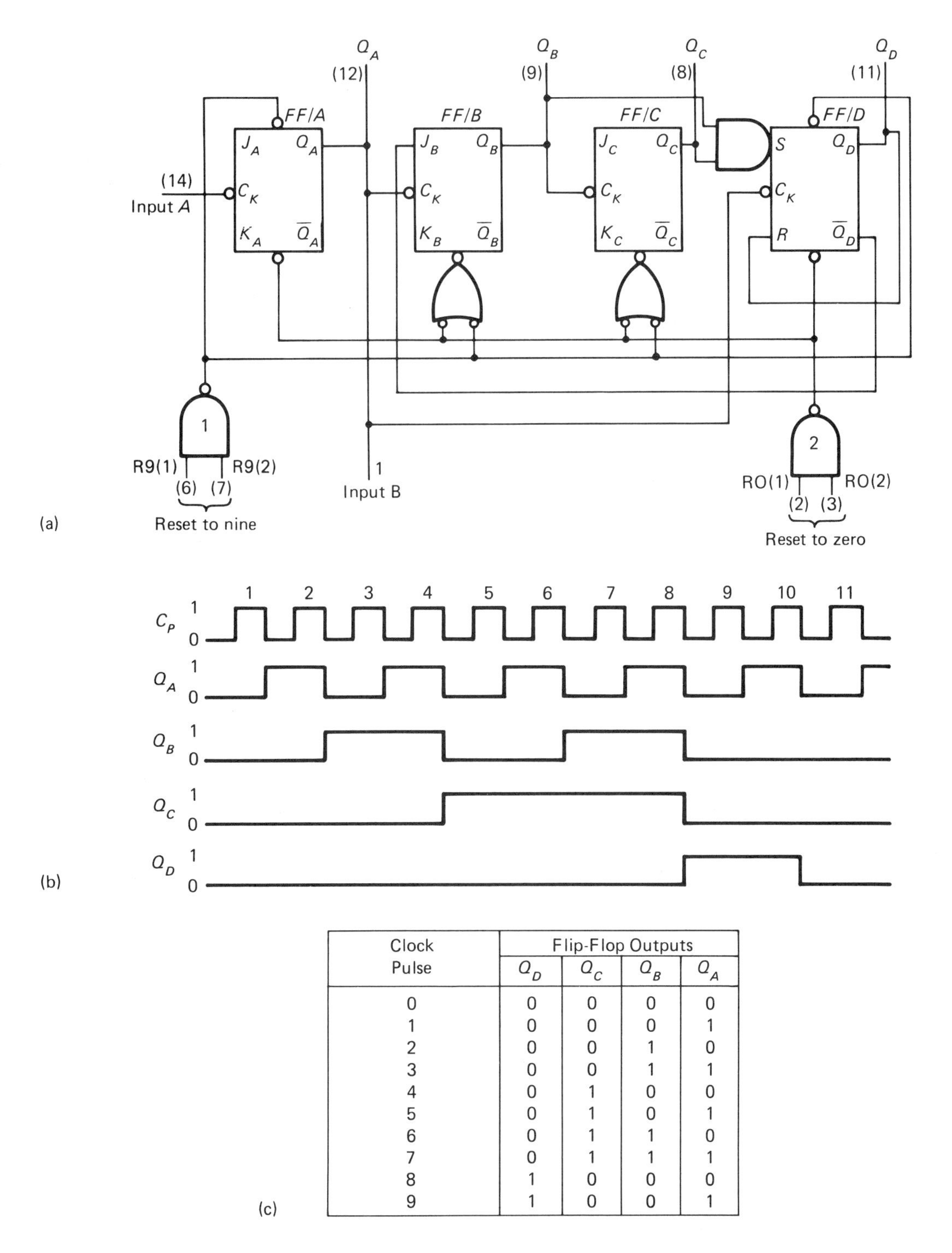

Clock Pulse	Flip-Flop Outputs			
	Q_D	Q_C	Q_B	Q_A
0	0	0	0	0
1	0	0	0	1
2	0	0	1	0
3	0	0	1	1
4	0	1	0	0
5	0	1	0	1
6	0	1	1	0
7	0	1	1	1
8	1	0	0	0
9	1	0	0	1

Figure 6-42. The 7490A BCD counter: (a) block diagram, (b) clock input and output waveforms, (c) count sequence table.

count is 0011. When pulse 4 is applied FFs A, B, and C toggle on the negative-going transition; the count is then 0100. When pulse 5 is applied only FF/A toggles and the count becomes 0101. When pulse 6 is applied FFs A and B toggle and the count is 0110. When pulse 7 is applied FF/A toggles, Q_B and Q_C and S are high but FF/D cannot change state until Q_A changes from a high to a low, and the count is then 0111. When pulse 8 is applied FFs A, B, and C toggle and since FF/D is an R-S FF, Q_D goes high; the count at this point is 1000. At the ninth pulse FF/A toggles and the count is 1001, Q_D is low and J_B is low. When pulse 10 is applied FF/A toggles $Q_A = 0$, FF/B is disabled, $Q_B = Q_C = 0$, and since the clock input to FF/D is $Q_A = 0$, $Q_D = 0$; that is, the counter has reset and the count is 0000.

As can be seen from the previous paragraphs a single BCD counter may be used to represent decimal numbers 0 through 9 or 10 states. Very often a requirement exists to count numbers greater than 9. This may be accomplished by cascading BCD counters with each BCD counter representing one decimal digit; the number of BCD counters in cascade is determined by the count requirement.

An example of cascaded BCD counters is illustrated in Figure 6-43, where four counters are connected in series to permit a maximum count of 9999. The most significant digit output Q_D of each counter is connected to the input of the next counter. As a result, as each counter counts ten pulses, it will recycle to the 0000 count, but the tenth pulse will initiate the count in the succeeding counter. In our example, the counter output represents 6521_{10}, or a total of 6521 pulses have been applied from the original reset condition. It should also be noted that each Q_D output has divided the input frequency by 10. This frequency division capability is extremely useful in conjunction with other mod-N counters in timing, frequency, and period measurement applications.

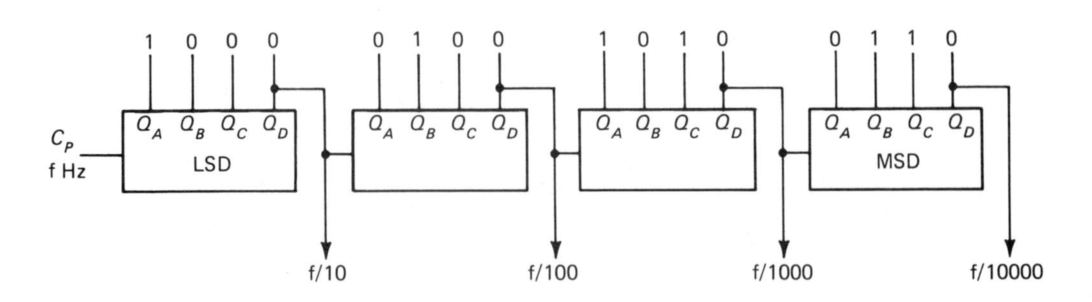

Figure 6-43. Cascaded BCD counters.

6-17-2 Modulo-N Counters

As we saw the binary counters could count in binary and also could divide the input signal frequency by factors of 2^n. The BCD counter, in addition to counting in decimal, can also divide by factors of 10^n. A divide-by-*N* counter is a counter whose output frequency is 1/*N* of the input frequency. Originally counters with this capability were constructed from *J-K* FFs with suitably modified inputs and feedback to achieve the desired count capability. Nowadays there are a large number of TTL MSI counters available which will provide the necessary divide-by-*N* capability or which can be modified to meet the requirements. A divide-by-*N* counter has *N* different count states and will, after *N* counts, return to its original state. A counter with *N* count states is called a modulus-*N* or mod-*N* counter. The four-bit binary counter has 16 count states and is termed a mod-16 counter; similarly the decade counter is a mod-10 counter.

6-17-2-1 Mod-3 Counter

The mod-3 counter can be constructed from a dual *J-K* FF as shown in Figure 6-44. This arrangement is synchronous since the same input pulse is applied to both FFs. The $\bar{Q}_B$ is also fed back into the *J* input of FF/*A*, and both *K* inputs are held high. Assuming initially that the counter is reset, both Q_A and Q_B are low, J_B is low, and J_A is high since $\bar{Q}_B$ is high. FF/A will toggle on the negative-going transition of pulse 1, Q_A will go high, and FF/*B* will not toggle. When pulse 2 is applied and going low, J_B is high, FF/*B* will toggle and set, and FF/*A* toggles and resets making J_A low. At the same time $\bar{Q}_B$ is low, which will make J_A low and disable FF/*A*. On the trailing edge of pulse 3, FF/*B* resets and the counter is back to its original state. As more input pulses are applied the counter will continue to cycle through this sequence repeatedly. As can be seen from the count sequence table, Q_A and Q_B are high once for every three input pulses; that is, the Q_A and Q_B outputs are divided by three outputs.

6-17-2-2 Mod-6 Counter

It is relatively easy to construct a mod-6 counter by cascading a mod-3 counter into another *J-K* FF, as shown in Figure 6-45. As can be seen from the waveforms and the count sequence table, the mod-3 counter cycles through its count with FF/*C* reset and then repeats the cycle with FF/*C* set. It should also be noted from the count sequence table that the counts are not binary or BCD. The count sequence is called an unweighted code. If it is desired the combination can be

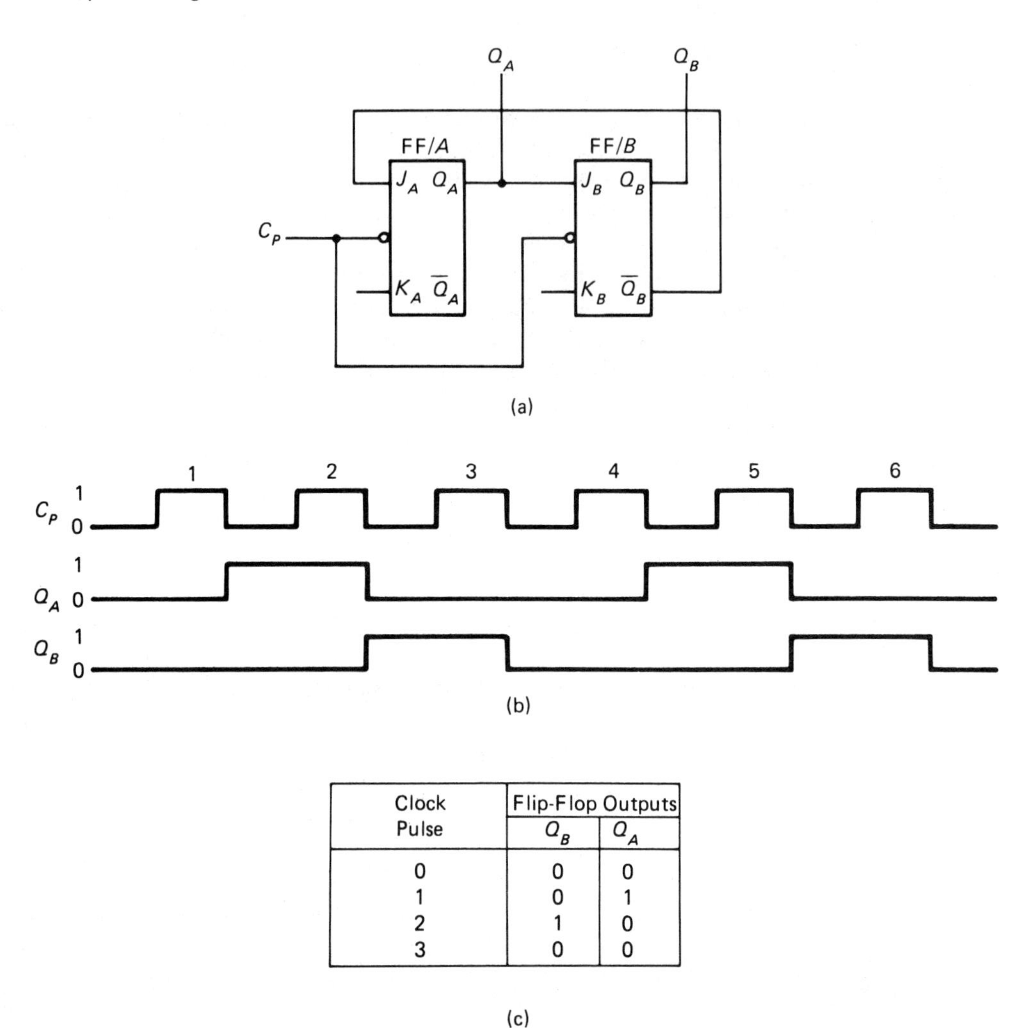

Clock Pulse	Flip-Flop Outputs	
	Q_B	Q_A
0	0	0
1	0	1
2	1	0
3	0	0

Figure 6-44. A mod-3 counter: (a) basic circuit, (b) input and output waveforms, (c) count sequence table.

extended to a mod-12 counter by adding yet another *J-K* FF in cascade with the others.

Normally mod-*N* counters are not formed from interconnected *J-K* FFs for mod-*N* counters above 4. The main reason is the wide selection of available MSI ICs such as the 7490, 7492, 7493 etc., which enable a wide range of divide-by-*N* counters to be formed.

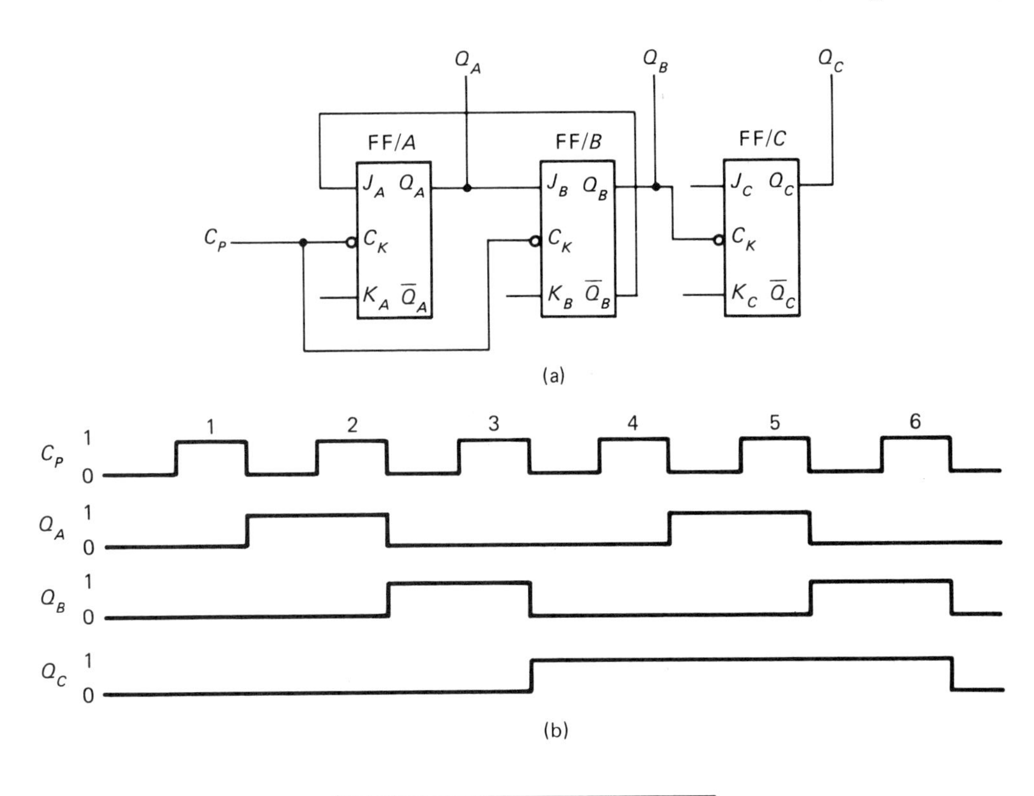

Clock Pulse	Flip-Flop Outputs		
	Q_C	Q_B	Q_A
0	0	0	0
1	0	0	1
2	0	1	0
3	1	0	0
4	1	0	1
5	1	1	0
6	0	0	0

(c)

Figure 6-45. A mod-6 counter: (a) basic circuit, (b) input and output waveforms, (c) count sequence table.

6-18 SHIFT REGISTERS

Shift registers are another widely used form of sequential logic and like the counter are made up of a number of FFs forming binary storage elements. The FFs permit information stored in the register to be moved to the left or right when input or clock pulses are applied to the register. This ability to shift and store data makes the shift register particularly valuable in arithmetic operations. In addition, they can be used to convert parallel data to serial data or vice versa, as well as being used as a buffer to change the transmission rate of serial data.

6-18-1 Four-Bit Shift Register

The most common type of shift register is formed by using four *J-K* FFs as shown in Figure 6-46. The serial input and its complement are applied to the *J-K* inputs of FF/*A*. The Q and $\overline{Q}$ outputs of each FF are fed into the *J-K* inputs of the following FF stage. Also, the clock input is supplied simultaneously to all the FFs, that is, since all the FFs are toggled together the shift register is a synchronous system. As can be seen from Figure 6-46(b) applying a low to the direct reset line will preset the register.

The operation of the register can be observed from the output waveforms. Prior to the receipt of pulse 1, the direct reset has made Q_A, Q_B, Q_C, and Q_D all low. Assuming that we are loading 1101_2 or 11_{10} into the register, as can be seen from the serial input waveform the LSB is loaded in first. That is, prior to clock pulse 1 the serial input is high, on the trailing edge of pulse 1, the binary 1 from the serial input is loaded into FF/*A*, since the *J-K* inputs are such that when the clock pulse is received the FF sets. Since clock pulse 1 is applied to the clock inputs of all FFs, the low stored by FF/*A* is transferred to FF/*B*, the low stored by FF/*B* is transferred to FF/*C*, and so on. That is, the original contents of the FFs in the shift register have been transferred one place to the right. On the second clock pulse the second binary 1 from the serial input is loaded into FF/*A*, the original high stored in FF/*A* is shifted into FF/*B*, the 0 state stored in FF/*B* had been transferred to FF/*C*, and similarly the 0 stored in FF/*C* has been transferred to FF/*D*.

With the third clock pulse, the binary 0 of the serial input is transferred into FF/*A*, the binary 1 from FF/*A* is transferred to FF/*B*, then the binary 1 from FF/*B* is transferred to FF/*C*, and FF/*D* still remains low. When the fourth clock pulse is applied, the binary 1 in the serial input is stored in FF/*A*, the 0 from FF/*A* is transferred to FF/*B*, the binary 1 of FF/*B* is transferred to FF/*C*, and the binary 1 of FF/*C* is shifted into FF/*D*. Therefore, after four clock pulses the serial input has been transferred into the shift register with Q_A representing the MSB and Q_D representing the LSB. This shift register is termed a shift-right register.

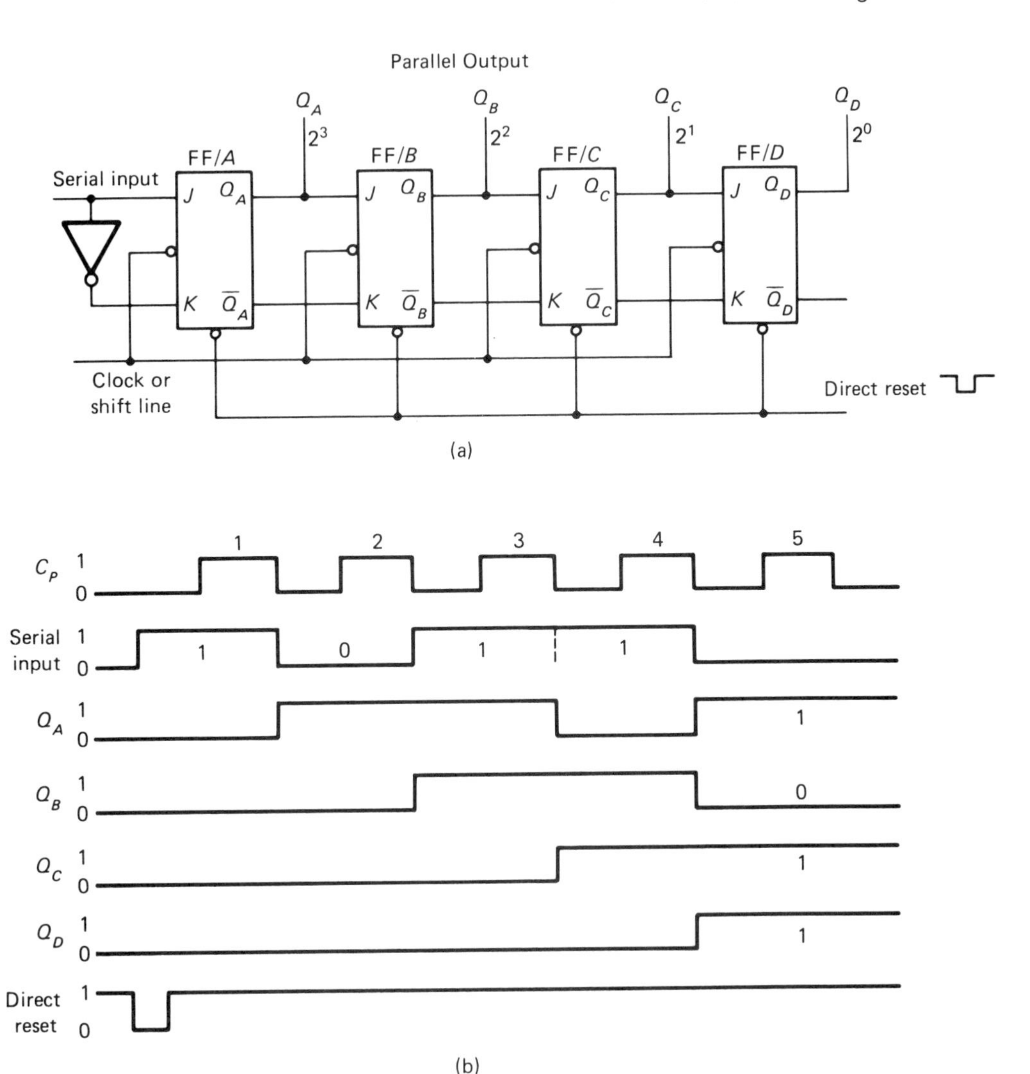

Figure 6-46. A 4-bit shift register using *J-K* flip-flop: (a) logic diagram, (b) input and output waveforms.

Shift registers are also classified in terms of the way data is loaded in and read out. For example, with the circuit shown in Figure 6-46(a), the data is loaded in serially; however, after four clock pulses, the data can be read as a parallel output at Q_A, Q_B, Q_C, and Q_D, and the shift register is termed as a serial in-parallel output shift register or SIPO. On the other hand if the data is read out at Q_D only then it is termed a serial in-serial output shift register or SISO. A SISO register will also introduce a time delay in data transmission; since there are N stages to the shift register the output serial data is delayed for N clock pulses.

Shift registers are available in various capacities; however, 4- and 8-bit shift registers are standard and are available with either D-type or J-K FFs in MSI or LSI ICs. Some examples are: parallel in–parallel out (PIPO) bidirectional; that is, shift right or shift left are available in D-type FFs as 54/74194 (4-bit) or 54/74198 (8-bit); serial in–parallel out (SIPO) are marketed for 8 bits with a gated D as 54/74164; parallel in–serial out (PISO) as an 8-bit shift-right register 54/74165; and serial in–serial out (SISO) are manufactured in shift-right 8-bit shift registers 54/7491A. Shift registers with 16 or more bits are available and usually are serial input–serial output.

6-19 RING COUNTERS

Another variation of the shift register is the **ring counter** or **sequencer** as it is sometimes called. In the case of a ring counter, the output of the last FF stage is coupled back to the input of the first FF, effectively arranging the FFs in a circle or ring; hence the term ring counter. An example of a six-step ring counter is illustrated in Figure 6-47. Initially, if the counter is reset, then since all the J's $= 0$ and the K's $= 1$, when the clock pulse makes a negative transition the FFs will reset, but since they are reset there will be no change in the output. For the ring counter to perform FF/A must be set, then when the clock pulse goes low the high from FF/A will be transferred to FF/B and FF/A will reset since the output $Q_F = 0$ is applied to the J input of FF/A, all the other FFs remaining low. Clock pulse 2 will move the high from FF/B to FF/C and FF/B will reset and all the other FFs remain low. With each succeeding clock pulse the high will be moved around the interconnected FFs. When FF/F is high, since Q_F is connected to the J input of FF/A, the next clock pulse will cause FF/A to go high and the cycle is repeated as long as clock pulses are applied.

The major disadvantage of the ring-counter arrangement shown in Figure 6-47 is that it must be preset to function as shown in Figure 6-47(b). To preset the ring counter, it is necessary to place a binary 1 into any one of the FFs, and leave all the other FFs low, as illustrated in Figure 6-48. The ring counter is preset by commoning the clear inputs of FFs B, C, D, E, and F and the preset input of FF/A. When a low is applied to the direct reset line, FF/A is set to binary 1 and all the other FFs are set to binary 0. The six-stage ring counter described above could be used combined with optocouplers to distribute the gate pulses to a six-pulse thyristor phase-controlled converter.

Another commonly used shift register counter is the Johnson or

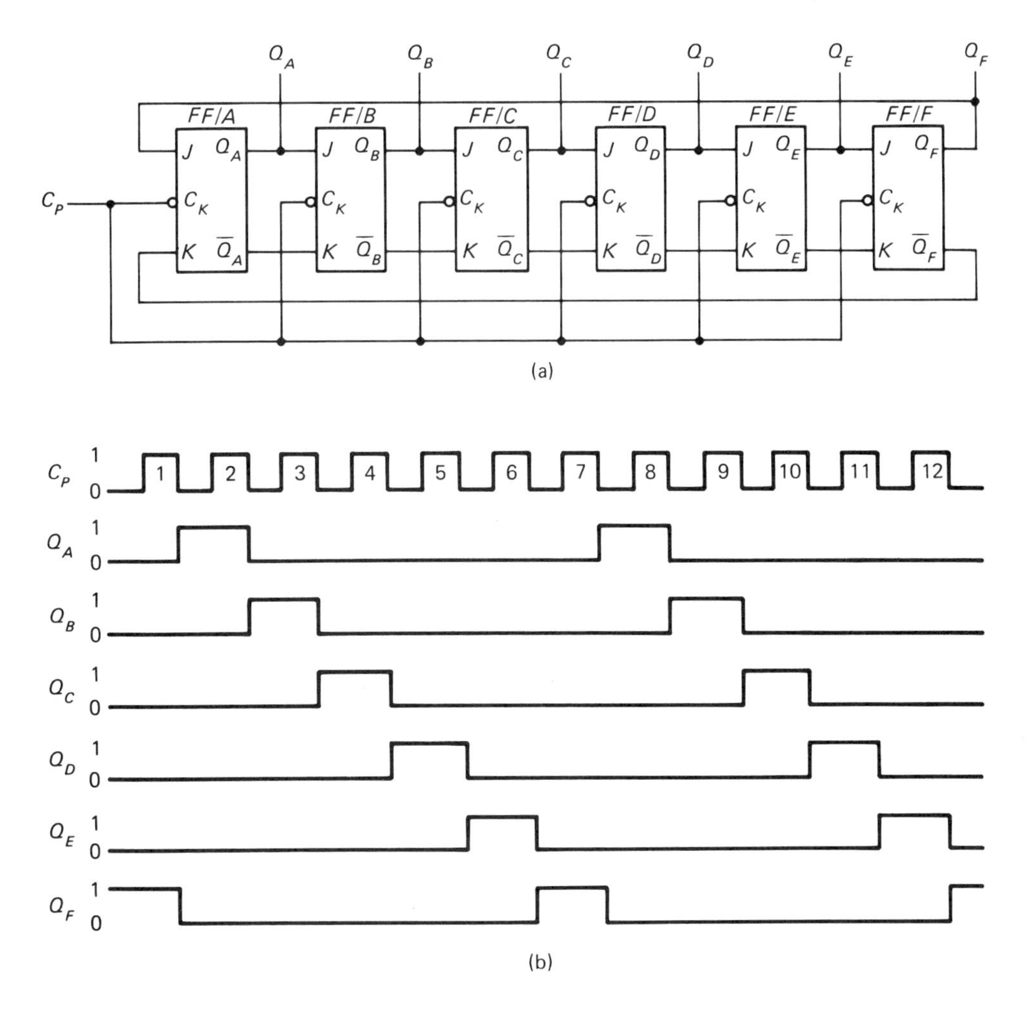

Figure 6-47. A 6-stage ring counter: (a) logic diagram, (b) output waveforms.

switch tail counter illustrated in Figure 6-49. A very common configuration is the five-stage counter which can be used as a decade counter. This counter is given the name switch tail because the Q_E output is connected to the K input of FF/A and the $\overline{Q}_E$ output is connected to the J input of FF/A. The counter is initialized by applying a low to the direct reset. As can be seen from the waveforms and the count sequence table, the initial clock pulse causes Q_A to be high and remain high for five clock pulses, the second clock pulse causes Q_B to be high and remain for five clock pulses, and so on. The shift register

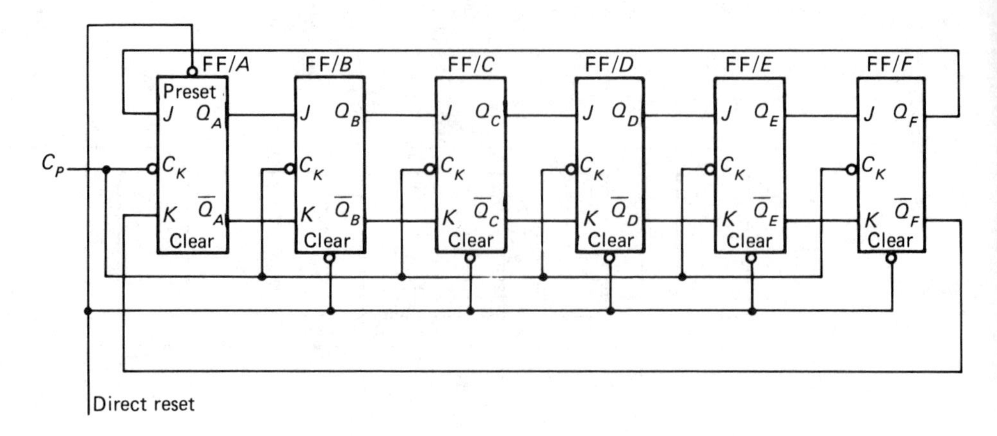

Figure 6-48. A 6-stage ring counter.

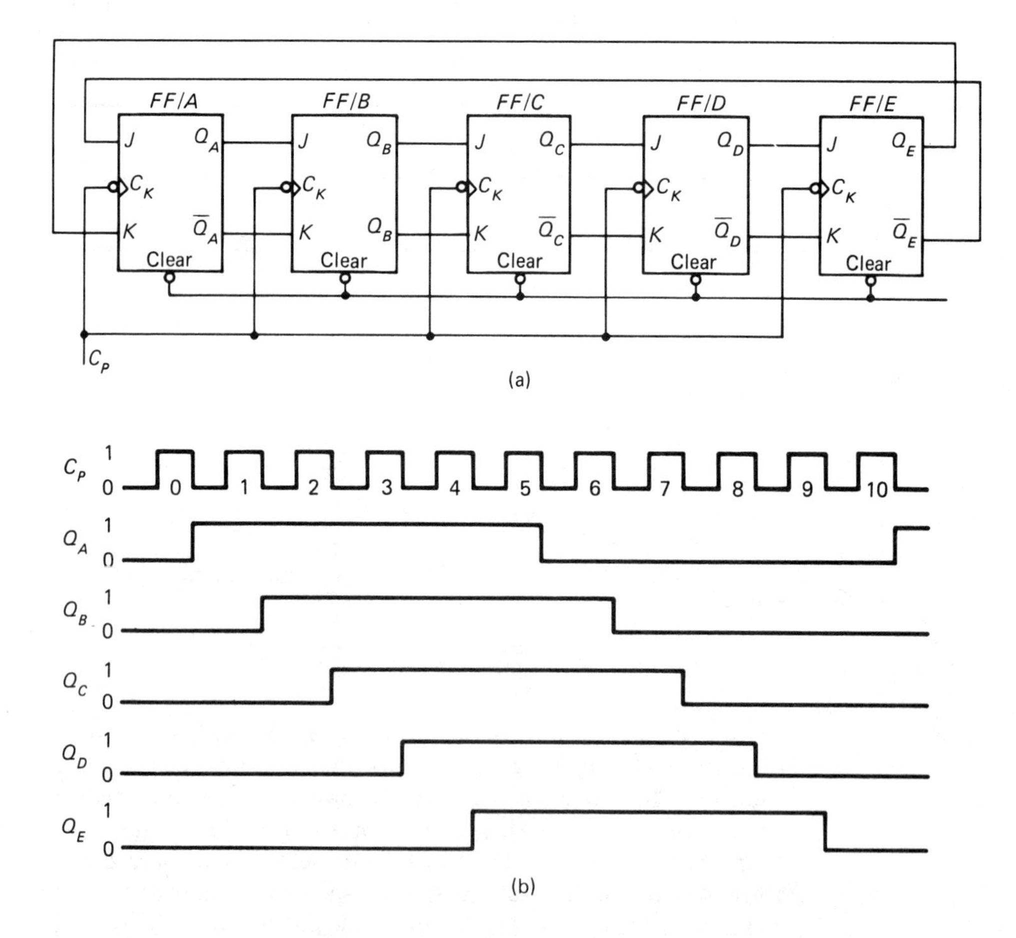

Figure 6-49. The 5-stage Johnson or switch tail counter: (a) logic diagram, (b) output waveforms, (c) count sequence table.

Clock Pulse	Flip-Flop Outputs Q_E	Q_D	Q_C	Q_B	Q_A	Binary Equivalent
0	0	0	0	0	0	0
1	0	0	0	0	1	1
2	0	0	0	1	1	3
3	0	0	1	1	1	7
4	0	1	1	1	1	15
5	1	1	1	1	1	31
6	1	1	1	1	0	30
7	1	1	1	0	0	28
8	1	1	0	0	0	24
9	1	0	0	0	0	16

(c)

Figure 6-49. continued

will fill up with 1's until $\overline{Q}_E$ goes low, making the *J* input of FF/*A* low; then the register will fill with 0's until $\overline{Q}_E$ goes high. At this point the counter will recycle.

It should be noted that there are 10 states which have been produced by five *J-K* FFs, this means that a Johnson counter has $2N$ different states, where N is the number of FFs forming the counter. As can be seen from the count sequence table, the output is a nonweighted binary which causes difficulty in counting applications.

6-20 COMBINATIONAL LOGIC APPLICATIONS

Combinational logic circuits, unlike sequential logic circuits, are combinations of logic gates such as the AND, OR, NAND, NOR, EXCLUSIVE-OR, and inverters, which are used to implement Boolean algebra equations defining a specific application. In addition to this type of application of the logic gates and inverters, combinational logic is used with decoders, encoders, multiplexers, comparators, and code converters.

6-20-1 Decoders

A **decoder** is a combinational logic circuit whose function is to detect and indicate a particular binary number from all the possible combinations of binary numbers present. Usually the input to a decoder is a parallel binary number, and the output is a binary number which shows the presence or absence of the desired binary number.

One of the most common decoding applications is the conversion of

BCD to decimal. The decoder input consists of the parallel 4-bit output of a BCD counter, whose outputs range from 0000 to 1001. These ten states represent the decimal numbers 0 through 9. To be able to decode the parallel 4-bit output of the BCD counter requires a total of ten NAND gates and eight inverters arranged as shown in Figure 6-50(a). The 4-bit BCD number *DCBA*, where *D* is the MSB and *A* is the LSB, is applied to the four inverter combinations whose outputs are the normal and complements of the respective inputs. If all the inputs to any one of the NAND gates are high, the output will be low; for any other combination of highs and lows at the input the output will be high.

As can be seen from Figure 6-50(a), when the appropriate BCD code is applied to the correct gate the output will be low. For example, when the input to gate 4 is 0011 $(AB\overline{C}\overline{D})$ the output will be low. In other words the inputs to a specific gate must be unique to obtain a low at the output. It should also be noted that the six invalid states will produce highs; therefore, the decoder will only convert 4-bit 8421 BCD numbers to decimal. The arrangement shown in Figure 6-50(a) can be assembled using SSI logic gates in single DIP packages. A possible combination could be five dual 4-input positive NAND gates (54/7420) and two hex inverters (54/7404). As can be seen, a total of seven ICs are required to assemble the BCD to decimal converter. A much more practical arrangement would be to use an MSI IC such as the 54/7442A IC which is available as a 16-pin DIP.

Another very common decoder application is the BCD-to-seven-segment decoder/driver used to control seven-segment LED displays used in such applications as **digital volt-ohm meters (DVOM)**, calculators, and digital clocks, to name a few uses.

The seven-segment numeric readout package is available in 10, 14, or 16 lead DIPs. The arrangement of a seven-segment display is shown in Figure 6-51(a) and permits the decimal numbers 0 to 9 inclusive to be represented by energizing the LED segments *a*, *b*, *c*, *d*, *e*, *f*, and *g*, as appropriate. The function table [Figure 6-51(b)] shows the energizing sequences of the segments for each of the 8421 BCD inputs to the decoder driver. A binary 0 indicates the respective segments that are illuminated and a binary 1 indicates the segments that are blanked out. A common BCD-to-seven-segment numeric display is the TIL 302 or TIL 303 numeric display, driven by a 54/7447A BCD-to-seven-segment decoder/driver.

The logic diagram of the 54/7447A decoder/driver is shown in Figure 6-52(a). The features of this circuit include, in addition to the four BCD inputs, a blanking input (BI) which is held at binary 1 or open if all 0 to 15 inclusive outputs are desired. The ripple blanking input (RBI) must be maintained at binary 1 or open if the blanking of a decimal zero is

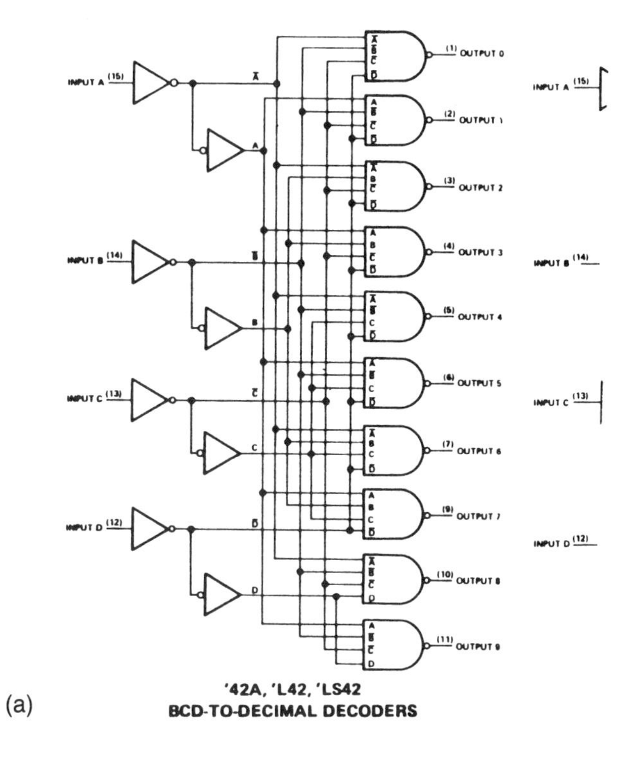

(a)

FUNCTION TABLE

NO.	'42A, 'L42, 'LS42 BCD INPUT				'43A, 'L43 EXCESS-3-INPUT				'44A, 'L44 EXCESS-3-GRAY INPUT				ALL TYPES DECIMAL OUTPUT									
	D	C	B	A	D	C	B	A	D	C	B	A	0	1	2	3	4	5	6	7	8	9
0	L	L	L	L	L	L	H	H	L	L	H	L	L	H	H	H	H	H	H	H	H	H
1	L	L	L	H	L	H	L	L	L	H	H	L	H	L	H	H	H	H	H	H	H	H
2	L	L	H	L	L	H	L	H	L	H	H	H	H	H	L	H	H	H	H	H	H	H
3	L	L	H	H	L	H	H	L	L	H	L	H	H	H	H	L	H	H	H	H	H	H
4	L	H	L	L	L	H	H	H	L	H	L	L	H	H	H	H	L	H	H	H	H	H
5	L	H	L	H	H	L	L	L	H	H	L	L	H	H	H	H	H	L	H	H	H	H
6	L	H	H	L	H	L	L	H	H	H	L	H	H	H	H	H	H	H	L	H	H	H
7	L	H	H	H	H	L	H	L	H	H	H	H	H	H	H	H	H	H	H	L	H	H
8	H	L	L	L	H	L	H	H	H	H	H	L	H	H	H	H	H	H	H	H	L	H
9	H	L	L	H	H	H	L	L	H	L	H	L	H	H	H	H	H	H	H	H	H	L
INVALID	H	L	H	L	H	H	L	H	H	L	H	H	H	H	H	H	H	H	H	H	H	H
	H	L	H	H	H	H	H	L	H	L	L	H	H	H	H	H	H	H	H	H	H	H
	H	H	L	L	H	H	H	H	H	L	L	L	H	H	H	H	H	H	H	H	H	H
	H	H	L	H	L	L	L	L	L	L	L	L	H	H	H	H	H	H	H	H	H	H
	H	H	H	L	L	L	L	H	L	L	L	H	H	H	H	H	H	H	H	H	H	H
	H	H	H	H	L	L	H	L	L	L	H	H	H	H	H	H	H	H	H	H	H	H

H = high level, L = low level

(b)

Figure 6-50. BCD-to-decimal, 4-line to 10-line decoder SN54/7442A: (a) logic diagram, (b) function table. (Courtesy of Texas Instruments Incorporated)

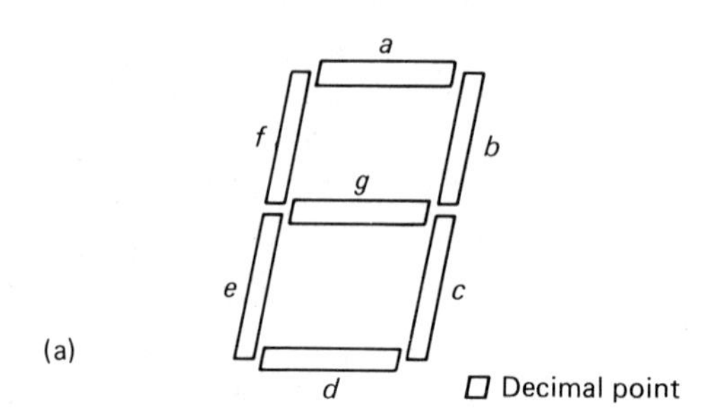

(b)

Decimal No.	B C D Outputs				Segments						
	D	*C*	*B*	*A*	*a*	*b*	*c*	*d*	*e*	*f*	*g*
0	0	0	0	0	0	0	0	0	0	0	1
1	0	0	0	1	1	0	0	1	1	1	1
2	0	0	1	0	0	0	1	0	0	1	0
3	0	0	1	1	0	0	0	0	1	1	0
4	0	1	0	0	1	0	0	1	1	0	0
5	0	1	0	1	0	1	0	0	1	0	0
6	0	1	1	0	1	1	0	0	0	0	0
7	0	1	1	1	0	0	0	1	1	1	1
8	1	0	0	0	0	0	0	0	0	0	0
9	1	0	0	1	0	0	0	1	1	0	0

Figure 6-51. BCD-to-seven segment decoder: (a) seven-segment decimal readout display, (b) function table.

not desired. When the blanking input is held at binary 0 all the segments of the display are turned off. This is a particularly desirable feature when using a number display to give a readout for multidigit numbers, since it will suppress any leading zeros. The display is tested to ensure all segments are operational by applying a binary 1 to the BI/RBO input and a binary 0 to the lamp test input.

Another useful feature is that the output light intensity of the display can be varied by applying a variable duty cycle pulse signal to the ripple blanking input without varying the applied voltage to the LEDs. The decoder decodes the BCD outputs and the driver actually controlls the segments of the LED display [see Figure 6-52(b)]. The driver is an *n-p-n* transistor which, when driven into saturation, causes the collector to go low and turn on the LED segment. The LED current is limited by means of the series-limiting resistor.

Similar decoder/drivers are available for hexadecimal displays. Numeric displays such as the TIL 306 or 307 can be obtained as a synchronous BCD counter, 4-bit latch, decoder/driver, and seven-

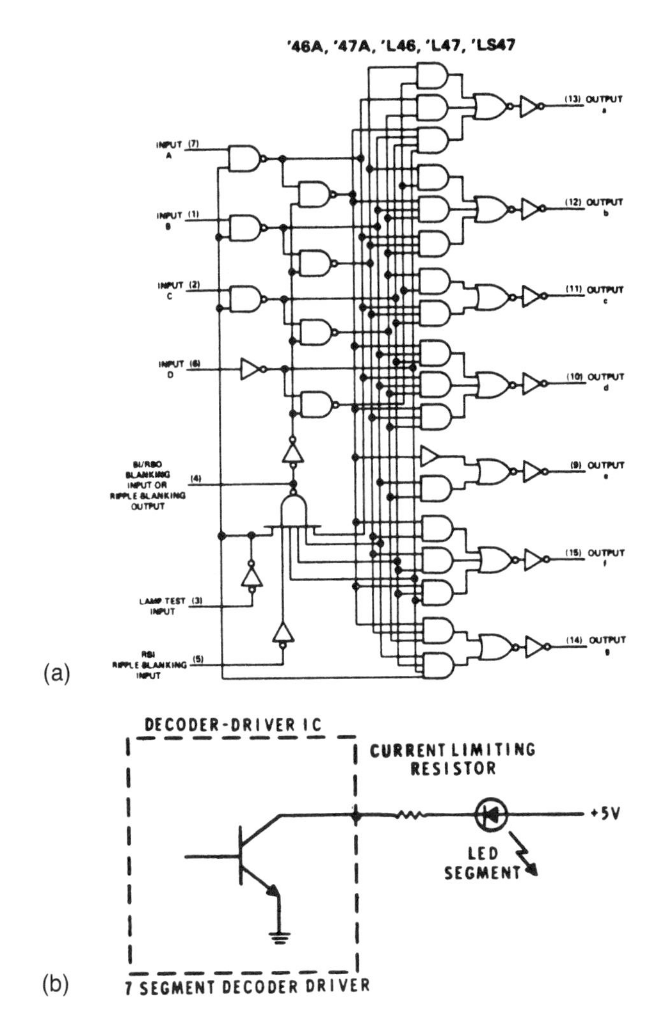

Figure 6-52. (a) BCD-to-seven segment decoder/driver SN54/7447A, (b) decoder-driver output circuit. (Courtesy of Texas Instruments, Incorporated)

segment LED display with left or right decimal points in a complete integrated package. These TTL MSI packages contain the equivalent of 86 gates and are TTL and DTL compatible.

6-20-2 Encoders

An **encoder** is another application of combinational logic where a number of inputs are supplied to the circuit and a multibit binary output is produced. A typical example is the Gray-to-binary encoder. It will be recalled that the Gray code has the characteristic that two adjacent Gray code numbers differ by only 1 bit. The 4-bit Gray code and the 4-bit binary equivalent are shown in Figure 6-53(a). The conversion from the 4-bit Gray code to the 4-bit binary code can be accomplished by the exclusive-OR logic gate combination shown in Figure 6-53(b).

Another very commonly used encoder is the decimal-to-BCD converter illustrated in Figure 6-54. This encoder can be used to convert a decimal keyboard input into 8421 BCD. If any one of the input lines is brought to binary zero then the corresponding 4-bit BCD code is generated at the output. For example, if the number 6 line is brought

Decimal	Gray				Binary			
	G_3	G_2	G_1	G_0	D	C	D	A
0	0	0	0	0	0	0	0	0
1	0	0	0	1	0	0	0	1
2	0	0	1	1	0	0	1	0
3	0	0	1	0	0	0	1	1
4	0	1	1	0	0	1	0	0
5	0	1	1	1	0	1	0	1
6	0	1	0	1	0	1	1	0
7	0	1	0	0	0	1	1	1
8	1	1	0	0	1	0	0	0
9	1	1	0	1	1	0	0	1
10	1	1	1	1	1	0	1	0
11	1	1	1	0	1	0	1	1
12	1	0	1	0	1	1	0	0
13	1	0	1	1	1	1	0	1
14	1	0	0	1	1	1	1	0
15	1	0	0	0	1	1	1	1

(a)

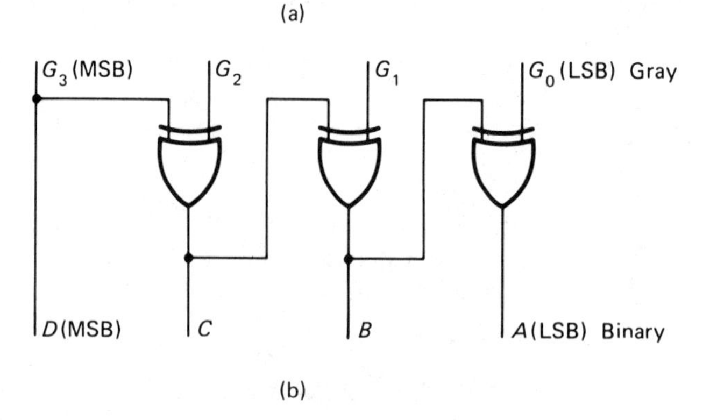

(b)

Figure 6-53. Gray-to-binary encoder: (a) truth table for a 4-bit Gray to 4-bit binary, (b) logic circuit.

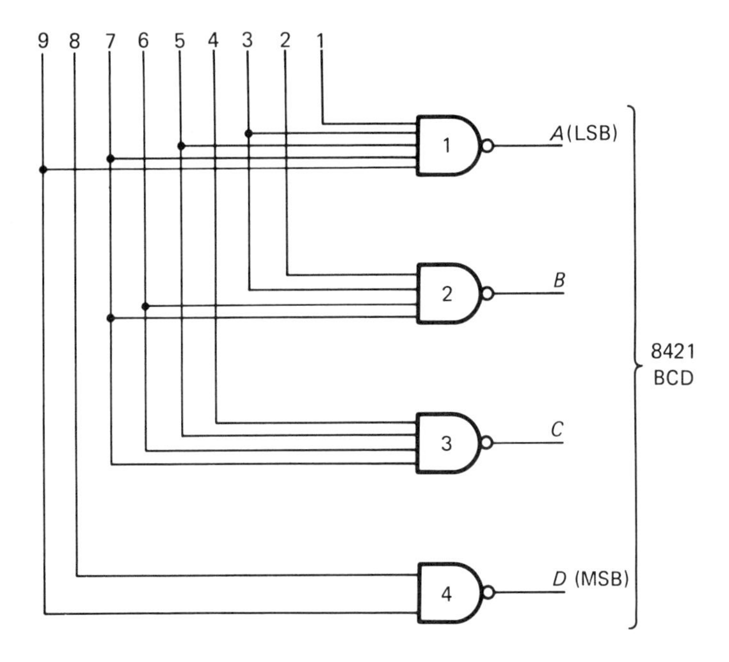

Figure 6-54. 10-line decimal to binary encoder.

to binary 0, then gates 1 and 4 are low and gates 2 and 3 are high, that is, 0110 which is the binary equivalent of 6_{10}.

Encoders are also available in TTL MSI, typical of which is the 54/74147 10-line to 4-line priority encoder shown in Figure 6-55(a). A priority encoder produces an output that will only encode the highest order data line, a feature which is particularly useful in converting parallel inputs of analog to digital (A/D) converters. The 54/74147 priority encoders convert nine data line inputs to a four-line BCD output. The 0 line is implied since a zero is encoded when all data lines are high. The function table for this encoder is shown in Figure 6-55(b).

6-20-3 Multiplexers and Demultiplexers

A **multiplexer (MUX)** is a combinational logic circuit that selects from a number of inputs and routes this input to a single output. Multiplexers are available with 2, 4, 8, and 16 inputs. A typical 8-input multiplexer is the 54/74LS151 data selector/multiplexer shown in Figure 6-56. The 54/74LS151 has a 3-bit address buffer which selects which of the data inputs *D*0 to *D*7 inclusive will appear at the output *Y*. The function of the strobe input is to enable the AND gates when it

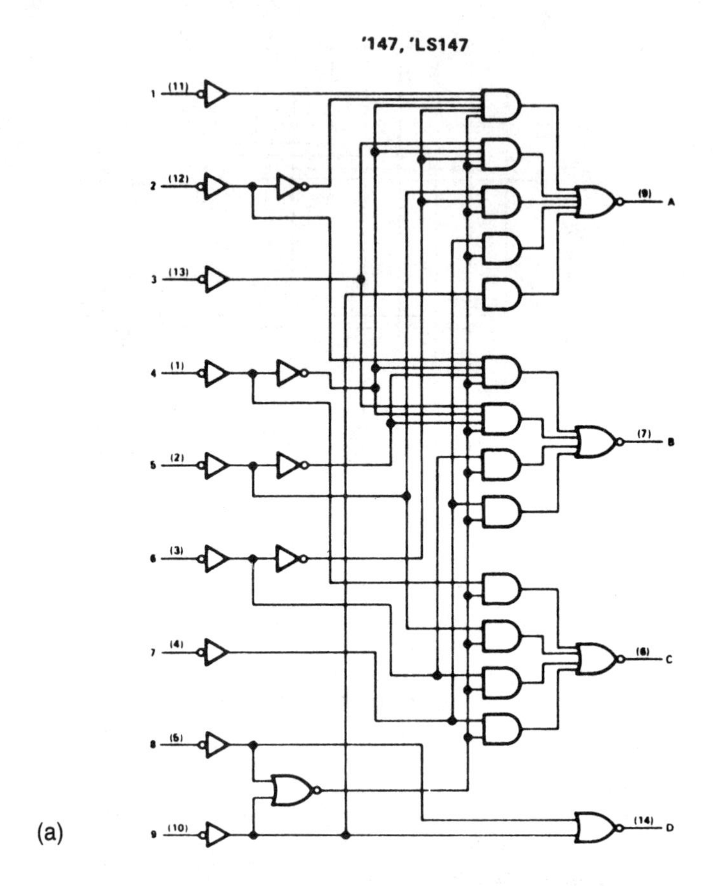

(a)

'147, 'LS147
FUNCTION TABLE

INPUTS									OUTPUTS			
1	2	3	4	5	6	7	8	9	D	C	B	A
H	H	H	H	H	H	H	H	H	H	H	H	H
X	X	X	X	X	X	X	X	L	L	H	H	L
X	X	X	X	X	X	X	L	H	L	H	H	H
X	X	X	X	X	X	L	H	H	H	L	L	L
X	X	X	X	X	L	H	H	H	H	L	L	H
X	X	X	X	L	H	H	H	H	H	L	H	L
X	X	X	L	H	H	H	H	H	H	L	H	H
X	X	L	H	H	H	H	H	H	H	H	L	L
X	L	H	H	H	H	H	H	H	H	H	L	H
L	H	H	H	H	H	H	H	H	H	H	H	L

(b)

H = high logic level, L = low logic level, X = irrelevant

Figure 6-55. 10-line to 4-line priority encoder type SN54/74147: (a) circuit schematic, (b) function table. (Courtesy of Texas Instruments, Incorporated)

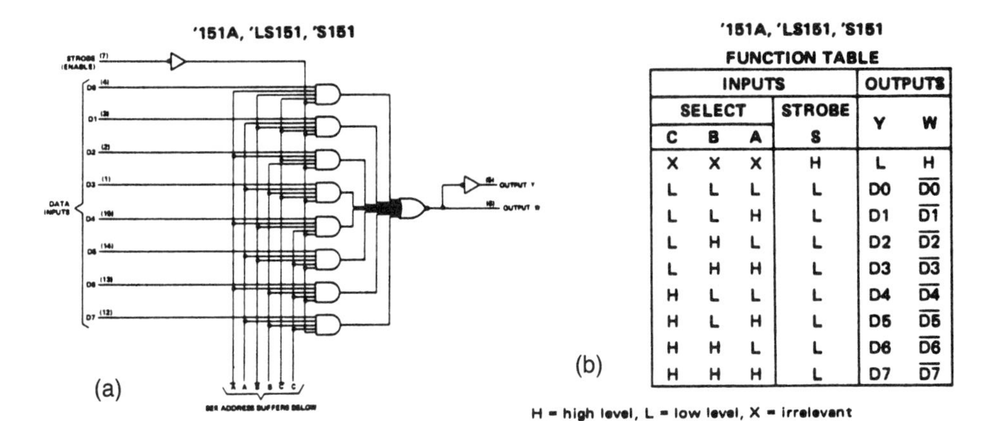

'151A, 'LS151, 'S151

FUNCTION TABLE

INPUTS				OUTPUTS	
SELECT			STROBE	Y	W
C	B	A	S		
X	X	X	H	L	H
L	L	L	L	D0	$\overline{D0}$
L	L	H	L	D1	$\overline{D1}$
L	H	L	L	D2	$\overline{D2}$
L	H	H	L	D3	$\overline{D3}$
H	L	L	L	D4	$\overline{D4}$
H	L	H	L	D5	$\overline{D5}$
H	H	L	L	D6	$\overline{D6}$
H	H	H	L	D7	$\overline{D7}$

H = high level, L = low level, X = irrelevant
$\overline{E0}$, $\overline{E1}$ . . . $\overline{E15}$ = the complement of the level of the respective E input
D0, D1 . . . D7 = the level of the D respective input

Figure 6-56. Data selectors/multiplexers type SN54/74LS151: (a) circuit schematic, (b) function table. (Courtesy of Texas Instruments, Incorporated)

is low. When the strobe input is low the Y output is low and the W output is high. The complete function table is shown in Figure 6-56(b). The 54/74LS151 is fully compatible with TTL and DTL and will perform parallel to serial conversion; in addition, it can also be used as a Boolean function generator.

The **demultiplexer (DMUX)** is the reverse of the multiplexer, that is, it accepts a single input and routes it to one of the multiple outputs under the control of the input binary address. A typical decoder/demultiplexer is the 4-line to 16-line 54/74154 which decodes a 4-bit 8421 BCD input into one of 16 initially exclusive outputs, or as a demultiplexer will distribute the data from one input line to any one of 16 outputs.

An example of the application of multiplexers and demultiplexers is the time-sharing of a single transmission channel by four signals. At the sending end the data is inputted into a multiplexer and transmitted over the single channel. At the receiving end the information is demultiplexed and the appropriate data is separated out to the correct output.

6-20-4 Exclusive-OR Circuits

Exclusive-OR circuits are used extensively to perform binary operations such as addition, subtraction, comparing, and parity generation/checking. The exclusive-OR gate has two inputs and one output. If both inputs are low or both are high the output is low. If the inputs are not the same the output is high.

6-20-4-1 Binary Addition

At this point it will pay to recall the rules of addition of two binary numbers:

$$0 + 0 = 0$$
$$0 + 1 = 1$$
$$1 + 0 = 1$$
$$1 + 1 = 10$$

↑
CARRY

These results are summarized in Table 6-9. The inputs A and B and the sum S form the truth table of the EXCLUSIVE-OR, so $S = A \oplus B$. The C column is the carry and $C = A \quad B$. The truth table shows that binary addition of two 1-bit numbers can be achieved by using an EXCLUSIVE-OR for the sum and an AND for the carry. This single bit binary adder shown in Figure 6-57 is called a **half-adder.** As can be seen from Figure 6-57(a), there is no provision for a carry-in from a previous binary addition.

Table 6-9 Binary Addition

A	*B*	*S*	*C*
0	0	0	0
0	1	1	0
1	0	1	0
1	1	0	1

When adding binary numbers with two or more bits, provision must be made for a carry-in from a previous operation as well as a carry-out. This can be accomplished by combining two half-adders, as shown in Figure 6-58, to form a full adder. The first half-adder is formed by the exclusive-OR gate 1 and the AND gate 2. This combination sums the two input bits A and B. To the output of gate 1, that is, to the sum of A

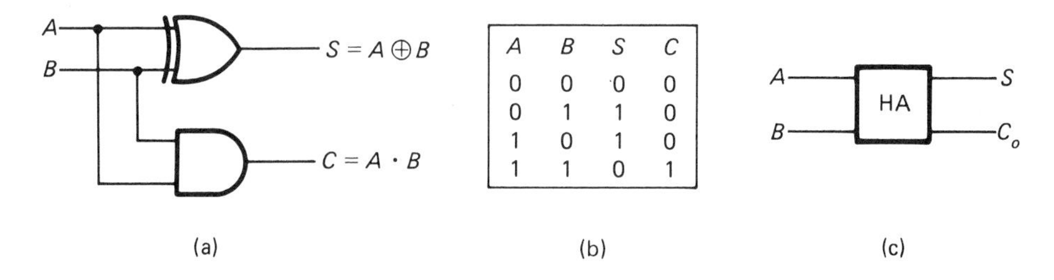

A	B	S	C
0	0	0	0
0	1	1	0
1	0	1	0
1	1	0	1

Figure 6-57. A half-adder: (a) logic circuit, (b) truth table, (c) block diagram.

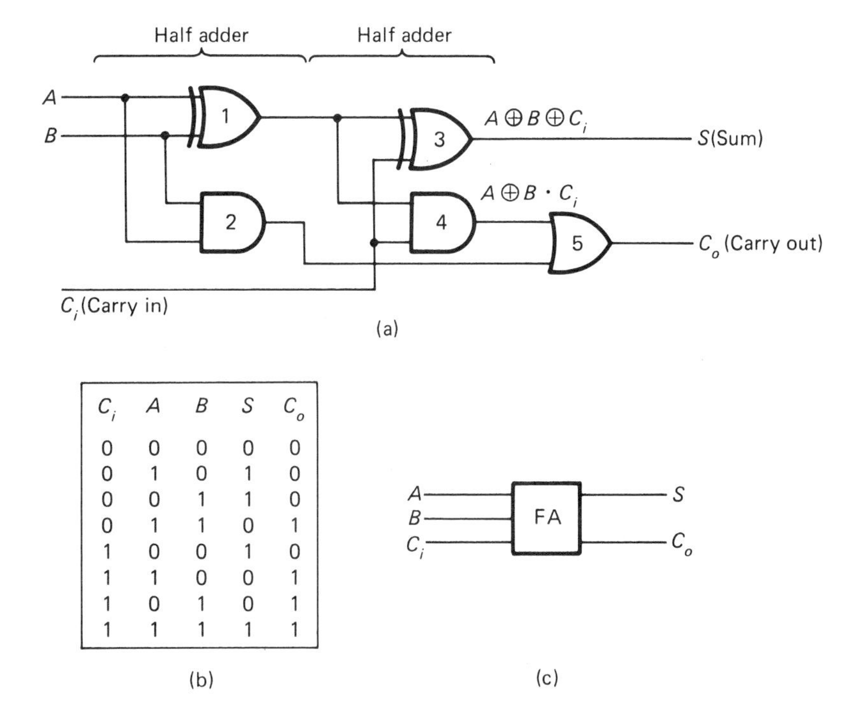

C_i	A	B	S	C_o
0	0	0	0	0
0	1	0	1	0
0	0	1	1	0
0	1	1	0	1
1	0	0	1	0
1	1	0	0	1
1	0	1	0	1
1	1	1	1	1

Figure 6-58. A full-adder: (a) logic circuit, (b) truth table, (c) block diagram.

and B, is added the carry-in C_i from a previous addition as inputs to the second exclusive-OR gate 3 of the second half-adder. The ouput of gate 3 is $A \oplus B \oplus C_i$ which is the sum S. The two carry-outs from gates 2 and 4 are ORed by gate 5 to form the carry-out of the full adder. As can be seen a full adder is required to add two bits and include the carry-in C_i from a previous lower weighted addition.

Larger binary numbers are added in a similar manner; for example, two 4-bit binary numbers can be added by a 4-bit parallel adder as shown in Figure 6-59. It is not normal practice to construct adders from exclusive-ORs since there are a number of TTL MSI integrated circuits such as the 54/7483A 4-bit binary full adders commercially available.

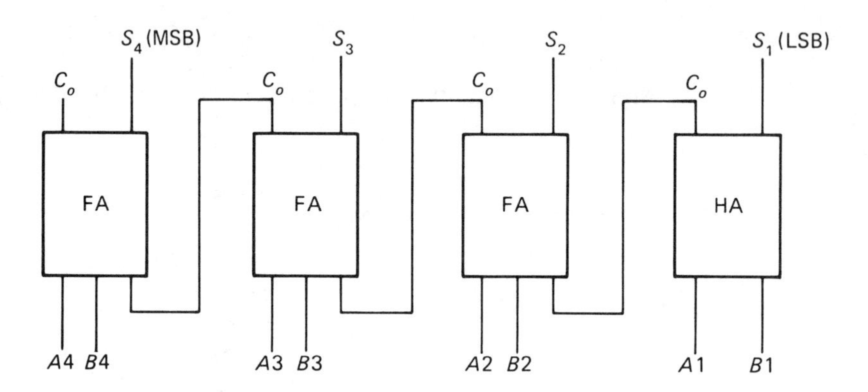

Figure 6-59. A 4-bit parallel adder.

6-20-4-2 Binary Subtraction

When subtracting two binary numbers, the smaller number must be subtracted from the larger. This requires that the circuit performing the subtraction operation, namely $A - B$, must be able to detect if $B > A$, which would result in the difference of $A - B$ being negative, and then carry out the operation $B - A$. You will recall that the rules for subtraction are:

$$0 - 0 = 0 \text{ (difference } D) \text{ (borrow } B_0) = 0$$
$$0 - 1 = 1 \text{ (difference } D) \text{ (borrow } B_0) = 1$$
$$1 - 0 = 1 \text{ (difference } D) \text{ (borrow } B_0) = 0$$
$$1 - 1 = 0 \text{ (difference } D) \text{ (borrow } B_0) = 0$$

These results are summarized in Table 6-10. The inputs A and B and the difference D form the truth table of an exclusive-OR, that is, $D = A \oplus B$. The borrow B_0 only occurs when $A = 0$ and $B = 1$ or $B > A$,

Table 6-10 Half-Subtractor

A	B	D	B
0	0	0	0
0	1	1	1
1	0	1	0
1	1	0	0

which is $B_0 = \overline{A}B$. These results suggest that a half-subtractor (HS) could be constructed from an exclusive-OR, an inverter, and an AND, as shown in Figure 6-60. This subtractor can only subtract two 1-bit binary numbers. To subtract multibit binary numbers requires the ability to accommodate the borrow from a previous operation. This is tackled in a similar manner to the full adder and is termed the borrow-in B_i. A full subtractor is shown in Figure 6-61 together with the truth table and the block diagram representation. If two 4-bit binary numbers are to be subtracted, a total of three full subtractors and a half- subtractor for the LSB would be required, as illustrated in Figure 6-62.

6-20-4-3 Binary Comparators

There is often the need to compare two binary numbers to see if they are the same, $A = B$, or different, $A \neq B$. One method of comparing two 4-bit binary numbers using exclusive-ORs is shown in Figure 6-63. As can be seen if the two binary numbers A and B are the same then the outputs of the respective exclusive-ORs will be binary 0 and the output of the 4-input OR will be binary 0. If the numbers are not equal the output of the OR will be binary 1. Comparators such as the 54/7485 4-bit comparators are available commerically. These comparators can

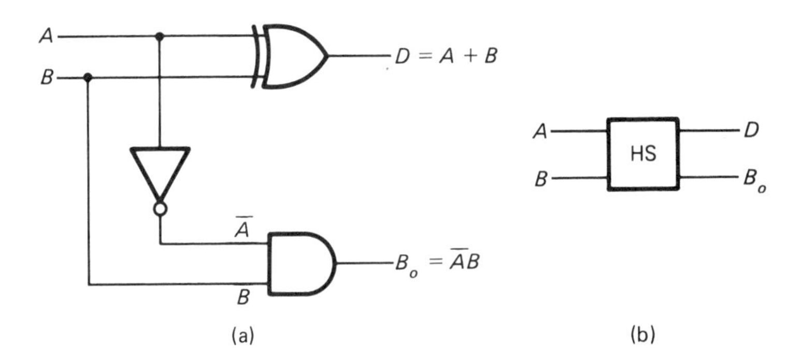

Figure 6-60. A half-subtractor: (a) logic circuit, (b) block diagram.

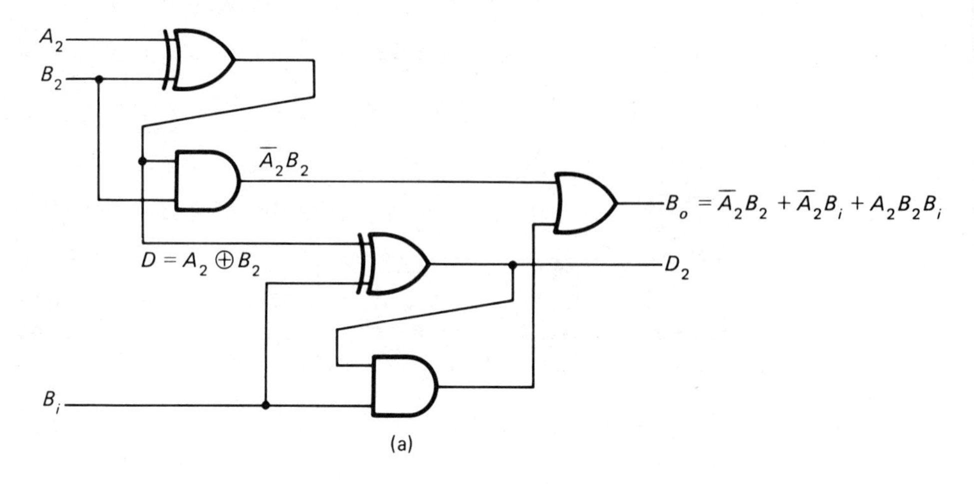

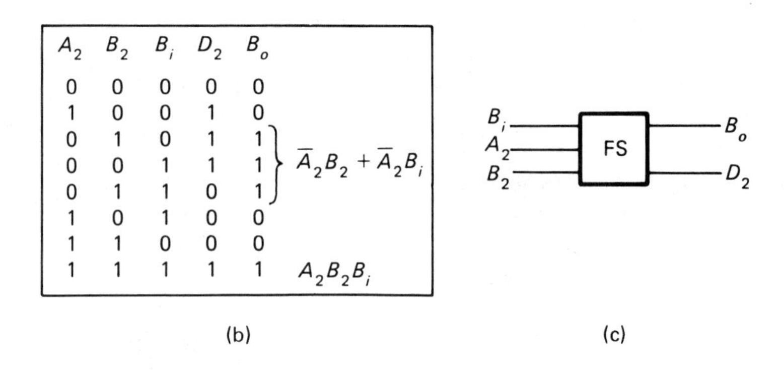

A_2	B_2	B_i	D_2	B_o	
0	0	0	0	0	
1	0	0	1	0	
0	1	0	1	1	
0	0	1	1	1	$\overline{A}_2B_2 + \overline{A}_2B_i$
0	1	1	0	1	
1	0	1	0	0	
1	1	0	0	0	
1	1	1	1	1	$A_2B_2B_i$

Figure 6-61. A full-subtractor: (a) logic diagram, (b) truth table, (c) block diagram.

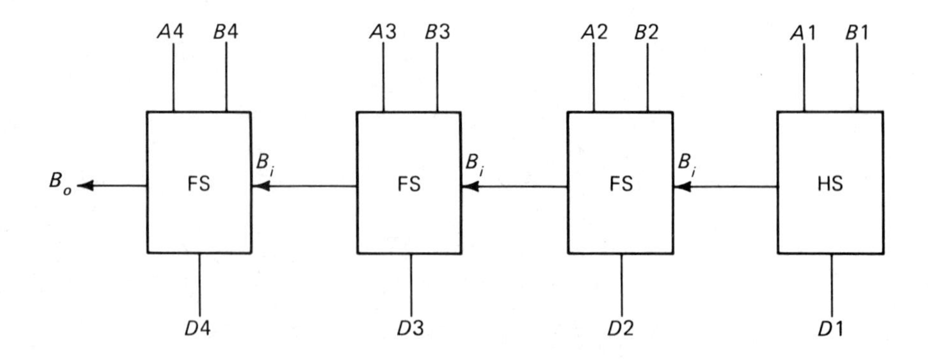

Figure 6-62. A 4-bit parallel binary subtractor.

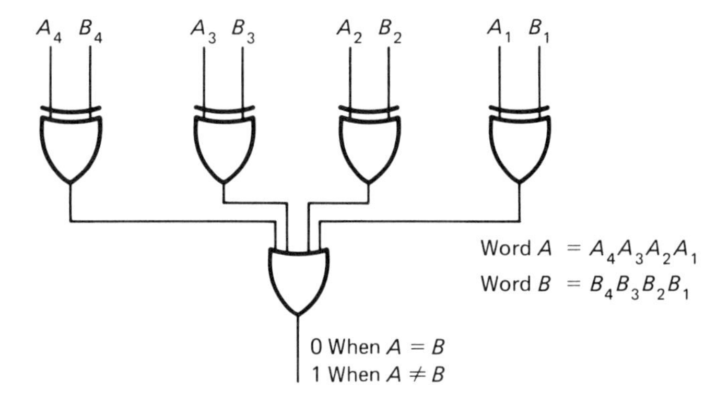

Figure 6-63. A 4-bit binary comparator.

compare both binary and 8421 BCD and determine whether $A > B$, $A < B$, or $A = B$. If binary numbers or words greater than 4-bits are to be compared the appropriate number of comparators are connected in casade.

6-20-4-4 Parity Generators/Checkers

In high-speed digital operations, a requirement exists to detect the introduction of errors in the data as it is being transmitted from one part of the system to another. The error usually takes the form of the inversion of one or more bits.

One method of checking the data is by the addition of an extra bit called the parity bit to each binary word being stored or transmitted, preferably adjacent to the MSB. The parity bit may be either a 1 or a 0 determined by whether odd or even parity is being used. In the odd parity system, the total number of 1's in the binary word or number including the parity bit is odd. Conversely, in the even parity system, the total number of 1's in the binary word including the parity bit is even. As a result, if one bit becomes inverted for either system, a parity error will be detected and a suitable indication will be given that an error has occurred. This system only responds when an odd number of errors occur but it will not detect any even number of errors. However, the odds are in favor of one error as against two or more errors occurring.

In order to be able to implement parity-checking, two circuits are required. First, a parity bit generator at the point where the binary word is generated and second, a parity checker at the point where the binary word is received. Both of these circuits can be implemented with exclusive-OR gates as shown in Figure 6-64. Figure 6-64 shows a parity generator for a 4-bit binary number; the inputs *A*, *B*, *C*, and *D* could be the normal outputs of a 4-bit shift register. Each pair of inputs is monitored by an exclusive-OR and in turn the outputs of these two exclusive-ORs are monitored by another one. This output together with the inverter produces the even and odd parity bit. At the receiving end an almost identical combinational logic circuit acts as the parity checker. The parity bit developed by the checker is compared against the transmitted parity bit, and assuming even parity the output of the last exclusive-OR will be binary 0 if the number has been transmitted correctly and a binary 1 if a parity error exists.

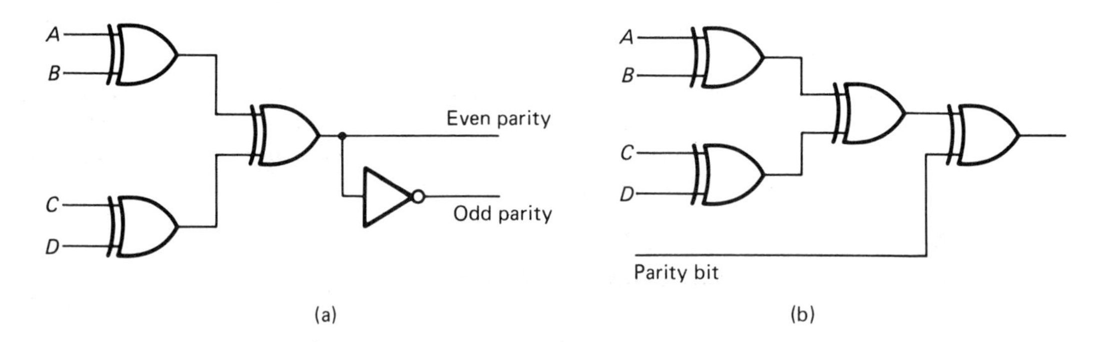

Figure 6-64. A 4-bit parity generator/checker: (a) parity generator, (b) parity checker.

Normally a parity generator/checker is used as a TTL MSI integrated circuit such as the 54/74180 9-bit odd/even parity generator/checker.

SUMMARY

In this chapter an overview has been made of a complex subject. The reader will most likely at this point wish to expand his or her knowledge. To this end a listing of recommended reference material is given at the end of the text.

GLOSSARY OF IMPORTANT TERMS

Number: A figure or a group of figures that represent a quantity.

Base or radix: Specifies the number of digits used to represent a number.

Number system: Represents quantities by summing successive powers of the base or radix.

Binary system: A number system that represents quantities by successive powers of 2.

Bit: A binary digit.

Word: The total number of bits representing a binary number.

Most significant bit (MSB): The left-hand bit of a binary number.

Least significant bit (LSB): The right-hand bit of a binary number.

Octal System: A number system that represents quantities by successive powers of 8.

Hexadecimal system: A number system that represents quantities by successive powers of 16. It uses the digits 0 through 9 and the letters A through F. It is commonly used with microprocessors.

Binary coded decimal (BCD): Represents the decimal digits 0 to 9 inclusive by a 4-bit binary code. The 4-bit equivalents of the decimal number 10 to 15 inclusive are not used. It also uses the standard weighting system of the binary number system, namely the 8421 positional weights.

Gray code: A non-weighted number system. Used as an error minimizing system since there is only a one bit change at a time from one number to the next.

Logic levels: The voltages assigned to the binary 0 and binary 1 states.

Propagation delay: The speed of operation of a logic circuit. It is the time required for the output of a digital device to respond to a change of logic level input.

Power dissipation: The measure of the power consumed by a logic gate or other digital circuit. Usually expressed in milliwatts per logic gate and is an average value of the power consumption in the binary 0 and 1 states.

Noise immunity: A measure of the response of a logic circuit to noise voltages in the inputs and outputs of logic circuits.

Noise margin: The dc noise margin is the potential difference between the input and output switching voltage levels.

Fan-out: Defines the number of standard logic loads that a logic gate output can supply and yet maintain correct operation at normal logic levels, speed, temperature, etc.

AND gate: A logic gate that produces a binary 1 at the output only if the inputs are at binary 1. The output is at binary 0 for any other combination of inputs.

OR gate: A logic gate whose output is a binary 1 if one or more of the inputs are at binary 1. The output is at binary 0 when the inputs are at binary 0.

NOT or inverter: The output is the complement or negation of the input.

NAND gate: A logic gate that produces a binary 0 at the output only if the inputs are at binary 1. The output is at binary 1 for any other combination of inputs.

NOR gate: A logic gate that produces a binary 1 at the output only if the inputs are at binary 0. The output is at binary 0 for any other combination of inputs.

Exclusive-OR: A logic gate whose output is binary 1 only when the inputs are different. When the inputs are the same, both binary 1 or binary 0, the output is binary 0. It is used in binary addition of two binary bits. It is also called a half-adder since there is no provision for the carry bit.

Exclusive-NOR: A logic gate whose output is binary 1 when both inputs are the same. If the inputs are different the output is binary 0. It is also called a comparator or coincidence function.

Boolean algebra: An algebraic system based entirely on the binary number system.

Karnaugh maps: A graphical technique designed to simplify Boolean expressions. The expressions must be expressed in the form of an OR function although the individual terms may be ANDed.

SSI: Small-scale integration; less than 10 transistors.

MSI: Medium-scale integration; between 10 and 100 transistors.

LSI: Large-scale integration; in excess of 100 transistors.

RTL: Resistor-transistor logic. Obsolete.

DTL: Diode-transistor logic.

TTL or T^2L: Transistor-transistor logic. The major logic family.

ECL: Emitter-coupled logic or current mode logic. A fast switching logic which is not compatible with other logic families.

PMOS: P-channel MOSFET.

NMOS: N-channel MOSFET.

CMOS: Complementary MOS.

Flip-flop or bistable multivibrator: A logic circuit with two stable states, that is, binary 1 or binary 0, at its outputs. There are two complementary outputs Q and Q. The FF acts as a 1-bit memory storage.

Set: The S or set input sets the FF so that the Q output is at binary 1 and the FF is storing a binary 1.

Reset: The R or reset input resets the FF so that the Q output is at a binary 0.

R-S flip-flop or latch: The simplest FF and is activated as soon as the set (S) or reset (R) conditions are initiated.

Clock pulses: Usually square voltage pulses generated by an astable multivibrator, a unijunction transistor, or a crystal oscillator. Used to transfer data from the input to the output of a FF.

Astable or free-running multivibrator: A modified FF, it has two unstable states and switches backwards and forwards between states at a rate determined by the time constants of the RC networks.

Clocked R-S FF: An *R-S* FF with the addition of two more gates to permit the use of a clock pulse. The data inputs S and R (double rail) are enabled when the clock pulse goes high, positive edge triggering.

Direct set (S_D): Overrides any other input, and the output changes immediately when the direct set input is applied.

Direct clear (C_D): Overrides any other input, and output changes immediately when the direct clear input is applied.

D-Type FF: Overcomes the ambiguous condition of the clocked *R-S* FF and transfers the data on the D input to the Q output on the leading edge of the next clock pulse.

Master-Slave R-S FF: Two clocked cascaded *R-S* FFs with the clock pulse to the slave inverted. The master stage is enabled and the slave stage disabled when the clock pulse goes high. When the clock pulse goes low the master stage is disabled and the slave stage enabled.

Toggle: The output state changes with each clock pulse.

T or toggle FF: A master-slave *R-S* FF with the S and R inputs of the master cross connected to the Q and Q outputs of the slave. It toggles on the negative going transition of the clock pulse.

J-K master-slave FF: Effectively a combination of the *R-S* and T FFs. Toggles when the J and K inputs are high and eliminates the indeterminant state of the *R-S* FF.

Monostable multivibrator: A multivibrator with one stable and one temporarily stable state. It remains in the stable state until an input pulse is applied, then the output goes high and will remain high for a period of time determined by an externally connected RC combination. It is used to delay, increase, and decrease pulse widths.

Schmitt trigger: A regenerative bistable multivibrator that converts slowly-changing input signal waveforms to a square wave output with fast rise and fall times.

Sequential logic: Usually formed from FFs and used to store or alter binary data in response to logic input signals, the sequence of operations being controlled by clock signals. The most commonly used applications are counters and shift registers.

Counters: One of J-K, toggle, or D-type FFs that count pulses.

Binary counter: A counter that counts in binary and divides by factors of 2^N.

Ripple or asynchronous counter: A counter in which only the first FF associated with the LSB is operated by the incoming clock. All the remaining FFs are toggled by the preceding FF.

Counting speed: The total time required from the application of the first pulse to the change of state of the MSB.

Up-counter: A binary counter whose output increases by one for each clock pulse.

Down-counter: A binary counter whose output decreases by one for each clock pulse.

Synchronous counter: A counter in which all the FFs are triggered simultaneously by the same clock pulse.

Modulus: The total number of states that can be assumed by a counter.

Shift register: Consists of a number of FFs acting as binary storage elements, which permit stored data to be moved to the left or right under the control of clock pulses.

Ring counter or sequencer or circulating register: Consists basically of a shift register with the outputs of the last FF connected back to the first FF, that is, Q to J and $\bar{Q}$ to K. Commonly used to produce repetitive signals one clock pulse period wide for control operations in or between digital circuits.

Decoder: A combinational logic circuit used to detect and indicate a particular binary number from all possible combinations of binary numbers present.

Encoder: A combinational logic circuit which accepts one or more inputs and produces a multibit binary output.

Multiplexer: A combinational logic circuit that selects from a number of inputs and routes this input to a single output.

Demultiplexer: The reverse of a multiplexer. It accepts a single input and routes it to one of the multiple outputs under the control of the input binary address.

Binary comparator: An arrangement of exclusive-ORs that determine whether two binary words are the same.

REVIEW QUESTIONS

6-1 Why is the binary number system used in digital circuits?

6-2 What are the advantages of the octal number system as compared to the binary number system?

6-3 What are the advantages of using the hexadecimal system?

6-4 What is the Gray code? What are the advantages and disadvantages of using the Gray code?

6-5 With the aid of a sketch explain positive and negative logic.

6-6 Define what is meant by propagation delay.

6-7 What is the significance of power dissipation of a logic gate or digital circuit?

6-8 What is meant by noise margin? What are the causes of noise and how may it be reduced?

6-9 Draw the circuit symbol and the truth tables for the following logic gates: (a) AND, (b) OR, (c) NAND, (d) NOR, (e) exclusive-OR, and (f) exclusive-NOR.

6-10 What is the standard TTL gate? What are the advantages to the use of TTL?

6-11 What is meant by totem-pole or active pull-up output and what are the benefits obtained by its use?

6-12 What are the variations of the basic TTL gate and what are the advantages and disadvantages of each type?

6-13 What are the advantages of using MOSFET IC logic gates?

6-14 What is meant by enhancement mode?

6-15 Sketch the basic circuit and explain the following circuits: (a) PMOS inverter, (b) NMOS NOR, (c) NMOS NAND.

6-16 What are the benefits of using CMOS?

6-17 Sketch the basic circuit and explain the following CMOS circuits: (a) inverter and (b) OR.

6-18 With the aid of a schematic, truth table, and timing waveforms explain the operation of an R-S FF.

6-19 Repeat Question 6-18 for a clocked R-S FF.

6-20 Repeat Question 6-18 for a D-type FF.

6-21 Repeat Question 6-18 for a master-slave R-S FF.

6-22 Repeat Question 6-18 for a T FF.

6-23 Repeat Question 6-18 for a J-K master-slave FF.

6-24 With the aid of a schematic explain the operation of an astable multivibrator formed from discrete components.

6-25 With the aid of a schematic explain the operation of an IC astable multivibrator.

6-26 With the aid of a diagram explain how a 555 timer may be used as a monostable.

6-27 With the aid of a sketch explain how the Schmitt trigger circuit of Figure 6-37 performs.

6-28 With the aid of a sketch, count sequence table, and timing waveforms explain how a 3-bit binary ripple up-counter performs.

6-29 Repeat Question 6-28 for a 3-bit binary ripple down-counter.

6-30 With the aid of a count sequence table explain the operation of the synchronous binary counter of Figure 6-41.

6-31 With the aid of a count sequence table and timing waveforms explain the operation of a 4-bit BCD counter.

6-32 Repeat Question 6-31 for a Mod-3 counter.

6-33 Sketch a four-stage shift register to accept a parallel input and produce a parallel output.

6-34 Sketch a six-stage Johnson counter.

6-35 Explain the function of a decoder and discuss an illustrative example of its use.

6-36 What is the function of a multiplexer?

6-37 What is the function of a demultiplexer?

6-38 Explain the operation of the following exclusive-OR circuits: (1) a half-adder, (2) a full adder, (3) a half-subtractor, and (4) a full subtractor.

6-39 Draw a circuit to compare two 6-bit binary numbers.

PROBLEMS

6-1 Convert the following decimal numbers to binary: (a) 255, (b) 33, (c) 121, (d) 17, and (e) 325.

6-2 Convert the following decimal numbers to binary: (a) 16.75, (b) 243.675, (c) 15.375, (d) 332.56, and (e) 13.575.

6-3 Add the following binary numbers: (a) 1110 + 101, (b) 110010 + 110110, (c) 1011111 + 1100111, (d) 1001 + 1101 + 1011, and (e) 110010 + 1011.

6-4 Add the following binary numbers: (a) 11010.101 + 101.11, (b) 11110.101 + 110.01, (c) 111.1101 + 1101.111, (d) 1011.011 + 101.11, and (e) 11.111 + 101.101.

6-5 Subtract the following numbers by binary subtraction and by complementing: (a) 101 − 11, (b) 1101 − 110, (c) 11010 − 10101, (d) 110111 − 101101, and (e) 10011 − 1101.

6-6 Subtract the following numbers by 1s and 2s complements: (a) 111000 − 11010, (b) 111 − 10001, (c) 1101 − 10111, (d) 110011 − 101010, and (e) 11001 − 1101.

6-7 Multiply the following binary numbers: (a) 1101 × 1101, (b) 11011 × 111, (c) 1101 × 1011, (d) 1001 × 1001 × 1001, and (e) 11.1 × 10.1.

6-8 Divide in binary the following numbers: (a) 1101 ÷ 101, (b) 11011 ÷ 111, (c) 11101 ÷ 1101, (d) 1100101 ÷ 1011, and (e) 100101 ÷ 1101.

6-9 Convert the following decimal numbers to octal: (a) 255, (b) 33, (c) 325, (d) 125.53125, and (e) 637.2782.

6-10 Convert the following octal numbers to decimal: (a) 271.47, (b) 175.32, (c) 2471, (d) 325, and (e) 111.

6-11 Convert the following binary numbers to octal: (a) 1101101.11, (b) 1001.011, (c) 111011, (d) 111101011, and (e) 110111011110.11011.

6-12 Convert the following decimal numbers to hexadecimal: (a) 122.63, (b) 44, (c) 35.125, (d) 19.0625, and (e) 28.1015625.

6-13 Convert the following binary numbers to hexadecimal: (a) 1001.1101, (b) 11010101.01110111, (c) 11110110.00011101, (d) 11101111.110101, and (e) 101.0101.

6-14 Convert the following hexadecimal numbers to binary: (a) AE7.C2, (b) 3C6.22F8, (c) 1B4.65D, (d) 8F.41, and (e) 176.4E.

6-15 Convert the following decimal numbers to BCD: (a) 22.68, (b) 3.1416, (c) 1.726, (d) 144, and (e) 225.45.

6-16 Convert the following BCD numbers to decimal: (a) 100101010010, (b) 111000101010011, (c) 110000010.0111, (d) 1010010000.00110101, and (e) 11010011001.

6-17 Prove by the rules of Boolean algebra the following:

(a) $A + AB = A$

(b) $A + \bar{A}B = A + B$

(c) $\bar{A}\bar{B}C + A\bar{B}C + AB\bar{C} + ABC = AB + \bar{B}C$

(d) $(A + B)(\bar{A} + \bar{B}) = A\bar{B} + \bar{A}B$

(e) $ABC + AB\bar{C} + A\bar{B}C + \bar{A}BC = AB + AC + BC$

(f) $\bar{A}B + \bar{B}C + \bar{C}A = A\bar{B} + B\bar{C} + \bar{A}C$

(g) $A\bar{B}C + A\bar{B}C + AB\bar{C} + \bar{A}\bar{B}C + \bar{A}B\bar{C} + A\bar{B}\bar{C} = A\bar{B} + B\bar{C} + \bar{A}C$

(h) $\overline{ABC}(A+B+C) = A\bar{B} + B\bar{C} + \bar{A}C$

(i) $\overline{\bar{A}B + A\bar{B}} = AB + \bar{A}\bar{B}$

(j) $A\bar{B}C + \bar{A}B + \bar{B}C + \bar{B}\bar{C} = \bar{A} + \bar{B}$

6-18 By the use of Karnaugh maps simplify the following expressions:

(a) $\bar{A}BC + A\bar{B}C + AB\bar{C} + \bar{A}\bar{B}C + \bar{A}B\bar{C} + A\bar{B}\bar{C}$

(b) $ABC + AB\bar{C} + A\bar{B}C + \bar{A}BC$

6-19 In Figure 6-28 $RB1$=22kΩ, $RB2$=22kΩ, (1) find the PRR when C1=C2=0.02μF and (2) find the *PRR* when C1=.0.02μF and C2=0.01μ F.

6-20 (1) What is the clock cycle time of a 2.5MHz clock? (2) If a device has a propagation delay time of 100ns. What is the maximum possible clock frequency?

6-21 A *J-K* FF has a propagation delay of 12ns to binary 1 from binary 0 after the clock pulse is applied, and a 15ns propagation delay to binary 0 from binary 1. A three-stage counter is in the 011 count state. After a clock pulse is applied: (1) what is the next count state? and (2) what is the time delay before the count state is 111?

6-22 A 1MHz signal is applied to a four-stage binary ripple counter. Determine: (1) the frequency at each of the Q outputs, (2) the count capability of the counter expressed in binary and decimal and (3) sketch the waveforms of the outputs at Q_A, Q_B, Q_C and Q_D assuming that the data is transferred on the trailing edge of the clock pulse.

6-23 A digital watch has a crystal controlled oscillator which produces a 32,768Hz signal. How many binary stages are required to produce an output pulse at 1s intervals?

6-24 A 256-bit shift register is to be used as a delay line. If the clock rate is 2MHz, what will be the delay introduced?

7

Analog and Digital Transducers

7-1 INTRODUCTION

In all industrial processes there is a requirement to measure accurately and reliably the value of the controlled variable. The most common controlled variables that require measurement are position, velocity, pressure, temperature, flow, and level. In all of these measurement requirements, a physical quantity is measured and usually an electrical output proportional to the physical quantity is derived. The final form of the electrical output is determined by the type of control means, that is, whether the system is **analog** or **computer controlled.** The major reasons for the conversion to an electrical output are ease of conveying information over relatively long distances, ease of amplification, filtering out unwanted information, and the ability to perform mathematical operations on the electrical output, such as differentiation (rate of change), integration (summation over time), addition, subtraction, etc.

7-2 TERMINOLOGY

A **transducer** by definition is a device that converts energy in one form to another, usually an electrical output. In instrumentation applications the function of the transducer is to measure the controlled variable with a minimum of energy being removed from the quantity being measured.

There are usually two elements in a transducer: first, the sensor which measures the controlled variable and, second, the element which converts the sensor signal into a useful electrical output.

A transducer must have two basic characteristics, namely, accuracy and speed of response.

7-2-1 Transducer Characteristics

Accuracy reflects the measurement capability of a transducer to approach the true value of the measured quantity. The true value is represented by internationally defined standards, which are in turn transferred to working standards and high-class calibration equipment. The transducer must be able to accurately reflect all changes that occur in the measured quantity be it pressure, velocity, temperature, level, flow, etc. In control applications the function of the transducer is to create a visual or audible signal to permit a manual change to be made or initiate a corrective control action. In either case the accuracy and speed of response of the transducer will reflect on the performance of the system.

7-2-1-1 Accuracy

There are four factors that affect the accuracy of a transducer: static error, dynamic error, reproducibility or repeatability, and dead time and dead zone.

Static error is defined as the deviation of the measured value of the physical quantity being measured from its actual value when the physical quantity is constant, and it is expressed as a percentage of the full-scale range of the transducer, not as a percentage of the actual reading.

Dynamic error is the error that occurs because of changes in the value of the physical quantity caused by either input or load changes. The measured quantity lags behind the actual value and the value of the dynamic error is continually changing and will be zero only when the measured quantity remains constant for a period of time. Dynamic error is present in all transducers and is additional to and independent of the static error, but it is present only when there are changes in the measured quantity.

Reproducibility or **repeatability** is the maximum deviation from the average of repeated readings when the measured quantity is constant. If a transducer has a good degree of repeatability but its readings are not accurate, it is usually possible to improve the accuracy; however, if the degree of repeatability is poor, it is impossible to improve the accuracy.

Two other factors that affect the accuracy of a transducer are dead time and dead zone. **Dead time** describes the time interval between a change occurring in the measured quantity and the change being noted at the transducer output. **Dead zone** defines the largest change occurring in the measured quantity without a detectable change in the transducer output. Dead zone introduces an initial time lag and thus reduces the speed of response of the transducer to changes in the measured quantity. Dead time on the other hand delays the reaction of the control system.

7-2-1-2 Speed of Response

Obviously the ideal transducer responds instantaneously to changes in the measured quantity. This condition is not attainable in practice because of dynamic error, dead time and dead zone effects.

The transducer may begin to respond immediately to the change; however, it takes time for the change to occur in the output. This time interval is known as the **lag,** the retardation of the transducer response compared to the changes in the measured quantity. As can be well appreciated, corrective action in industrial processes cannot take place until the change in the measured quantity has been detected by the transducer. As a result, the faster the response of the transducer the faster the controlled variable will be corrected.

The speed of response of a transducer depends upon a number of factors and can be best illustrated by considering temperature sensing using a thermocouple. The first factor, thermal resistance, is affected by the nature of the medium (liquid or gaseous) being sensed and the rate of flow of the medium. The second factor depends upon the capacitance of the thermocouple and its protective shield and casing. The third factor is the thermal mass of the transducer which causes a delay in heat transfer. The last factor is the transmission lag, that is, the time interval from the thermocouple responding to the change in temperature until it is detected at the measuring point.

In general, these factors are appropriate to all transducers and their combined effect determines the speed of response of the transducer.

7-3 SENSORS

As was previously mentioned, a transducer usually consists of two parts, namely, the sensor and the signal conditioner which modifies the sensor output to a useful form. In this section we will concentrate on the sensors, and in a later section we will review the methods and problems of signal conditioning. Sensors can be grouped in terms of the process variable being measured, that is, motion, velocity, temperature, etc.

7-3-1 Linear Motion Sensors

Linear motion sensors are used for the measurement of linear displacement and, as will be seen later, are also used in such applications as the measurement of pressure.

7-3-1-1 Linear Motion Potentiometer

A linear motion potentiometer probably employs the simplest method of linear motion measurement. The basic principle, illustrated in Figure 7-1(a), depends upon the movement of a slider connected to a moving process causing variations in the output resistance of the potentiometer. The change in output resistance can be measured by a resistance bridge calibrated in terms of the linear movement, or a varying current output signal can be obtained by placing a constant voltage source in series with the potentiometer.

Alternatively, a varying voltage output can be obtained by using a constant current source. In each case an output signal is derived which is proportional to the change in resistance and in turn is proportional to the linear motion of the process being measured. The increase or decrease of the resistance determines the direction of motion. The resolution of the resistance readings are determined by whether the potentiometer is wire-wound or a film-type element. The advantage of this type of sensor is its relatively low cost. However, it has a number of limitations: first the range of travel is usually between 1.25 and 6 in; second, the life expectancy is relatively low because of mechanical wear between the slider and resistance element; and last, it can introduce electrical noise caused by the wiping action into the system. Common applications are in the measurement of the movement of pneumatic and hydraulic actuators and linkage systems.

7-3-1-2 Linear Motion Variable Inductor

In outward appearance the linear motion variable inductor resembles the linear motion variable potentiometer and may be used in similar applications. The principle of operation depends upon the variation of the coil inductance by changes in the reluctance of the magnetic circuit caused by variations in the position of the moveable ferromagnetic core (see Figure 7-2). Inductive transducers always require an external power supply, usually ac, and form part of a voltage-divider or variable-frequency control circuit. The variations in inductance directly related to the movement of the ferromagnetic core can be measured by an impedance bridge and recorded by a bridge-type recorder. The sensor coil can be made to be part of an oscillator tank

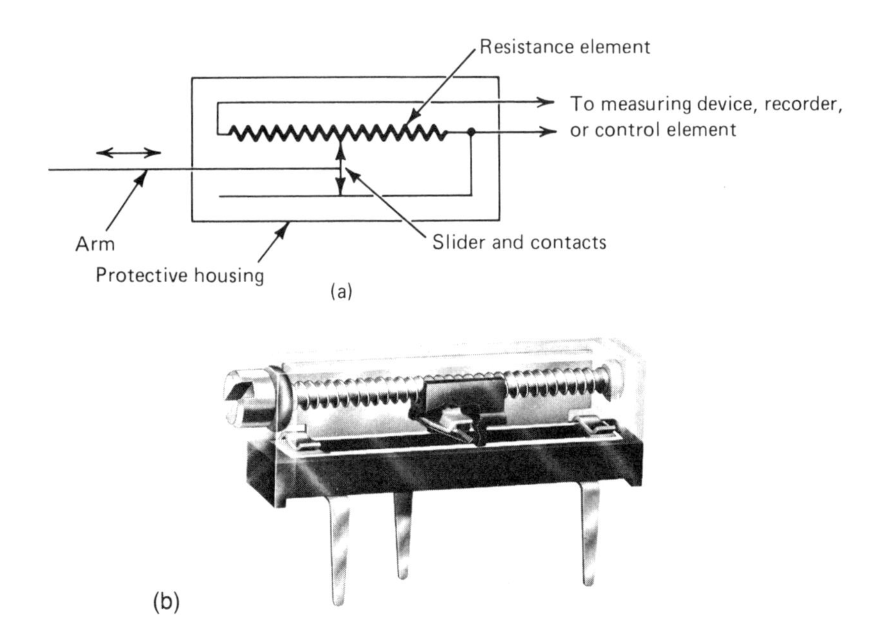

Figure 7-1. Linear motion potentiometer: (a) schematic, (b) commercial unit.

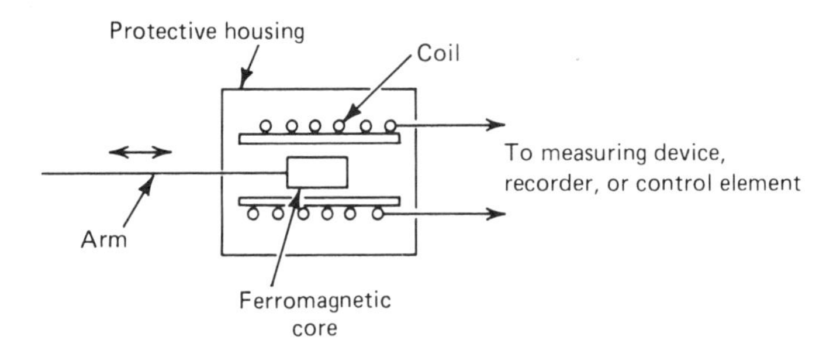

Figure 7-2. Schematic representation of a linear motion variable inductor.

circuit where changes in inductance change the frequency of the oscillator.

The linear motion variable inductor has the advantage, compared to the linear motion potentiometer, of eliminating the sliding contact and thus eliminating wear and electrical noise. In addition, there is electrical isolation between the input and output, and the whole device is more rugged and less subject to errors caused by vibration, shock, and hostile environments.

7-3-1-3 Linear Variable Differential Transformer (LVDT)

The LVDT is more commonly encountered than the linear motion variable inductor. While it is more expensive, it has the advantage of greater sensitivity and can produce polarity sensitive dc voltages or a phase sensitive ac voltage output. The amplitude and sense of these voltages is proportional to the amount of core movement and direction. LVDTs are available in a number of types to meet application requirements such as: standard with stroke ranges usually between ± 0.04 to ± 2.00 in; miniature with stroke ranges ± 0.005 to ± 1.5 in; dc-to-dc long and short stroke with an integral solid-state oscillator-demodulator; and high-temperature capability up to 427 °C (800 °F).

The principle of operation is basically the same as a transformer; it consists of a primary winding supplied typically with 10 V at 60 or 400 Hz and two identical and symmetrical secondary windings connected in opposition, see Figure 7-3(a), with both windings placed on a cylindrical former. The moveable nickel-iron core is positioned by the element being measured. When the core is in the center or null position, the voltages produced by each secondary winding are equal and opposite and the output voltage is zero. As the core is moved from the null position, the voltage in the secondary surrounding the core increases because of the increased flux linkage, the voltage in the other secondary decreases, and the output voltage increases, its phase with respect to the primary input determined by the direction of travel.

In some applications a dc output is required for use with dc servomotors, recorders, indicators, and controllers. This is accomplished by using a demodulator, see Figure 7-3(b), to produce a dc output voltage proportional to the displacement of the core with its polarity determined by the direction of the movement of the core from the null. At the null position the output voltage is zero.

The LVDT is used to directly measure linear displacement, or as a component in more complex transducers to measure force, thickness, weight, pressure, torque, speed, horsepower, flow, level, and viscosity, to name a few applications. The sensitivity of the LVDT is usually expressed in millivolts per mil, a typical value being 2.0 mV/mil (1 mil = 0.001 in), it also possesses a good resolution with a highly linear output.

7-3-1-4 Linear Motion Variable Capacitor

The operation of the linear motion variable capacitor depends on the fact that its capacitance is inversely proportional to the distance between the two plates. That is, as the moveable plate approaches the stationary plate the capacitance increases, and vice versa. The change

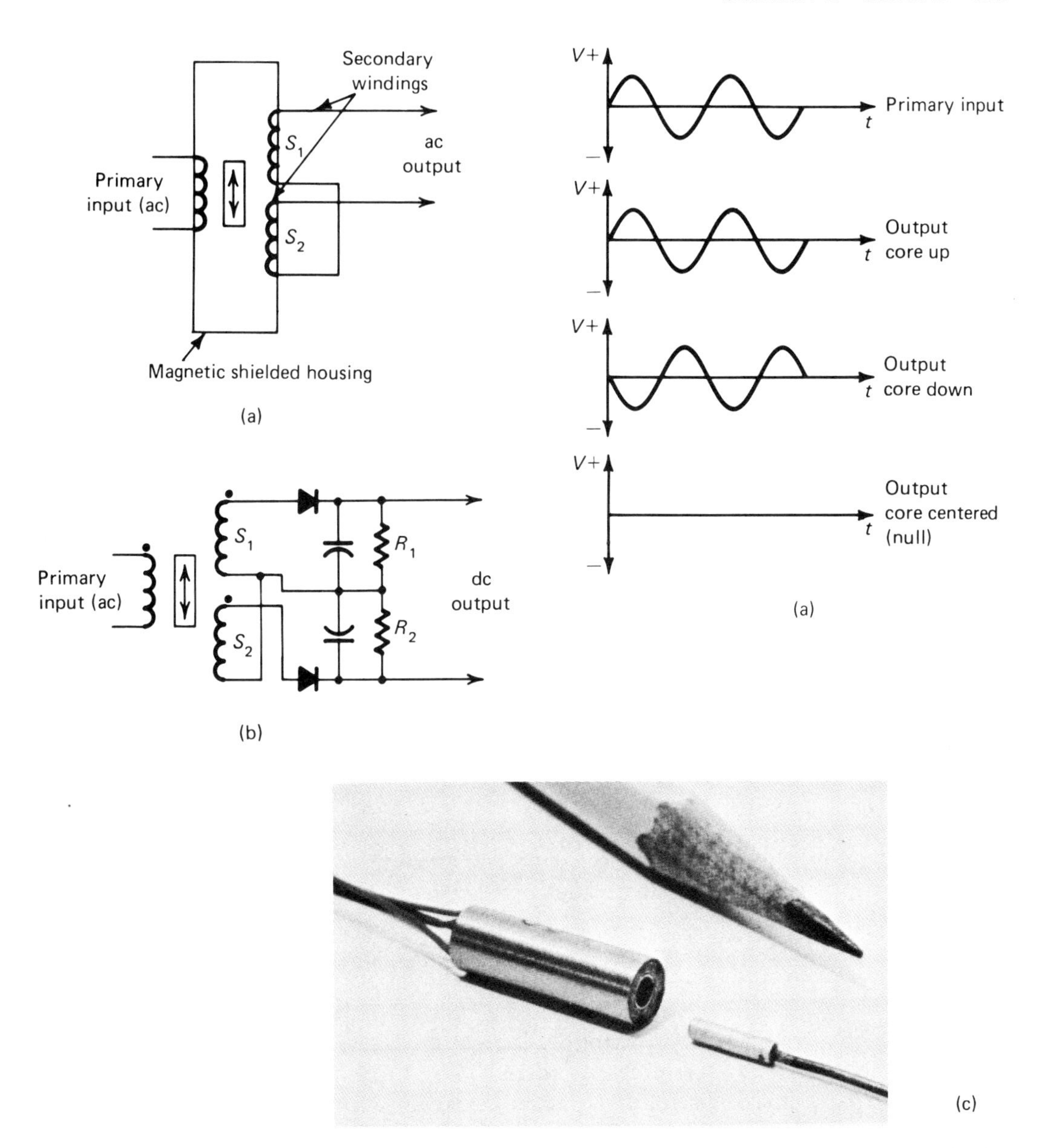

Figure 7-3. Linear variable differential transformer: (a) ac output, (b) dc output, (c) typical LVDT transducers.

in capacitance is proportional to the amount of movement and the direction of movement is determined by sensing the change of capacitance. The capacitance can be measured directly by an impedance bridge or the capacitor can be used as the variable component in an *LC* oscillator and the changes in frequency used to measure the amount of movement in a manner similar to the linear motion variable inductor (see Figure 7-4).

The variable frequency output signal can be fed into a frequency converter and a counter to produce a digital output. The major limitation of this transducer is the rather small range of movement of the sensing plate; however, this is offset by the ability to measure linear movement with high accuracy down to as little as one microinch. It also has good resolution and linearity.

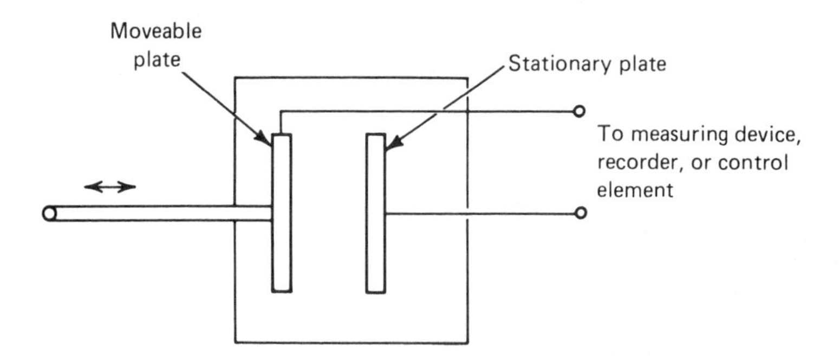

Figure. 7-4. Linear motion variable capacitor.

7-3-1-5 Linear Encoders

So far in our discussion the output of the sensor has been continuous, that is, an analog output. Nowadays, with the rapid introduction of numerically controlled machines, a requirement exists for sensors with a digital output that is acceptable to digital computers, digital displays and, digital readouts.

Digital position sensors are called **encoders**, which in turn are broken down into two groups, **incremental** and **absolute**. Incremental encoders determine the instantaneous position relative to a reference point or datum. In turn, the incremental encoder may be unidirectional or bidirectional. Absolute encoders give a direct digital readout of position and therefore do not require a reference or datum.

The incremental encoder output consists of a series of pulses, each pulse representing an increment of movement. If the pulses are counted with respect to the zero or reference position, the amount of movement is known but not the direction. The speed of movement can also be determined by comparing the pulse rate to time. For the encoder to be able to determine whether to add or subtract pulses, that is, be bidirectional, it is necessary to produce a second train of pulses displaced 90° from the first pulse train. The phase relationship of these quadrature signals, that is, leading or lagging, depends upon the direction of motion, and interfacing to suitable logic determines the direction of motion. The required accuracy determines the resolution of the encoder scale; for example, if the required measurement capability is 0.001 in, then 1000 pulses must be produced per inch of travel.

Incremental linear encoders usually consist of a scanning head and a long scale rigidly attached to the device whose motion is being measured. The scale may be transparent if a transmitted light encoder is used or a steel scale if a reflected light encoder is used. In the first case the scale consists of alternate transparent and opaque areas, and in the second case the scale has a grating etched on its surface. There are usually at least two tracks and a zero reference marked on the scales, the scales are coded in either binary or BCD. In either case the scale moves past the scanning head, which contains a light source and usually photovoltaic sensors (see Figure 7-5). Most manufacturers supply incremental encoders with a built-in electronic package which outputs two square waves phase shifted 90° with respect to each other and a zero or reference pulse.

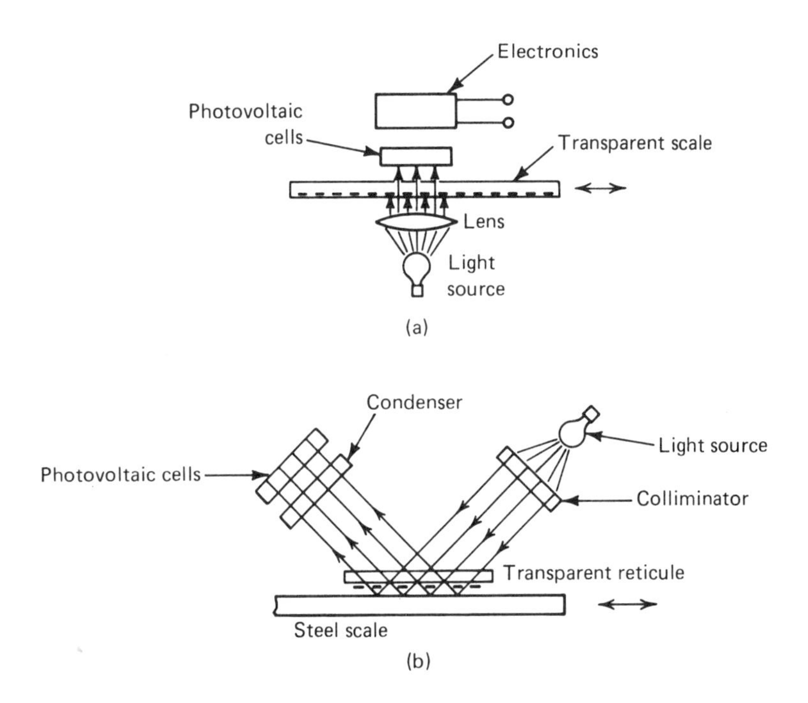

Figure 7-5. Linear incremental encoders: (a) transmitted light type, (b) reflected light type.

7-3-2 Angular Motion Sensors

Angular motion sensors are devices whose angular movement is usually less than 300°; however, there are some devices such as multiturn potentiometers (2-, 5-, and 10-turn), shaft encoders and synchros whose angular movement may be greater than 360°.

7-3-2-1 Angular Motion Potentiometers

Angular motion potentiometer devices have a circular wire-wound or film-type resistance element. The slider is attached to a shaft and sweeps over the resistance element as the shaft is rotated by the device whose motion is being measured (see Figure 7-6). In both the linear and angular motion potentiometers, the wire-wound types have a low resistance temperature coefficient and are little affected by humidity. They possess good resolution and are relatively inexpensive, but are limited to the order of 2 to 3 million cycles of operation at a relatively low speed of operation. The film type has a very high degree of resolution and accuracy and can be operated at relatively high speeds. They do have a higher resistance temperature coefficient, are sensitive to humidity, and can be expected to give up to 100 million cycles of operation.

An important advantage possessed by all potentiometers is the ease with which they can be modified to generate nonlinear outputs by using a shaped card, varying the wire spacing and changing wire size and material in the case of wire-wound potentiometers, or shaping the resistance element in the core of film-type potentiometers.

Even though potentiometers are probably the simplest method of

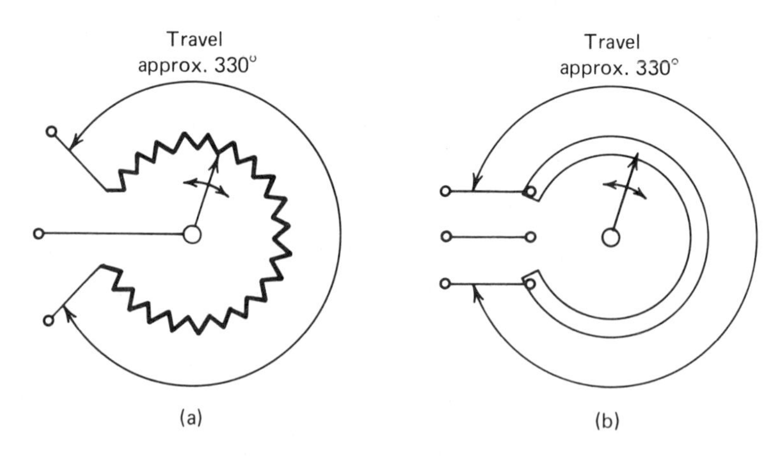

Figure 7-6. Angular motion potentiometer: (a) wire wound, (b) film type.

measuring linear and angular motion they are also the cheapest sensor available.

7-3-2-2 Rotary Variable Differential Transformer (RVDT)

As can be seen from Figure 7-7, the basic principle of operation of a RVDT is identical to the LVDT except that the core rotates, its travel usually being limited to ±45° from the null position. The output voltage will be zero when the cone-shaped core is in the null position as shown. Offsetting the core in either direction will produce a differential voltage whose amplitude increases with the angular offset and whose phase, with respect to the ac input voltage, will change depending upon the direction of angular movement from the null position.

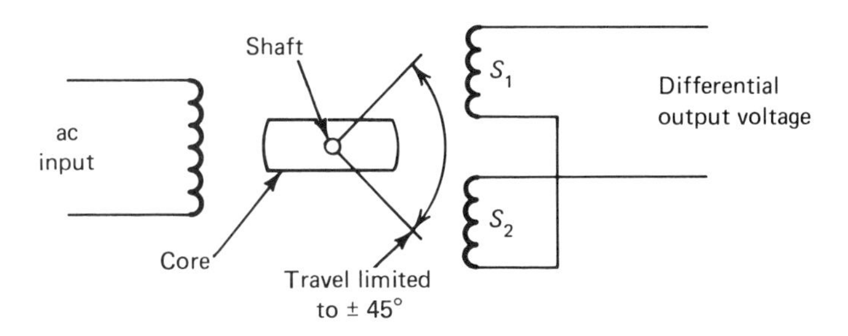

Figure 7-7. Rotary variable differential transformer.

7-3-3 Synchros

So far only limited angular motion sensors have been discussed. There are many requirements for the measurement of unlimited angular motion, with at times the need to add or subtract angular information, accept digital inputs or provide either analog or digital outputs, and be able to perform rectangular-to-polar or polar-to-rectangular conversions. These requirements are met by a class of rotary transformers called **synchros** and **resolvers.**

Synchros are somewhat similar to an ac motor, in that the stator windings are star- or delta-connected three-phase windings and convert an angular position to an electrical signal, or vice versa. There are two classes of synchros, **torque** and **control**. The torque units transmit angular data directly without amplifiers and may be used to drive light loads; for example, pointers and remote displays; however, because the angular error depends upon loading as well as dynamic errors caused by internal friction and inertia, an overall error as great as 1.5° may be produced. The control unit, on the other hand, is only transmitting data and can achieve an accuracy of 10 min of arc. All synchros are subjected to errors caused by lack of uniformity of the three-phase windings, irregularities in the magnetic circuit, choice of slot combinations, uneven winding distribution, and rotor pole structures, to name a few causes.

Synchro stators are basically similar in construction to a three-phase induction motor stator, with high-permeability high-nickel alloy laminations with skewed slots to reduce "cogging" and thus remove a source of angular error. If the stator slots are not skewed, the rotor laminations are skewed. The stator windings are distributed in the slots to produce windings 120° apart with particular attention given to the uniformity and balance of the windings. These windings are either star- or delta-connected and form the secondary winding of the synchro acting as a rotary transformer.

The difference in the functions of the various types of synchros lies in the differences in rotor design. The two basic rotor types are the salient pole type, consisting of a laminated H-type rotor energized by a single-phase ac source via brushes and slip rings, and the three-phase rotor, similar in construction to the rotor of a three-phase wound rotor induction motor with power being supplied via three brushes and slip rings.

The usual operating voltages and frequencies of standard synchros are 26 V at 400 Hz and 115 V at 60 or 400 Hz. Synchros are manufactured in various sizes and mounting arrangements. It should be remembered, prior to discussing the operation and functions of the individual types, that the input for all types is mechanical shaft rotation and the output is an ac voltage.

7-3-3-1 Synchro Transmitter (CX)

The synchro transmitter has an emf induced in the three-phase stator windings that is proportional to the angular position of the ac energized single-phase rotor [Figure 7-8(a), (b)]. It can be seen from the waveforms at the stator terminals that there is a unique relationship between the magnitudes and polarities of the induced voltages. The general construction of the synchro can be seen from Figure 7-8(c).

7-3-3-2 Synchro Receiver (CR)

The construction of the synchro receiver is identical to that of the transmitter except that the receiver rotor has a flywheel oscillation damper mounted on the shaft that is free to move up to approximately 45° in each direction. The function of this damper is to provide inertia to reduce oscillations of the rotor during alignment and prevent the

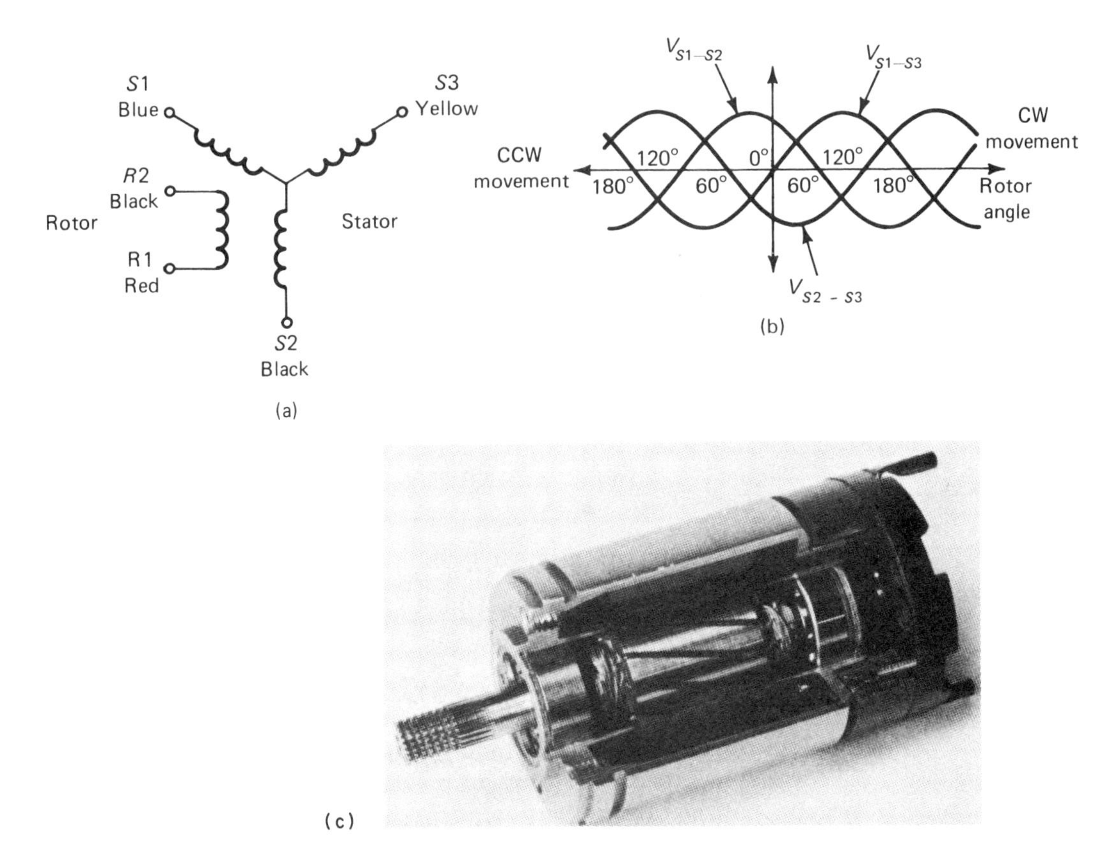

Figure 7-8. Synchro transmitter: (a) schematic diagram, (b) RMS stator terminal voltages versus rotor angle, (c) exploded view of synchro transmitter.

rotor from overspeeding in a continuous rotation situation. In general, 60 Hz receivers may also be used as transmitters, but a transmitter may not be used as a receiver since it does not possess an oscillation damper.

The simplest possible configuration of a transmitter and receiver is shown in Figure 7-9. The magnitude of the voltages induced in the stator windings of the transmitter is determined by the rotor position. These stator voltages are transmitted over the connecting lines between the transmitter and receiver and cause currents to flow in the stator windings of the receiver. In turn, the currents in the receiver stator windings create a magnetic field in the stator bore, and since the receiver rotor is energized from the same single-phase supply as the

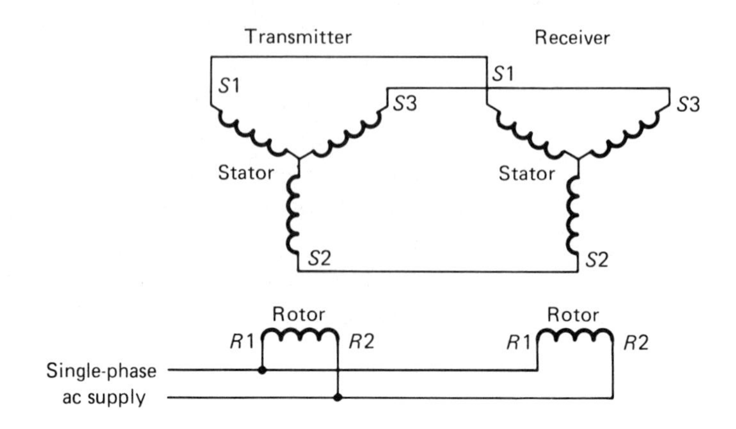

Figure 7-9. Synchro transmitter and receiver connections.

transmitter there is an interaction between the stator and rotor magnetic fields which creates a torque to align the rotor field with the stator field. It should be noted that when the rotor field is aligned with the stator field there is no current flow in the interconnecting lines between the transmitter and receiver stators.

7-3-3-3 Synchro Differential Transmitter (CDT)

A synchro differential transmitter is connected between a synchro transmitter and receiver whenever it is necessary to add or subtract angular information in the system.

The stator of the differential transmitter is identical to those of the synchro transmitter and receiver; however, the rotor is cylindrical and has a three-phase star- or delta-connected winding which is connected via slip rings to the receiver stator. Normally the stator and rotor windings have a 1:1 transformation ratio. The electrical connections of a differential transmitter system are shown in Figure 7-10.

The stator of the synchro transmitter is directly connected to the differential transmitter stator and the rotor connections of the differential transmitter are connected to be the stator connections of the receiver. As can be seen, the output voltages of the synchro transmitter applied to the differential transmitter stator produce a magnetic field whose orientation is determined by the angular position of the transmitter rotor. In turn, these voltages induce, by transformer action, voltages in the differential transmitter stator whose relative amplitudes depend upon the rotor position. These voltages are in turn applied to the receiver stator and are equal to the sum or difference of the angles represented by the mechanical position of the two inputting

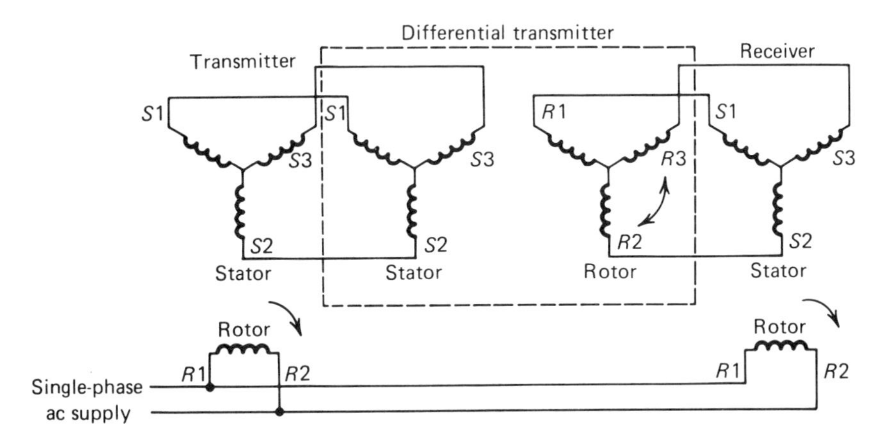

Figure 7-10. Synchro differential transmitter system: schematic symmetrical connection.

unit rotors. The electrical information used at the receiver to produce a mechanical positioning of the rotor is dependent upon:

1. The directions of rotor movement of the transmitter and differential transmitter,
2. The arrangement of the interconnecting wiring between the transmitter and differential transmitter stators and,
3. The arrangement of the interconnecting wiring between the differential transmitter rotor and the receiver stator.

The connections of the arrangement shown in Figure 7-10 are known as the symmetrical connection, and

$$\text{receiver angle output} = \text{transmitter angle} - \text{differential transmitter angle.}$$

If the sum of the two angles is desired, two of the connections between the transmitter stators are reversed; that is, $S1$ to $S3$ and $S3$ to $S1$, as well as reversing two of the connections between the differential transmitter rotor and the receiver stator, that is, $R1$ to $S3$ and $R3$ to $S1$. The angles are then additive, that is,

$$\text{receiver angle} = \text{transmitter angle} + \text{differential transmitter angle.}$$

If it is desired to subtract the synchro transmitter angle two of the connections between the two transmitter stators are reversed, that is,

$S1$ to $S3$ and $S3$ to $S1$. Then

receiver angle = − (transmitter angle + differential transmitter angle).

A final variation involves the reversal of two of the connections between the differential transmitter rotor and the receiver stator, that is, $R1$ to $S3$ and $R3$ to $S1$. Then

receiver angle = differential transmitter angle − transmitter angle.

7-3-3-4 Synchro Differential Receiver (CDR)

The differential receiver is similar to the differential transmitter, the difference being that an oscillation damper is mounted on the shaft to prevent rotor oscillations. The function of the differential receiver is to produce a mechanical output at the rotor shaft, that is, the sum or difference of two synchro transmitters; the stator of one transmitter is connected to the stator of the differential receiver, and the stator of the other is connected to the rotor.

7-3-3-5 Synchro Control Transformer (CCT)

The synchro control transformer is designed to produce an electrical output at the rotor terminals proportional to the angular offset of the rotor with respect to the stator magnetic field. While the stator is similar in construction to the other synchro units its impedance is significantly greater, with the effect of reducing the excitation current. The rotor is cylindrical, has a high impedance as compared to other units, and should never be connected to a low impedance load, although this is normally not a problem since it is usually outputted to a high input impedance amplifier.

The normal electrical connection arrangement is as shown in Figure 7-11. The stator of the control transformer is supplied from the synchro transmitter and the rotor is mechanically coupled to the servomechanism whose position is being controlled (for example, the azimuth or elevation members of a radio telescope). The currents supplied to the control transformer stator establish a magnetic field whose orientation is the same as the rotor of the transmitter. This magnetic field will induce a voltage in the rotor of the control synchro, the amplitude of which is proportional to the angle of offset between the rotor and stator magnetic fields. This error voltage is usually used to develop corrective action to reduce the error signal to zero.

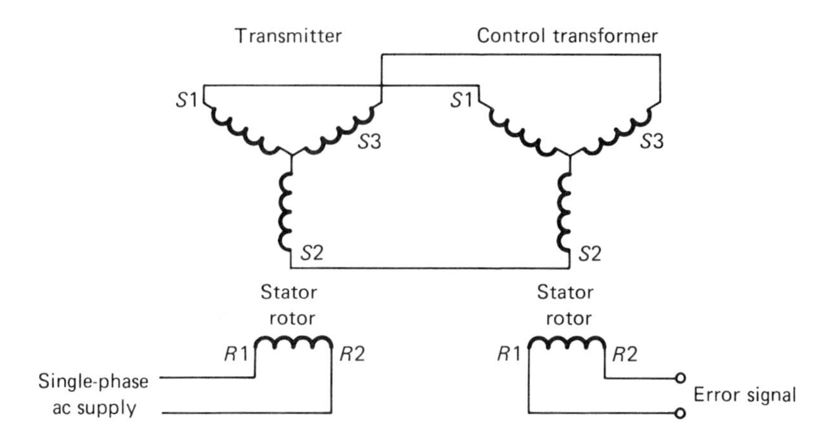

Figure 7-11. Synchro control transformer schematic.

7-3-3-6 Synchro Capacitors

When using control transformers, differential transmitters, and receivers, it is common practice to connect matched sets of capacitors in either a star or delta configuration across the lines connecting the stators. The function of these capacitors is to inject a substantial leading power-factor component to offset the lagging power-factor effect of the excitation currents and as a result reduce the current and the heating effects.

7-3-3-7 Zeroing

To ensure that the angular indication between transmitters, receivers, and control transformers corresponds, it is necessary to have an accurate alignment procedure. All of the units have an electrical zero position which is determined by the angular position of the rotor with respect to the stator windings. The procedure is essentially electrical and requires a voltmeter. The outer stator housing of all synchro units may be racked around by means of a pinion and gear to achieve alignment which is measured by obtaining a minimum voltage reading. The electrical connections for zeroing 115 V synchro units are shown in Figure 7-12.

7-3-3-8 Dual-Speed Synchro Systems

When there is a need for highly accurate angular information to be transmitted (for example, a fraction of a minute of arc), a dual-speed synchro system using geared synchros is commonly used (see Figure 7-

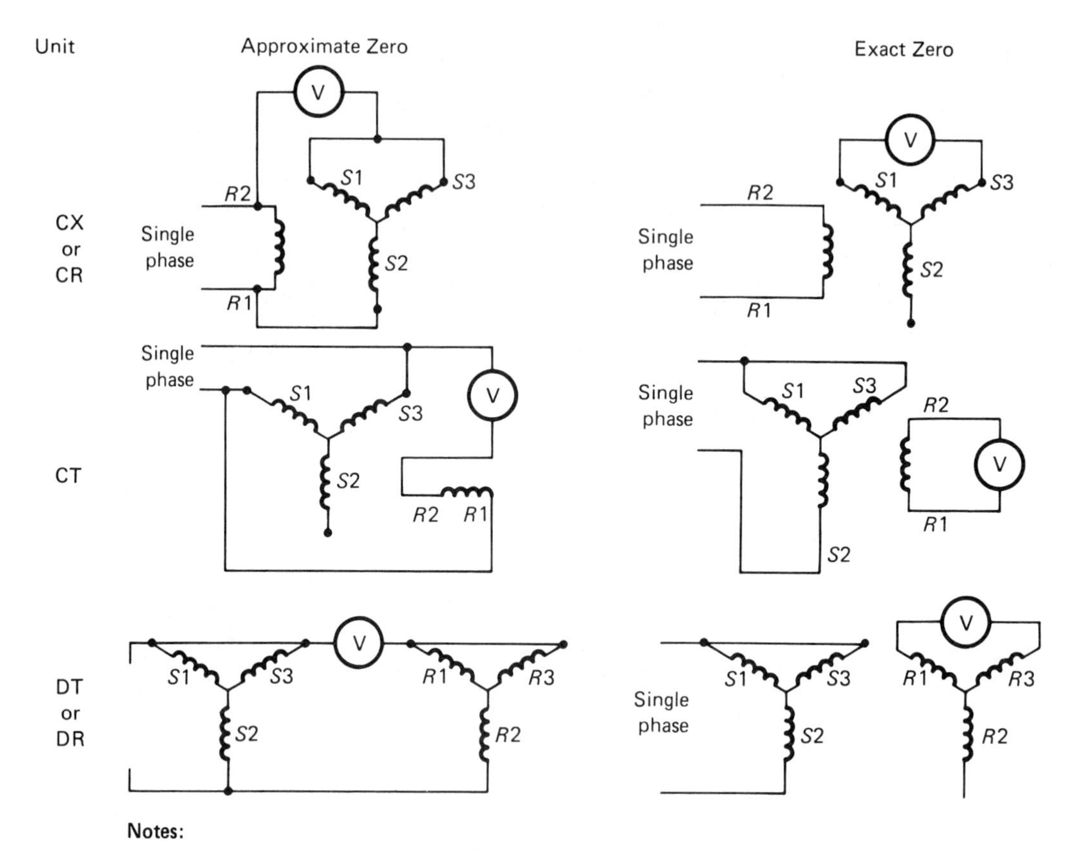

Notes:

1. First, find the approximate zero by turning the shaft to obtain the minimum reading on the voltmeter.
2. Then find the exact zero by turning the shaft through a small angle to obtain the minimum voltmeter reading.

Figure 7-12. Electrical zeroing techniques for synchro units.

13) to overcome the inherent inaccuracy of a synchro at its null position. The concept is best illustrated by considering an azimuth control system for a radio telescope.

Basically the system consists of a coarse (1-speed, 1X) synchro transmitter and a fine (36-speed, 36X) synchro transmitter with their shafts coupled together via a 36:1 gearbox, with the desired angular information fed in by turning the rotor shafts. Similarly the rotors of a coarse and fine pair of synchro control transformers are coupled together via a 36:1 gearbox, their shafts turned by the azimuth motion of the radio telescope.

The system operates in such a manner that, when the angular error between desired and actual output angles is large, the coarse or 1-

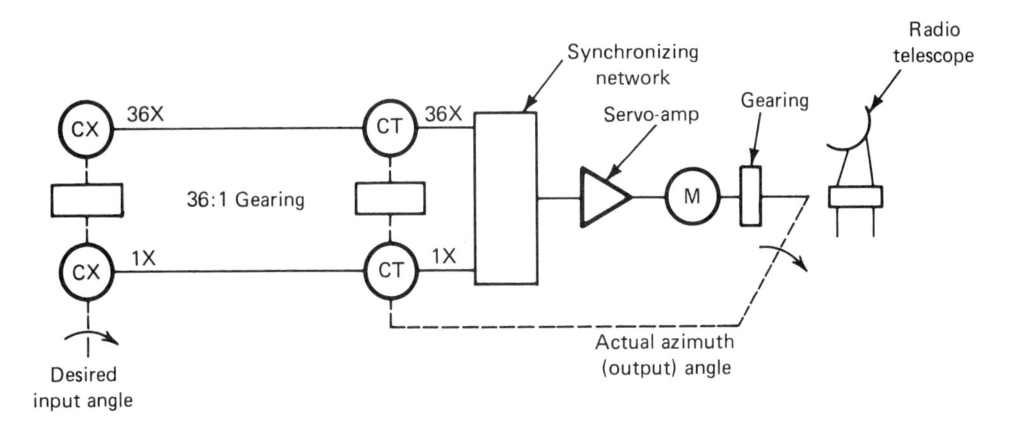

Figure 7-13. Simplified dual-speed synchro system: schematic of 36:1 dual-speed system controlling azimuth motion of a radio telescope.

speed system takes over control and drives the system to reduce the error to a small value. At this point the fine or 36-speed system takes over and reduces the error to the acceptable accuracy.

In the radio telescope system shown, the desired azimuth angle is fed into the system by positioning the synchro transmitter rotor. The electrical analog of this angle is represented by the magnetic fields produced by the 1-speed and 36-speed control transformer stators. The rotors of the control transformers are positioned by a mechanical feedback showing the actual azimuth bearing of the radio telescope. If there is a large error, the synchronizing network will operate under the control of the 1-speed system producing an error signal to the servoamplifier which will create an output signal and cause the motor to drive rapidly in the direction required to reduce the error. When the error has been reduced, the synchronizing network transfers control to the 36-speed system and the servoamplifier and motor bring the system into alignment.

Commonly used gear ratios are 16:1, 20:1, 25:1, and 36:1; however, it must be appreciated that the gearing has backlash which in itself will introduce an error into the system.

7-3-3-9 Resolvers

Resolvers are members of the synchro family and are used extensively to perform: (1) rectangular-to-polar and polar-to-rectangular conversions; (2) coordinate transformation, that is, rotate coordinate axes; and (3) phase shifting.

The resolver is a variable transformer with a coupling coefficient that is proportional to the sine or cosine of the rotor angle. There are usually two sets of windings mounted at right angles to each other to minimize interaction between each pair. These windings are placed in pairs in the stator and rotor. The ends of the rotor windings are brought out to four slip rings. The resolver is represented schematically in Figure 7-14.

The output of the rotor or secondary windings depends upon the voltages applied to the stator or primary windings. Since the windings are at right angles to each other on both the stator and rotor, the induced rotor output voltages represent sine and cosine functions of the angle of offset of the rotor windings. Simultaneous application of voltages to the stator windings establishes magnetic fields at right angles to each other. The resultant magnetic field is the phasor sum of

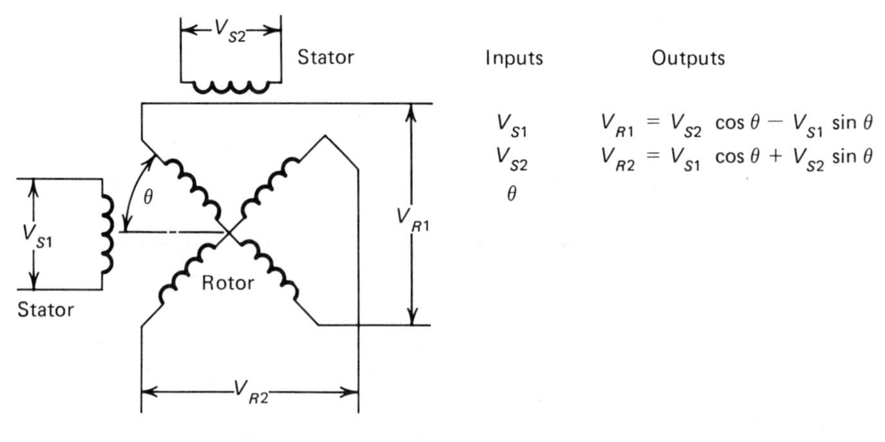

(a) Both stator windings energized

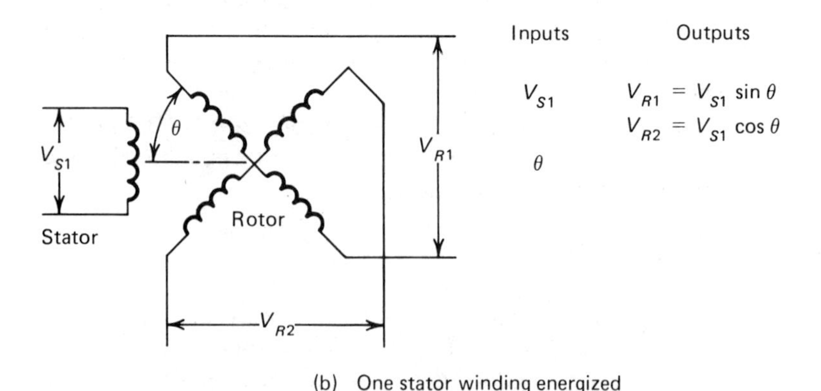

(b) One stator winding energized

Figure 7-14. Resolver, schematic, and voltage relationships: (a) both stator windings energized, (b) one stator winding energized.

the individual fields, whose amplitude is determined by the instantaneous magnitude of the stator currents and its direction by the arc tangent of the ratio of the two stator currents. The rotor-induced voltages are proportional to the resultant flux component which is at right angles to the respective coil.

The resolver is particularly suited for the measurement of small angular movement, usually between $\pm 45°$.

7-3-3-10 Scott-Tee Transformers

While not a synchro, the Scott-Tee transformer arrangement permits conversion from a two-phase system to a three-phase system, or vice versa. This means that they are particularly suited for resolver-to-synchro and synchro-to-resolver conversions. The principle is illustrated in Figure 7-15.

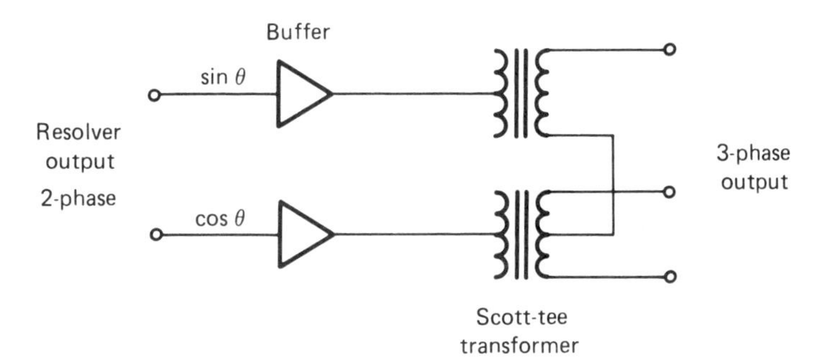

Figure 7-15. Scott-Tee transformer converting 2-phase to 3-phase or vice versa.

7-3-4 Synchro Converters

The major advantages of synchro systems were their ability to provide accurate and reliable angular information between remotely separated points under a wide range of operating conditions. With the ever increasing emphasis on digital techniques, it is logical that a marriage between synchro (analog) and digital techniques should take place. As a result, digital-to-synchro or synchro-to-digital conversions are becoming more and more common as a means of interfacing between analog positioning systems and microprocessors of digital computers.

Computer Conversions Corporation currently is marketing a range of synchro-to-digital converters that accept 3-wire synchro or 4-wire resolver inputs and produce 10-, 12-, or 14-bit outputs with the capability of supplying up to 5 standard TTL loads at input angular rates up to 36000°/sec. Multi-speed synchro-to-digital and digital-to-synchro converters, as well as digital-to-synchro or resolver converters, are also available.

Natel Engineering Co., Inc. is marketing a synchro/resolver-to-digital converter in a 36-pin DDIP hybrid configuration. It operates from a single 5 V dc power supply and draws 20 mA with a power dissipation of 100 mW. It has the ability to track up to 3600°/sec in the 14-bit mode with an accuracy of ±1.3 arc min. The logic inputs and outputs are TTL and CMOS compatible, and it has the ability to drive one 54/74 gate load or four 54LS/74LS gate loads, as well as being compatible with 8- and 16-bit microprocessors.

7-3-5 Shaft-Angle Encoders

The increasing use of computer-controlled systems in industry has led to the development of shaft-angle encoders. These encoders convert an analog signal in the form of a shaft angle or position into a digital code such as binary, BCD, Gray and BCD-excess-3 for interfacing with a control computer and for a direct display.

There are a number of choices available such as contact (brush) or noncontact (optical) encoders, incremental or absolute, and single-turn or multiturn types. It is necessary to choose a unit whose code and output voltage levels are computer compatible.

As previously discussed in the case of a linear encoder, incremental encoders produce an output train of pulses which are a function of position. To know the exact position requires that the pulses be counted with respect to a zero or reference point. In the case of a shaft encoder, it is also necessary to generate a second train of pulses in quadrature with the first train in order to determine the direction of travel. The lead-lag information sensed may then be fed via an up-down counter to a decade counter for output to the desired termination.

On the other hand, the absolute shaft encoder produces a unique coded digital representation for each position of the encoder. This is achieved by using coded patterns on the surface of the disc. As a result, if there is a power failure, the exact positional information is retained and as an added benefit data are not destroyed by electrical noise or transients [see Figure 7-16(a)].

Absolute shaft encoders are either directly connected to the unit whose angular position is being measured or coupled by a gearing system. There are basically two types of encoders, contact and non-

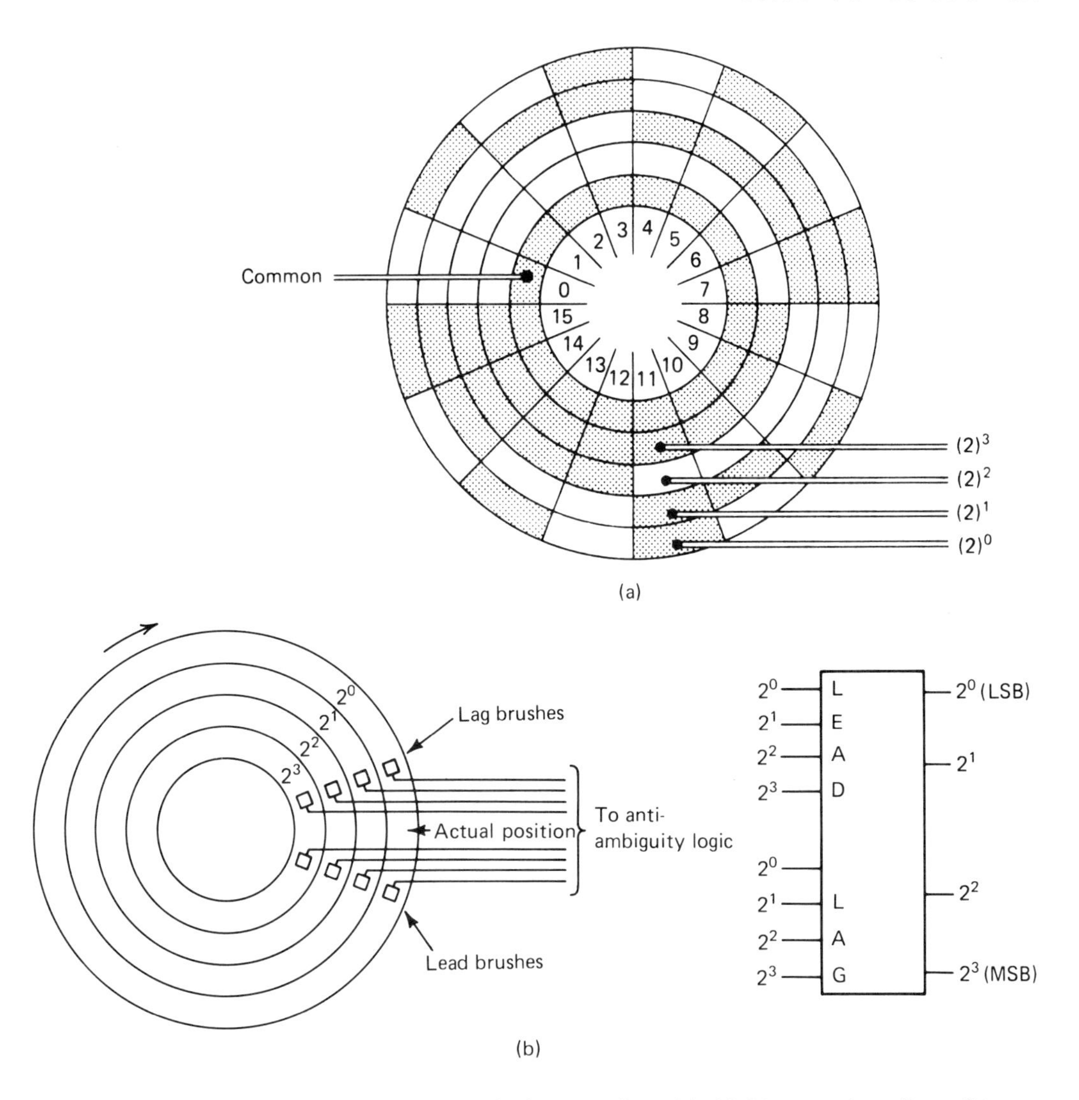

Figure 7-16. Absolute shaft encoder: (a) 22.5° encoder disc, (b) antiambiguity sensing.

contact. In the case of the contact type, it normally consists of a conducting metal disc etched with the digital pattern, the nonconducting areas usually being insulated with an epoxy material. Contact brushes, usually of the same material as the disc, make contact with the disc, and depending on whether the contact fingers are making contact with the conducting or insulated areas a combination of highs and lows are produced. For reliable reading the contacts must lie along a radii of the disc. This precise alignment is often quite difficult to achieve in

Table 7-1 Resolution Equivalents

Bits	Bits/Turn	Degrees/Bit	Minutes/Bit	Seconds/Bit
0	1	360.0	21600.0	1296000.0
1	2	180.0	10800.0	648000.0
2	4	90.0	5400.0	324000.0
3	8	45.0	2700.0	162000.0
4	16	22.5	1350.0	81000.0
5	32	11.25	675.0	40500.0
6	64	5.625	337.5	20250.0
7	128	2.8125	168.75	10125.0
8	256	1.40625	84.375	5062.5
9	512	0.70313	42.1875	2531.25
10	1024	0.35756	21.09375	1265.625
11	2048	0.17578	10.54687	632.8125
12	4096	0.08789	5.27344	316.40625
13	8192	0.04394	2.63672	158.20315
14	16384	0.02197	1.31836	79.10156
15	32768	0.01099	0.65918	39.55178
16	65536	0.00549	0.32959	19.77589
17	131072	0.00275	0.16480	9.88794
18	262144	0.00137	0.08239	4.94397
19	524888	0.0006866	0.04120	2.47199
20	1048576	0.0003433	0.0206	1.23599

practice. To overcome this problem it is usual to position a second set of contact brushes displaced by some mechanical angle from the first set. These brush arrangements are usually referred to as the lead and lag brushes and their outputs are fed into an antiambiguity logic unit that outputs the correct code [see Figure 7-16 (b)].

The need for antiambiguity logic could of course be eliminated by using the Gray code, since for each position only one bit changes. However, since the Gray code is usally not compatible with most systems used in industry, conversion to binary or BCD would be required.

The second type of absolute encoder consists of a transparent disc with opaque areas representing the code pattern photodeposited on the surface. A narrow radial light source is mounted on one side of the disc, and optoelectronic sensors in the same radial line are mounted on the opposite side of the disc. Just as before antiambiguity logic is required to ensure an accurate digital output.

The advantages lie with the noncontacting or optical absolute encoders; namely, they require less torque to turn, can be operated at much greater angular velocities, and have a much greater life expectancy since there are no physical contacts.

The resolution requirements are obviously determined by the process requirements, the number of tracks (bits) determines the resolution. The relationship between bits and bits per turn and the corresponding angular measurements are shown in Table 7-1.

Absolute encoder applications include control rod positioning in nuclear reactors, screwdown and side guide positioning in rolling mills, automated machine control, and the feedback element in dc motor position control, to give a few examples.

7-4 ROTARY SPEED MEASUREMENT

The requirement to be able to accurately measure angular velocity is very important, and a number of analog and digital speed transducers have been developed. Analog speed transducers depend upon the output voltage being proportional to the speed of rotation, that is, linear over the normal range of operation. Digital speed transducers depend upon the frequency being proportional to angular velocity, or a pulse train being produced so that the pulse count over a sampling period represents the speed of rotation.

7-4-1 Rotary Analog Speed Transducers

7-4-1-1 dc Permanent-Magnet Tachometers

dc tachometers usually consist of a permanent-magnet field structure and a conventional armature with commutator segments. The output dc voltage is directly proportional to angular velocity and is usually of the order of 10 to 20 V per 1000 rpm. Another advantage is that the polarity of the dc output voltage reverses with reversal of the direction of rotation. There are a number of disadvantages to their use: a voltage ripple which can be minimized by increasing the number of coils and commutator segments; brush arcing creating radio frequency interference (RFI), which can be reduced by ensuring that the commutator is not eccentric and the brushes are correctly selected and properly fitted in close-fitting brushholders; and the maintenance problems resulting from commutator and brush wear.

In closed-loop speed-control applications it is also essential that the coupling between the load and the tachogenerator be torsionally stiff to prevent oscillations being introduced into the system. It is reasonable to expect an accuracy of the order of 0.1% to 0.25% at the maximum designed speed rating.

7-4-1-2 ac Permanent-Magnet Tachometers

The ac permanent-magnet tachogenerator consists of a permanent-magnet rotor and a polyphase stator winding. The output voltage and frequency are proportional to the angular velocity of the rotor. The major advantage is the elimination of RFI since the commutator and brushes of the dc permanent-magnet tachometer have been eliminated, which also reduces the torque requirement and maintenance. Unfortunately, performance at low angular velocities is poor. Since the output voltage is ac it is not sensitive to changes in the direction of rotation, and it must also be rectified because most control inputs are dc. An alternative method of determining the angular velocity is to measure the frequency and, by means of frequency-to-voltage conversion techniques, produce a compatible output representing angular velocity.

7-4-1-3 ac Induction Tachometer

The ac induction tachometer has a laminated two-phase stator with two separate windings at right angles to each other; one is called the excitation winding and the other the output winding. The rotor usually consists of a copper or aluminum cup or cylinder, which has also given the name drag cup generator to the induction tachometer. See Figure 7-17(a). The rotor revolves between the pole pieces. The principle of operation [Figure 7-17(b)] depends upon the production of an eddy current flux in the drag cup. When the drag cup is stationary, an eddy current flux is induced by the excitation winding which is at right angles to the output winding, and as a result there is no output voltage. However, when the drag cup rotor is turned, an eddy current flux is developed at right angles to the excitation winding flux, and the flux linking the output winding is the resultant of the two fluxes. The output voltage is at the same frequency as that supplied to the excitation winding and is proportional in amplitude to the angular velocity. When the direction of rotation is changed, the eddy current flux changes direction by 180° and the phase of the output voltage with respect to the excitation voltage changes phase by 180°. As a result, by using a phase-sensitive rectifier, a dc voltage is obtained whose amplitude is proportional to the angular velocity and whose polarity is determined by the direction of rotation. The major disadvantages of the drag cup tachometer are: two airgaps are required in the magnetic circuit, which is offset by the low inertia of the unit, and a well-regulated excitation voltage source operating at constant frequency is required.

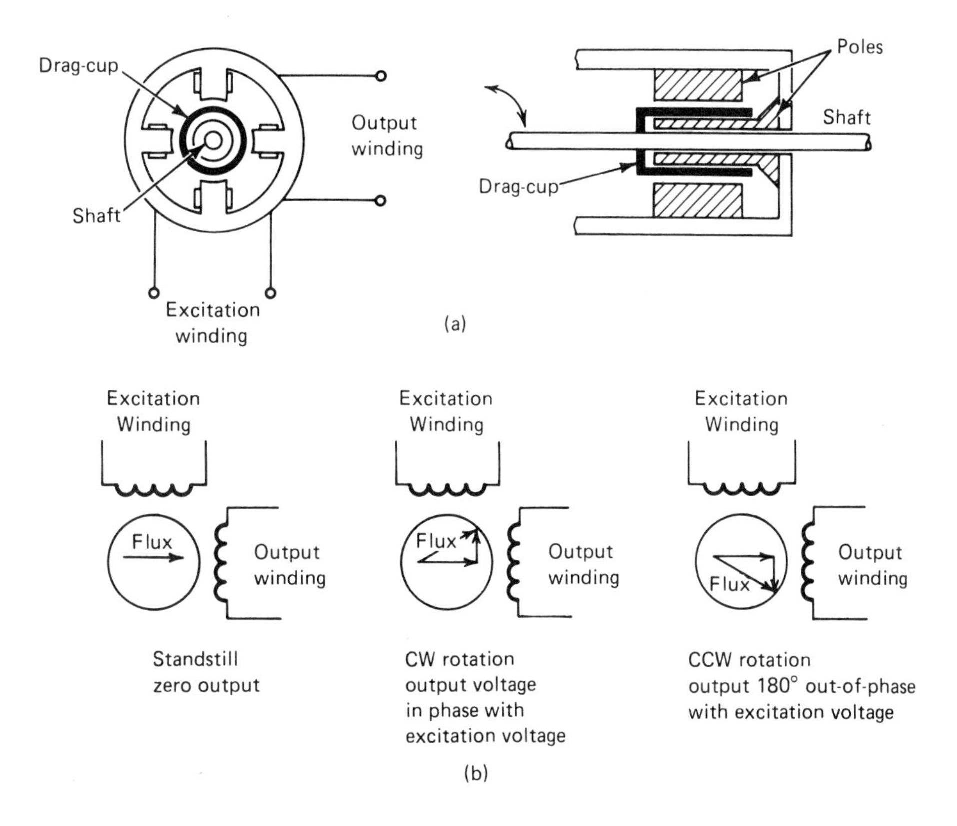

Figure 7-17. ac induction tachometer (drag cup generator): (a) sectionalized view, (b) principle of operation.

7-4-2 Rotary Digital Speed Tachometers

Digital tachometers produce voltage pulses. These pulses in turn are counted over a period of time (the counter being used directly in the control process) and displayed as a readout or converted to an analog dc voltage. Digital tachometers are essentially an angle sensor and the speed is determined by the relationship

$$S = \frac{\text{count} \times \text{angle/count}}{\text{time of count}}. \tag{7-1}$$

This speed is an average value over the counting interval. The accuracy of the reading can be improved by increasing the count in a given counting interval, which is achieved by increasing the number of voltage pulses per revolution, thus increasing the resolution capability.

If the tachometer is to be used in an application requiring a fast response, then the counting interval must be decreased, which in turn implies for a high-accuracy fast-response system that there must be a large number of voltage pulses per revolution. The limiting factors are obviously the number of pulses that can be generated per revolution and the ability of the associated electronics to accurately respond. An illustration of the problem is shown in Table 7-2. Assume that a closed-loop motor speed-control system is to operate with a bandwidth of 10 Hz. Then the count time should be less than the periodic time, that is, less than 0.1 sec, assuming 0.01 sec and a count of 1000 in the counting interval. From Table 7-2 it can be seen that, as the speed decreases, to maintain the count of 1000 in an interval of 0.01 sec, the number of pulses generated must increase.

This increased count requirement very quickly exceeds the ability of the sensors. The two most common types of rotary digital speed tachometers are the magnetic type and the optical type.

Table 7-2

Speed	Revolutions In 0.010	Counts/Revolution
6,000	1	1,000
600	0.1	10,000
60	0.01	100,000

7-4-2-1 Magnetic Proximity Digital Tachometer

The magnetic proximity digital tachometer consists of a magnetized toothed wheel rotating past a magnetic pick-up. The magnetic pick-up consists of a permanent-magnet core over which is wound a pick-up coil. A voltage is induced in the coil each time a tooth of the rotor passes by the pick-up (see Figure 7-18). The output waveform is approximately sinusoidal, and its amplitude depends upon the air gap between the toothed wheel and the permanent magnet of the pick-up coil and the width of the teeth and the angular velocity of the toothed wheel.

7-4-2-2 Optical Digital Tachometer

The principle of the optical digital tachometer is illustrated in Figure 7-19. The periphery of the rotating disc is either slotted or has a series

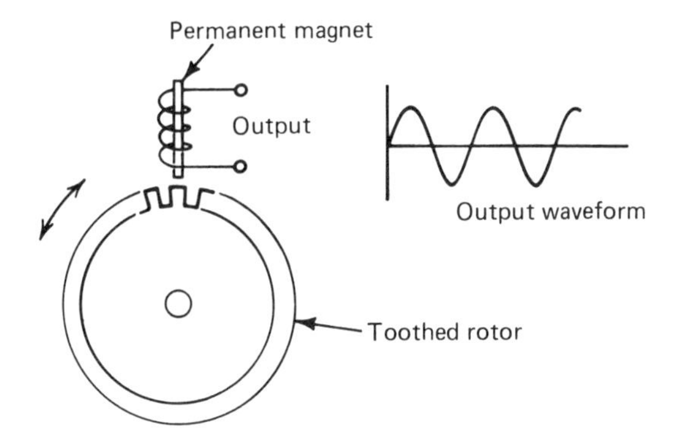

Figure 7-18. Magnetic proximity digital tachometer.

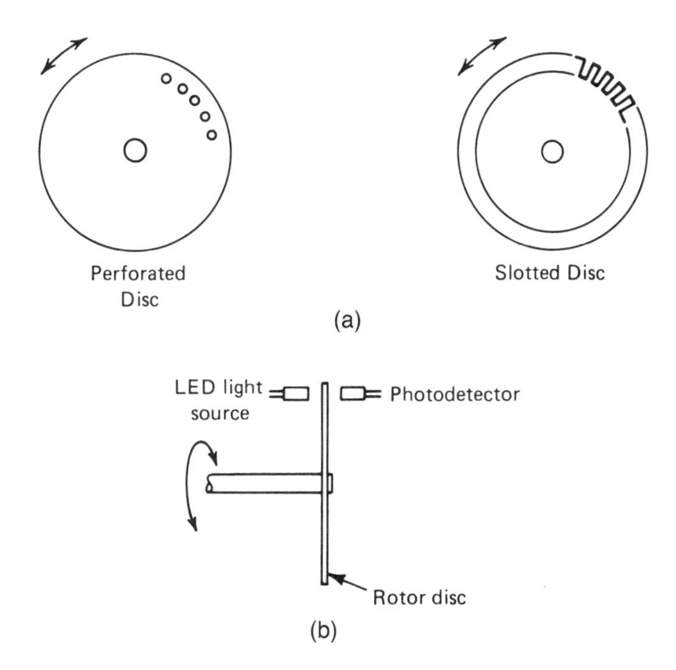

Figure 7-19. Optical digital tachometer: (a) rotor configurations, (b) principle of operation.

of holes. The disc rotates between a light source such as a LED and a photodetector such as a photodiode or phototransistor. The light from the light source is sensed each time a hole or slot passes between it and the photodetector.

7-5 FORCE SENSORS

Strain, force, load, torque, acceleration, and vibration can all be determined by measuring the displacement of predefined points on a surface. The amount of the displacement is usually of the order of several thousandths of an inch. Since this is a much smaller movement than has been considered to date, new techniques are required to measure the displacement. Probably the most important transducer that will be encountered is the **resistance strain gauge** which is used in many applications (such as load cells, pressure gauges, flowmeters, etc.) under quite often adverse operating conditions. There are two basic types of strain gauges, **bonded** and **unbonded.** The bonded resistance strain gauge normally consists of a resistance element bonded to a flexible membrane and in turn the membrane is attached to the member being tested by adhesives. The strain occurring in the element under test is then transferred directly to the resistance element where it is converted into changes of resistance. The unbonded strain gauge is used to measure small displacements which are transferred to it by a mechanical linkage, that is, the force is acting directly on the strain gauge element, whereas with the bonded strain gauge it measures the strain occurring at a point.

7-5-1 Bonded Resistance Strain Gauges

Recall that the resistance of a metallic conductor operating at constant temperature varies directly with length and inversely with area; that is,

$$R = \frac{\rho l}{A} \ \Omega, \tag{7-2}$$

where ρ = specific resistance or resistivity

l = conductor length

A = cross-sectional area of conductor

From Equation (7-2) it can be seen that when the conductor is under tension, l increases, A decreases, and R increases. Similarly, when the conductor is under compression, l decreases, A increases, and R decreases.

It can be shown that the strain $\varepsilon = \Delta l/l$ μ in/in (microinches/inch) where Δl is the change in length. The gauge sensitivity factor K is

$$K = \frac{\Delta R/R}{\Delta l/l} = \frac{\Delta R/R}{\varepsilon} \tag{7-3}$$

where ΔR is the change in resistance. This relationship is the unit change of resistance per unit strain and is dimensionless. It represents the sensitivity of the strain gauge; typical values range from 2 to 3.5.

The bonded resistance strain gauge is illustrated in Figure 7-20. Typical conductor materials are copper-nickel alloy and nickel-chrome alloy. Strain gauges are also subjected to two temperature effects. First, as the temperature of the member to which the strain gauge is bonded changes, strain will be applied to the gauge. This effect may be minimized by selecting a gauge material whose coefficient of expansion is as close as possible to that of the member being tested. The second temperature effect is the change in resistance caused by the resistance temperature coefficient of the resistance element. This may be minimized by selecting a low-resistance temperature coefficient material such as Constantan.

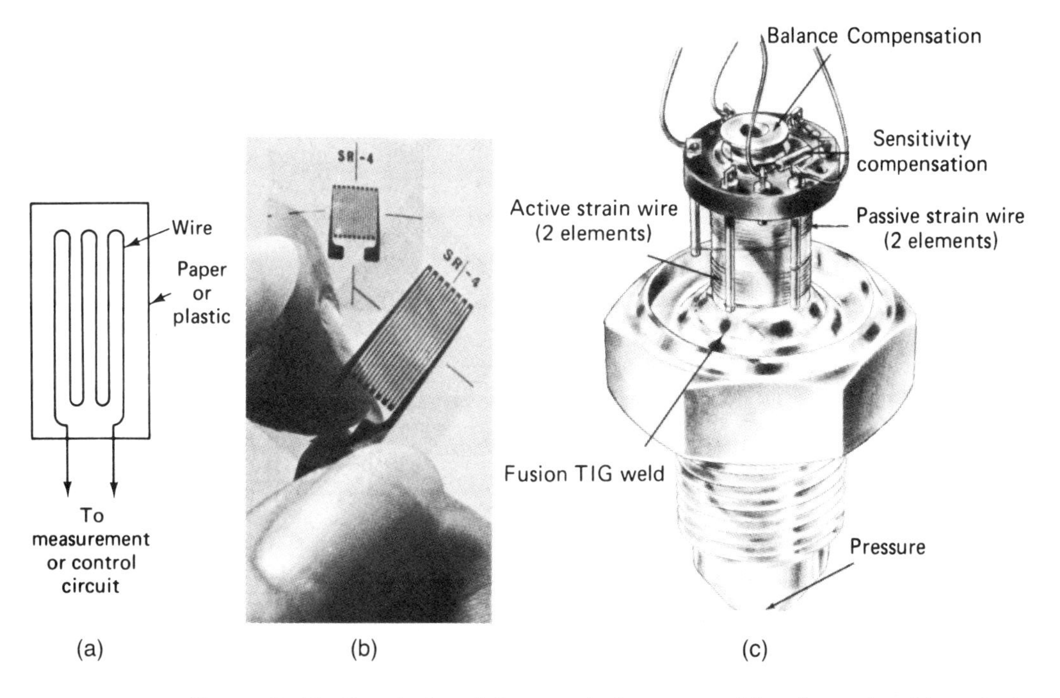

Figure 7-20. Bonded resistance strain gauge: (a) schematic, (b) typical commercial strain gauge, (c) pressure transducer using a bonded wire strain gauge.

Bonded resistance strain gauges may be used at elevated temperatures up to 1400°F (760°C) using nickel-chrome or platinum alloy resistance elements combined with a suitable membrane base and adhesives. Moisture effects such as shrinkage of the membrane and changes in resistance can be combatted by a silicon grease or lacquer coating.

The measurement of the variation of resistance of the strain gauge element requires a Wheatstone bridge to measure the small variations of resistance (see Figure 7-21). In its simplest form, Figure 7-21(a), the strain gauge forms one arm of the Wheatstone bridge; however, there is no compensation for the effects of temperature changes on the strain gauge, thus providing a source of error. Temperature compensation can be easily arranged by using a second identical but unstressed strain gauge mounted in the same temperature environment as the first gauge.

A typical load cell measuring compressive forces is shown in Figure 7-22(a). Cells of this type have been built to measure up to a compressive load of 8×10^6 lbs (3.63×10^6 kg).

7-5-2 Unbonded Resistance Strain Gauge

When relatively small forces require measurement, the unbonded strain gauge shown in Figure 7-23 is sometimes used. The resistance wire is mounted under tension to the limited movement strain sensitive elements, so that when in operation both compressive and tensile forces may be measured. The resistance elements may be stressed repeatedly up to approximately 2000 μ in/in. Care must be exercised in

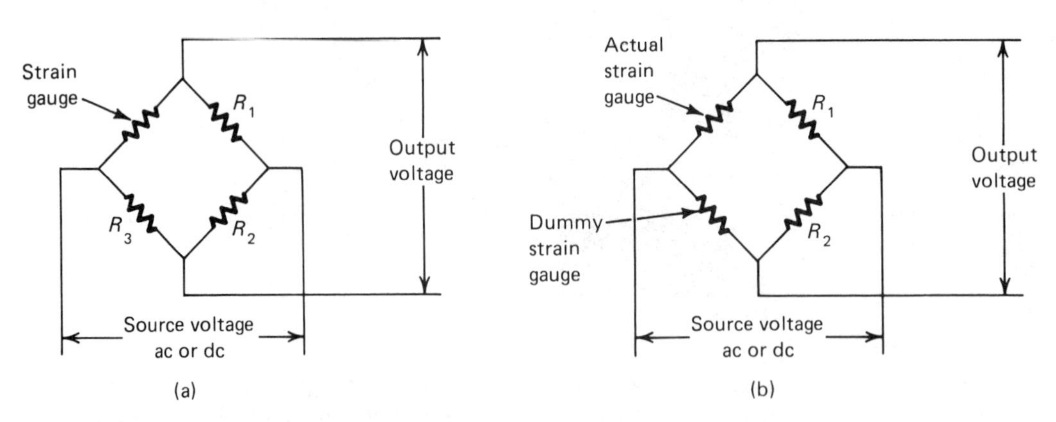

Figure 7-21. Strain gauge bridge: (a) uncompensated, (b) temperature compensated.

the application of this type of transducer to ensure that this range of movement is not exceeded or that it is not subjected to torsional forces.

7-5-3 Semiconductor Strain Gauges

Semiconductor strain gauges depend upon the piezoresistance effect, that is, the changes in specific resistance that occur under stress.

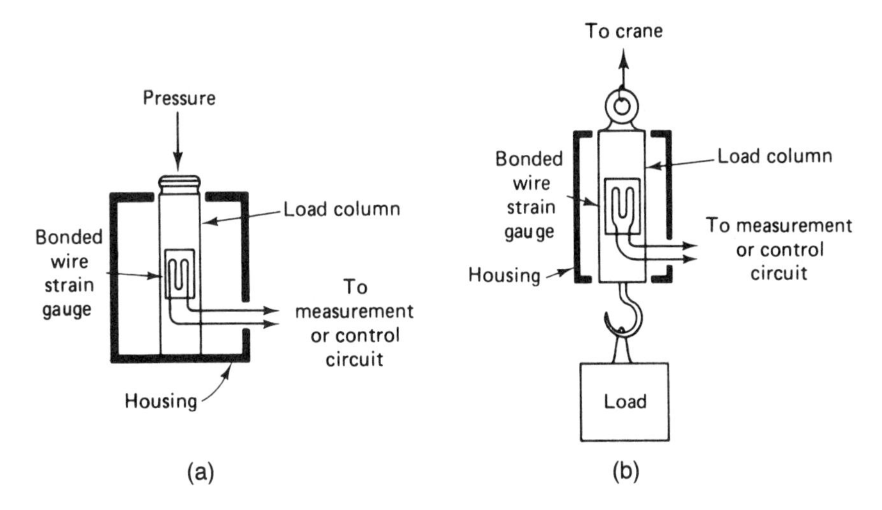

Figure 7-22. Bonded strain gauge applications: (a) load cell, (b) crane scale.

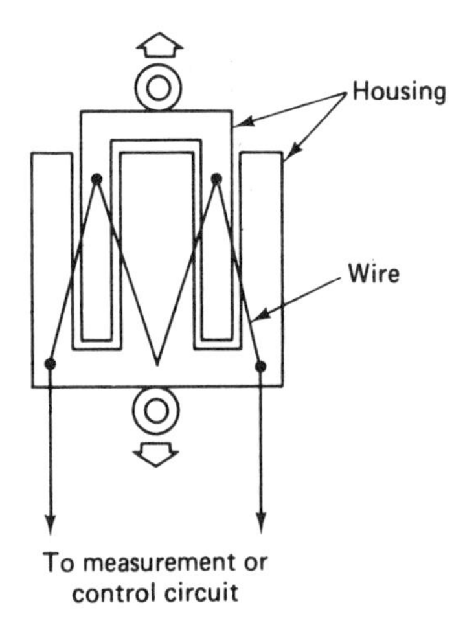

Figure 7-23. An unbonded strain gauge.

This characteristic is present in all materials to a greater or lesser extent; however, materials such as pure silicon have an almost perfect elasticity up to approximately 1000 °F (538 °C). An additional benefit is that the gauge sensitivity factors of semiconductor strain gauges approach 200 which permits the elimination of strain gauge amplifiers required for the other types of strain gauges. Also, because of their nature, thermistors may be built into the cell to compensate for errors arising from temperature changes. However, a disadvantage of semiconductor strain gauges is that they are expensive and sensitive to vibration and shock.

7-5-4 Acceleration and Vibration Transducers

There are two forms of acceleration that must be considered in the industrial environment, namely, a slowly changing force causing a body to accelerate or decelerate and a limited but rapidly changing backward and forward motion commonly called vibration. The devices that measure acceleration are called **accelerometers** and those that measure vibration are called **vibration sensors**.

The basic principle of operation depends upon the fact that if a mass is at rest it tends to remain at rest because of inertia. It will remain at rest until a force is applied, then the force will move the mass in the same direction as the force. If the mass is in motion, the force will accelerate the mass if it is applied in the same direction as the motion, or it will decelerate the mass if applied in the opposite direction. Since $F = ma$, the mass is known, the force can be measured, and then the acceleration can be determined. Normally accelerometers are designed to have a natural frequency of oscillation as high as possible and vibration transducers with a low natural frequency. Otherwise the vibration transducer will tend to operate as an accelerometer at frequencies below its natural frequency. The principles of some of the more common types of accelerometers and vibration transducers are described in the following sections.

7-5-4-1 Piezoelectric Accelerometer

When certain crystals or ceramic materials are subjected to a compressive force, a voltage is generated across opposite faces of the crystal. Conversely if a piezoelectric crystal is subjected to a varying electric field the crystal will expand and contract. Typical piezoelectric crystals are quartz, Rochelle salt, tourmaline, cadmium sulfide, and zinc oxide. Piezoelectric crystals are used in a number of applications such as accelerometers, ultrasonic transducers, crystal and ceramic microphones, phonograph pickups, and heart pacemakers, to name a

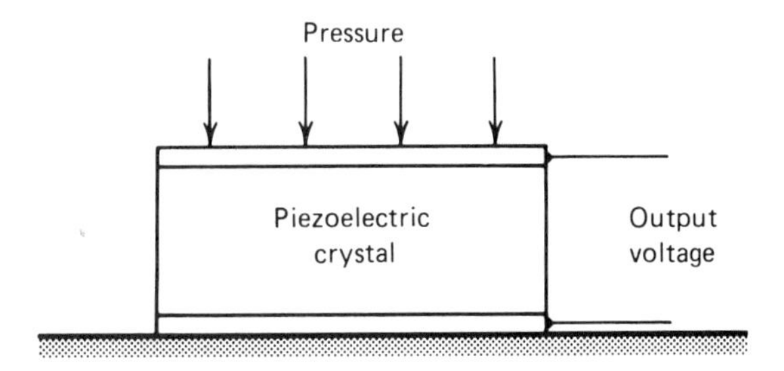

Figure 7-24. A piezoelectric crystal accelerometer.

few applications. The output voltage, which is proportional to the rate of change of the applied force, means that the transducer will only respond under dynamic conditions (ranges from 1 to 100 mV); the polarity of the voltage is dependent upon the direction of the force. The basic construction of a piezoelectric accelerometer is shown in Figure 7-24. The output voltage may be measured by a high impedance voltmeter which is usually calibrated in terms of acceleration.

7-5-4-2 Inductive Vibration Transducer

Vibration transducers are designed to detect and measure rapid to-and-fro motion. Vibration measurement is very important in industry since the detection and location of vibration prevents damage to expensive equipment and is commonly used as part of preventive maintenance programs. The inductive vibration transducer (Figure 7-25) has a number of advantages such as a low output impedance, which

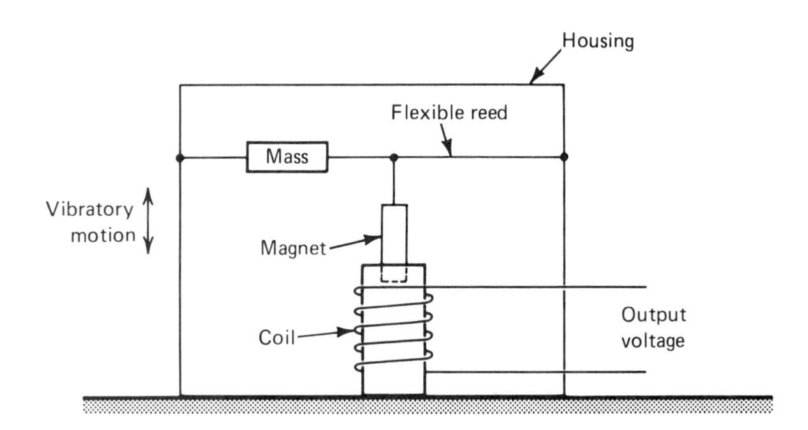

Figure 7-25. An inductive vibration sensor.

permits the indicating unit to be located at some distance from the measuring unit and relatively high output voltages at low frequencies thus reducing the need for preamplification. Among the disadvantages are a high mechanical impedance and a low-resonant frequency which limits use to frequencies below 2000 Hz. The principle of operation is very simple. The motion transmitted to the permanent magnet causes the flux linkages with the coil to change and, as a result, induces a voltage in the coil whose frequency is determined by the oscillations of the permanent magnet; the amplitude represents the vibrational movement.

7-5-4-3 Capacitive Vibration Transducer

Since capacitance in a given capacitor is inversely proportional to the distance between the plates, the variations in capacitance are a measure of the amplitude of the vibration. The basic construction of the transducer is shown in Figure 7-26. The variations of capacitance may be used to frequency modulate an oscillator and, if the output of the oscillator is demodulated and supplied to a suitable indicator, a direct read-out of the vibration amplitude can be obtained.

7-5-4-4 Linear Variable Differential Transformer Vibration Transducer

As will be seen, the LVDT is a very versatile device and may be used in a number of applications to provide an output. One such application is illustrated in Figure 7-27, where the mass of the transducer is the core of the LVDT. Vibration causes the core to be displaced either side

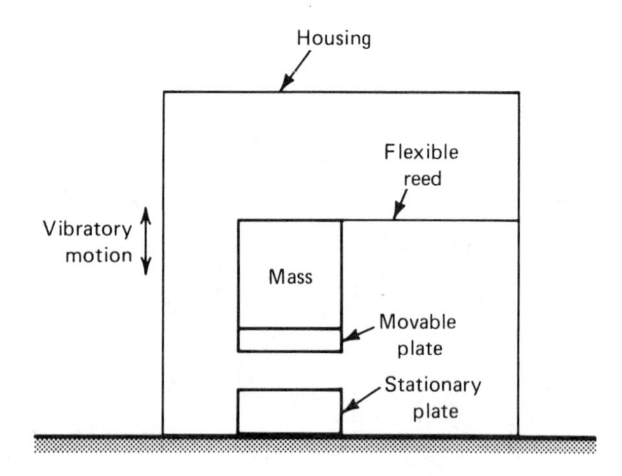

Figure 7-26. A capacitive vibration transducer.

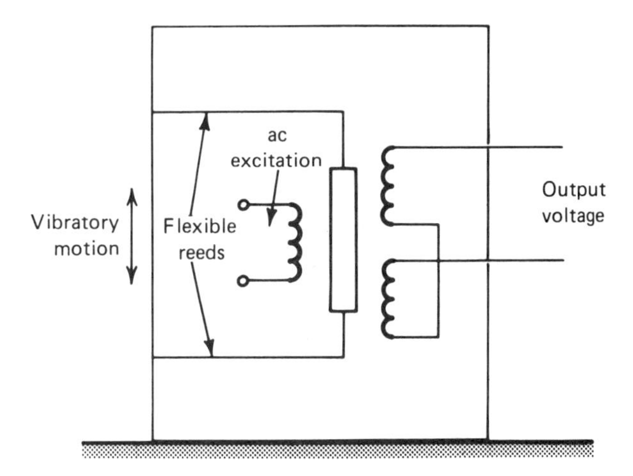

Figure 7-27. A linear variable differential transformer vibration transducer.

of the null position, the amplitude of the output voltage represents the amplitude of the vibration; its phase, with respect to the source voltage, determines the direction of motion.

These vibration transducers and accelerometers are representative of the techniques used; however, numerous methods such as potentiometer, bonded strain gauge, semiconductor, and optical transducers are also available.

7-6 FLUID MEASUREMENT

Liquids and gases are classified as fluids. Fluid pressure acting over an area creates a force. The techniques of measuring fluid pressure depend upon the production of a force in a mechanical element, and then in turn measuring the force.

There are three groups of fluid measurement: (1) pressure common to both fluids and liquids; (2) flow common to both, and; (3) level, liquids only.

7-6-1 Pressure Transducers

There are a number of devices used to convert fluid pressure to a force; the most common are the bellows, the diaphragm, and the Bourdon tube. The bellows basically consists of a very thin brass, phosphor-bronze or stainless steel corrugated walled cylinder sealed at one end. When pressure is applied to the inside of the cylinder it stretches; conversely, pressure applied to the outside causes it to contract. In either case the motion is opposed by the springiness of the metal walls of the cylinder. Bellows generally are used in low pressure applications, however, higher pressures can be accommodated by using an external spring to prevent excessive movement. See Figure 7-28(a).

The diaphragm is available in two basic forms: the flat disc and corrugated disc types. The circular disc, usually made of thin springy phosphor-bronze, is mounted on the end of a stiff cylinder. The disc is corrugated to amplify the motion when pressure is applied to the cylinder [see Figure 7-28(b)].

The Bourdon tube consists of a length of flattened oval cross-section phosphor-bronze tubing formed into either a flat-spiral, helical-spiral, or an arc of constant radius and strain tubes. The principle of operation relies on pressure being applied to the inside of the tube causing the tube to deform, the amount of deformation begin proportional to the applied pressure [see Figure 7-28(c)]. It may be used in applications up to several thousand pounds per square inch.

Pressure-to-motion converters, in these and other forms, will be encountered in instrumentation. They will be used to convert changes in fluid pressure to an electrical analog in conjunction with potentiometers, capacitors, LVDTs, strain gauges, etc. We will review the basic principle of operation of some typical applications.

7-6-1-1 Bellows Pressure Transducer

The bellows provides an accurate and reliable pressure sensing system. The motion of the output when coupled, for example, to a potentiometer produces an output voltage or current which is directly proportional to the applied pressure (see Figure 7-29).

7-6-1-2 Differential Bellows Pressure Transducer

A requirement exists in a number of applications to sense the difference in pressure between two fluids or between different sections of a process. A differential bellows pressure transducer is shown in Fig. 7-30(a); the basic principle of operation is illustrated in Figure 7-30(b). Application of pressure to the left hand bellows will cause the connecting rod to move to the right; similarly, applying pressure to the right hand bellows will move the connecting rod to the left. The movement of the connecting rod is proportional to the difference in pressures applied to the two bellows units. This movement, in turn, can be applied to a linear potentiometer, whose output will then be proportional to the pressure difference.

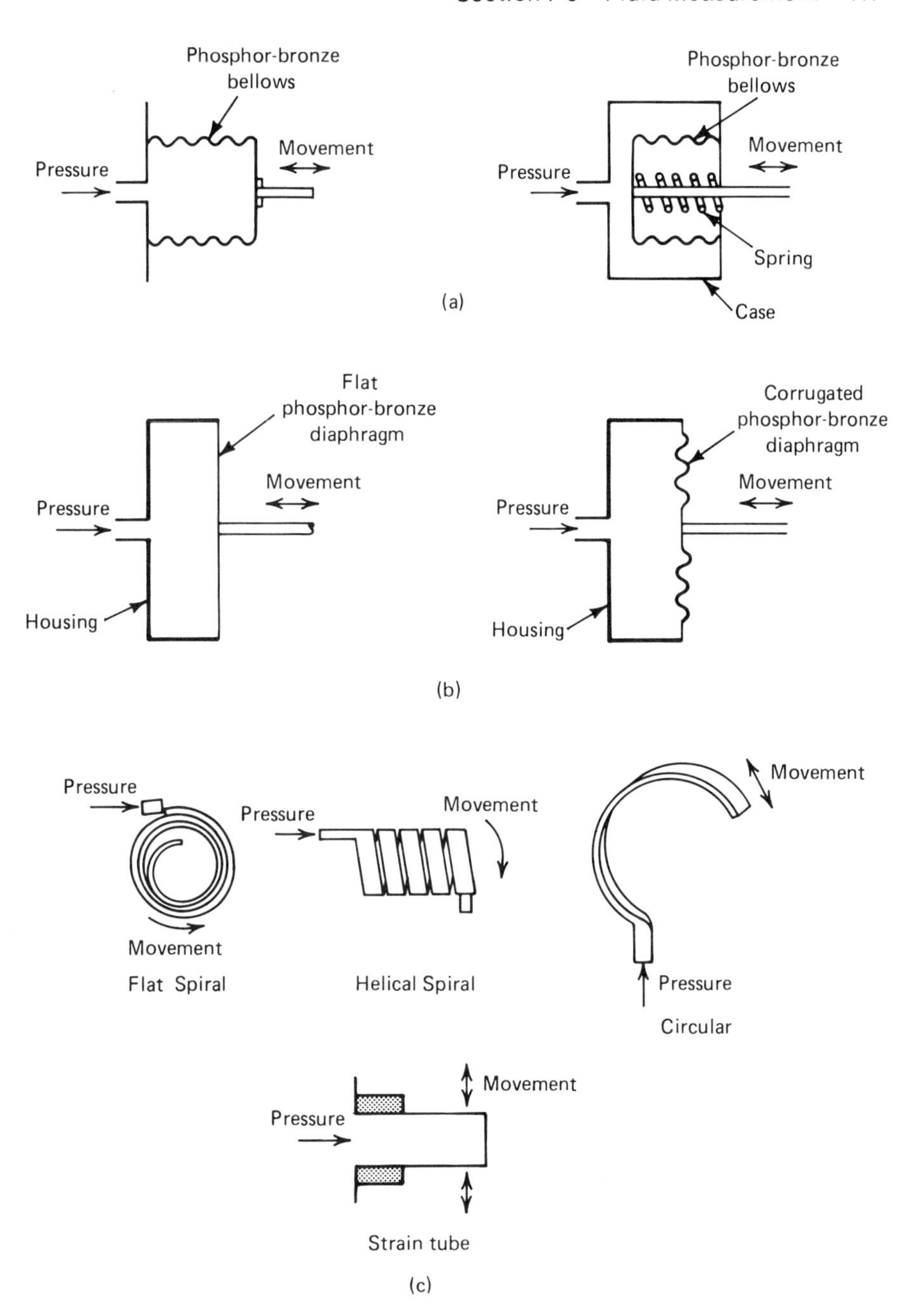

Figure 7-28. Fluid pressure to motion converters: (a) bellows, (b) diaphragms, (c) Bourdon tubes.

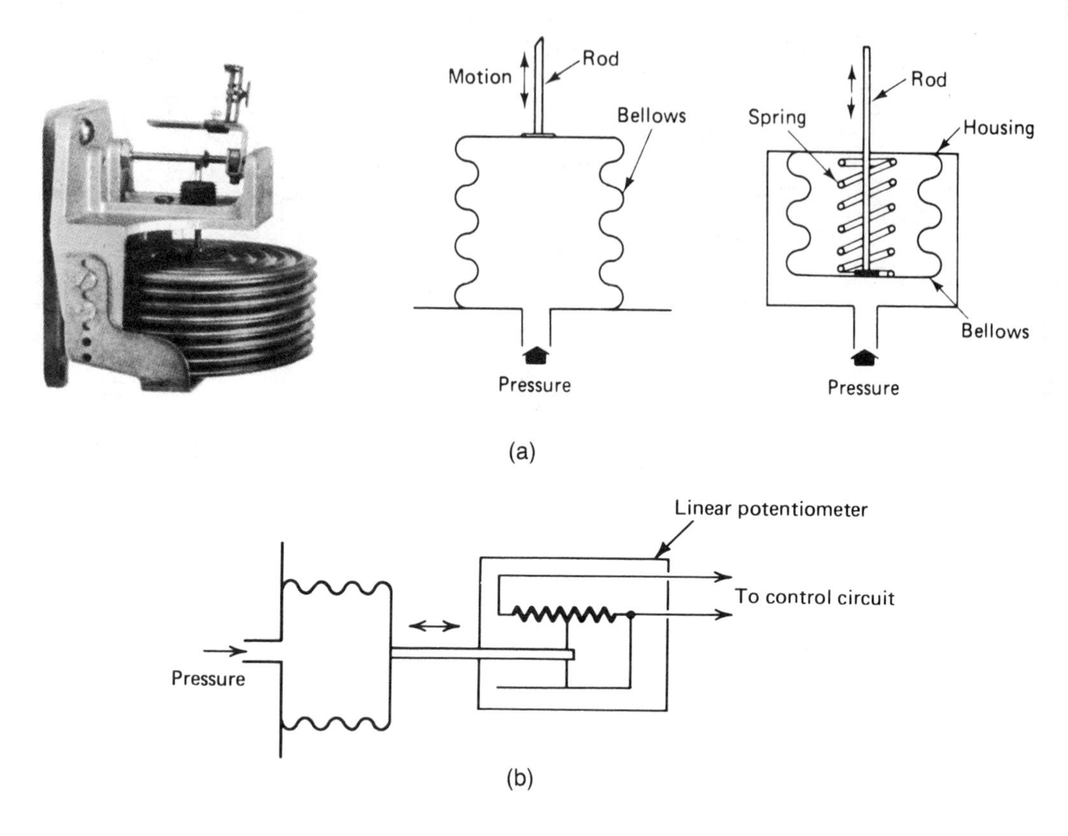

Figure 7-29. Bellows transducer: (a) bellows element, (b) schematic.

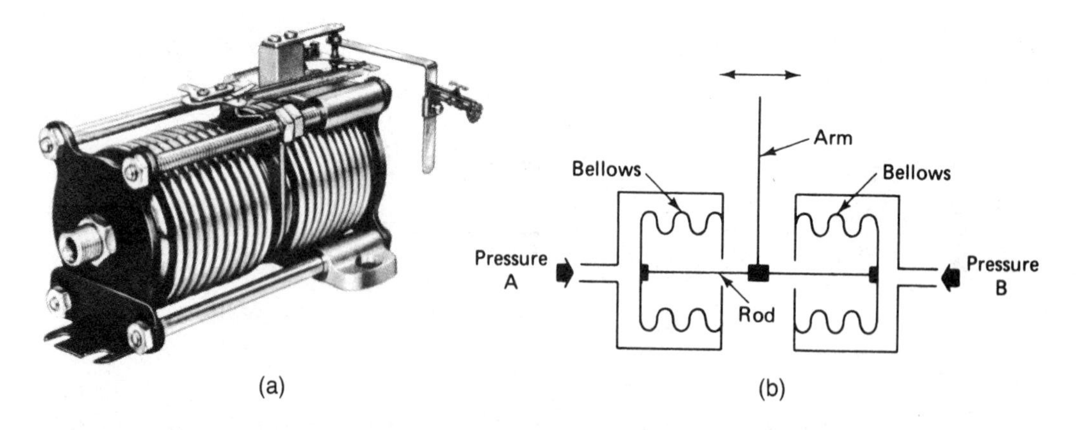

Figure 7-30. Bellows type differential pressure sensor: (a) industrial unit; (b) schematic.

The bellows sensors are more expensive to produce than the Bourdon tube sensors, so as a result the Bourdon tube sensor is used in many high pressure applications.

7-6-1-3 Bourdon Tube Pressure Transducers

The various configurations of the Bourdon tube readily lend themselves to being easily adapted for use with electrical output devices. Figure 7-31(a) illustrates a transducer using a helical Bourdon tube with an angular motion potentiometer mounted directly on the helix. Figure 7-31(b) uses a constant arc tube with the free end directly attached to the core of an LVDT. The constant arc Bourdon tube is also frequently encountered with a mechanical linkage system hooked to a pointer to give a direct read-out of pressure.

The last type of transducer utilizes a twisted Bourdon tube and a variable inductance pickup or E-type transformer. The physical arrangement is shown in Figure 7-31(c) and the principle of the variable reluctance pickup is illustrated in Figure 7-31(d). When the pivoted armature is in the position shown, the reluctances of the two airgaps are equal and the inductances of the differentially connected windings on the outer legs are also equal. As a result, the output voltage of the center leg winding is zero. As the armature pivots in either direction, the output voltage increases with the angle and the phase relationship of the output voltage, with respect to the excitation voltage, determines the direction of movement.

The helical Bourdon tube has been modified using a contact type angular absolute encoder to provide an 8421 BCD output.

7-6-1-4 Diaphragm Pressure Transducers

The diaphragm sensor lends itself readily to a number of applications. However, we will only consider one; namely, a differential pressure arrangement as illustrated in Figure 7-32. The principle is shown in Figure 7-32(a). The difference in pressure on both sides of the diaphragm determines the direction of movement of the pivoted arm which, in turn, can be readily connected to an electrical transducer such as a linear potentiometer.

7-6-1-5 Bonded Strain Gauge Pressure Transducer

A number of applications, such as hydraulic systems, exist where high pressures require measurement. An adaption of the Bourdon tube, called a strain tube, is used. Bonded strain gauges are bonded with adhesives to the outside of the strain tube which expands and

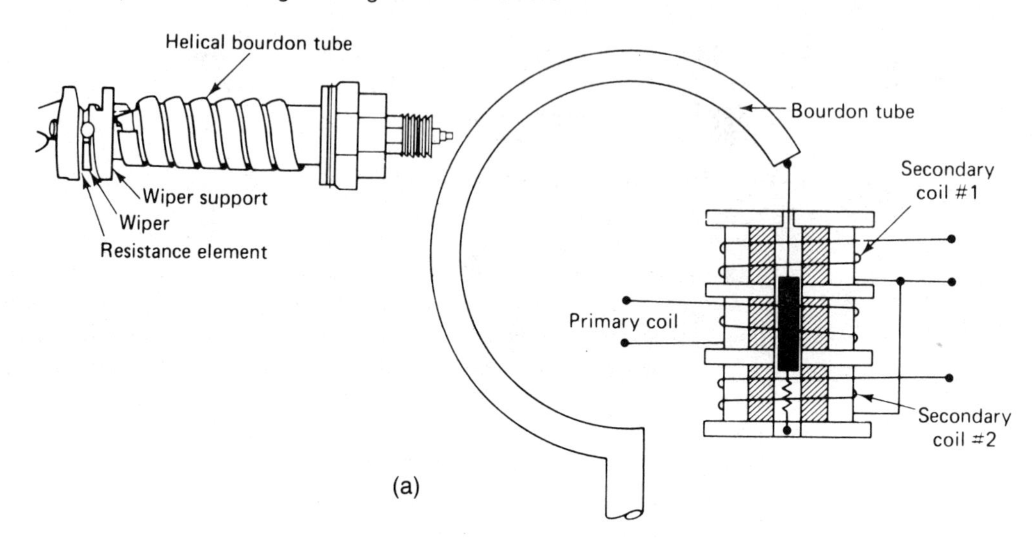

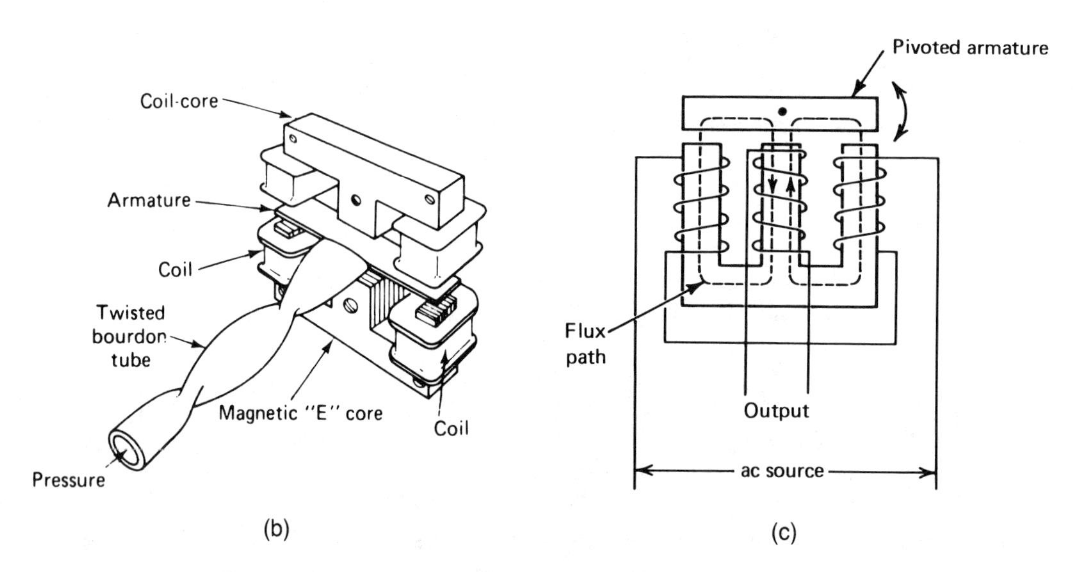

Figure 7-31. Bourdon tube transducers: (a) helical Bourdon tube with integral potentiometer, (b) constant arc Bourdon tube with LVDT, (c) twisted tube with variable inductance *E*-type transformer, (d) principle of *E*-type transformer.

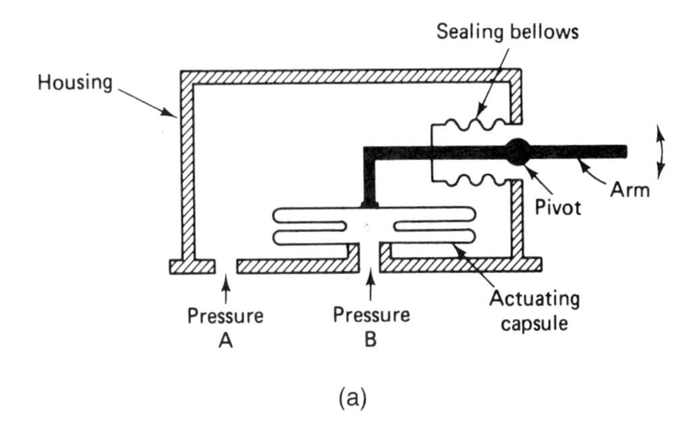

(a)

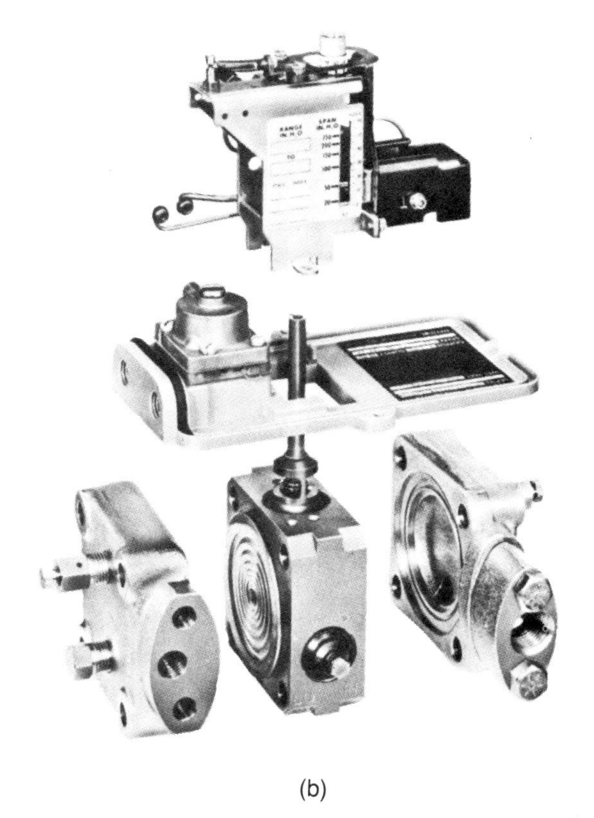

(b)

Figure 7-32. Differential pressure diaphragm sensor: (a) schematic, (b) industrial unit.

contracts with pressure changes, the resistance changes being proportional to pressure changes. These types of transducers are available in ranges up to 10,000 psi, and two types are illustrated in Figure 7-33.

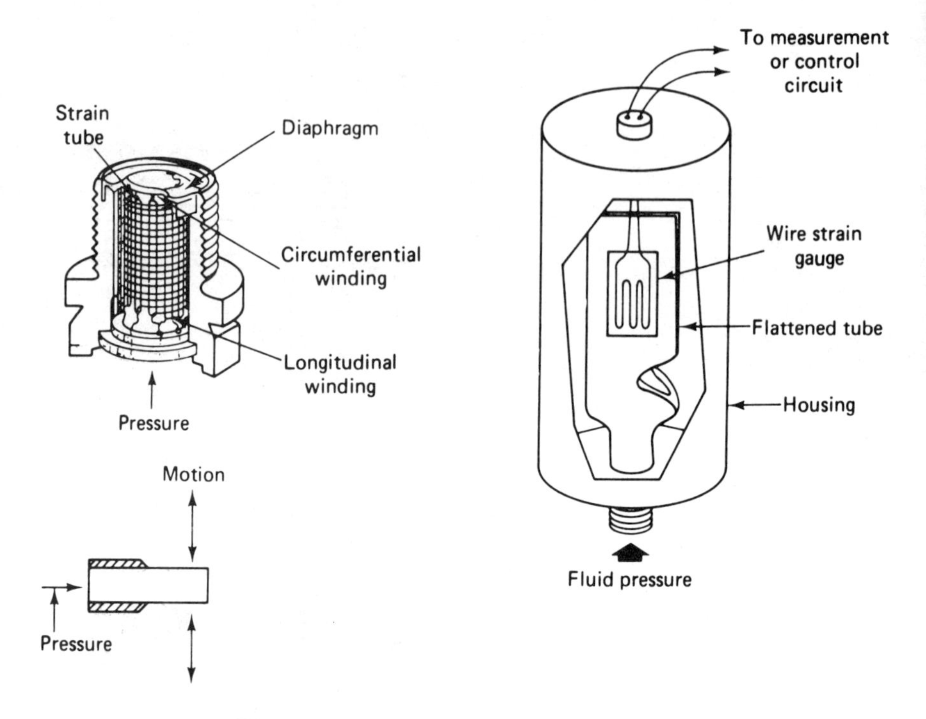

Figure 7-33. Strain tube pressure sensors.

7-7 FLUID FLOW SENSORS

Very often a critical measurement in industrial processes is that of fluid flow. Measurement of fluid flow may be broken down into a number of different methods. The method selected is dependent upon the type of fluid, the volume, and the desired accuracy. The most common methods used in industrial processes are head flowmeters, electromagnetic flowmeters, mass flowmeters, variable area flowmeters, and positive displacement flowmeters.

7-7-1 Head Flowmeters

Head flowmeters are probably the most common type of device used to measure the flow of liquids, gases, and slurries. In each case the device depends upon the development of a pressure differential across a restriction in the flow path. The basic theory of their operation depends upon Bernoulli's theorem which states that

$$\frac{\rho v^2}{2g} + p = \text{constant} \tag{7-4}$$

where ρ = density of the fluid

v = velocity of the fluid

p = pressure, and

g = acceleration due to gravity

A pressure differential is created in a piping system by introducing a restriction to flow. In the case of a liquid which is incompressible, a head h is developed so that

$$h = \frac{p_1 - p_2}{\gamma} = \frac{v^2_2 - v^2_1}{2g} \qquad (7\text{-}5)$$

where p_1 = the pressure on the upstream side of the restriction

p_2 = the pressure on the downstream side of the restriction

v_1 = the velocity on the downstream side of the restriction, and

v_2 = the velocity on the downstream side of the restriction

γ = the specific weight of the fluid

The restriction to the flow may take a number of forms (see Figure 7-34). Figure 7-34(a) shows the orifice plate method used widely to measure flow rates because of its simplicity, low cost, and ease of installation. It consists of a circular plate (with a circular hole at its center) with sufficient rigidity that it will not be distorted by the developed pressure differential. As can be seen, the velocity of the fluid is less on the upstream side of the orifice plate where the pressure is the greatest. Conversely, the velocity is at its greatest and the pressure is at its least on the downstream side of the orifice plate. This type of head flowmeter is most satisfactory for flow rate (gal/min, l/min, ft^3/min, etc.) measurements of liquids and gases that do not contain solids in suspension. When measuring liquids containing suspended solids it is necessary to use a venturi tube. See Figure 7-34(b). The venturi tube is far more expensive than the orifice plate. Basically it consists of a specially fabricated section of pipe that narrows down to an accurately machined throat section and then enlarges back up to the normal pipe diameter. A simpler and less expensive device is the flow nozzle illustrated in Figure 7-34(c).

In each of these devices, the pressure differential may be determined by a variety of measuring means ranging from the U-tube manometer to the differential pressure transducer previously described. These methods of head flow measurement measure the flow rate in terms of the square root of the differential pressure where

$$Q = K\sqrt{h} \tag{7-6}$$

where Q = flow rate, and

K = constant determined for each installation

As a result, when it is desired to use strip chart recorders it is normal practice to use a square root converter to produce a linear output.

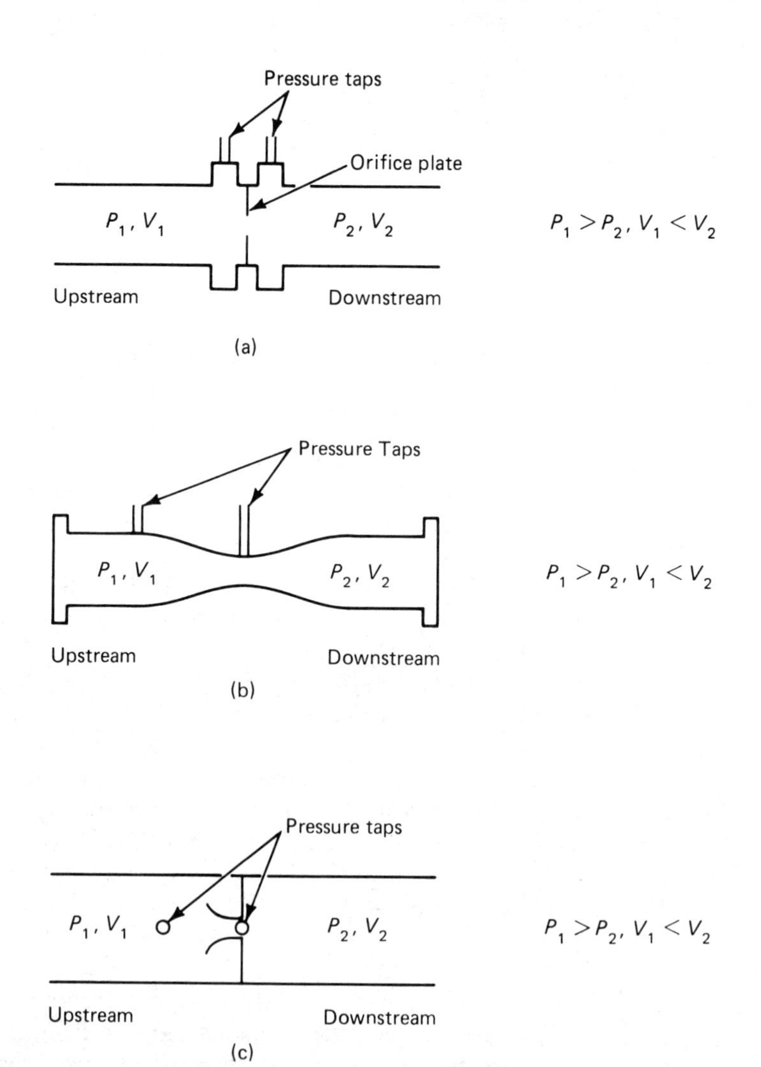

Figure 7-34. Head flowmeters: (a) orifice plate, (b) venturi tube, (c) flow nozzle.

7-7-2 Electromagnetic Flowmeters

The head type flowmeter extracts energy from the moving liquid. The electromagnetic flowmeter, since it has no moving parts, does not extract energy from the moving liquid. The principle of the electromagnetic flowmeter is illustrated in Figure 7-35. A nonconducting pipe section is placed in a strong magnetic field with a flux density B Wb/m^2. Two electrodes are inserted into the pipe at right angles to the magnetic field. The liquid whose flow rate is being measured cuts the magnetic field and a voltage is induced in the conducting liquid (Faraday's Law). This voltage is detected by the electrodes and is

$$V_{OUT} = BDV, \qquad (7\text{-}7)$$

where V_{OUT} = the induced emf (V)

B = magnetic flux density (Wb/m^2) (T)

D = inside pipe diameter (m)

V = fluid velocity (m/sec)

From Equation (7-7) it can be seen that the output voltage V_{OUT}, with a constant flux density B, is proportional to the velocity of the fluid through the flowmeter. This type of flowmeter is satisfactory for any

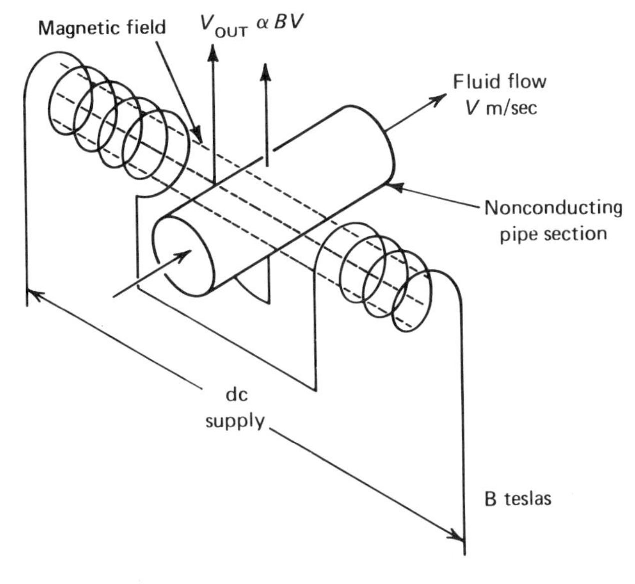

Figure 7-35. Electromagnetic flowmeter.

fluid with reasonable conductivity and is unaffected by variations in consistency, viscosity, temperature, and turbulence. One disadvantage is that, if the conductivity of the fluid is too low, polarization of the contacting area of the electrodes will occur with a dc energized field coil system. Polarization can be overcome by using an ac excited field, however, this will give rise to additional problems. First, the dielectric constant of the fluid will have a shunting effect upon the induced voltage and second, there will be an additional induced voltage caused by the alternating magnetic field. The major disadvantages of this type of flowmeter are: (a) since the output voltages are usually in the microvolt range, amplifiers are required to produce a useful output and (b) the cost of the unit is much greater than other types of flowmeters.

7-7-3 Turbine Flowmeters

The constructional details of the turbine flowmeter are shown in Figure 7-36. As liquid flows through the device, the impeller rotates; the speed of rotation is determined by the velocity of the liquid. As the impeller rotates, a ceramic magnet passes by a stationary pick-up coil and produces a sine-shaped pulse at the coil terminals. The frequency of the pulses is directly proportional to the flow rate of the liquid or gas being measured. In turn these pulses can be relayed to suitable signal conditioning circuits to give a direct readout, or be used in the control of the process. Turbine flowmeters, which must only be used with liquids or gases free of solids, are available for liquid flow rates varying from less than a gal/min to as high as 15,000 gal/min (4.55 to 68,250 l/min) and for gas flow rates from about 20 to 10,000 ft^3/min. (0.57 to 284 m^3/min).

The major disadvantages are initial and maintenance costs and loss of accuracy due to turbulence. Turbulence, however, may be reduced by the use of flow straightening vanes. The major advantages are the ability to operate over wide temperature ranges [$-450\,°F$ ($-267.8\,°C$) to $1000\,°F$ ($537.8\,°C$)] and pressures up to 50,000 lb/in^2 (344,500 kN/m^2).

7-7-4 Mass Flowmeters

All the devices discussed so far measure the rate of flow. However, there are many industrial and commercial applications where it is necessary to measure the total quantity of liquid flow, for example, the gasoline pump at the local gas station.

Usually this type of measurement is made using a positive displacement flowmeter. In general, these devices trap a known volume of

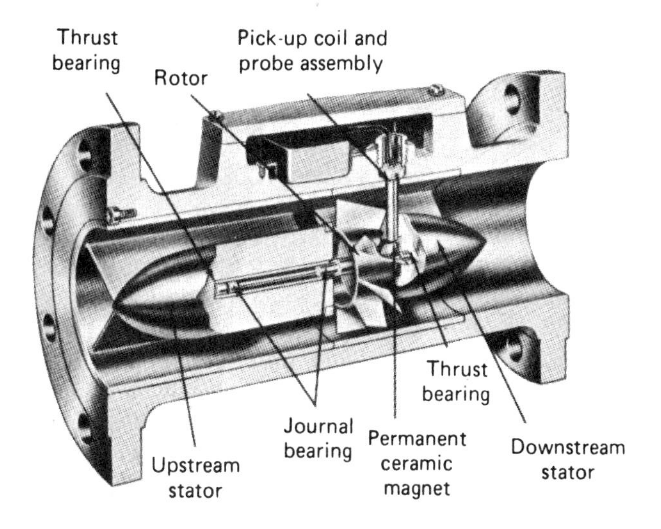

Figure 7-36. Turbine type flowmeter.

liquid and then release it by a pumping action. Probably the most common flowmeter of this type is the nutating piston pump used in most houses to meter the water supplied by the local utility.

The principle of the nutating or disc pump is illustrated in Figure 7-37. It consists of a sealed cylindrical measuring chamber. Inside the chamber is a flat circular disc or piston which divides the chamber into two equal parts. The disc, the only moving part in the flowmeter, is mounted on an off-center pivot so that one side of the disc is making a sealed contact with the top of the chamber and the other side seals off the bottom of the chamber.

This arrangement prevents direct liquid flow from the inlet to the outlet. With the disc in the position shown in Figure 7-37, liquid enters from the right and fills the chamber above the disc. The pressure exerted by the incoming fluid is applied across the whole upper surface of the disc and, since it is mounted off-center, there is a greater area on the left of the disc which causes the disc to tilt and dump the liquid into the left-hand chamber. The liquid then flows out through the outlet connection on the left. The disc then tilts back to its original position and the process is repeated. A shaft projects from the top of the disc and engages a counting mechanism. Since the volume of the chamber is known, the product of the volume and the number of oscillations permits the total quantity of liquid flowing to be measured.

This type of mass flowmeter gives reliable measurement from nearly zero flow rate up to the maximum designed capacity of the unit.

Figure 7-37. Nutating type positive displacement mass flowmeter.

7-7-5 Variable Area Flowmeter

The variable area flowmeter is extremely simple and basically consists of two parts: (a) a vertically mounted tapered circular glass tube and (b) a metering float (see Figure 7-38). When there is zero liquid flow the metering float will sit at the bottom of the tapered glass tube. As the flow rate increases, the float is forced up until there is sufficient area between the periphery of the circular float and the tapered glass tube. The float will come to rest at a particular height which is dependent upon the velocity of the incoming fluid. The rate of flow can be determined in a number of ways ranging from the simplest method, consisting of comparing the float position against an engraved scale on the side of the tube, to more sophisticated methods such as coupling the float to the core of a linear variable differential transformer or to the arm of a potentiometer. The advantage of the last two methods is that the electrical output with suitable signal conditioning may be used to control the process.

7-8 LEVEL SENSORS

A number of industrial processes require the measurement of a liquid, slurry, or some type of dry material. The problem of level measurement may also be compounded because the liquid or material to be measured may be under high pressure, corrosive, at elevated temperatures, or radioactive to name a few possibilities. Level measurement basically consists of measuring the actual height of the liquid visually

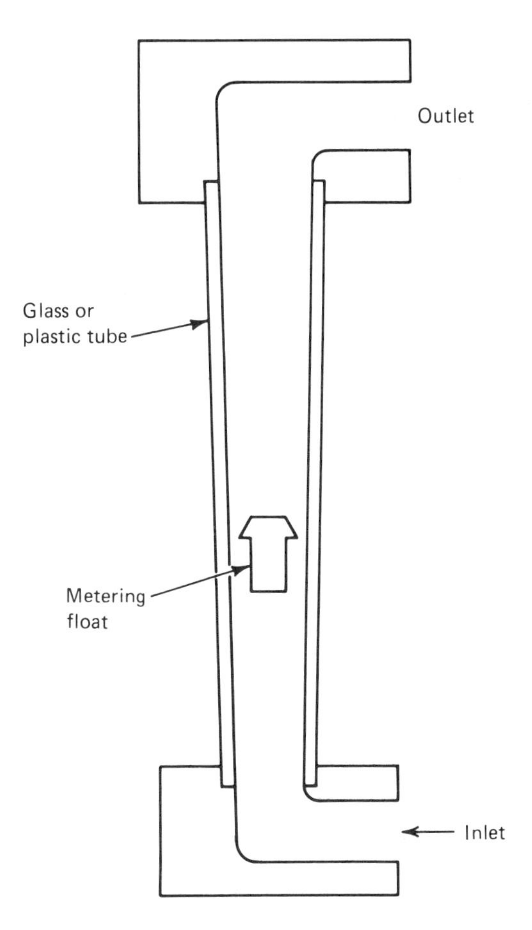

Figure 7-38. Variable area flowmeter.

or electrically, and in turn maintaining the liquid between two preset levels (low and high points desired in the container).

There are many methods that may be used to measure level. A simple method involves sensing the hydrostatic pressure of the fluid by using a pressure sensor mounted in the bottom of the tank as shown in Figure 7-39(a). This method is quite reliable for measurement of constant density liquids, since the height of the liquid is equal to the pressure divided by the density of the liquid. In turn the pressure sensor can be made to produce an electrical output which may be used to provide a direct analog or digital readout, or may be used to control flow in and out of the tank.

Another very simple method is to use a float-actuated rheostat [shown in Figure 7-39(b)] where the rheostat resistance is inversely proportional to the height of the fluid.

Two forms of capacitive liquid level sensing are shown in Figures 7-39(c) and (d). In Figure 7-39(c), an insulated metal electrode is mounted close to and parallel to the wall of a metal tank. Assuming that the liquid is nonconductive, it forms the dielectric of a capacitor consisting of the electrode and the tank wall. Alternatively, if the liquid is conductive and the tank is nonconductive, then two insulated electrodes are used and the liquid acts as the dielectric between them [see Figure 7-39(d)]. Since the capacitance C is proportional to the wetted area between the electrodes, the capacitance will be directly proportional to the level of the liquid.

Another simple liquid level sensing system depends upon the liquid being conductive, and the conductivity between two parallel electrodes is proportional to the liquid height [see Figure 7-39(e)]. There are several disadvantages to this method. First, it should not be used with any liquid that will leave a deposit on the electrodes and thus introduce resistance and, second, it cannot be used with any flammable liquids.

Another method of measuring liquid level depends upon placing a radioactive source at the bottom of the tank and using a Geiger-Muller tube mounted at the top of the tank to sense the intensity of the gamma rays emitted by the radioactive source. The radiation sensed will decrease as the liquid level rises. See Figure 7-39(f).

The methods briefly described above only cover a limited number of the possible ways of sensing level. In general, the techniques utilized in industrial applications are determined by the required accuracy, maintainability, and economics. The more common methods of level measurement of liquids and dry materials are made by mechanical means, pressure changes, weighing, ultrasonics, electrical, and electronic devices.

7-8-1 Level Limit

Another form of level sensing requires the ability to determine when the liquid or dry material entering a tank has reached a predetermined upper level so that, when this level is reached, the flow of the liquid or material may be stopped.

A number of approaches to the problem of level limit control are shown in Figure 7-40. These methods will not be discussed since the principle of operation is considered to be self-explanatory.

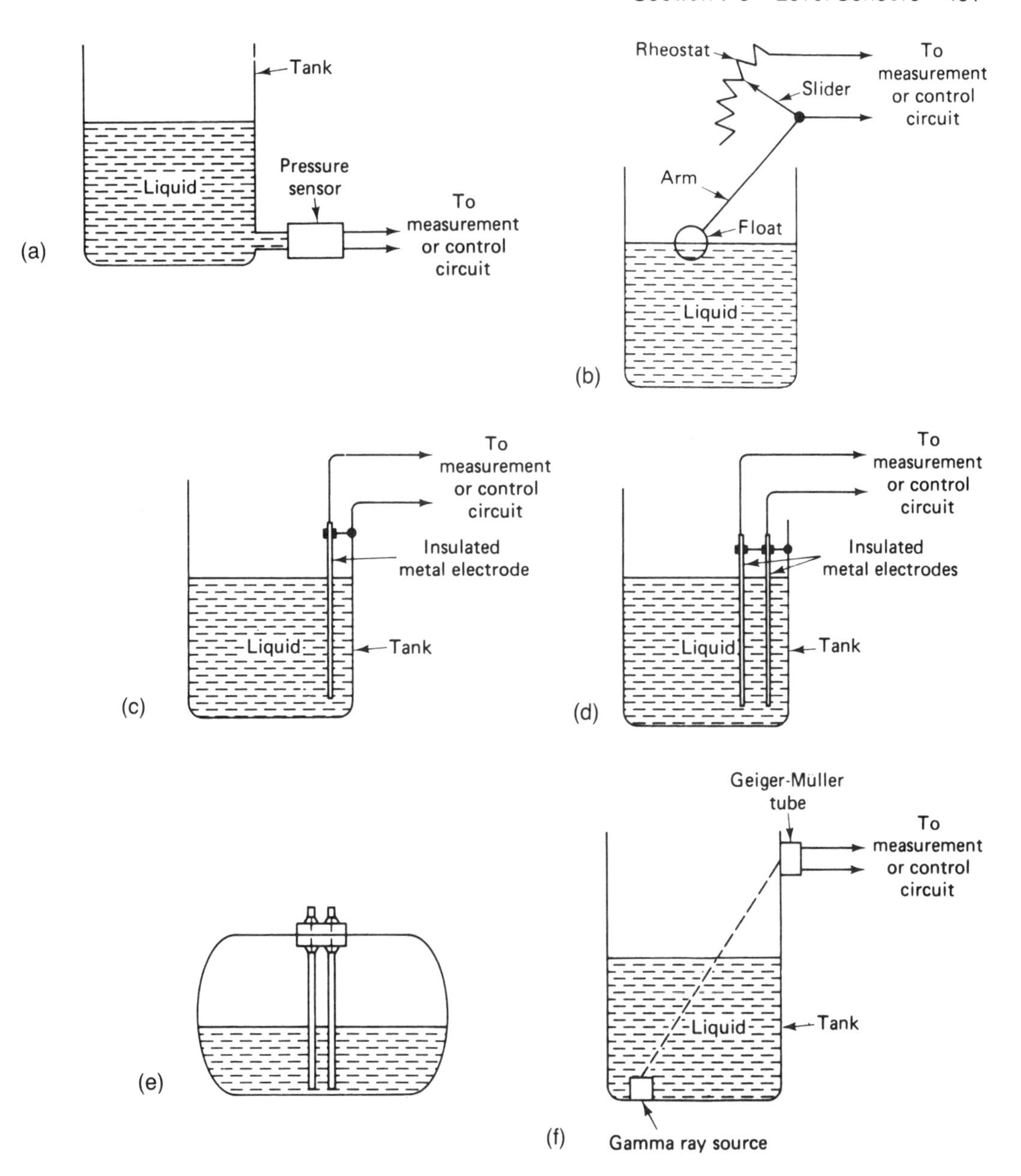

Figure 7-39. Liquid level sensing: (a) pressure sensing, (b) float operated rheostat, (c) capacitive sensing metallic tank, (d) capacitive sensing nonmetallic tank, (e) conductivity sensing nonmetallic tank, (f) gamma ray level sensing.

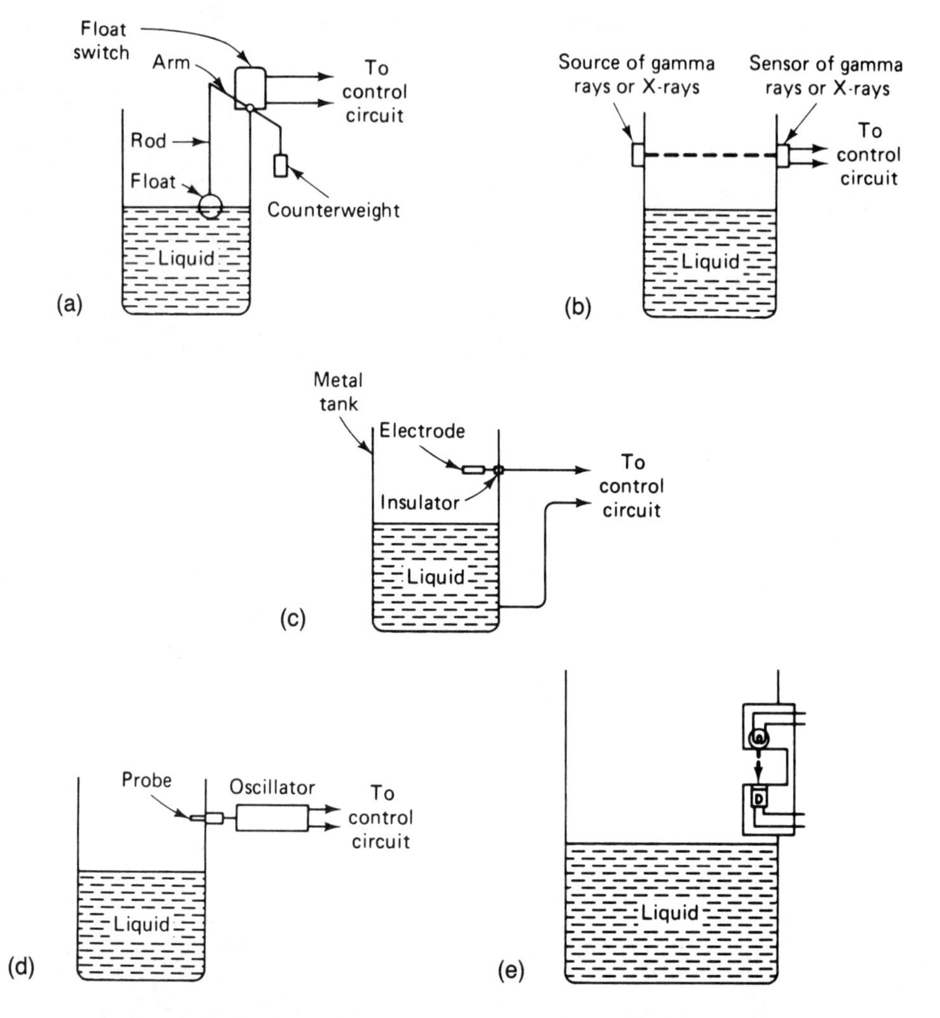

Figure 7-40. Level limit methods: (a) float switch, (b) gamma or X-ray, (c) conductivity, (d) ultrasonic, (e) optical sensing.

7-9 TEMPERATURE SENSORS

In process control instrumentation, temperatures ranging from −470 °F (−267.8 °C) to +7500 °F (4150 °C) require to be measured. To accomplish this task requires a variety of temperature sensors. The types of sensors and typical ranges for each type are shown in Table 7-3. As can be seen, it requires a variety of devices to provide for the entire range of possible temperature measurement.

A number of factors determine the selection of a suitable temperature sensor. First, sensitivity which may be defined as the rate of change of the output to the change of input. The sensitivity of the whole temperature measuring system is more important than just the device, for example, the sensitivity of the resistance temperature detector and the associated resistance bridge measuring system. Second, accuracy which is the ratio of the error in the indicated value to the indicated value. Accuracy is usually expressed as a percentage of the to-

Table 7-3 Types of Temperature Sensors and Ranges

Sensor	Temperature Range			
	Min.		Max.	
	°F	°C	°F	°C
Thermocouples				
Copper-Constantan	− 300	(− 184)	+ 700	(371)
Iron-Constantan	− 300	(− 184)	+ 1,400	(760)
Chromel-Alumel	+ 32	(0)	+ 2,300	(1260)
Iridium/Rhodium-Iridium	+ 1,425	(774)	+ 3,600	(1982)
Resistance Temperature Detectors				
Platinum	− 325	(− 198)	+ 1,000	(538)
Nickel	− 40	(− 40)	+ 400	(204)
Copper	− 330	(− 201)	+ 250	(121)
Radiation Pyrometers				
Low range	+ 100	(38)	+ 700	(371)
Intermediate range	+ 1,000	(538)	+ 3,400	(1871)
High range	+ 1,700	(927)	+ 7,000	(3871)
Optical Pyrometers				
Disappearing filament	+ 1,400	(760)	+ 7,500	(4149)
Thermistors	− 75	(− 59)	+ 750	(399)
Filled Thermometer Systems				
Liquid-filled	− 300	(− 184)	+ 600	(316)
Mercury-filled	− 38	(39)	+ 1,000	(538)
Vapor-filled	− 432	(− 258)	+ 600	(316)
Gas-filled	− 450	(− 268)	+ 1,400	(760)
Bimetallic Thermometers	− 300	(− 184)	+ 1,000	(538)

tal scale span. Accuracy is a more important parameter than sensitivity. The third parameter, response speed or response time, is the time required for the sensor to attain a specified percentage of the total change in response to a step input change of temperature. Response time is determined by the characteristics of the sensor and the thermal time constant of the protecting tube or head.

7-9-1 Thermocouples

The thermocouple depends upon the Seebeck effect which states that if two dissimilar metals are joined together at one end and heated (hot junction) then a small voltage (millivolts) will be produced at the other end (cold junction). A wide range of metal combinations may be used; the choice of the combinations also determines the practical range of temperatures that may be measured. Figure 7-41(a) shows the principle of operation and Figure 7-41(b) shows the temperature-voltage curves for iron-constantan and chromel-alumel thermocouples. Figure 7-41(c) shows a selection of typical industrial thermocouples.

The major advantages of thermocouples are: (a) relatively inexpensive, (b) available commercially in a wide variety of configurations, (c) the electrical output simplifies interfacing with measuring and control systems, and (d) they have a fairly fast rate of response. The principal limitations are: (a) the output is not linear, (b) not suitable for temperature measurement of moving objects, and (c) relatively poor accuracy as compared to resistance temperature detectors.

7-9-2 Resistance Temperature Detectors (RTDs)

The resistance temperature detector depends upon the property of conductors to experience resistance changes with changes of temperature. In general, the resistance of most metallic conductors increases with increases in temperature, and vice versa. The basic construction of the RTD is shown in Figure 7-42(a). The resistance temperature sensing element is in the form of a coil of fine wire, usually copper, nickel or platinum wound on a mica support and enclosed in a metal tube for protection. Examples of commercially available RTDs are shown in Figure 7-42(b).

The main advantages of the RTD as a temperature sensor are: (a) the resistance-temperature characteristics are easily reproduced, unlike the thermocouple and (b) are very accurate and retain their accuracy over many years. The major disadvantages are: (a) they are relatively expensive as compared to a thermocouple, (b) they have an upper temperature limit of 1400 °F (760 °C), (c) possess a relatively long response

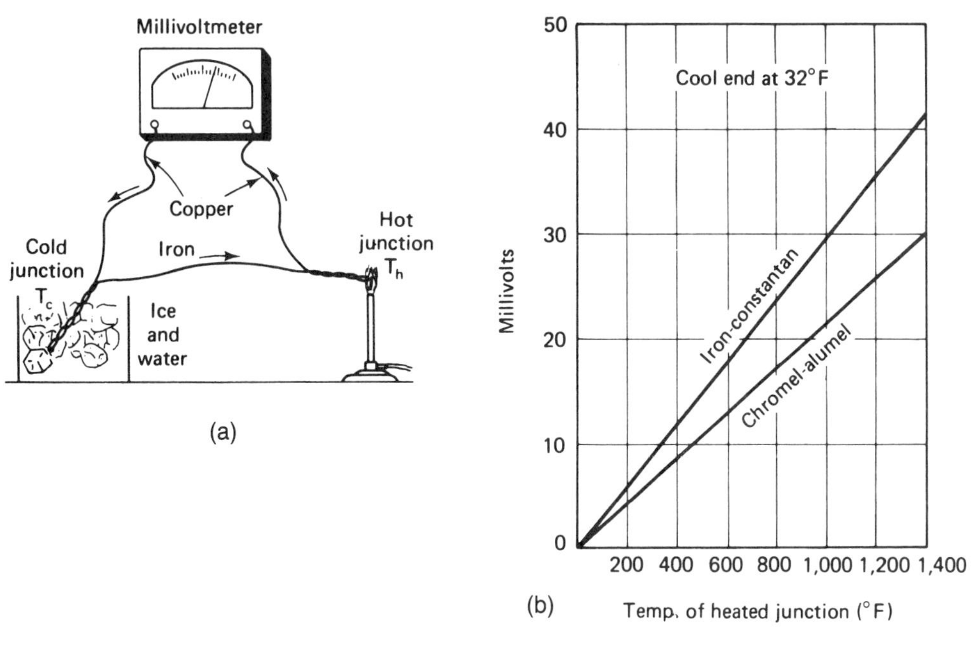

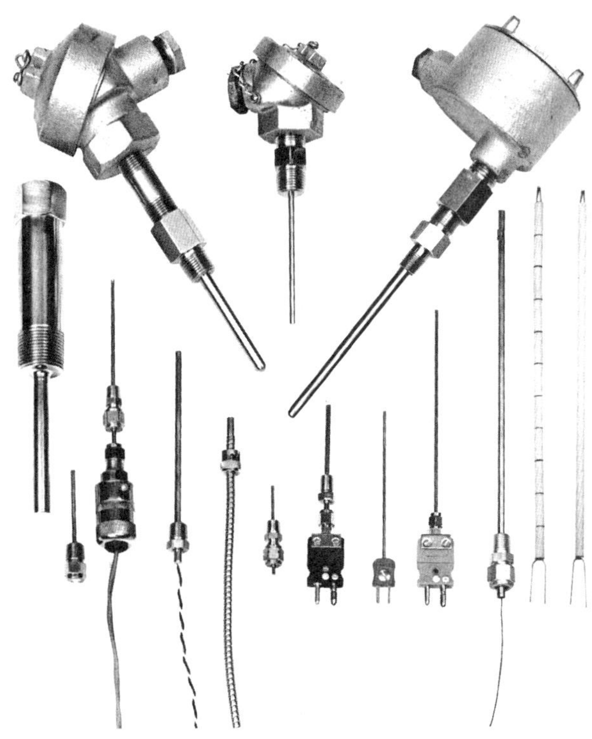

Figure 7-41. Thermocouples: (a) schematic, (b) representative voltage-temperature curves, (c) typical industrial thermocouple assemblies.

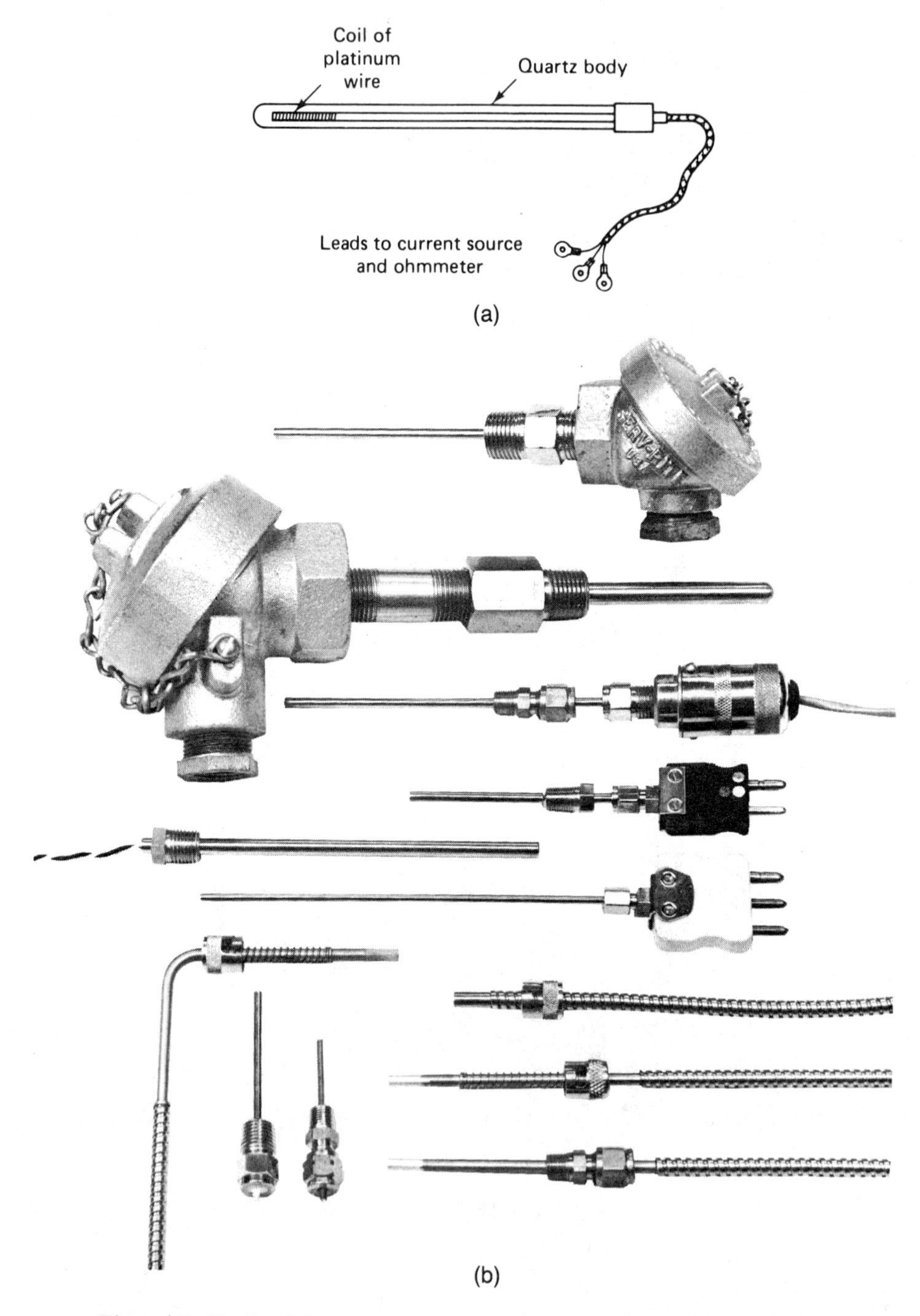

Figure 7-42. Resistance temperature detectors: (a) schematic, (b) typical industrial RTD assemblies.

time, and (d) are not as durable as a thermocouple in situations where there is a fair amount of vibration present.

7-9-3 Radiation and Optical Pyrometers

The prime advantage of using radiation pyrometers is that temperature may be measured without the measuring device making contact with the process. High temperature sources such as molten cast iron radiate energy in the visible and infrared portions of the electromagnetic spectrum. If this radiant energy is focused onto a group of thermocouples connected in a series aiding configuration, called a **thermopile**, a millivolt output which represents temperature is obtained. The output voltage is monitored by a millivoltmeter calibrated in °F or °C, as appropriate. If a continuous record is required, the output must be amplified by a dc amplifier in order to drive the pen of a strip chart or circular chart recorder. The radiation pyrometer is illustrated in Figure 7-43.

Another method of measuring high temperatures without making contact uses a photovoltaic pyrometer. The principle of operation is similar to that of the total radiation pyrometer except that the spectral response is usually confined to a specific wavelength. The choice of the specific wavelengths, usually in the visible spectrum, is determined by the need to maximize the emission from the hot body.

Optical pyrometers compare the radiated energy at a specific wavelength from the hot body to that of an internal reference source. The internal source is usually a standard tungsten lamp. The brightness of the lamp increases with temperature and is controlled by varying the current flow through the lamp until the filament appears to match the image of the hot body. Once again a millivoltmeter calibrated in °F or °C, as appropriate, is used to record the temperature.

The advantages of pyrometers are: (a) they are the only devices that can measure high temperatures, (b) they do not make contact with the body or material whose temperature is being measured, and (c) have fairly high response speeds. The principal disadvantages are: (a) they are subject to errors caused by smoke and dust and (b) are relatively expensive.

7-9-4 Thermistors

A **thermistor** is a semiconductor device usually made by sintering various mixtures of metallic oxides such as nickel, manganese, cobalt, copper, iron, and uranium. A thermistor, because it is covalently bonded, will release free electrons as the temperature increases; that is,

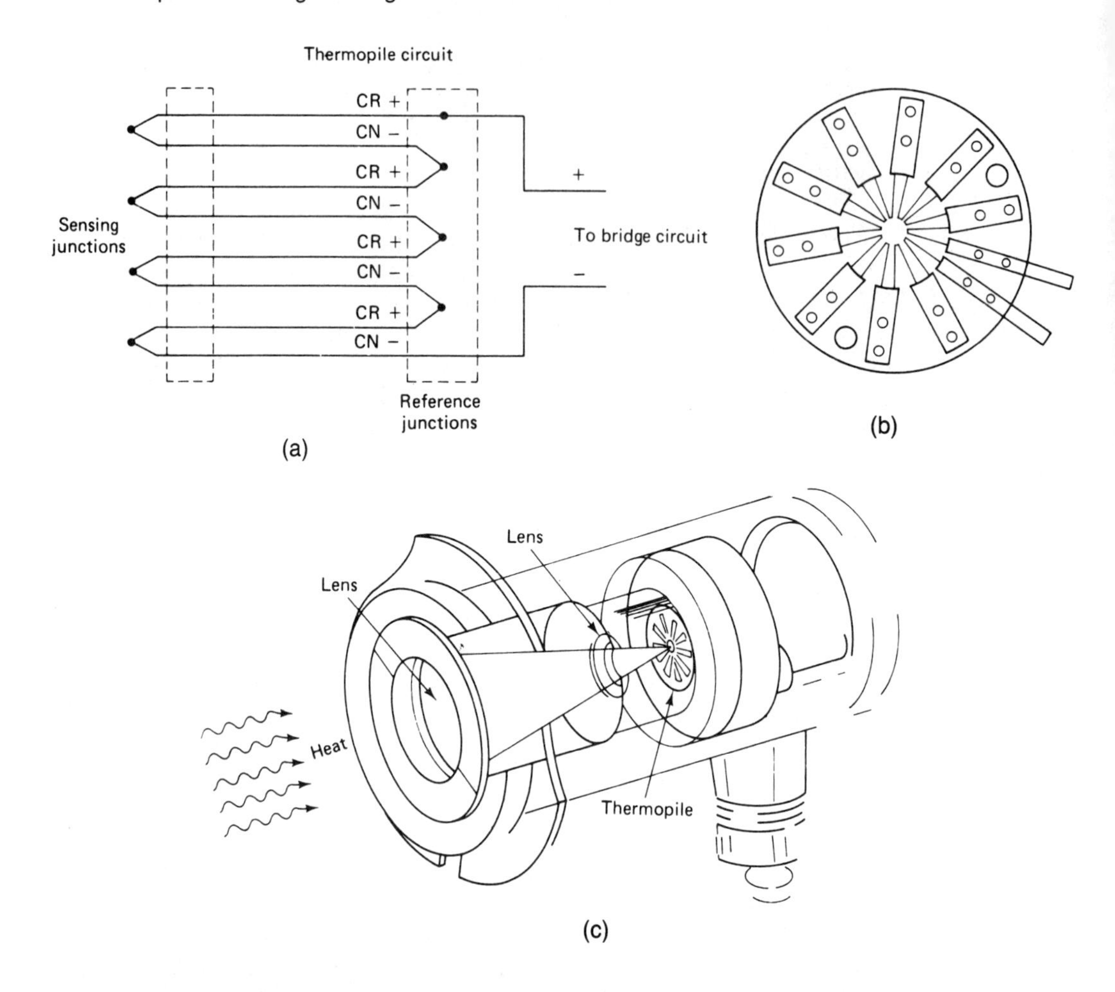

Figure 7-43. Total radiation pyrometer: (a) series-aiding thermocouples, (b) thermopile, (c) sectional view of a radiation pyrometer.

it has a negative resistance temperature coefficient. Thermistors may be placed in one leg of a Wheatstone bridge, see Figure 7-44 (a), or be used to develop a voltage drop across a fixed resistor, see Figure 7-44(b). Commercial thermistors are available in a wide variety of shapes and sizes and are manufactured with tolerances of $\pm 0.5\%$.

The main advantages of thermistors are: (a) very sensitive, (b) available in very small packages, (c) fast response, (d) low cost, and (e) readily adapted to electrical readouts. The major disadvantages are: (a) less stable than other sensors, (b) limited temperature range, and (c) currently limited in their application to process control.

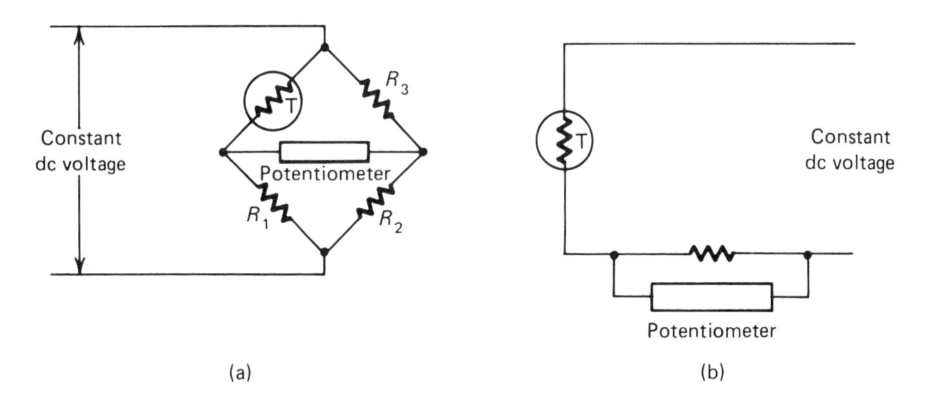

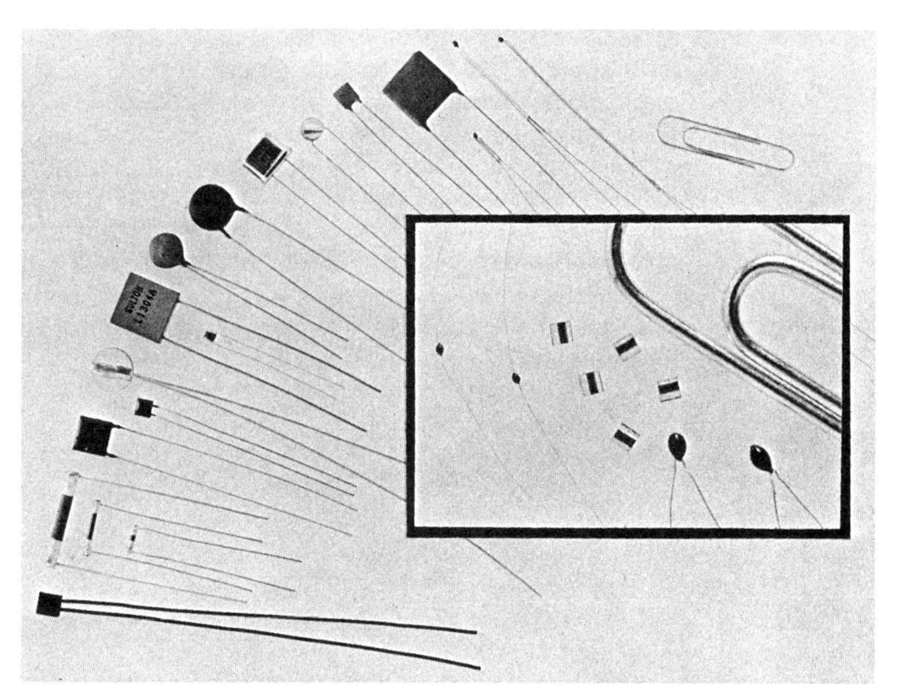

Figure 7-44. Thermistors: (a) Wheatstone bridge arrangement, (b) voltage drop arrangement, (c) typical industrial thermistors.

7-9-5 Filled Thermal Systems

Filled thermal systems depend upon the volumetric expansion of liquids such as, alcohol, mercury, gases, and vapors with increasing temperatures. They are used to cover measurement of temperatures from −450°F (−268°C) to 1400°F (760°C).

They are classified into three different classes. Class I liquid expansion sensors are characterized by a uniform scale and can be used to measure temperatures as low as −300°F (−184°C) and as high as +600°F (316°C), although the usual span of a single device is not sufficient to cover the entire range. Class II are vapor pressure devices with the capability of providing measurements between −432°F (−258°C) and +600°F (316°C), although the span of an individual device can vary from a minimum of 18°F (7.8°C) to 400°F (204°C). Class III are gas pressure devices and provide measurement capabilities over the range of −450°F (−268°C) to +1400°F (760°C) with a minimum span of 400°F (204°C) to a maximum of 1400°F (760°C).

Classes I, II, and III systems can provide remote indication usually up to 25 ft (7.62 m) by using a Bourdon tube pressure sensor as shown in Figure 7-45. An electrical output can be obtained by connecting the mechanical pointer system to an electrical device such as an LVDT or linear potentiometer.

The major advantages of filled systems are: (a) rugged construction, (b) simple, (c) low cost, and (d) totally self-contained. The principal disadvantages are: (a) failure of the bulb or capillary tubing requires replacement of the entire unit, (b) if remote indication at some distance is required, the output must be converted to an electrical output, (c) less sensitive and accurate compared to comparable electric systems, for example, thermocouples and RTDs, and (d) not suitable for usage in temperature applications in excess of 1400°F (760°C).

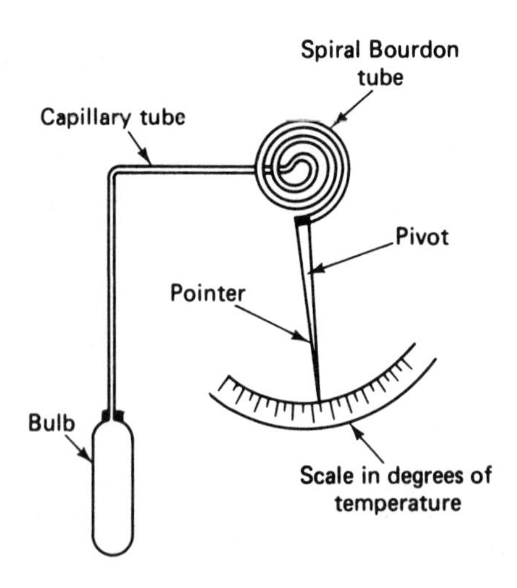

Figure 7-45. Filled thermal system.

7-9-6 Bimetallic Temperature Sensors

Bimetallic temperature sensors depend on the coefficient of expansion of metals when heated for their ability to measure temperature. If two materials with widely different coefficients of expansion are bonded together as shown in Figure 7-46, the bonded element will distort when heated. The most commonly used materials are invar (which has a very low coefficient of expansion), brass (which has a relatively high coefficient of expansion), or a nickel alloy for very high temperatures.

These devices are low cost, possess a reasonably wide temperature range $-300°F$ ($-184°C$) to $1000°F$ ($538°C$), are easy to install, and provide reasonable accuracy. Their main disadvantages are they are limited to indicating, except in the case of thermostats and thermal overloads, possess only a fair sensitivity and accuracy, are limited to local mounting, and do not provide remote indication.

7-10 MISCELLANEOUS SENSORS

There are a number of applications of sensors to particular tasks such as thickness and density measurement. It is not intended to discuss light sensors at this point. Light sensors and their applications will be considered in greater detail in Chapter 8.

7-10-1 Thickness Measurement

There are a number of different ways of measuring the thickness of a material. Among the methods in most common use are inductive, capacitive, ultrasonic, and gamma or X-ray devices.

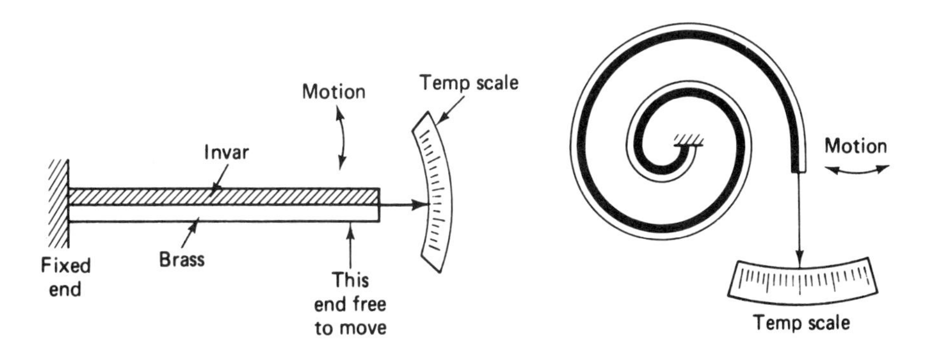

Figure 7-46. Bimetallic temperature sensors.

7-10-1-1 Inductive Sensor

An inductive sensor is most appropriate when measuring the thickness of magnetic materials such as sheet steel. The principle is shown in Figure 7-47 (a). The sensor consists of a square U-shaped ferromagnetic laminated core with a multiturn coil wound on it. When the sensor is in physical contact with the magnetic material to be measured, it completes the magnetic circuit and the inductance of the coil is related to the thickness of the magnetic material. In turn the inductance is measured and equated to the thickness of the material under test.

7-10-1-2 Capacitive Sensor

A capacitive sensor is particularly suitable for the measurement of nonconducting materials such as plastic sheet. The material under test is passed between two metal plates. Since capacitance is proportional to the area and the dielectric constant of the material and inversely proportional to the distance between the plates, if the capacitance is measured it can be equated to the thickness of the material being tested. See Figure 7-47 (b).

7-10-1-3 Ultrasonic Sensor

Ultrasonic transducers depend upon either of two methods of producing ultrasonic sound waves. The first method is the piezo-electric transducer which consists of a specially cut crystal of quartz sandwiched between metal plates. When this combination is excited by a high-frequency ac voltage, it alternately expands and contracts at the same frequency, producing compressions and rarefractions of the adjacent air. The second method is the magnetostrictive transducer which depends upon a ferromagnetic material such as transformer steel expanding and contracting under the influence of a high-frequency alternating magnetic field, thus producing ultrasonic sound. Normally ultrasonic transducers operate at frequencies in the range of 20 to 100 kHz.

Ultrasonic transducers measure thickness by being placed in contact with the material and measuring the time it takes a pulse of utrasonic sound to travel in the material. The system, in its simplest form, consists of one transducer which acts as a transmitter and receiver, see Figure 7-47 (c). The system is pulsed and a sound wave is transmitted and travels through the thickness of the material to the interface between the material and the bottom surface, where it is reflected back. At this point in time, the transmitter has been muted and is now

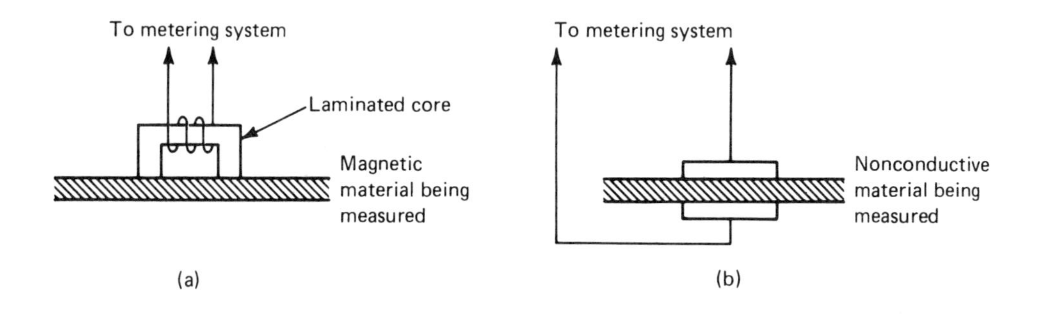

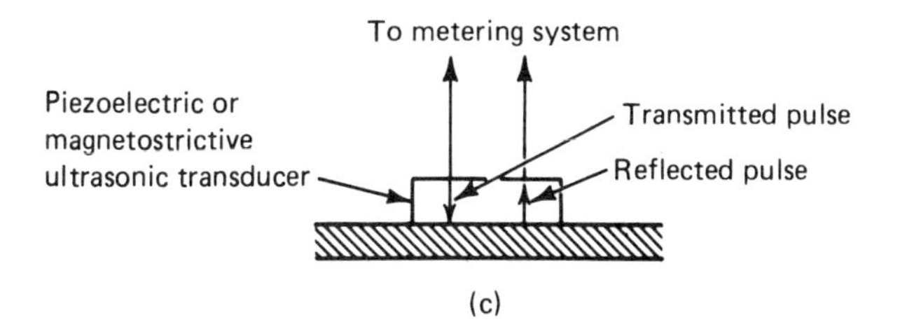

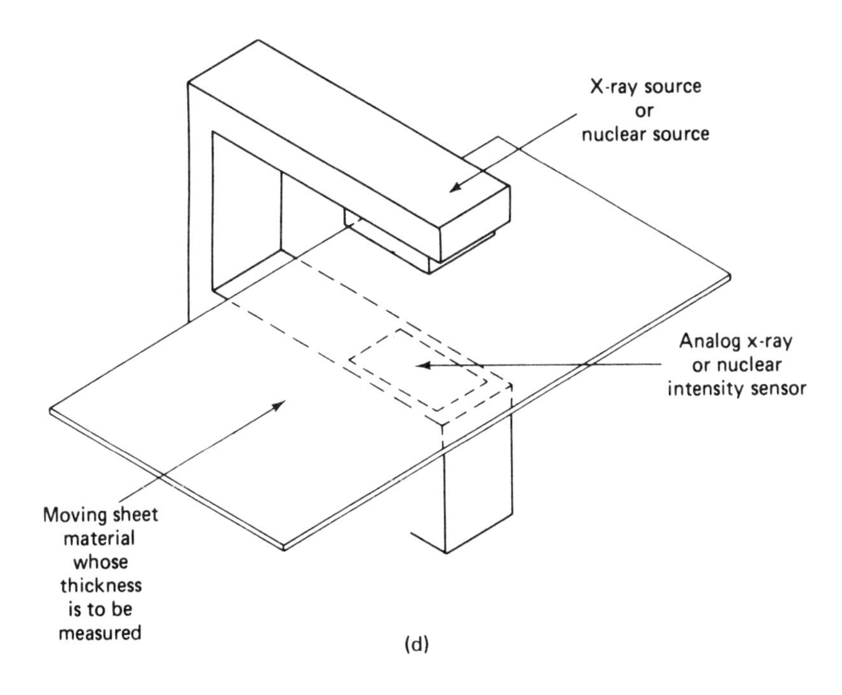

Figure 7-47. Thickness measurement: (a) by change of inductance, (b) by change of capacitance, (c) by ultrasonics, (d) by X-rays.

in the receiving mode. The time interval between the transmitted and received pulse is measured and equated to the thickness of the material.

This method can also be extended to nondestructive testing where it is used to detect cracks, slag inclusions, etc., as well as thickness measurement.

7-10-1-4 X-Ray Thickness Sensor

The X-ray thickness gauge is particulary suitable in measuring the thickness of hot moving metal sheets, and can achieve an accuracy of ± 0.001 in (± 0.0254 mm).

The device is illustrated in Figure 7-47 (d). A continuous stream of X-rays, at a known intensity level, is transmitted from the source in the lower arm and passes through the material being measured. A number of the rays are absorbed in the material as it passes between the source and the sensing unit; the amount of the absorption is dependent upon the type of material and its thickness. The sensor consists of a sodium iodide or zinc sulfide cell which will produce a bright flash of light each time it is struck by a gamma ray. A scintillation counter is used to count the flashes, which in turn equates the count to thickness.

7-10-2 Density Sensing

From our studies of physics we know that the density of a substance is the mass or weight per unit volume. Normally density is expressed in terms of specific gravity, where specific gravity is the ratio between the mass of a substance and the mass of an equal volume of water.

There are many industrial applications where continuous monitoring of the density of a liquid is required. A simple but effective method utilizes both the hydrometer and the LVDT. This system is shown in Figure 7-48. The core of the LVDT forms an integral part of the hydrometer, the excitation and sensing coils being mounted on the outside of the glass tube. The principle of operation depends upon the hydrometer rising in the tube as the density of the fluid increases, and dropping as the density decreases. In turn the differential output voltage will indicate the amount of the change of density and whether it is increasing or decreasing.

SUMMARY

The purpose of the preceding review of sensors has been to give an overview of the range of measuring devices and the basic principles of

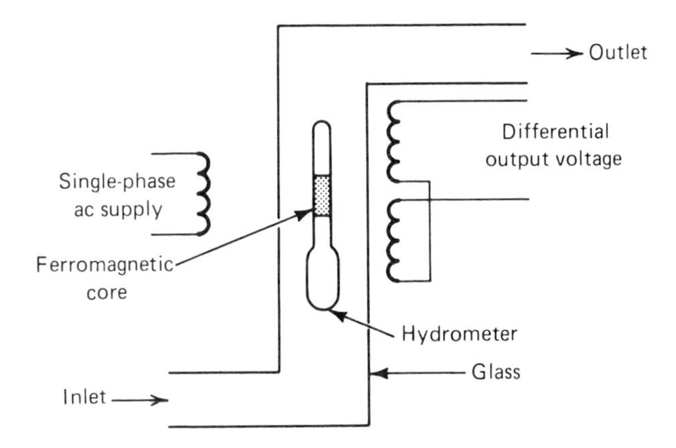

Figure 7-48. Liquid density sensing using an LVDT.

their operation. All industrial processes depend to lesser or greater degree upon sensors to provide information that will permit effective control of the process to be maintained. In subsequent chapters a number of these sensors will be applied to illustrative applications.

GLOSSARY OF IMPORTANT TERMS

Transducer: A device that converts energy in one form to another form.

Controlled variable: The output variable of a controlled process.

Sensor: The device that measures the controlled variable.

Accuracy: The measurement capability of a transducer to approach the true value of the measured quantity. It is affected by static error, dynamic error, reproducibility or repeatability, dead time, and dead zone.

Static error: The deviation of the measured value of the physical quantity being measured from its actual value when the physical quantity is constant. It is expressed as a percentage of the full-scale range of the transducer.

Dynamic error: The error that occurs because of changes in the value of the physical quantity caused by either input or load changes.

Reproductibility or repeatability: The maximum deviation from the average of repeated readings when the measured quantity is constant.

Dead time: The time interval between a change in the measured quantity and the change being noted at the transducer output.

Dead zone: The largest change that can occur in a measured quantity without it's being detected at the transducer output.

Speed of response: The time interval between a change in the measured quantity occurring and the time the change is detected at the transducer output.

Linear motion sensors: Sensors that measure linear displacement.

Angular motion sensors: Sensors that measure angular displacement usually limited to less than 300°.

Synchros: Unlimited angular motion sensors, used in control and signaling applications to transmit angular information by electrical means.

Resolvers: Members of the synchro family which perform: (1) rectangular-to-polar and polar-to-rectangular conversions, (2) coordinate transformation, and (3) phase shifting.

Scott-Tee transformers: Convert two-phase systems to three-phase or vice versa.

Synchro converters: Perform digital-to-synchro or synchro-to-digital conversions and are used to interface between analog positioning systems and microprocessors or digital computers.

Encoders: A device that assigns digital values to linear or angular motion.

Angular velocity transducers: Devices that measure angular velocity.

Analog speed transducers: Produce an output voltage directly proportional to angular velocity.

Digital speed transducers: Depend upon the output frequency being proportional to angular velocity, or a pulse train being produced so that the pulse count over the sampling interval is proportional to the angular velocity.

Force sensors: A family of sensors that measure the displacement caused by the effects of strain, force, load, torque, acceleration, and vibration.

Fluid pressure sensors: Liquids and gases under pressure create a force over an area. The displacement created by this force is used to measure pressure.

Fluid flow sensors: Sensors that measure either the rate of flow of a fluid, or the total quantity of liquid flow.

Level sensors: Sensors that measure the height of a liquid, slurry, or dry materials.

Level limit: A specific application of level sensors to control the maximum or minimum levels of a liquid etc., in a process.

Temperature sensors: Sensors that measure temperature and provide an indication of temperature, by visual means, by change of resistance, by change of pressure, by producing a voltage or, by differences in the coefficient of expansion (bimetallic).

REVIEW QUESTIONS

7-1 Explain what is meant by accuracy and discuss the factors that affect the accuracy of a transducer.

7-2 Define speed of response of a transducer and discuss the factors that affect the speed of response.

7-3 Explain with the aid of sketches the principle of operation of an LVDT and how the circuit may be used to provide a dc output. List typical applications.

7-4 What is a linear encoder? Explain what is meant by (1) an incremental and (2) an absolute encoder.

7-5 With the aid of a sketch explain the principle of operation of an incremental encoder.

7-6 Explain the differences in construction between the following synchro devices: (1) transmitter, (2) receiver, (3) differential transmitter, (4) differential receiver, and (5) control transformer.

7-7 With the aid of sketches discuss the zeroing procedure for synchro units.

7-8 What is the function of a dual-speed synchro system?

7-9 What is a resolver? How is it constructed? What function does it perform?

7-10 What is a synchro converter?

7-11 What is a shaft angle encoder? Discuss (1) the incremental encoder and (2) the absolute encoder. Give typical applications.

7-12 Explain the principle of operation of a dc permanent magnet tachogenerator. What are the advantages and disadvantages?

7-13 Explain the principle of operation of an ac induction tachometer.

7-14 Discuss the advantages and disadvantages of using a digital tachometer. Explain the principle of operation of (1) a magnetic proximity digital tachometer and (2) an optical digital tachometer.

7-15 Explain the principle of operation of a resistance strain gauge and explain the differences between a bonded and unbonded strain gauge. What precautions must be taken to prevent errors resulting from temperature effects. List typical applications of each type.

7-16 What is the difference between an accelerometer and a vibration transducer?

7-17 With the aid of sketches explain the principle of operation of (1) a piezoelectric accelerometer, (2) an inductive vibration transducer, and (3) an LVDT vibration transducer.

7-18 Explain the basic principle of operation and construction of: (1) a Bourdon tube, (2) a diaphragm, and (3) bellows type pressure to motion transducers.

7-19 Discuss the types of electrical outputs available from pressure to motion transducers.

7-20 What are the commonly used types of head flowmeters? Explain the principle of operation of each type. What are their limitations?

7-21 Explain the principle of operation and discuss the advantages and disadvantages of electromagnetic flowmeters.

7-22 Repeat Question 7-21 for a turbine flow meter.

7-23 Explain the principle of operation of a variable area flowmeter.

7-24 With the aid of sketches explain the principle of operation and limits of the following level sensors: (1) capacitive (both types), (2) conductivity, and (3) gamma ray.

7-25 What factors determine the selection of a suitable temperature sensor?

7-26 With the aid of sketches explain the principle of operation and the advantages and limitations of the following types of temperature sensors: (1) thermocouples, (2) resistance temperature detectors, (3) radiation and optical pyrometers, and (4) thermistors.

7-27 Explain the principle of operation of filled thermal temperature sensors. How can their responses be converted to an electrical output? What are their advantages and limitations?

7-28 Explain the principle of operation of the following types of ultrasonic transducers: (1) piezoelectric, and (2) magnetostrictive. What are the principle commercial applications of these devices?

7-29 Explain the principle of operation of an X-ray thickness gauge.

8

Optoelectronics

8-1 INTRODUCTION

The field of optoelectronics is the result of the marriage of two previously independent technologies, namely optics and electronics. Because of the rapid rate of introduction of these devices in the industrial field, it is essential that a basic understanding of their principles and applications be obtained. Currently the devices are being used in such applications as level sensing, smoke detection, speed and position measurement, high-voltage isolation, fiber optics, logic circuits, data communication, etc., the only limitations being the engineer's imagination.

By definition, an **optoelectronic device** is one which:

1. Detects light or radiation, or
2. Emits or modulates coherent and noncoherent light, or
3. Depends internally for its operation on light or radiation

Light is electromagnetic radiation occurring in the electromagnetic spectrum from 0.3 μ (1 x 10^{15} Hz) to 30 μ (1 $\times$ 10^{13} Hz), and inlcudes the range from 0.38 to 0.78 μ, the portion visible to the human eye, as well as portions of the ultraviolet and infrared sections of the spectrum invisible to the human eye to which many optoelectronic devices are responsive (see Figure 8-1).

The interrelationship between light and electricity was first observed by Becquerel in 1839, while in 1873 Smith noted a change in the resistance of a selenium conductor when exposed to sunlight. In 1907 the emission of yellow light was observed by Round when approximately 10 V was applied to a silicon carbide crystal. Initially these methods of producing and sensing light were basically ignored. It was

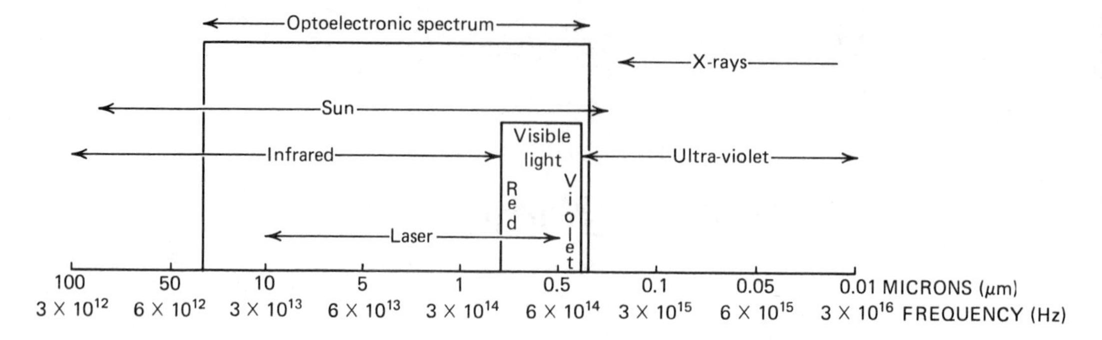

Figure 8-1. Portion of the electromagnetic spectrum applicable to optoelectronic devices.

only with the introduction of the semiconductor diode and transistor, which were found to be sensitive to light, that interest was rekindled in the earlier experiments. This renewed interest has in turn led to the rapid and continuing development of modern optoelectronic devices.

8-2 DEFINITIONS AND UNITS

In our study of optoelectronics it is essential that we differentiate between photometric and radiometric sources of light. **Photometric quantities** refer to the visual effects as seen by the human eye, while **radiometric quantities** refer to the total energy output of a light source and include the ultraviolet and infrared as well as the visible part of the electromagnetic spectrum.

8-2-1 Wavelength

In most industrial power applications we are used to dealing with alternating quantities in terms of frequency. At the higher frequencies encountered in optoelectronics it is more convenient to use wavelengths (λ) instead of frequency. The relationship between frequency and wavelength is expressed by

$$\lambda = \frac{c}{f}, \qquad (8\text{-}1)$$

where λ = wavelength of one complete cycle in meters (m),
c = velocity of light
= 3×10^8 m/sec, and
f = frequency, (Hz).

This relationship is shown graphically in Figure 8-2.

The common units for wavelengths encountered in optoelectronics are microns (μ), millimicrons (mμ), or Angstroms (Å), where

$$1 \text{ micron} = 1 \times 10^{-6} \text{ meters} = 1\ \mu,$$

$$1 \text{ millimicron} = 0.001 \times 10^{-6} \text{ meters} = 1 \text{ m}\mu,$$

$$1 \text{ nanometer} = 1 \times 10^{-9} \text{ meters} = 1 \text{ nm} = 1 \text{ m}\mu, \text{ and}$$

$$1 \text{ Angstrom} = 0.1 \times 10^{-9} \text{ meters} = 1\ \text{Å}.$$

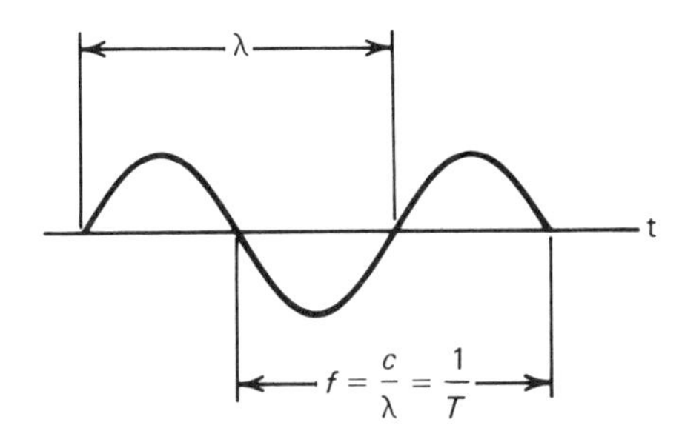

Figure 8-2. Relationship between wavelength and frequency.

Example 8-1

Express the following as wavelengths: (1) 20 kHz; (2) 1×10^{15} Hz.

Solution:

(1) From Equation (8-1)

$$\lambda = \frac{c}{f} = \frac{3 \times 10^8 \text{ m/sec}}{20 \times 10^3 \text{ Hz}} = 1.5 \times 10^4 \text{ m}$$

$$= 1.5 \times 10^4 \text{ m} \times \frac{1\ \mu}{1 \times 10^{-6} \text{ m}}$$

$$= 1.5 \times 10^{10}\ \mu$$

$$= 1.5 \times 10^4 \times \frac{1 \text{ m}\mu}{1 \times 10^{-9} \text{ m}}$$

$$= 1.5 \times 10^{13} \text{ m}\mu$$

$$= 1.5 \times 10^4 \times \frac{1\ \text{Å}}{1 \times 10^{-10} \text{ m}}$$

$$= 1.5 \times 10^{14}\ \text{Å}.$$

(2) From Equation (8-1)

$$\lambda = \frac{c}{f} = \frac{3 \times 10^{8}\ \text{m/sec}}{1 \times 10^{15}\ \text{Hz}} = 3 \times 10^{-7}\ \text{m}$$
$$= 3.0 \times 10^{-1}\mu$$
$$= 3.0 \times 10^{2}\ \text{m}\mu$$
$$= 3.0 \times 10^{3}\ \text{Å}.$$

As can be seen from this example, the lower the frequency the greater the wavelength and the greater the frequency the shorter the wavelength.

8-3 SPECTRAL RESPONSE OF THE HUMAN EYE

Spectral response of an optoelectronic device or the human eye is the ratio of the energy detected to the power transmitted by the source at a specific wavelength. Usually the readings are plotted as shown in Figure 8-3 to produce a spectral response curve. As can be seen from the figure, the human eye is most responsive to yellow green and least responsive to ultraviolet and infrared. The threshold wavelength λ_o is the maximum wavelength or minimum frequency that produces a detectable response in the device.

Optoelectronic phenomena are based on the quantum theory. Planck postulated that light or radiant energy consisted of discrete bundles of concentrated energy called quanta or photons. Each photon possesses a quantum of energy, E, whose magnitude depends upon the frequency of the radiation

$$E = fh, \tag{8-2}$$

where E = photon energy in joules (J),

f = frequency (Hz), and

h = Planck's constant

$= 6.626 \times 10^{-34}$ joule-sec.

Photons appear to possess both wave-like and particle-like properties, i.e., the color of a light source depends upon the wavelength and the intensity depends upon the number of incident photons per unit time. Equation (8-2) also shows that the photon energy increases with frequency which explains why high frequency radiations such as X-rays and gamma rays are able to penetrate dense materials such as welds.

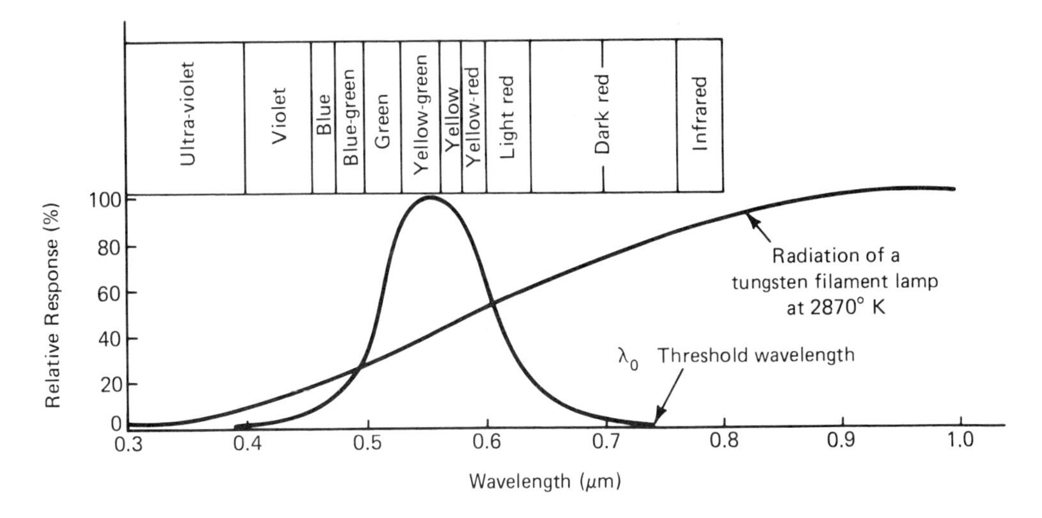

Figure 8-3. Spectral response of the human eye and the radiation characteristic of a tungsten filament lamp at 2780 °K.

8-4 LIGHT PRODUCTION

Light is produced by either of two methods, namely, incandescence or luminescence. The tungsten filament lamp is the most common method of light generation by **incandescence.** It consists of a tungsten wire filament in an evacuated glass bulb which is heated by the flow of an electric current. As the current is increased the filament temperature increases until, at approximately 1255 °K (1800 °F), light can be detected. As the filament temperature is increased, the light output increases, i.e., the color becomes whiter, and the efficiency or the ratio of light output to input power improves. Since in most applications the color or wavelength of the light is extremely important in the selection and application of detectors, tungsten filament lamps are designed to operate at a standard operating temperature of 2870 °K. At this temperature the peak output occurs at approximately 1 μ, and from Figure 8-3, it can be seen that most of the lamp output occurs at wavelengths beyond the visible range of the human eye.

Luminescence, or cool light, is light that is produced without relying upon the temperature of a material. There are two categories of luminescence which are of particular interest in industrial applications: first, electroluminescence, or the production of light by electron-hole recombinations, the most important examples of which are the LED and the laser and, second, photoluminescence or the production of light by the absorption of radiations such as ultraviolet and X-ray.

8-5 THE PHOTOMETRIC AND RADIOMETRIC SYSTEMS

The energy that is transmitted by a light source is proportional to the square of the amplitude, and the color is proportional to the wavelength or frequency. These two characteristics immediately lead to two methods of specifying the characteristics of light, namely, the illumination system and the radiation system. The photometric system is concerned only with that portion of the electromagnetic spectrum that is visible to the human eye and is thus a physiological concept.

The radiometric system, while it can cover the whole electromagnetic spectrum, is normally used in the measurement of radiated energy in the wavelengths or frequencies outside the range visible to the human eye. The two systems are required because most light sources produce emissions that are in the regions that are visible and invisible to the eye, as well as a number of photodetectors responding to both the visible and invisible portions of the electromagnetic spectrum.

8-5-1 Photometric

The modern standard of luminous intensity *(I)* is the candela or cd (formerly the candle). The SI definition of the **candela** is the luminous intensity of 1/600,000 of a square meter of projected area of a black body radiator operating at the temperature of solidification of platinum under a pressure of 101,325 N/m^2.

The amount of luminous flux radiating out from a uniform point source of 1 cd through 1 unit solid angle (1 steradian) is 1 **lumen** (lm). Therefore 1 lm is the amount of luminous flux radiating from a 1 cd source perpendicular to a spherical surface of area 1 ft^2 at a distance of 1 ft from the source, or upon a spherical area of 1 m^2 at a distance of 1 m.

There are 4 π unit solid angles (4 π steradians) surrounding a point source; then the total luminous flux *(F)* emitted by a point source of luminous intensity *(I)* is

$$F\,(\text{lm}) = 4\,\pi\,(\text{lm/cd}) \times I\,(\text{cd}), \tag{8-3}$$

where F = total luminous flux, lm and

I = luminous intensity, cd

from which it can be seen that 1 cd radiates 4 πlm.

Example 8-2

A 60 W incandescent lamp has a luminous intensity I of 70 cd. What is the total luminous flux F radiated by the lamp and the luminous efficiency of the lamp?

Solution:

$$
\begin{aligned}
F &= 4\pi I \\
&= 4\pi \text{ lm/cd} \times 70 \text{ cd} \\
&= 879.65 \text{ lm.} \\
\text{Luminous efficiency} &= \text{lm/W} \\
&= 879.65/60 \\
&= 14.66 \text{ lm/W.}
\end{aligned}
$$

The illuminance E of a surface is the luminous flux F incident upon a unit area of a surface; therefore

$$
\begin{aligned}
E &= \frac{F}{A} \\
&= \frac{4\pi I}{4\pi r^2} = \frac{I}{r^2}.
\end{aligned}
\tag{8-4}
$$

Example 8-3

What is the illuminance E on a perpendicular surface 100 cm from an incandescent lamp with a luminous intensity I of 70 cd?

Solution:

$$E = \frac{I}{r^2} = \frac{70}{(1.0 \text{ m})^2} = 70 \text{ lm/m}^2.$$

Typical luminous flux values for tungsten filament incandescent lamps range from 440 lm for a 25 W lamp to 1740 lm for a 100 W lamp, while a T12, 40 W fluorescent lamp has an output of 3200 lumens.

8-5-2 Radiometric

The radiometric system is normally used to measure the radiated energy outside the frequencies or wavelengths that are visible to the eye. Irradiance H is the radiant flux density incident on a surface, i.e., it is the ratio of radiated power to the area of irradiated surface and is represented by

$$H = P/A \tag{8-5}$$

where H = irradiance or radiant flux density (W/m^2)

P = total radiated power (W)

A = area (m^2)

Example 8-4

Calculate the irradiance H at a distance of 2m from a source radiating 40 W of energy.

Solution:

$$H = P/A$$
$$= 40/(4\pi\,(2\text{ m})^2)$$
$$= 0.796\text{ W/m}^2$$
$$= 7.96 \times 10^{-4}\text{ m W/m}^2$$

It should be noted that although Equation (8-5) is normally used outside the visible portion of the electromagnetic spectrum, it is equally valid in the visible range.

8-6 PHOTODETECTORS

Photodetectors are devices that convert radiant energy to an electrical output by an interaction between the radiant energy and a semiconductor material. They are classified into two groups, bulk type devices which do not have a p-n junction and junction type devices with p-n junctions.

To understand the basic principle of operation of both types it is necessary to recall the energy band diagram. This diagram, Figure 8-4(a), has three regions; the valence band, the forbidden region, and the conduction band. In a semiconductor material the majority of electrons are tightly held in the valence band.

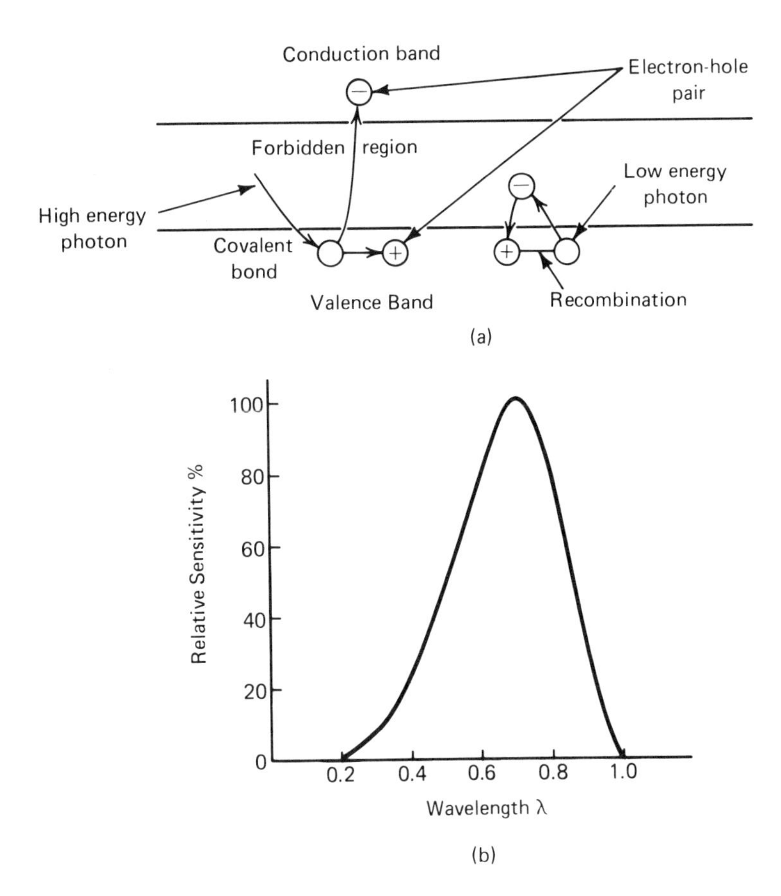

Figure 8-4. Radiation effect on a homogeneous semiconductor: (a) energy band diagram, (b) spectral response curve.

When a photon strikes the semiconductor material then dependent upon the energy and wavelength of the photon, two things will occur. First, if the energy possessed by the photon is sufficient, i.e., greater than the energy gap between the valence and conduction bands, a covalent bond in the valence band will be broken and an electron will rise to the conduction band where it becomes a charge carrier and leaves a hole in the valence band. Thus the photon has created an electron-hole pair. Second, if the wavelength of the incident photon is not of the correct wavelength for the material, insufficient energy will be imparted to the valence band electron to cause it to cross the forbidden gap. As a result the electron will dissipate its energy in the forbidden gap and fall back to the valence band where it recombines with a hole.

If a photon with sufficient energy and of the correct wavelength strikes the semiconductor material, it produces an electron-hole pair with an electron possessing sufficient energy to remain in the conduction band. Thus the conductivity of the semiconductor material has increased; however, because the distribution of electrons in the valence and conduction bands is not uniform, the spectral response of the semiconductor material to incident radiant energy is also not uniform, Figure 8-4(b). It is also slightly skewed because of absorption, recombination, and thermal generation of electron-hole pairs.

8-6-1 Bulk Type Photoconductive Sensors

Bulk type photoconductive sensors are also frequently called **photoresistors** and depend upon the variation of the resistance of the semiconductor material when exposed to varying levels of incident radiant energy. The most commonly used photoresistor devices are either cadmium sulphide (CdS) or cadmium selenide (CdSe). These types of cells will operate with either a dc or ac source; a cell consists of a thin sheet of either CdS or CdSe with metallic electrodes deposited on a ceramic substrate. They are available in TO5 cases or in chip form. A typical CdS or CdSe cell is shown in Figure 8-5. It is to be noted that an external source of emf V is required to provide current flow.

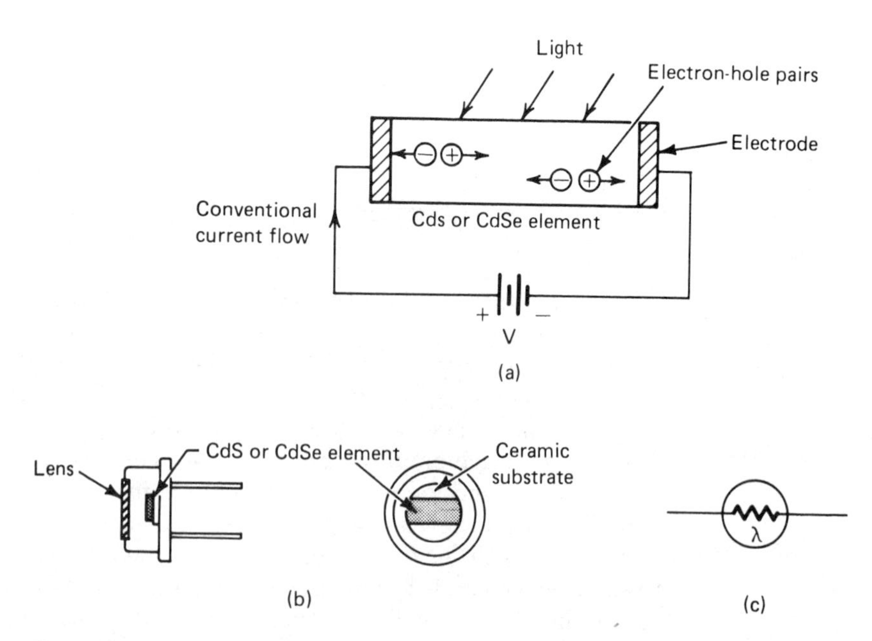

Figure 8–5. Typical photoconductive cell: (a) schematic, (b) T05 structure, (c) circuit symbol.

The essential characteristics of CdS and CdSe cells are shown in Table 8-1.

CdS cells are used primarily in light measurement applications, for example, aperture control of a camera, light meters, etc. CdSe cells are typically used as counters because of their fast response time as compared to the CdS cell.

The major advantages of photoresistor cells are as follows: low cost; ruggedness; high sensitivity; a high dark-resistance-to-light-resistance ratio typically of the order of 100:1; they are most responsive in the visible light region; with the use of a heat sink they possess a high power dissipation; and they may be operated with relatively high voltages, typically 100 V with some devices having the capability to safely handle 300 V, which permits their direct application in relay and subfractional horsepower (kW) motor control circuits. A major disadvantage of photoresistor cells is slow response to the application of light (typically 30 to 50 ms for a CdS cell and 1 to 3 ms for a CdSe cell) which in turn limits the frequency response to approximately 300 to 1000 Hz. Another important limitation is hysteresis or the ability to respond to rapidly changing light levels. This effect may be minimized

Table 8-1 Typical Photoresistor Parameters

MAX. VOLTAGE	300 V
MAX. TEMP.	65 °C
MAX. CURRENT	100 mA
POWER DISSIPATION	500 mW
RISE TIME TO 65% OF LIGHT RESISTANCE	1 ms to 100 ms
FALL TIME	Usually slightly longer than the rise time
SPECTRAL RESPONSE	400 to 800 nm
OPERATING LIGHT LEVEL RANGE	0.01 to 11,000 lm/m^2
FREQUENCY CAPABILITY	300 to 1,000 Hz
LONG TERM STABILITY	Fair to good
SENSITIVE AREA	18 mm^2

NOTES

1. 1 foot candle (fc) = 10.764 lm/m^2
2. 1 lux = 1 lm/m^2
3. The rise time of a CdSe cell is typically 1 to 3ms, while a CdS cell has a rise time of the order of 30 to 50 ms.

by ensuring that the cell is not subjected to widely varying light levels or by using a CdSe cell.

Another parameter of interest is the **cell sensitivity**, the cell current for a given voltage at a specified illumination level. Current sensitivity is affected by the color of the light, its intensity, the type of material, and the geometry of the active area. Photoresistors are affected in different ways by temperature; the CdS cell resistance remains relatively unchanged by temperature changes while, on the other hand, the CdSe cell experiences significant resistance changes with temperature changes.

The resistance or illumination characteristic of a photocell provides important data for cell application. A typical resistance characteristic for a photoresistor is shown in Figure 8-6. It should be noted that the graph of call resistance versus illumination is a log-log plot and remains linear up to approximately 1000 lm/m^2.

The spectral responses of the CdS and CdSe cells and the human eye are compared in Figure 8-7. It should be noted that the CdS cell responds to visible light more closely to the eye than the CdSe cell which is slightly more sensitive to infrared. The peak spectral response of the cells may be altered by doping the photocondutive material.

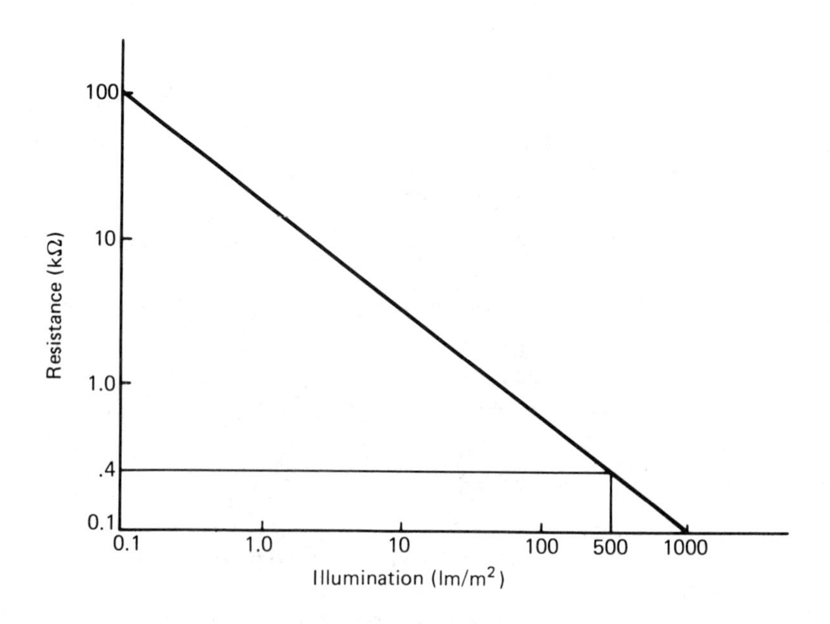

Figure 8-6. Resistance characteristic of a photoconductive cell.

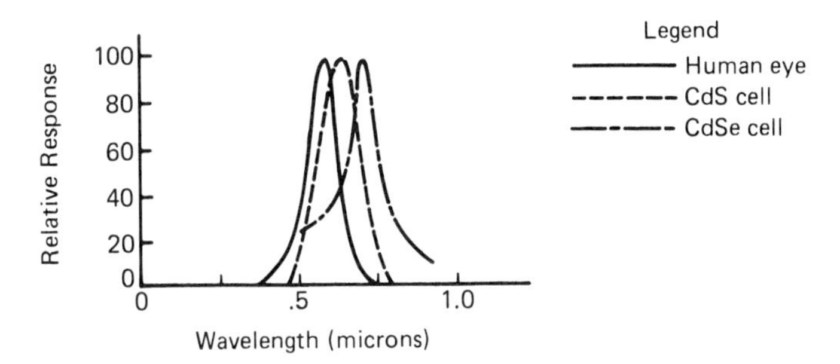

Figure 8-7. Spectral response curves of the eye, CdS and CdSe cells.

Example 8-5

A photoconductive cell with the resistance characteristic shown in Figure 8-6 is used to control a relay. The relay is supplied from a 24 V source and requires 10 mA to pick-up when the illumination level is 500 lm/m^2 and must remain deenergized when the cell is dark. Draw the circuit and calculate the value of the necessary series resistance and the dark current.

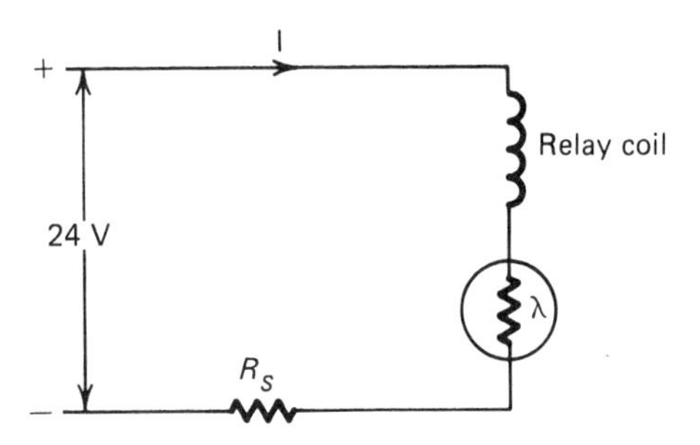

Figure 8-8. Relay controlled by a photoconductive cell.

Solution:

$$I = \frac{24\text{ V}}{R_s + (\text{cell resistance})}.$$

From Figure 8-6 the cell resistance at 500 lm/m^2 is 400 Ω. Then

$$R_s = \frac{24\text{ V}}{10\text{ mA}} - 400\,\Omega = 2\text{ k}\Omega$$

The dark cell resistance is 100 kΩ.
Therefore, the dark cell current is

$$\frac{24\text{ V}}{2\text{k}\Omega + 100\text{ k}\Omega} = 0.235\text{ mA}.$$

Other combinations of materials such as silicon (Si), germanium (Ge), indium antimonide (InSb), and indium arsenide (InAs) are used as infrared detectors in the 2 to 20 μ region of the electromagnetic spectrum. When cooled by means of a Dewar flask, this type of sensor is widely used as a temperature sensor in steel rolling mills.

8-6-1-1 Photoconductive Cell Applications

Some representative applications of bulk type photoconductors are shown in Figure 8-9. A common application is the protective circuit of an oil furnace [Figure 8-9(a)]. If the flame is extinguished, the dark resistance of the photoconductor causes the *n-p-n* transistor to be forward biased. This in turn causes the relay to pick up, which shuts down the oil burner pump motor and prevents a dangerous accumulation of fuel. Under normal operating conditions the light resistance of the cell causes the *n-p-n* transistor to be reverse biased which prevents the relay picking up.

Figure 8-9(b) illustrates the use of a CdS cell when applied as a light meter. The microammeter scale is calibrated in foot-candles or lm/m^2. This configuration permits the meter to be used in both low and high light-level applications. In low-level lighting measurement $R2$ is at a maximum so that the majority of current is passed through the meter movement. In high-level light situations $R2$ will be at a minimum shunting current from the meter. The meter scale is a dual-range scale.

Figure 8-9(c) illustrates the basic principle of a street light control. When the ambient light level diminishes, the increase in the cell resistance forward biases the *n-p-n* transistor which causes the relay to pick-up and switch on the street light. When the ambient light level increases, the cell resistance decreases and the transistor is reverse biased, the relay deenergizes, and the street light is turned off.

8-6-2 The p-n Junction Photodiode

The basic construction of a photodiode is the same as the normal *p-n* junction diode, except that a window or lens is formed in the diode casing to permit radiation to fall on the *p-n* junction. The *p-n* junction photodiode is operated under reverse-bias conditions which increases the

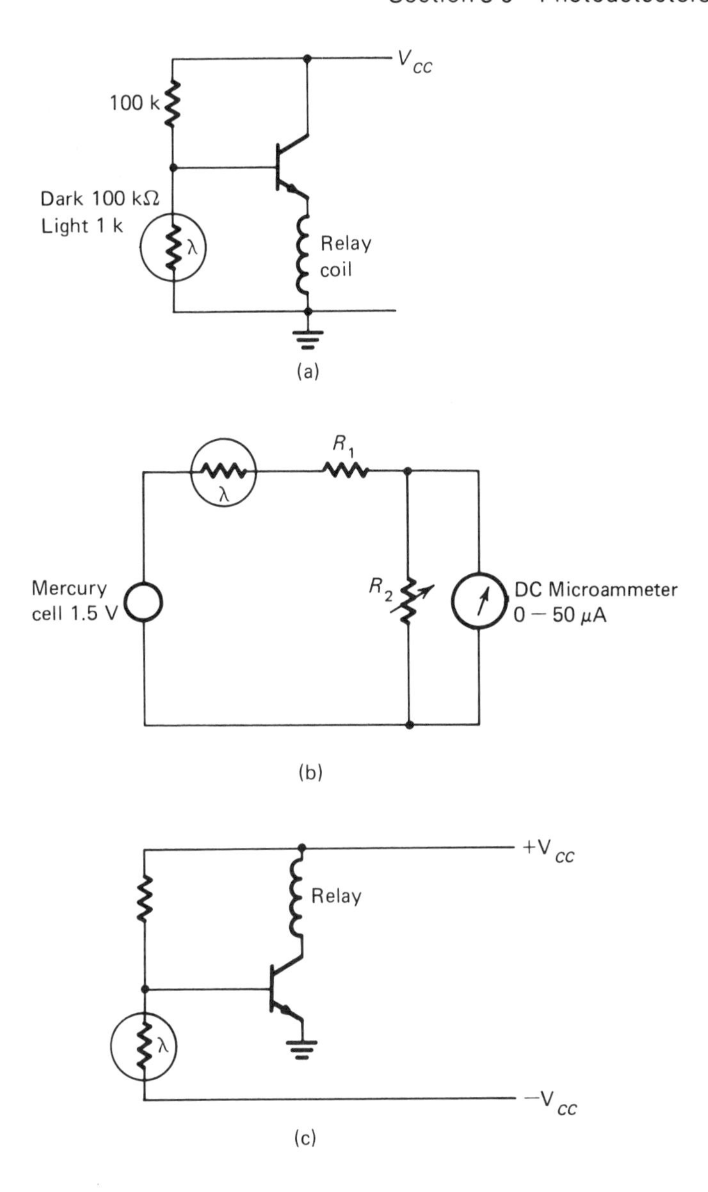

Figure 8-9. Photoconductive cell applications: (a) oil burner flame out protection, (b) light meter, (c) street light control.

width of the space charge region. When incident radiation falls on the *p-n* junction, electron-hole pairs are created which increase the availability of minority carriers; that is, electrons in the *p* material and holes in the *n* material. The minority carriers under reverse-bias conditions cause a significant increase in the leakage current I_L [see Figure

8-10(a)]. It should be noted that if the photodiode was forward biased, the increase in the majority carriers would cause a negligible increase in the forward current. In general, the photodiode is characterized by a linear relationship between the incident radiation and the output current.

For a type TIL 81, it can be seen from Figure 8-10(b) that when $H = 0$, the dark current I_D is 0.01 μA with $V_R = 10$V. When $H = 20$ mW/cm^2, the light current I_L is typically 170 μA with $V_R = 0$ to 50 V. The light current rise time t_r is 350 ns and the fall time t_f is 500 ns.The peak spectral response is approximately 0.9μ and the operating temperature range from $-65\,°C$ to $+125\,°C$. Usually these devices are mounted in TO5 or TO18 packages. Typical applications are encoders, tape and card readers, optical tachometers, and optical character readers operating at frequencies up to 1 MHz depending upon the manufacturer. The spectral response may be varied by controlling the depth of the junction. For shorter wavelength radiation, the electron-hole pairs are created closer to the surface of the silicon; conversely, for longer wavelength radiation, the electron-hole pairs are created further in from the surface. It is common practice to combine a photodiode and an amplifier on the same silicon chip.

8-6-3 The Avalanche Photodiode (APD)

The **avalanche photodiode** is a device whose photocurrent gain has been increased in the range of 25 to 100 times by operating in the reverse breakdown or avalanche region. As the reverse-bias voltage is increased, electron-hole pairs are created by absorbed photons, which in turn collide with the substrate atoms and generate additional electron-hole pairs. The magnitude of the photocurrent gain is dependent upon the magnitude of the reverse voltage. The TIED 55-56 silicon avalanche photodiodes have a dark current, typically 0.8 nA, with a photocurrent gain at avalanche greater than 600 and has been designed for high-speed detection in the near infrared (0.9μ) region. Other advantages are an excellent signal-to-noise ratio with rise times typically of the order of 15 ns; the major disadvantage is cost.

8-6-4 The pin Photodiode

The speed of response of the silicon photodiode may be greatly enhanced by sandwiching a layer of intrinsic silicon (i.e., pure undoped silicon) approximately 2 to 3μ thick between the *p*- and *n*-type materials. The high resistance of the intrinsic silicon produces two major advantages: (1) a decrease in the capacitance between the *p*- and *n*-type

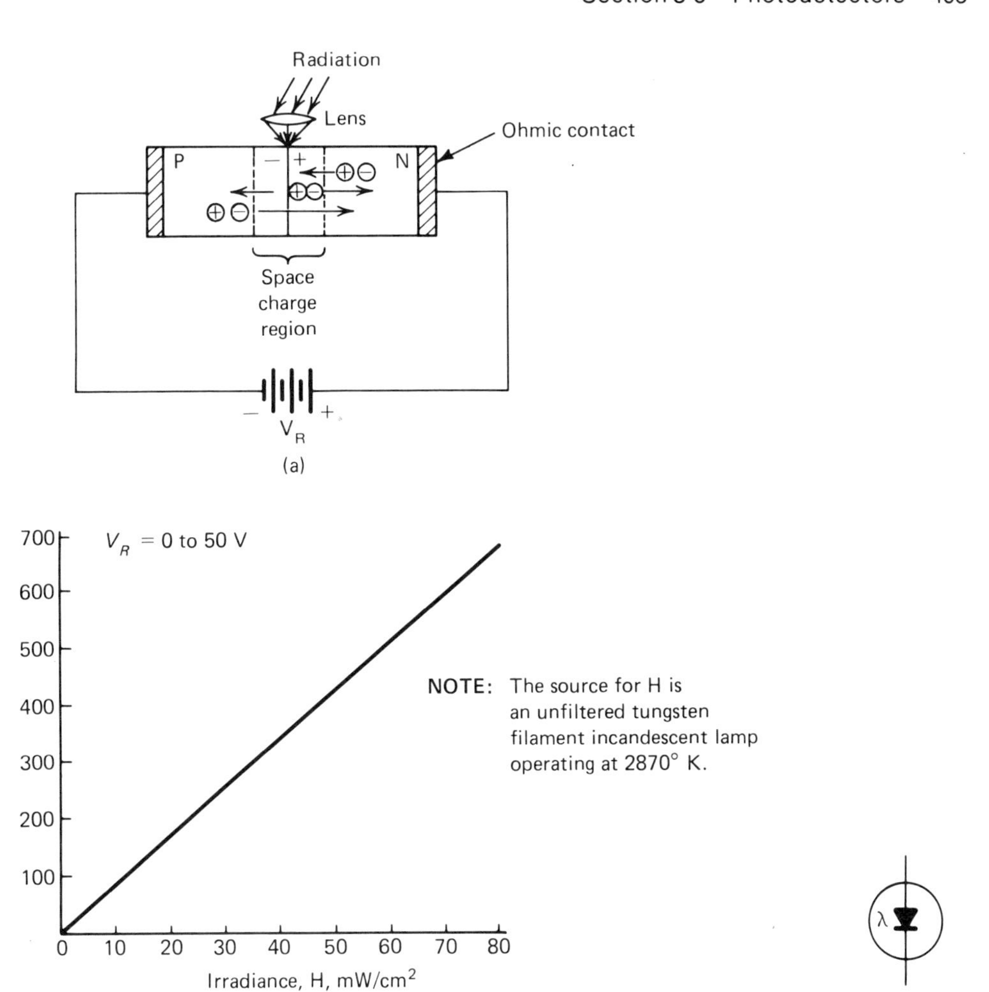

Figure 8-10. *p-n* junction photodiode: (a) schematic, (b) light current versus irradiance, (c) circuit symbol.

materials, typically 2.5 pF, which reduces the transit time of the photoinduced electron-hole pairs and thus increases the frequency response to optical signals (typical response times are less than 1 ns) and (2) permits greater reverse-bias voltages to be applied, which results in an increased production of minority carriers and high radiation sensitivity S_R, typically 3.0 μA/mW/cm², with $V_R = 20$ V. By comparison, for a standard silicon *p-n* photodiode, S_R ranges from 0.4 to 4 μA/mW/cm². The irradiated voltage-current characteristic of a Motorola MRD 500 pin silicon photodiode shows in Figure 8-11 that light current I_L is substantially constant over a wide range of reverse voltage V_R, i.e., it acts as a current generator. Pin photodiodes are used in laser detection, detection of visible and near-infrared light-emitting diodes, shaft and position encoders, and switching and logic currents, to name a few applications.

Pin photodiodes are replacing the photomultiplier tube because of their ability to respond to low radiant energy levels, compactness, broad spectral response, ruggedness, and low cost.

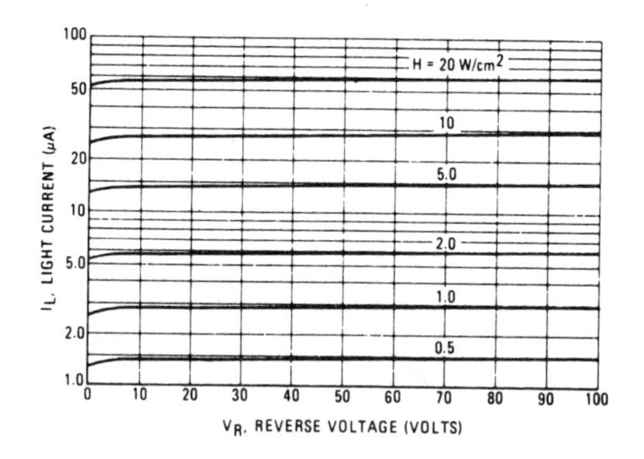

Figure 8-11. Irradiated voltage-current characteristics for MRD500 pin photodiode. (Courtesy of Motorola Inc.)

8-6-5 The n-p-n Phototransistor

The current flow of a phototransistor is controlled by the application of incident radiation of the correct wavelength. The basic construction of the phototransistor is similar to the normal bipolar transistor, except that a lens or window in the protective casing permits radiant energy to be focused on the collector-base junction. The phototransistor can be either a two- or three-terminal device. In the three-terminal configuration, the base connection is available to permit operation as a normal bipolar transistor with or without responding to incident radiant energy. The most common configuration is the the two-terminal version.

Figure 8-12 shows the connections of the *n-p-n* phototransistor, with the emitter-base region forward biased and the collector-base region reverse biased. As in the p-n diode the reverse-biased junction creates a leakage current, which in the normal bipolar transistor is negligible. However, in the phototransistor, the radiant energy incident upon the

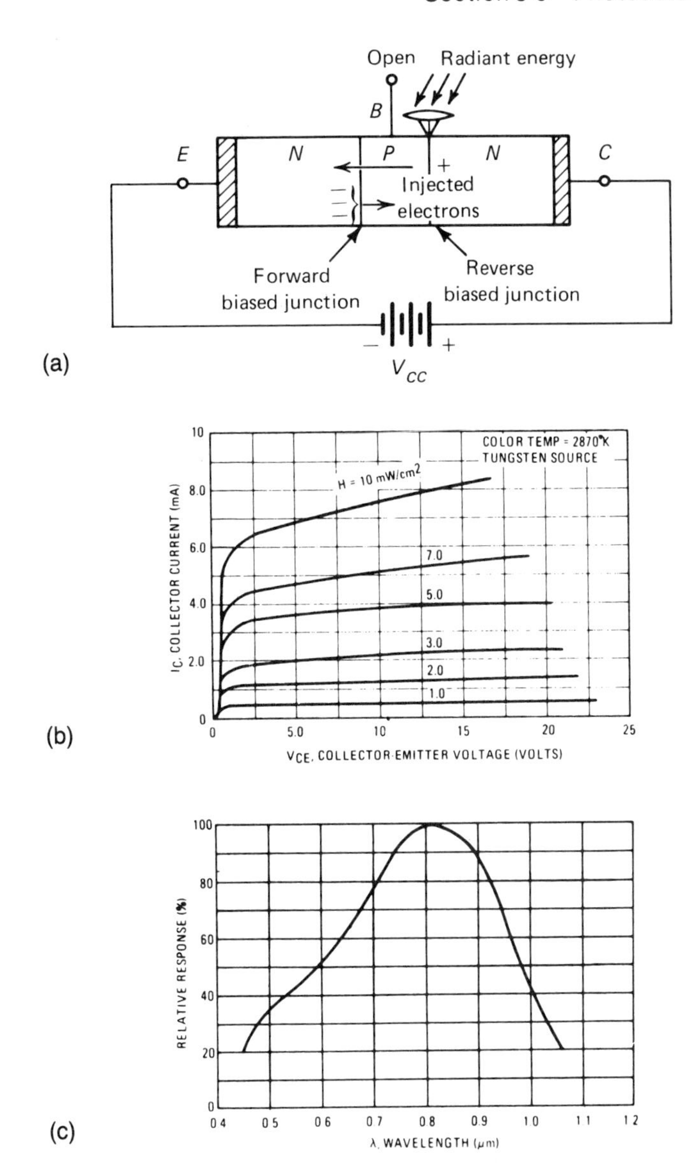

Figure 8-12. The *n-p-n* phototransistor: (a) schematic, (b) collector-emitter characteristics, (c) constant energy spectral response. (Courtesy of Motorola Inc.)

collector-base junction creates electron-hole pairs in both the collector and base regions. The holes created in the base remain there, and the holes created in the collector move to the base under the influence of the reverse-bias. From the base they migrate to the emitter-base junction and move to the emitter under the influence of the forward-bias voltage. This results in electrons in turn being injected by the emitter into the base. The emitter injection efficiency far exceeds the base injection efficiency, so as a result each injected hole results in a large number of injected electrons. The creation of electron-hole pairs at the collector-base junction depends upon the amount of incident radiation received; more minority carriers are created by increasing as far as is practical the target area.

The electrical characteristics of an MRD450 silicon phototransistor are shown in Figure 8-12(b). This phototransistor has a minimum sensitivity of 0.2 mA/mW/cm², with a photocurrent rise time t_r of 2.5 μs and a fall time of 4.0 μs. The dark current at $T_A = 85\,°C$ is typically 5.0 μA and a dissipation capability of 100 mW at $T_A = 25\,°C$. Figure 8-12(c) shows the normalized spectral response which peaks at 0.8 μ. The maximum collector current is approximately 25 mA. Typical applications of the phototransistor are counters, sorters, switching and logic circuits, and process and industrial control where radiation sensitivity and stable performance are required.

Example 8-6

A Motorola MRD450 phototransistor with the collector-emitter characteristics shown in Figure 8-13 has $V_{CC} = 20$ V and a collector load resistance of 2.5 kΩ. Determine the output voltage V_{CE} when the radiation level H is (1) zero, (2) 5.0 mW/cm², and (3) 10 mW/cm².

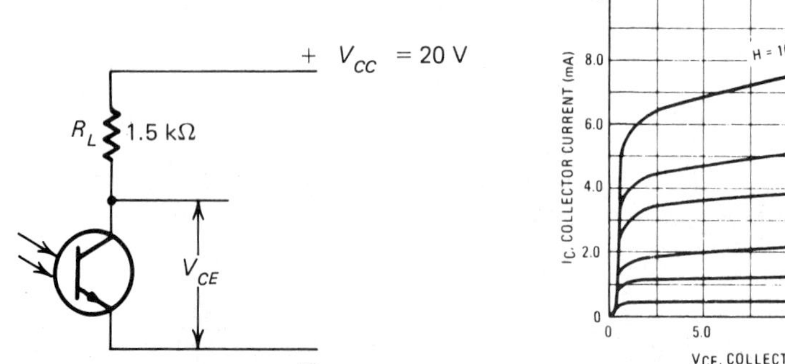

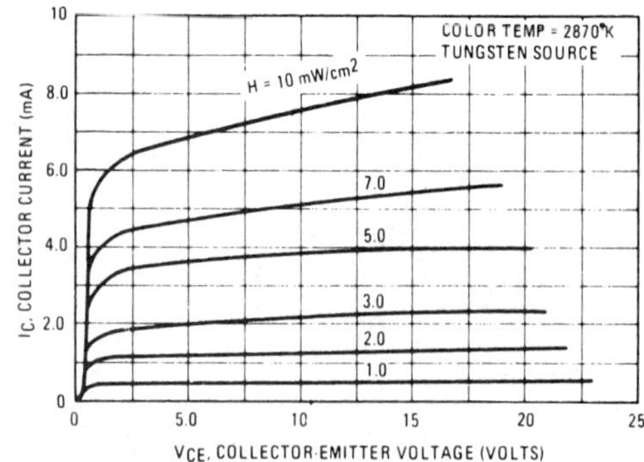

Figure 8-13. Determination of the dc load and V_{CE}. (Courtesy of Motorola Inc.)

Solution:

The load line is determined as follows: when $I_C = 0$, $V_{CE} = V_{CC} = 20$ V, which establishes point A on the graph. When $V_{CE} = 0$, $I_C = V_{CC}/R_L = 20/2.5\ \text{k}\Omega = 8.0$ mA, which establishes point B at $V_{CE} = 0$, $I_C = 8.0$ mA. Next the load line is drawn from point A to point B.

From the intersections of the characteristics and the load line:

when $H = 0$, $V_{CE} = V_{CC} = 20$ V;

when $H = 5.0\ \text{mW/cm}^2$, $V_{CE} = 10.7$ V; and

when $H = 10.0\ \text{mW/cm}^2$, $V_{CE} = 3.5$ V

8-6-6 The n-p-n Photodarlington

High sensitivity at low light levels may also be obtained by using a phototransistor to drive the base of an *n-p-n* silicon transistor to obtain a high collector current. This combination is called a **photodarlington amplifier** (see Figure 8-14) and provides a higher radiation sensitivity than the phototransistor through the visible and near infrared regions of the electromagnetic spectrum. Typical figures for the MRD370 are a maximum output current of 250 mA and a maximum power dissipation of 250 mW. The increased output, however, is achieved at the expense of increased switching times, typically $t_r = 15\ \mu$s and $t_f = 25\ \mu$s, with $I_L = 1$ mA. The photodarlington can be found in such applications as process control, counters, sorters, and switching and logic circuits where a high radiation sensitivity at low illumination levels is required.

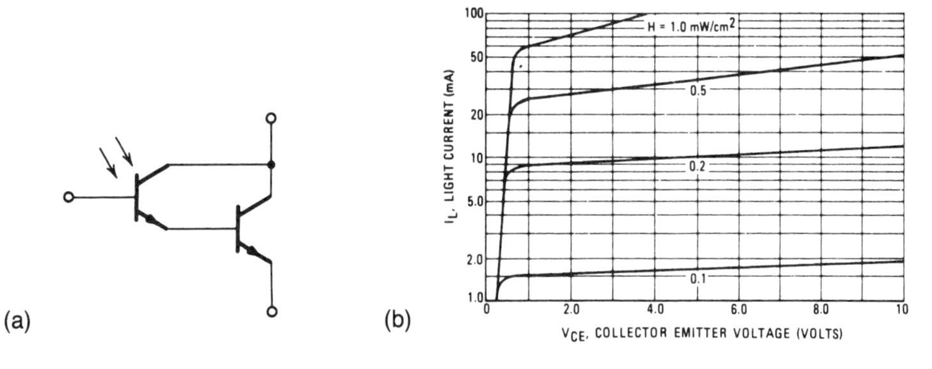

Figure 8-14. The *n-p-n* photodarlington: (a) schematic, (b) collector characteristics. (Courtesy of Motorola Inc.)

8-6-7 The n-Channel Photo-FET

The circuit representation of the *n*-channel junction FET is shown in Figure 8-15(a). The gate-source leakage current I_{GSS} is controlled by the reverse-biased *p-n* junction and typically is a few nanoamperes, the drain current I_D is controlled by the amplitude of the gate-to-source voltage V_{GS}, I_D decreasing as V_{GS} is made more negative. If incident radiation is permitted to fall upon the normally reverse-biased drain-to-gate junction, then electron-hole pairs are generated which produce an increase in the gate-source leakage current.

In the photo-FET the gate-source leakage current is termed the gate current λI_g and I_{GSS} is the dark gate-leakage current. Similarly, the drain current in the photo-FET is designated λI_d. Placing a resistor R_G in series with the gate connection results in the production of a gate voltage $V_{GS} = g_{fs} R_G \lambda I_G$ which is amplified by the photo-FET, where g_{fs} is the forward transductance and is equal to $\Delta I_D / \Delta V_{GS}$ with V_{DS} held constant. The radiation sensitivity is directly proportional to R_G, gains of the order of 10^4 can be achieved by the photoinduced current, and the sensitivity may be varied by as much as 6 orders of magnitude. Unlike the phototransistor the sensitivity decreases with increases of the device temperature. For example, if the device is cooled down to the temperature of liquid nitrogen ($-198\,°C$), g_{fs} increases and the sensitivity increases. The photo-FET is characterized by a high power gain and a very fast rise time.

8-6-8 The Photothyristor or Light-Activated SCR (LASCR)

Although a *p-n-p-n* device, the **light-activated silicon-controlled rectifier (LASCR)** may be considered as a member of the junction type photoconductor family. The LASCR is basically a conventional SCR that may be turned on by incident radiation falling upon the normally reverse-biased J_2 junction which generates electron-hole pairs and the resulting photoinduced current initiates turn-on of the SCR (see Figure 8-16). Normally LASCRs are arranged to permit normal gate-signal control as well as radiation initiated turn-on. The LASCR responds to radiation from the visible to infrared wavelengths; the incident radiation falls on the J_2 junction through a lens in the protective casing. Turn-on time is directly related to the irradiance H. To provide the sensitivity needed to achieve turn-on, the silicon pellet must of necessity be small and thin, which immediately limits the LASCR to low-current applications; typical currents are up to 1.6 A rms with a forward blocking voltage of 200 V. Typical applications of the LASCR are logic circuits and optocouplers.

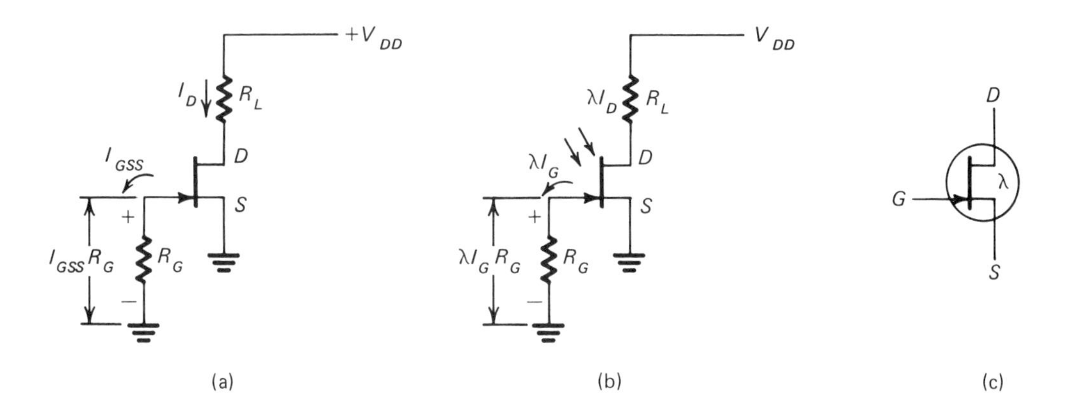

Figure 8-15. The JFET and photo-FET: (a) JFET, (b) photo-FET, (c) circuit symbol of a photo-FET.

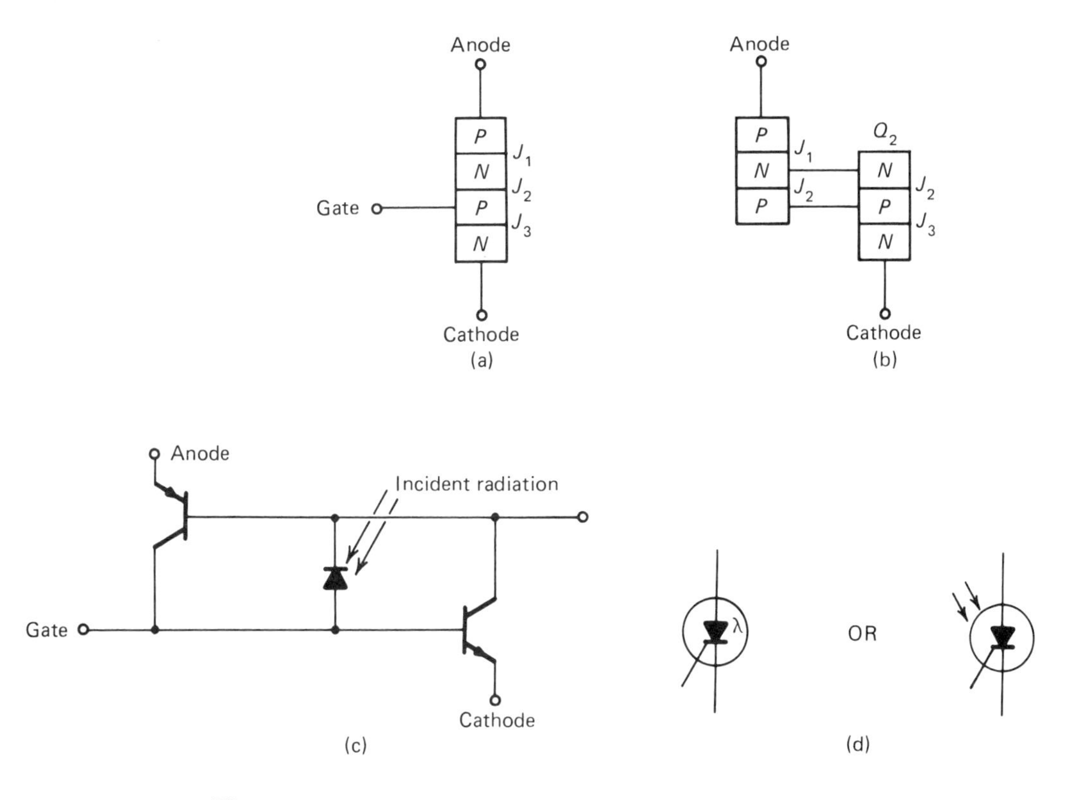

Figure 8-16. Photothyristor or LASCR: (a) SCR basic structure, (b) as two complementary transistors, (c) two-transistor equivalent circuit of the LASCR, (d) circuit symbol.

8-6-9 The Phototriac

The **phototriac** is a TRIAC with a radiation sensitive gate which is triggered by a low radiation density (0.5 mW/cm^2 for an MRD 3011 TRIAC driver). It is designed for applications requiring light or infrared LED TRIAC triggering provided at low cost in a TO18 package. The phototriac and typical applications are illustrated in Figure 8-17.

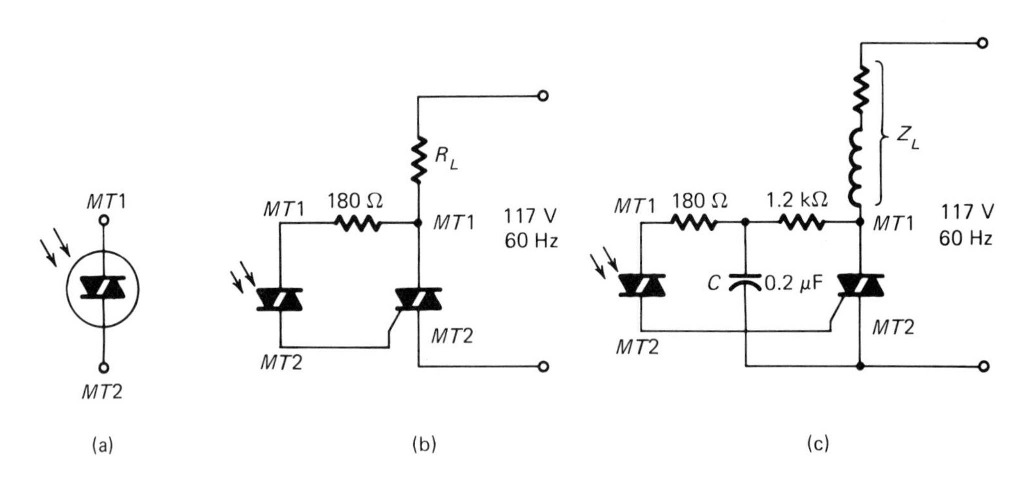

Figure 8-17. Phototraic: (a) circuit symbol, (b) connection for a pure resistive load, (c) connection for an inductive load.

8-7 PHOTOVOLTAIC DEVICES

When a voltage is developed due to radiation falling on a *p-n* junction, it is known as a **photovoltaic effect** and the device is called a **photovoltaic cell.** Modern photovoltaic cells consist of a silicon crystal with the *p-n* junction very close to the surface. The active area or *n*-type silicon is usually about 1 μ thick and permits radiant energy of the correct wavelength to pass through to the *p-n* junction with very little loss of energy. Electron-hole pairs are created at the *p-n* junction. Charge diffusion takes place, i.e., minority charge carriers are swept across the junction increasing the number of electrons in the *n*-type material and the number of holes in the *p*-type material. The space charge region, or barrier potential, prevents the majority carriers (electrons for the *n*-type material and holes for the *p*-type material) from crossing the junction; in other words, the minority charge carrier flow reduces the potential barrier.

When the cell is connected to an external load, current will flow in the external circuit and the device is polarized with the p-type material effectively forming the positive terminal and the n-type material forming the negative terminal of the battery. See Figure 8-18(a). As can be seen from Figure 8-18(b), when a reverse-bias is applied to the device, it acts as a junction photodiode, and when the bias is removed, it acts as a photovoltaic cell; i.e., depending upon the presence or absence of a reverse-bias, the same junction photodiode can operate in the photoconductive mode or in the photovoltaic mode.

Typical characteristics for a silicon photovoltaic cell are 0.5 V on open-circuit and 1 mA short-circuit current, an overall conversion efficiency of approximately 14%, and a peak spectral response at approximately 1μ, which is beyond the spectral response of the eye. When a spectral response is required in the visual range, for example, an automatic aperture control for a single-lens reflex camera, then a selenium photovoltaic cell, is used but at the expense of output power. Typical applications of photovoltaic cells are light meters, optical equipment, and the detection of low-level radiation.

8-8 SOLAR CELLS

Solar cells are basically photovoltaic cells with an increased active area. In our current climate of diminishing hydrocarbon reserves, namely oil and natural gas, combined with our concerns with respect to atmosphere pollution, it is only logical that a considerable amount of attention should be devoted to the development of an efficient solar energy conversion system.

Currently solar cell power is not economically competitive with fossil fuel and nuclear power. The major aims of research organizations are first directed to improving cell efficiency from approximately 14% in 1970 to of the order of 22% currently, with some indications that efficiencies as high as 28% may be achieved for single-cell arrangements. There are indications that by using tandem cell techniques, which will more effectively utilize the wavelengths present in sunlight, higher efficiencies may be obtained. Secondly, an equally important factor is the reduction in the cost per watt from the $500/W in the early 1970s to the current $8 to $15/W to, hopefully, a cost of less than $1/W by the mid-1980's.

Another important factor is the cost of energy storage which currently is by means of storage batteries, although there are alternative storage methods such as hydrogen, where the output of the solar cells is used to electrolyze water into hydrogen and oxygen with the hydrogen later being used directly or in fuel cells.

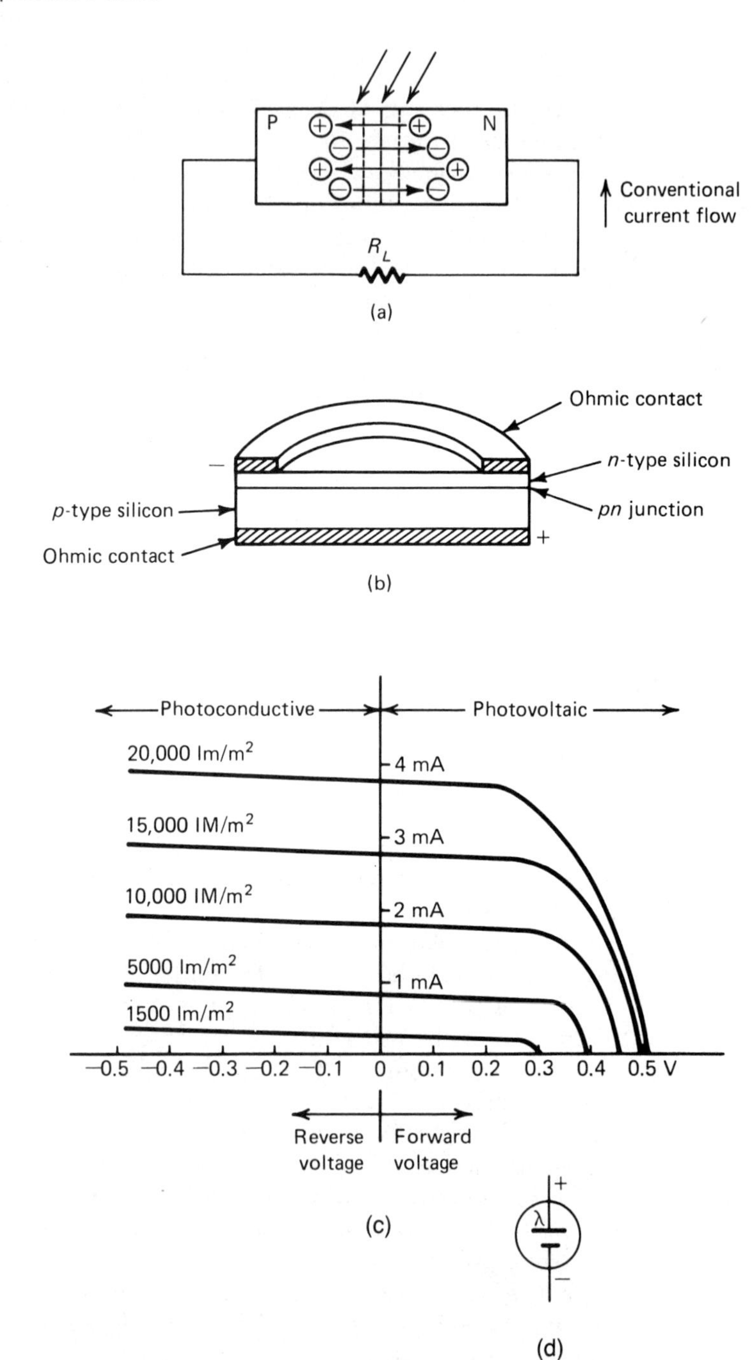

Figure 8-18. Silicon photovoltaic cell: (a) schematic, (b) cell structure, (c) illumination characteristics, (d) circuit symbol.

Currently, solar cell arrays are being developed and placed in operation in applications such as remote navigational aids, microwave repeater stations, meteorological platforms, seismic monitors, and UHF/VHF radio repeaters, to name a few. In view of the potential increase in usage of solar cells in the near future it is appropriate to briefly survey the improved techniques that are being currently researched.

8-8-1 Silicon Photovoltaic Cells

The basic principle of silicon photovoltaic cells has been described; however, the active cell area has been increased. The major thrust of research is to significantly reduced the cost by producing single crystal sheets of silicon by extruding molten silicon into a thin ribbon through a narrow slit, as compared to the slicing of the silicon discs from large single crystal ingots produced by the Czochralski process, the conventional method used in the production of semiconductor devices. Also, it has been determined that the cells do not have to be a single crystal, which has permitted the development of polycrystalline silicon sheets resulting in significant cost savings. In addition, the ability to produce rectangular cells which cover an array area more effectively than circular cells has increased the output capability of an array.

Improved methods of forming the semiconductor junction by ion implantation have also reduced costs as compared to the conventional diffusion process.

8-8-2 Copper Sulphide–Cadmium Sulphide Cells

Copper sulphide–cadmium sulphide cells consist of a thin film of copper sulphide deposited on a layer of cadmium sulphide, the interface between the copper sulphide and cadmium sulphide forming the *p-n* junction (see Figure 8-19). The major advantage of this type is that the active materials are deposited on a metal or glass substrate by spraying or evaporation, a process which is relatively unexpensive. The major disadvantages are low efficiency, typically 5%, and a reduction in cell output with aging. The copper sulphide–cadmium sulphide is classified as a heterojunction device since the *p-n* junction is formed between two dissimilar materials.

8-8-3 Gallium Arsenide Cells

When constructed from a single crystal, **gallium arsenide cells** perform with an efficiency of approximately 22%. A cheaper but less

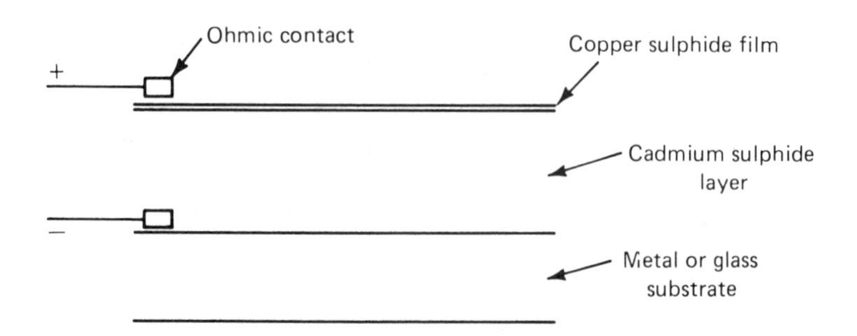

Figure 8-19. Copper sulphide—cadmium sulphide photovoltaic cell.

efficient form (approximately 7%) has been constructed from polycrystalline gallium arsenide.

8-8-4 Amorphous Silicon Cells

Amorphous silicon cells are produced by passing silane (SiH_4) under low vacuum over a noncrystalline silicon substrate in the presence of a radio frequency discharge. Although to date they have a low efficiency, they are cheap and easy to produce.

8-8-5 Electrochemical Solar Cells

Bell Laboratories has developed a semiconductor-liquid junction cell. The cell is constructed from an etched gallium arsenide crystal whose surface has been treated with ruthenium to increase efficiency, the whole crystal being immersed in a selenium compound solution. The overall efficiency is of the order of 12% and production is simple and cheap.

Texas Instrument Inc. is developing an electrochemical cell which consists of a silicon *p-n* junction immersed in an acid solution. Incident radiation on the junction produces a photoinduced current which breaks down the acid and hydrogen is given off as a by-product.

The above gives a brief insight into some of the methods that are currently being researched. In the US expenditures of $1.5 billion have been authorized by the government to make photovoltaic technology competitive with more conventional power sources. However, an equally great amount of research must also be devoted to storage methods to ensure that a viable alternative to the present electrical production techniques is available.

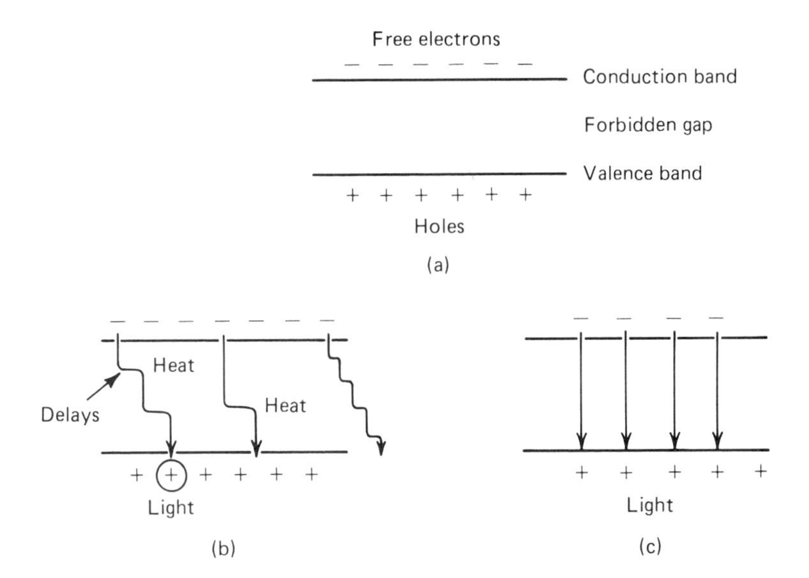

Figure 8-20. Light-emitting diode: (a) energy levels, (b) indirect gap diode, (c) direct gap diode.

8-9 LIGHT-EMITTING DIODES (LED)

When a *p-n* junction diode is forward biased, free electrons in the conduction band of the *n*-type material are injected into the *p*-type material, where they recombine with holes in the valence band. Since the electrons are at a higher energy level than the holes, as the electron falls to the lower energy valence band, the excess energy is dissipated as light and/or heat. In the case of germanium and silicon diodes, the falling electron may be delayed one or more times in the forbidden gap. On the occasion of each delay, energy is given up as heat; consequently, by the time the electron has reached the valence band, there is very little energy left to be given up as light energy. Germanium and silicon diodes are known as indirect gap diodes (see Figure 8-20).

In the case of light-emitting diodes, which are classified as direct-gap diodes, the falling electron is not impeded in its descent to the valence band, and as a result, the excess energy during recombination is converted to light. However, the actual light that is available has been reduced by absorption in its passage to the surface and further reduced by reflection at the surface back into the diode, which is known as **reabsorption.**

Reabsorption can be minimized by using a domed plastic lens. Direct gap diodes have a higher threshold voltage (usually in the range of 1.0

to 1.6 V) than germanium and silicon diodes and are fabricated from materials that peak at different wavelengths. For example: gallium phosphide/gallium phosphide (GaP/GaP) emits at 565 nm (green); gallium arsenide phosphide/gallium phosphide (GaAsP/GaP) emits at 585 nm (yellow); gallium arsenide phosphide/gallium phosphide (GaAsP/GaP) emits at 635 nm (red); gallium arsenide phosphide/gallium arsenide (GaAsP/GaAs) emits at 655 nm (red); gallium phosphide/gallium phosphide (GaP/GaP) emits at 697 nm (red); and gallium arsenide (GaAs) emits at 940 nm (infrared). These emitters form a sample of the range currently available. Emitting diodes produce light by a process known as electrolumunescence and do not operate under vacuum conditions as is required by incandescent lamps.

As can be seen from above, the LED may be classified as a visible emitter, i.e., those devices with emissions in the range of 0.38 to 0.78 μm and infrared emitters (0.940 μm) which are invisible to the naked eye. They have very fast rise and fall times, typically 300 and 200 ns, respectively, and a high life expectancy combined with a low power consumption. The infrared emitters are spectrally matched to silicon photodetectors, while the visible emitters are spectrally matched to CdS and CdSe photoconductors. Among a wide range of applications, emitters are used in solid-state relays, smoke detectors, optical encoders, high-voltage isolators, optical switches, security systems, level indication, and remote control to name a few.

8-10 ALPHANUMERIC DISPLAYS

There are several methods of producing an alphanumeric display on a monolithic structure. One method is to use a seven-segment display shown in Figure 8-21(a) which permits the display of any numeral from 0 to 9 inclusive. From Figure 8-21(b) it can be seen that the light emission from the LED is enlarged and brought to the face of the encapsulated device by light pipes.

Normally the seven-segment display is controlled by a decoder/driver. In Figure 8-22 the functional block diagram of a TIL 302 numeric display is shown, together with the truth table which illustrates the operation of the device with a BCD input.

The second method of producing an alphanumeric display is by the use of a 5×7 array of LEDs plus one for the decimal point. By logic control of the columns and rows, a total of 64 characters can be displayed; the characters include the alphabet, all numbers from 0 to 9, and the arithmetic signs. The use of TIL 504 as a single-character display is illustrated in Figure 8-23. The character that is displayed is

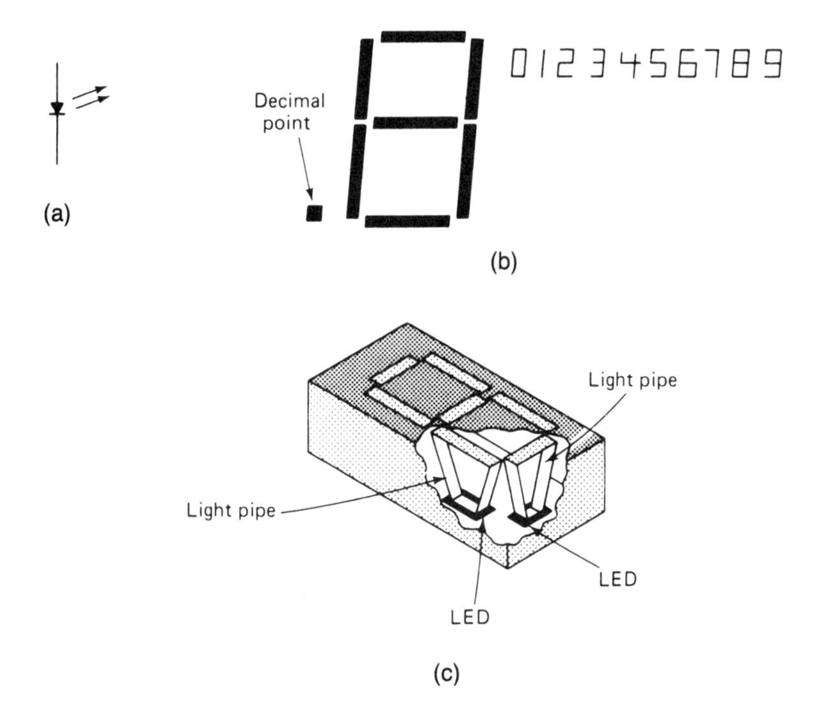

Figure 8-21. Seven-segment numeric display: (a) symbol, (b) arrangement, (c) construction.

a function of the logic input, lines 1 to 7 inclusive, and the blanking input. The display is inhibited when a low logic-level signal is applied to the blanking input. The five columns of the TIL 504 5 × 7 alphanumeric display are scanned in turn, the sequencing being controlled by the UJT oscillator, the SN7496 shift register, and one of the SN7416 hex inverter/buffer drivers that invert and feed the outputs back to the serial input to form a ring counter.

The outputs of the ring counter drive the column drivers (A5T2907's) and the column select inputs of the read-only memory TMS4100 after being inverted through another SN7416 hex inverter/buffer. The logic inputs 1 to 7 inclusive are also inverted by a SN7416 to make the inputs compatible with positive logic, and series 54/74-levels LED-type alphanumeric displays are designed to accept various inputs such as BCD, USASC11, and EBCDIC and may or may not contain internal logic to produce alphanumberic, numeric, or hexadecimal displays. These units are compatible with most TTL and DTL circuits and have a wide viewing angle and a high luminous intensity.

TYPICAL APPLICATION DATA

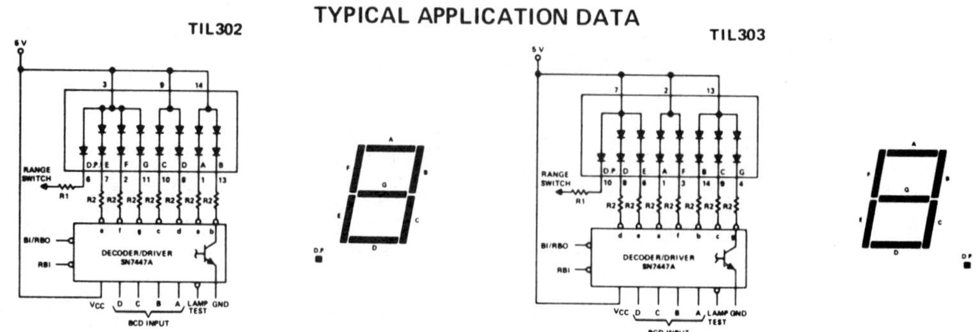

NOTES: A. R1 and R2 are selected for desired brightness.

B. SN74L47 may be used in place of SN7447A in applications where segment forward current will not exceed 20 mA.

FUNCTION TABLE
SN7447A

DECIMAL OR FUNCTION	INPUTS						BI/RBO†	SEGMENTS							NOTE
	LT	RBI	D	C	B	A		a	b	c	d	e	f	g	
0	H	H	L	L	L	L	H	ON	ON	ON	ON	ON	ON	OFF	1
1	H	X	L	L	L	H	H	OFF	ON	ON	OFF	OFF	OFF	OFF	1
2	H	X	L	L	H	L	H	ON	ON	OFF	ON	ON	OFF	ON	1
3	H	X	L	L	H	H	H	ON	ON	ON	ON	OFF	OFF	ON	1
4	H	X	L	H	L	L	H	OFF	ON	ON	OFF	OFF	ON	ON	1
5	H	X	L	H	L	H	H	ON	OFF	ON	ON	OFF	ON	ON	1
6	H	X	L	H	H	L	H	OFF	OFF	ON	ON	ON	ON	ON	1
7	H	X	L	H	H	H	H	ON	ON	ON	OFF	OFF	OFF	OFF	1
8	H	X	H	L	L	L	H	ON	ON	ON	ON	ON	ON	ON	1
9	H	X	H	L	L	H	H	ON	ON	ON	OFF	OFF	ON	ON	1
10	H	X	H	L	H	L	H	OFF	OFF	OFF	ON	ON	OFF	ON	1
11	H	X	H	L	H	H	H	OFF	OFF	ON	ON	OFF	OFF	ON	1
12	H	X	H	H	L	L	H	OFF	ON	OFF	OFF	OFF	ON	ON	1
13	H	X	H	H	L	H	H	ON	OFF	OFF	ON	OFF	ON	ON	1
14	H	X	H	H	H	L	H	OFF	OFF	OFF	ON	ON	ON	ON	1
15	H	X	H	H	H	H	H	OFF	OFF	OFF	OFF	OFF	OFF	OFF	1
BI	X	X	X	X	X	X	L	OFF	OFF	OFF	OFF	OFF	OFF	OFF	2
RBI	H	L	L	L	L	L	L	OFF	OFF	OFF	OFF	OFF	OFF	OFF	3
LT	L	X	X	X	X	X	H	ON	ON	ON	ON	ON	ON	ON	4

H = high level (logic 1 in positive logic), L = low level (logic 0 in positive logic), X = irrelevant.

†BI/RBO is wire-AND logic serving as blanking input (BI) and/or ripple-blanking output (RBO).

NOTES: 1. The blanking input (BI) must be open or held at a high logic level when output functions 0 through 15 are desired. The ripple-blanking input (RBI) must be open or high if blanking of a decimal zero is not desired.

2. When a low logic level is applied directly to the blanking input (BI), all segment outputs are off regardless of any other input.

3. When the ripple-blanking input (RBI) and inputs A, B, C, and D are at a low logic level with the lamp test input high, all segment outputs are off and the ripple-blanking output (RBO) of the decoder goes to a low level (response condition).

4. When the blanking input/ripple blanking output (BI/RBO) is open or held high and a low is applied to the lamp-test input, all segments are illuminated.

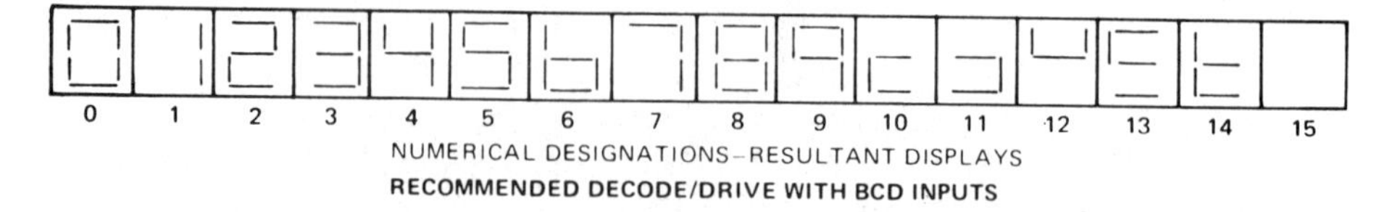

NUMERICAL DESIGNATIONS—RESULTANT DISPLAYS

RECOMMENDED DECODE/DRIVE WITH BCD INPUTS

Figure 8-22. Application of a TIL302 numeric display: (a) decoder/driver, (b) function table. (Courtesy Texas Instruments Incorporated)

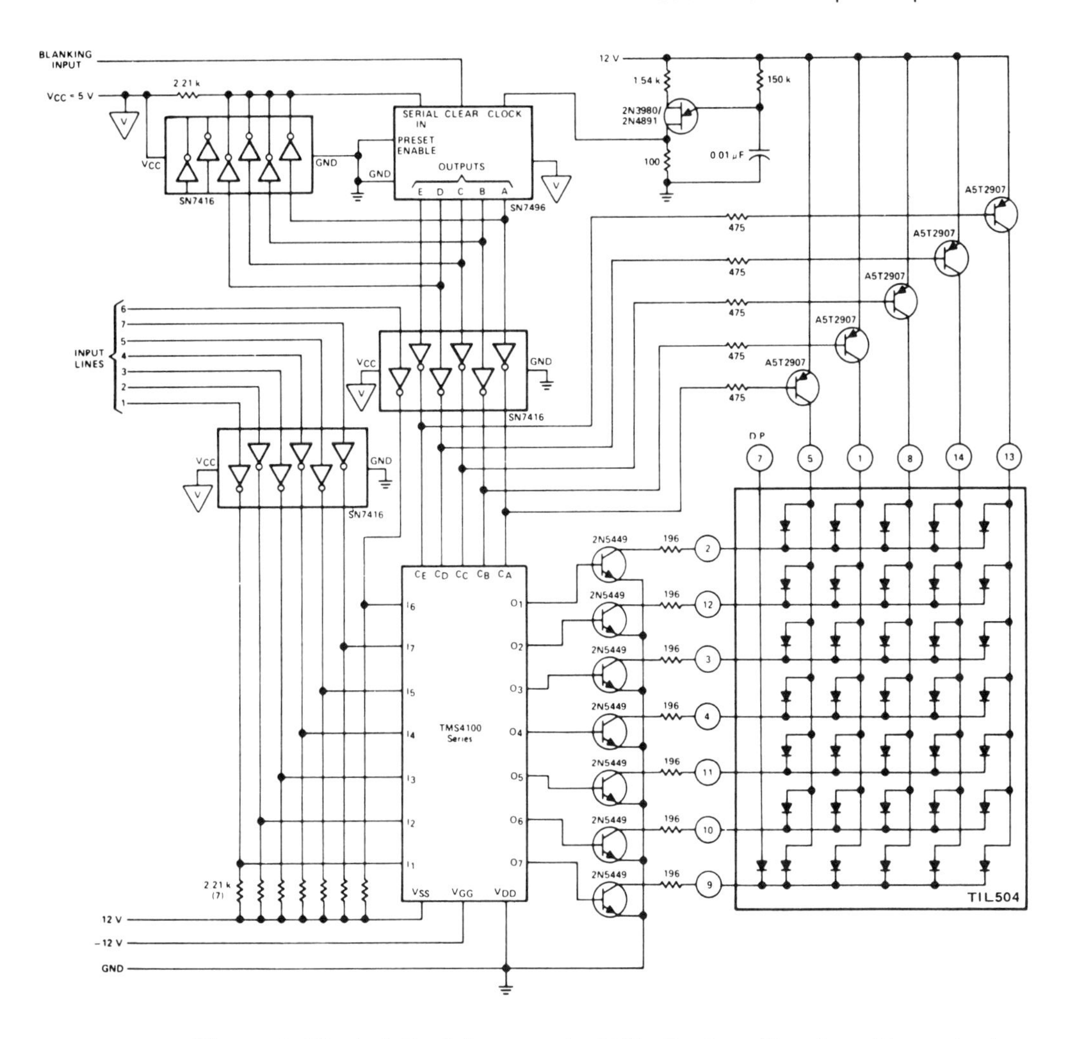

Figure 8-23. A 5x7 alphanumeric LED display. (Courtesy Texas Instruments Incorporated)

8-11 OPTOCOUPLERS

Optocouplers, also called photocouplers and optoisolators, usually consist of a gallium arsenide (GaAs) infrared LED and a photodetector encapsulated in a metal can or plastic DIP. The radiant energy from the infrared light-emitting diode (IRED) is coupled optically by an insulating optical path to a silicon photodetector. The signal transfer is unidirectional because the emitter and detector cannot reverse their roles. The input capacitance combined with the dielectric characteristics makes the optocoupler relatively unresponsive to common mode input signals as well as providing complete isolation between the input and output circuits. The degree of electrical isolation is dependent upon the material in the light path and the physical distance between the emitter and detector; however, the greater the physical distance the smaller the current transfer ratio (CTR), which is the ratio of the detector current to the emitter current.

The CTR also depends upon the radiation efficiency of the IRED, the area and sensitivity of the detector, and the amplifying gain of the detector. Typical CTRs range from a minimum of 25% to 400%. The major advantages of optocouplers are as follows: no moving parts or contacts; compactness, typically 0.35 in (8.9 mm) by 0.26 in (6.6 mm) for a DIP; high voltage isolation between input and output, typically a minimum 7,500 V ac peak for all Motorola devices with in excess of 12,000 V ac peak on a repeatable basis; high CTR; high speed and excellent frequency response; wide operating temperature range, $-55\,°C$ to $100\,°C$; and compatibility with DTL and TTL integrated circuits.

Optocouplers are used to couple analog or digital signals to a system and are unaffected by electrical transients as well as being free of problems associated with circulating grounds. They may be operated in either the switching or linear modes. In the industrial environment, optocouplers are used in gate-firing circuits of phase-controlled converters, as SCR and TRIAC triggers, counters, sorters, and in switching and logic circuits.

Since the optocoupler depends upon the detection of emitted radiation, while the emitter is usually an IRED, the detector may be a photovoltaic cell, a photoconductor, a photodiode, a phototransistor, a photodarlington, a phototriac, or a photo SCR. The more common optocoupler configurations are shown in Figure 8-24.

8-11-1 Optocoupler Applications

Figure 8-25 illustrates a few applications of optocouplers. Figure 8-25(a) illustrates the coupling between a gate-pulse generator in a thyristor phase-controlled converter via an optocoupler consisting of an IRED and a photodarlington-amplifying detector, the output being used as the gate signal to trigger the SCR.

Figure 8-25(b) shows the application of an IRED-phototransistor optocoupler in the switch mode. The output pulses are in phase with the input if taken from the emitter and 180° out of phase if taken from the collector.

Figure 8-25(c) shows a method of coupling an ac input signal to an operational amplifier. In order to reduce distortion produced by the

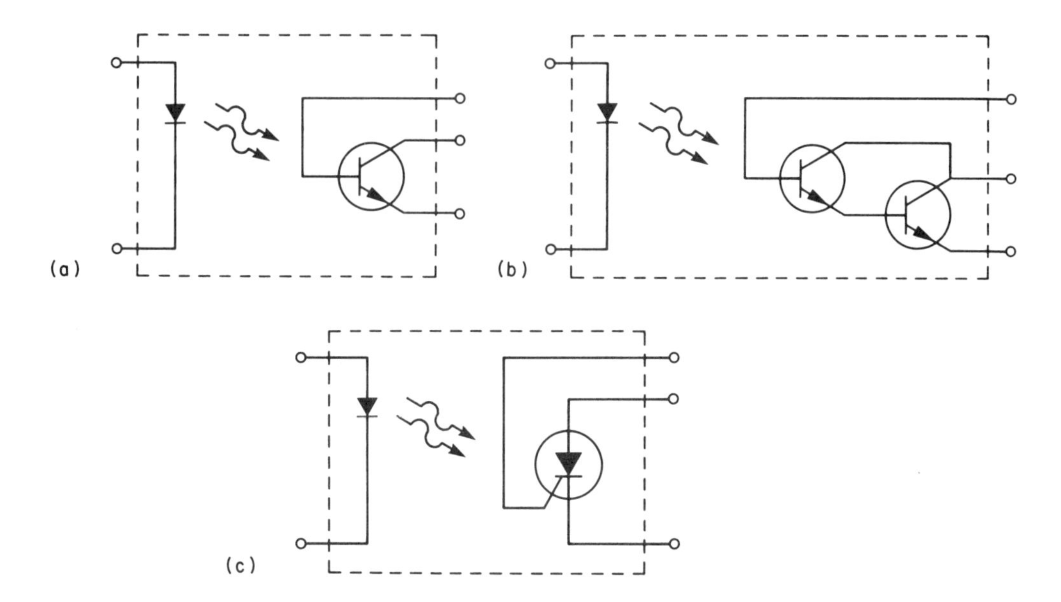

Figure 8-24. Optocoupler configurations: (a) phototransistor, (b) photodarlington, (c) photo SCR.

coupler, the IRED is biased with a dc current and the ac input signal is summed so that the output of the optocoupler will have ac and dc components. By capacitively coupling the optocoupler to the operational amplifier only an ac output will be present at the output terminal.

Figure 8-25(d) illustrates the use of an optocoupler as part of a Schmitt trigger. Transistor $Q1$, the photodetector of the optocoupler, is one of the Schmitt trigger transistors. When the base voltage of $Q1$ exceeds $V_E + V_{BE}$, $Q1$ turns on, which in turn causes $Q2$ to turn on and the output goes high. When the input signal to the IRED is removed, Q2 turns off and the output is low. Since the impedance in the base of $Q1$ is high, the turn-off delay time is approximately 6 μs. This configuration can be used as a limit switch, counter, chopper, position sensor, and in conjunction with optical encoders. The Spectronics division of Honeywell has produced the Spectronics SPD8600 Opto-Schmitt Detector which is a single-chip IC photodetector consisting of a photodiode, amplifier, voltage regulator, Schmitt trigger, and an output stage.

8-12 LASERS

Laser, which stands for light amplification by stimulated emission of radiation, is a device which emits radiation at one wavelength or

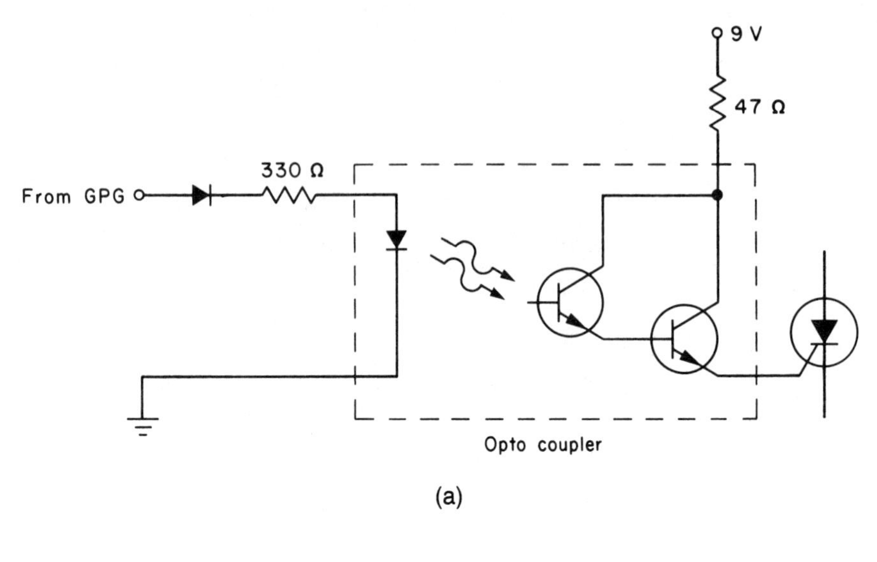

(a)

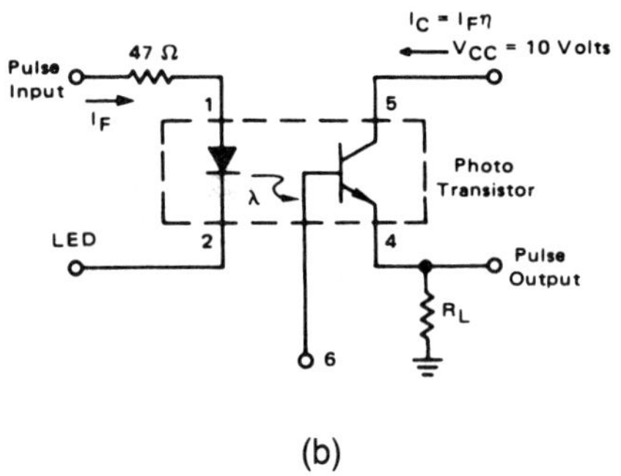

(b)

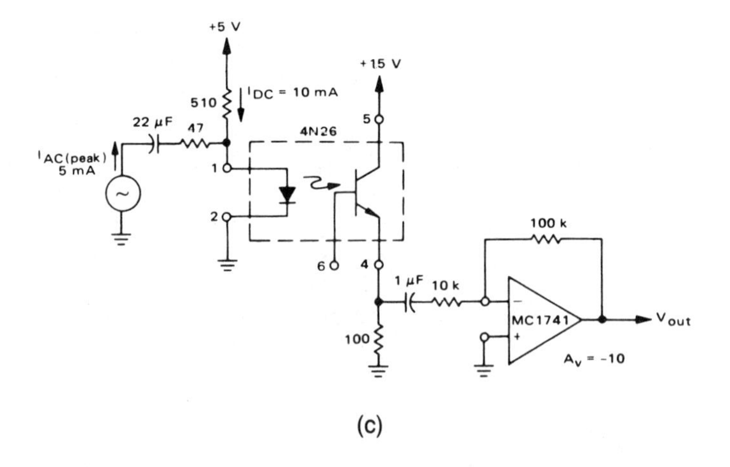

(c)

Figure 8-25. Optocoupler applications: (a) gate pulse isolation, (b) pulse mode circuit, (c) coupling an ac signal to an operational amplifier, (d) optically coupled Schmitt trigger. (b, c, and d Courtesy of Motorola Inc.)

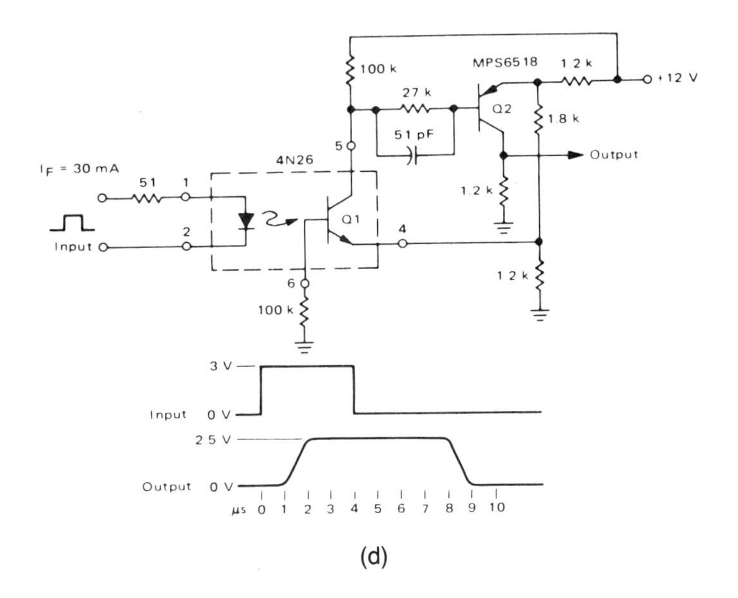

(d)

Figure 8-25. continued

frequency, i.e., the output is monochromatic or one color. The light output is termed coherent as against the light output of, for example, an incandescent lamp which consists of a large number of different wavelengths and is in turn termed incoherent light.

A major property of the light beam produced by a laser is that it is very narrow and diverges very little. There are many applications in modern society of laser technology such as cancer treatment and cauterizing of hemorrhages in the human eye, metal welding, precision leveling and range finding, fiber optical communications (audio, video, and data), and even potentially as weapons, to name a few examples.

8-12-1 The Injection Laser Diode

Even though the GaAs LED has a very fast response time, usually in the 2–5ns region and a spectral response that is compatible with silicon detectors, the limitation for optical communication applications has been the relatively low optical power output, typically up to 200 mW for operation at 25°C and up to 1.5 W when operated at low temperature.

The development of the GaAs injection laser diode increased the optical power output a 1,000 times, but with the restriction that it required special operating conditions. It was found that by operating the diode under cryogenic conditions, for example, at the temperature of liquid nitrogen, the drive requirement was greatly reduced and the duty cycle could be increased from typically 2% to 8% without overheating.

The basic construction of a GaAs injection diode is shown in Figure 8-26(a). The physical length of the junction (L) determines the wavelength of the emitted light. The ends of the diode are polished to form reflecting surfaces, with the end at which emission occurs being a partial reflector. The function of the reflecting ends is to reflect internally generated light back into the *p-n* junction.

Under conditions of forward bias, with increasing forward current an increased quantity of charge carriers enter the depletion region and strike and excite the atoms. The excited atoms randomly emit photons; eventually some of the photons strike the reflective surfaces at right angles. These photons are reflected back along the junction, in turn exciting more atoms and rapidly increase the number of photons of radiated energy. Eventually, the amplifcation effect causes a coherent beam of light to be radiated through the partial reflector. See Figure 8-26(b).

GaAs laser diodes operate with drive currents ranging from 100 mA up; under low drive conditions, for example 10mA, they operate as a conventional IRED. Beyond the threshold current the radiation intensity increases rapidly and the bandwidth of the output radiation decreases. Normally the injection diode laser is operated under pulsed conditions to prevent overheating. A note of caution should be made at this point: because of the high energy intensity of the radiation, protective eye glasses must be worn at all times when working with laser diodes.

8–13 APPLICATION FUNDAMENTALS OF OPTOELECTRONIC CONTROLS

The usual optoelectronic control system consists of a light-source (emitter) photodetector combination that produces an input signal to an amplifying and conditioning circuit which transforms the output to a useful electrical signal.

There are two basic types of control: self-contained and modular. The **self-contained** unit consists of an emitter and photodetector in one package, and in turn can be further subdivided into interrupters and reflectors [see Figure 8-27(a), (b)]. The advantages of the self-contained modules are the ease of alignment and reduction of wiring. The **modular** type consists of two units, the emitter and the photodetector. See Figure 8–27(c). The major advantage of the modular type is that the scanning distance between units can be as much as 100 ft (30.48 m), but the units are usually not further than 10 ft (3 m) apart.

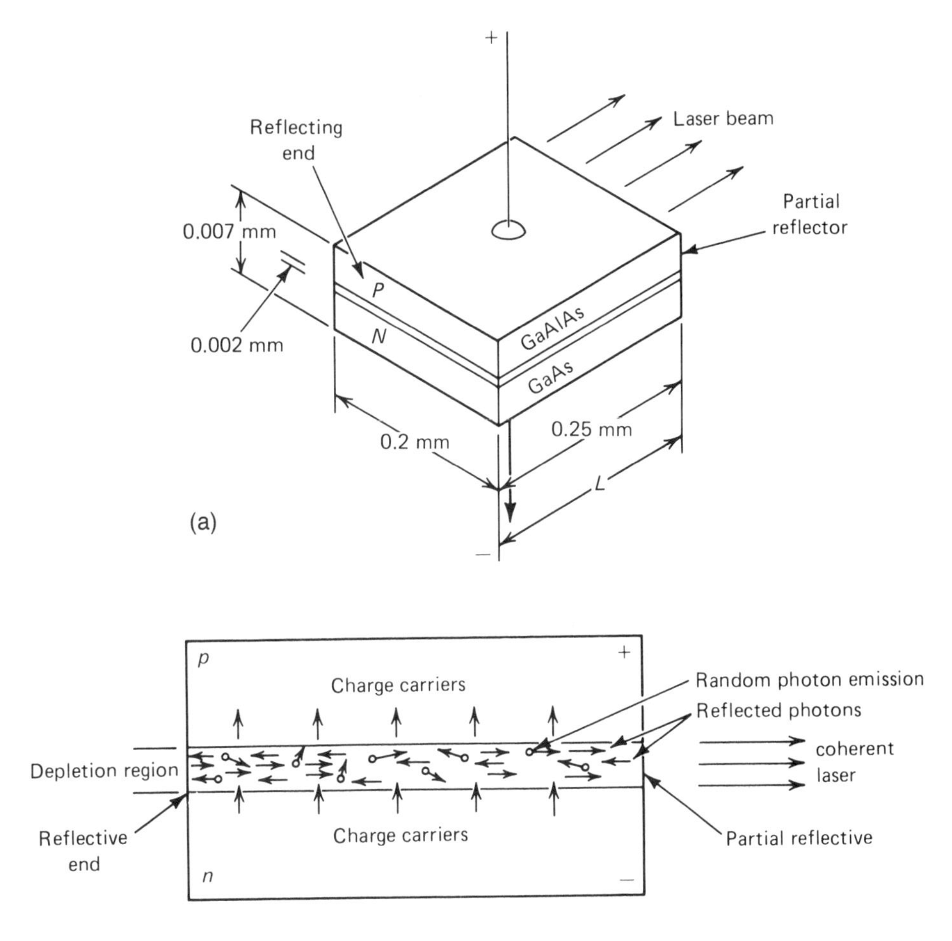

Figure 8-26. Injection diode: (a) construction, (b) principle of photon amplification.

Usually the emitter is a LED, and LEDs may be operated in the modulated or unmodulated modes. Modulated LED systems are frequency modulated, with a phototransistor and receiver circuit designed to respond to a narrow frequency spectrum on either side of the LED pulse frequency. Modulated systems permit operation under high ambient light conditions as well as poor scanning conditions such as fog, dust, and with transluscent materials. The unmodulated system must be protected against high ambient light conditions to prevent incorrect responses. Optoelectronic controls can, with suitable modifications, be made to respond to radiation levels above or below a preselected threshold. In addition, they can be made to respond to the rate of change of radiation as against radiation intensity.

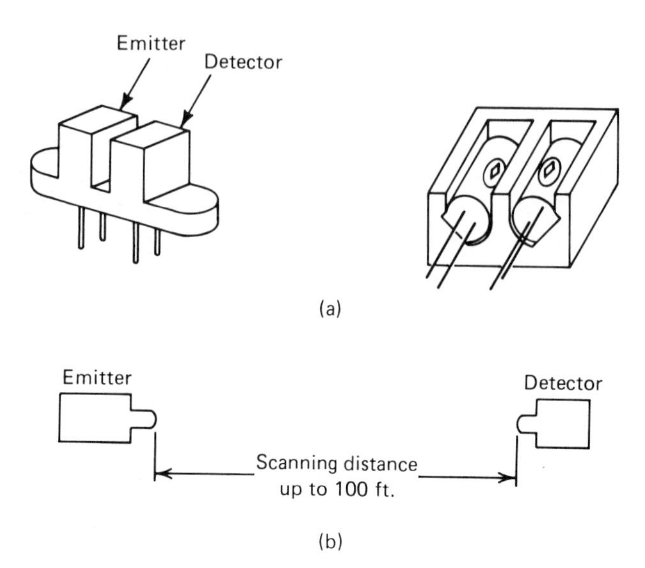

Figure 8-27. Optoelectronic controls, self-contained: (a) interrupter, (b) reflector, (c) individual modules.

Optoelectronic controls respond most easily to changes in shape, size, contrast, finish, opaqueness, position, and direction of travel, the response determined by the radiant energy reflected or transmitted by the object being sensed.

There are two scanning techniques in common use. The **through** or **direct scan** method, where the emitter and sensor are directly mounted opposite each other with the object to be detected passing between them [Figure 8-28(a)] is most effective for opaque objects; however, the object should be at least 50% or greater than the diameter of the detector lens. Direct scanning permits greater separation between the emitter and detector since no light is lost at a reflecting surface. Transluscent objects reduce the transmitted light energy and as a result require greater care in the adjustment of the sensitivity of the control system.

The second type of scanning technique is the **reflective scan**. This method is used when finish inspections are being made or a code or registration mark is being sensed. There are three reflective scanning techniques: diffuse, specular, and retroreflective. The first two are most commonly used. **Diffuse scanning** is used with materials such as wood, cardboard, fabrics, and paper. The incident light is reflected or scattered in nearly all directions, and as a result only a small percentage returns to be detected by the sensor. Normally the sensor is mounted quite close to the scanned surface and at an angle of 45°. This

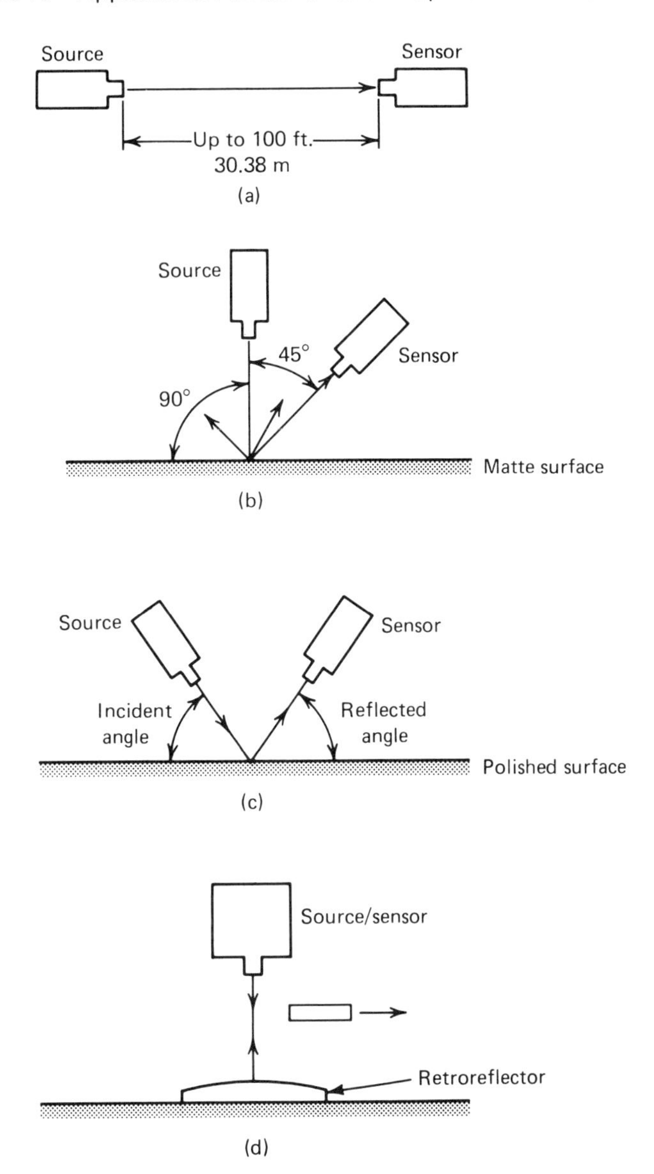

Figure 8-28. Scanning techniques: (a) direct scan, (b) diffuse scan, (c) specular scan, (d) retroreflective scan.

technique is also used where there is a slight flutter to the scanned material. Normally the alignment of the source and sensor with each other and with the scanned material is not very critical. See Figure 8-28(b).

In the case of **specular scanning**, since the scanned material is highly polished (such as shiny plastic, a mirror, or polished metal), the positioning of the source and sensor is critical. First, the size of the angle between the source and sensor determines the depth of the scanning field; with a narrow angle the depth of the field is greater, with a wide angle the depth of the field is smaller. Second, the incident angle and the reflected angle must be equal. See Figure 8-28(c).

The **retroreflective scan** uses a combined source/sensor, and the emitted light travels to a retroreflective target, where the light beam is reflected back along the same path as it was transmitted. A high proportion of the transmitted light is reflected, and alignment may be as much as 15° to either side of the perpendicular to the target. A further advantage is that this arrangement permits single-side wiring, thus lowering maintenance and installation costs. Also, because of the high percentage of reflected light, the distance between the source/sensor and the retroreflective target can be as great as 30 ft (9 m). This type of scanning is especially effective for the detection of translucent objects. See Figure 8-28(d).

Among the many industrial uses of optoelectronic controls are counting, conveyor control, liquid level control, elevator levelling control, etc. Typical applications are illustrated in Figures 8-29 and 8-30.

Photoelectric controls are found in many different kinds of applications. They satisfy a wide range of control needs because they respond to the presence or absence of either opaque or translucent materials at distances from a fraction of an inch, up to 100 or even 700 feet. These controls need no physical contact to trigger them, and are, therefore, safe to use where sensed objects must remain untouched; whether the application involves food processing, counting tiny parts which must not be detained, or detecting freshly painted or delicate objects. Photoelectric controls respond rapidly to parts moving quickly and in varying positions along a conveyor, yet operate dependably if activated only infrequently during inspection procedures or security surveillance. There are controls for indoor use, for varying ambient light conditions, for high vibration, severe environments, and even for explosive locations. In short, there is a photoelectric control to satsify almost any need.

MICRO SWITCH photoelectric controls bring solid state dependability and long life to all applications. Although some applications demand a particular control, most can be handled easily and efficiently using any one of several controls and scanning techniques. MICRO SWITCH offers a broad line of controls with an equally broad range of logic capabilities to handle routine or unique applications.

- *COUNTING*
- *LABELING*
- *CONVEYOR CONTROL*
- *BIN LEVEL CONTROL*
- *INSPECTION*
- *FEED AND/OR FILL CONTROL*
- *MAIL AND PACKAGE HANDLING*

Figure 8-29. Typical photoelectric control applications. (Courtesy of MICRO SWITCH, a Honeywell Division)

- *THREAD BREAK DETECTION*
- *EDGE GUIDE*
- *WEB BREAK DETECTION*
- *REGISTRATION CONTROL*
- *FOOD PROCESSING*
- *PARTS EJECTION MONITORING*
- *BATCH COUNTING*
- *SEQUENTIAL COUNTING*
- *SECURITY SURVEILLANCE*
- *ELEVATOR CONTROL*
- *RAMP MONITORING*
- *DETECTING TRANSLUCENT OBJECTS*

The photoelectric controls pictured count, inspect, detect conveyor line stops (A), brewery operations (B), shotgun shell manufacturing (C), bottle cap detection (D), pharmaceutical production (E) and disposable bag processing (F).

A.

B.

C.

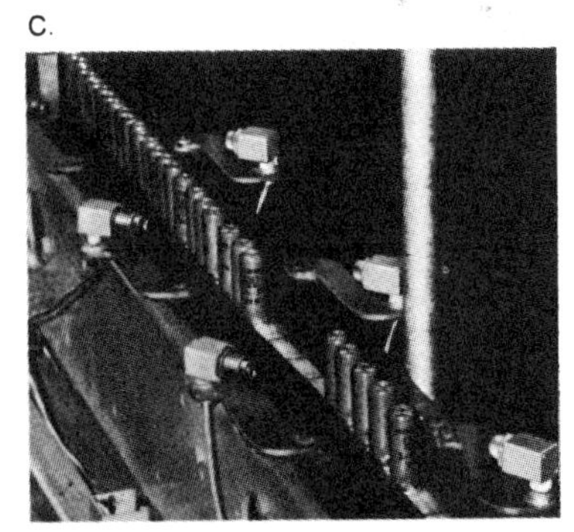

D.

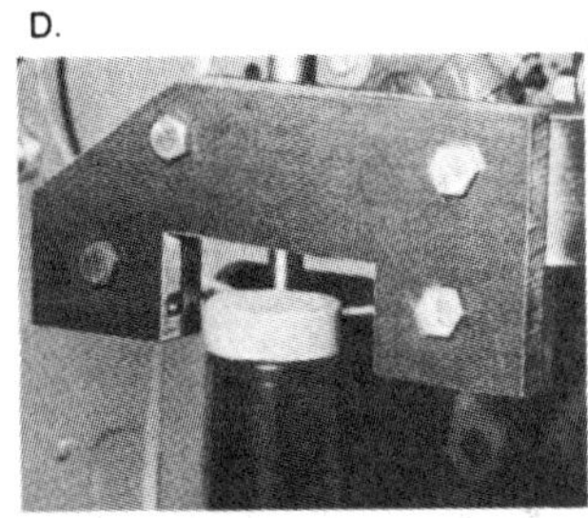

E.

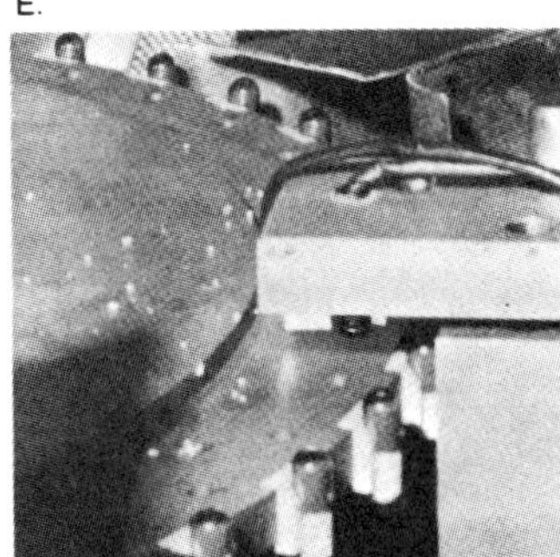

F.

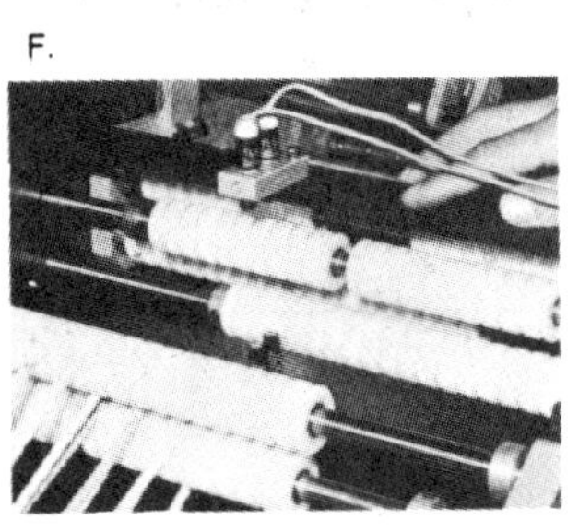

Figure 8-29. continued

Typical applications

The drawings here give you only a hint of the variety and potential for photoelectric control applications.

A. Two light source-photoreceiver pairs are used to keep hopper fill level between high and low limits.

B. Counting products is a common application of photoelectric controls. Counting batches or groups of cans or other items prior to packaging or group processing is also common.

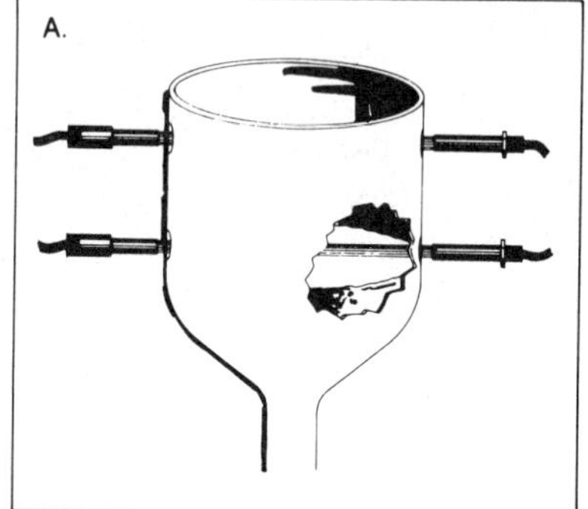

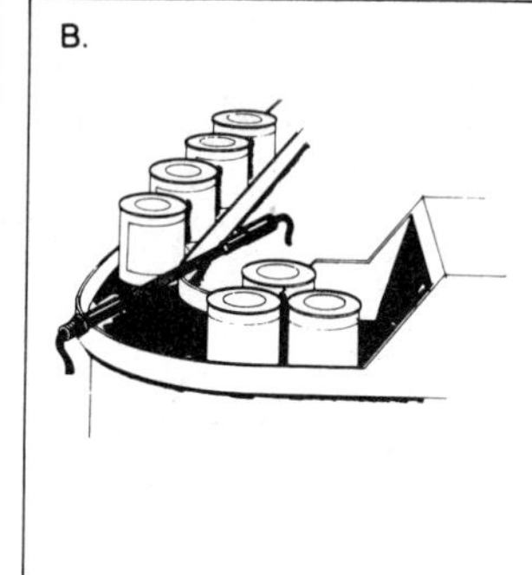

C. A photoelectric control operating on reflected light is a simple way to detect a web break. An alternative is to put a light source above the web, and a photoreceiver below.

D. Dark caps are checked for white liners by a photoelectric scanner. The scanner activates a mechanism that rejects caps without the liners.

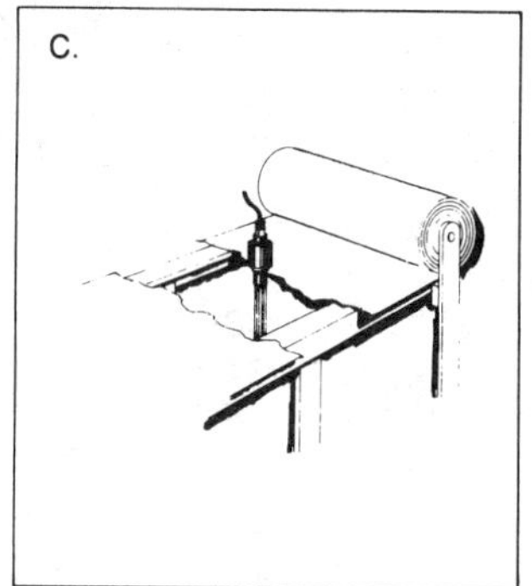

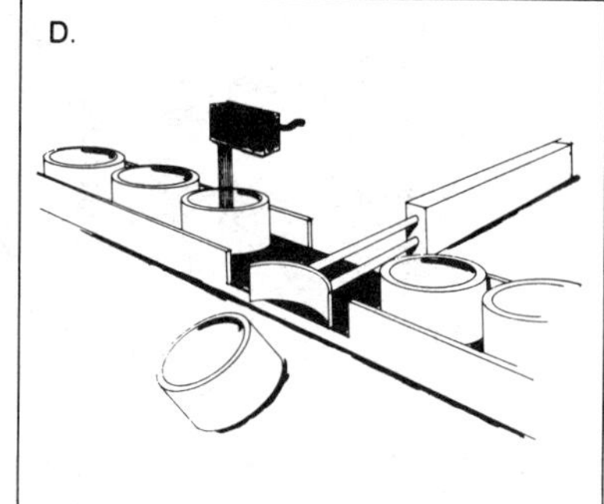

E. To prevent collisions where two conveyors merge, each conveyor is monitored by a control that powers the other conveyor when its own is cleared.

F. A tubular light source and photoreceiver in a specially designed bracket detect registration marks to initiate any related operation, such as printing, cutoff, or folding.

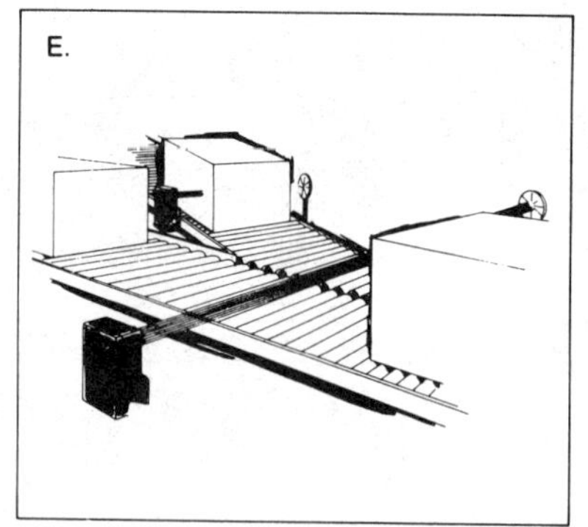

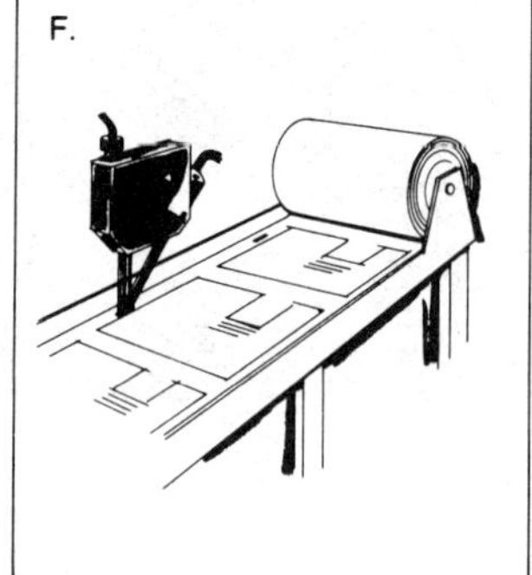

Figure 8-30. Further photoelectric control applications. (Courtesy of MICRO SWITCH, a Honeywell Division)

G. Gluing, buffing, or flattening can be done efficiently by controlling the pressure rollers or buffer with a photoelectric light source and photoreceiver that detect the product to be processed.

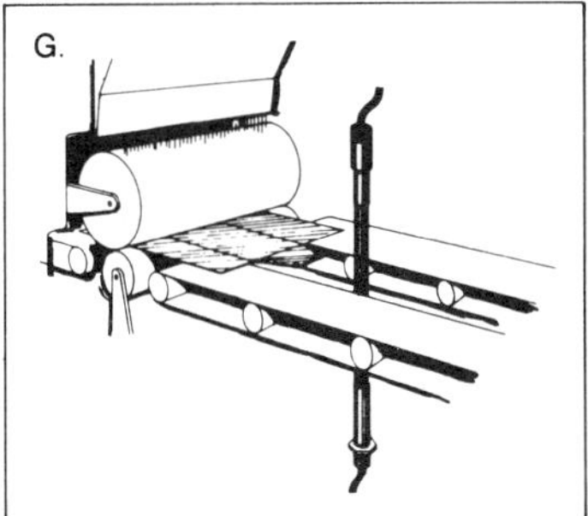

H. Using logic for one-shot pulse output, a photoelectric control slows a conveyor and fills the carton which has interrupted the light beam.

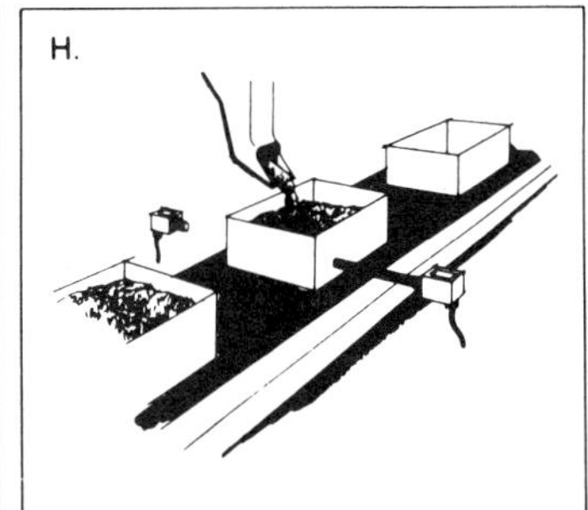

I. Two light source-photoreceiver pairs work together to check fill level. The box-detecting pair turns on, or enables, the fill inspection pair — thereby preventing the inspection pair from mistaking the space between boxes as an "improper fill".

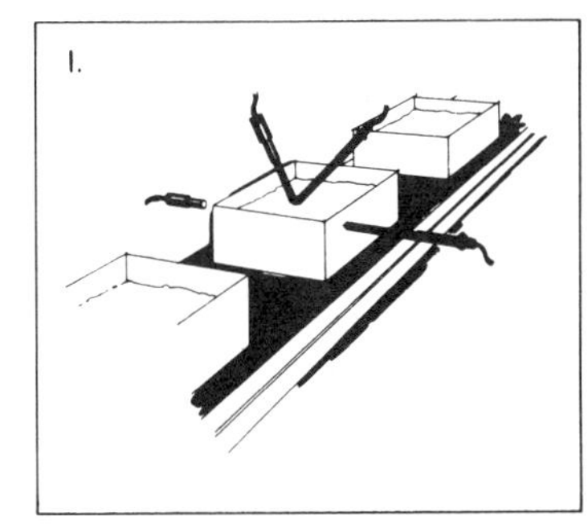

J. Light source and photoreceiver placed near a guillotine are used to detect products and operate the blade for cutting the link between products.

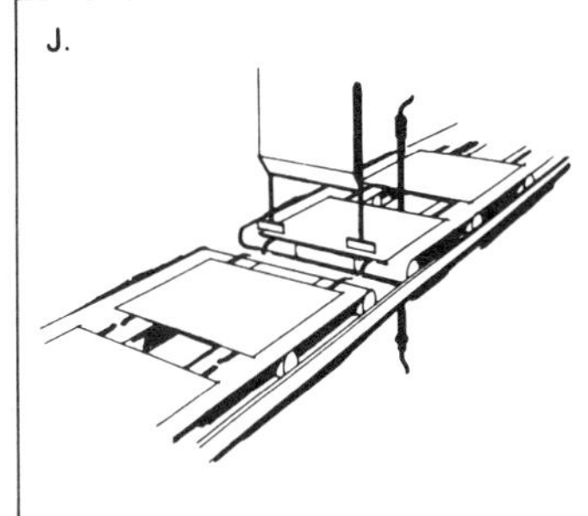

K. Thread break detection is easy when the photoelectric beam is interrupted by a lightweight flag riding on the taut thread.

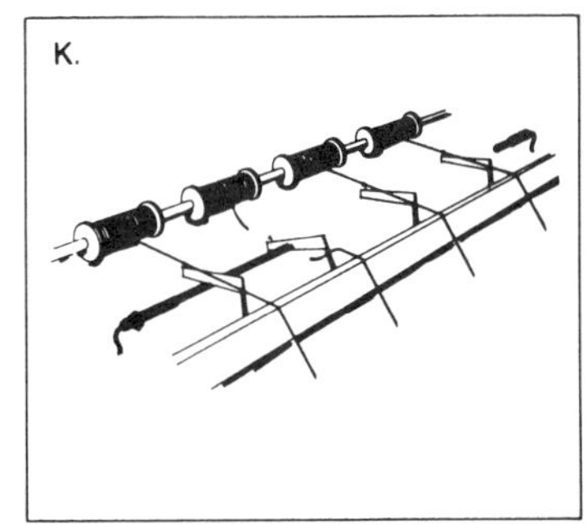

L. The size of a paper or fabric roll can be controlled by positioning a light source and a photoreceiver so the roll diameter blocks the beam.

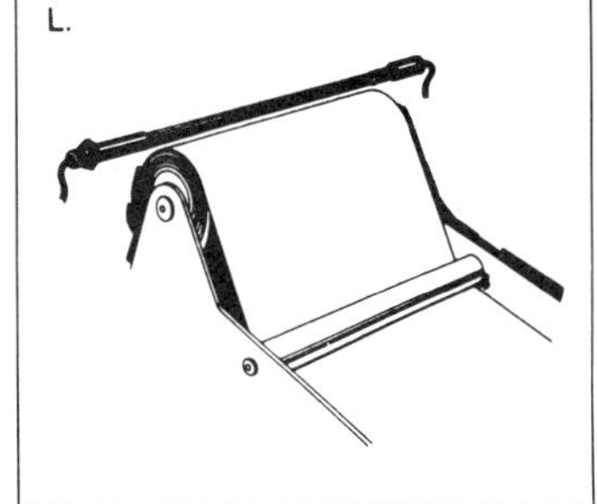

M. Turning a glue nozzle on and off is an example of process control easily effected with a photoelectric control.

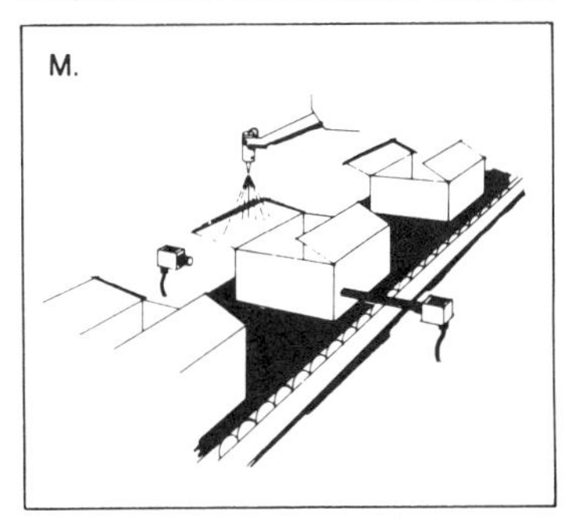

Figure 8-30 continued.

8-14 FIBER OPTICS

Fiber optics is a rapidly emerging technology based upon the transmission of light through ultrathin strands of very pure glass or plastic. These strands or filaments are called optical fibers and are used in conjunction with a modulated light source, usually a LED, to transmit digital bit data or analog signals to a detector, usually a photodiode, at the receiving end. The major applications that exist in the industrial field are data transmission between computers, programmable controllers, and control signals to and from equipments. There are a number of advantages for using fiber optic systems from performance and cost terms over the more conventional electrical methods such as coaxial or hand-wired transmission systems:

1. Accepts data in any form from dc to 10 Mb/sec.
2. Electrical isolation from transmitter to receiver.
3. Electromagnetic interference (EMI) and radio frequency interference (RFI) immune.
4. No signal or noise emission.
5. Safe to use in an explosive environment.
6. Freedom from short-circuits, since there is no current flow.
7. Secure transmission, i.e., cannot be tapped.
8. Lightweight, since there is no metallic conductor.
9. Does not require lightning or transient voltage protection.
10. Relatively low cost as compared to other methods.

There are several disadvantages:

1. Signals are subject to attenuation, which limits transmission distance.
2. The signal is subject to distortion which limits the maximum data rate that can be transmitted.
3. Requires highly skilled personnel to make connections, splices, etc.

8-14-1 Basic Physics of Light

In free space electromagnetic energy including light travels at 3×10^8 m/sec; however, in denser mediums, such as glass, plastic, water, etc., the speed of light is reduced. The reduction in speed as light passes from free space into a denser medium results in the light being bent as it enters the denser medium. This process is known as refraction. In addition, the amount of bending is also dependent upon the wavelength of the incident light (see Figure 8-31). The laws of refrac-

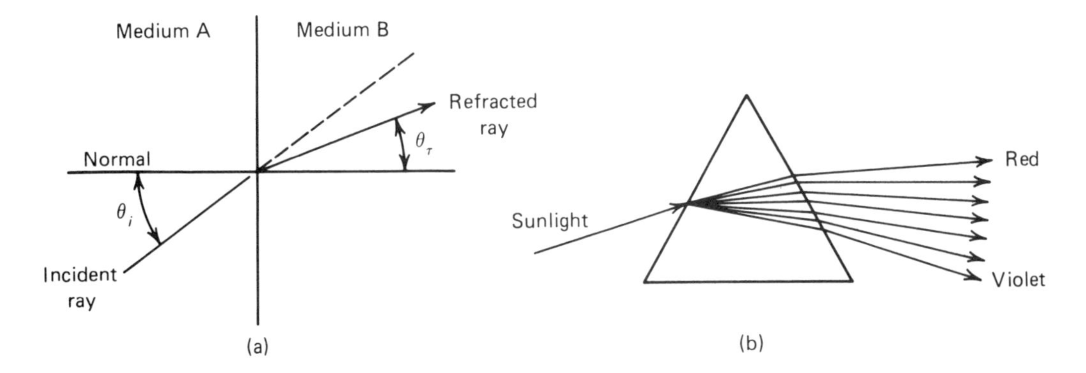

Figure 8-31. Light refraction: (a) monochromatic refraction, (b) refraction of sunlight by a prism.

tion are called **Snell's Law**, which states that if θ_i is the angle of incidence of a light ray and θ_r is the angle of refraction of the ray,

$$\frac{\sin \theta_i}{\sin \theta_r} = \frac{\text{speed of light in medium } A}{\text{speed of light in medium } B} = \text{a constant} \quad (8\text{-}6)$$

The constant is called the index of refraction (n). Typical values of n are: for a vacuum $n = 1$; for air $n = 1.0003 \simeq 1$; for glass $n = 1.5$; and for silicon $n = 3.4$.

Usually we are interested in the effect on a light ray when the ray is incident on the interface between two materials of different refractive indices n_1 and n_2. Then

$$n_1 \sin \theta_i = n_2 \sin \theta_r \quad (8\text{-}7)$$

where n_1 and n_2 are the ratios of the speed of light in free space to the speed of light in the denser medium.

The speed of light in a medium is expressed as

$$\text{speed of wave} = \text{frequency} \times \text{wavelength}. \quad (8\text{-}8)$$

However, although the index of refraction n is a function of the wavelength of the light, the variation is sufficiently small to be neglected. From Equation (8-7) it is possible to calculate the angle of refraction θ_r:

$$\sin \theta_r = \frac{n_1}{n_2} \sin \theta_i \quad (8\text{-}9)$$

From Equation (8-9) it can be seen that if $n_1 < n_2$, then $\theta_r < \theta_i$, i.e., the light ray is bent away from the normal. It should also be noted that the intensities of reflected and refracted light rays at the surface of a transparent medium are dependent upon the angle of incidence θ_i. When the angle of incidence is small, i.e., the incident light is perpendicular or nearly perpendicular to the surface, the percentage of light transmitted is large and the percentage of light reflected is small. As the angle of incidence increases the percentage of light transmitted decreases and the percentage of light reflected increases.

In the case of light originating in a medium with a higher refractive index than the refractive index of the medium into which it is being transmitted, the angle of refraction θ_r is greater than the angle of incidence. There is a maximum angle of incidence where the angle of refraction is 90°. This angle is called the critical angle of incidence (θ_C). If the angle of incidence exceeds the critical angle, the incident light is reflected at the interface. See Figure 8-32(a). For θ_i approaching θ_c, all the light is transmitted, and when $\theta_i = \theta_c$, the light is all reflected, as is the case in practice. As θ_i approaches θ_c, less and less light is transmitted and more and more light is reflected until, when $\theta_i = \theta_c$, all light is reflected. This characteristic is a very important factor in fiber optics. The critical angle θ_c can be obtained from Equation (8-9) as follows:

$$\sin\theta_r = \frac{n_1}{n_2}\sin\theta_i$$

When $\theta_r = 90°$, $\sin\theta_r = 1$, and $\theta_i < \theta_c$

$$1 = \frac{n_1}{n_2}\sin\theta_c$$

$$\sin\theta_c = \frac{n_2}{n_1} = \frac{1}{n_1} \qquad \text{if } n_2 \text{ is air} \qquad (8\text{-}10)$$

Example 8-7

The refractive index of glass is 1.5. What is the critical angle θ_c of light passing from the glass into air? (Assume the refractive index of air is 1.)

Solution:

$$\sin\theta_c = \frac{1}{n_1} = \frac{1}{1.5} = 0.67$$

and

$$\theta_c = 41.81°$$

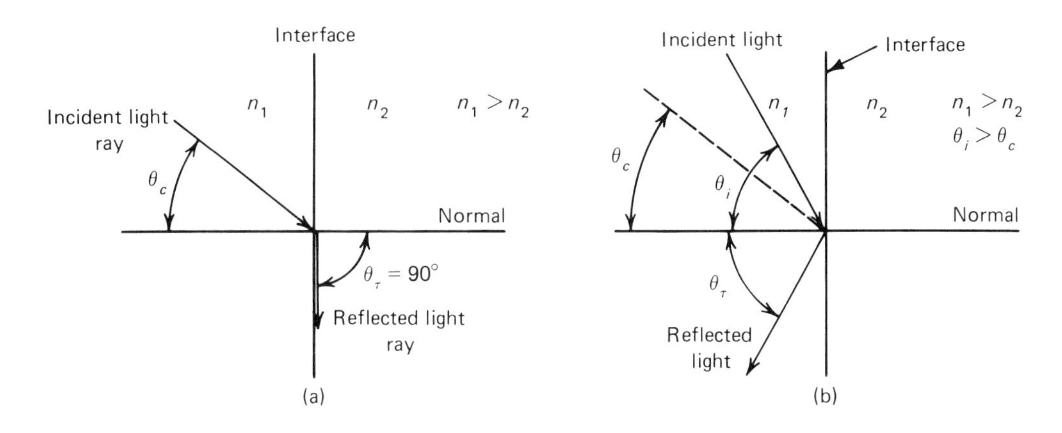

Figure 8-32. (a) Critical angle of incidence, (b) total reflection.

When the angle of incidence is greater than the critical angle, the angle of the reflected ray is equal to the angle of the incident ray, see Figure 8-32(b).

8-14-2 Optical Fibers

The basic construction of an optical fiber is illustrated in Figure 8-33(a). The actual propagation of light takes place in the central glass core. Surrounding this core is the cladding layer which is also usually glass but may be plastic. During manufacture the refractive indices of the core and cladding are adjusted to ensure that the refractive index of the cladding is less than that of the core. Finally, to provide mechanical protection, a rubber or plastic protective jacket is extruded over the fiber.

Prior to discussing the various types of optical fibers, it is appropriate at this point to discuss the meaning of **numerical aperture (NA)** which is the measure of how far the incident light beam may be off the center axis of the core while still being accepted by the core. A low value of NA indicates that the light source and the core must be carefully aligned for the core to accept and transmit light. On the other hand, a high NA indicates that the fiber will accept and transmit light over a wide angle and that the alignment of the source and core is not as critical as is the case for a low NA.

There are three classes of optical fibers: single-mode, multi-mode, and graded-index. The **single-mode fiber** usually has a core with a diameter about 3μ, together with a cladding that brings the overall diameter up to between 70 and 150μ, and it has a small numerical aperture. The single-mode fiber, as its name implies, will only accept a single ray or mode directly along its axis, has a very wide bandwidth (frequency range), and can transmit about 10 Gb/sec/km. The major problem is that they are difficult to couple into a splice and hardware alignment is very critical.

The **multi-mode fiber** has a core diameter, typically 50μ with a cladding thickness of 1 to 50 μ . However, since they have a relatively large numerical aperture, the entering light rays have two paths, one directly along the axis and the other a series of reflections from the core-cladding interface. See Figure 8-33(b). This type of construction is termed a **stepped-index fiber** since the glass core is surrounded by a cladding which has a slightly lower refractive index. It is obvious that the light rays entering the core off the center axis reflect from the core-cladding interface a large number of times, and since the velocity of light rays in both paths is the same, the rays following the zig-zag path have a longer transit time. As a result, if the input is a square pulse, the output pulse tends to be trapezoidal. This property, called **pulse dispersion** or **modal dispersion,** increases with increased numerical aperture, and places a limit on the bandwidth capabilities of the fiber. The data rate is limited to the range of 25 to 30 Mb/sec/km. However, the multi-mode fiber is less sensitive to misalignment of the interfacing hardware.

Graded-index fiber cores have a core with a varying refractive index which is highest at the center of the fiber. The effect of varying the refractive index is to reduce pulse dispersion (by slowing down light traveling in the center of the core) and as a result increases the bandwidth and the data transmission rate, which usually lies between the rates for single- and multi-mode fibers. The graded-index fiber core is relatively large being of the order of 200 μ and is self-focusing, thus minimizing hardware alignment problems. See Figure 8-33(c).

Signal loss in the fiber is only one of the signal loss factors, since the total loss of a system is the sum of the losses in all parts. However, in industrial applications the fiber runs are relatively short which minimizes the light loss problem. The major loss occurs at the ends of the fibers, and glass fiber core ends must be cut precisely and polished. In the case of plastic cores, precision cutting and polishing is relatively easy with suitable equipment.

There are two competing solid-state devices for use as transmitters: the LED and the injection laser. At the moment the LED is leading with its low cost and transmission rate capability of up to 50 Mb/sec, combined with low power consumption and long life. The injection laser, with high optical power and more satisfactory coupling to the fiber because of its directional properties, offers strong competition for long-distance transmission and high-transmission rates up to 1

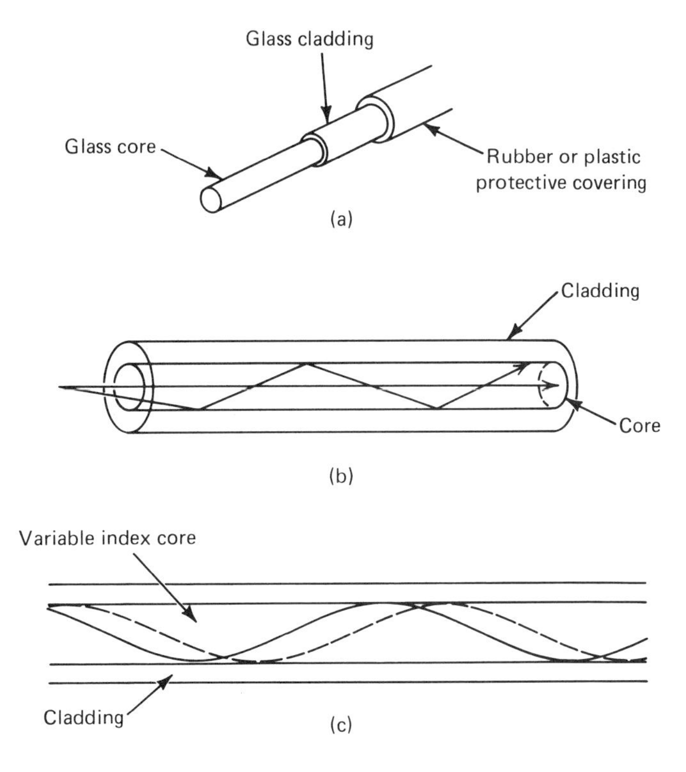

Figure 8-33. Optical fibers: (a) single fiber construction, (b) multimode stepped index fiber, (c) graded index fiber.

Gb/sec. However, it is relatively short lived, usually less than 10,000 hours, and is temperature sensitive.

At the receiving end there are once again two competitors: the avalanche photodiode and the pin photodiode. At the moment the pin photodiode is in the lead because of its low noise and capability to operate in bandwidths up to 50 MHz combined with relatively low cost, typically about one dollar. The avalanche photodiode is far more sensitive than the pin photodiode because of internal amplification. However, it has temperature sensitivity problems as well as requiring high reverse-bias voltages up to 200 V, combined with a cost in the range of $25 to $35.

The task of selecting and replacing emitters and detectors has been simplified by the joint efforts of AMP, Inc. and Motorola with their "Straight Shooter" line of mechanically matched connectors, emitters, and detectors (see Figure 8-34).

The complete fiber optic link in schematic form is illustrated in Figure 8-35.

The fiber optic emitters and detectors are in the new and unique ferrule package and in the standard lensed TO-18 type package. This ferrule package was developed to provide maximum coupling of light between the die and the fiber. The package is small, rugged and producable in volume. The ferrule mates with the AMP ferrule connector #227240-1 for easy assembly into systems and precise fiber-to-fiber alignment. This assembly permits the efficient coupling of semiconductor-to-fiber cable and allows the use of any fiber type or diameter.

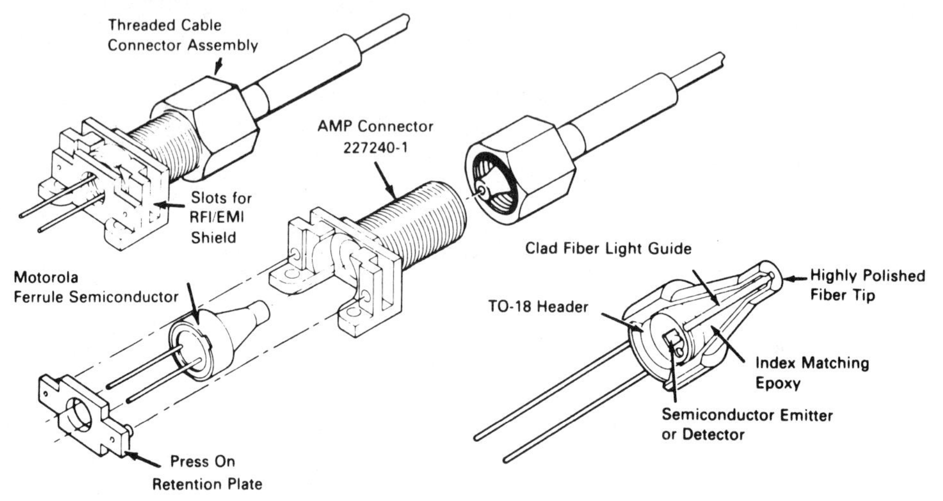

Figure 8-34. Standard ferrule package. (Courtesy of Motorola Inc.)

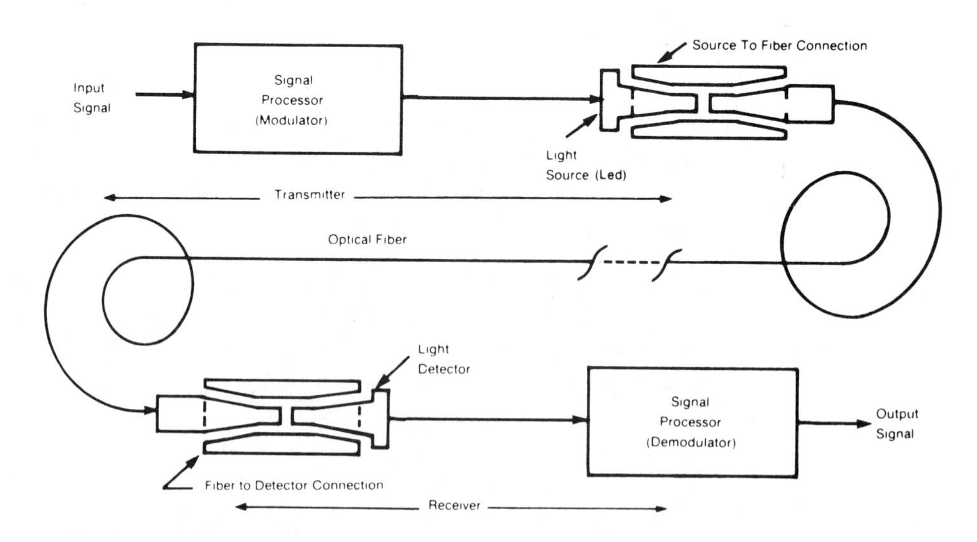

Figure 8-35. The basic fiber optic link. (Courtesy of Motorola Inc.)

In conclusion, industrial fiber optics are emerging as the major means of data and instrumentation communications where their electrical isolation and freedom from electromagnetic interference and noise offer a tremendous advantage. In addition, there is a continuing move toward complex and high-density communications that can only be satisfied economically by fiber optics. Additionally, fiber optic sensors are being introduced to sense variables such as pressure, temperature, viscosity, humidity, fluid level, fluid flow, and motion.

GLOSSARY OF IMPORTANT TERMS

Optoelectronic device: An electronic device that detects light or radiation, or emits or modulates coherent and noncoherent light or radiation, or may also depend entirely upon light or radiation for its operation.

Light: Electromagnetic radiation occurring in the electromagnetic spectrum from 0.3μ (1×10^{15}Hz) to 30μ (1×10^{13}Hz).

Visible light: The portion of the electromagnetic radiation visible to the human eye, that is, $0.38\,\mu$ to $0.78\,\mu$.

Photometric quantities: Refer to the visual effects as seen by the human eye.

Radiometric quantities: Refer to the total energy output of a light source including the ultraviolet and infrared as well as the visible portion of the electromagnetic spectrum.

Wavelength (λ): The shortest distance between two successive points in an electromagnetic wave that are in phase.

Spectral response: Of the human eye or of an optoelectronic device, the ratio of the energy detected to the energy transmitted at a specific wavelength.

Incandescence: The production of light by raising the temperature of a material.

Luminescence: The production of light without being dependent upon the temperature of a material.

Candela (cd): The unit of luminous intensity (I). SI definition is the luminous intensity of 1/600,000 of a square meter of projected area of a black body radiator operating at the temperature of solidification of platinum under a pressure of $101{,}325 N/m^2$.

Lumen (lm): The amount of luminous flux radiating out from a uniform point source of lcd through 1 unit solid angle (1 steradian).

Luminous flux (F): The total luminous flux radiated from a point source of luminous intensity I.

Irradiance (H): The radiant flux per unit area incident on a surface.

Photodetector: Devices that convert radiant energy to an electrical output.

Photoconductors: Devices that change their resistance when exposed to radiant energy.

Spectral response: Curves showing the relative response of a photodetector or the human eye to varying wavelengths of radiated energy.

Photodiode: A *p-n* device operated under reverse-bias conditions which experiences an increase in leakage current when radiant energy falls on the *p-n* junction.

Avalanche photodiode (APD): A photodiode whose gain has been increased by operating in the reverse or avalanche breakdown region.

pin photodiode: A silicon photodiode with a very thin layer of intrinsic silicon sandwiched between the *p*- and *n*- type materials to improve the speed of response.

Phototransistor: Similar in construction to the normal bipolar transistor except that radiant energy is focused on the collector-base junction. Leakage current is increased by the current gain of the transistor.

Photodarlington: A phototransistor connected in combination with another transistor to form a darlington amplifier.

Photo-FET: A JFET in which I_D varies when radiant energy is applied to the gate-channel junction.

Photovoltaic device: A device in which a voltage is produced when radiant energy falls on a *p-n* junction.

Light emitting diode (LED): A *p-n* junction diode in which charge carrier recombinations produce visible light.

Optocoupler: Usually consists of an infrared LED and a photodetector.

Laser: Light amplification by stimulated emission of radiation. The device emits radiation at one frequency or wavelength.

Fiber optics: A general term covering the conversion of electrical energy to optical energy and then transmitting to other locations by means of ultrathin strands of glass or plastic and converting the optical energy back to electrical energy.

Numerical aperture (NA): The measure of how far the incident light beam may be off the center axis of the core and still be accepted by the core.

REVIEW QUESTIONS

8-1 Why are high frequency radiations such as X-rays and gamma rays capable of penetrating dense materials?

8-2 Discuss incandescence and luminescence.

8-3 Discuss the photometric and radiometric systems.

8-4 What is meant by a bulk type photoconductive sensor? Discuss the construction and characteristics of two photoconductive sensors.

8-5 With the aid of a sketch explain how a street light may be controlled by a photoconductive cell.

8-6 With the aid of a sketch describe the operation of a photodiode. How may the spectral response be varied? List typical applications.

8-7 Explain the principle of operation of an avalanche photodiode. Discuss its advantages and disadvantages.

8-8 Repeat Question 8-7 for the pin photodiode.

8-9 Repeat Question 8-7 for the phototransistor.

8-10 Repeat Question 8-7 for the photodarlington.

8-11 Repeat Question 8-7 for a photo-FET.

8-12 Repeat Question 8-7 for a phototriac.

8-13 Discuss the principle of operation of a photovoltaic cell.

8-14 Briefly discuss the principle of operation and construction of the currently available solar cells.

8-15 Explain the principle of the light emitting diode and explain what is meant by direct and indirect gap diodes. What is reabsorption?

8-16 Describe two types of alphanumeric displays.

8-17 Explain the operation of a TIL 504 numeric display as illustrated in Figure 8-22.

8-18 Discuss the different types of optocouplers, listing the advantages and typical applications of each type.

8-19 Discuss the principle of operation of an injection diode.

8-20 Discuss the self-contained and modular types of optoelectronic controls including scanning techniques, advantages, and limitations. Illustrate your answer with typical industrial applications.

8-21 Discuss the advantages and disadvantages of using fiber optic systems as against the use of conventional electrical systems.

8-22 With the aid of sketches explain the principle of operation of a fiber optic fiber. What is meant by the critical angle?

8-23 Discuss the construction of optical fibers.

8-24 What is meant by pulse or modal dispersion? How may its effects be reduced?

8-25 With the aid of a sketch discuss a typical fiber optic link with specific attention being given to the transmitting and receiving end devices.

PROBLEMS

8-1 Express the following frequencies in wavelengths: (1) 150kHz, (2) 2GHz, and (3) 4.2×10^{14}Hz.

8-2 Express the frequencies of Question 1 in: (1) microns, (2) millimicrons, and (3) Angstroms.

8-3 A 200 W incandescent lamp has a luminous intensity of 300 candela. Calculate (1) the total luminous flux radiated, (2) the luminous efficiency, and (3) the illuminance on a perpendicular surface 150 cm from the lamp.

8-4 A bulk type photoconductor having a resistance characteristic as shown in Figure 8-6 is used to control a relay. The relay is supplied from a 15V source and requires 15mA to pick-up when the illumination level is 100 lm/m^2. The relay remains deenergized for all illumination levels less than 100 lm/m^2. Design and draw the circuit to accomplish this task. What is the value of the series resistance and the dark current?

8-5 A photodiode with the illumination characteristic of Figure 8-18(c) is connected in series with a 100Ω resistance and is reverse biased by a 0.4V supply. Draw the dc load line and calculate the diode voltages and currents for illumination levels of 5,000, 10,000, and 15,000 lm/m^2.

8-6 Determine the diode resistance of the diode in Question 8-5 for illumination levels of 5,000, 10,000 and 15,000 lm/m^2.

8-7 An MRD 450 phototransistor having the characteristics shown in Figure 8-13 has a supply voltage of 15V and a collector load resistance of 1.5kΩ. Determine the output voltage V_{CE} when the illumination level is (1) zero, (2) 7.0mW/cm^2, and (3) 3.0mW/cm^2.

9

Amplifiers And Control Elements

9-1 INTRODUCTION

An **amplifier** is a device in which an input signal is used to control a source of power so that the amplified output possesses a definite relationship to the input signal. The term "amplifier" has by common usage been applied exclusively to electronic applications. However, in industrial control applications the term "amplifier" must be extended to include pneumatic, hydraulic, rotating amplifiers such as the Amplidyne and Rototrol, and thyristor phase-controlled converters.

9-2 THE OPERATIONAL AMPLIFIER

The **operational amplifier** or op-amp as it is more commonly known, is a high-gain direct-coupled amplifier with the capability of amplifying signals ranging from dc (0 Hz) to the MHz range. Originally, it was constructed using vacuum tubes to perform mathematical operations such as addition, multiplication, division, subtraction, integration, and differentiation in analog computers. Unfortunately, the original op-amps were unstable, unreliable, and expensive. With the development of IC technology, stable, reliable and, inexpensive op-amps became readily available and now form the basic building block of electronic circuits in instrumentation, control, signal conditioning, filtering, impedance transformation, and mathematical operations to name a few of the many uses to which the op-amp has been applied.

The main advantages of op-amps are: (a) small and light, (b) low power consumption, (c) low heat dissipation, (d) highly stable, and (e) highly reliable.

9-2-1 Characteristics Of An Ideal Operational Amplifier

The ideal operational amplifier should possess the following characteristics:

1. An infinite input impedance; permits any source to be used without loading effects being present, that is, no power is consumed.
2. A zero output impedance; enables the op-amp to drive any connected load.
3. Infinite gain; a very small change in the input signal should produce a very large change in the output signal.
4. Infinite bandwidth; the range of input signal frequencies over which the device may operate without the output being distorted.
5. A zero response time; that is, the changes in output occur at exactly the same time as the changes in the input signal.
6. A zero offset voltage; a zero output voltage when there is a zero input voltage.
7. Insensitive to variations in supply voltage and temperature.

As defined by these characteristics, the ideal op-amp has yet to be developed. However, with the advanced state of IC technology the modern commercially available op-amp closely approaches many of these characteristics.

Even though the op-amp is an IC our understanding of the operation will be greatly improved if we have some basic understanding of the major stages inside the IC.

In its simplest form the op-amp consists of three stages: a differential amplifier, a voltage amplifier, and an output amplifier (see Figure 9-1).

9-2-1-1 The Differential Amplifier

To clearly understand the operation of a differential amplifier it is best to develop the concept in stages. The simplest form of the differential amplifier is shown in Figure 9-2(a), and is known as the single-input single-output arrangement.

When a sinusoidal signal v_1 is applied to the base of $Q1$ with the base of $Q2$ grounded, $Q1$ will act as an emitter follower and $Q2$ acts as a common-base amplifier. The voltage swings of the input signal v_1 are

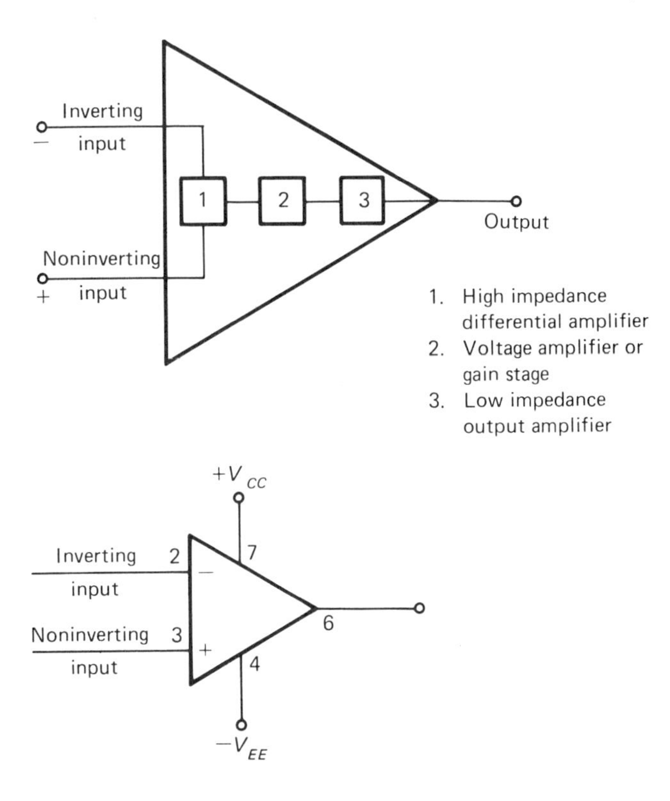

Figure 9-1. Basic arrangement of an op-amp and circuit symbol with pin numbers.

also followed at the emitter of $Q1$. In turn, this signal is applied to the emitter of $Q2$, and since $Q2$ is acting as a common-base amplifier an amplified output voltage in phase with v_1 will be present at V_{02}. It should be noted that, since the two transistors are directly coupled, the amplifier will be equally effective with dc or ac input signals.

The single-input, differential-output circuit shown in Figure 9-2(b) is almost identical to that of Figure 9-2(a). The only difference is that a second output V_{01} has been brought out at the collector of $Q1$. As before the voltage at V_{02} is in phase with the input v_1 applied to the base of $Q1$. Since $Q1$ acts as a common-emitter amplifier, the output voltage at V_{01} will be 180° out of phase with the input signal v_1. The two output signals V_{01} and V_{02} will be equal in amplitude but opposite in phase to each other.

The differential-input, differential-output configuration is shown in Figure 9-2(c). Usually the two input signals v_1 and v_2 are identical, but out of phase with each other, and produce an output signal v_o which is

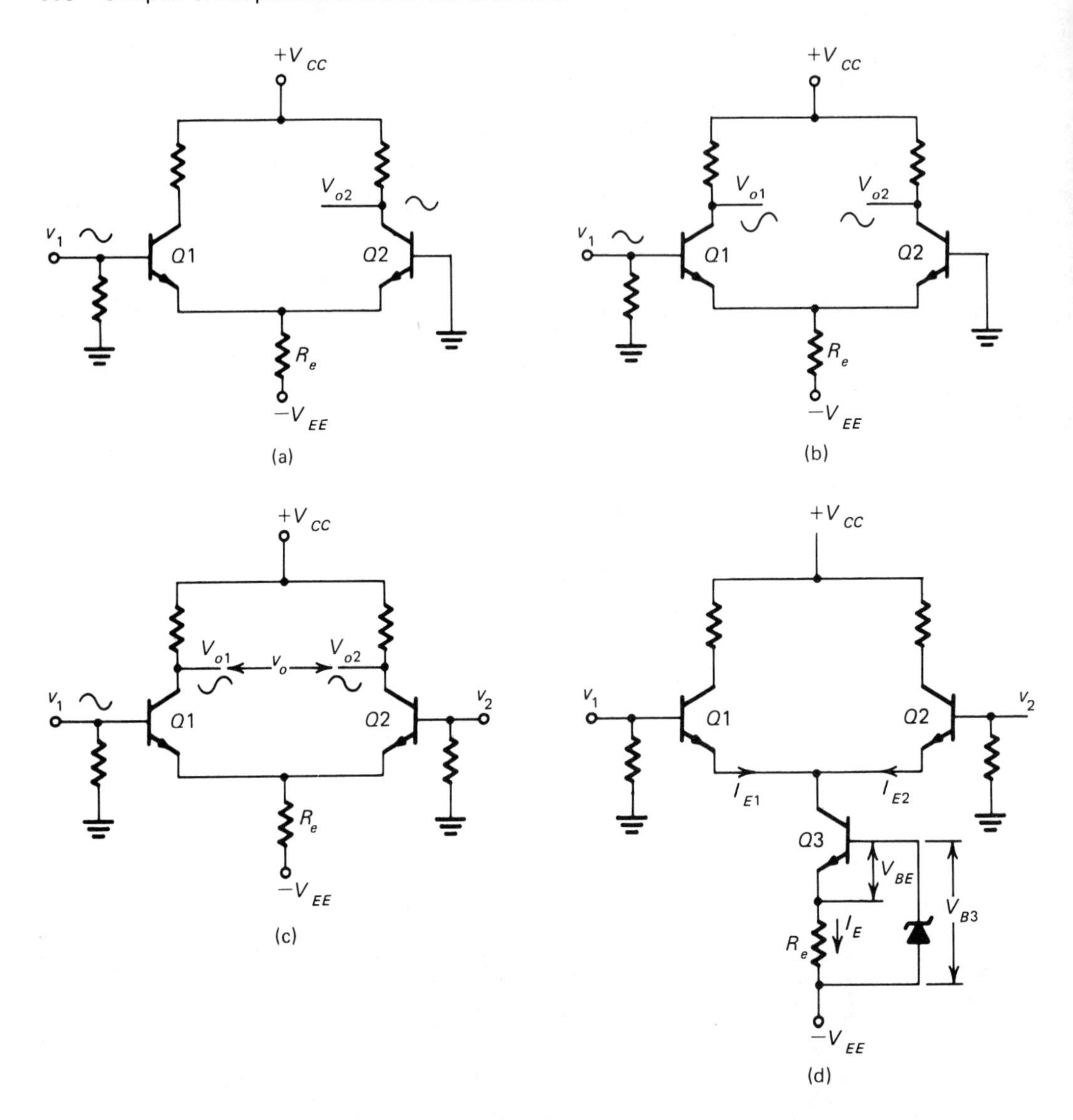

Figure 9-2. Basic differential amplifier circuits: (a) single-input, single-output, (b) single-input, differential-output, (c) differential-input, differential-output, (d) differential amplifier with a constant current tail.

obtained by connecting the load between V_{01} and V_{02}. The output signal is given by

$$v_0 = A_d(v_1 - v_2) \tag{9-1}$$

where A_d is the gain of the differential amplifier.

The characteristics of $Q1$ and $Q2$ and the associated collector resistances are matched during the manufacturing process. If two equal in phase input voltages v_1 and v_2 are simultaneously applied to the inputs this is termed a common-mode input. The differential amplifier should not respond to common-mode signals. In order to reject common-mode signals, it is necessary to ensure that the common-mode gain is minimized. This may be achieved by increasing the value of R_e, but, as R_e is increased, the collector and emitter currents of $Q1$ and $Q2$ will decrease. The solution is to use a constant current source [see Figure 9-2(d)]. As a result,

$$I_E = I_{E1} + I_{E2} \tag{9-2}$$

$$= (V_{B3} - V_{BE})/R_e \tag{9-3}$$

Consequently, I_E will remain constant irrespective of the changes in v_1 and v_2, and as a result common-mode signals will be rejected.

The differential amplifier will reject common-mode signals where

$$v_c = \tfrac{1}{2}(v_1 + v_2) \tag{9-4}$$

and amplify difference signals, where

$$v_d = v_1 - v_2 \tag{9-5}$$

The ability of the differential amplifier to amplify differential signals and reject common-mode signals is commonly expressed as the common-mode rejection ratio (CMRR), the ratio of the differential gain to the common-mode gain.

The differential gain A_d is the change of output voltage Δv_o to the change of differential input voltage Δv_d, therefore,

$$A_d = \frac{\Delta v_o}{\Delta v_d} \tag{9-6}$$

The common-mode gain A_c is the change of output voltage Δv_o to the change of common-mode voltage Δv_c, therefore,

$$A_c = \frac{\Delta v_o}{\Delta v_d} \tag{9-7}$$

Therefore,

$$\text{CMRR} = \frac{A_d}{A_c} \tag{9-8}$$

9-2-1-2 The Voltage Amplifier

The output of the differential amplifier is fed into a high-gain voltage amplifier. The arrangement differs with different manufacturers, however, usually this stage consists of Darlington pairs which achieve gains in excess of 200,000.

9-2-1-3 The Output Amplifier

The output stage is usually a common-collector amplifier or emitter follower, which is characterized by a low output impedance. This feature combined with the high input impedance of the op-amp provides an excellent method of impedance matching between a high impedance source and a low impedance load.

9-2-2 Electrical Characteristics

The ideal op-amp cannot be manufactured. However, the practical op-amp characteristics are acceptable in most applications. The following is a review of the most important characteristics and an explanation of the information normally included in the manufacturers data sheets:

1. Input Resistance (R_i). This is the equivalent resistance seen by an external source between the inverting and noninverting input terminals. Usually R_i ranges between 10 kΩ and 1 MΩ.

2. Open-loop Gain (A_d). This is the ratio of the output voltage to the input voltage. Typical values of A_d range up to 200,000 or greater. Another way of expressing the gain is as the ratio of output volts to input millivolts. For example, with a gain of 200,000 this may be expressed as 200 V/mV. From Figure 9-3(a) it can be seen that any nonzero values of input voltage will drive the output voltage into saturation.

3. Input Offset Voltage (V_{io}). This is the dc voltage that must be applied to the input terminals to produce a zero output dc voltage. See Figure 9-3(b).

4. Output Resistance (R_o). This is the ratio of output voltage change to input current change. Typical values range between 50 to 200 Ω. It should be noted that when external feedback circuits are used the output resistance can approach zero.

5. Common-Mode Rejection Ratio (CMRR). This is the ratio of open-loop gain to the common-mode gain. A CMRR of 30,000 or higher is

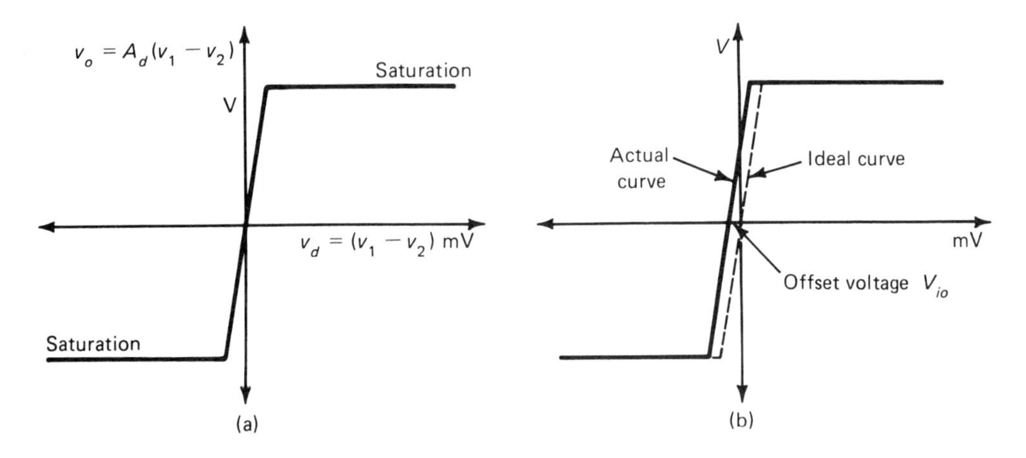

Figure 9-3. Transfer curves: (a) ideal, (b) with offset.

common. CMRR may also be expressed in terms of dB; a CMRR of 90 dB is typical for a 741 op-amp.

6. Input Offset Current (I_{io}). This is the difference between the base currents of the input transistors when the output voltage is zero. A maximum of 500 nA for a 741 op-amp.

7. Input Bias Current (I_{ib}). The average of the input currents with a zero volts output. Typically 1.5 μ A for a 741 op-amp.

8. Input Capacitance (C_i). The capacitance between input terminals with one terminal grounded. In high-frequency applications input capacitance may be very important. Usually $C_i < 2$pF.

9. Slew Rate (SR). This is the time rate of change of the output voltage of a closed-loop amplifier in response to a step input signal. It is usually expressed in terms of volts per microsecond (V/μs).

There are a number of additional characteristics defined in the data sheets, but in all cases these are self-explanatory and do not require further explanation.

9-2-3 Closed-Loop Operations

The normal methods of using op-amps all involve the use of negative or degenerative feedback. The effects of negative feedback reduce the circuit gain, improve stability, and increase the bandwidth. In **closed-loop** operations the output signal is fed back to either the inverting or noninverting input, and as a result the feedback signal acts in opposition to the input signal.

9-2-3-1 Inverting Amplifier

The inverting op-amp is an amplifier whose output voltage is opposite in polarity or phase to the input voltage. See Figure 9-4(a). The input signal V_{in} is applied to the input resistance R_{in} and the feedback resistance R_f is connected between the output and the inverting input (−). The noninverting input (+) is connected to ground, and since the differential voltage between the two inputs is very small, it places the inverting input at very nearly ground potential.

The inverting input terminal forms a virtual ground, since when an input signal is applied it initially moves the potential away from ground. However, the negative feedback signal from the amplifier output opposes this effect and returns the potential to ground potential.

The input voltage and current to the amplifier are V_a and I_a and the output voltage is V_o. The output voltage V_o is opposite in polarity (phase) to V_{in} and is given by

$$V_o = -V_{in}\left(\frac{R_f}{R_{in}}\right) \tag{9-9}$$

This relationship may be proved as follows. At the summing junction

$$I_{in} + I_f + I_d = 0$$

or

$$I_{in} + I_f + V_d/R_i = 0$$

where R_i is the input resistance of the op-amp.

Normally V_d is very small because of the high amplifier gain and R_i is high by the design of the op-amp. As a result, V_d/R_i is very small and may be ignored, therefore,

$$I_{in} = -I_f$$

and by Ohm's Law, it follows that

$$\frac{V_{in} - V_d}{R_{in}} = -\frac{(V_o - V_d)}{R_f}$$

Once again since $V_d << V_{in}$ and V_o then

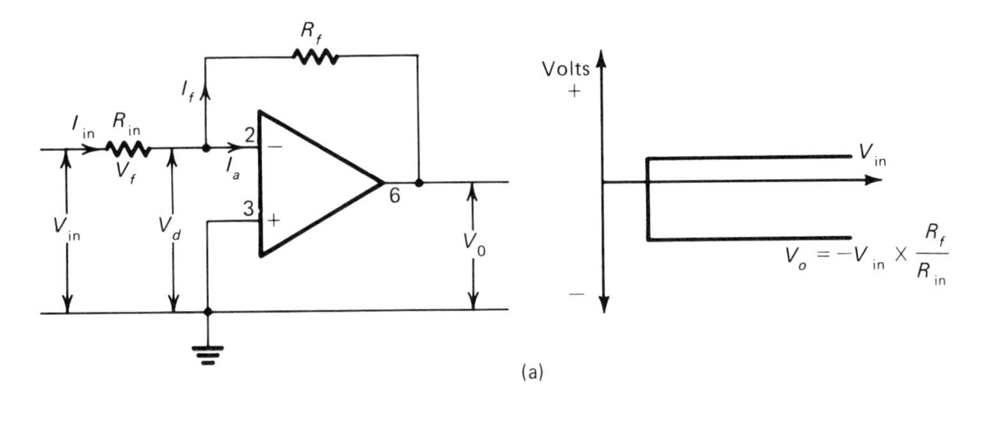

(a)

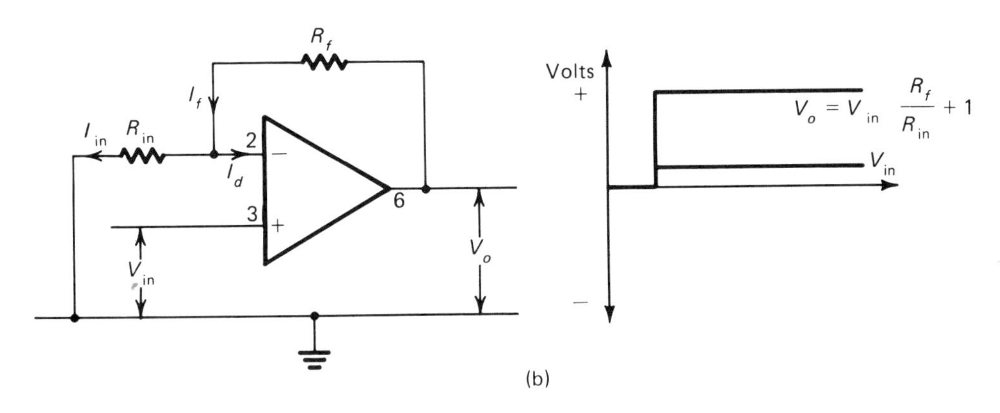

(b)

Figure 9-4. Basic amplification: (a) inverting amplifier, (b) noninverting amplifier.

$$\frac{V_{in}}{R_{in}} = \frac{-V_o}{R_f}$$

or

$$V_o = -V_{in}\left(\frac{R_f}{R_{in}}\right)$$

If we rearrange this equation, then

$$\frac{V_o}{V_{in}} = -\left(\frac{R_f}{R_{in}}\right) \tag{9-10}$$

and by definition the closed-loop gain A_{CL} is

$$A_{CL} = \frac{V_o}{V_{in}} \tag{9-11}$$

Therefore,

$$V_o = -V_{in}A_{CL} \tag{9-12}$$

This means that the closed-loop gain A_{CL} is solely dependent upon the values of the input and feedback resistors.

Also the saturation output voltage V_{sat} of most op-amps is approximately ±80% of the supply voltages.

Example 9-1

The inverting amplifier shown in Figure 9-4(a) has $R_f = 1\ \text{M}\Omega$ and $R_{in} = 100\ \text{k}\Omega$ and an open-loop gain $A_d = 50{,}000$. The supply voltages are ±15 V. Determine: (1) the closed-loop gain; (2) the output voltage if $V_{in} = 500$ mV; and (3) what input voltage will drive the output into negative saturation?

Solution:

(1) $$A_{CL} = \frac{V_o}{V_{in}} = \frac{-R_f}{R_{in}} = \frac{-1\ \text{M}\Omega}{100\ \text{k}\Omega} = -10$$

(2) $$V_o = (-10)\ (500\ \text{mV}) = -5\ \text{V}$$

(3) The saturation voltage is ±80% of 15 V, that is, ±12 V, therefore,

$$V_{in} = \frac{V_o}{A_{CL}} = \frac{-12\ \text{V}}{-10} = 1.20\ \text{V}$$

It should be noted that the circuit of Figure 9-4(a) can be modified to make the op-amp act as a straight inverter, that is, a unity gain inverting amplifier by making $R_{in} = R_f$. This circuit is also known as a phase inverter.

9-2-3-2 Noninverting Amplifier

The same op-amp can also be used as a noninverting amplifier when connected as shown in Figure 9-4(b). Since the differential voltage $V_d \simeq 0$ V, the voltage at the inverting input is approximately equal to V_{in}. Also the feedback resistor is connected between the inverting input and the output. Since R_{in} is grounded, the output voltage V_o is divided between R_{in} and R_f. V_o is in phase with V_{in} and, since the voltage appearing at the inverting input is very nearly equal to V_{in}, the voltage drop across $R_{in} \simeq V_{in} = V_f$.

The feedback action can be best described as follows. When V_{in} goes positive the amplified output V_o also goes positive. Since R_{in} and R_f are acting as a voltage divider, as V_o increases V_f also increases. The increase in V_f partially offsets the increase in V_{in}. The op-amp responds to the difference voltage between the terminals. By the voltage divider rule

$$V_f = \frac{R_{in}}{R_f + R_{in}} (V_o)$$

dividing both sides by V_o, then

$$\frac{V_f}{V_o} = \frac{R_{in}}{R_f + R_{in}}$$

and

$$\frac{V_o}{V_f} = \frac{R_f + R_{in}}{R_{in}} = \frac{R_f}{R_{in}} + \frac{R_{in}}{R_{in}}$$

$$= \frac{R_f}{R_{in}} + 1$$

However, since $V_f \simeq V_{in}$ and $A_{CL} = \frac{V_o}{V_{in}}$, then

$$A_{CL} = \frac{R_f}{R_{in}} + 1 \qquad (9\text{-}13)$$

This relationship once again shows that the closed-loop gain A_{CL} is determined solely by the external resistances.

9-2-3-3 Offset

As was mentioned in Section 9-2-2, the offset resulted in the op-amp output V_o being at some value other than zero when there was a zero input signal. The effect of offset may be corrected by applying an input offset voltage V_{io} to ensure that $V_o = 0$ when $V_{in} = 0$. Figure 9-5 illustrates three common methods of obtaining offset compensation.

Figure 9-5(a) illustrates a method of providing external compensation for an inverting amplifier by supplying V_{io} from a 10 kΩ potentiometer connected across the op-amp's supply voltages. A similar arrangement applicable to a noninverting amplifier is shown in Figure 9-5(b).

Finally the arrangement shown in Figure 9-5(c) depends upon the individual op-amp. Some manufacturers provide pin connections so that a 10 kΩ potentiometer may be connected across the terminals with the slider connected to the negative supply voltage V_{EE}.

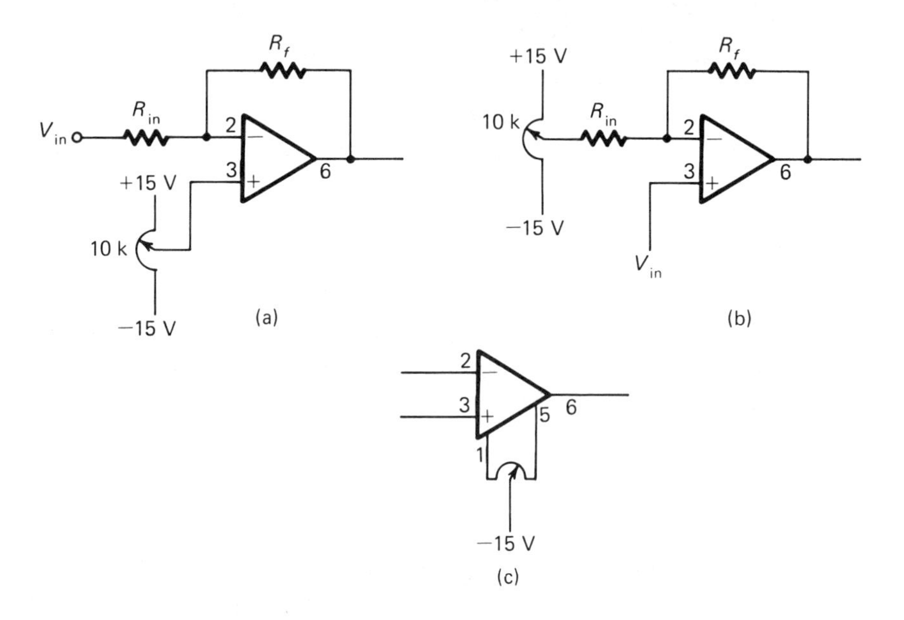

Figure 9-5. Offset compensation techniques.

9-2-3-4 The Voltage Follower

The voltage followers shown in Figure 9-6 are special forms of the noninverting amplifier. Figure 9-6(a) shows a unity gain dc voltage follower. From Equation (9-13) the closed-loop gain was

$$A_{CL} = \frac{R_f}{R_{in}} + 1$$

but $R_f = 0$, therefore $A_{CL} = 1$. This means that, since V_{in} is connected to the noninverting input, V_o will be an exact duplicate of V_{in}. Specially designed op-amps for voltage follower applications such as the LM 102 produced by National Semiconductor Corporation, have very high input resistances and low output resistances, typically 10,000 MΩ and approaching zero ohms, respectively.

The major application of the ac voltage follower shown in Figure 9-6(b) is to provide an impedance buffer between a high-impedance signal source and a low-impedance load. Capacitors $C1$ and $C2$ are effectively shorted at all input frequencies. Resistors $R1$ and $R2$ provide RC coupling and a path for dc input current to the non-inverting input. The closed-loop gain A_{CL} is approximately 1. The op-amp input resistance as seen by the source is $R1/(1 - A_{CL})$, that is, the input resistance is very high.

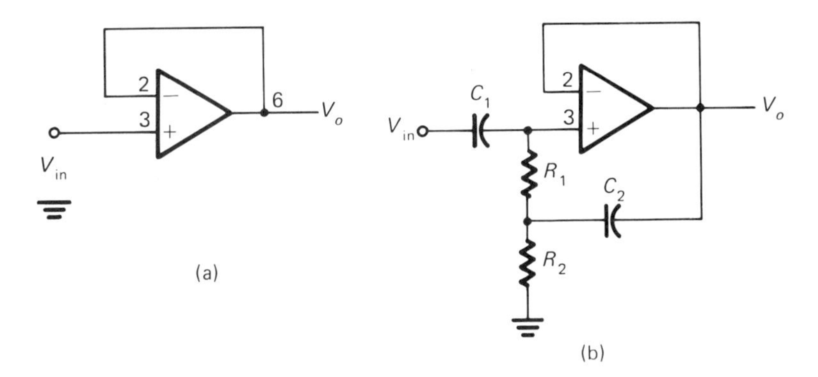

Figure 9-6. Voltage follower: (a) dc unity gain voltage follower, (b) ac voltage follower.

9-2-3-5 The Differential Amplifier Or Difference Amplifier

There are a number of applications where it is necessary to amplify the difference in voltage between two inputs neither of which is grounded. Figure 9-7(a) shows the basic arrangement of the differential amplifier. The closed-loop gain of the circuit is determined by the values of the four resistances. If all resistances are equal in value then

$$V_o = V_2 - V_1 \tag{9-14}$$

and the difference amplifier is acting as a subtractor. However, in most cases the input resistances $R1$ and $R2$ are equal, as are R_f and $R3$. Therefore, with $R1 = R2$ and $R_f = R3$ the output voltage is

$$V_o = \frac{R_f}{R_1}(V_2 - V_1) \tag{9-15}$$

Equation (9-15) tells us that the op-amp has amplified the difference voltage and that the closed-loop gain is solely dependent upon the externally connected resistances.

A word of caution should be injected at this point so that care will be exercised to ensure that the maximum differential input voltages specified in the appropriate data sheet is not exceeded. Failure to follow these maximum values will result in the destruction of the op-amp.

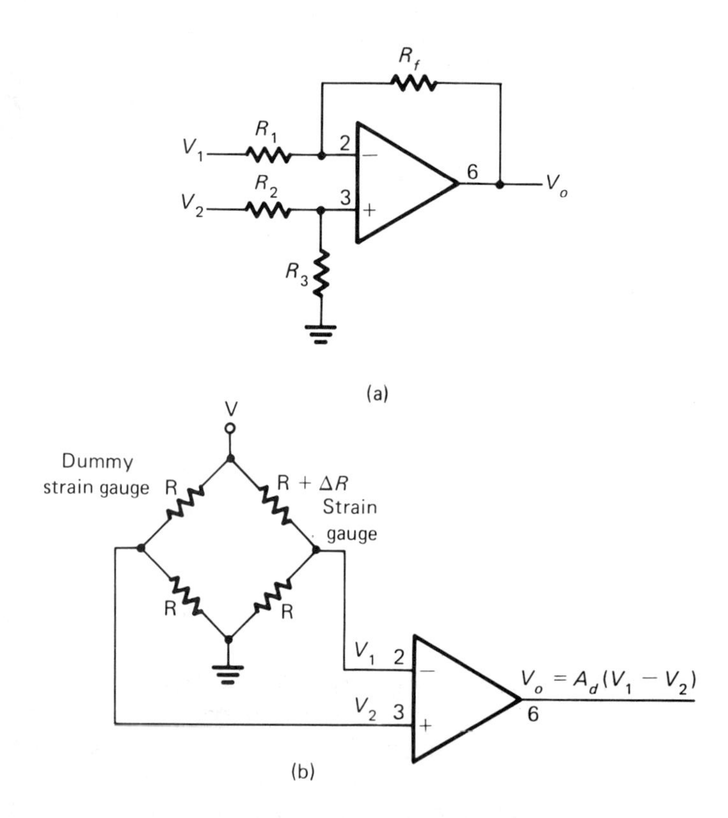

Figure 9-7. Differential amplifier: (a) basic circuit, (b) applied to a strain gauge bridge.

The circuit illustrated in Figure 9-7(b) may be used to amplify the output from a strain gauge bridge. The bridge is arranged so that the four resistances R are all equal in value. Assuming that the strain gauge is under tension, its resistance becomes $R + \Delta R$. The fractional change of resistance $\delta = \Delta R/R$. It can be shown that the output voltage V_o is

$$V_o = -\frac{A_d V}{4} \bullet \frac{\delta}{1 + \delta/2} \tag{9-16}$$

Since the change in the strain gauge resistance is usually very small, δ will be very small and

$$V_o = -\frac{A_d V \delta}{4} \tag{9-17}$$

This basic circuit can also be used to measure other transducer outputs such as thermocouples.

9-2-3-6 Adder or Summing Amplifier

More than one input may be fed into an op-amp and the output V_o will be proportional to the sum of the inputs; this arrangement is known as an **adder** or **summing amplifier** (see Figure 9-8).

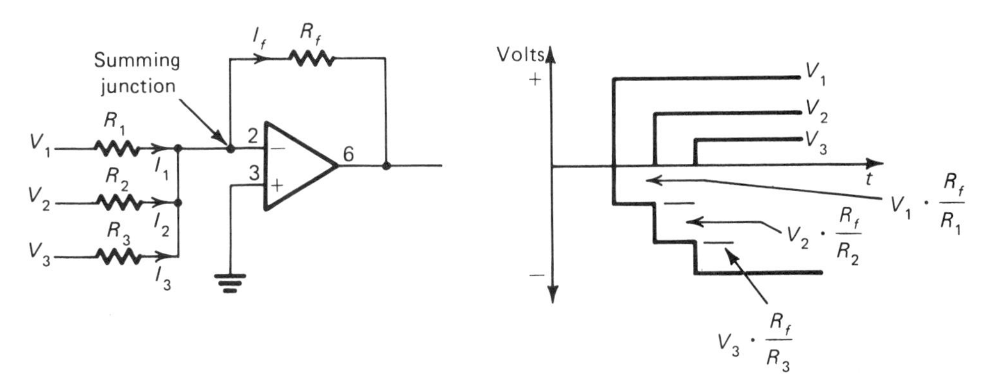

Figure 9-8. Summer or adder.

Consider the currents at the summing junction

$$I_1 + I_2 + I_3 + I_f = 0.$$

Therefore,

$$\frac{V_1}{R_1} + \frac{V_2}{R_2} + \frac{V_3}{R_3} = \frac{V_o}{R_f}$$

and

$$-V_o = V_1\left(\frac{R_f}{R_1}\right) + V_2\left(\frac{R_f}{R_2}\right) + V_3\left(\frac{R_f}{R_3}\right) \tag{9-18}$$

As can be seen, the output V_o is proportional to the sum of the inputs V_1, V_2, and V_3.

Example 9-2

In the circuit of Figure 9-8, $R_1 = 5\ \text{k}\Omega$, $R_2 = 7.5\ \text{k}\Omega$, $R_3 = 10\ \text{k}\Omega$, and $R_f = 20\ \text{k}\Omega$. The input voltages are $V_1 = 1.25$ V, $V_2 = -0.5$ V, and $V_3 = 0.75$ V. Calculate V_o.

Solution:

From Equation (9-18)

$$-V_o = V_1\left(\frac{R_f}{R_1}\right) + V_2\left(\frac{R_f}{R_2}\right) + V_3\left(\frac{R_f}{R_3}\right)$$

$$= 1.25\left(\frac{20\ \text{k}\Omega}{5\ \text{k}\Omega}\right) - 0.5\left(\frac{20\ \text{k}\Omega}{7.5\ \text{k}\Omega}\right) + 0.75\left(\frac{20\ \text{k}\Omega}{10\ \text{k}\Omega}\right)$$

$$= 5.0 - 1.33 + 1.50\ \text{V}$$

$$-V_o = 5.17\ \text{V}$$

$$V_o = -5.17\ \text{V}$$

From this example it can be seen that the weight of the input is determined by R_f/R_n, that is, the smaller the ohmic value of the input resistor the greater the amplification or weighting factor applied to the input voltage signal.

If all the resistances are equal in value then

$$-V_o = V_1 + V_2 + V_3 \tag{9-19}$$

The operation of the summer may be checked by applying the input signals V_1, V_2, and V_3 in turn and checking that $V_o/V_1 = -R_f/R_1$, etc.

A typical application of this circuit is the comparison of the desired speed voltage against the actual speed voltage in a closed-loop motor speed control system. The output V_o would then be the difference or error signal.

9-2-3-7 Voltage-To-Current Converter

The voltage-to-current converter is a specialized application of the inverting amplifier. As can be seen in Figure 9-9, a load resistance R_L is connected in place of the feedback resistor R_f. As a result, the load current I_L is completely independent of the load resistance R_L and the op-amp acts as a current generator. As in the inverting amplifier, the inverting input forms a virtual ground and does not draw any current. The load current I_L as a result is equal to the input current, that is,

$$I_L = I_1 = \frac{-V_{in}}{R_1} \tag{9-20}$$

The load current I_L is independent of the load resistance, and both the voltage source and op-amp must be capable of supplying the necessary current.

9-2-3-8 Current-To-Voltage Converters

Devices such as photocells and photomultipliers are current generators with a very high output impedance and produce an output current which is independent of the load. The current-to-voltage converter or current amplifier is shown in Figure 9-10. Since the inverting input forms a virtual ground there will be no current flow through R_s and nearly all the input current I_s will flow through the

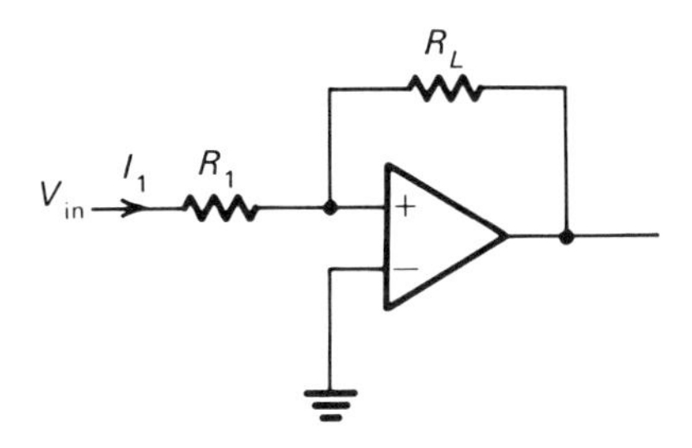

Figure 9-9. Voltage-to-current converter.

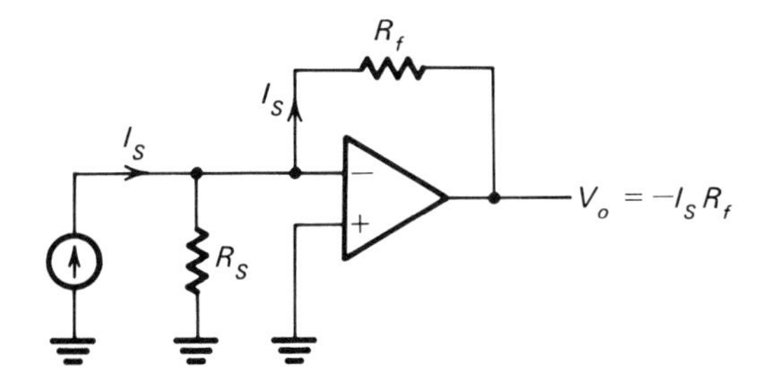

Figure 9-10. Current-to-voltage converter.

feedback resistor R_f. As a result, the output voltage of the op-amp is

$$V_o = -I_s R_f \tag{9-21}$$

The minimum current that can be converted to a voltage is determined by the input bias current I_{ib}. Very often a capacitor is connected in parallel with the feedback resistor R_f to reduce the effect of high-frequency noise.

9-2-3-9 Integrators

The op-amp has been demonstrated in applications such as addition, subtraction, and multiplication. The op-amp can also perform more complex mathematical operations such as integration and differentiation. The basic op-amp integrator configuration is shown in Figure 9-11(a). As can be seen, the resistive feedback element has been replaced by a capacitor. As a result, the output voltage V_o will increase at a rate determined by V_{in}, R_{in}, and C. The inverting input forms a virtual ground. As a result, an input current $I_{in} = V_{in}/R_{in}$ will flow in the capacitor feedback circuit and the capacitor begins to charge. Initially $V_o = 0$ since the feedback impedance at the instant charging commences is zero. As the capacitor charges, a voltage is developed across its terminals which opposes current flow. To maintain the charging current constant the amplifier output voltage V_o must increase at a constant rate until V_{in} is removed or until the op-amp has saturated. The charge applied to the capacitor is

$$q = CV_o = I_{in} \times t = \frac{V_{in}}{R_{in}} \times t \tag{9-22}$$

Therefore,

$$-V_o = \frac{t}{R_{in}C} = V_{in} \tag{9-23}$$

An alternative way of considering the circuit is based upon

$$i_c = C\frac{dV}{dt}$$

but

$$i_c = i_{R_{in}}$$

Therefore,

$$\frac{V_{in}}{R_{in}} = -C\frac{dV_o}{dt} \tag{9-24}$$

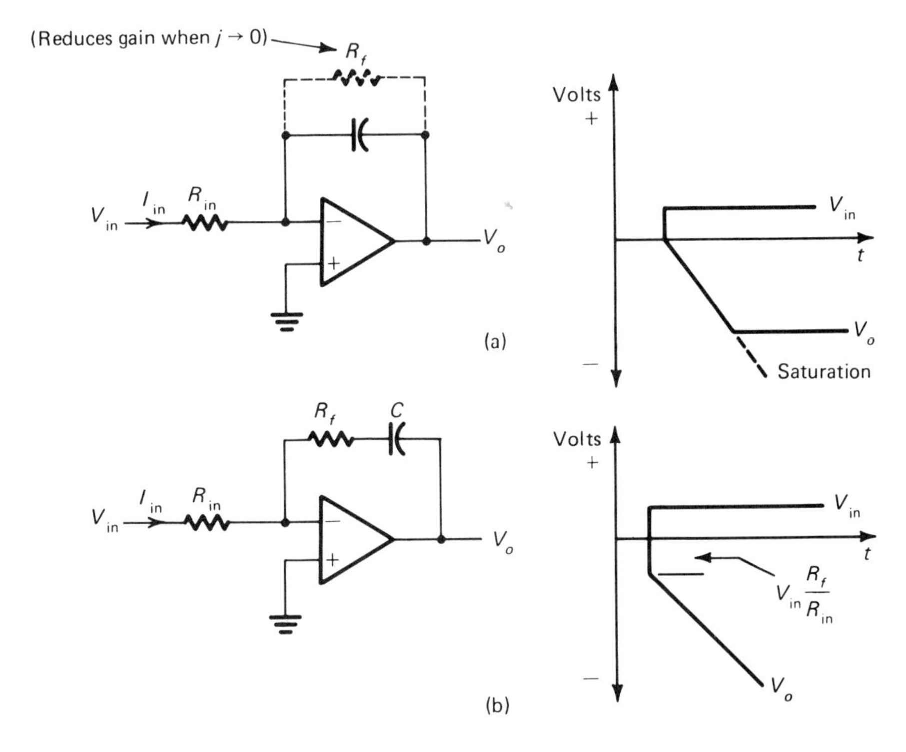

Figure 9-11. Op-amp integrator: (a) basic circuit, (b) integrator with lead time constant.

and

$$V_o = -\frac{1}{R_{in}C} \int V_{in}dt + C \tag{9-25}$$

Equation 9-23 shows that, with a constant positive (negative) V_{in}, the output voltage V_o decreases (increases) linearly with time. Equation 9-25 shows that the integrator produces an output voltage V_o proportional to the integral of V_{in}.

The effects of the frequency on the input signal must also be considered. Recall

$$\frac{V_o}{V_{in}} = \frac{R_f}{R_{in}} = A_{CL}$$

This may then be modified so that

$$\frac{V_o}{V_{in}} = \frac{1/j2\pi fC}{R_{in}} \angle 180°$$

$$A_{CL} = \frac{1}{2\pi fCR_{in}} \angle 90° \qquad (9\text{-}26)$$

Equation 9-26 shows that as the frequency decreases the closed-loop gain increases, or vice versa. When f = 0 Hz, that is, direct current, the gain approaches infinity; in actual practice it approaches the open-loop gain. This creates a problem since any small unbalances in the input differential amplifier or small input currents produce a voltage unbalance which when multiplied by the open-loop gain will drive the op-amp into saturation. This effect may be reduced by connecting a feedback resistor R_f in parallel with the capacitor to reduce the gain when the frequency approaches zero. The value of R_f is selected to reduce the gain to approximately 100.

Example 9-3

The input voltage V_{in} of the integrator shown in Figure 9-11(a) changes from 0 to 1.5 V at time $t = 0$, $R_{in} = 5$ kΩ, and $C = 1\mu$F. Calculate: (1) V_o after 2.0 ms; (2) the rate of change of V_o; and (3) how long after V_{in} changes will the amplifier limit if $V_{CC} = \pm 18$ V and limiting occurs at 80% of V_{CC}.

Solution:

(1) From Equation (9-23)

$$-V_o = \frac{t}{R_{in}C} V_{in}$$

$$= \frac{2.0 \times 10^{-3} \times 1.5}{5 \times 10^3 \times 1 \times 10^{-6}} = 0.60 \text{ V}$$

(2) After 1 ms

$$-V_o = \frac{1.0 \times 10^{-3} \times 1.5}{5 \times 10^3 \times 1 \times 10^{-6}} = 0.30 \text{ V}$$

This means that the output voltage is changing at the rate of -0.20 V/ms or

$$\frac{-0.20}{1 \times 10^{-3}} = -200 \text{ V/sec.}$$

(3) $0.8 \times -18 = -14.40$ V. From part (2) the voltage is changing at -0.20 V/ms, therefore, the time t required for the integrator to limit is

$$t = \frac{-14.40 \text{ V}}{-0.20 \text{ V/ms}} = 72 \text{ ms}$$

The integrator op-amp configuration is widely used in industrial and power electronic circuits as a waveform generator, a ramp generator, a phase shifter, etc.

A deficiency of the op-amp integrator is that the output drifts as a result of offset voltage and input bias current. This effect can be minimized by using a general purpose FET op-amp such as the TL081C or LF355 which have very low input currents and offset voltages.

Another form of the integrator is shown in Figure 9-11(b). This circuit is sometimes called an integrator with a lead time constant and is commonly used to stabilize a control loop. When an input V_{in} is applied, an input current flows and the output voltage V_o will immediately step to $-V_o = V_{in}R_f/R_{in}$. As the capacitor charges, the output voltage will ramp from this value up to saturation.

As can be seen, as long as a fixed V_{in} is applied to an integrating op-amp the output voltage will ramp up into saturation. As a result, it is often desirable to limit the maximum value of V_o. The principles of limiting V_o are illustrated in Figure 9-12. Figure 9-12(a) illustrates a method of providing an adjustable limit on V_o. Assume a constant positive V_{in}, then V_o will ramp negatively. Initially diode $D1$ will be reverse biased by the positive voltage at the wiper arm of poten-

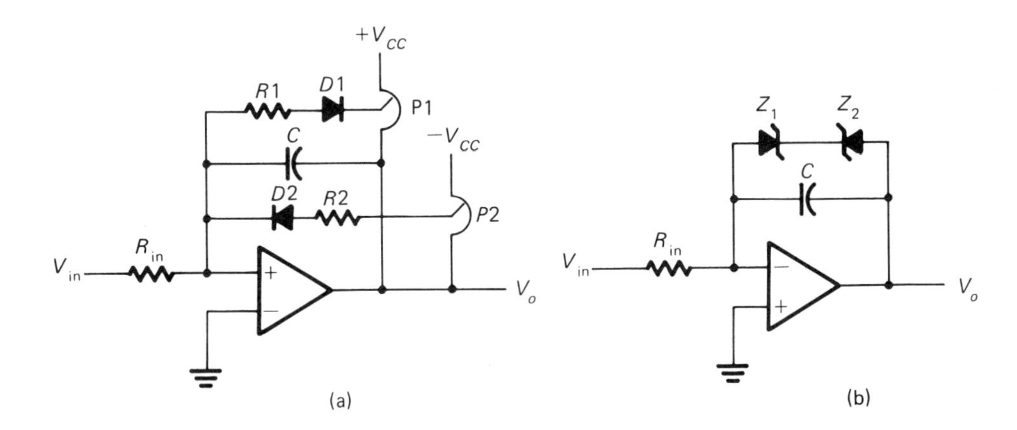

Figure 9-12. Integrator output voltage limiting: (a) variable limit, (b) fixed limiting.

tiometer $P1$. As V_o becomes more negative, the potential at the wiper arm of $P1$ becomes less positive. There will be some value of V_o at which the reverse bias applied to $D1$ is removed. At this value of V_o, $D1$ conducts and forms a low resistance path through $R1$ to the summing junction at the inverting input. As a result, V_o will be maintained at any desired value depending upon the adjustment of $P1$. Similarly $P2$, $D2$, and $R2$ will provide the same control when V_{in} is negative.

If it is not required to have an adjustable limit on V_o, then a fixed limit can be established by connecting a pair of back-to-back zener diodes in parallel with the feedback capacitor [see Figure 9-12(b)]. The zener diodes are selected so that their breakdown voltage is less than the saturation voltage of the integrator.

9-2-3-10 Differentiators

Differentiators are similar to the integrator except that the relative positions of the resistor and capacitor have been reversed (see Figure 9-13). When the input signal V_{in} is constant, current will not flow in the capacitor circuit. As a result, there will not be any current in the feedback circuit and the output voltage V_o will be zero. However, if the input signal voltage V_{in} changes at a uniform rate there will be a steady current flow I_c in the capacitor circuit. It follows that there will be a current flow in the feedback path since $I_f = I_c$ and there will be a constant output voltage V_o.

Since

$$I_c = C\frac{dV_{in}}{dt}$$

and

$$I_f = \frac{V_o}{R_f}$$

Therefore,

$$C\frac{dV_{in}}{dt} = \frac{V_o}{R_f}$$

and

$$-V_o = CR_f\frac{dV_{in}}{dt} \tag{9-27}$$

This means that the output voltage V_o is proportional to the time derivative of the input signal V_{in}.

The effects of frequency upon the gain are best described by

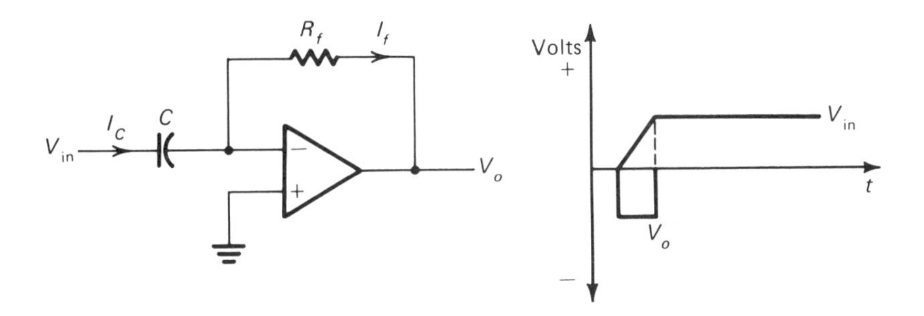

Figure 9-13. Op-amp differentiator.

$$\frac{V_o}{V_{in}} = \frac{R_f}{1/j2\pi fC} \angle 180^\circ$$

$$= 2\pi fCR_f \angle -90^\circ \qquad (9\text{-}28)$$

This means that the gain of a differentiator increases as the frequency increases. This relationship illustrates two major shortcomings of the differentiator; it will amplify the high frequency components of any random noise signals or transient voltage signals.

9-2-3-11 Comparators

In all the applications that have been considered to date, the op-amp has always been used with negative feedback. There is one interesting application of an op-amp used without feedback. The voltage comparator shown in Figure 9-14 is an open-loop op-amp arrangement, and is being used to compare a triangular wave input V_1 against a sinusoidal input V_2. Assuming that the open-loop gain of the op-amp is 15,000 and that the op-amp limits at 80% of $V_{CC} = \pm 15$ V. If V_1 is more positive than V_2, for example 1 mV, then $V_o = -(1\times10^{-3} \times 15\times10^3) = -15$ V. Since the amplifier limits at 80% of 15 V = 12 V, then $V_o = -12$ V. Similarly if V_2 is more positive than V_1 then $V_o = +12$ V. These relationships are shown in the waveforms of Figure 9-14(b) and (c).

9-3 POWER AMPLIFIERS

The major difference in the design of a **power amplifier**, as compared to any other amplifier, is that the emphasis is on the maximum amount of power delivered to the load. The load of a power amplifier is usually a transducer, a servomotor, or a current operated load. As a

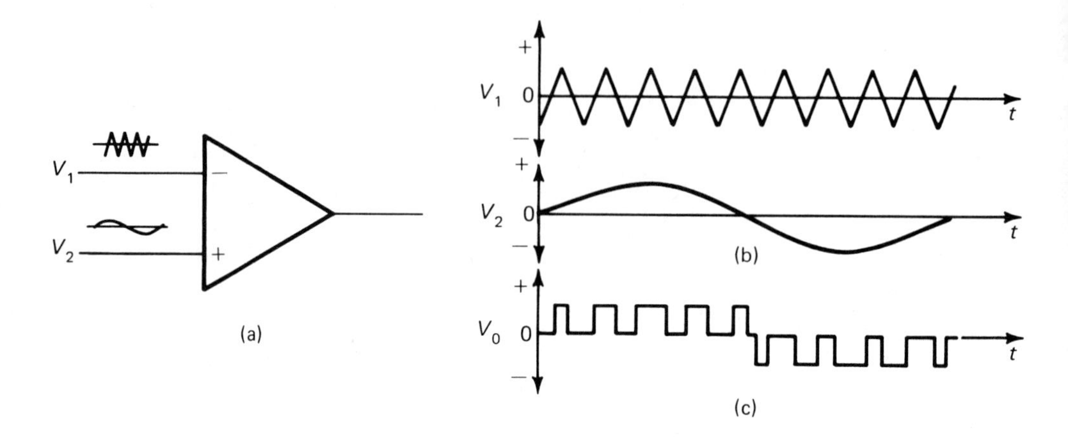

Figure 9-14. Voltage comparator: (a) basic circuit, (b) input waveforms, (c) output waveform.

result, relatively high currents and voltages are common in power amplifier circuits; therefore, power transistors are normally used. Power transistors are physically larger than normal transistors and require special heat dissipation arrangements. Most power transistors have the collector grounded to the case to lower the thermal impedance. Gain and high-frequency characteristics of power transistors are secondary to the need to achieve a high power dissipation capability.

Power amplifiers are classified according to their mode of operation. The main classes are:

1. Class A: The amplifier is biased so that collector current flows for the entire duration of the input signal.

2. Class B: The amplifier is biased so that collector current flows for 180°, or one half-cycle of the input signal.

3. Class C: The amplifier is biased so that collector current will flow for less than 180° of the input signal. Class C amplifiers are mainly used to boost the power level of radio frequency signals in transmitter applications. As a result, we will not discuss their operating characteristics.

9-3-1 Class A Power Amplifiers

In class A power amplifiers, the load is either directly connected into the collector circuit or coupled into the collector circuit by a transformer. Transformer coupling is the preferred method since the dc power loss is minimized because of the low resistance of the primary winding. Also the use of a transformer permits impedance matching and therefore permits maximum power transfer (see Figure 9-15).

It can be shown that the theoretical maximum efficiency of a directly coupled class A single-ended power amplifier is 25%, while in the case of the transformer coupled power amplifier the theoretical maximum efficiency is 50%, although in actual practice 35% would be a more realistic figure. The reason that the transformer coupled amplifier is more efficient is because the power loss in the load in the collector circuit has been eliminated assuming a resistanceless transformer primary winding. However, the collector power dissipation is the same in both connections for the same power output. This means that the collector dissipation rating should be at least twice the projected maximum amplifier power output. In other words, a class A single-ended amplifier with a power output of 1 W would require a power transistor with a maximum collector power dissipation of at least 2 W. In addition, with transformer coupling the collector voltage varies about the operating point. As a result, the collector voltage can be as great as $2V_{CC}$, which means that V_{CC} must not be greater than one-half of the maximum collector emitter breakdown voltage V_{CEO}.

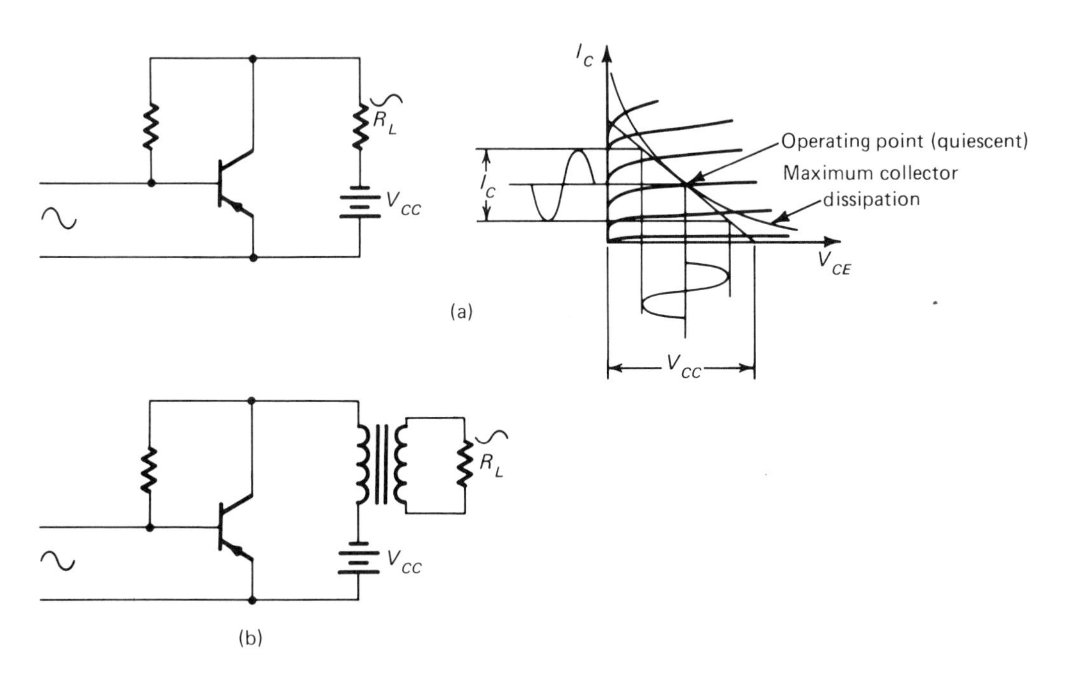

Figure 9-15. Single-ended class A power amplifiers: (a) directly coupled, (b) transformer coupled.

In order to increase the power output capability the push-pull configuration is quite often used [see Figure 9-16(a)]. Compared to the single-ended class A amplifier the major benefits are: (a) there is no dc saturation of the output transformer core and (b) there are no signal frequency currents flowing in the power circuit. The push-pull arrangement requires the use of two matched power transistors.

Transistors $Q1$ and $Q2$ are supplied with base drive signals 180° out of phase with each other, and their outputs are combined by the center-tapped output transformer. It is normal practice to use an input transformer. If an input transformer is not used, RC coupling is used with a phase inversion stage. Apart from increasing the power output capability the push-pull amplifier is not responsive to power supply hum occurring in the amplifier, since the in-phase ripple currents will cancel at the output. It should be noted that ripple frequencies in the input signal will be amplified at the output.

Class A amplifiers provide a linear output but have an inherently low efficiency caused by the quiescent loss and, even when used in the push-pull configuration, will not produce sufficient power output for use as a servo amplifier. For servo amplifier requirements a class B amplifier is usually used.

9-3-2 Class B Power Amplifiers

The conversion efficiency is improved in the class B power amplifier, but it is usually operated in the push-pull configuration to minimize harmonic production. There is one disadvantage of operating a class B amplifier with a zero bias, which is the ideal situation. Operation at this point introduces crossover distortion which occurs at the crossover between the positive and negative half-cycles of the output. This problem is usually remedied by applying a small forward bias to the transistors; the resulting output is then very nearly linear.

It can be proved that the maximum output efficiency is 78%, attained at maximum power output. However, in most applications, the output efficiency is in the region of 60 to 65%.

The class B push-pull amplifier circuit is shown in Figure 9-16(b). Since the transistors $Q1$ and $Q2$ require signals 180° out of phase with each other this can be easily obtained by using an input transformer. An alternative method would be to use the outputs from an emitter-coupled differential amplifier. Collector currents will only flow when the base-emitter junctions of $Q1$ and $Q2$ are forward biased, and since the base signals are 180° out of phase with each other only one transistor at a time will be in operation. The collector currents are directly proportional to the amplitude of the base voltages. As a result, the base power must be relatively large when maximum power output

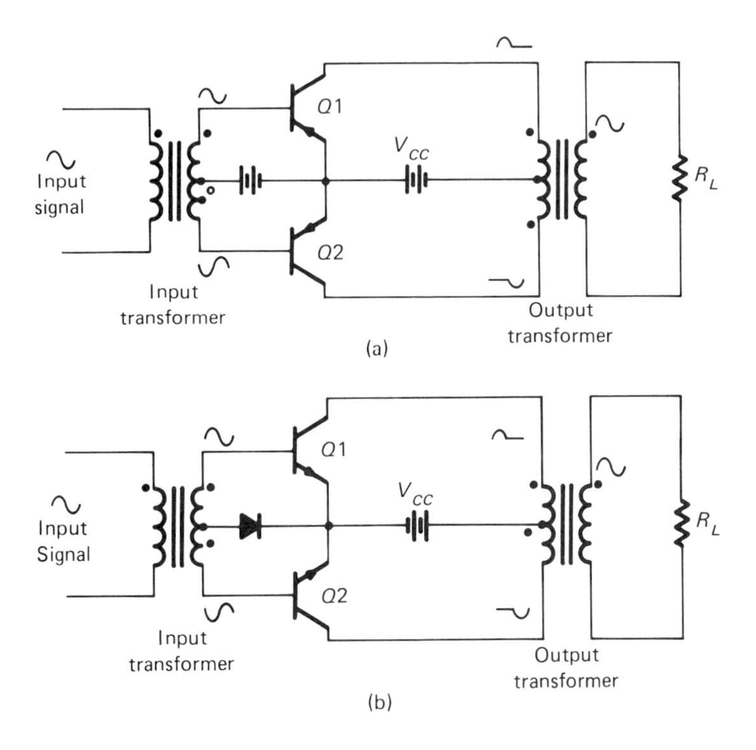

Figure 9-16. Push-pull power amplifier: (a) class *A*, (b) class *B*.

is desired. At the same time, the collector currents will also be large, which in turn demands that the V_{CC} power supply must have a good voltage regulation. However, unlike the class A amplifier, the collector dissipation is low when the input signal is small.

The use of center-tapped transformers is an easy way of producing equal out-of-phase signals; however, they are heavy and expensive, and reduce the frequency response of the amplifiers. The transformers may be eliminated by the use of a pair of complementary matched transistors (see Figure 9-17).

The complementary *p-n-p* ($Q1$) and *n-p-n* ($Q2$) transistors are driven by a common input signal applied to their bases. When the input signal is positive, $Q1$ is cut off and $Q2$ conducts. Point A will be positive with respect to point B. During the negative half-cycle, $Q2$ is cut off and $Q1$ conducts making point B positive with respect to point A. The resulting output waveform seen across the load R_L is an amplified version of the input signal. In order to eliminate crossover distortion and dead zone, which occurs when neither $Q1$ or $Q2$ is conducting, it is usual practice to apply a small amount of forward bias to the transistors.

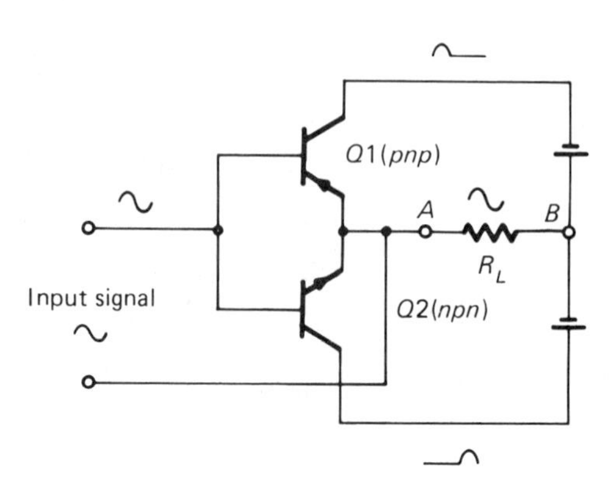

Figure 9-17. Complementary symmetry class *B* power amplifier.

9-4 FEEDBACK AMPLIFIER PRINCIPLES

So far with the exception of the op-amp circuits previously discussed, all the amplifier circuits have been open-loop amplifiers, the term open-loop being used to define an amplifier in which the output is not fed back to modify the amplifier performance. When the output signal fed back is in phase with the input signal this is known as positive or regenerative feedback and is used in oscillator circuits. When the fed back signal is antiphase with the input signal this is known as negative or degenerative feedback. Negative feedback is used to improve the performance of a system by: (a) increasing the bandwidth, (b) improving the stability, (c) reducing distortion, (d) reducing noise, (e) reducing the gain, (f) increasing the input impedance and decreasing the output impedance, and (g) reducing time lags.

There are a number of ways of obtaining the feedback signal. They are:

1. Series feedback where the output signal, or a desired fraction of it, is connected in series with the input signal.
2. Shunt feedback where the input and output signals are converted to current, summed and converted back to a voltage signal and then applied to the amplifier.

Any amplifier can be considered as a four-terminal device consisting of two pairs of terminals, the input and output terminals. In turn the circuits between them can be represented by the Thévenin and Norton equivalents (see Figure 9-18). It should be remembered that in both representations the amplifiers do not produce a voltage or current at the input terminals. Therefore, the only quantity left is the input

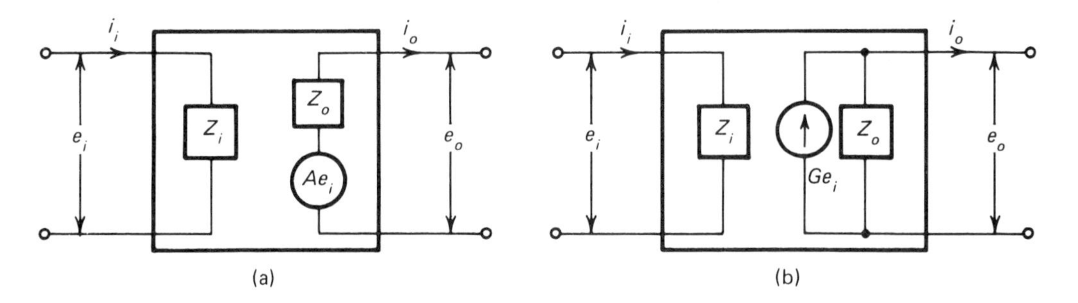

Figure 9-18. Amplifier representation: (a) by Thévenin's theorem, (b) by Norton's theorem.

impedance Z_i. As far as the output terminals are concerned, it is more convenient to use the Thévenin equivalent circuit, see Figure 9-18(a), when considering output voltages; and the Norton equivalent circuit, see Figure 9-18(b), when considering output currents. However, in either case the output impedance Z_o is the same. In Figure 9-18(a) the open circuit output voltage is e_o, the open-loop gain of the amplifier $A = e_o/e_i$, and the voltage generator produces a voltage $Ae_i = e_o$. In Figure 9-18(b) the current generator output is Ge_i, where $G = i_o/e_i$ mhos (siemens).

9-4-1 Voltage Feedback

Figure 9-19(a) shows the application of negative series feedback to an amplifier. The feedback signal is proportional to the output voltage and is subtracted from the input signal by connecting it in series with it. The Thévenin equivalent circuit of the closed-loop amplifier is shown in Figure 9-19(b) where Z_i' is the input impedance, Z_o' is the output impedance, and the no-load gain is A'. The feedback factor β is the ratio of the feedback voltage to the output voltage. The β network is usually a voltage divider and $\beta < 1$. The circuit may be analyzed in the following manner.

From Figure 9-19(a)

$$e_o' = Ae \tag{9-29}$$

and

$$e = e_i - \beta e_o' \tag{9-30}$$

Therefore,

$$e_o' = \frac{Ae_i}{1 + \beta A} \tag{9-31}$$

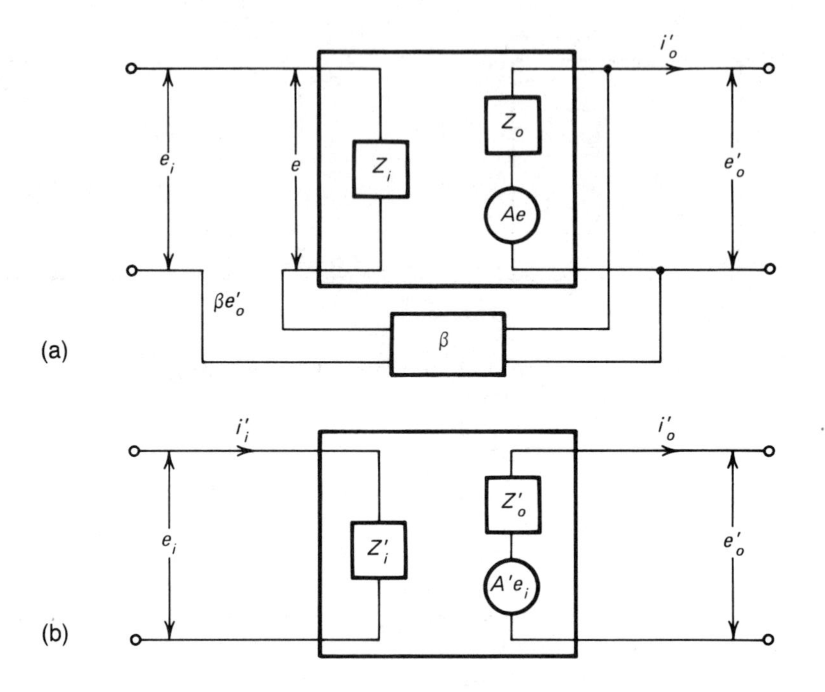

Figure 9-19. Negative series voltage feedback: (a) schematic, (b) Thévenin equivalent.

Since this is the open circuit voltage, the no-load gain A' is

$$A' = \frac{A}{1 + \beta A} \tag{9-32}$$

This means that the effect of the negative series voltage feedback has been to reduce the amplifier gain, since $A' < A$.

The input impedance is

$$Z_i' = \frac{e_i}{i_i'}$$

but

$$i_i' = \frac{e_i - \beta e_o'}{Z_i}$$

$$= \frac{e_i - \beta (Ae_i/(1 + \beta A)}{Z_i}$$

$$= \frac{e_i}{Z_i(1 + \beta A)}$$

Therefore,

$$Z_i' = Z_i(1 + \beta A) \tag{9-33}$$

This shows that the input impedance has been increased.

If we apply a short circuit to the output terminals of Figure 9-19(a) there will be no feedback since the output voltage e_o' is zero, and the input error signal is e_i. Therefore,

$$\text{short circuit current} = Ae_i/Z_o \tag{9-34}$$

Combining Equations (9-31) and (9-34) results in

$$Z_o' = \frac{\text{open circuit voltage}}{\text{short circuit current}}$$

$$= \frac{Z_o}{1 + \beta A} \tag{9-35}$$

which means that the negative feedback has reduced the output impedance by the factor $(1 + \beta A)$.

Example 9-4

A closed-loop amplifier has an open-loop gain $A = 150$, $\beta = 0.2$, $Z_i = 95\,\Omega$, and $Z_o = 10\,\Omega$. Calculate: (1) the no-load gain A'; (2) the input impedance Z_i'; and (3) the output impedance Z_o'.

Solution:

$$1 + \beta A = 1 + 0.2 \times 150$$
$$= 31$$

(1) From Equation (9-32)

$$A' = \frac{A}{1 + \beta A} = \frac{150}{31}$$
$$= 4.84$$

(2) From Equation (9-33)

$$Z_i' = Z_i(1 + \beta A)$$
$$= 95 \times 31$$
$$= 2945\,\Omega$$

(3) From Equation (9-35)

$$Z_o' = \frac{Z_o}{1 + \beta A}$$
$$= \frac{10}{31}$$
$$= 0.32\,\Omega$$

9-4-2 Noise

In addition to the desired output signal, all amplifiers produce spurious voltage signals (noise) that are developed in circuit components, interconnections, etc. The best way to consider the effects of noise is by assuming that the amplifier is perfect and that there is a noise generator e_s in series with the input impedance which will produce an output voltage Ae_s. Then

$$e_o' = A(e + e_s)$$
$$e = e_i - \beta e_o' \qquad (9\text{-}36)$$

Therefore, the output voltage with negative feedback is

$$e_o' = \frac{Ae_i}{1 + \beta A} + \frac{Ae_s}{1 + \beta A}, \qquad (9\text{-}37)$$

as compared to the output voltage e_o without feedback

$$e_o = Ae_i + Ae_s \qquad (9\text{-}38)$$

Comparing Equations (9-37) and (9-38) shows that if we compare the ratio of the two, that is, the signal-to-noise ratio, the effect of the negative feedback is to reduce both the desired output and the noise signal by the same amount. This leads to the conclusion that an improvement in the signal-to-noise ratio can only be achieved by increasing the desired signal without increasing the noise signal.

9-4-3 Loading Effects

Our discussion would not be complete if the effects of load on the amplifier paramenters were not considered. Assuming that a load impedance Z_L is connected to the output terminals of the amplifer shown in Figure 9-19(b), then

$$i_o' = \frac{A'e_i}{Z_o' + Z_L}$$

and since

$$e_o' = i_o'Z_L$$

then

$$e_o' = \frac{A'e_iZ_L}{Z_o' + Z_L}$$

$$= \frac{\dfrac{(Ae_iZ_L)}{(1 + \beta A)}}{\dfrac{Z_o'}{(1 + \beta A)} + Z_L}$$

Therefore,

$$e_o' = \frac{AZ_Le_i}{Z_o' + Z_L(1 + \beta A)} \tag{9-39}$$

The effects on load voltage caused by changes in load impedance can be found by differentiating Equation (9-39) with respect to Z_L, therefore,

$$\frac{de_o'}{dZ_L} = \frac{AZ_o'e_i}{[Z_o' + Z_L\ (1 + \beta A)]^2} \tag{9-40}$$

In the case of the open-loop amplifier the equivalent relationship is

$$e_o = \frac{AZ_LV_L}{(Z_o + Z_L)}$$

Therefore,

$$\frac{de_o}{dZ_L} = \frac{AZ_0e_i}{(Z_o + Z_L)^2} \tag{9-41}$$

Comparing Equations (9-40) and (9-41) shows that the change in output voltage is less when negative feedback is applied to the amplifier than when the amplifier was operated without feedback. In general, the results of loading the amplifier has been a reduction in the output voltage and a reduction in the closed-loop gain A'. Since originally the increase in input impedance was a result of negative series voltage feedback, a decrease in the output voltage will cause a greater input current to be drawn from the source. The result of a greater input current will be a reduction in input impedance, which in the case of a short circuit $Z_L = 0$ will result in $Z_i = Z_i'$; that is, the input impedance of the feedback amplifier will equal the open-loop impedance.

9-4-4 Reduction of Time Lags

In our previous studies, the concept of time constants in *RL* and *RC* circuits was considered. In the case of an *RL* circuit the build-up of current was delayed after the application of a voltage to the circuit. In the case of any device, depending upon the effects of current flow such as relays, dc motors and generators, etc., this delay in current build-up introduces a time delay or time lag in the response of the device to a control signal. Apart from the nuisance value of the time delay, these delays can also result in instability in a control system. In order to avoid a complex mathematical explanation, we will use an illustrative example to show how feedback reduces the time delay effects in a system. Probably the easiest and most readily understandable example would be the case of the control of a separately excited dc generator (see Figure 9-20). Assume that the armature is rotating at constant speed and that all the inductance is concentrated in the field circuit, neglecting armature circuit inductance. Then when a step input signal e_i is initially applied, since the output voltage e_o' is zero, there will not be a feedback voltage, and as a result the full input voltage e_i is applied to the field. Initially the current is zero and then builds up exponentially at a rate determined by the time constant L_f/R_f of the field circuit. Under steady-state conditions the error voltage e is

$$e = e_i - \beta e'_o = e_i - \beta A e \qquad (9\text{-}42)$$

Therefore,

$$e_i = e + \beta A e = e(1+\beta A) \qquad (9\text{-}43)$$

When e_i is applied and $e_o' = 0$, the initial transient error voltage e_t is

$$e_t = e_i = e(1 + \beta A) \qquad (9\text{-}44)$$

The voltage directly applied to the field circuit is $(1 + \beta A)$ times the steady-state error voltage, and the field current will rise $(1 + \beta A)$ times as fast as when the steady-state error voltage is applied. This means that the time constant of the field circuit $\tau = L_f/R_f$ is reduced by the factor $1/(1 + \beta A)$, or in other words the time lag has been reduced.

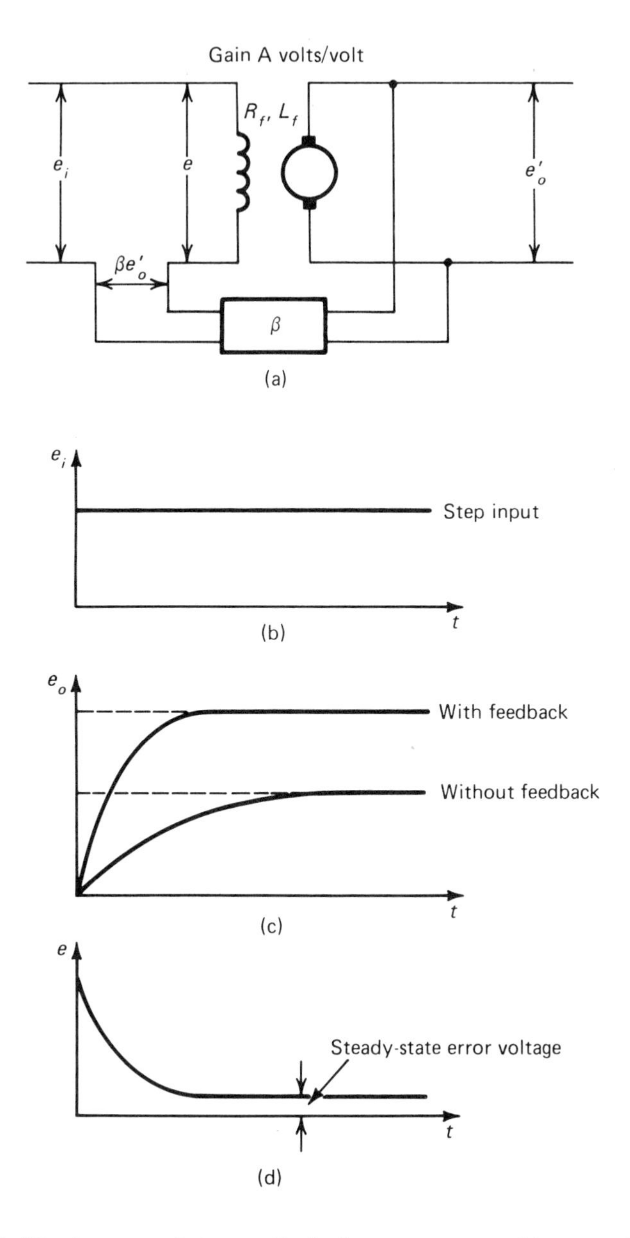

Figure 9-20. A separately excited dc generator with negative series voltage feedback: (a) schematic, (b) step input signal, (c) output voltage response with and without feedback, (d) error signal.

9-4-5 Current Feedback

Figure 9-21 shows the application of negative series current feedback. In this case the feedback signal is derived as a voltage developed across the resistance r which is equal to $i_o'r$ and proportional to the output current.

Under open circuit conditions, there is no current flow through r and the output of the current generator is Ge_i. This current must all flow through Z_o, therefore,

$$\text{open circuit voltage} = Ge_iZ_o \tag{9-45}$$

When the output terminals are short-circuited, assuming that $r << Z_o$, the output current i_o' will all flow through r; therefore,

$$i'_o = Ge \tag{9-46}$$

and

$$e = e_i - i_o'r \tag{9-47}$$

from which

$$i_o' = \frac{Ge_i}{(1 + rG)} \tag{9-48}$$

This is also the short-circuit current in Figure 9-21(b), then

$$G' = \frac{G}{(1 + rG)} \tag{9-49}$$

where G' is the closed-loop gain. As before the gain has been reduced with negative feedback.

If we combine Equations (9-45) and (9-48) we find that the closed-loop output impedance is

$$Z_o' = Z_o(1 + rG) \tag{9-50}$$

which means that the output impedance has been increased and is consistent with the requirement that the output impedance of a current generator must be high.

The no-load input impedance Z_i' is

$$Z_i' = \frac{e_i}{i_i'}$$

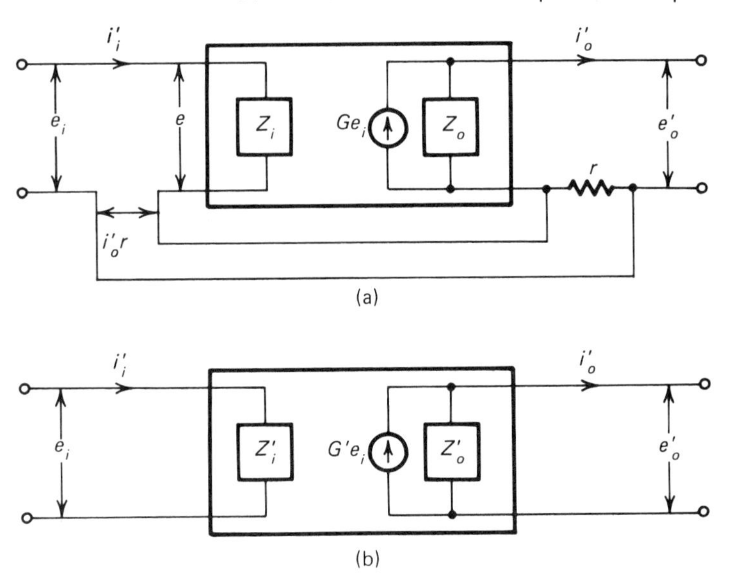

Figure 9-21. Negative series current feedback: (a) schematic, (b) Norton equivalent.

but

$$i'_i = \frac{e}{Z_i} = \frac{e_i - i'_o r}{Z_i} \tag{9-51}$$

$$\frac{i'_o}{e_i} = \frac{G}{(1 + rG)} \tag{9-52}$$

therefore,

$$i'_o = e_i \left(1 - \frac{Gr}{(1 + Gr)}\right) \Big/ Z_i$$

but

$$Z_i' = \frac{e_i}{i'_i} = Z_i\,(1 + rG) \tag{9-53}$$

As was the case with negative series voltage feedback, the input impedance has been increased. This effect can be explained by the fact that the voltage applied to the amplifier has been reduced; consequently, the input current is reduced and the input impedance has effectively been increased.

Under load conditions the output current will change as will the amplifier gain. By the current divider rule, when a load impedance Z_L is connected across the output terminals of Figure 9-21(b),

$$i'_o = Ge \frac{Z_o}{(Z_L + r + Z_o)} \tag{9-54}$$

and since

$$e = e_i - i'_o r$$

then

$$\frac{i'_o(Z_L + r + Z_o)}{GZ_o} = e_i - i'_o r$$

$$e_i = \frac{i'_o\left[Z_L + Z_o + r(1 + GZ_o)\right]}{GZ_o}$$

But the closed-loop gain is

$$G' = \frac{i'_o}{e_i}$$

$$= \frac{GZ_o}{\left[Z_L + Z_o + r(1 + GZ_o)\right]} \tag{9-55}$$

which shows that the effect of loading the amplifier reduces the closed-loop gain. It can also be shown that the input impedance Z'_i also decreases as Z_L increases and will be $Z'_i = (Z_i + r)$ when $Z_L = \infty$, or the amplifier is open circuited.

This review of the principles of negative feedback amplifiers has been designed to show the basic principles and really only applies to steady-state operation. A full analysis under dynamic load conditions is beyond the scope of this text. However, the principles developed will permit us to appreciate the operation of complex systems.

9-5 SERVOMOTORS

There has been a considerable upsurge in the use of **servomotors** as part of electromechanical servomechanisms with the increased usage of industrial robots and computerized numeric control systems in industry.

The two basic servomotors are the two-phase ac squirrel cage induction motor and the dc motor. Initially, the first choice was the ac servomotor because of its reliability, simplicity, and low cost. The dc servomotor was relegated to second choice because of brush problems and radio frequency interference (RFI). However, as a result of considerable research and improved materials, the magnetic circuits have been improved as well as the resolution of the brush problems and a reduction in RFI.

Usually servomechanisms under 100 W are classified as instrument servo systems and it is in this field that the two-phase induction motor finds its main applications. dc servomotors are usually found in applications where higher outputs are required, mainly because dc control systems are simpler and because the modern dc servomotor has a higher efficiency, faster response, and a greater peak power capability.

9-5-1 dc Servomotors

The usual forms of dc servomotors are: (a) wound field types (separately, shunt, series, and compound wound); (b) the moving coil motor; (c) the printed circuit motor; (d) the dc torquer; and (e) the stepper motor.

At this point we will not discuss the details of dc motor speed control. Suffice to say that the steady-state speed of dc motors can be controlled by either varying the voltage applied to the field circuit (speed is inversely proportional to the field pole flux) or by varying the voltage applied to the armature circuit. Nearly all dc servomotors used in closed-loop control systems have their speed controlled by armature voltage control.

9-5-1-1 Separately Excited dc Servomotor

For the separately excited dc servomotor, Figure 9-22(a), the relationships betwen steady-state speed and the applied armature voltage V_a at constant torque are shown in Figure 9-22(b). This curve shows that under conditions of constant torque, the case when the dc servomotor is operated under armature voltage control, the shaft speed is proportional to V_a. In normal operation, a dc servomotor does not operate against a constant torque load. The curves shown in Figure 9-22(c) illustrate the actual operating conditions. As can be seen, the greatest torque output occurs when the motor is at standstill or at low operating speeds. This high torque output will rapidly accelerate the motor and, as a result, rapidly positions the connected servomechanism. As the actual position of the servomechanism approaches the desired position, the error signal decreases, as does the armature voltage V_a.

However, the inertia of the motor and the connected load will cause the servomechanism to overshoot the desired position. At this point the servomotor acts as a generator and produces a counter or reverse torque which damps out the oscillations.

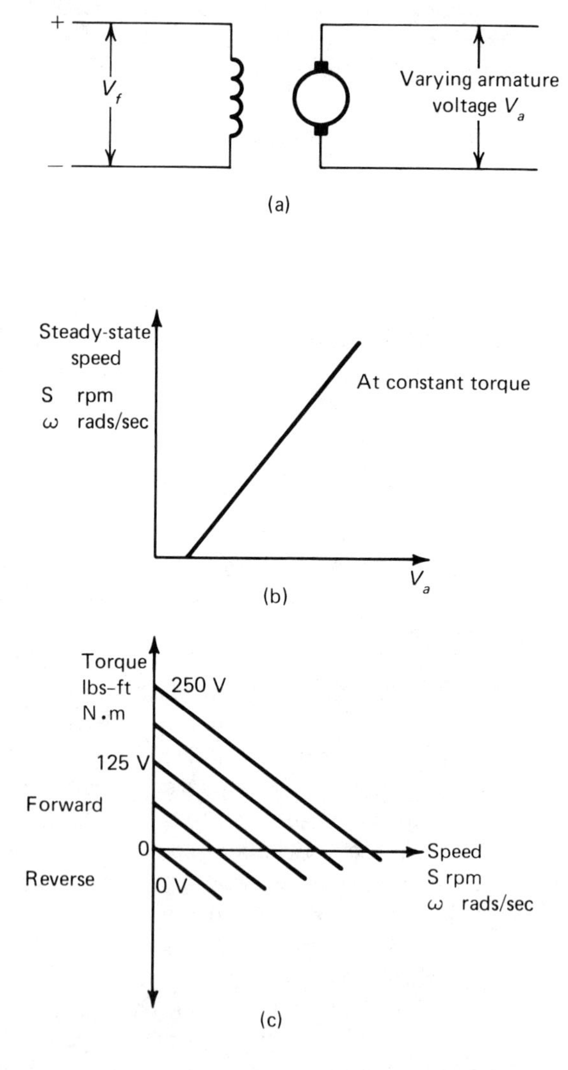

Figure 9-22. Separately excited dc servomotor: (a) schematic, (b) speed versus armature voltage at constant torque, (c) torque versus speed for varying armature voltages.

9-5-1-2 Moving-Coil Servomotors

Modern servomotors use permanent-magnet field structures. These magnet structures can be molded into any desired shape and are usually either alnico or ceramic magnets, although there is a third family of magnetic materials known as rare-earth magnets. The most common member of this family is the samarium-cobalt magnet. However, it is very expensive. A slightly cheaper and lower performance material is the misch metal magnet. The major benefits derived by using permanent-magnet field pole structures are: (a) the motor is designed to optimize the magnetic field, (b) the motor is more efficient since there are no field copper losses, (c) more responsive since there is no field time constant, and (d) is cheaper. The disadvantages are: (a) cannot be speed controlled by using field control methods, and (b) may exhibit commutation problems since there are no interpoles.

The requirements of high performance systems, such as magnetic tape drives where very high acceleration and deceleration rates are required, can only be met by using low inertia ironless armatures. The armature contains no iron and consists of the armature windings forming a cylindrical shell which is reinforced with glass tape and the whole structure epoxyed. Since the resulting structure is small in diameter, the mechanical inertia is minimized; also, because of the lack of iron, the armature inductance is minimal. As a result, the armature circuit is extremely responsive to armature voltage control signals giving the moving-coil motor the highest acceleration capability of any dc motor with accelerations in excess of 700,000 rad/sec^2; yet having the capability of completing several thousand stop-start cycles per second. This latter feature also permits their use in incremental motion as well as continuous motion applications.

A major disadvantage of moving-coil motors is their low thermal time constant and low thermal capacity, that is, they will heat up rapidly in fast response applications. Normally these drives are driven by transistorized current source amplifiers using pulse width modulation techniques.

9-5-1-3 Printed Circuit Motors

The printed circuit motor is effectively a disc moving-coil type motor characterized by a relatively large diameter but a very short axial length. The armature and commutator are printed on a thin copper disc. This arrangement results in a very low armature inductance with a low mechanical time constant, low inertia, and good heat dissipation. Very often the moving-coil and printed circuit motors are also supplied with integral analog tachometers or optical incremental encoders, and/or electromagnetic brakes.

9-5-1-4 Direct-Drive Torque Motors or Torquers

The direct-drive dc torque motor is a servo actuator that is directly connected to the load it drives. They are particularly suited to servo system applications where size, power, weight, and response times must be minimized and yet accurate positional and rate accuracies must be maintained. They normally have a wound armature and a permanent-magnet field structure and are usually frameless. They are relatively thin compared to the diameter and quite often have relatively large shaft holes in the armature for easy connection to bosses, hubs, and shafts.

By being directly connected to the load they eliminate the need for gearing and as a result eliminate backlash, therefore ensuring better resolution.

9-5-2 ac or Two-Phase Servomotors

Two-phase ac servomotors have a high-resistance low-inertia rotor. Although the high resistance rotor is essential for linear performance, it contributes to low efficiency which is not compatible with the normal requirement of a high output. The basic winding connections, performance, and response curves are shown in Figure 9-23. The basic principle of operation is the same as any polyphase induction motor, that is, a rotating magnetic field is created by the two-phase stator winding, Figure 9-23(a); or if only a single-phase supply is available, connecting a capacitor in series with the reference or fixed stator winding will produce the desired 90° phase shift, see Figure 9-23(b). The rotating magnetic field induces rotor emfs and currents in the rotor circuit, and the interaction between the rotating magnetic field and the rotor field produces rotation.

The two-phase ac servomotor differs from the conventional polyphase induction motor because the magnitude of the rotating field, and thus the torque and speed, is controlled by the magnitude of the control winding voltage. The direction of rotation is determined by the phase relationship between the fixed and control winding voltages. Figure 9-23(c) shows the relationship between speed and torque for varying values of control phase voltages. It should be noted that the greater the negative slope of these curves the greater the inherent damping capability of the servomotor. Usually the control phase winding is connected to the output of a servoamplifier. An alternative method of supplying the two-phase power is by means of an SCR amplifier connected in each phase so that the power requirement at zero speed is approximately zero. The stall torque varies as the product of the fixed and control phase voltages, which means that the torque versus error curve tends to be a square law curve.

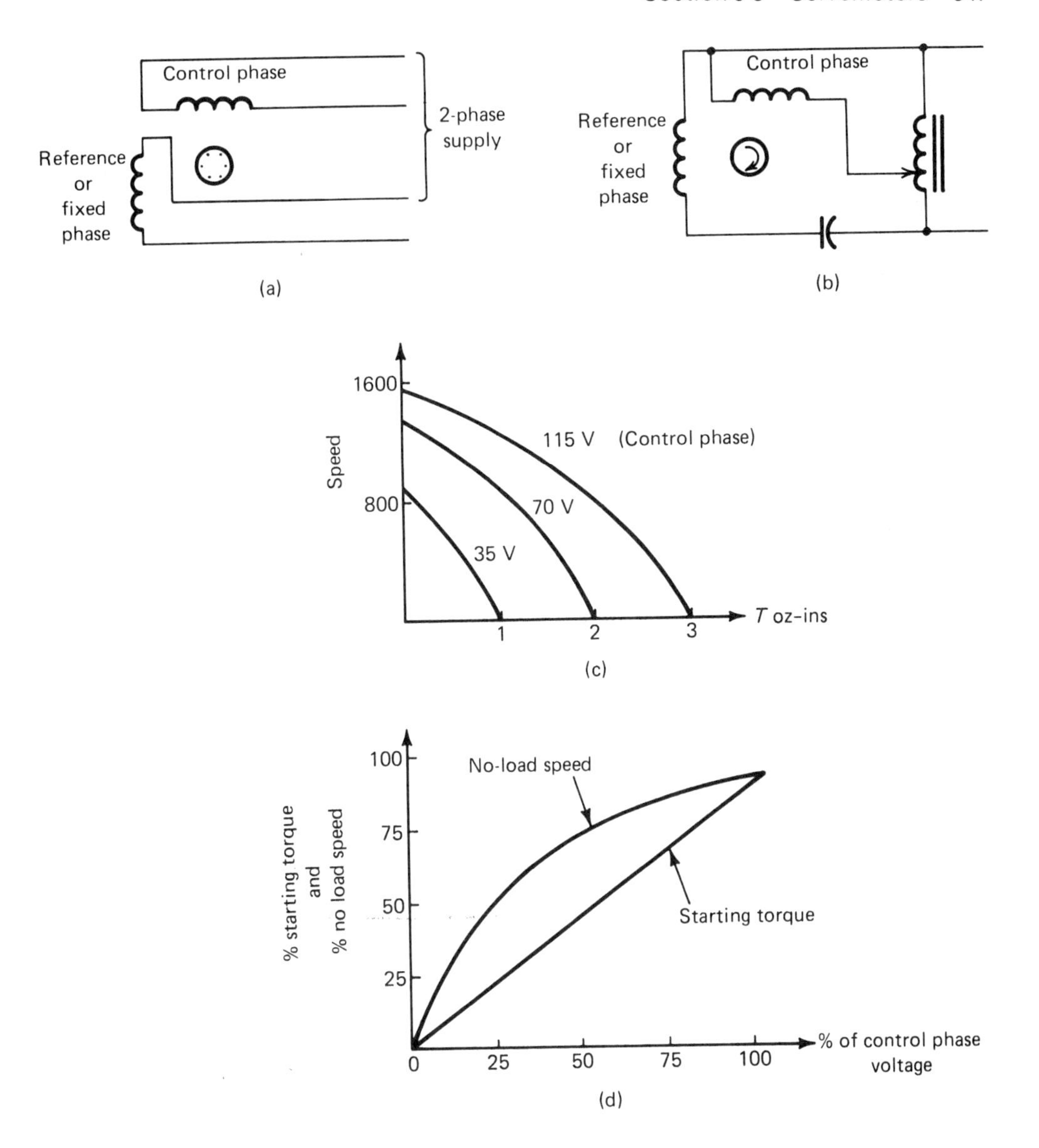

Figure 9-23. Two-phase ac servomotor: (a) schematic, (b) schematic with single-phase supply, (c) speed versus torque, (d) response curves.

9-5-3 Stepper Motors

Stepper motors are dc excited devices in which a pulsed dc supply is switched to each winding in turn so as to cause the rotor to turn in a series of equal steps. There is a one-to-one correspondence between the input pulses and the angular steps. The direction of rotation is determined by applying the pulses to either the clockwise or counterclockwise drive circuits. The rotational speed is determined by the frequency of the applied input pulses.

Stepper motors are being used in industrial applications such as machine tool, computerized numerical control, and process control drives. They are also very extensively used in computer systems for tape and printer drives.

The main advantages of using stepper motors are: (a) easily interfaced with minicomputers and microprocessors; (b) fast response, up to 10,000 full steps (1.8°) or 20,000 half steps (0.9°) per second; and (c) may be used in open-loop or closed-loop systems.

The two most common stepper motors are the variable reluctance and permanent-magnet types.

9-5-3-1 Variable Reluctance Stepper Motors

The basic principle of operation for a variable reluctance stepper motor is illustrated in Figure 9-24. The rotor is a soft iron-toothed rotor. When the stator coils are energized, the rotor teeth will align with the energized stator poles; that is, they move to a position of minimum reluctance. In Figure 9-24 each of the three-phase windings (1, 2, and 3) are wound on stator poles that are located 90° apart around the stator. For example, if phase 1 is energized, four north poles are produced at 90° intervals around the stator. Rotor teeth marked **a** will align themselves with these poles. The nonexcited poles between the energized poles will provide the return magnetic circuit, that is, they will be induced south poles. If we next energize the phase 2 winding, the rotor teeth marked **b** will move counterclockwise (CCW) to align themselves with the energized stator poles. Similarly, if we energize the phase 3 winding, the rotor teeth marked **a** will align themselves with the energized poles. In other words, for each pulse applied to the stator windings in a 1, 2, and 3 sequence the rotor turns one step CCW. Similarly, if the pulses were applied in a 1, 3, and 2 sequence to the stator windings, the rotor turns one step in a clockwise direction.

The relationship between the number of step sequences (SS) and the number of rotor teeth (T_r) gives the steps per revolution (SPR); this relationship is

$$\mathrm{SPR} = (\mathrm{SS})\,T_r. \tag{9-56}$$

In our example SS = 3 and $T_r = 8$; therefore SPR = $3 \times 8 = 24$ SPR. That is, the step angle is 15°.

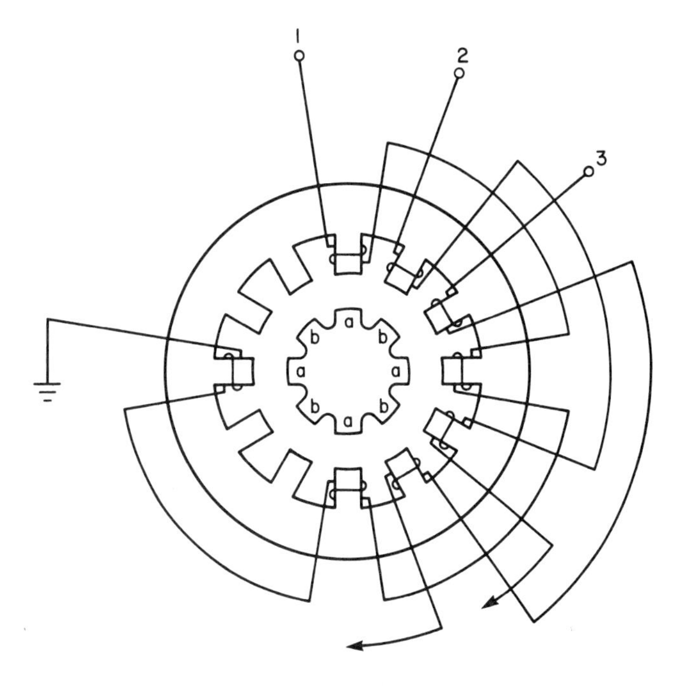

Figure 9-24. A three-winding variable reluctance stepper motor.

9-5-3-2 Permanent-Magnet Stepper Motors

Permanent-magnet stepper motors usually have a polyphase wound stator and a permanent-magnet rotor. The permanent-magnet stepper motor shown in Figure 9-25, represents a particular design commonly used in digital drive applications where small step angles, for example 1.8°, are required. Both the stator field pole faces and the rotor periphery are toothed. The teeth on the rotor surface and the stator pole faces are offset so that there will only be a limited number of rotor teeth aligning themselves with an energized stator pole. Applying the right hand grasp rule, it can be seen that when currents I_1 and I_3 flow as shown the associated stator poles will become N poles. Similarly, when I_2 and I_4 are energized, the stator poles become S poles. The windings are arranged on the stator poles so that when the coil currents are applied in the sequence I_1, I_2, I_3, and I_4 the rotor will turn clockwise at a stepping rate determined by the rate of applying the current pulses. If the sequence of pulses was changed to I_1, I_3, I_2, and I_4 the rotor would turn CCW. This particular type of construction is designed so that very small stepping angles are obtained. Because of the permanent-magnet rotor and the tooth arrangements there will be a holding torque when the windings are deenergized.

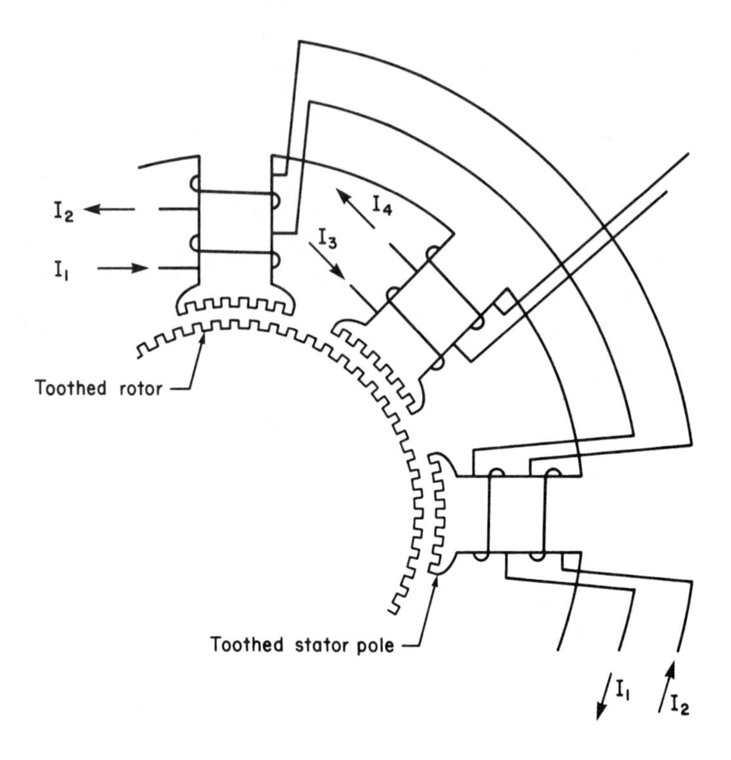

Figure 9-25. A four-winding permanent-magnet stepper motor.

The step angle is determined by the tooth arrangements on the rotor and stator. The magnitude of the input pulse signals will have no effect on the step angle or the rate of rotor movement. However, if the stator current increases the torque produced will increase. The angular velocity of the rotor is solely determined by the step angle and the number of pulses per second. Therefore the angular velocity is

$$\omega = \frac{2\pi\,(\mathrm{PPS})\,(\mathrm{SA})}{360}\ \text{rads/sec} \tag{9-57}$$

$$= \frac{2\pi\,(\mathrm{PPS})}{(\mathrm{SPR})}\ \text{rads/sec} \tag{9-58}$$

where ω = angular velocity rads/sec

PPS = pulses per second

SA = step angle (degrees), and

SPR = steps per revolution

9-5-3-3 Stepper Motor Characteristics

When a stepper motor has a steady dc signal applied to one winding, the rotor will be lined up with that stator field. If the shaft is turned by an external force, a restoring torque will be developed. The maximum value of this torque will occur when the rotor has been offset ± 1 step angle. This maximum value is called the holding torque and is usually measured in ounce-inches (oz-ins) and will usually be specified in the manufacturer's data sheet. An important feature of a stepper motor is that it can maintain the holding torque indefinitely when the rotation of the rotor is stopped, for example when used in a universal milling machine and the table is stalled against the stop. Obviously, the running torque or pull-out torque must be less than the holding torque or the rotor will not turn. The pull-out torque is the true indication of the torque output capability of the motor; this torque also varies with the stepping rate.

Under dynamic conditions the rotor is subject to oscillation in response to an input pulse. The amplitude and duration of the oscillation is determined by: (a) the load torque T_L, (b) by the load inertia, and (c) the current rise time in the stator winding. The oscillation will be at its greatest under light load conditions and decreases as the load increases. As would be expected, the amplitude and duration of oscillation increases with increasing load inertia. The effects of load inertia are especially noticeable at start and stop. The inductance of the stator windings delays the build-up of current and the build-up of the magnetic field. Since torque is the interaction of two magnetic fields, there will be a delay in torque build-up. This delay in torque build-up will also slow down the response of the motor to an input pulse.

Normally the current in an inductive circuit is assumed to have attained a steady-state value in five time constants; therefore,

$$\text{current rise time} = 5T \tag{9-59}$$

where

$$T = L/R \text{ sec,} \tag{9-60}$$

L = stator inductance per phase (henrys), and

R = stator resistance per phase (ohms)

The time constant can be decreased by increasing the denominator, that is, by adding extra resistance (R_s) in series with the motor winding then

$$\text{current rise time} = \frac{5L}{R + R_s} \tag{9-61}$$

To maintain the current at the rated value, the applied voltage must also be increased.

There is a maximum response rate which is the maximum input pulse rate that an unloaded motor can respond to from a standing start. As the load inertia increases, the maximum response rate decreases. However, for a given motor, it can repeatedly be stopped and started provided the maximum input pulse rate is not exceeded. However, once the motor is running, the maximum response rate can be exceeded and the motor is then operating in the slew range, where it will stay in synchronism with the input pulses up to the maximum designed input pulse rate.

To operate the motor in the slew range from standstill it is necessary to apply input pulses to bring the motor up to a stepping rate in the start-stop range, for example, a velocity ω, as shown in Figure 9-26. At this point the input pulse rate is ramped up to the desired stepping rate in the slew range. The rotor will accelerate up to this stepping rate provided the motor is developing sufficient torque to overcome the load torque and inertia and the motor inertia. The motor is stepped at the desired rate until it is necessary to stop it. At this point the motor is ramped down to the start-stop range and then stopped.

In order to drive the stepper motor it is necesary to have a drive control system which can: (a) determine the required direction of rotation, (b) correctly sequence the windings, (c) have sufficient power capacity to accelerate the motor under load conditions, and (d) provide rapid switching control of each phase winding.

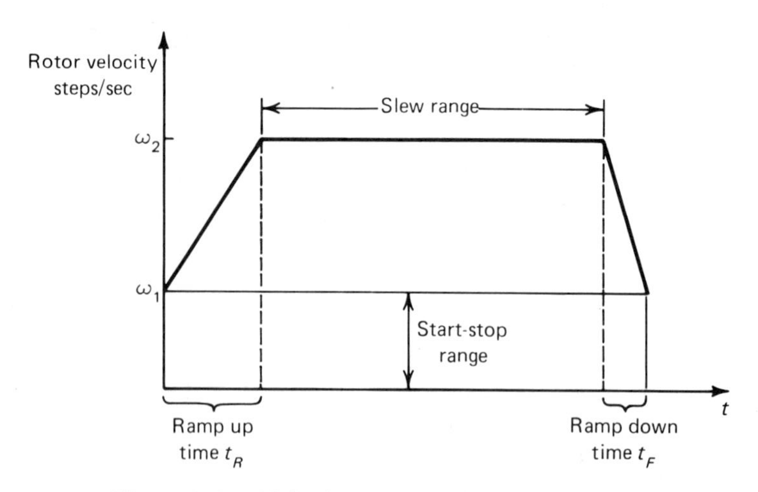

Figure 9-26. Velocity control of a stepper motor.

First of all, in looking at the excitation requirements, we assume that we are controlling the permanent-magnet motor shown in Figure 9-25. In order to have a clockwise rotation, it is necessary to energize the windings in the sequence 1, 3, 2, and 4. Figure 9-27 shows the winding connections and required energization sequence for a single-phase four winding sequence. The stepper motor windings are each energized once in every four input pulses, that is, the pulse train for each winding has a 25% duty cycle.

The control logic must perform two functions: (a) control direction of rotation and (b) control the stepping rate. Figure 9-28 shows a control logic scheme using ANDs, NORs, inverters, and J-K FFs, with a clock input from an external microprocessor, minicomputer, programmable controller, or any other digital source. The basic sequence of operation is as follows:

1. Assuming that a clockwise rotation is desired, a high (1) is applied to the directional inverter.
2. Flip-flop 1 (FF1) has been preset so that $J = 1$, $K = 0$; and the outputs $Q = 0$ and $\overline{Q} = 1$.
3. Flip-flop 2 (FF2) has also been preset so that the inputs are $J = 0$ and $K = 1$; and the resulting outputs are $Q = 0$ and $\overline{Q} = 1$.
4. Both inputs to AND gate 1 are high and the output is high, thus energizing phase 1 of the motor.
5. On the negative going transition of the first input clock pulse,

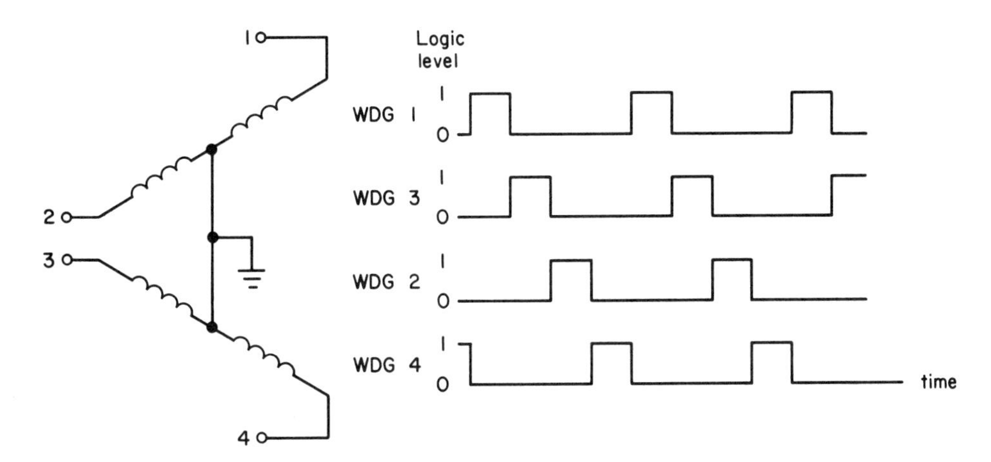

Figure 9-27. A single-phase four-winding stepper motor: (a) winding schematic, (b) pulse sequence.

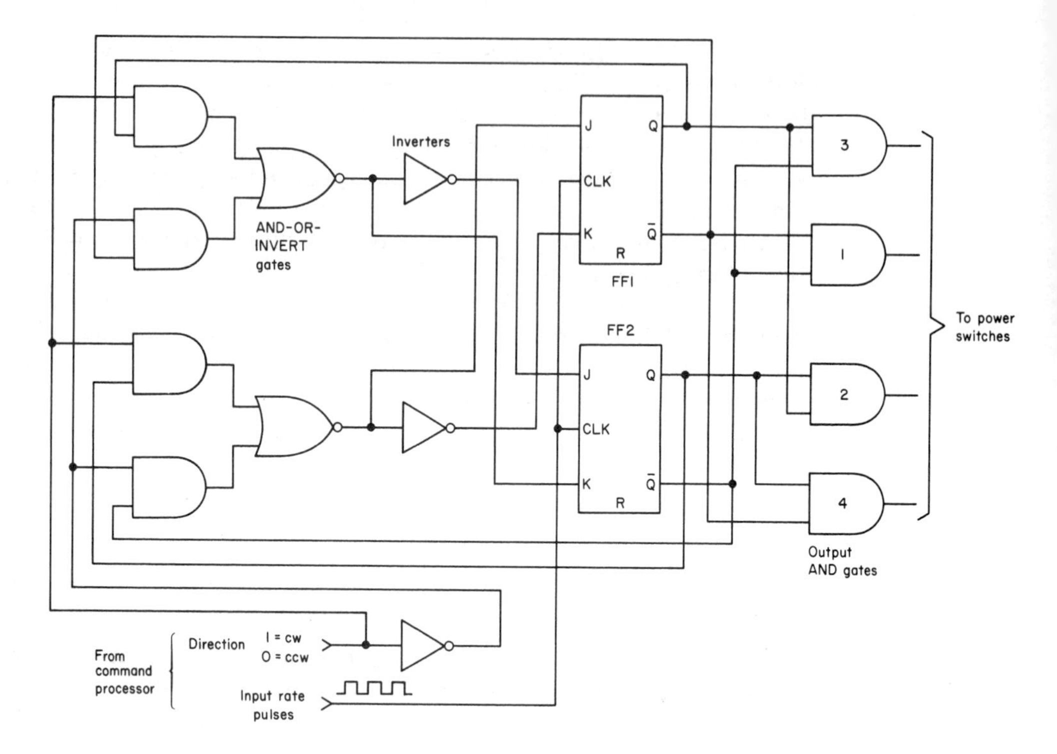

Figure 9-28. A single-phase sequencing logic control.

FF1 toggles and $Q = 1$ and $\overline{Q} = 0$; FF2 will not be affected since it has been reset, then $Q = 0$ and $\overline{Q} = 1$.

6. Both inputs to AND gate 3 are high and the output is high, therefore, phase 3 of the motor is energized. FF1 will remain set, that is, $J = 1$ and $K = 0$; however, FF2 will set, that is, $J = 1$ and $K = 0$.

7. The negative going transition of the second input clock pulse will leave the output state of FF1 unchanged, but FF2 will toggle so that $Q = 1$ and $\overline{Q} = 0$.

8. This makes both inputs to AND gate 2 high and phase 2 of the motor will be energized.

9. The negative going transition of the third input clock pulse will cause FF1 to reset, that is, $J = 0$ and $K = 1$; $Q = 0$ and $\overline{Q} = 1$. Also FF2 will remain unchanged.

10. Both inputs to AND gate 4 will be high and phase 4 of the motor will be energized.

11. The negative going transition of the fourth input clock pulse will cause the sequence to repeat. If a low (0) is applied to the directional

inverter, a 4, 3, 2, and 1 sequence of outputs results and the motor reverses.

To meet industrial requirements one of two methods of control are normally used: (a) the translator method, where the translator changes pulses from a minicomputer, microprocessor, programmable controller or, some other source of digital pulses defining step rates, direction, and even number of steps into the sequencing and switching logic for bidirectional control of the stepper motor; and (b) the preset indexer method which provides bidirectional control; however, the direction, stepping rate, and number of steps are usually set in manually by an operator.

Philips has developed the SAA1027 IC driver which is used to control four-phase stepper motors without the need of discrete power stages. This chip has the capability of driving stepper motors which draw less than 350 mA per phase with two phases on from a 12 V dc power supply.

9-6 THE AMPLIDYNE

The normal dc generator is a single-stage rotating amplifier. For the particular case of the separately excited generator, there are two amplification factors: the voltage amplification factor (VAF) and the power amplification factor (PAF) defined as follows:

$$\text{VAF} = \frac{\text{open circuit armature voltage}}{\text{field voltage}} = \frac{V_{a_{oc}}}{V_f} \tag{9-62}$$

and

$$\text{PAF} = \frac{\text{armature power}}{\text{field power}} = \frac{V_a I_a}{V_f I_f} \tag{9-63}$$

If two separately excited dc generators are connected in cascade as shown in Figure 9-29, they form a two-stage power amplifier whose amplification factor is the product of the factors of each stage. This arrangement has a number of disadvantages of which the most important is the fact that there are two separate magnetic circuits.

The amplidyne developed by General Electric uses the quadrature axis armature reaction flux as the separately excited field of the second stage, and utilizes one magnetic circuit for both stages (see Figure 9-30). The control field establishes a flux Φ_f along the direct axis and produces an emf V_{a1} between the short-circuited brushes on the quadrature axis. The resulting short-circuited current I_{a1} produces a quadrature axis armature reaction flux Φ_q. The rotating armature conductors cut this flux and produce the second-stage armature voltage V_{a2} at the brushes mounted on the direct axis. Assuming that the output is connected to an external load, the resulting second stage armature reaction flux Φ_d produced by I_{a2} will act along the direct axis and directly oppose the control field flux Φ_f, causing the airgap flux to collapse. The second-stage flux Φ_d must be eliminated by an equal but opposite flux. This flux Φ_c is produced by the series-connected compensating field which, since it is carrying the load current, will always be equal and opposite to Φ_d.

To further improve performance as well as the power amplification factor, the amplidyne also usually has commutating poles in series with the load current and a series-excited field connected in series with the path between the short-circuited brushes. The amplidyne is manufactured with a polyphase squirrel cage induction motor driving the armature, the whole being integrated into one package. Power amplification factors of 40,000 can be obtained with output power capacities up to 250 kW. The amplidyne is principally used as a high-gain power amplifier and provides high accuracy control. Their major applications are in the areas of voltage regulation and motor speed and position control.

In Figure 9-30 it should be noted that the main field poles contain a number of separate control windings which are used to provide feedback, biasing control, etc. The effect of these independently connected windings is to produce a resultant air gap flux which controls the output of the amplidyne.

Figure 9-31 illustrates the use of voltage and current feedback techniques as applied to the amplidyne. Figure 9-31(a) shows the basic connection for negative voltage feedback. A high resistance potentiometer is connected across the output and some fraction of the output voltage βV is fed back to the control winding in opposition to the reference voltage V_1. As a result, the control field voltage $V_f = V_1 - \beta V$.

Figure 9-31(b) shows a similar arrangement using current feedback. A low resistance R is connected in series with the output. The voltage developed across this resistance under load conditions is IR and opposes the reference voltage V_1. The voltage across the control winding is $V_f = V_1 - IR$.

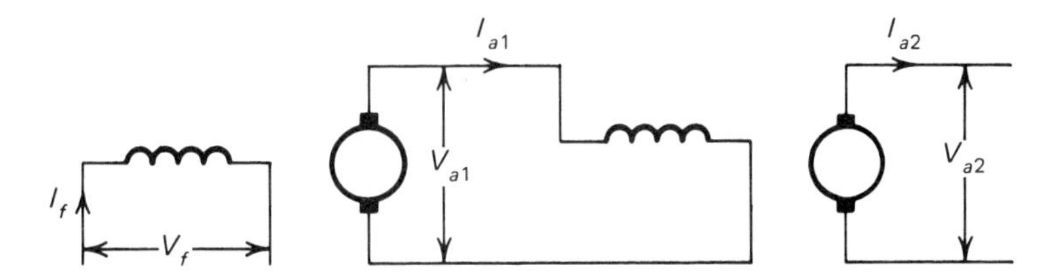

Figure 9-29. Cascaded separately excited dc generators.

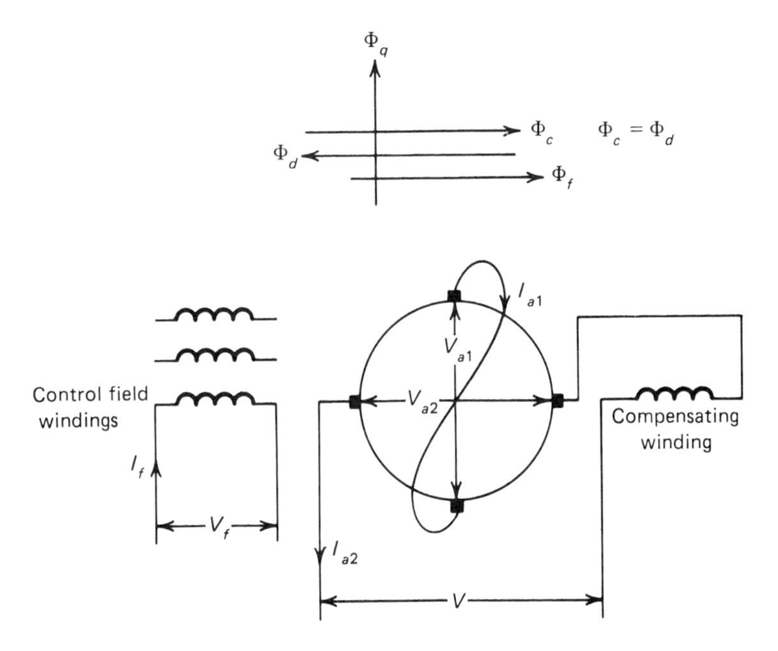

Figure 9-30. Basic principle of the amplidyne.

The amplidyne, although in common use in industry, is being replaced in newer installations by the thyristor phase-controlled converter, which has the advantage of lower maintenance costs since there are no rotating parts and increased efficiency. However, the amplidyne is still performing well in many applications in the steel and paper-making industries, electrical power generation, and electric transportation.

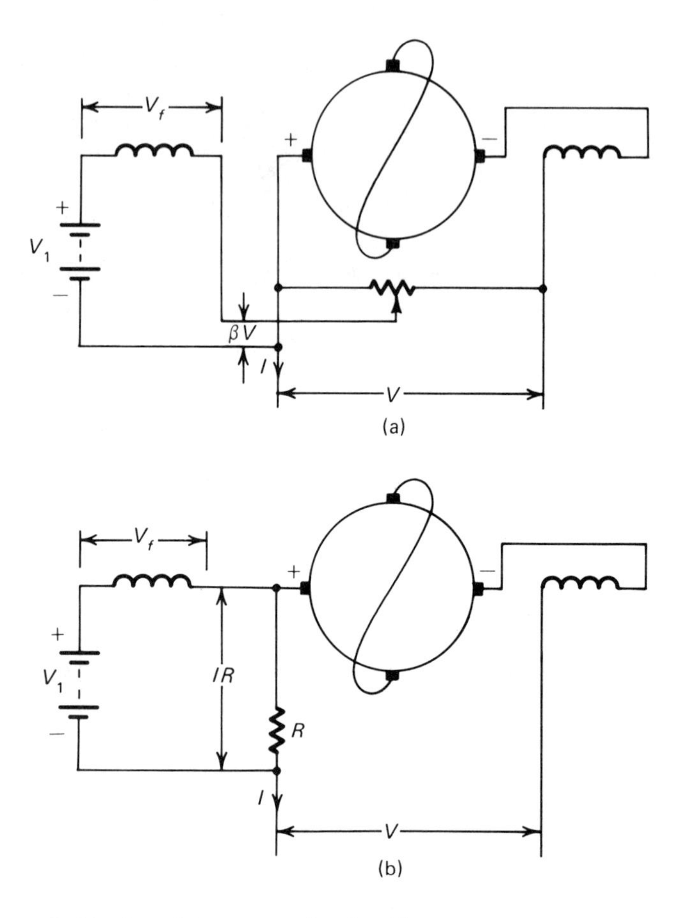

Figure 9-31. Amplidyne negative feedback: (a) voltage, (b) current.

GLOSSARY OF IMPORTANT TERMS

Amplifier: A device in which an input signal is used to control a source of power so that the amplified output possesses a definite relationship to the input signal. The term is applied to pneumatic, hydraulic, electrical, magnetic, and electronic amplifiers.

Operational amplifier (Op-Amp): A high-gain direct-coupled amplifier with a differential input capable of amplifying signals ranging from dc to many megahertz. It forms the basic building block of electronic circuits in industrial electronics, instrumentation, signal conditioning, etc.

Common mode rejection ratio (CMRR): The ratio of open-loop gain to the common mode gain.

Slew rate: The time rate of change of the output voltage of a closed-loop amplifier in response to a step input signal.

Inverting amplifier: An amplifier whose output voltage is opposite in polarity or phase to the input voltage.

Voltage follower: A unity gain noninverting amplifier with a very high input impedance and a low output impedance and acts as a buffer.

Difference amplifier: An amplifier that amplifies the difference in voltage applied to the input terminals.

Summing amplifier: An amplifier whose output is proportional to the sum of the inputs.

Integrator: An op-amp that has a capacitor in the feedback loop and sums the input signal over a period of time.

Differentiator: An op-amp that has a capacitor in series with the input and produces an output as long as the input is changing.

Comparator: An op-amp configuration that determines which input is more positive or negative than the other.

Power amplifier: An amplifier designed to deliver the maximum power to the load.

Class A power amplifier: An amplifier that is biased so that collector current flows for the entire duration of the input signal.

Class B power amplifier: An amplifier that is biased so that collector current flows for 180°, or one half-cycle of the input signal.

Feedback amplifier: An amplifier in which a portion of the output is fed back to modify the amplifier performance. If the fed back signal is in phase with the input signal this is known as positive or regenerative feedback. When the fed back signal is antiphase with the input signal this is known as negative or degenerative feedback.

Negative series voltage feedback: A feedback signal proportional to the output voltage is subtracted from the input signal by connecting it in series with it.

Negative series current feedback: A voltage signal proportional to the output current is subtracted from the input voltage signal.

Servomotors: Two-phase ac squirrel cage induction and dc motors used in instrument control systems and electromechanical servomechanisms.

Stepper motor: A dc excited device in which rotation is a series of steps controlled by a pulsed dc supply.

Amplidyne: A high gain high power output fast response generator utilizing armature reaction flux.

REVIEW QUESTIONS

9-1 What characteristics should an ideal op-amp possess?

9-2 What is meant by the term common mode rejection ratio?

9-3 With the aid of a sketch explain the operation of an inverting op-amp.

9-4 Explain the term "virtual ground."

9-5 Explain the feedback action of a noninverting op-amp.

9-6 Explain the term "offset" and discuss methods of compensating for offset.

9-7 What is a voltage follower? Sketch and explain a dc unity gain and an ac voltage follower.

9-8 Show how a difference amplifier may be used to amplify the output from a strain gauge bridge.

9.9 What is meant by a summing amplifier; illustrate your answer with a sketch.

9-10 With the aid of a sketch explain the principle of operation of a voltage-to-current converter.

9-11 With the aid of a sketch explain the principle of operation of a current-to-voltage converter.

9-12 With the aid of a sketch explain the principle of operation of an integrator. What precautions should be taken as the input signal frequency approaches zero?

9-13 With the aid of a schematic explain how an adjustable limit on the output voltage of an integrator may be obtained.

9-14 With the aid of a schematic explain the principle of a differentiator. What problems can be experienced with a differentiator?

9-15 Explain the principle of a comparator.

9-16 What is meant by a power amplifier? What are the main classes and their characteristics?

9-17 Why is a transformer coupled Class A power amplifier preferred to the directly coupled configuration?

9-18 With the aid of a sketch explain the operation of a Class A push-pull power amplifier. What are the advantages of using the push-pull configuration?

9-19 Repeat Question 9-18 for a Class B push-pull amplifier.

9-20 Repeat Question 9-19 for a complementary symmetry Class B power amplifier.

9-21 What are the benefits derived from using negative series voltage feedback?

9-22 How may negative feedback be used to reduce time lags in a separately excited dc generator control system?

9-23 Explain with the aid of a schematic and graphs the principle of a separately excited dc servomotor.

9-24 Explain the principle of a moving-coil servomotor. What are the advantages and disadvantages of this type of motor?

9-25 What are the advantages and disadgantages of using permanent magnet field structures in dc servomotors?

9-26 Briefly explain the construction of a printed circuit motor.

9-27 What is a torquer? Where would it be used?

9-28 Explain the principle of a two-phase ac servomotor. Show how the direction and speed of rotation is controlled.

9-29 What is a stepper motor and where is it used?

9-30 Explain the principle of operation of (1) a variable reluctance stepper motor and (2) a permanent magnet stepper motor. Discuss the characteristics of each type.

9-31 Discuss a control logic scheme that will control the direction and the stepping rate of a stepper motor.

9-32 What is meant by the translator and preset index methods of stepper motor control?

9-33 Explain with the aid of a schematic the principle of an amplidyne generator.

9-34 With the aid of schematics explain how negative voltage and current feedback may be applied to an aplidyne generator.

PROBLEMS

9-1 An inverting amplifier has $R_f = 1.5\text{M}\Omega$, $R_{in} = 200\text{k}\Omega$ and an open loop gain $A_d = 60{,}000$. The supply voltages are $\pm 15\text{V}$. Determine (1) the closed-loop gain, (2) the output voltage if $V_{in} = 100\text{mV}$, and (3) what input voltage will drive the output into negative saturation.

9-2 A noninverting op-amp has $R_f = 22\text{k}\Omega$, $R_{in} = 10\text{k}\Omega$ and $V_{in} = 200\text{mV}$. Calculate (1) the closed-loop gain, and (2) the output voltage.

9-3 An operational amplifier is being used as a summer (see Figure 9-8) where $R_1 = 15\text{k}\Omega$, $R_2 = 10\text{k}\Omega$, $R_3 = 20\text{k}\Omega$ and $R_f = 50\text{k}\Omega$.

The input voltages are $V_1 = 1.2V$, $V_2 = 0.5V$, and $V_3 = -0.85V$. Calculate V_O.

9-4 Design a weighted summer such that $-V_O = 2V_1 + 3.5V_2 + V_3$.

9-5 A step input voltage V_{in} which varies from 0 to + 2V is applied to an integrator. When $R_{in} = 10k\Omega$ and $C = 1.0\mu F$, calculate (1) V_0 after 4ms, (2) the rate of change of V_0, and (3) how long after V_{in} changes will the amplifier limit if $V_{CC} = \pm 22V$ and limiting occurs at 85% of V_{CC}.

9-6 A closed-loop amplifier has an open-loop gain A = 900. Calculate the closed-loop gain when $\beta = 0.09$. As a result of component aging the open-loop gain is degraded to 600. What is the percentage reduction in gain with feedback?

9-7 A closed-loop amplifier has an open-loop gain $A = 200$, $= 0.25$, $Z_i = 100\,\Omega$, and $Z_o = 5\,\Omega$. Calculate (1) the no-load gain A', (2) the input impedance Z_i', (3) the output impedance Z_o', and (4) the no-load gain A'.

9-8 The open-loop gain of an amplifier is 50 and $Z_i = 325\,\Omega$ and $Z_o = 0.45\,\Omega$ without feedback. If 25% of the output voltage is fed back, calculate: (1) Z_i', and (2) Z_o'.

9-9 An amplifier employing negative series voltage feedback has a constant 3.5V input signal applied and produces a constant 40V output. The output voltage rises to 50V when the feedback is removed. Calculate the percentage of the output voltage that was fed back.

10

Closed-Loop Control Principles

10-1 INTRODUCTION

Over the centuries man has learned to use and develop sources of energy to perform tasks that were far beyond his physical capabilities. Beginning with the lever and the wheel he progressed to the waterwheel and windmill. With the discovery of the steam engine, he applied the principles of closed-loop control with Watt's use of the flyball governor in 1788. These advances set the way for the Industrial Revolution which was further aided by the first railway locomotive in 1824 and the steamship. These developments, combined with the rapidly developing use of electrical energy in all its applications, have completely changed our manufacturing concepts and the role of the worker.

Initially the machines were simple, but as the complexity of the machines and processes increased, it became obvious that it was necessary to replace human beings as the operator and controller. This need for sophisticated control has developed at a phenomenal rate since the early 1950's until whole processes and systems are currently being automated in every aspect of our commercial and industrial activities.

The major advantages of automatic control are:

1. The ability to produce a product with repeatable accuracy.
2. The more effective use of plant facilities.
3. The release of the operator to more skilled and productive work.
4. The reduction of boredom and the elimination of workers being exposed to hazardous operations.
5. Higher productivity.

A major change is rapidly developing in the industrial scene, namely, in order to use energy effectively it is necessary to be able to control and regulate it. There is an ever increasing need for highly skilled personnel in industry with highly developed skills in industrial electronics, power electronics, heavy rotating machines, control systems, and computer systems who are able to understand current systems and develop and service future systems.

10-2 OPEN-LOOP AND CLOSED-LOOP CONTROL SYSTEMS

A control system that does not sense the output and make corrections to the process is called an **open-loop** or **open-cycle control system.** For example, a dc separately excited motor has an inherent drooping speed characteristic and will decrease its speed as necessary to meet changing torque demands, but it will not run at a constant speed irrespective of the connected load.

A closed-loop or closed-cycle system, on the other hand, is a system that senses the output, compares the output against the desired condition, and corrects the system to achieve the desired output. Referring back to our dc separately excited motor if a dc tachogenerator is mechanically connected to the shaft a speed voltage directly proportional to the shaft speed is produced. If this speed voltage is fed back and compared against an input voltage representing the desired or set speed, an error signal will be produced whenever there is a difference between desired and actual speeds. In turn this error signal suitably amplified can be used to control the firing delay angle of a thyristor phase-controlled converter, which in turn can return the shaft speed to its desired value by varying the armature circuit voltage.

A closed-loop control system can be described as an assembly of components used to maintain a desired output by controlling the energy input. Closed-loop control systems regulate the energy supplied to the process. An open-loop control system does not have a feedback and the control element has no reference data about the variable being controlled.

10-3 BLOCK DIAGRAMS

Control systems are usually diagrammatically represented by a series of interconnected blocks. The blocks represent the individual functions of the system and are interconnected by lines, which are used to represent the variable or quantity involved with directional arrows showing the direction of information flow. Figure 10-1 illustrates the basic concepts of open- and closed-loop control systems by means of block diagrams.

A closed-loop system can be broken down into four basic operations:

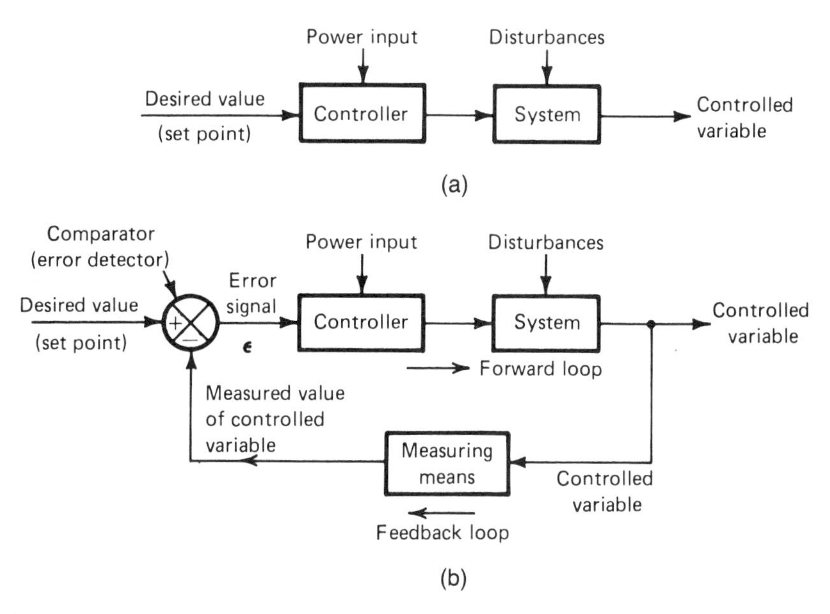

Figure 10-1. Block diagram representation of (a) an open-loop system and (b) a closed-loop system.

1. Measurement of the controlled variable C_m which may be temperature, position, velocity, thickness, etc. This measurement is accomplished using a sensor (see Chapter 7) and by means of a transducer converting the measurement made by the sensor into a useful signal, usually a voltage or current, representing the controlled variable.

2. Determination of the error, usually accomplished by a comparator or error detector. The comparator compares the measured value of the controlled variable against the desired value or setpoint (SP) and generates an error signal. Error signal = Set Point − measured value of controlled variable

$$\varepsilon = SP - C_m, \tag{10-1}$$

3. The use of the error signal to generate the appropriate control action.

4. The use of the control action to reduce the error signal by driving the controller, and thus the final control element, so that the actual value of the controlled variable approaches the desired value or setpoint. It should be noted that closed-loop control systems are error actuated, that is, an error must be present for the system to initiate

corrective action. The final control element in the dc variable speed drive would be the thyristor phase-controlled converter.

10-3-1 Characteristics of a Closed-Loop Control System

Intuitively it would appear that the final measure of success of a closed-loop control system would be the degree of closeness between the desired value (setpoint) and the measured value of the controlled variable, regardless of the frequency and magnitude of load changes or setpoint changes. As was mentioned previously, the system is error actuated, that is, there must be a change in the measured value of the controlled variable before corrective action can occur. The quality of the system determines the maximum error that must be present before corrective action can take place.

Another extremely important characteristic is the speed of response or settling time, that is, the time interval between the detection of the error and the completion of the corrective action.

The last, but by no means least, characteristic is the offset, residual error or steady-state error. The residual error is the final difference between the desired value and the measured value of the controlled variable, after the system has responded to a large change in the desired value or controlled variable.

These three characteristics of a good control system are illustrated in Figure 10–2(a), and tend to be mutually exclusive. The residual error should respond to an increase in gain of the controller, but increasing the gain makes the system more sensitive, and as a result may increase the maximum value of the error as well as increasing the settling time. Another effect of increasing the controller gain is to change the type of damping of the system in response to disturbances, see Figure 10–2(b). The type of damping is best described in terms of the controller gain as follows:

1. Overdamped: Low gain; the dynamic or transient response is very slow and a large residual error may be present.

2. Critically Damped: Low to medium gain; the least amount of damping that produces an output without any overshoot or oscillation.

3. Underdamped: Gain has been increased, and the output overshoots and oscillates with a diminishing amplitude response. Any further increase in gain will result in the system becoming unstable.

There are a number of stabilizing methods which may be used to increase the damping effect and at the same time permit an increase in the gain of the controller in order to reduce the settling time and ampli-

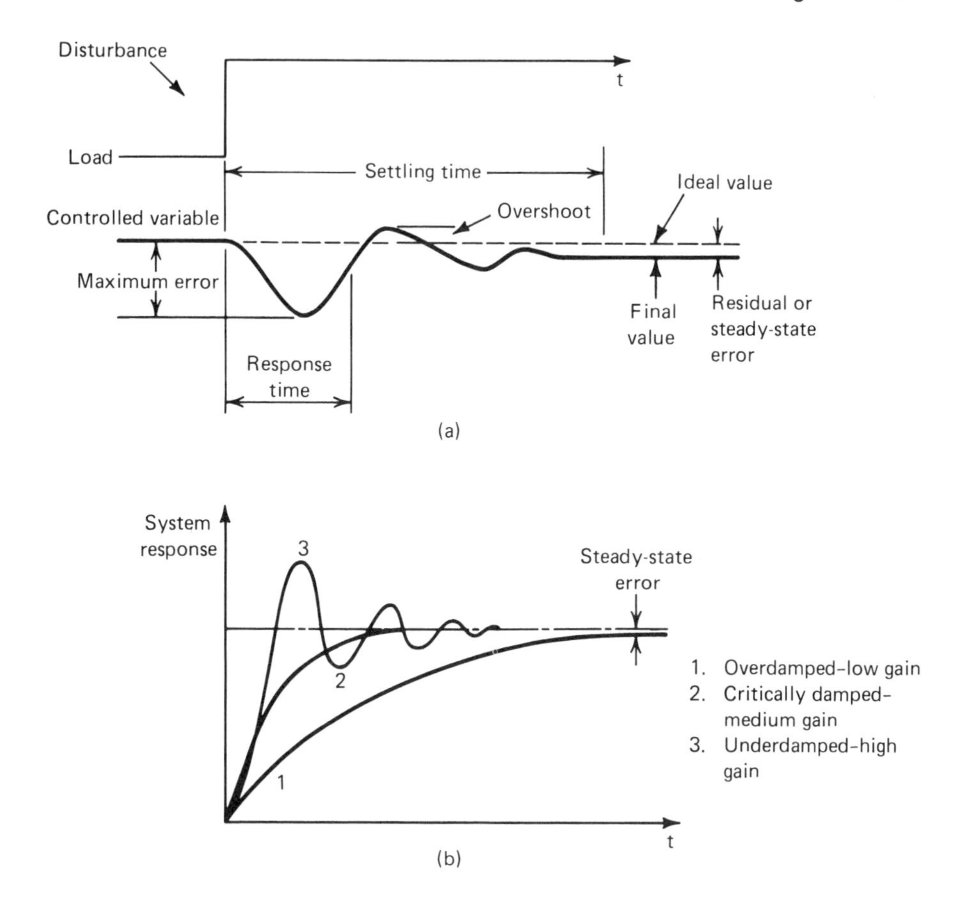

Figure 10-2. Closed-loop system characteristics: (a) system objectives, (b) system responses.

tude of the residual or steady-state error. These methods depend upon developing a force or signal to oppose changes in the controlled variable.

Electromechanical systems, especially position control systems, can be damped by developing friction forces that oppose oscillatory motion by extracting energy from the system. These methods are known as **viscous damping** and are obtained by using devices such as fluid, magnetic particle, or eddy current brakes. These methods are preferable to using mechanical brakes, which will reduce the positioning accuracy of a position control system. These systems are usually restricted to relatively small systems and have the disadvantages of increasing the cost, weight, and size.

Viscous damping techniques are not adaptable to pure electrical and electronic systems. There are several commonly used techniques that are frequently employed in these types of damping applications: output derivative damping, transient velocity damping, error rate damping, and integral damping.

10-3-1-1 Output Derivative Damping

Since damping is only required during transient conditions when changes in the controlled variable are occurring, it may be achieved by developing a signal that opposes changes in the controlled variable. Output derivative damping or rate feedback depends upon the production of a signal that represents the rate of change or derivative of the controlled variable, and then subtracting the derivative signal from the error signal. This technique is illustrated in Figure 10–3(a). In this case, a change of the setpoint will introduce a large error which is compounded by time lags within the system. Initially the controlled variable will change slowly, however, the error signal input to the amplifier is large and the controller output will also be large. As the error signal decreases, the rate of change of the controlled variable increases, and will continue increasing, until the error and output derivative signals are equal. At this point, the error input signal to the amplifier is zero and the controller output is zero, but the controlled variable continues to increase because of stored energy in the system. The output derivative signal is now the greater signal, and the polarity of the amplifier input signal has reversed; the controller output is reversed and acts to reduce the rate of change of the controlled variable, rapidly bringing the controlled variable to a steady-state condition.

When used in continuous motion position control systems, in order for the output to rotate at a constant speed a constant input signal must be applied to the amplifier. Since the output derivative signal is subtracted from the error signal, the error signal voltage must be great enough to provide a controller output to overcome the friction effects of the system, in addition to a voltage equal to the output derivative voltage. The result is that the output drive will rotate at a constant speed, but there will be an error between the input and output positions. This is known as velocity lag.

10-3-1-2 Transient Velocity Damping

An error will exist if there is an output derivative signal present when the system is operating under steady-state conditions. This condition mainly occurs in position control systems. If we differentiate the output (controlled variable) twice, the result is a signal proportion-

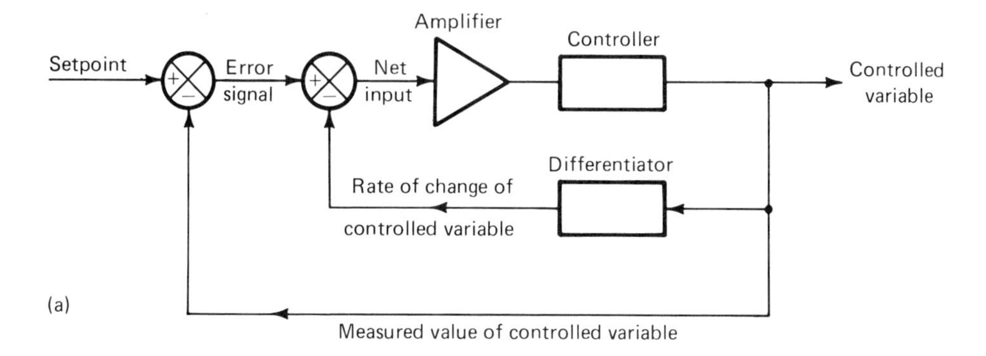

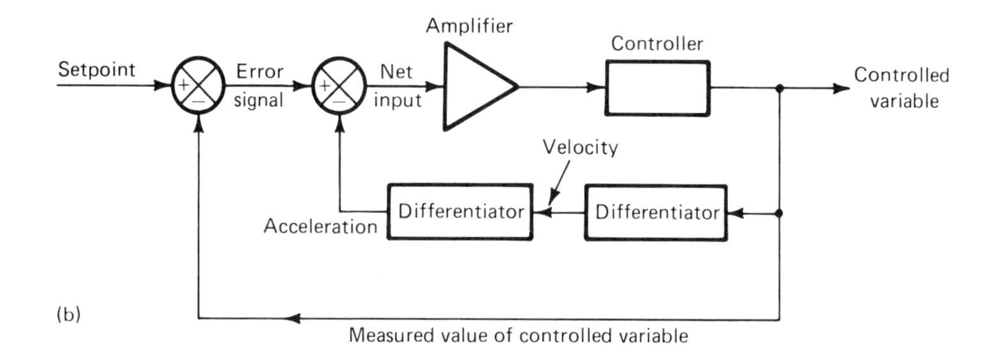

Figure 10-3. Damping: (a) output derivative damping, (b) transient velocity damping.

al to the output acceleration. The output from the second differentiator will be zero when the output is operating at a constant speed. As a result, the error voltage need only be great enough to overcome friction and load torque. The stabilizing loop will introduce a signal only when the system is accelerating or decelerating, which only occurs when there is a change in output velocity. See Figure 10-3(b).

10-3-1-3 Error Rate Damping

Error rate damping does not introduce a velocity error. This is accomplished by adding the derivative of the error signal to the error signal. The circuit shown in Figure 10-4 is a simple method of achieving this result. First, disregarding the capacitor C, $R1$ and $R2$ form a simple voltage divider and the voltage seen across $R2$ is proportional to the error signal. Second, disregarding $R1$, the voltage developed across $R2$ is the derivative of the error signal. With both components

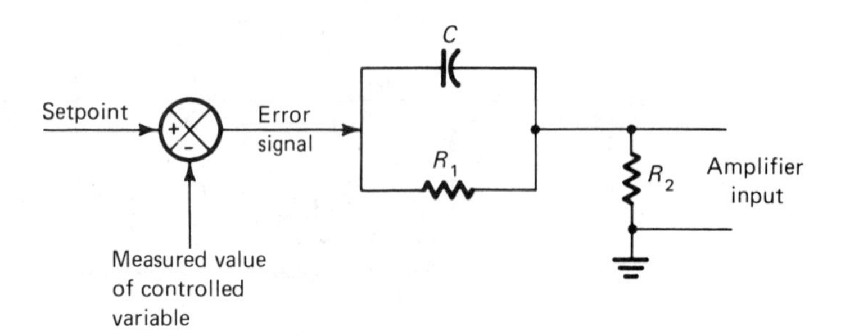

Figure 10-4. Error rate damping.

in parallel the input voltage at the amplifier terminals is then the error signal plus the derivative of the error signal.

10-3-1-4 Integral Damping

All of the stabilizing techniques discussed to date have left the problem of the residual or steady-state error unresolved. In speed regulators the speed will be less than specified because of the effects of friction, windage, and load torque; position controllers will have a velocity lag caused by output derivative damping and friction when operating at constant speed; and voltage regulators will produce an output voltage less than desired because of internal voltage drops.

Integral damping is basically an error plus the integral of error stabilizing system. In the case of a speed control system with a sudden change of setpoint, the immediate effect will be a sudden increase in the error voltage and the integrator output will be zero; the sum of the error voltage and the integrator output are applied to the amplifier. As time progresses, the error signal decreases but the integrator output increases. When the error signal is zero, the integrator output is constant and its output will be sufficient to eliminate the steady-state error.

Summary

Any control system can be stabilized by any of the methods described above. It should be remembered that output derivative damping introduces a steady-state error. Transient velocity stabilization and error rate damping improve the performance under transient conditions but do not eliminate the residual or steady-state error. Integral damping at a minimum significantly reduces the steady-state error, but may introduce instability to the system because of additional time lags that are introduced.

10-4 TYPES OF CONTROL SYSTEMS

Before proceeding any further, it is desirable that the classification of control systems be discussed. Since open-loop control systems are infrequently used, our discussion will be concentrated on closed-loop systems.

Closed-loop control systems can be divided into two main groups:

1. Regulator systems. These are feedback control systems in which the setpoint is rarely adjusted; for example, the automatic voltage regulator in your car or the thermostat in home heating systems, be they oil, gas, or electric. They function so as to maintain the controlled variable constant irrespective of large load changes.

2. Follow-up systems. These are feedback control systems where the setpoint is frequently changed. Typical examples would be reversible and nonreversible dc motor drive systems associated with rolling mills, mine hoists, etc.

In turn these systems can be further subdivided into:

1. Process control. Process control is a term commonly applied to the control of variables in manufacturing processes. Process control applications maintain such variables as temperature, flow rates, pressure, viscosity, density, levels etc., in chemical plants, oil refineries, food processing, blast furnace operations, automobile production lines, and nuclear power plants to name a few illustrative examples. Depending upon the complexity of the process there may be as many as 100 or more control loops involved in the complete process. Modern practice is to bring all the individual control functions under the control of a system such as the Honeywell TDC 2000. A system such as the TDC 2000 supervises the process by comparing actual process conditions against the optimum required performance and determining and implementing the changes necessary to achieve optimum performance. Process control systems usually involve manufacturing processes which are essentially slow processes.

2. Servomechanism. The term servomechanism has been traditionally reserved for position control closed-loop systems. As compared to process control systems, servomechanisms are relatively fast. Typical examples of servomechanisms are the elevation and azimuth controls of a radio telescope.

3. Sequential control. A sequential control system is a system that performs a series of operations in a prescribed sequence. The simplest example that comes to mind is the automatic washing machine found

in most homes. Other examples are automated engine block machining in the automobile industry and automatic winding of polyphase ac motor stators.

4. Numeric control. There are many manufacturing operations involving punching, drilling, milling, tapping, welding, etc., which must be performed accurately and on a repetitive basis. Numeric control (N/C) uses a set of predetermined instructions to control the necessary sequence of manufacturing operations. These instructions are converted from the manufacturing blueprints into a symbolic program and stored on some readily accessible storage medium such as paper tape, punched cards, magnetic tape, or floppy discs. When a specific part is to be manufactured, the appropriate stored program is fed into the machine control unit and the machine performs all the programmed operations under the control of the stored program. The term numeric control was assigned to the process since nearly all the instructions such as direction, position, velocity, cutting speeds etc., are defined mathematically. Numeric control is not designed for production line operation, but is usually limited to small runs of the desired parts. A major factor in selecting N/C equipment is cost. For example, a computerized lathe costs in the region of $200,000 as compared to $30,000 for a conventional lathe of the same capability. For this reason only about 5% of the metal working machinery in the U.S. is numerically controlled.

5. CAD/CAM. CAD/CAM stands for computer aided design/computer aided manufacture. The main benefits that are derived by the user of CAD/CAM include:

(a) A significant reduction in drafting office time. By using computer graphics, for example, a drawing revision which takes two days by conventional drafting office procedures can be produced in two hours.
(b) Improvements and corrections are easily made since the original design stored in the computer memory may be recalled to the screen and amended by a light pen.
(c) Assists engineers in their designs by utilizing stored programs to create the finished design. The major users of CAD/CAM are the U.S. auto industry and aircraft manufacturers such as McDonnell-Douglas who probably have the most sophisticated CAD/CAM facility in the world at St. Louis, Missouri. An example of the use of CAD/CAM by McDonnell-Douglas is in the control of numerical equipment at their Toronto, Ontario plant for the manufacture of DC-9 and DC-10 aircraft assemblies via land line from St. Louis.

CAD/CAM potential is estimated at 50,000 systems with half

of them to be installed in the U.S. The major drawback, apart from the huge capital investment, is the dearth of software to convert the computer image into the full set of instructions for automatically controlling the entire manufacturing process, which will be the factory of the future. However, probably within five years this will be a fact.

10-5 PROCESS RESPONSE

Before discussing the functions and control methods used in automatic controllers, it is necessary to have some knowledge of the behavior of the process. Each control system can be analyzed from the overall block diagram, where each block represents an important part of the process, for example, an amplifier, a measuring means, etc.

Each block receives an input signal and produces an output signal. The signals can be in many forms, for example, current, voltage, pressure, temperature, velocity, position, etc. The signal paths may also be wiring, mechanical linkages, pneumatic or hydraulic lines; that is, any medium that is capable of passing information.

The relationship between the input and output of a component or block is known as the **transfer function**. The component or block may affect the transfer of information in two ways: gain and time. Gain is defined as the change in output per unit change in input.

$$\text{Gain} = \frac{\text{change in output}}{\text{change in input}} \tag{10-2}$$

Usually the input signals have different units. For example, in a dc tachogenerator the input is rpm and the output is voltage, the gain is expressed as V/1000rpm. Some typical examples are shown in Figure 10-5. Also in each of these examples there is a time delay introduced, that is, there is a time difference between the application of the input and the result occurring at the output. Consider the case of the armature voltage-controlled dc motor. The effect of the application of the armature voltage is delayed by the time constant of the armature RL circuit and also by the mechanical inertia of the motor and the connected load. This effect is illustrated in Figure 10-6.

In addition, the measurement accuracy of the measuring means is affected by static error, dynamic error, reproducibility, and dead zone. **Static error** is the deviation of the indication of the measuring device from the true value of the measured variable. Static error is present in all measuring devices and obviously large values of static error are not desirable; however, it is not a major problem in a process where a variable is to be maintained at a constant value.

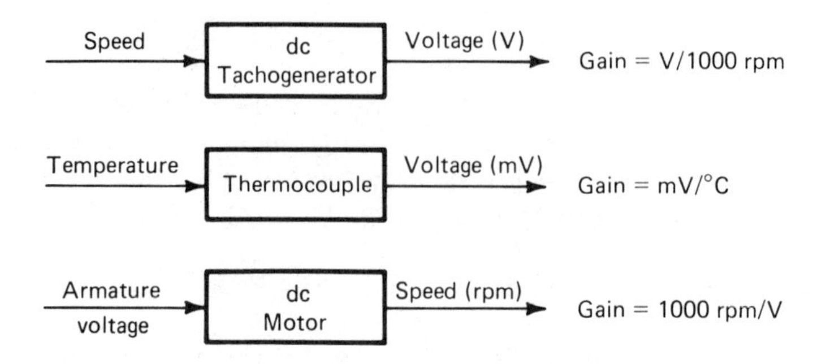

Figure 10-5. Block diagrams of transducers.

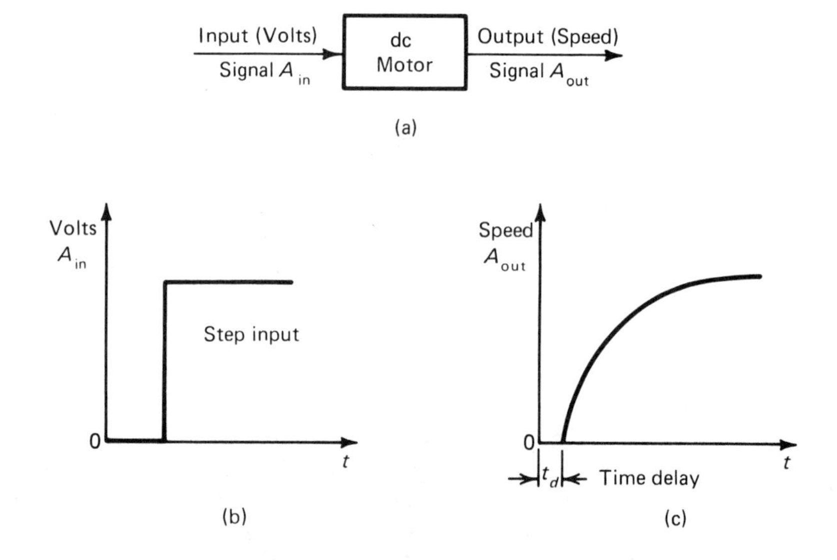

Figure 10-6. Response of a dc motor to a step input signal: (a) block diagram, (b) step input of armature voltage, (c) output response showing time delay.

Dynamic error is caused by a time lag in the measuring device and is only present under changing conditions and is independent of static error. If both the dynamic error and the static errors are known, the system can be adjusted to operate within accepted tolerances.

Reproducibility, or the ability to measure the measured value repeatedly within close limits, is actually more important than accuracy. **Dead zone** is the amount of change that occurs in a variable before the change is detected and, as a result, introduces a time delay called **dead**

time. The effect is not to change the characteristics of the process, but to delay them.

A system in which the controlled variable responds immediately to a change in the final control element is said to have little or no resistance to the energy changes involved. If initially the response of the controlled variable is rapid and then decreases, this is known as the capacity of the process. The rate of change is called the **process reaction rate**. The usual effect of more than one capacity in a system is to increase the dead time. Figure 10-7(a) is the process reaction curve of a process to a step input change. Initially the resistance of the process is small as can be seen from the high rate of change, but the rate of change decreases rapidly indicating capacitance in the system as energy is stored. Figure 10-7(b) represents a system with resistance and two or more capacities in response to a step input change. Initially there is no response of the controlled variable to the input signal, this interval is the dead time. When the controlled variable changes, it changes quite slowly and then increases until it approaches the specified final value at which point the reaction rate decreases. This situation is the result of energy being transferred between capacities, referred to as the **transfer lag**. It should be emphasised that dead time

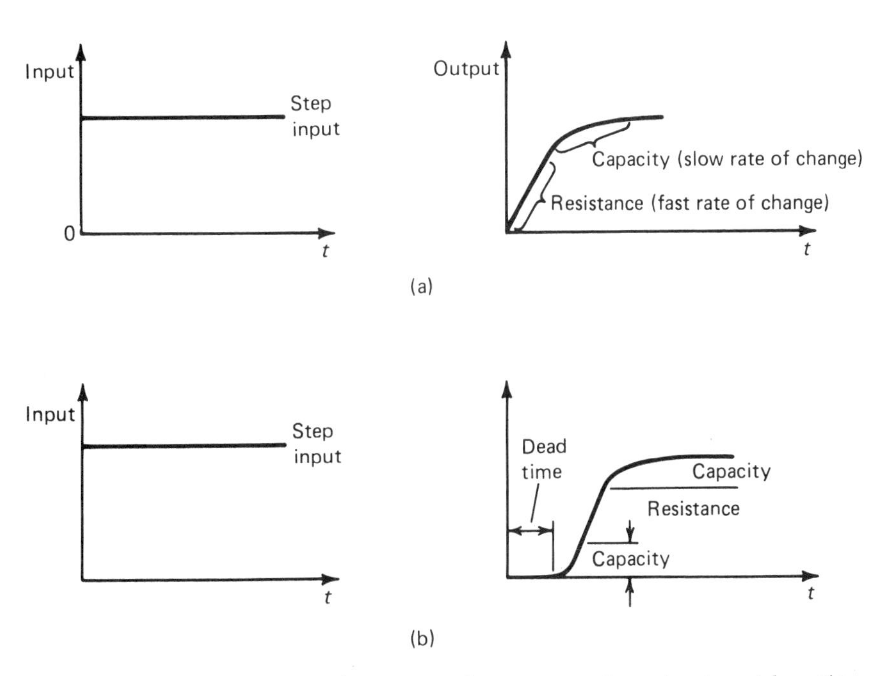

Figure 10-7. Process reaction curves in response to a step input function: (a) simple resistance and capacity system, (b) complex resistance and capacity system.

occurs whenever energy or mass is transported between two points. Typical causes of dead time are signal transmission speeds and flow rates of liquids and solids. It should be noted that dead time, or **transport time**, or **pure delay**, or **distance-velocity lag** as it is also sometimes called, is usually not a problem in a fast process such as a motor speed control system, but definitely can become a very significant factor in a process control system.

The purpose of this section is to illustrate that it is necessary to consider all factors in a control system prior to selecting a controller to meet the process requirements.

10-6 BASIC CONTROL MODES

Controllers used in closed-loop control systems are usually driven electrically, pneumatically, or hydraulically. The function of an automatic controller is to respond to an error signal and produce an output which will reduce the error to zero.

Automatic controllers are classified in accordance with their mode of control. There are six basic control modes; they are:

1. On-off or two-position.
2. Proportional.
3. Integral.
4. Proportional plus integral.
5. Proportional plus derivative.
6. Proportional plus integral plus derivative.

10-6-1 On-Off or Two-Position Control

An on-off or two-position system, as its name implies, has two states, on and off. This type of control is relatively inexpensive and simple, and is very commonly used in industrial, commercial, and residential applications for heating control. A good example of an on-off controller is the thermostat. This usually contains a bimetallic sensing element which moves as a result of a temperature change. This movement can be used to directly operate a switch, for example, a mercury switch or open or close a set of contacts. A typical residential electric heat thermostat will control up to 5 kW of heating elements; that is, 22 A at 240 V, with a control range of 41 °F (5 °C) to 86 °F (30 °C) and a differential of 1.8 °F (1 °C), where the differential is the difference in temperature between the make and break points of the contacts during normal operation. Figure 10-8 illustrates the operating characteristics of the system. Fig. 10-8(a) shows the differential action of the bimetallic element. Figure 10-8(b) shows the variations of room temperature

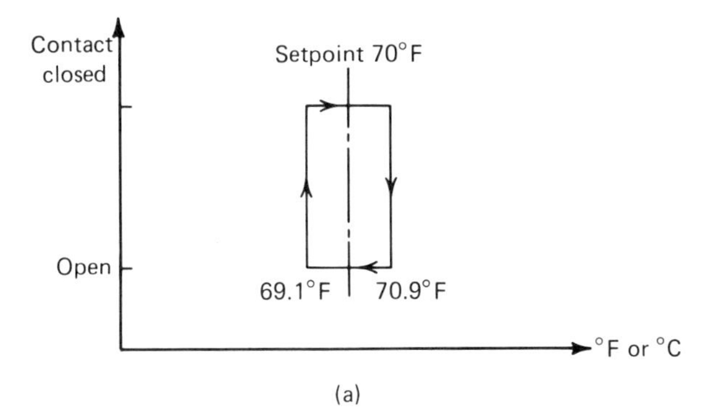

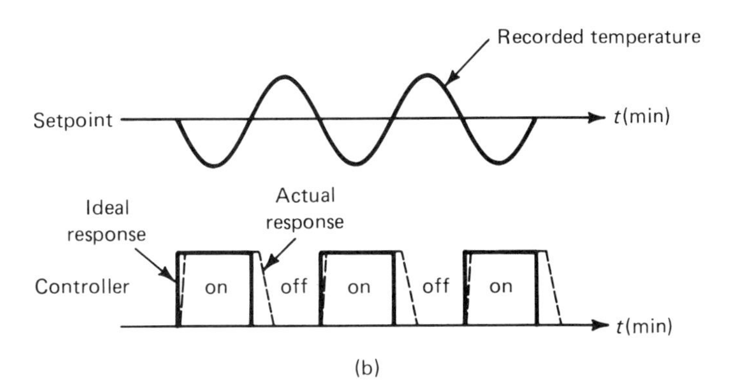

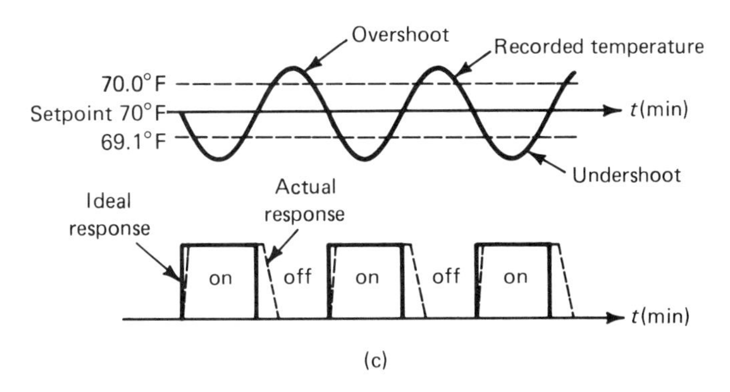

Figure 10-8. On-off or two-position control: (a) differential action, (b) without differential, (c) with differential.

about the setpoint without differential action. It should be noted that the heating element has a thermal time constant which results in the trapezoidal shape shown for the heating element output; this is transfer lag. The actual room temperature rises above and falls below the setpoint temperature because of transfer lag. If transfer lag were not present, the amplitude of the temperature oscillations would be reduced but the frequency would be increased. With a differential the heating element is not switched on or off until the setpoint has been passed. As can be seen from Figure 10-8(c), the duration of the on-off periods of the contactor are the same as before, but the angle of lag is less than 180° as compared to the actual temperature changes. If the differential is increased, the frequency of contact operation is reduced but the amplitude of the temperature variations will increase. If this condition is acceptable it will result in reduced contact wear and a greater life expectancy. If the differential is decreased, the frequency of operation increases and the amplitude of temperature variations decreases and contact wear increases.

The on-off type of control is most suited to applications where the reaction rate is slow, with little or no dead time and where transfer lag is a minimum.

10-6-2 Proportional Control

In on-off control the contactor was either on or off. If the control device had been a valve, the valve would have been closed or fully open irrespective of the magnitude of the deviation of the controlled variable from the setpoint. Consider a temperature control system as illustrated in Figure 10-9(a), the cold water enters at approximately a constant temperature and exits at a higher but constant temperature. It is assumed that the motor driven valve-controlled steam heating coil system does not introduce dead time or transfer lag. The proportional controller will drive the valve so that the amount of opening is proportional to the difference in temperature between actual and setpoint values. It is usually arranged so that the valve is 50% open when the system is operating normally [see Figure 10-9(b)]. The proportional band is usually defined as the range of values of the controlled variable that causes the correcting element (motor driven valve) to vary over its full range. In proportional controllers the proportional band is adjustable over a reasonably wide range. Mathematically proportional control action is defined as

$$A_{\mathrm{OUT}} = K_p \varepsilon \tag{10-3}$$

where A_{OUT} = controller output

K_p = proportional gain of the controller, and

ε = error signal

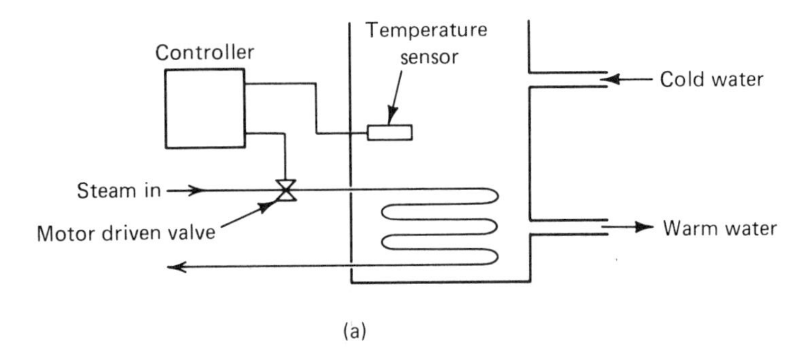

(a)

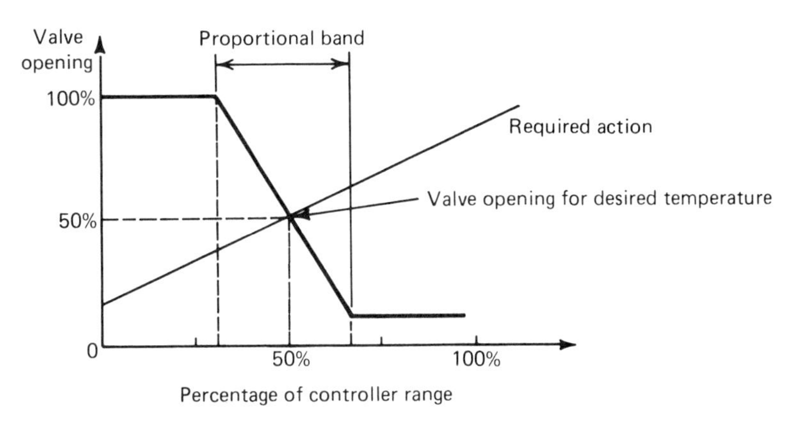

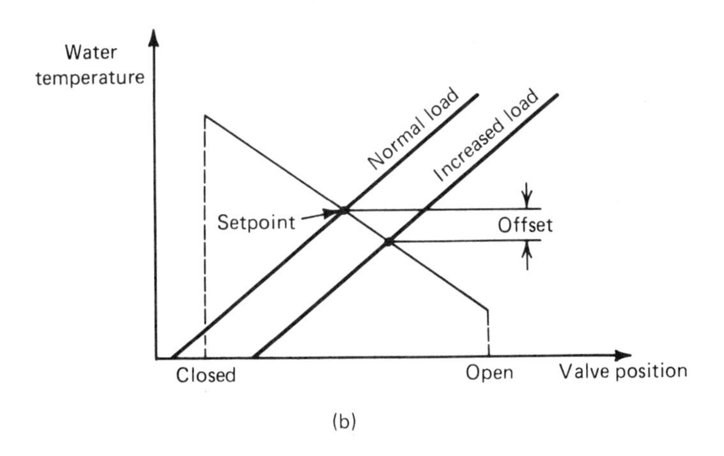

(b)

Figure 10-9. Proportional control: (a) temperature control system, (b) proportional band.

This relationship is accurate for steady-state conditions. However, since all controllers have time constants, the dynamic operations must be considered.

So far we have only considered the situation where the flow rate of the water is constant. If there is a large increase in the inflow and outflow rates, then the controller will open the valve to increase the steam flow to bring the water temperature back to the setpoint. As the temperature increases the valve opening decreases. However, the only way that sufficient steam can be supplied to the heating coils is for the valve to remain open at a larger setting at a new lower temperature; this action is shown in Figure 10-9(c). The difference between the original equilibrium temperature and the new temperature is called **offset**. This condition applies equally well to speed and position control systems.

If the proportional band is narrowed, the controller acts as an on-off controller. At the other extreme, if the proportional band is too wide the system is sluggish and will not respond fast enough to follow load changes or disturbances. If transfer lags and dead times are present, the gain or sensitivity K_p must be reduced but this results in larger offsets.

Proportional control is most effective in systems where:

1. Large load changed are not present.
2. Only small transfer lags and dead times are present.
3. The process reaction rate is slow.

10-6-3 Integral Control (Reset Control)

Integral control is continuous and the output of the control element changes at a rate proportional to the magnitude and duration of the error signal. Using the same heating system illustrated in Figure 10-9(a), the rate of the valve movement is proportional to the error signal; that is, the valve position A_{OUT} is the time integral of the error, or

$$\frac{dA_{OUT}}{dt} = K_i\varepsilon \qquad (10\text{-}4)$$

or

$$A_{OUT} = K_i \int \varepsilon dt$$

where

A_{OUT} = controller output

K_i = integral gain, and

ε = error signal

The action of an integral controller is shown in Figure 10-10. When the error signal is zero, $A_{OUT} = 0$. When an error is present the output action is proportional to the error.

The integral control action is most effective in systems where:

1. The transfer lags and dead times are small.
2. The measuring means and controller lags are small.
3. The range of setpoint variation is large.
4. The process tends to be self regulating.

Integral control is very rarely used by itself, but is usually used in conjunction with proportional control, where it eliminates offset.

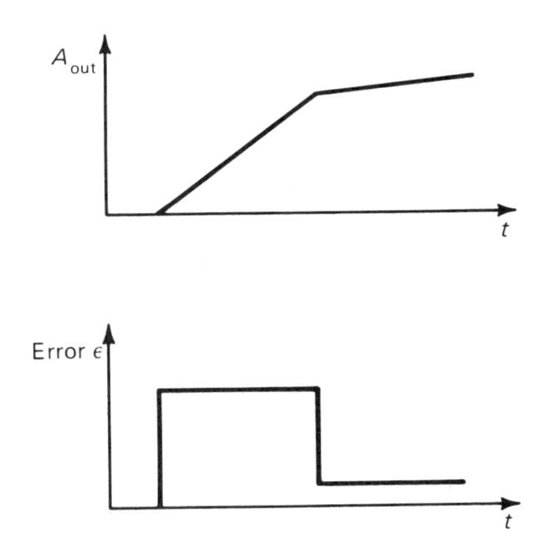

Figure 10-10. Integral control action.

10-6-4 Proportional Plus Integral Control

The control action of a proportional plus integral controller is defined mathematically by

$$A_{OUT} = K_p\varepsilon + K_i \int \varepsilon dt \tag{10-5}$$

This is actually Equation (10-3) plus Equation (10-4).

The action of the controller may be interpreted to mean that as long as an offset is present the correcting element (motor driven valve) from our hot water heating system will open or close at a rate proportional to the deviation from the setpoint until the setpoint condition is attained.

However, this gives rise to two problems: (a) the settling time of the system is greater than with proportional control only and (b) with a negative error the integral action will move the proportional band above the setpoint until the controlled variable attains the setpoint condition. In other words the control element (valve) will not start to close until the system has overshot the setpoint.

Usually a proportional plus integral controller is capable of having both actions independently adjustable. The major portion of the control action is achieved by the proportional element; the integral element is relatively slow moving as it eliminates offset. For best performance, the gain K_p of the proportional control action should be low to obtain good stability. The control action of a proportional plus integral controller is illustrated in Figure 10-11. It should be noted that the best results are obtained when: (a) the load changes are small and slow and (b) transfer lags and dead times are small.

10-6-5 Proportional Plus Derivative

Proportional plus derivative control is used in applications where it is desired to reduce overshoot during sudden load changes; that is, the proportional element produces an output proportional to the error and the derivative element produces an output proportional to the rate of change of the error. Mathematically this relationship is

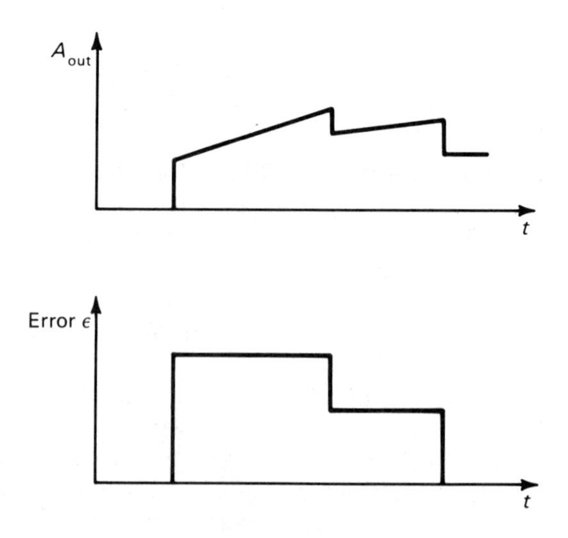

Figure 10-11. Proportional plus integral control action.

$$A_{OUT} = K_p \varepsilon + K_d \frac{d\varepsilon}{dt} \qquad (10\text{-}6)$$

where K_d = derivative gain.

In the case of the motor driven valve temperature control system, the valve is positioned by extra amounts as the temperature increases or decreases. The major advantage of proportional plus derivative control is that it anticipates the required action. However, especially in electrical/electronic systems, it responds to noise and may saturate the amplifier. Derivative control is never used in isolation since it is only responsive during transient conditions. The action of a proportional plus derivative controller is shown in Figure 10-12.

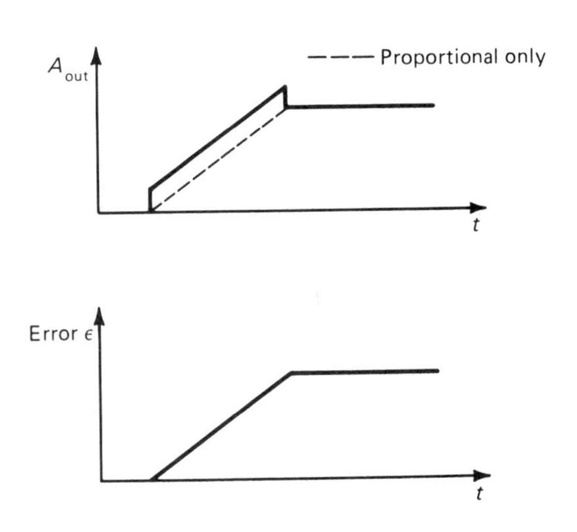

Figure 10-12. Proportional plus derivative action.

10-6-6 Proportional Plus Integral Plus Derivative Control (PID)

For satisfactory control of processes with large transfer lags and/or large dead times which may also be subjected to sudden large load changes, a combination of proportional plus integral plus derivative control is required. Mathematically this is represented by

$$A_{OUT} = K_p \varepsilon + K_d \frac{d\varepsilon}{dt} + K_i \int \varepsilon \, dt \qquad (10\text{-}7)$$

The effect of adding derivative action to a proportional plus integral controller is to reduce the amplitude of the maximum error and reduce the cycle time. It also gives the greatest benefit when transfer lags are present, but is more commonly used to reduce the effects of dead time. The control action of a proportional plus integral plus derivative controller in response to an error signal is shown in Figure 10-13. It should be noted that proportional plus integral plus derivative, or three-term control action, is almost exclusively applied in the process control field.

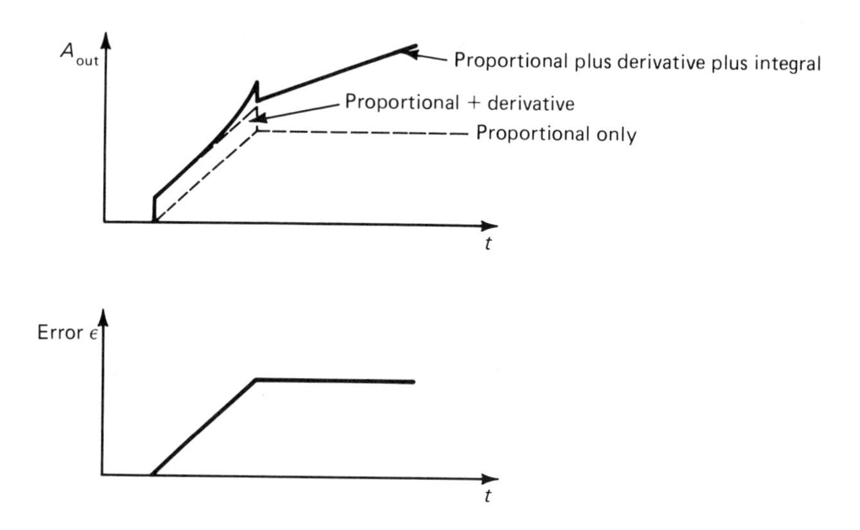

Figure 10-13. Proportional plus integral plus derivative control action.

SUMMARY

In general, the application and performance of the various control modes can be summarized as follows:

1. On-off control is best suited for applications where the process reaction rate is slow and transfer lags and dead times are small. In addition, the magnitude of load changes should be small and the cycle time long.

2. Proportional control is best suited for applications where there are small, slow load changes. However, it can accommodate small to moderate transfer lags and dead times with moderate to long process reaction rates.

3. Integral control is very rarely used by itself.

4. Proportional plus integral can handle varied reaction rates with a

wide range of slowly changing load disturbances. However, it will have a long settling time if transfer lags and dead times are too great.

5. Proportional plus derivative is most suitable for applications where the reaction rates are not small, transfer lags and dead times are moderate, and load disturbances are small irrespective of their frequency.

6. Proportional plus integral plus derivative. This is the most effective control mode, since it is responsive to a wide range of reaction rates, transfer lags, dead times, and amplitude and frequency of load disturbances.

Obviously when making a selection of a suitable controller, initial and maintenance costs, ease of maintenance, and reliability are all factors that must be considered. A reasonably safe rule is to select the simplest control mode that will meet the installation and process requirements.

GLOSSARY OF IMPORTANT TERMS

Open-loop control: Any system in which there is not a comparison between the actual result and desired result.

Closed-loop control: Any system which compares the actual result with the desired result and uses the difference to initiate a corrective action to move the actual result toward the desired result.

Settling time or speed of response: The time interval between the detection of an error and the completion of the corrective action.

Residual or steady-state error: The final difference between the desired and measured values of the controlled variable.

Viscous damping: Frictional forces that oppose oscillatory motion by extracting energy from the system.

Output derivative or output rate feedback: The simplest method of obtaining damping is achieved by subtracting a feedback signal proportional to the rate or velocity of the controlled variable from the error signal, but introduces a steady-state error.

Transient velocity damping: A method of damping that adds a signal to the error signal when the controlled variable is changing. Improves performance but does not eliminate steady-state error.

Error-rate damping: Eliminates velocity error by adding the derivative of the error signal to the error signal. Improves performance but does not eliminate steady-state error.

Integral damping: Eliminates steady-state error by monitoring the error signal and as long as an error signal is present adds the integral of the error signal to the error signal.

Regulator system: A feedback control system in which the setpoint is rarely adjusted.

Follow-up system: A feedback control system in which the setpoint is frequently adjusted.

Process control: The control of variables in a manufacturing process.

Servomechanism: Traditionally used to describe closed-loop position control systems.

Sequential control: A control system that performs operations in a prescribed sequence.

Numeric control: A control system using a set of predetermined instructions to control a sequence of manufacturing operations.

Transfer function: The relationship between the input and output of a component or system and defines the gain and time relationships between the input and output signals.

Static error: The deviation between the indication and true value of a measured variable.

Dynamic error: The error resulting between the change of a measured variable and the indication of the change.

Reproducibility: The ability to repeatedly measure the measured variable within close limits.

Dead zone: The amount of change in the measured variable before the change is detected.

Dead time: The time delay introduced by dead zone.

On-off or two-position control: An on-off energy control system used when the process capacitance is greater to reduce the cycling frequency.

Proportional control: A controller whose output is directly proportional to the error signal.

Integral or reset control: A controller whose output is proportional to the magnitude and direction of the error signal.

Proportional plus integral control: A controller whose output is a linear combination of both proportional and integral control actions.

Proportional plus derivative control: A controller whose output is a linear combination of both proportional and derivative control actions.

Proportional plus integral plus derivative control: A controller whose output is a linear combination of proportional, integral, and derivative control actions.

REVIEW QUESTIONS

10-1 Discuss the advantages of automatic control.

10-2 Discuss the four basic operations performed by a closed-loop control system.

10-3 Discuss the characteristics of a closed-loop control system.

10-4 What is meant by viscous damping? Give examples of its use and discuss the advantages and disadvantages.

10-5 Explain the basic principles and discuss the limitations of output derivative, transient velocity, error rate, and integral damping techniques.

10-6 Explain the difference between regulator and follow-up control systems.

10-7 Explain what is meant by the term transfer function.

10-8 Explain what is meant by process reaction rate.

10-9 What are the six basic control modes?

10-10 What is meant by on-off or two-position control? What is meant by differential action? What is the effect of increasing the differential? What are the advantages and disadvantages of this type of control?

10-11 What is meant by proportional control? What is meant by proportional band and offset? What are the effects of changing the width of the proportional band? What are the advantages and disadvantages of this type of control?

10-12 What is meant by integral control? When should it be used?

10-13 What is meant by proportional plus integral control? When should such a system be used and what are the disadvantages to this type of controller?

10-14 What is meant by proportional plus derivative control? What is the advantage of using this type of control action? What are the disadvantages?

10-15 Why is derivative control action never used by itself?

10-16 What is meant by proportional plus integral plus derivative control? When should this type of controller be used?

10-17 What is meant by transfer lag?

11

dc And ac Motor Speed Control

11-1 dc DRIVES

The dc motor has been used for many years in industrial and transportation applications. The major advantages for using dc motors are:

1. A wide range of speed control above and below base speed can be obtained.
2. They can be operated under variable or constant torque conditions.
3. They can be rapidly accelerated or decelerated and reversed. These qualities are particularly advantageous in steel making, mine hoist control, and electric traction applications.
4. They can be operated in closed-loop control systems to provide accurate speed or position control.
5. They can be operated under regenerative conditions.

As a result, the dc motor is still widely used in all applications where variable speed control is required. However, the newer drives for heavy rotating equipment such as steel mills, seamless tube mills, paper making machines, urban transit systems, excavators, etc., are now controlled by thyristor phase-controlled converters or choppers. These solid-state drive systems reduce initial procurement costs and maintenance costs as well as operating at higher efficiencies. Thyristor phase-controlled converters are used to control dc motors ranging from fractional horsepower (fractional kW) to as high as 10,000 hp (7,500 kW).

The ac motor, on the other hand, has always been considered as a constant speed machine. It is only a relatively short time since the modern variable-frequency ac drive has become commercially available, and even though the odds appear to be in favor of the ac variable-frequency drive, the overall cost and complexity have resulted in their slow acceptance. However, in the past three or four years this situation has started to change rapidly.

11-2 dc MOTOR CONTROL

It is not necessary to have a detailed understanding of the dc motor in order to understand their control by thyristor phase-controlled converters. The most commonly used motors for variable-speed applications are the separately-excited and series-excited dc motors. Since series motors are almost exclusively used in electric traction and crane applications, we will not study their usage but concentrate instead on the separately-excited dc motor.

11-2-1 dc Motor Construction

Variable-speed drive motors usually differ from standard design motors in that: (a) they have a physically larger commutator to provide extra insulation capability because of the large and rapid voltage variations; (b) they have laminated main and interpoles, as well as a laminated yoke to reduce the effects of eddy currents; and (c) they have low inertia armatures to improve the response. Usually the larger motors will also have series-connected compensating windings let into the main pole faces to minimize the effects of armature reaction.

11-2-2 dc Motor Characteristics

The performance of a dc motor is readily developed based on the following assumptions:

1. The motor is operating under steady-state load conditions.
2. The magnetic circuit is not saturated.
3. The rotational and stray power losses can be neglected.
4. The armature circuit resistance R_a is negligible.

Considering the separately-excited motor connected as shown in Figure 11-1 to a source of armature voltage V_a, then the armature circuit equation is

$$V_a = E_c + I_a R_a \tag{11-1}$$

where E_c is the counter-emf. Then assuming that R_a is small enough to be neglected, the armature voltage drop $I_a R_a$ is also small, then

$$V_a \simeq E_c \tag{11-2}$$

The steady-state operation of the motor is determined by the fact that the counter emf E_c is slightly less than the applied armature voltage V_a. The counter emf E_c is proportional to the speed of rotation of the armature and the field pole flux and can be represented by

$$E_c = K\Phi S \tag{11-3E}$$

where $K = (ZP/60a) \times 10^{-8}$

$= \text{constant of proportionality}$

$S = \text{speed (rpm)}$

or

$$E_c = k\phi\omega \tag{11-3SI}$$

where $k = (ZP/2\pi a)$

$= \text{constant of proportionality, and}$

$\omega = \text{angular velocity (rad/sec).}$

Then, solving for speed, we have

$$S = \frac{E_c}{K\phi} \simeq \frac{V_a}{K\phi} \tag{11-4E}$$

or

$$\omega = \frac{E_c}{k\phi} \simeq \frac{V_a}{k\phi} \tag{11-4SI}$$

From Equations (11-4E) and (11-4SI) it can be seen that the speed of a dc motor is directly proportional to E_c, and since $E_c \simeq V_a$, directly proportional to the applied armature voltage V_a, and inversely proportional to the field pole flux Φ (or ϕ). See Figure 11-1(b), (c).

The torque developed by a dc motor is directly proportional to the armature current I_a and the field pole flux Φ (or ϕ); that is,

$$T = KI_a\Phi \text{ ft-lb} \tag{11-5E}$$

or

$$T = kI_a\phi \text{ Nm} \tag{11-5SI}$$

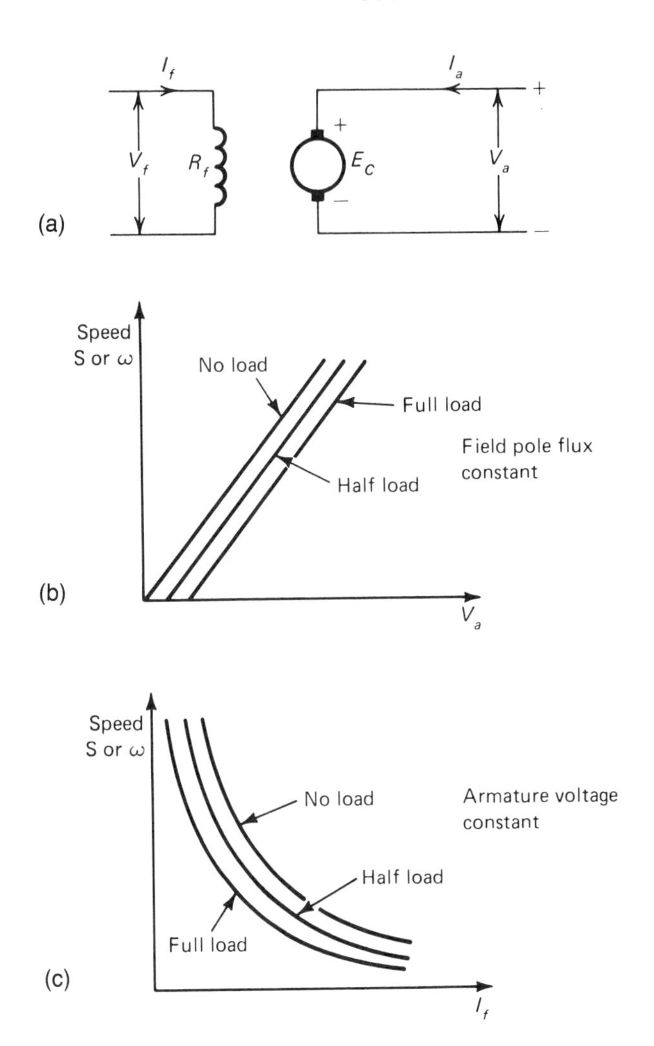

Figure 11-1. Separately excited dc motor: (a) schematic, (b) speed versus armature voltage, (c) speed versus field current.

The horsepower (kW) output is proportional to both torque and speed, therefore,

$$\text{hp} = KTS \tag{11-6E}$$

or

$$\text{kW} = kT\omega \tag{11-6SI}$$

Equations (11-5E and SI) and (11-6E and SI) show that if the armature current I_a remains constant, the following holds:

1. By varying the applied armature voltage V_a, with the field pole flux remaining constant, the output torque remains constant, and the output horsepower (kW) is directly proportional to speed. This is known as the **constant torque mode.**

2. By applying rated armature voltage, and weakening the field pole flux, the output horsepower (kW) remains constant and the output torque decreases. This is known as the **constant horsepower (kW) mode.**

These relationships are illustrated in Figure 11-2. The base speed of a dc motor is defined as the speed at which the motor operates with rated armature voltage and rated field current.

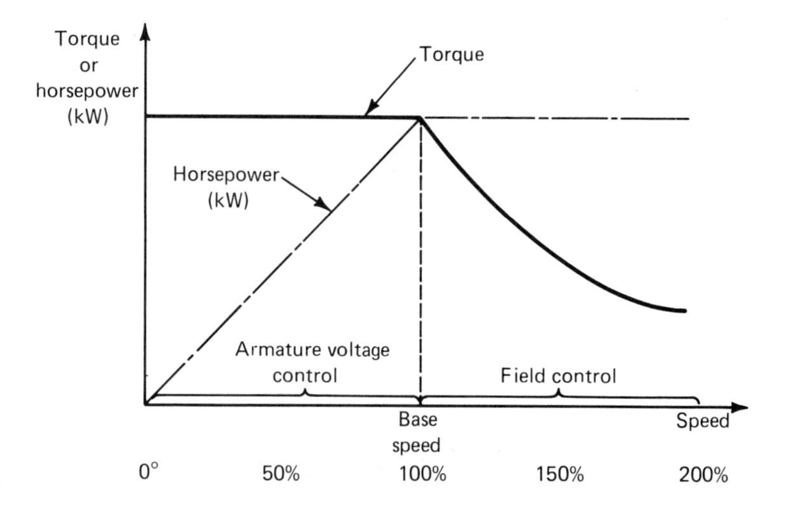

Figure 11-2. Torque and horsepower (kW) versus speed characteristics of a separately excited motor.

11-2-3 Four-Quadrant Control or Dual Converters

When a one-quadrant converter is used for dc motor control the power flow is from the ac source to the motor, and power cannot be returned from the dc motor to the ac source; that is, the converter cannot operate in the synchronous inversion mode. From Figure 11-3, it can be seen that a one-quadrant converter can only operate in quadrant 1 (forward motoring) or in quadrant 3 (reverse motoring). Reversal can be achieved by one of two methods: (a) reversing the field connections;

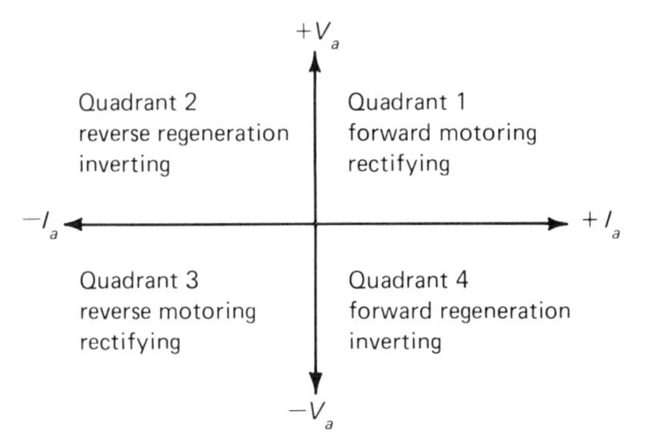

Figure 11-3. Four-quadrant operation of a dc motor.

however, this method is not commonly used because of the high time constant of the field circuit or (b) reversing the armature circuit; however, since thyristors only permit current flow in one direction, this method requires the use of a reversing contactor in the armature circuit.

A two-quadrant converter permits the machine to operate in the motoring or regenerative modes, that is, it can operate in quadrant 1 (forward motoring) and quadrant 4 (forward regeneration), or quadrant 3 (reverse motoring) and quadrant 1 (reverse regeneration). A two-quadrant drive can be reversed by reversing the armature circuit after the applied armature voltage has been reduced to zero by retarding the firing pulses.

There are many industrial applications where reversible drive operation is required, for example, a mine hoist or a reversing steel mill drive. In these applications the use of a single two-quadrant converter with a reversing contactor in the armature circuit is not very practical. A far more efficient arrangement is to use two two-quadrant converters connected back-to-back to supply the armature circuit. This arrangement permits reversing control and regeneration without switching either the field or armature circuits and has been previously discussed in more detail in Section 4-5.

11-2-4 dc Motor Speed Control

A typical closed-loop armature voltage dc motor speed control system is illustrated in Figure 11-4. There are a number of requirements that must be met by the system: (a) limit inrush currents during starting, (b) maintain the output speed at set speed, and (c) provide overcurrent protection.

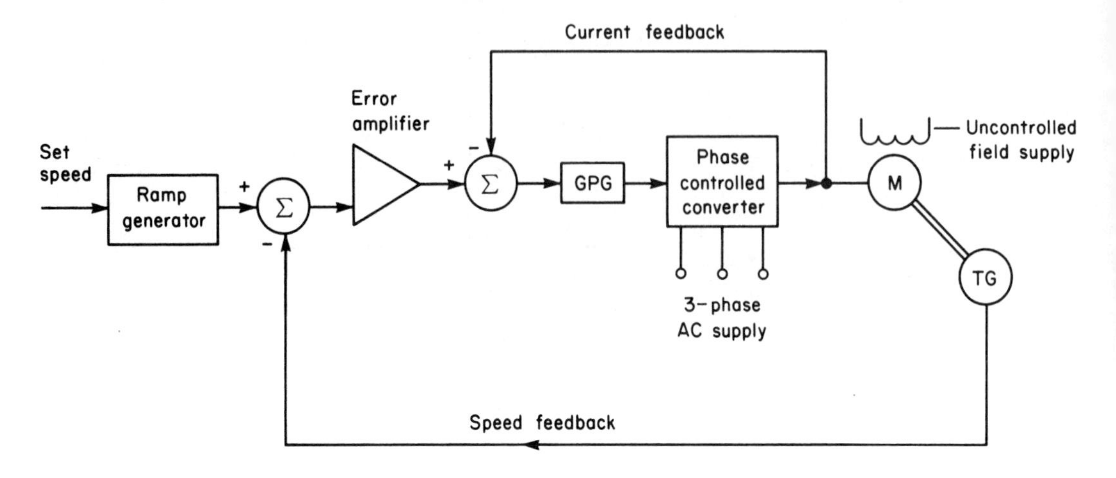

Figure 11-4. Block diagram of a phase-controlled converter dc motor speed control system.

11-2-4-1 Starting

It can be shown that the armature current I_a at any instant is

$$I_a = \frac{V_a - E_c}{R_a} \tag{11-7}$$

Under starting conditions $E_c = 0$. Since R_a is usually very small, in order to limit I_a to an acceptable maximum, usually 150% to 175% of the normal full-load current, it is necessary to reduce V_a. In electromagnetic control systems additional resistance was added in series with the armature to limit inrush current during start. However, in thyristor phase-controlled converters this additional resistance is unnecessary since it is only needed to move the firing pulses to the maximum retard position and then advance the firing pulses from the fully retarded position under the control of a ramp generator. It is also necessary to ensure that the rate of advance of the firing pulse does not result in excessive armature current. This can be assured by monitoring the load current, either on the dc side by means of a shunt or on the ac side by means of current transformers, and rectifying the output. In either case, a dc voltage signal is produced representing actual current which is compared against a voltage representing the maximum permissible current. If an overload current is present, an inhibit signal is used to increase the firing delay angle until the motor comes out of the current limit condition.

11-2-4-2 Ramp Generator

A simple ramp generator circuit is shown in Figure 11-5. A reference dc input voltage, usually in the range of 0 to 10 V, is applied to the inverting input of *U*1. The diodes *D*1 and *D*2, together with the potentiometer *P*1 and *P*2, form a variable gain amplifier; the function of *P*1 and *P*2 is to respectively control the acceleration and deceleration rates of the drive. The inverted output of *U*1 is connected to the integrating op-amp *U*2 whose output is a positive going ramp. This signal is used to control the firing point of the thyristors in the converter section. A negative feedback signal is also taken from the output of *U*2 and applied to the inverting input of the unity gain op-amp *U*3. The output of *U*3 is a negative going ramp which is compared at a summing junction with the input dc reference voltage; when the two signals are equal and opposite, their sum is zero, and the output voltage of the ramp generator will remain constant at a value determined by the set speed signal. This in turn maintains the output voltage constant by controlling the position of the firing pulses.

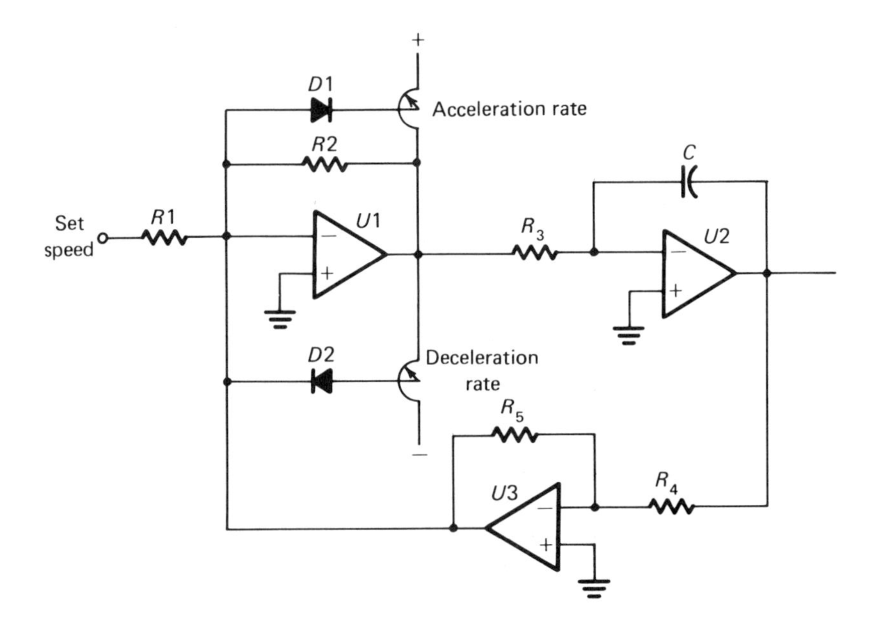

Figure 11-5. A simple ramp generator.

11-2-4-3 Speed Control

To be able to maintain set speed it is necessary to generate a dc signal that represents actual speed. There are two basic methods: (a) by using the counter emf E_c of the armature circuit, since $E_c \propto \omega$ assuming a constant field and reduce the voltage level by a voltage divider network and (b) by using a tachogenerator coupled directly to the shaft to ensure torsional stiffness. Normally, an ac tachogenerator is used and the output rectified and attenuated. In either case, this negative feedback voltage is compared against the output voltage of the ramp generator. If the motor speed is below the desired speed, the resulting error signal will cause the gate pulse generator to advance the firing pulses and increase the converter output voltage, and the motor speed will increase. However, because of the inertia of the motor armature and the connected load, the motor will accelerate above the desired speed. Once again, the resulting error signal will cause the gate pulse generator to retard the firing pulses; that is, increase the firing delay angle and slow down the drive.

As can be seen, the motor armature will never remain constantly at the set speed, but will vary slightly about the desired speed; the amount of the variation is a measure of the speed regulating capability of the drive control system. Armature voltage sensing usually produces a speed regulating capability of approximately $\pm 2\%$, whereas an ac or dc tachogenerator will produce a speed regulating capability of the order of $\pm 0.1\%$, which is usually more than adequate for most large horsepower (kW) drives.

11-2-4-4 A Three-Phase Industrial dc Drive

Figure 11-6 shows a typical three-phase, variable-speed dc motor drive converter, the Veritron® Type ASD6 series produced by Brown Boveri, with output ratings ranging from 4.5 kW to 154 kW. These converters are designed for infinite speed control of dc motors for one-quadrant operation in such applications as fans, pumps, calenders, extruders, conveyors, agitators, grinders, crushers, machine tools, etc. However, since they are fully controlled converters, that is, all active devices are thyristors, they may be used as two-quadrant converters; with the addition of torque reversal logic, a single converter can be made to operate as a four-quadrant converter, thus providing driving and regenerative braking in both directions.

The control scheme illustrated in Figure 11-6 is applicable to both one- or two-quadrant operation. The principle of operation is as follows:

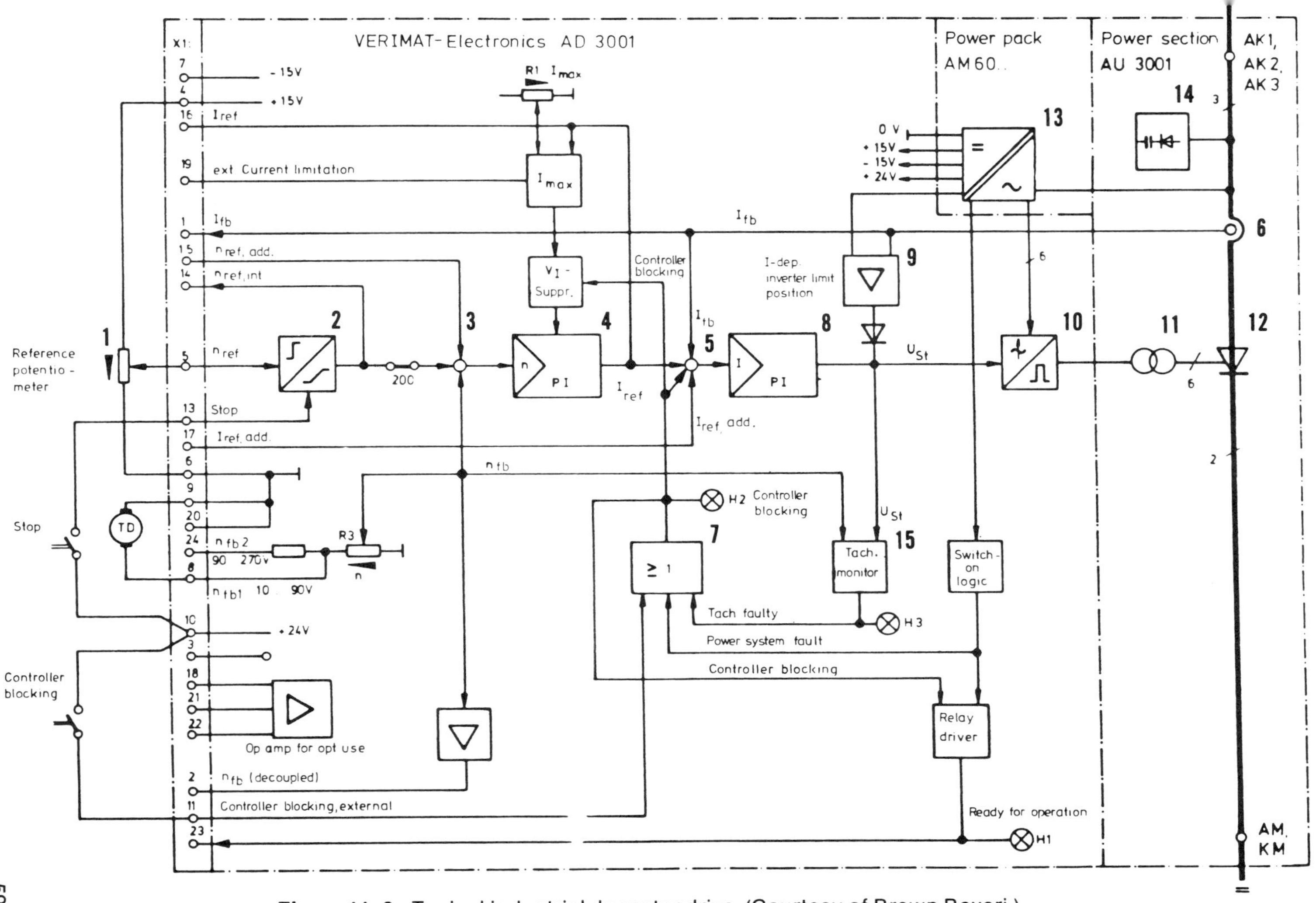

Figure 11-6. Typical industrial dc motor drive. (Courtesy of Brown Boveri.)

1. A voltage n_{ref}, representing the desired speed, is established by the reference potentiometer (1).

2. In turn, n_{ref} is applied to the acceleration/deceleration ramp generator (2) which is adjustable and controls the acceleration and deceleration rates of the drive.

3. The output of the ramp generator is summed at (3) with a voltage signal from the tachogenerator *TD* suitably attenuated by *R*3 and a current limit signal.

4. The resulting error voltage signal is applied to the inverting input of the speed amplifier (4). The speed amplifier is a proportional integral op-amp whose time constant is matched to the motor time costant.

5. The output of the speed amplifier is a voltage proportional to current signal. Recall $T \propto I_a \Phi$ and, assuming Φ is constant, then the drive is essentially current controlled.

6. The output signal of the speed amplifier is summed at (5) as follows: (a) The current feedback signal I_{fb} is derived from two current transformers (6) connected in series with the three-phase ac supply lines. The current transformer output is rectified by a diode and converted to a voltage signal representing actual line current. (b) An output from (7) functions as an OR gate and produces an inhibit signal as a result of a tachogenerator feedback signal failure or a power system fault such as a phase loss, under-voltage, reversed phase rotation, or an external controller blocking signal.

7. The resulting error signal output from the summing point is then applied to the inverting input of the current amplifier (8). This amplifier is also a proportional integral amplifier.

8. The output signal of the current amplifier is a dc voltage which tracks the drive error signal. This variable dc signal is compared against a sine wave timing signal, modified by end stop control (9) to produce a correctly positioned trigger pulse to the appropriate gate pulse generator (10). In turn the output of the gate pulse generator is amplified and electrically isolated by the correct gate pulse amplifier (11) before being applied to the appropriate thyristor (12) in the converter.

Other features of the drive include the power pack AM60 (13) which provides ± 15 and $+24$ V for the converter control system as well as ac phasing signals to the gate pulse generator; transient suppression (14) across the incoming three-phase supply; the tachogenerator monitor (15) which is basically a NAND gate whose function is to detect the loss of the tachogenerator feedback.

The gate pulse generator produces two output pulses spaced 60°

apart (double pulsing) to ensure thyristor turn-on when supplying inductive loads.

11-2-5 Chopper or dc-dc Control

The chopper or dc-dc converter is a method of control where a constant voltage dc source is switched rapidly on and off by a thyristor to obtain a variable dc voltage at the output terminals of the converter (see Figure 11-7).

The mean load voltage V_{do} can be varied in one of the following ways:

1. t_{ON} variable, t_{OFF} variable, and the periodic time T constant; this is known as **pulse width modulation.**

2. t_{ON} constant and t_{OFF} variable; this is known as **pulse rate modulation** or **frequency modulation.**

The preferred method of operating choppers is the pulse width modulation or constant frequency scheme as it is sometimes called.

The mean load voltage over a cycle is given by

$$V_{do} = \frac{t_{ON}}{t_{ON} + t_{OFF}} V_d = \frac{t_{ON}}{T} V_d \tag{11-8}$$

where T is the periodic time.

Choppers are used in a wide range of industrial and transportation applications where a constant dc voltage source is available. Typical applications are dc traction motor control in subways, cars, trolley buses, streetcars, switching power supplies or battery operated equipment such as forklift trucks, etc. The advantages are: (a) they eliminate rheostatic control for dc drive motor applications, (b) they eliminate the need for low efficiency motor-generator sets since the fixed dc voltage is usually obtained from a diode rectifier, (c) they present a high power factor to the ac supply for the rectifier, (d) they may be operated in four-quadrant applications, and (e) they are available up to 3,000 hp (2,240 kW).

11-2-5-1 The Basic Step-Down Chopper

The simplest form of the chopper is illustrated in Figure 11-7(a). This circuit uses one thyristor SCR1. When SCR1 is turned on, the dc source voltage V_d is applied to the load and diode $D1$ is reverse biased, the voltage across it being V_d. As can be seen from Figure 11-7(c), the load current will rise exponentially due to the inductive nature of the

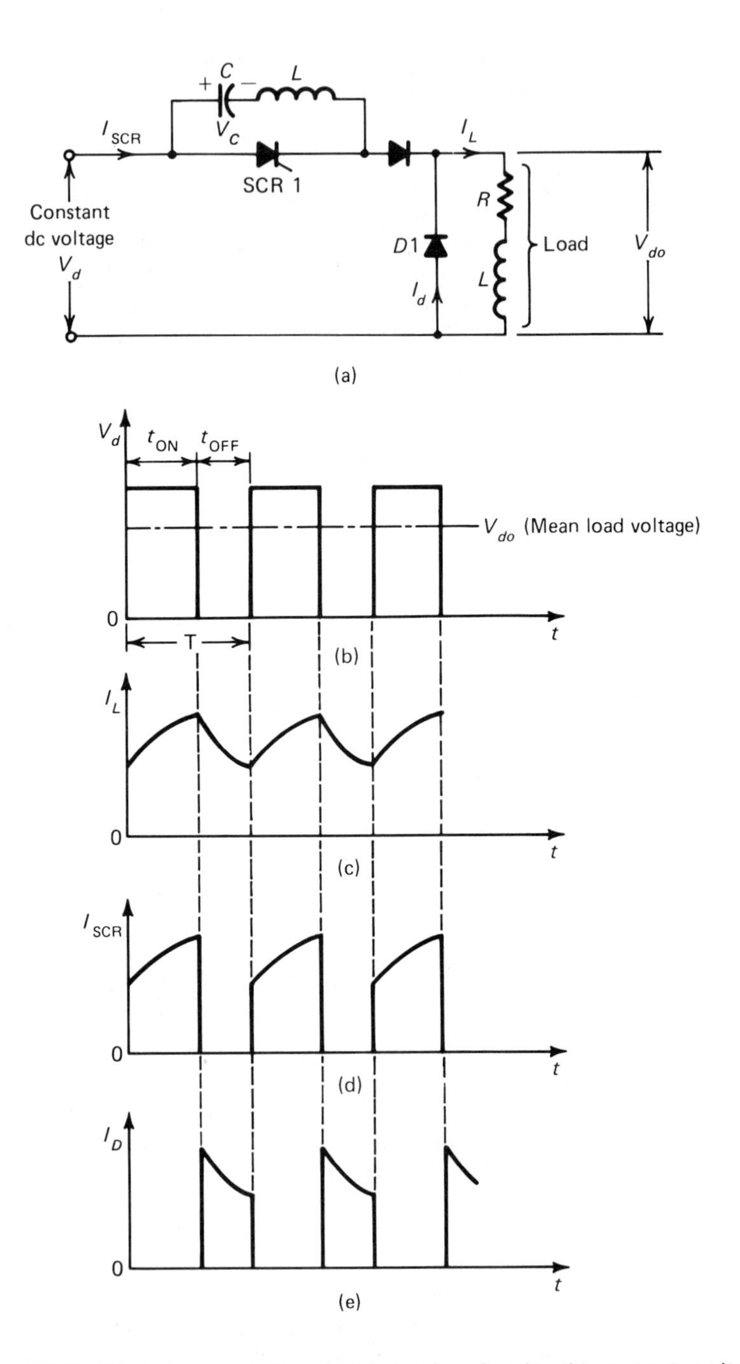

Figure 11-7. Step-down chopper: (a) basic circuit, (b) output voltage waveform, (c) load current waveform, (d) thyristor current waveform, (e) freewheel diode current waveform.

load, and the capacitor C, which was originally precharged with the lefthand plate positive as shown, will have the charge reversed by a counterclockwise current flowing through SCR1, L, and C. At this point the capacitor current reverses and the capacitor begins to discharge and replaces the load current flow through SCR1 initiating turn-off (assuming that there is not a gate signal present). At the instant that the capacitor starts to discharge, the voltage applied to the load is $V_{do} = V_d + V_c$. Assuming a constant load current I_L, the capacitor voltage V_c will decay linearly until $V_{do} = 0$. At this point the load current will be supplied from the stored energy in the inductance of the load via the freewheel diode $D1$. The function of diode $D2$ is to ensure that $V_c > V_d$.

This circuit, while it illustrates the basic step-down chopper principle, also introduces the concept of forced commutation; however, the commutation circuit has its drawbacks. The conducting period t_{ON} is partially determined by the values of L and C, but it is also modified by the magnitude and nature of the load current. As a result, this method of forced commutation is rarely used.

11-2-5-2 The Basic Step-Up Chopper

The step-down chopper always produces an output voltage V_{do} less than the input voltage V_d. However, a chopper can also be made to produce an output voltage greater than the input voltage. The basic step-up chopper circuit is shown in Figure 11-8. When SCR1 is gated on, current will flow through the inductance L and energy is stored in the magnetic field. When SCR1 is commutated off, the stored energy is returned to the circuit and the voltage applied to the load is $V_{do} = V_d + e_L$.

Assuming that the inductance L is quite large and that the load cur-

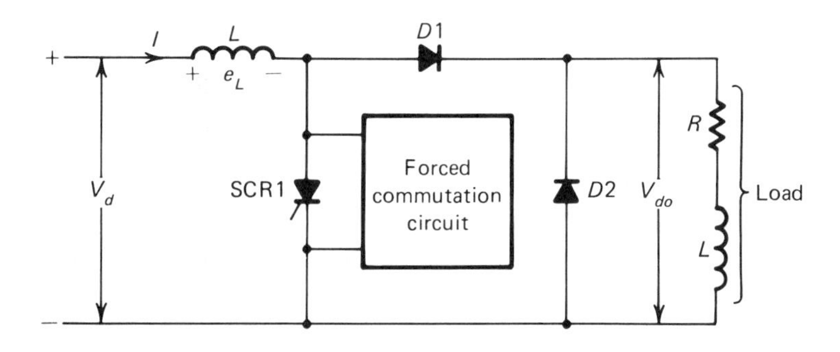

Figure 11-8. Step-up chopper.

rent I remains practically constant, during the time interval that SCR1 is conducting t_{ON} the energy stored in the inductor is

$$W_{in} = V_d I t_{ON} \ . \tag{11-9}$$

During the time that SCR1 is off t_{OFF}, the energy released to the load is

$$W_{out} = (V_{do} - V_d) I t_{OFF} \tag{11-10}$$

Assuming zero losses then

$$V_d I t_{ON} = (V_{do} - V_d) I t_{OFF} \tag{11-11}$$

therefore,

$$\begin{aligned} V_{do} &= V_d \frac{t_{ON} + t_{OFF}}{t_{OFF}} , \\ &= V_d \frac{T}{T - t_{ON}} \ . \end{aligned} \tag{11-12}$$

As can be seen, as t_{ON} approaches zero V_{do} approaches V_d, and as t_{ON} approaches the periodic time T, V_{do} approaches infinity.

Normally the chopping frequently is limited by the thyristor and the forced commutation circuitry to the range of 200 to 400 Hz. The chopper frequency may be raised by using power transistors or gate turn-off thyristors (GTOs). However, their use is currently limited by their power handling capability.

11-2-5-3 The Jones Circuit

The Jones circuit shown in Figure 11-9 utilizes pulse width modulation control to vary the mean load voltage V_{do}. Figure 11-9 illustrates the application of the Jones circuit to the control of a dc series motor. The major advantage of the Jones circuit is that the commutating capacitor C does not need to be precharged prior to initiating the load current pulse.

Assuming that C is discharged when SCR1 is fired, load current flows through $L1$ and the load, and since $L1$ and $L2$ are closely coupled, capacitor C will be charged with the lower plate positive by the induced voltage produced in $L2$ by the high di/dt in $L1$. The charge on capacitor C is trapped by diode $D1$ until it is released by turning on SCR2, at which point SCR1 is reverse biased and commutates off, and the charge on C is reversed via C, SCR2, $L1$ and the load.

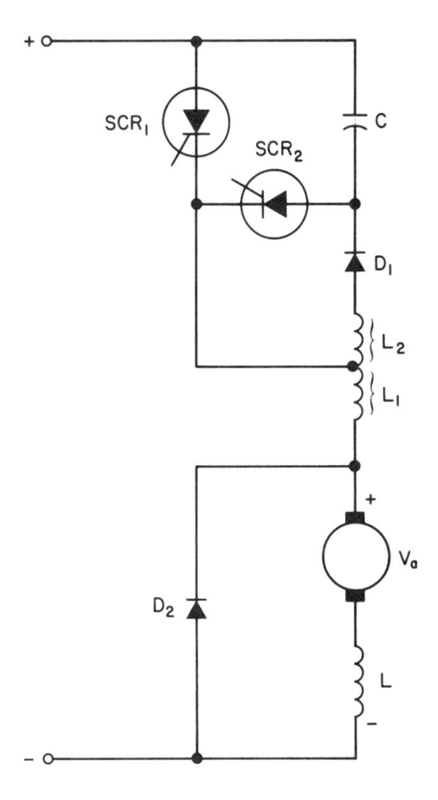

Figure 11-9. Series motor control using the Jones circuit.

Each time SCR1 is turned on, C is charged. The magnitude of the charge depends upon the magnitude of the load current. As a result, under heavy load current conditions, the charge on C will be at its greatest, which in turn aids commutation by decreasing the turn-off time of SCR1 under heavy loads.

The most common use of chopper control has been in the electric traction field. For example, they are being used to control trains in the BART (Bay Area Rapid Transit) system in the San Francisco area and subway cars in the Montreal and Toronto subway systems. Choppers are especially effective in battery operated vehicles since there is no wasted energy in rheostatic control and both regenerative and dynamic braking techniques can be utilized. The major benefits of chopper controlled drives are: smooth acceleration and deceleration; high efficiency; fast response; and, when supplied from an uncontrolled rectifier, the voltage applied to the load may be varied without the power factor penalty incurred when using thyristor phase-controlled converters. The minimum output voltage is determined by the turn-off times of the main and commutating thyristors which immediately determines the switching frequency.

11-2-6 Digital Speed Control

This section is based on an article presented in the *ASEA Journal* 1976, Volume 49, Number 5.

In most modern industrial plants there are a large number of speed-controlled dc motor drives, both for the control of the motor and also to control other variables in the process. For example, modern high speed paper machines require very precise control systems to ensure that the number of web breaks are kept to a minimum. Web breaks result from speed differences between speed controlled sectional drives. Canadian International Paper has recently installed a paper machine designed to run at 3,000 ft/min (914 m/min). The machine consists of 11 sections, each individual section being driven by a direct digitally controlled dc motor using ASEA's programmable microprocessor based DS-8 System.

ASEA has replaced the conventional analog tachogenerator by a pulse transmitter directly coupled to the drive motor shaft to ensure torsional stiffness used in conjunction with a frequency/digital converter. The major advantages of this arrangement are vastly improved accuracy and speed of response as well as improved linearity.

The key to the performance of the system is the frequency/digital converter. Past practice in digital speed measurement has been to count the number of pulses over a given time interval. The disadvantage of this method has been that it is necessary to have a count of several thousand pulses per revolution in order to achieve a satisfactory accuracy for the control system. This requirement is very difficult to achieve.

ASEA's patented frequency/digital converter uses a pulse transmitter with a relatively low number of pulses per revolution. This arrangement is possible because the previous speed reading is also used in the computation process. The basic principle is shown in Figure 11-10. *M* and *N* are numbers which affect the resolution and range of the converter.

A fixed frequency pulse train f_c is produced when a crystal controlled clock pulse generator 4 is fed to a digital frequency multiplier 5. The output of the multiplier is a pulse train f_R whose frequency is $(F/M)f_c$, where F is a binary number representing the measured value and stored in the result register 6 and M is the capacity of the result register. The pulse train with frequency f_x is produced by the pulse transmitter 2 and is supplied to the control circuit 3, which in turn controls

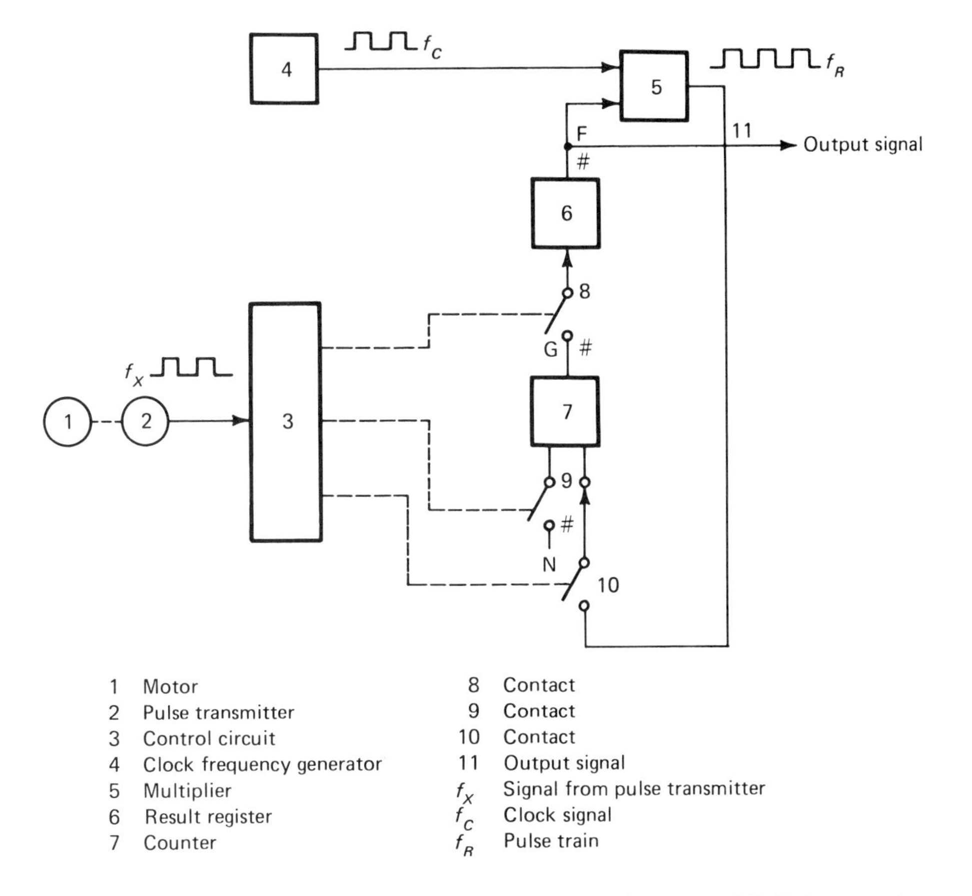

Figure 11-10. Schematic diagram of ASEA's frequency/digital converter. (Courtesy of ASEA Journal.)

the three contacts 8, 9, and 10. The following sequence of events occurs for each cycle of the unknown frequency f_x.

At the beginning of the cycle, contact 10 is closed and the contents of the counter 7 are successively counted down by one unit for each pulse in the pulse train f_R. During one cycle the number of subtracted pulses is

$$f_R/f_x = (F/M)(f_c/f_x) \tag{11-13}$$

At the end of the cycle, contact 10 opens, contact 9 closes, and a number N is added to the contents of the counter 7. Then contact 8 closes and the contents G of the counter are entered into the result register 6. The contents of the result register are now

$$F_{new} = F_{old} - (F_{old}/M)(f_c/f_x) + N \quad (11\text{-}14)$$

Contacts 8 and 9 open and contact 10 closes, the whole sequence of contacts opening and closing taking place in less than one clock cycle. The next calculation cycle takes place, and the whole procedure is repeated. After a number of cycles, the contents of the result register F will stabilize at

$$F = F_{new} = F_{old} = (M \cdot N/f_c)f_x \quad (11\text{-}15)$$

Figure 11-11 shows the dynamic response of the converter as a result of a change in f_x. If f_x suddenly increases at time t_1, the contents F of the result register will change from F_1 to F_2. The changeover from the old to the new values basically follow an exponential relationship. A time constant for this curve of the order of 2.5 ms may be obtained with a suitable choice of the parameters (f_c, M, and N).

Consider a conventional digital system for the following requirement:

Speed range	4 to 1000 rpm
Desired resolution in speed measurement	0.025%
Desired time constant of speed measurement	5 ms

A pulse transmitter producing 48,000 pulses per revolution with a pulse frequency between the pulse transmitter and counter of 800 kHz would be required. The method just described would only need 750 pulses per revolution and the maximum frequency between the pulse transmitter and the frequency/digital converter would be 12.5 kHz.

The block diagram of a typical closed-loop speed control utilizing this system is illustrated in Figure 11-12.

11-3 ac MOTOR SPEED CONTROL

The vast majority of polyphase ac motors used in industrial, commercial, and residential applications run at fixed speeds. Variable-speed operation of polyphase ac motors has long been the goal of engineers, and many interesting schemes have been developed. The increasing use of the thyristor, power transistor, and gate turn-off thyristor (GTO) have made variable-speed operation of ac motors not only possible but economically viable because of the steadily decreasing cost of solid-state circuitry.

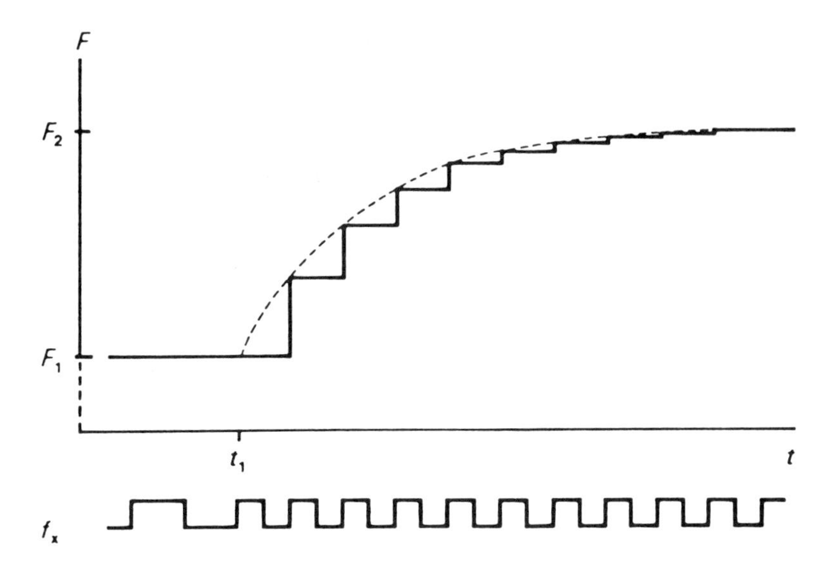

Figure 11-11. Dynamic response of the converter to a change in speed. (Courtesy of ASEA Journal.)

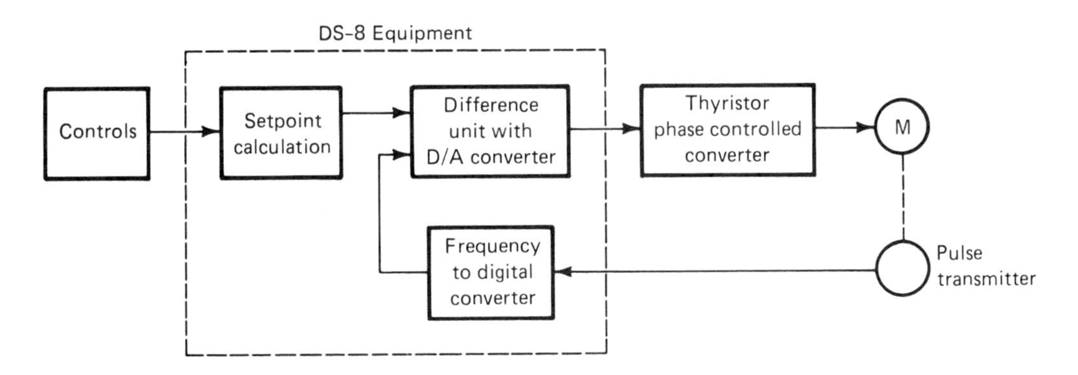

Figure 11-12. Digital speed control system using the DS-8 system. (Courtesy of ASEA Journal.)

The main types of polyphase ac motors that readily lend themselves to variable speed control are the polyphase squirrel cage induction motor (SCIM), the wound rotor induction motor (WRIM), and the synchronous and synchronous-reluctance motor.

The normal methods of speed control are all based upon

$$S = \frac{120f}{P} \quad \text{rpm} \tag{11-16E}$$

or

$$\omega = \frac{4\pi f}{P} \text{ rad/sec} \tag{11-16SI}$$

where S or ω = synchronous speed of the rotating magnetic field

f = frequency (Hz), and

P = number of stator poles/phase

Synchronous machines such as the synchronous motor and the synchronous-reluctance motor rotate at synchronous speed. In the case of asynchronous motors, such as the SCIM and WRIM, the actual rotor speed (S_R or ω_R) is load dependent and is slightly less than the synchronous speed. Since the rotor torque is produced by the interaction of the rotor magnetic field produced by the induced rotor currents and the polyphase rotating magnetic field, the speed difference is termed the slip, and is

$$s = \frac{S - S_R}{S} \tag{11-17E}$$

or

$$s = \frac{\omega - \omega_R}{\omega} \tag{11-17SI}$$

From Equations (11-16E), (11-16SI), (11-17E), and (11-17SI) there are two possible ways of varying the speed of polyphase motors. The first method, which applies to all polyphase motors, is by controlling the frequency, since S or ω is proportional to frequency, or changing the number of poles/phase. The second method, which applies primarily to WRIMs, is speed controlling by varying the slip; however, if rheostatic methods are used, this method results in very high losses and low efficiencies.

The preferred method is by variable-frequency control. As was previously developed in Chapter 5 [Equation (5-14)], the internal torque developed by a polyphase induction motor is

$$T = \frac{mP}{4\pi} \cdot \left[\frac{E_1}{f}\right]^2 \cdot \frac{f_r + r_2}{\left[r_2^2 + s^2 x_2^2\right]}$$

from which it can be seen that the internal torque developed is proportional to E_1/f. The air-gap flux must be maintained constant at all frequencies over the range where a constant torque is required. This means that the ratio E_1/f must also be constant, in order to produce a constant air-gap flux. Assuming that the stator leakage flux is small, E_1 is approximately equal to the applied stator voltage V_1, and the air-gap flux will also be constant as long as the ratio of V_1/f is constant.

However, at low frequencies the phase resistance is the major component of the phase impedance. This condition will result in a decrease of air-gap flux. Normally, the V_1/f ratio is slightly boosted at low frequencies in order to maintain the torque, even though this may result in the stator iron becoming saturated.

The maximum or breakdown torque is constant for a given machine at all frequencies up to the rated or motor nameplate frequency and is proportional to the air-gap flux squared and inversely proportional to the rotor leakage reactance. The only effect of rotor resistance is to determine the slip at which breakdown torque is developed. Under constant air-gap flux conditions, the starting torque is greater at all frequencies than when supplied at rated frequency and voltage (see Figure 11-13).

In summary, operating a polyphase induction motor, either a squirrel cage or wound rotor, with a constant V/Hz ratio will result in higher starting and breakdown torques, and with the same full load slip the torque is greater at the higher frequencies. Additionally, the horsepower (kW) output and efficiency are greater at the higher frequencies. Failure to maintain the constant V/Hz ratio will affect the constant torque output by the square of the air-gap flux density, or will cause the stator current to increase and overheat the motor.

In most variable-frequency drives the constant V/Hz ratio is maintained up to the rated frequency of the motor and then the applied stator voltage is maintained at its rated value as the frequency is increased. As a result, the motor operates in a constant torque mode up to rated frequency (usually 60 Hz in North America and 50 Hz in Europe), and then operates in a constant horsepower (kW) mode above rated frequency (see Figure 11-14).

11-3-1 Variable Frequency Control (Six-Step)

A typical commercial six-step variable-frequency dc link converter is shown in Figure 11-16. This converter is the Veritron® RWK 00831 Frequency Converter manufactured by Brown Boveri Corporation. The vital statistics are: three-phase 380 V, 60 Hz; normal rated output 8.5 kVA, short time maximum rated output 13.0 kVA; input line current 13 A; output frequency range 5 to 150 Hz; output voltage at rated current 30 to 360 V, three-phase; rated output line current 13 A; and maximum rated line current 20 A.

This converter is a dc link converter using an uncontrolled three-phase diode rectifier. The constant dc output voltage is smoothed by a

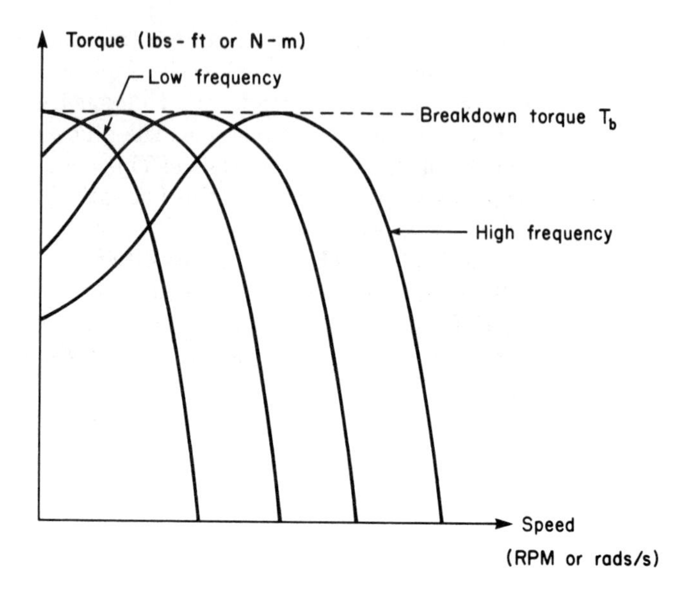

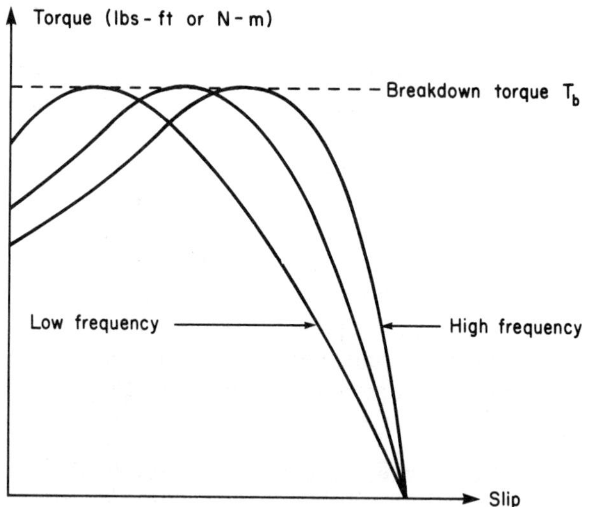

Figure 11-13. Torque-speed and torque-slip curves of a polyphase induction motor with a variable-frequency supply.

series inductance and parallel capacitor arrangement at dc link 1. In order to meet the *V*/f requirement, the input dc voltage to the inverter is controlled by the two thyristor chopper shown in Figure 11-15. Assuming that capacitor *C*1 is precharged to approximately V_d, the basic principle of operation is as follows. When SCR1 is turned on it connects the dc source to the inverter and load current will flow. At the

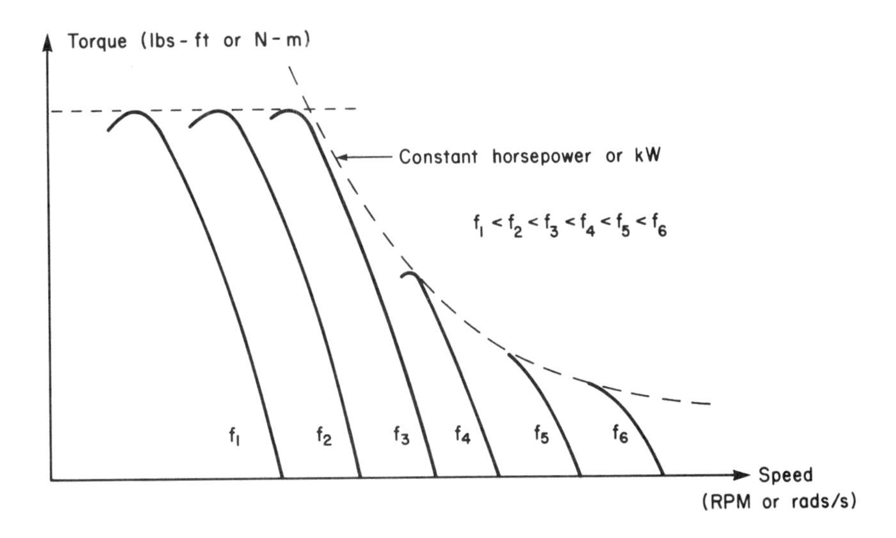

Figure 11-14. Torque-speed curves for variable-frequency operation in constant torque and constant horsepower (kW) modes.

same time, the charge on capacitor $C1$ is reversed and $C1$ is charged via $L1$, $D1$, and SCR1. When the capacitor charging current has decayed to zero, the charge is retained by the reverse bias on $D1$. Load current will continue to flow through SCR1 until turn-off is initiated by turning on SCR2. The load current is now supplied via SCR2 and $C1$ and SCR1 commutates off, and the load current is transferred to the freewheel path formed by $D3$ and $D4$. When SCR1 is gated on at the beginning of the next ON cycle, $D3$ is reverse biased and load current again flows throught SCR1. The action of the chopper is to produce a variable dc input voltage suitably smoothed by dc link 2 to the inverter.

The inverter section is a conventional three-phase thyristor inverter producing a six-step output with three thyristors in conduction at any one time, as previously described in Section 5-9-1. However, the commutation technique is unique in that all conducting thyristors are commutated off together by the collective turn-off circuit consisting of SCR3, $L3$, and SCR4 (see Figure 11-15). This circuit, when turned on, pulls the positive dc bus down to zero, forcing all conducting SCRs to turn off; when SCRs 3 and 4 block, the desired SCRs in the inverter are triggered. This action is repeated each time an SCR requires to be commutated off.

The collective turn-off circuit operation is as follows. Capacitor $C2$ charges with the top plate positive from the positive dc bus through $D2$, $L2$, and $L4$. When SCRs 3 and 4 are turned on, $C2$ discharges through SCR4 and recharges in the opposite direction and then at-

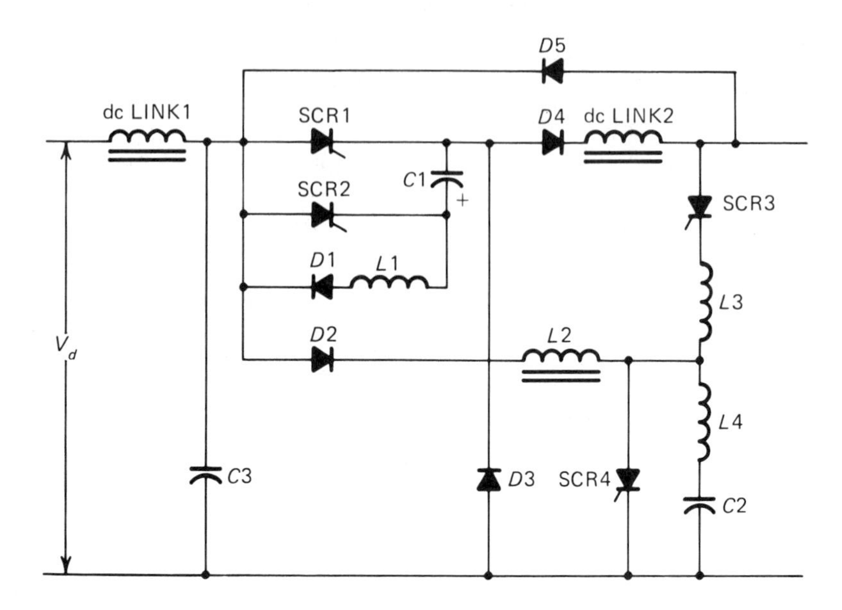

Figure 11-15. dc voltage control and group commutation for the Veritron RWK 00831 frequency converter.

tempts to discharge again through SCR4 and commutates it off. $C2$ then recharges again through $D2$, $L2$, and $L4$, ready for the next collective turn-off of the thyristors in the inverter. With this commutation circuit, the commutation energy is constant irrespective of the inverter load current. However, if the motor load is in excess of the current ability of the commutation circuit, the group turn-off action will fail and dc current will be applied to the motor causing it to brake.

A diode is connected in antiparallel with every thyristor in the inverter to provide a feedback path for the returning reactive energy from the motor. The function of diode $D5$ is to provide a path for this returned energy to bypass the chopper, the excess energy being stored in capacitor $C3$, which acts to prevent a voltage build-up across the inverter.

Referring to Figure 11-16, the overall operation of the inverter is as follows. Initially relay $K1$ is energized to precharge the capacitor connected directly across the output of the uncontrolled bridge converter. A current signal I_{ref}, designating the desired operating frequency (speed), is applied to the ramp generator (1). The output of the ramp generator establishes: (a) the V/Hz ratio by means of the voltage control (2), and (b) the frequency transmitter (3) generates an output signal six times the desired switching frequency. The voltage control output is applied to the chopper control (4) which directly establishes the control signals to the chopper SCRs to produce a variable dc output

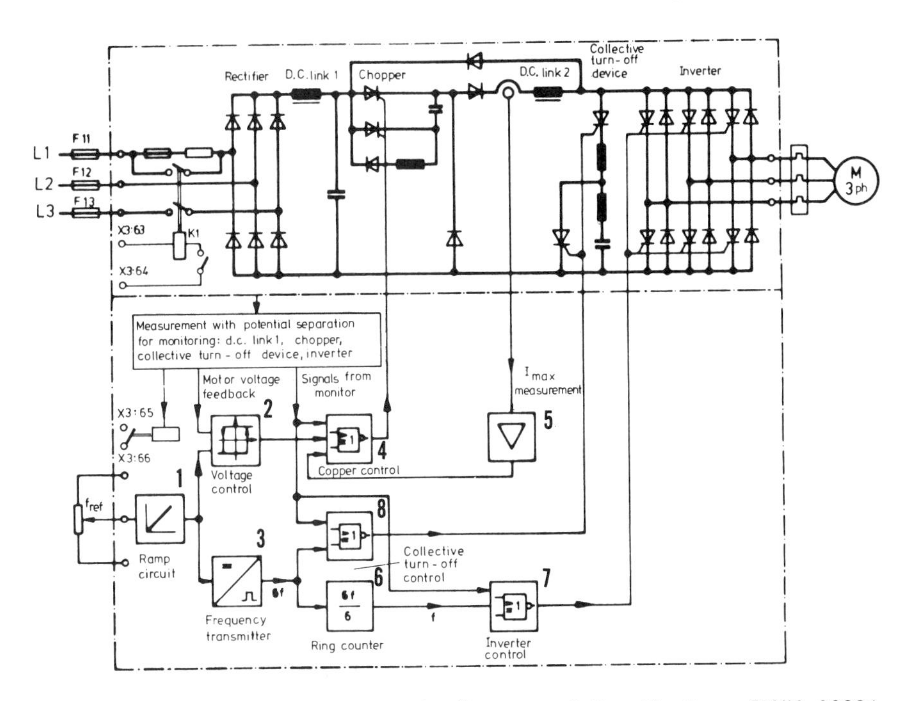

Figure 11-16. Elementary circuit diagram of the Veritron RWK 00831 frequency converter. (Courtesy of Brown Boveri.)

voltage. This voltage is smoothed by the inductance dc link 2, before being applied to the inverter section. The current is monitored by the I_{max} measurement (5) unit and, if it exceeds a prespecified value, it will apply an inhibit signal to the chopper control module (4) which will cause the duty cycle of the chopper to be reduced. The frequency transmitter output is applied to a six-stage ring counter (6) whose output is applied to the inverter control (7). The inverter control, provided an inhibit signal is not received from the monitoring section, will then apply correctly sequenced gating pulses to the inverter SCRs at 60° intervals. The frequency transmitter (3) also produces a 6f signal to the collective turn-off control (8) which is ANDed with a signal from the monitoring section. The output signal from the collective turn-off control directly controls the SCRs in the electronic crowbar to force commutate the conducting SCRs in the inverter section.

11-3-2 Variable-Frequency Control (Pulse Width Modulation)

Modern dc link converters can be programmed to provide the constant V/Hz ratio for constant torque control by controlling the amplitude of the output ac voltage by voltage control in the inverter using pulse width or pulse width modulation control (see Section 5-7-2). The preferred method is pulse width modulation (PWM) which achieves a significant reduction in the harmonic content of the output waveform. However, a pulse width modulated system will also produce high frequency currents in the motor circuit as a result of the rapid switching requirements. Even though these currents are small in amplitude due to the reactance presented by the motor windings, they will produce vibration and noise in the iron of the magnetic circuit as a result of magnetostriction effects. An additional problem is the effect of a train of high-voltage pulses on the motor winding which will impose a high-voltage stress on the insulation of the first few turns of each phase winding of the motor. As a result, the life expectancy of the winding insulation must be derated when using full-voltage PWM control.

Lovejoy Electronics Inc., has developed the Love AC MPR-IV microcomputer controlled inverter, which is available in a range of outputs from 1 hp (1.64 kVA) to 25 hp (29 kVA). A typical MPR-IV inverter is shown in Figure 11-17.

Figure 11-17. The Lovejoy MPR IV inverter. (Courtesy of Lovejoy Electronics, Inc.)

The conventional six-step inverter creates 5th, 7th, 11th, and 13th harmonics which are relatively large in amplitude and produce motor heating. Unfortunately, these harmonics cannot be eliminated. PWM inverters decrease the harmonic amplitudes by increasing the number of pulses per cycle. The limitation on the number of pulses per cycle has been the SCR which is limited by its switching and recovery times. Lovejoy utilizes gate turn-off thyristors (GTOs) with switching times of approximately 200 ns. As a result, the carrier frequency can be significantly increased, and in conjunction with the dedicated microprocessor which calculates the optimum number of pulses per cycle for each speed setting produces an accurately sine-coded output at all frequencies. For example, for a 1 to 3 Hz output the GTOs are switched at 1200 pulses per cycle.

The block diagram of the Lovejoy MPR-IV inverter is shown in Figure 11–18. The inverter is supplied from a 230 or 460 V three-phase source which is then converted by a three-phase diode bridge to a constant dc output voltage, filtered and applied to the static inverter via a *di/dt* current limit sensor. The board power supplies are also directly supplied from the three-phase ac input. The setpoint speed voltage signal is applied to the buffer and ramp generator where the acceleration and deceleration rates and minimum and maximum speeds are determined. The output of the ramp circuit is applied to the voltage-controlled oscillator (VCO). The VCO produces a variable-frequency pulse output which represents the set speed voltage signal modified as appropriate by the ramp generator. This signal is used as a clock signal by the microprocessor and the PWM control. The PWM control section establishes the constant V/Hz relationship and provides an automatic V/Hz boost for starting the motor against a high load torque. The logic section consists of a dedicated 48-pin microprocessor based on the Intel 8080 designed and manufactured by Lovejoy. The microprocessor logic scheme with sine-coded, double-edged modulation, coupled by optocouplers to the GTOs in the power section, results in the production of a very close approximation to a sine wave output to the motor load.

Other features of the MPR-IV inverter include the following.

1. Linear acceleration from 0 V 0 Hz to the set speed over an adjustable time range of 2 to 120 sec. This feature eliminates conventional motor starting equipment and limits the starting current.

2. The provision of jog control, adjustable between 0 and 40 Hz, provides position control of the load.

3. The motor may be reversed at any speed without damage to the inverter; the inverter dynamically brakes the motor to a complete stop

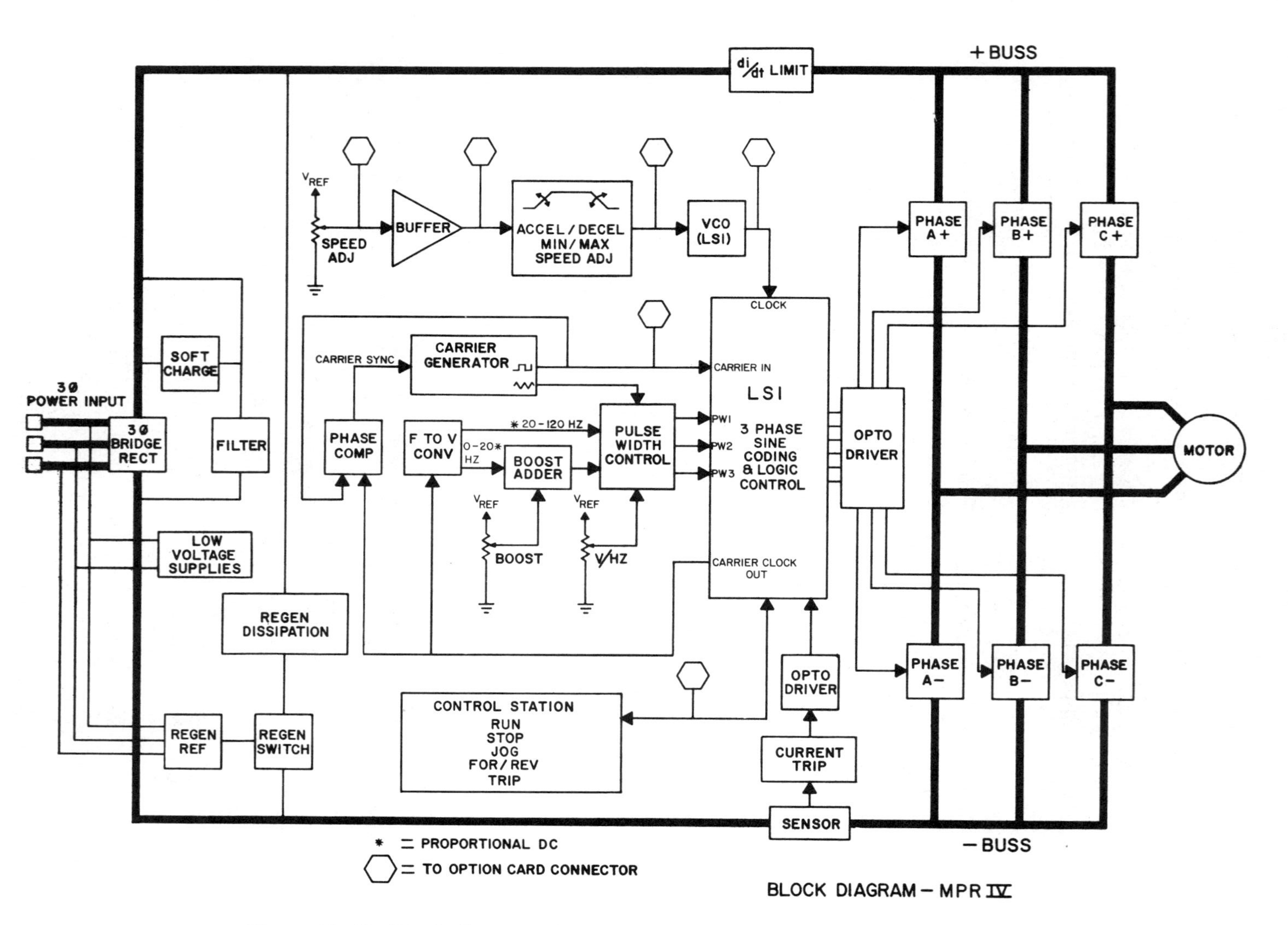

Figure 11-18. Block diagram of the Lovejoy MPR IV Inverter. (Courtesy of Lovejoy Electronics, Inc.)

and then accelerates it back to the original speed in the reverse direction while at the same time limiting the output current.

4. The inverter can also accept the returned energy under overhauling load conditions up to 150% of the continuous current rating for up to 12% of any one-minute duty cycle. With an add-on option this capability can be extended up to 100% of the duty cycle.

5. A built-in control ride through circuit compensates for power interruptions of six cycles or less (0.1 sec) without damage to the unit.

Adjustable frequency speed control of synchronous motors produces one of the most efficient process drives. A variable frequency inverter may be used to control several motors in parallel across its output, or simultaneous control of a number of inverters can be achieved by using a master oscillator or digital ratio controller.

Because of the essentially constant speed characteristics of polyphase induction motors, unless very precise speed control is required, it is normal practice to use open-loop control techniques and rely on the stability of the VCO.

When a constant speed is essential, a speed signal representing the actual shaft speed must be brought back to the VCO to modify the output frequency of the inverter. Alternatively, in a constant speed application a synchronous or synchronous-reluctance motor should be used.

11-3-3 Wound-Rotor Motor Control—Slip Power Recovery

The rotor speed of a polyphase wound rotor induction motor (WRIM) can be controlled by either varying the frequency of the three-phase supply to the stator or by controlling the power flow in the rotor circuit.

The most common method of obtaining variable speed control of a WRIM was by the use of an adjustable external resistance connected to the rotor circuit to obtain changes in the torque-speed characteristics of the drive. This arrangement was particularly suitable in applications requiring a high starting torque with a low inrush current, or smooth acceleration. Typical examples of these applications are elevators, forced draft fans, water and sewage pumps, cranes, and conveyor systems. The major disadvantage of rotor resistance control is that the energy in the rotor is dissipated as heat in the controlling resistance. As a result, the motor efficiency drops significantly as speed is reduced. In general, with this method the speed is usually never reduced below 50% of the synchronous speed because the overall efficiency will also be less than 50%.

In this age of increasing energy costs, it only makes good sense to convert, by solid-state means, the traditional rheostatic method of speed control to a regenerative control system. This type of system is known as slip power recovery, and a typical system outlined in Figure 11-19, can result in up to 98% of the previously wasted energy being recovered. The variable-voltage variable-frequency rotor voltage is converted to dc by the three-phase diode rectifier. The dc voltage varies from approximately zero at full speed to a maximum at standstill. The dc voltage is supplied to a three-phase inverter whose SCRs are gated in synchronism with the stator supply voltage. By varying the inverter firing angle, the amount of power extracted from the rotor circuit and returned to the three-phase supply can be controlled, and thus the rotor speed can be varied. To operate as a closed-loop system with variable torque and speed capabilities requires that input signals representing the desired speed and torque be compared with the actual speed signal derived from the tachogenerator coupled to the motor shaft. The actual torque derived is the voltage drop across the resistor in the dc link circuit, the resulting error signals being used to control the inverter firing angle. As the inverter firing angle decreases, the amount of power returned to the three-phase supply increases, and the motor speed decreases, or vice versa.

At first glance the cost of the control system would not appear to be justified; however, with power costs escalating as they are at present and in the foreseeable future, the improved efficiency results in a relatively short payback time. For a new installation, the overall cost of the whole drive is less than for an equivalent adjustable speed dc drive. In addition, there is a significant reduction in overall maintenance costs. Also, because of the total control capability, starting inrush currents are eliminated, which immediately removes the limitation on the number of starts before the machine must be allowed to cool. The major disadvantage of the system is its inherently poor power factor, which is of the order of 0.5 at maximum speed and full load and decreases as the speed is reduced. This problem can be reduced by connecting power factor correcting capacitors in delta across the stator terminals.

11-3-4 Variable Stator Voltage Speed Control of Induction Motors

The variable-frequency control schemes already discussed permit operation at the best efficiency while permitting the motor to operate at a speed approaching synchronous speed for the source frequency. However, there are a number of applications where the critical factor is the output torque not the motor speed. Recall that the maximum torque developed by a polyphase induction motor is proportional to

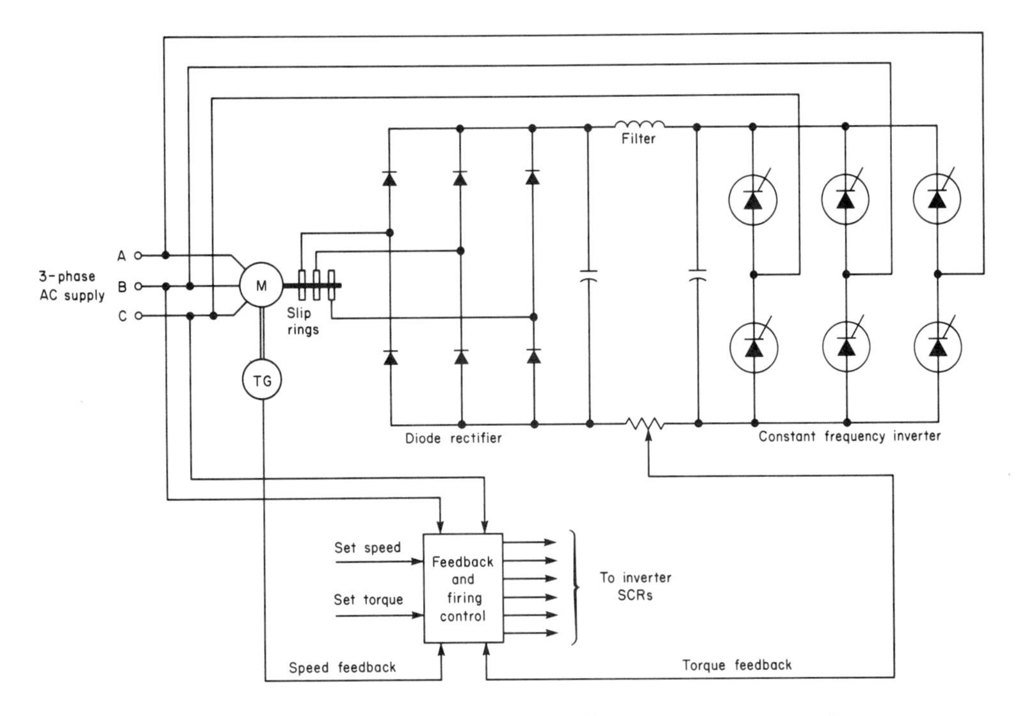

Figure 11-19. Wound rotor slip power recovery system.

the square of the air-gap flux, which in turn is controlled by the applied stator voltage.

As a result, the output torque of the motor is proportional to the square of the applied stator voltage; therefore, under steady-state load conditions, the motor torque adjusts itself to supply the load torque. As the stator voltage is decreased, the rotor speed decreases; however, the power dissipated by the rotor circuit must increase in order to maintain a constant torque output. With a normal rotor (NEMA Class A or B) and a constant load torque, this method of speed control will result in excessive losses and poor efficiency. The use of a high resistance rotor, NEMA Class D, will reduce these losses and give a stable speed control system. The system is most suitable in controlling fan and pump loads as encountered in heating, ventilating, and air conditioning (HVAC) where variable air flow is required. A typical system is shown in block diagram form in Figure 11-20.

The actual speed signal produced by the tachogenerator coupled to the motor is compared with the desired speed signal, and the output of the error amplifier is used to control the firing delay of the inverse-parallel connected SCRs in series with the motor stator. If there is an

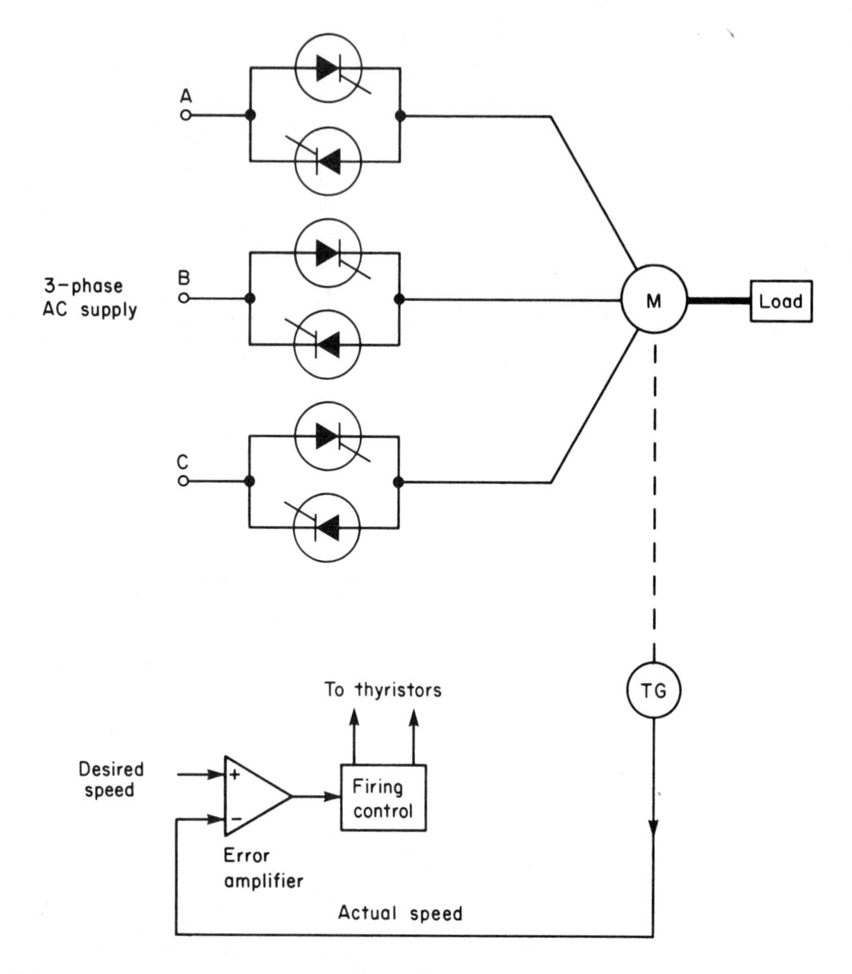

Figure 11-20. Polyphase induction motor speed control by means of variable stator voltage control.

increase in the load torque, the resulting speed decrease is detected by the tachogenerator. The resulting error signal will cause the firing control to advance the firing pulses so as to increase the stator voltage and speed, and vice versa. This circuit can also be modified to give a reduced voltage starting capability by the addition of a ramp generator, which will give a linear timed acceleration starting voltage to ensure a breakaway torque is established. As the motor accelerates, the tachogenerator feedback signal will cause the stator current to be reduced to a value that will give smooth acceleration up to the set speed. A disadvantage of this system is that the stator voltage waveforms are distorted and the current flow is not continuous. This will result in the production of harmonics in the stator and rotor circuits which will produce harmonics and increase the eddy current heating effects, that is, the motor will run at a higher operating temperature and lower performance as compared to its operation with a sinusoidal supply.

When used in continuous motor operation applications, the performance may be improved by using a power factor controller invented by Frank J. Nola in 1977 and patented by the National Aeronautics and Space Administration. The Nola power factor controller senses the power factor of the line supplying the motor and adjusts the applied stator voltage so that the motor is always operating at a fully loaded state which in turn improves the power factor. Nola type reduced voltage controllers can be applied to existing motor applications without serious matching problems, the advantages being that the overall power factor is improved, the motors operate at lower temperatures, and as a result, the insulation life expectancy has been extended as well as operating at lower vibration levels. Probably the most important benefit is the reduction of energy consumption.

11-3-5 Summary

Solid-state drive systems, either ac or dc, are making tremendous inroads into every area of industry. Initially, dc drives were developed more rapidly than the ac drives because of the reduced complexity of the control circuitry, initial cost, and because closely regulated single or multi-motor dc drive systems are simpler than the equivalent ac system. However, the extremely rapid advances that have been made in the area of variable-frequency control combined, with the lower initial and subsequent maintenance costs of the polyphase induction motor, are resulting in the variable-frequency ac drives challenging the traditional dc drive.

The continuing decline in the cost of thyristors, together with increased power capabilities and high efficiency, is seeing the thyristor making significant inroads into all areas of control which used to be dominated by electromagnetic methods. The rapid increase in the capability of the power transistor has extended its use into high horsepower (kW) ac motor speed control applications where it has the advantage that forced commutation circuitry is unnecessary. Another competitor that is beginning to emerge is the GTO, which has yet to completely prove itself, but has very good potential for PWM variable-frequency ac drives.

11-4 UNIVERSAL MOTORS

The series universal motor is used extensively in portable tools, in household appliances such as blenders, and in small high-speed applications. Solid-state control circuits for the control of universal motors are simpler and cheaper than those associated with dc motors and polyphase ac motors. In general, they are classified as half-wave or full-wave controls with or without feedback.

11-4-1 The R-C Phase Shift and Extended R-C Phase Shift Circuit

Most series universal motor controls use standard firing circuits which are based on the RC phase-shift principle. The basic block diagram of a phase-shift is shown in Figure 11-21. The phase-shift control is synchronized to the ac source and generates the firing delay angle. The function of the trigger is to produce a strong enough pulse to trigger a wide range of thyristors without modifying the circuit design.

The simplest possible RC phase-shift circuit is shown in Figure 11-22. This depends upon the fact that the voltage drop across the resistance V_R leads the voltage drop across the capacitator, V_C or V_{OUT}, by 90°, the phasor sum of V_R and V_C at all times being equal to V_{ac}. The phase shift between V_{OUT} and V_{ac} is the firing delay angle α, as shown in Figure 11-22(b). Increasing R will increase V_R and increase α, and reducing R will reduce V_R and α. This circuit theoretically permits α to be varied from 0 to 90°; however, in actual practice a variation from 10 to 80° is more realistic.

From Figure 11-22 (b)

$$\tan \alpha = \frac{V_R}{V_{\text{OUT}}} = \frac{IR}{Ix_c} \tag{11-18}$$

where I is the current flowing through the series R-C circuit. Therefore

$$\tan \alpha = \frac{R}{x_c} = \frac{R}{1/\omega C} = R\omega C \tag{11-19}$$

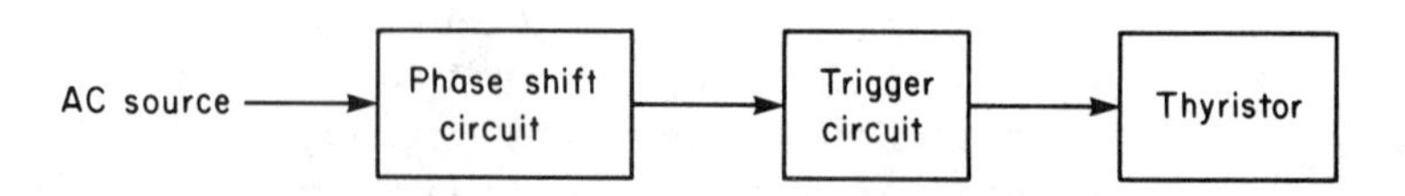

Figure 11-21. Block diagram of a complete phase shift control of a thyristor.

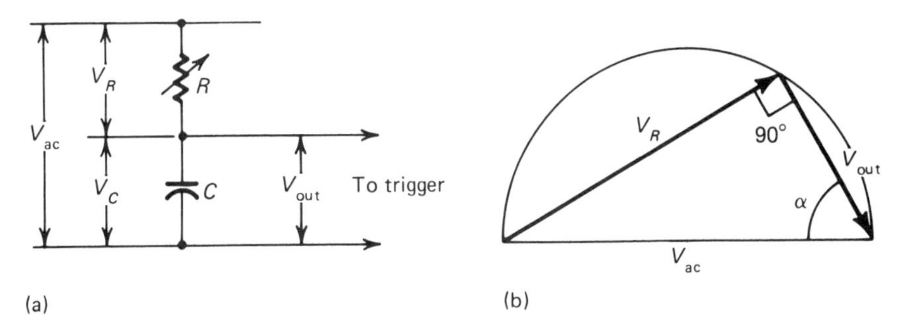

Figure 11-22. Basic *RC* phase shift: (a) basic circuit; (b) phasor diagram interrelating V_R, V_{OUT}, and α.

where R is the resistance, ohms

ω is the angular velocity, rad/s

C is the capacitance, farads

Example 11-1

In Figure 11-22(a), when $R = 10\text{k}\Omega$, $C = 0.22\mu\text{F}$, and $V_{ac} = 115\sin 377t$, what is the firing delay angle α?

Solution:

From Equation (11-19)

$$\tan\alpha = R\omega C = 10\text{x}10^3\text{x}377\text{x}0.22\text{x}10^{-6}$$

$$= 0.8294$$

and

$$\alpha = 39.67^0$$

To increase the range of firing angle control the basic *R-C* phase shift is modified as shown in Figure 11-23. The circuit consists of two equal value resistances $R1$ and $R2$ connected across the ac source to form one arm of a bridge. The other arm of the bridge is formed by the variable resistance R and the capacitor C. The output voltage V_{OUT} is taken from points A and B. As can be seen from the phasor diagram, as R is increased to infinity, α approaches 180°, and as R is decreased to zero, α decreases to 0°. In triangle ABC it should be noted that V_{R1}, V_{R2}, and V_{OUT} are equal. From triangle ABC

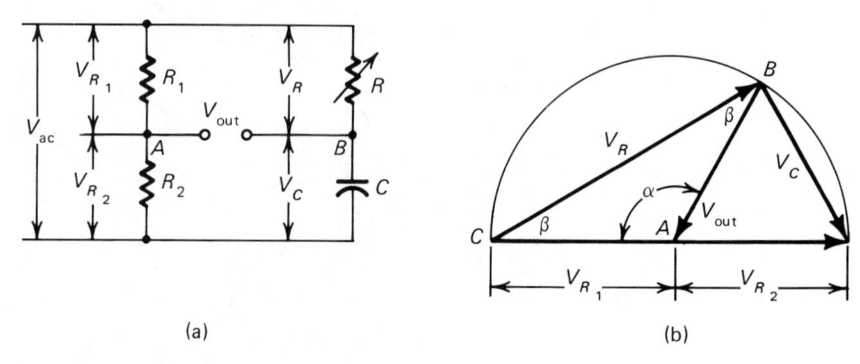

Figure 11-23. Extended *RC* phase-shift control: (a) basic circuit; (b) phasor diagram.

$$\alpha + 2\beta = 180° \tag{11-20}$$

$$\alpha = 180° - 2\beta \tag{11-21}$$

$$= (90° - \beta) \tag{11-22}$$

Since

$$\tan\beta = \frac{V_c}{V_R} = \frac{IX_c}{IR} = \frac{X_c}{R} = \frac{1/\omega C}{R} = \frac{1}{R\omega C}$$

and

$$\tan(90°-\beta) = \cot\beta = \frac{1}{\tan\beta}$$

then

$$\tan\alpha/2 = \cot\beta = R\omega C \tag{11-23}$$

Example 11-2

In Figure 11-23(a), when R=10kΩ, C=0.22μF and V_{ac}=115sin377t, what is the firing delay angle α?

Solution:

From Equation (11-23)

$$\tan\alpha/2 = R\omega C$$
$$= 10\text{x}10^3\text{x}377\text{x}0.22\text{x}10^{-6}$$
$$= 0.8294$$

and

$$\alpha/2 = 39.67°$$

therefore $\alpha = 79.34°$

This means that the delay angle α is exactly twice the value calculated in Example 11-1 when exactly the same component values were used in both circuits.

11-4-1-1 Phase-Shift Circuit Design

In order to select the most suitable values of R and C it is necessary to analyze the application and then determine the component values. The average power delivered to a load is

$$P = V^2/R \tag{11-24}$$

where V is the rms or effective voltage

In terms of the firing delay angle α it can be shown that

$$V = \left\{ \frac{1}{2\pi} \int_{\alpha}^{\pi} V^2_{max} \sin^2\omega t \omega t \right\}^{1/2} \tag{11-25}$$

Fortunately it is not necessary to solve this equation since *rms* and average relationships versus delay angle α curves have been developed to permit graphical solution techniques. Figure 11-24(a) is used for half-wave and SCR applications and Figure 11-24(b) is used for full-wave and TRIAC applications.

The use of these curves is best illustrated by an example.

Example 11-3

An electric heater with a resistance of 15Ω is supplied from a 208V 60Hz source. It is necessary to maintain the power supplied to the heater between 1000W and 1400W when using a phase shift controlled SCR. Determine (1) the required range of the firing delay α, and (2) when $C = 1\mu F$ determine the required range of R.

Solution:

The maximum power that can be delivered to the load is

(1) $$P_{max} = \frac{V^2}{R} = \frac{208^2}{15} = 2884\text{W}$$

The ratio of $\frac{P}{P_{max}}$ is calculated for the required range.

$$\frac{P_{\alpha(min)}}{P_{max}} = \frac{1400}{2884} = 0.49$$

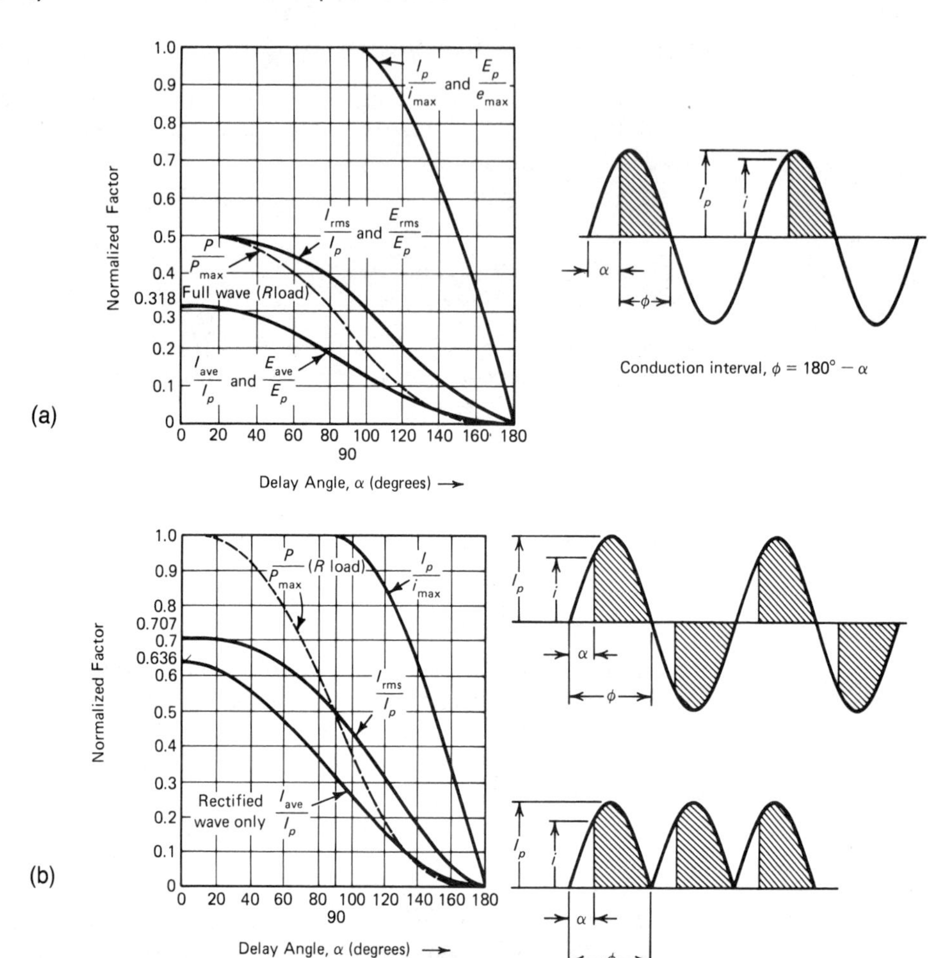

Figure 11-24. RMS and average relationships for (a) half-wave and SCR applications; and (b) full-wave and TRIAC applications. (Courtesy of Westinghouse Electric Corp., Semiconductor Div.)

and
$$\frac{P_{\alpha(\max)}}{P_{\max}} = \frac{1000}{2884} = 0.35$$

From Figure 11-24(a) for

$$\frac{P_{\alpha(\min)}}{P_{\max}} = 0.49 \qquad \alpha_{(\min)} = 25°$$

and
$$\frac{P_{\alpha(\max)}}{P_{\max}} = 0.35 \qquad \alpha_{(\max)} = 70°$$

(2) Since the maximum firing delay angle is 70° the basic R-C phase shift circuit may be used. Then

$$\tan \alpha = R\omega C$$

and

$$R = \frac{\tan \alpha}{\omega C}$$

therefore

$$R_{min} = \frac{\tan \alpha_{(min)}}{\omega C}$$

$$= \frac{\tan 25°}{2\pi \times 60 \times 1 \times 10^{-6}}$$

$$= 1237\Omega$$

$$R_{max} = \frac{\tan \alpha_{(max)}}{\omega C}$$

$$= \frac{\tan 70°}{2\pi \times 60 \times 1 \times 10^{-6}}$$

$$= 7288\Omega$$

The practical circuit would most likely be a 1kΩ fixed resistor in series with a 10kΩ potentiometer.

Example 11-4

Repeat Example 11-3 using a TRIAC.

Solution:

(1)

$$\frac{P_{\alpha(min)}}{P_{max}} = 0.49$$

and

$$\frac{P_{\alpha(max)}}{P_{max}} = 0.35$$

From Figure 11-24(b) for

$$\frac{P_{\alpha(min)}}{P_{max}} = 0.49 \qquad \alpha_{(min)} = 92°$$

and

$$\frac{P_{\alpha(max)}}{P_{max}} = 0.35 \qquad \alpha_{(max)} = 112°$$

(2) Since $\tan \alpha/2 = R\omega C$

then

$$R_{min} = \frac{\tan \alpha_{(min)}/2}{\omega C}$$

$$= \frac{\tan 46°}{2\pi \times 60 \times 1 \times 10^{-6}}$$

$$= 2747\Omega$$

and

$$R_{max} = \frac{\tan 56°}{2\pi \times 60 \times 1 \times 10^{-6}}$$

$$= 3933\Omega$$

Since α ranges between 92° and 112°, the extended phase shift circuit must be used and the practical resistor arrangement would be a 2kΩ fixed resistor in series with a 2kΩ potentiometer.

11-4-1-2 Trigger Devices

The function of a trigger device is to match the *R-C* phase shift circuit to the thyristor without having to redesign the *R-C* circuit for each different thyristor. The trigger device rapidly transfers the stored energy of the capacitor to the thyristor gate. The most commonly used devices all have negative resistance characteristics and at the breakdown point act as a low resistance switch. Typical devices are the unijunction transistor (UJT), the bidirectional diode thyristor (DIAC), the programmable unijunction transistor (PUT), and the silicon controlled switch (SCS).

11-4-2 R-C Phase Shift Circuits

The most commonly used configurations, shown in Figure 11-25, are used for half-wave and TRIAC control. In the case of an inverse-parallel SCR arrangement, firing control in both halves of the cycle can be obtained by modifying the UJT relaxation oscillator as shown in Figure 11-26.

The major disadvantage of the phase shift control circuits illustrated in Figures 11-25 and 11-26 is that relatively large variations of the variable resistance *R* are required to change the load voltage; in addition, the response is not linear. The transfer characteristic of the system can be improved, that is, the variation of load voltage can be

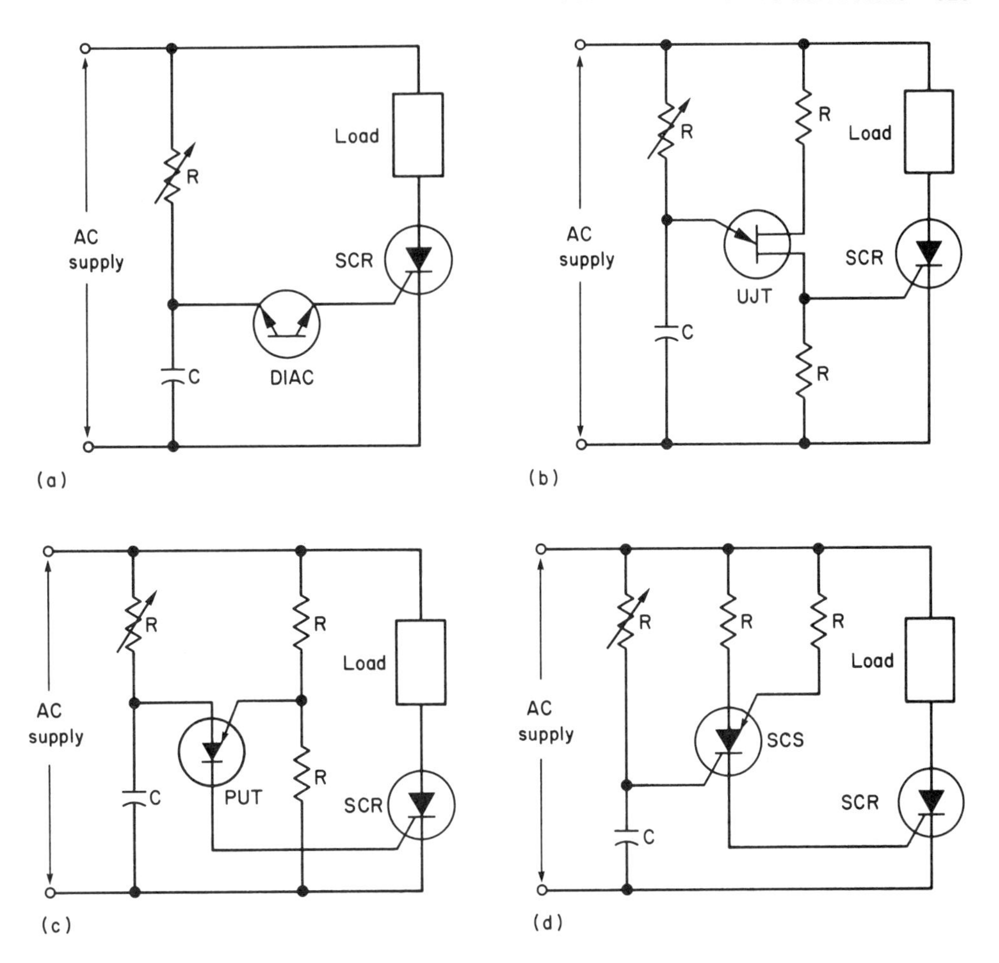

Figure 11-25. *RC* phase shift control circuits using as trigger devices (a) a DIAC, (b) a UJT, (c) a PUT, and (d) an SCS.

made more proportional to the firing delay angle, by using a ramp and pedestal control.

11-4-2-1 Ramp and Pedestal Control

The basic circuit of a ramp and pedestal control is shown in Figure 11-27. The capacitor C will charge very rapidly through the diode D to V_C, and this establishes the pedestal voltage. The capacitor will continue charging exponentially at a rate determined by the time constant $R2C$. When the potential at point B rises above the potential at point A, the diode becomes reverse biased, and the capacitor continues

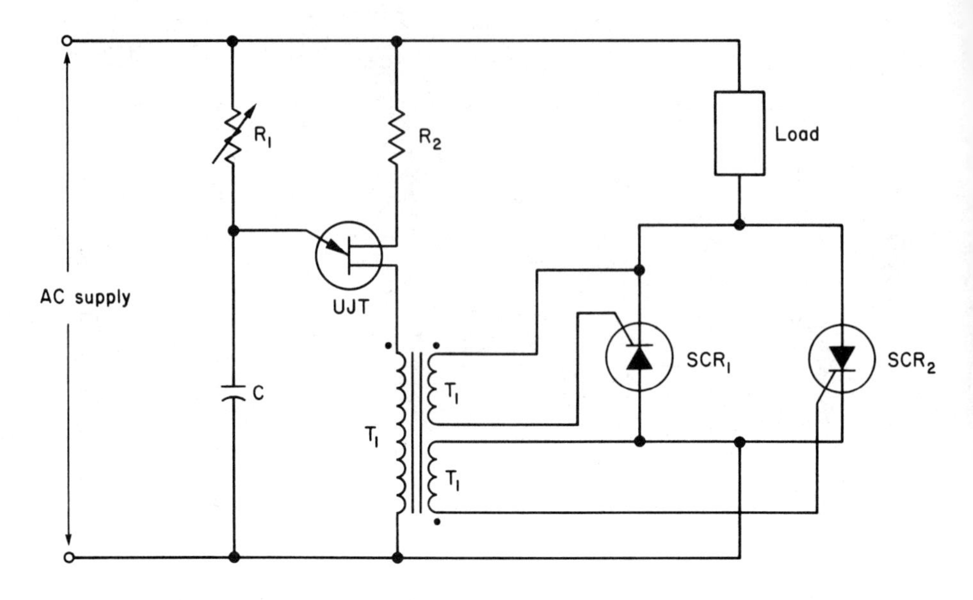

Figure 11-26. Phase control firing circuit for inverse parallel SCRs.

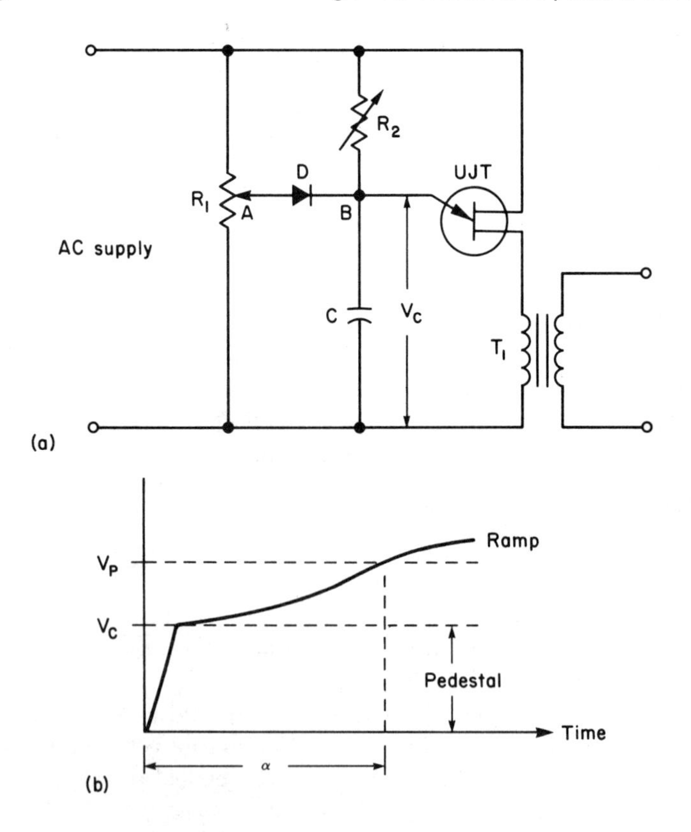

Figure 11-27. Ramp and pedestal control: (a) basic circuit; (b) ramp and pedestal voltages.

charging along the ramp until it reaches the triggering voltage V_p of the UJT. The ramp is not linear, since it is supplied from the sinusoidal ac source. The slope of the ramp is determined by the adjustment of $R2$; the firing delay angle is varied by varying $R1$ and thus the voltage V_c across the capacitor. Varying $R1$ to increase the pedestal voltage decreases α, or vice versa.

The linearity of the ramp may be improved by using the circuit of Figure 11-28. In this circuit the capacitor C is supplied directly from the ac line; this adds a cosine component to the nonlinear ramp, with the result that the ramp becomes nearly linear and the transfer characteristic is also nearly linear, and the gain is almost constant.

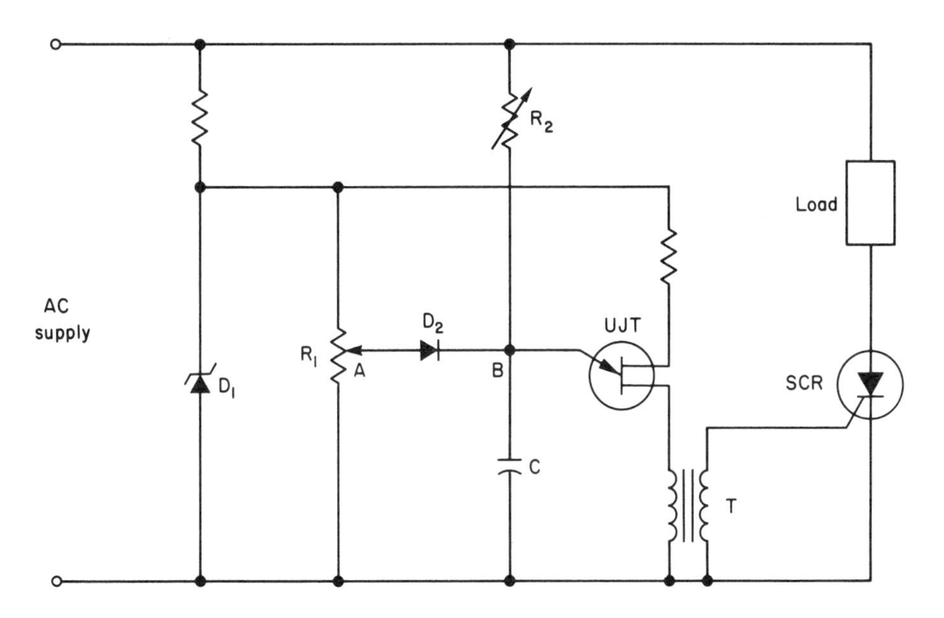

Figure 11-28. Cosine modified ramp and pedestal control.

11-4-3 Nonfeedback Universal Motor Control Circuits

Since the series motor produces positive torque for either half-cycle of the ac supply, it is capable of operation under half- or full-wave control.

The cheapest and simplest half-wave control circuit is shown in Figure 11-29, which utilizes an R-C phase shift circuit to vary the firing delay angle and a neon bulb as the trigger device to supply the gate pulse signal to the SCR. The major disadvantages of this circuit are,

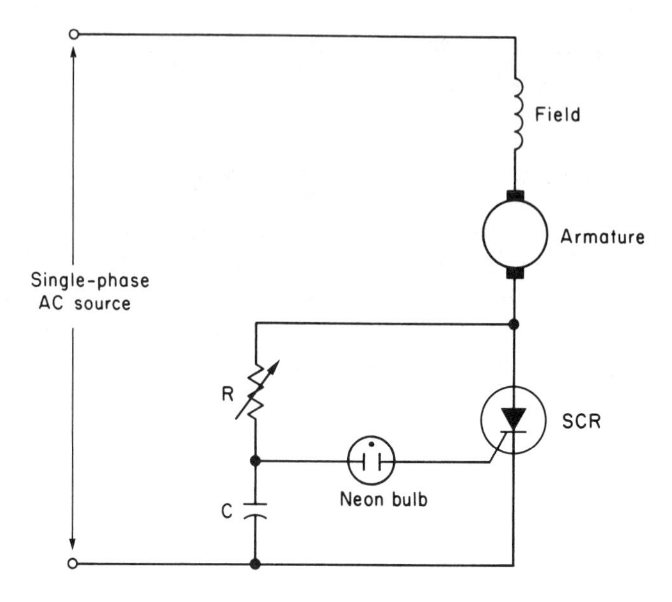

Figure 11-29. Half-wave, nonfeedback, universal motor control.

first, there is only one torque pulse per cycle, and, second, the speed regulation is very poor.

The torque output of the universal motor can be greatly improved by using a TRIAC instead of the SCR in Figure 11-29 to obtain full-wave phase control. A typical circuit is shown in Figure 11-30. As before, the phase shift control is obtained by a simple *R-C* phase shift network, using a DIAC as the trigger device. The firing delay angle α increases as R increases, or vice versa.

The major disadvantage of both these circuits is that the speed regulation is very poor. Speed regulation can be improved by using feedback circuits.

11-4-4 Feedback Universal Motor Control Circuits

The simplest form of feedback circuit is the half-wave Momberg circuit shown in Figure 11-31. A voltage signal V_d representing the desired speed is obtained by varying the speed regulating potentiometer $R2$. The voltage V_a represents the armature counter-emf which is proportional to the motor speed. The diode D will conduct only when $V_d > V_a$, if V_d has been established for the desired speed. As long as the motor is running at the desired speed, D will block the gate signal to the SCR. If the motor speed drops, the armature feedback signal V_a decreases, and since $V_d > V_a$ a gate signal will be applied to the

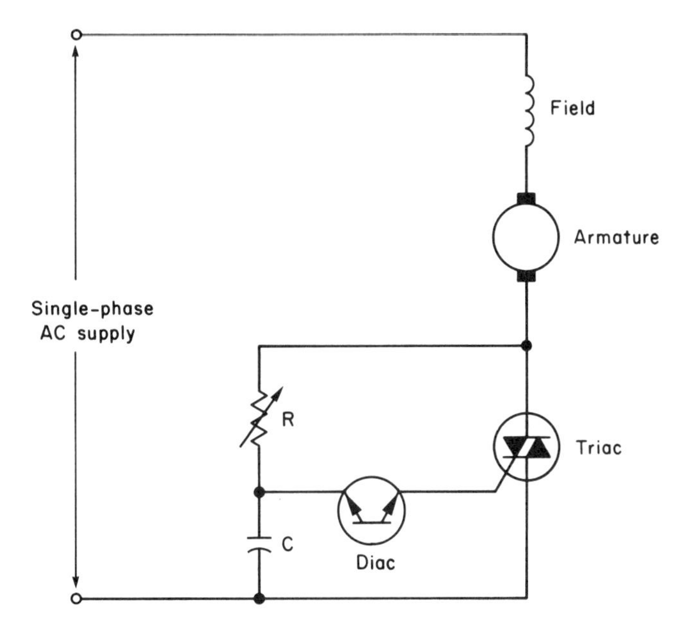

Figure 11-30. Full-wave, nonfeedback, universal motor control.

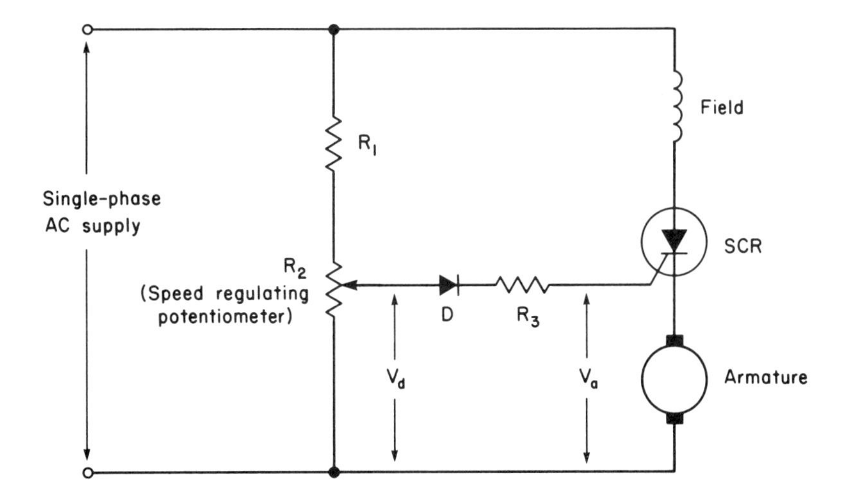

Figure 11-31. The Momberg half-wave feedback circuit.

SCR. The greater the drop in the motor speed, the earlier D will conduct in each positive half-cycle, and the greater will be the mean dc voltage applied to the motor armature, which in turn will result in the armature accelerating toward the desired speed.

The Momberg circuit has two major disadvantages: first, the final speed will be slightly less than the set speed. Second, at low-speed settings the range of speed control is greatly reduced. An improved feedback circuit that overcomes some of the disadvantrages of the Momberg circuit is the Gutzwiller half-wave feedback circuit shown in Figure 11-32. In this circuit the voltage across the speed regulating potentiometer $R2$ is established by the zener diode $Z1$. The voltage across the capacitor V_c is determined by the setting of $R2$; the blocking diode $D2$ will apply a gate pulse to the SCR whenever $V_c > V_a$. The greater the difference between V_c and V_a, the sooner in the positive half-cycle will the gate signal be applied to the SCR, and as a result the greater will be the mean dc voltage applied to the armature and the greater the shaft speed.

The Gutzwiller circuit improves speed regulation, the range of speed control, and the stability of the motor, but at an increased cost of the control circuit components.

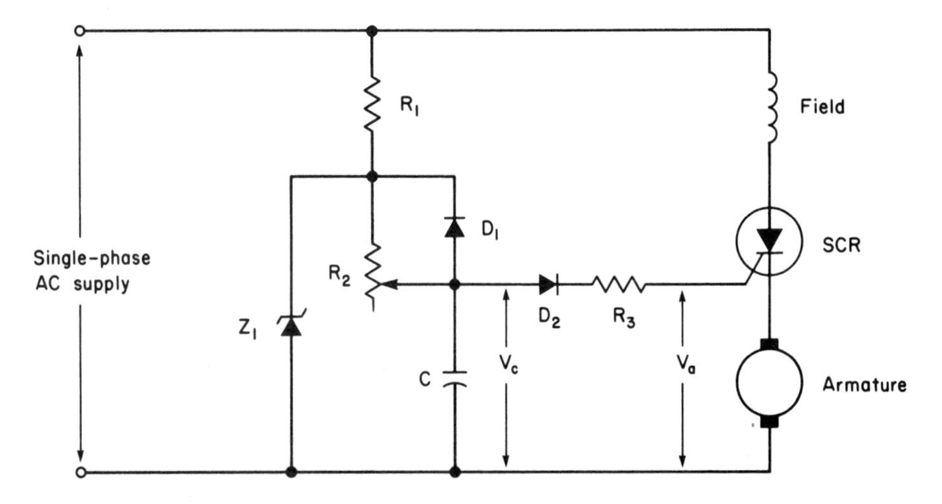

Figure 11-32. The Gutzwiller half-wave feedback circuit.

GLOSSARY OF IMPORTANT TERMS

Base speed: Normally the rated nameplate speed. In the case of a dc motor it is the speed achieved with rated armature voltage and rated field current. In the case of an ac motor it is the speed achieved when the stator is supplied at rated frequency.

Constant-torque drive: A drive operating under varying armature (or stator) voltage and constant airgap flux conditions. The drive supplies a constant torque output over the speed range from zero to base speed.

Constant horsepower (kW) drive: A drive operating under constant armature (or stator) voltage and varying airgap flux conditions. The drive supplies a constant horsepower (kW) output (torque × speed) over a speed range whose minimum value is the base speed.

Four-quadrant or dual converter: A drive system that operates in all four quadrants.

Torque limit: Basically current limiting since $T \alpha I_a \Phi$.

Current limit: Limiting armature (stator) current to prevent excessive motor and thyristor currents.

Chopper: A force commutated dc to dc converter. It is used to obtain a variable voltage dc output by rapidly switching on and off a thyristor which is supplied from a constant voltage dc source.

Pulse width modulation: As applied to choppers, t_{ON} and t_{OFF} are variable and the periodic time is constant. Also called the constant frequency system.

Pulse rate modulation: Sometimes used by choppers, where t_{ON} is constant and t_{OFF} is variable. Also sometimes known as frequency modulation.

REVIEW QUESTIONS

11-1 Explain what is meant when a dc motor drive is said to be operating in (1) the constant-torque mode and (2) the constant horsepower (kW) mode. How are those conditions achieved with a separately excited dc motor drive?

11-2 Explain what is meant by a four-quadrant converter and discuss the various methods that may be used to achieve this result. Give typical applications.

11-3 With the aid of a block diagram explain a typical closed-loop thyristor phase-controlled converter dc motor speed control system.

11-4 Why are starting resistances not required with a thyristor phase-controlled converter dc motor drive?

11-5 What is the purpose of a ramp generator in a dc motor drive control?

11-6 Explain how the output speed of a dc drive is maintained constant. Discuss the methods of obtaining an actual speed signal.

11-7 Discuss the principle of current limiting and how it may be applied to a dc motor drive.

11-8 Discuss with the aid of a schematic diagram a typical three-phase variable-speed dc motor drive converter.

11-9 What is meant by a chopper? List typical applications.

11-10 Explain the principle of a step-down chopper.

11-11 Explain the principle of a step-up chopper.

11-12 With the aid of a sketch explain the principle of the Jones circuit.

11-13 With the aid of sketches and block diagrams explain the principle of digital dc motor speed control.

11-14 Discuss the basic requirements that must be met by a variable-frequency ac motor speed control system.

11-15 With the aid of sketches and a block diagram discuss a six-step dc link variable frequency converter, using chopper voltage control.

11-16 Explain the principle of dc voltage control using a two thyristor chopper.

11-17 What problems may be encountered with a standard polyphase ac motor when it is supplied from a pulse width modulated variable frequency inverter?

11-18 With the aid of a block diagram describe a microprocessor based inverter for polyphase ac motor speed control.

11-19 With the aid of a schematic explain the principle of wound rotor motor speed control using a slip power recovery scheme. What are the advantages and disadvantages of such a system?

11-20 Discuss the principle of polyphase ac motor speed control using variable stator voltage control.

11-21 What is the principle of a Nola power factor controller and what are the benefits of using this system?

11-22 With the aid of sketches and a phasor diagram explain the principle of the basic *RC* phase shift circuit, and discuss its limitations.

11-23 Repeat Questions 11-22 for the extended *RC* phase shift circuit.

11-24 What is the function and characteristics of a trigger device? List commonly used trigger devices.

11-25 With the aid of schematics explain the operation of phase shift control schemes using the following as trigger devices: (1) a DIAC, (2) a UJT, (3) a PUT, and (4) an SCS.

11-26 What is the underlying purpose behind using a ramp and pedestal control?

11-27 With the aid of a sketch explain the operation of a ramp and pedestal control system.

11-28 With the aid of a sketch explain the operation of a cosine modified ramp and pedestal control.

11-29 What are the disadvantages of using a nonfeedback half-wave system to control a universal motor?

11-30 What are the advantages to using a feedback control system?

11-31 With the aid of a sketch explain the Momberg circuit. Discuss the advantages and disadvantages of this circuit.

11-32 Repeat Question 11-31 for the Gutzwiller circuit.

PROBLEMS

11-1 In the circuit of Figure 11-22, $V_{ac}=163\sin 377t$, $R=10\text{k}\Omega$ and $C=0.1\mu\text{F}$. Calculate (1) the firing delay angle α and (2) the maximum voltage developed across the capacitor.

11-2 In the circuit of Figure 11-23, $V_{ac}=294\sin 377t$, $R=20\text{k}\Omega$ and $C=1.0\mu\text{F}$. What is the firing delay angle α?

11-3 A 15Ω resistive load is supplied from a 230V, 60 Hz ac source. It is necessary to maintain the power supplied to the load between 750W and 1500W when using a phase shift controlled SCR. Calculate (1) the required range of firing delay angle control and (2) the required range of R when $C=0.1\mu\text{F}$.

11-4 Repeat Question 11-3 using a TRIAC.

12

Industrial Applications

12-1 INTRODUCTION

In this chapter we will look at a number of applications that consolidate material previously presented in other chapters.

12-2 PROGRAMMABLE CONTROLLERS (PCs)

Programmable controllers (PCs) were first introduced in 1969 in the automobile manufacturing industry as a means of replacing electromagnetic relay logic by electronic control systems. The major benefits achieved were the elimination of costly wiring changes for each model year retooling, reduced change-over time and space requirements, as well as dramatically increased flexibility in production line control. Even today the automotive industry is the major user of the programmable controller, although the use of PCs now extends into nearly all industrial activities.

The original PCs were limited to performing the functions normally carried out by the electromagnetic relays they replaced, for example, sequencing operations, timing, etc. Since their introduction, PCs have experienced a tremendous increase in their capabilities. The modern PC has a wide range of capabilities, such as, arithmetic calculation, data acquisition and storage, report generation, distributed control capabilities, analog and digital signal handling, computer and peripheral interfaces, intercommunication capability between PCs, self-diagnostics, system troubleshooting, and control of motors and systems.

The main reasons for the rapid increase in PC applications are: (a) ease of programming, (b) relatively low cost, (c) reliability, (d) low maintenance cost, and (e) ruggedness combined with their ability to operate on site in industrial applications.

12-2-1 A Typical Industrial Programmable Controller

As an example of a typical low cost PC, the Allen-Bradley Mini-PLC-2 provides an excellent illustration of versatility at a price that is well within the budget of most industrial users.

The Mini-PLC-2 is a microprocessor based controller which can monitor and control up to 128 input/output (I/O) channels. The main elements are:

1. The central processing unit (CPU) and memory are contained in the Mini-Processor Module. The Mini-Processor memory is available in two sizes: (a) 512 words (½k) or (b) 1024 words (1k), both of which accept 16-bit data. The function of the Mini-Processor is to store the user program, the current I/O status and carry out switching sequences, control internal counters and timers, and perform two function arithmetic. The memory is a Complementary Metal Oxide Semiconductor (CMOS) read/write memory whose contents are protected against loss during a power failure by a backup battery.
2. The I/O modules permit the Mini-Processor to monitor and control externally connected equipment and systems. The input modules sense voltage levels and the status of input devices such as push buttons, overloads, pressure switches, limit switches, etc. The output modules produce output signals which energize motors and motor starters, solenoids, electronic circuits, etc. The I/O modules feature optoelectronic isolation between the processor and input or output circuits. In addition, the input modules have filters to protect against circuit transients, and most of the output modules are fused. A wide range of I/O modules are available ranging from a 120V ac/dc and 220/240V ac/dc inputs, 8-bit and 12-bit analog inputs, TTL inputs, encoder/counter inputs, thermocouple inputs, and an 8-bit absolute encoder input to name a few of the available input modules. There are also a corresponding range of output modules.
3. The system power supply which can be operated from 120V or 220/240V ac provides logic level power for the Mini-Processor and the I/O modules. It has built-in under and overvoltage, and overcurrent protection as well as an automatic power up sequence.

The capability of the Mini-PLC-2 controller is enhanced by being interfaced to a number of peripheral devices, such as the Industrial Terminal or Program Panel. The former incorporates both a CRT display and a keyboard and enables programs to be entered, modified, or monitored; the CRT displays the logic rungs as the program is entered. The Program Panel permits program entry, as well as monitoring or editing, and the display shows the individual logic element status. A digital cassette recorder can be used to store programs as a backup to the system. Since the programming terminals are RS-232-C compatible, it is possible to interface with data terminals for hardcopy printouts of the controller program as well as reports defining the current operating status, warning messages and diagnostic messages. The general functional relationship of the components of a typical PC are shown in Figure 12-1.

In large industrial applications a requirement often exists to extend the capabilities of the production process by better utilization of new and existing equipment by improved communication and control. This can be accomplished by means of a data highway interconnecting a number of PCs, computers, input terminals, printers, etc., so that they operate as an integrated system.

Most PCs permit the programmer to carry out ladder diagram programming, a technique which is very similar to the conventional relay ladder diagram. Depending upon the system, real-time or off-line programming changes can be made. With real-time programming, the PC will continue to operate while it is being programmed; in the case of off-line programming, the PC is usually halted prior to the program being entered.

The advantage of real-time programming is that the PC exercises control, while a portion of the program is being entered or modified; adjustments can be made to a process and their effects observed.

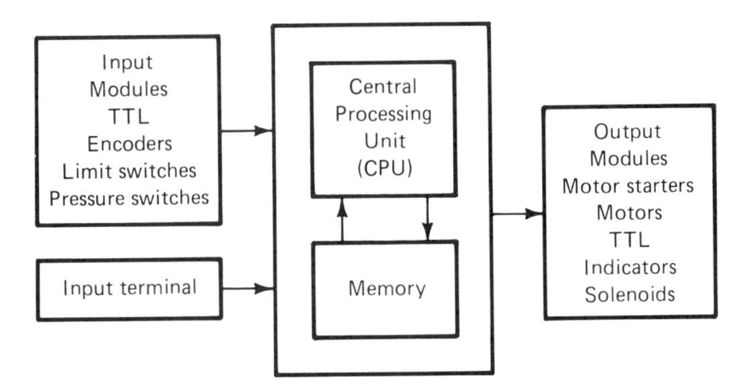

Figure 12-1. Functional block diagram of a typical programmable controller.

12-2-2 Programming Concepts

Prior to writing a program there are two basic steps that must be followed. They are:

1. Determine the operating sequence and conditions that must be met to complete the task.
2. Assign an address to each input and output device.

Example 12-1

The example of Figure 12–2 illustrates how the relay logic of Figure 12–2(a) is converted into ladder diagram rungs. When the START push button is pressed, the control relay CR is energized through the normally closed overload contacts. At the same time, the CR auxiliary contacts seal in the control relay and energize the on delay timer TD. After 5 secs the on delay timer picks up and energizes the M contactor and the M on light. Figure 12–2(b) illustrates the inputs and outputs of the I/O modules. Figure 12–2(c) is the ladder rung diagram, the addresses of the example are as follows:

Input	Address	Output	Address
START	11401	CR	01211
STOP	11402	TD	03012
OL	11403	M	01213
		M ON	01214

12-3 UNINTERRUPTIBLE POWER SUPPLIES (UPS)

There are an ever increasing number of complex electrical and electronic systems in the industrial environment which require an uninterrupted power supply to prevent a shut down or damage to critical equipment and systems in the event of a power failure. Typical systems requiring protection include computers, on-line data processors, processing control computers, boiler controls, flame failure systems, fuel processing and control systems, critical lighting, important instrumentation and recording equipment, data loggers, etc.

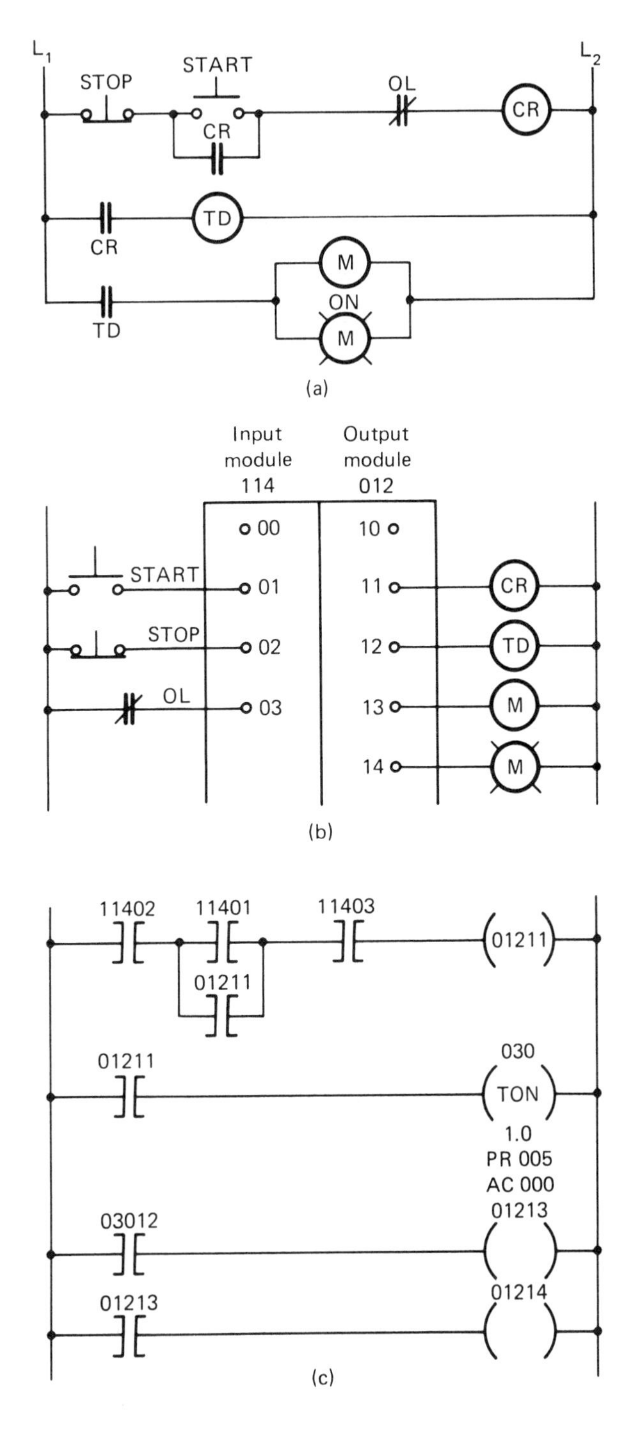

Figure 12-2. Programming example: (a) relay logic, (b) I/O module, (c) PC ladder diagram.

12-3-1 UPS Systems

There are basically two types of alternative power systems. The first type or transfer system is used in applications where a switch-over time of several cycles is acceptable, usually in the range of 4 to 12 cycles. The principle of this type is illustrated in Figure 12–3(a). As can be seen, the load is normally supplied from the preferred ac source via an automatic bus transfer switch to the load, with the battery being charged via the rectifier battery charger. When power fails the automatic bus transfer switch connects the inverter which is being operated from the battery supply to the load. This arrangement is satisfactory for applications where a short power interruption is not critical.

The continuous or float type UPS shown in Figure 12–3(b) is simple, relatively inexpensive, and ensures that there is no interruption of power to the load. Under normal operating conditions, the ac input is rectified to dc and maintains the charge on the battery bank connected across the inverter input terminals. The ac output of the inverter is usually filtered by a ferro-resonant transformer so that the output waveform is approximately sinusoidal. In the event of an ac power failure, the inverter is supplied by the battery and there is no interruption of power to the load. Since the ability of the battery to support load is limited, usually between 10 to 60 minutes, a standby generator set may be automatically started up after a predetermined time interval and connected to the rectifier by means of an automatic change over switch. The standby generator has effectively replaced the ac source and maintains the battery charge. When ac power is restored, the standby generator is automatically disconnected. The function of the synchronizing line is to ensure that the inverter output frequency and phase relationships are synchronized to the ac source.

There are three basic types of continuous UPS systems:

1. The forward system in which the load is supplied directly from the ac line via a static or solid-state transfer switch. See Figure 12–4(a). In the event of a power failure the static transfer switch automatically transfers the ac load to the inverter. When power is restored, the load is transferred back to the ac source. The disadvantage of this system is that there is no isolation between the source and load.

2. The reverse system in which the load is supplied from the inverter and is only transferred to the ac source as a result of an inverter failure. The advantage of this system is that there is electrical isolation between the source and load, and the output of the inverter can maintain a good voltage regulation and a stable frequency. See Figure 12–4(b).

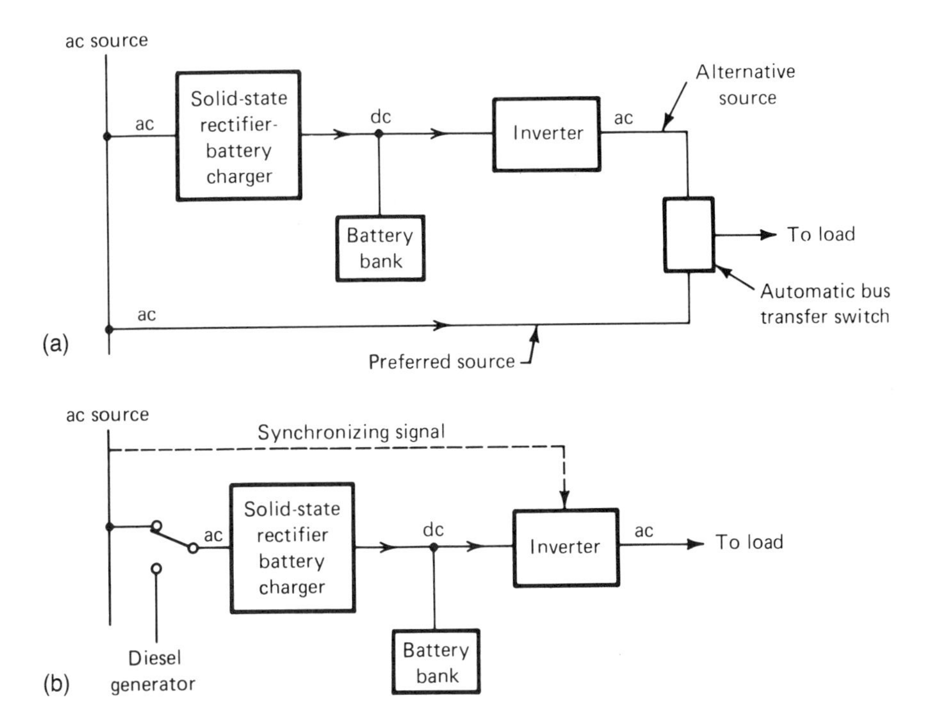

Figure 12-3. Uninterruptible power supplies: (a) changeover type, (b) continuous-float type.

3. A combination of the reverse system and a standby generator effectively gives double protection in the event of an inverter failure.

There is a further classification of UPS systems; namely, nonredundant and parallel redundant. The systems that we have discussed so far are categorized as nonredundant and are simple, low cost, and reliable.

The parallel redundant system consists of two or more nonredundant systems connected in parallel with each capable of being isolated. This system which is expensive and provides increased reliability in the event of an inverter failure, would only be used in applications where it is absolutely essential that there is no interruption of power to the load.

12-4 SWITCHED-MODE POWER SUPPLIES

There has been a considerable movement toward the use of switched-mode power supplies (SMPS). The major advantages are increased effi-

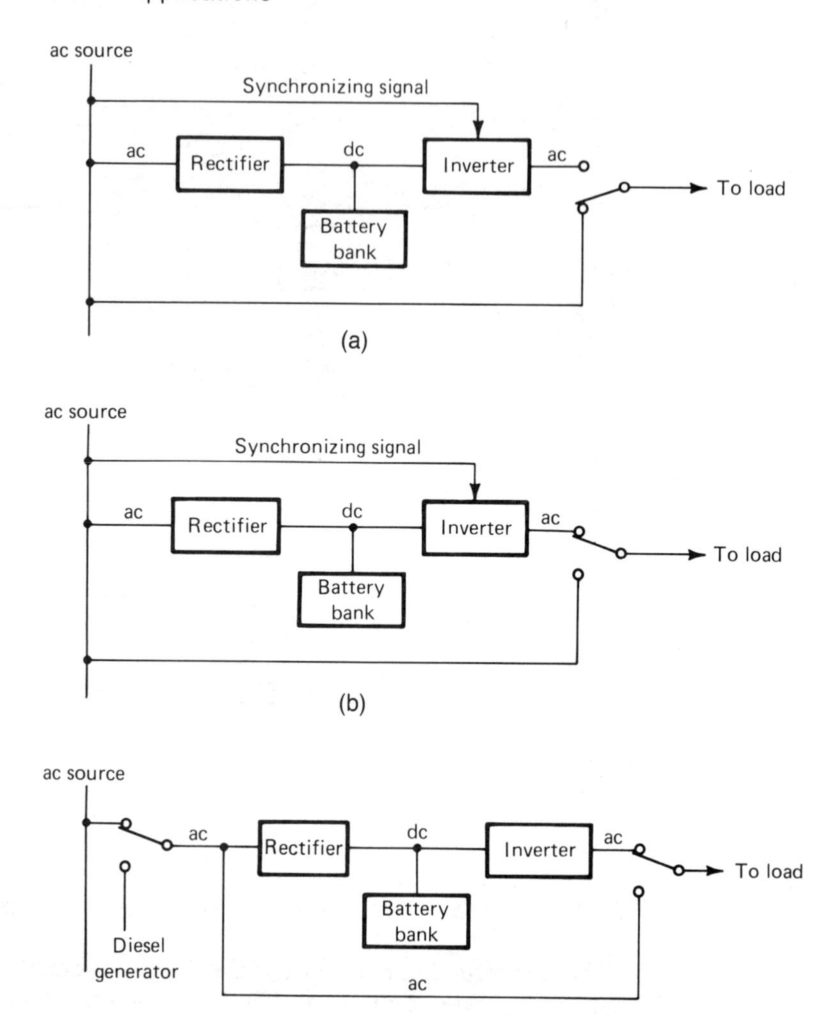

Figure 12-4. Types of continuous UPS systems: (a) forward, (b) reverse, (c) reverse plus standby generator.

ciency, reduced power losses, less heat dissipation, and a considerable reduction in component sizes and weights. The main disadvantages are self-induced noise and radio frequency interference (RFI) which, with proper design or input filters, can be reduced to acceptable levels.

In order to appreciate the benefits of the SMPS, we should briefly review the basic linear power supply shown in Figure 12-5. Electrical isolation is achieved by means of the 60-400Hz transformer which, unfortunately, must be relatively large in size because of the limitations of the maximum flux density imposed by normally available

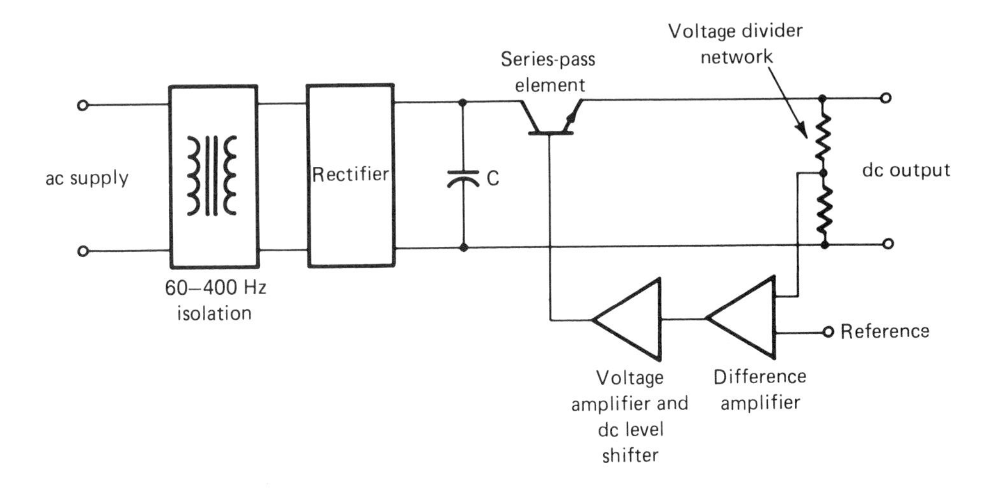

Figure 12-5. Linear regulated power supply.

materials; although, if the system is operated at 400Hz, there will be some reduction in the transformer size. The rectifier is usually composed of high surge rated silicon diodes which eliminate the need for a choke input filter. The series-pass element maintains the output dc voltage constant by changing its effective resistance in response to the signal applied to the base. This signal is derived by the negative feedback loop consisting of the voltage divider network, the difference amplifier and voltage amplifier, and dc level shifter.

The basic operation of the negative feedback loop is that the derived signal from the voltage divider is compared to the reference voltage. The difference voltage is amplified and further amplified and dc level shifted before being applied to the base of the series-pass transistor. The problem with this method of voltage regulation is that the load current flows through the series-pass element, and as a result the greater the difference between the input dc voltage and the output dc voltage the greater is the power dissipated by the series-pass transistor. This results in a low overall efficiency for this type of power supply.

The switched-mode power supply, on the other hand, utilizes a low loss switching element instead of the series-pass transistor. The switching element may be a bipolar power transistor or a MOSFET. When these devices are operated in the switching mode, the only energy dissipated occurs during switching on and off. The operation of the step down SMPS shown in Figure 12-6 is similar to that of the linear power supply. A signal is derived from the voltage-divider network across the output and compared to the reference voltage; the difference voltage is amplified and applied to the voltage controlled pulse width

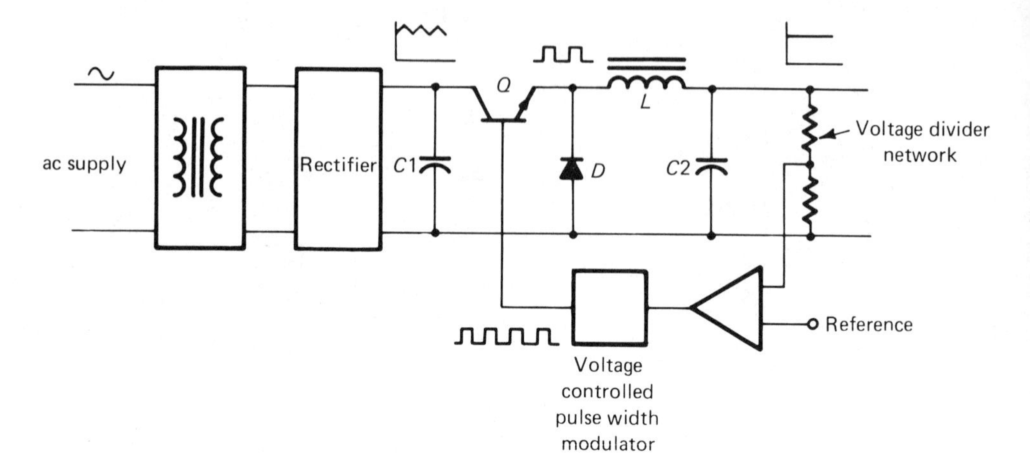

Figure 12-6. Switched-mode power supply.

modulator. As the voltage applied to the pulse width modulator increases or decreases, the ratio of t_{ON} to the periodic time T decreases or increases; the corresponding switching occurs at the switching transistor Q, and the output voltage remains constant. Normally the SMPS operates at a constant switching frequency which may be anywhere in the range of 3 to 50 kHz, although switching rates about 20 kHz seem to be the standard. Diode D acts as a freewheel diode, maintains current through inductor L, and limits the induced voltage in L due to high di/dt when the switching transistor turns off. Diode D must be capable of turning off in less than one microsecond; therefore, a fast recovery diode must be used. Inductor L and capacitor $C2$ are relatively small since they are operating at a switching frequency of 20 kHz. It should be noted that only high quality electrolytic capacitors are suitable. A problem experienced with SMPSs is that fast switching of high load currents produces noise spikes which may cause problems in other sections of the system as a result of capacitive and inductive coupling. These spikes may be reduced by shielding and filtering. The SMPS is also relatively slow in responding to rapid changes in load current.

In general, since the input ripple is less of a problem in the SMPS as compared to the linear power supply, a smaller input filter capacitor can be used. With an overall efficiency in excess of 90%, there is less heat dissipation. As a result, smaller heat sinks and cooling fans are required. Combined with the absence of resistors and physically smaller inductances and capacitors, a relatively small overall packaging arrangement results.

12-5 ROBOTS

The modern industrial robot is best described by CAM-I (Computer Aided Manufacturing International) as "a device that performs functions ordinarily ascribed to human beings, or operates with what appears to be almost human intelligence."

The modern industrial robot is a programmable manipulator that performs useful work without the aid of a human operator. The tasks that are being assigned to industrial robots are: material handling, loading, unloading, sorting and stacking, spot welding, paint spraying, casting, grinding, sanding, etc. In general, they are being assigned to tasks that humans find unhealthy, distasteful, and monotonous.

Industrial robots or manipulators fall into two categories:

1. Limited sequence type. This type has arms with limited axes of movement which move from point to point in sequence. The arms are usually controlled either electrically or by pneumatically controlled servomotors. The work sequence is programmed into the robot's memory. They are relatively low cost and suitable for pick and place or simple repeated operations involving light payloads.

2. Computer controlled type. This type usually has electrically or hydraulically servo controlled arms which may have as many as six axes of movement. They are capable of being programmed to operate over any range of points within the sphere of movement of the manipulator's arm(s). However, they must be walked through the sequence of tasks by an operator in order that they may be "taught" how to do the prescribed tasks.

These types of industrial robots are restricted in their capabilities by being rigidly programmed and requiring accurate placement of the parts being processed. In addition, in the absence of a means of visual sensing, they do not have the ability to sort and assemble.

A second generation of industrial robots are now making their appearance. These robots are using microprocessor controlled ac servomotor drives, with adaptive control systems which enable them to make two- and three-dimensional searches. In addition, in many cases TV cameras are being used to feed information into the computer control system in which, for example, the pattern or shape of the objects being processed has been memorized. This seeing capability enables the robot to compare objects against the memory and then position the robot arm to pick up, inspect, and assemble units.

Unimation Inc., which is probably the largest manufacturer of industrial robots in the world, has developed a five axis Programmable Universal Machine for Assembly (PUMA). The PUMA possesses both vision and the ability to make decisions and is under the control of three small computers; one responds to the video input, one controls the movement of PUMA's arm, and the third acts as a supervisory computer overviewing the whole automatic operation.

Cincinnati Milacron developed the software package that enables the computer to interpret the video data from the camera. The basic principle of operation is that two lights are focused to form a bright line across a conveyor. The video camera scans an assortment of parts crossing the line and transmits 128 slices of data to the computer where this information is compared against stored data representing the exact shape(s) of the parts to be assembled into an assembly, and will ignore all parts not previously identified. Once the parts have been identified the x and y coordinates, the part identification, and the orientation of the part are transferred to the supervisory computer. The next step is to teach the robot what to do with each part by using a special training box to control the movement of the hand, this information in turn being stored in the supervisory computer memory. Experience has shown that once the PUMA has been trained, it can correctly select and pick up the correct parts even though they may be arranged in any order. A number of companies, such as Westinghouse and General Motors, are experimenting with PUMA robots for batch type assembly operations.

There is no doubt that the industrial robot is making a very significant impact in manufacturing industries, especially the automotive where it is helping to reduce costs and improve productivity and quality. The National Bureau of Standards is working towards a robotic metal working shop by 1988. There would be a limited requirement for human staff to mainly design new parts or assemblies using CAD/CAM, order new material for inventory, and last, but probably the most important, maintain and service the robots.

12-6 MICROCOMPUTER BASED POSITION REGULATOR

The system about to be discussed was developed by the Electrical and Automation Systems Department of Algoma Steel Corporation Ltd., Sault Ste. Marie, Ontario, Canada.

The position regulator system was designed to provide a standardized general purpose regulator that was capable of replacing or upgrading existing position control systems associated with the production and rolling of steel.

The general requirements of the system are:

1. The capability to position up to three variable-speed (VS) or discrete-speed (CP) systems using one Z-80 microcomputer,

2. The positioning system must be critically damped, that is, achieve the specified position in the minimum time without overshoot,
3. That it be capable of monitoring and changing all regular constants and variables.

The control system includes the capability of interfacing with constant and variable voltage dc drives, hydraulic systems, variable frequency or variable voltage converters, RTL, DTL and TTL logic, programmable controllers, mini or microcomputers, and analog or digital position transducers.

In order to develop the capability to operate with new and existing systems the control logic is carried out external to the Z-80 microcomputer. The information is supplied via control lines to I/O conversion modules where appropriate signal conditioning takes place. This concept is illustrated in Figure 12-7, where the inputs are as follows:

1. Digital position reference provided as:
   (a) 15-bit absolute binary and a sign bit, or
   (b) 8421 BCD to 19999 and a sign bit.
2. Digital position feedback provided as:
   (a) 15-bit absolute binary and a sign bit, or
   (b) 8421 BCD to 19999 and a sign bit, or
   (c) 16-bit 2s complement binary.

3. A bipolar analog input voltage 0 to ± 10 Vdc which is used when enabled to provide a dc offset to the digital position reference.

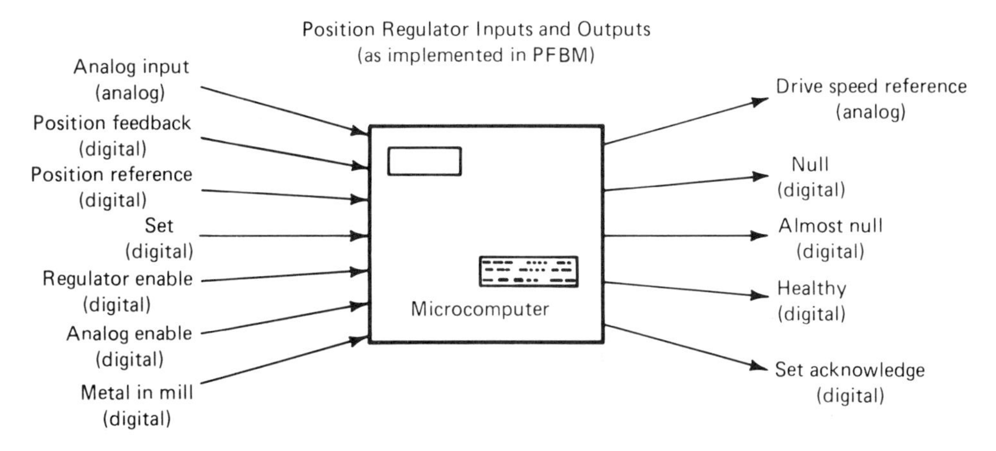

Figure 12-7. Position regulator inputs and outputs. (Courtesy of the Algoma Steel Corporation)

4. A "set" signal which when true will cause the digital position reference to be read.

5. "Inhibit forward" and "inhibit reverse" signals which if true prevent any further motion in the desired direction(s).

6. A "metal in stand" signal which when true buffers any position reference until the signal becomes false. The regulator will substitute the present position reference with a buffered position reference. This prevents roll position changes when there is metal in the stand.

7. "Analog enable" which if true permits the bipolar analog input voltage to be read and added to the position reference.

8. "Regulator enable" which if true permits the regulator to function normally. If false, a 0 Vdc reference signal is supplied to all associated variable voltage (VV) drives or a stop signal is supplied to all discrete-speed (CP) drives.

The microcomputer outputs are as follows:

1. A $\pm$10Vdc bipolar analog voltage which provides the speed reference to the associated variable-speed drives.

2. Three-speed points selects, forward, reverse, and stop to the associated discrete-speed drives.

3. "Set acknowledge" digital signal when the "set" signal is true and the digital position reference is correctly formatted and within prescribed limits.

4. "Almost null" and "null" digital signals when the position error is within the almost null and null band limits.

5. A "healthy" digital signal which toggles high and low to indicate correct operation of the regulator. The following conditions will cause the healthy signal to stop toggling:
   (a) An 8421 BCD position feedback in which there is an illegal character, or
   (b) The position feedback remains unaltered for a predetermined time interval in spite of the fact that the position error is greater than the null value even though there is no motion inhibit for the desired direction and the regulator is enabled.

6. A bipolar analog voltage output 0 to $\pm$10Vdc for a recorder drive.

7. Alarms indicating the presence of the conditions of paragraph 5(a) and (b) above as well as:
   (a) A position reference less than the specified minimum.
   (b) An 8421 BCD position reference in which there is an illegal character, and

(c) Time out on analog-to-digital (A/D) conversion when analog offsets are being read. These alarms are usually presented in the form of messages specifying the regulator number outputted to a video display terminal or printer.

8. Instructions are also provided to the operator to permit the following:

(a) Display and change any regulator constant or variable.
(b) Display any regulator variable or constant and update the display with a change.
(c) Record any regulator variable or constant by means of a strip chart recorder.

The operator must supply the instructions complete with data address and data type, that is, floating point, 8-bit or 16-bit hexadecimal.

The operation of the regulator depends upon three control algorithm programs written in FORTRAN and assembly languages. These programs provide: (a) multiplex control of up to four variable-speed drives, (b) automatic strip gauge control of the rolling mills, and (c) fast accurate control of constant potential torque drives.

The operation of the position regulator is best illustrated by means of a flowchart, see Figure 12-8.

When the system is powered up the tune-up constants which are stored in a programmable read only memory (PROM) are read into a random access memory (RAM). The tune-up software which is run as a background schedule is interrupted every millisecond by a clock pulse which permits sheduling and timekeeping. In the case of a single drive system the tune-up software is also interrupted by the change of state of the LSB of the digital position feedback so that the drive speed may be measured.

Once every 20ms in the case of a single drive system, or in the case of the four multiple drive regulators, one of the drive regulators is serviced every 10ms. First the set input is checked, and if true a digital position reference is read into the regulator and converted to scaled floating point. In the event of a conversion error an error message is outputted; if there is not a conversion error the scaled floating point reference is checked against the limits. If the reference is not within the limits it is rejected and an error message is outputted; if within the limits it is saved in the RAM as the next working reference and acknowledged to the external logic. Then provided the set input is true and there is no metal in the stand, the next working reference is saved as the working reference, and the system is checked to determine if an overshoot constant is needed. If it is required then the overshoot constant is added to the reference and the overshoot word is set to true.

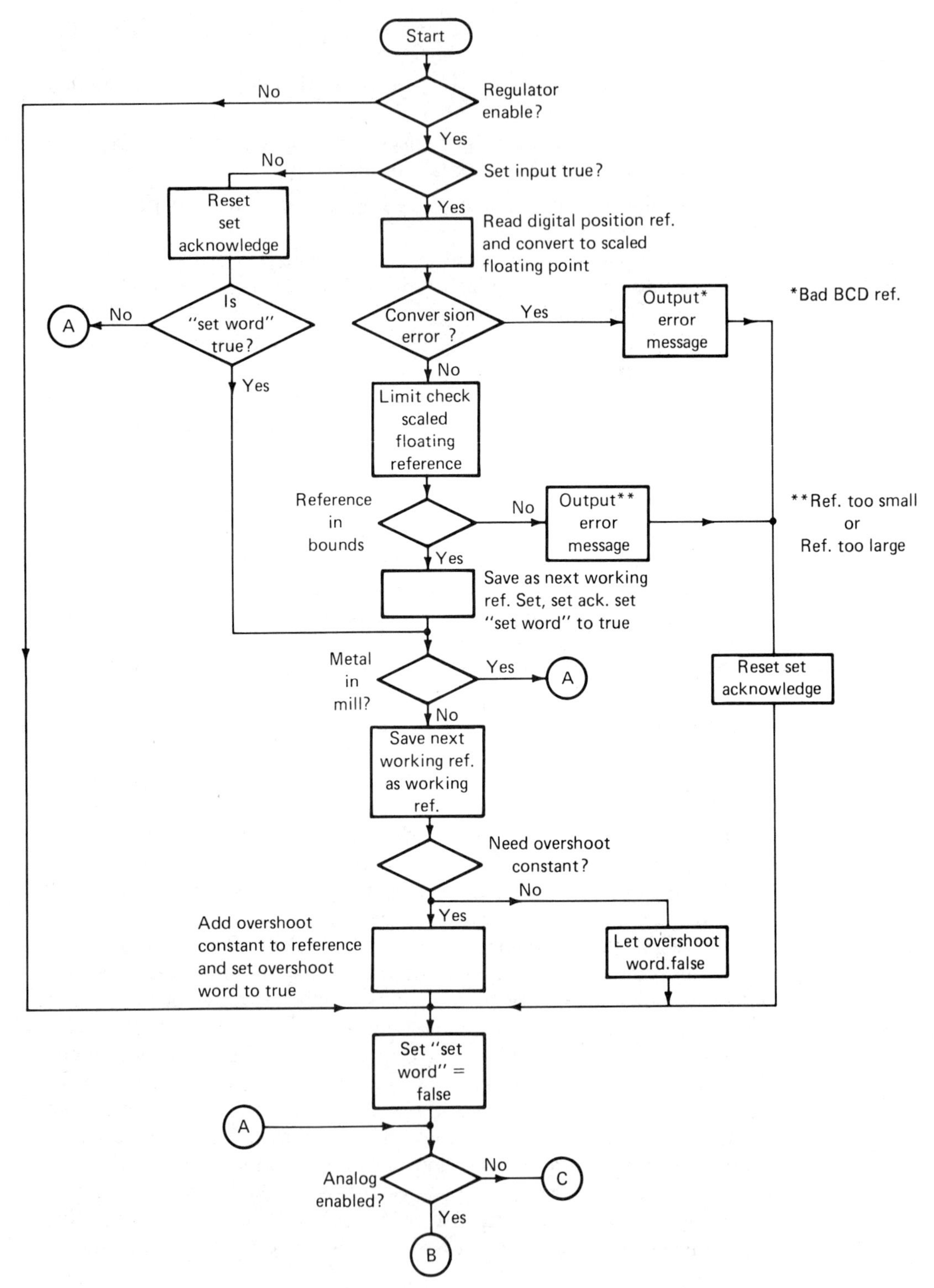

Figure 12-8. Microcomputer position regulator. (Courtesy of the Algoma Steel Corporation)

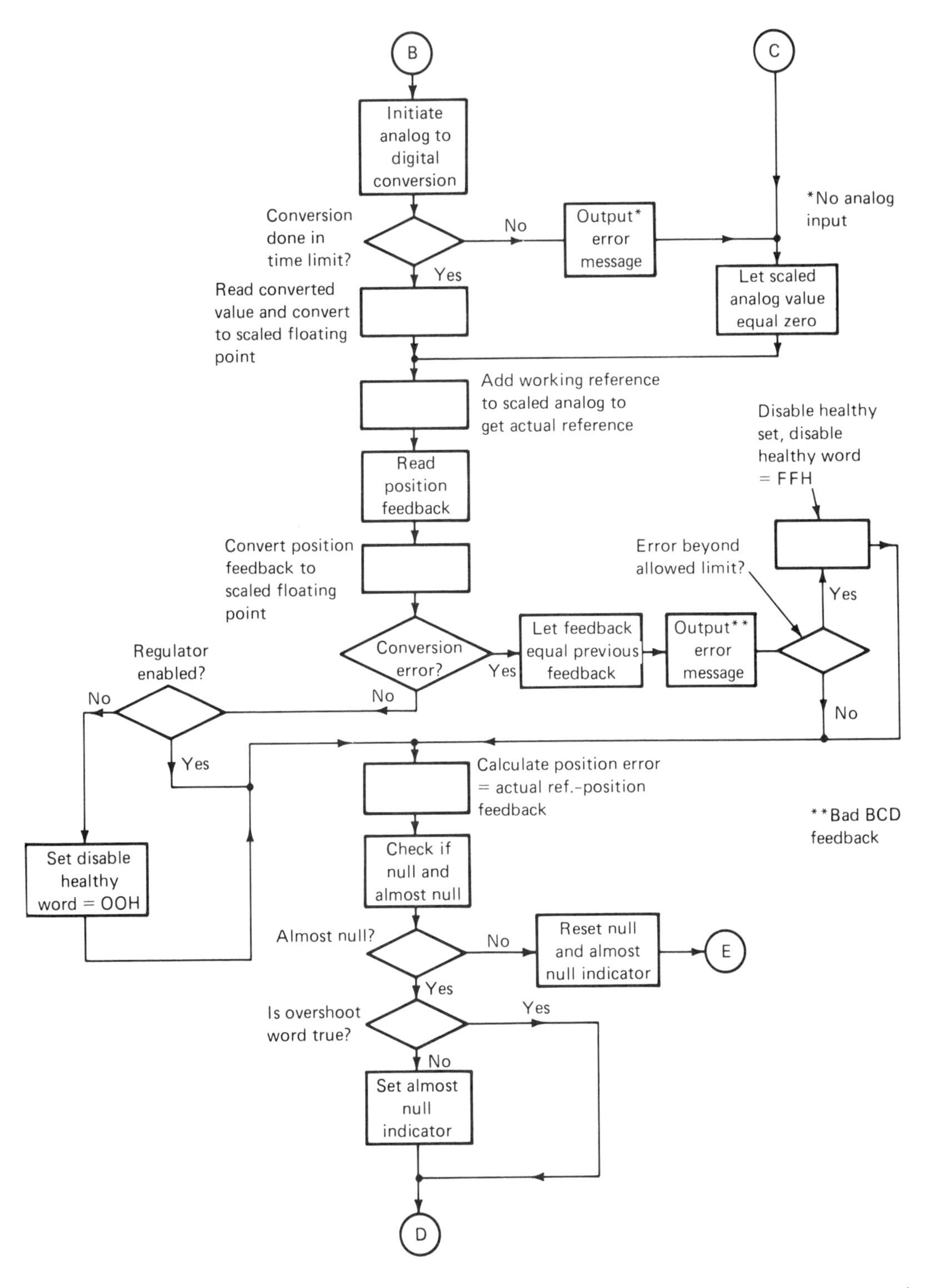

Figure 12-8 continued.

(Continued on next page)

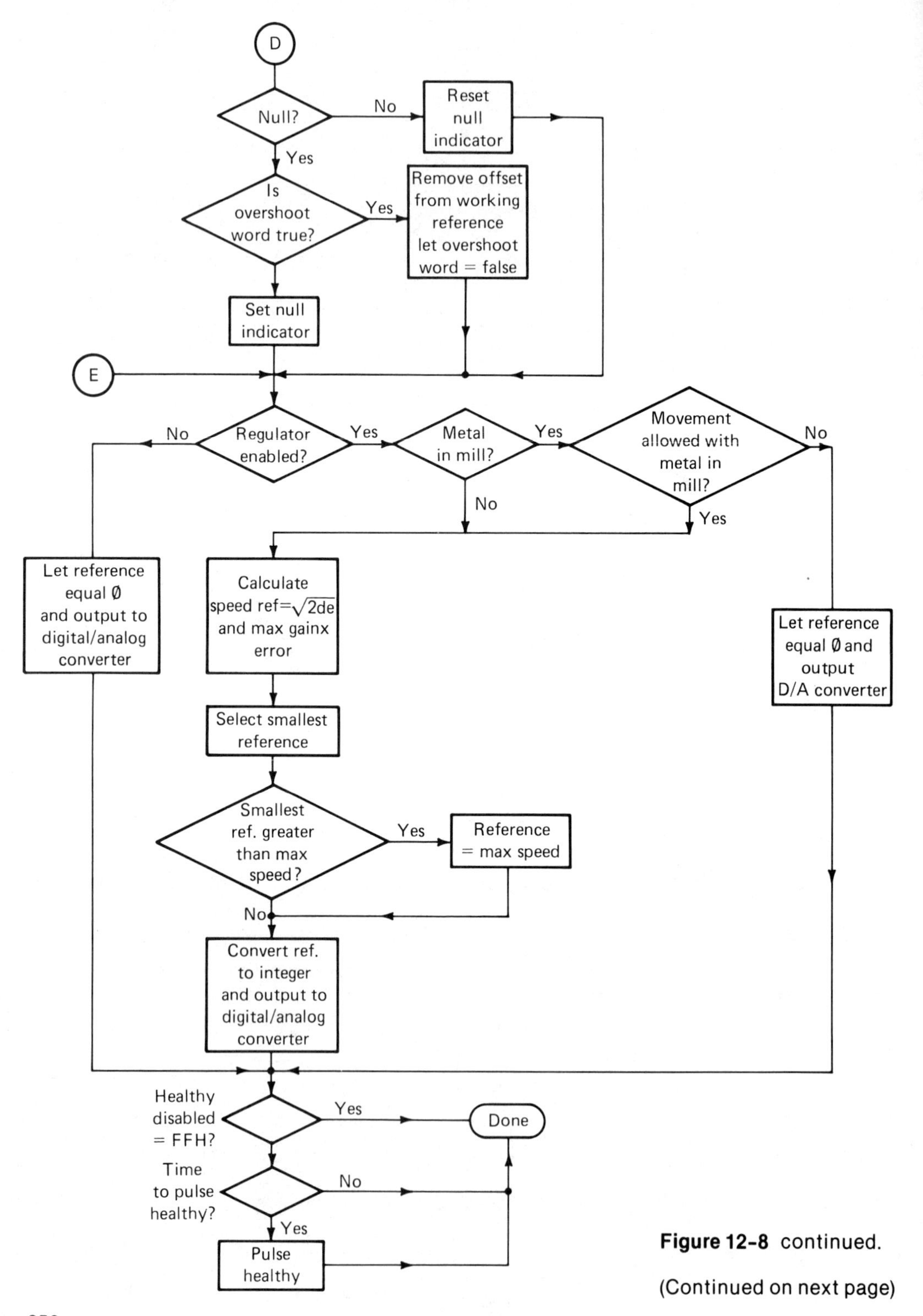

Figure 12-8 continued.

(Continued on next page)

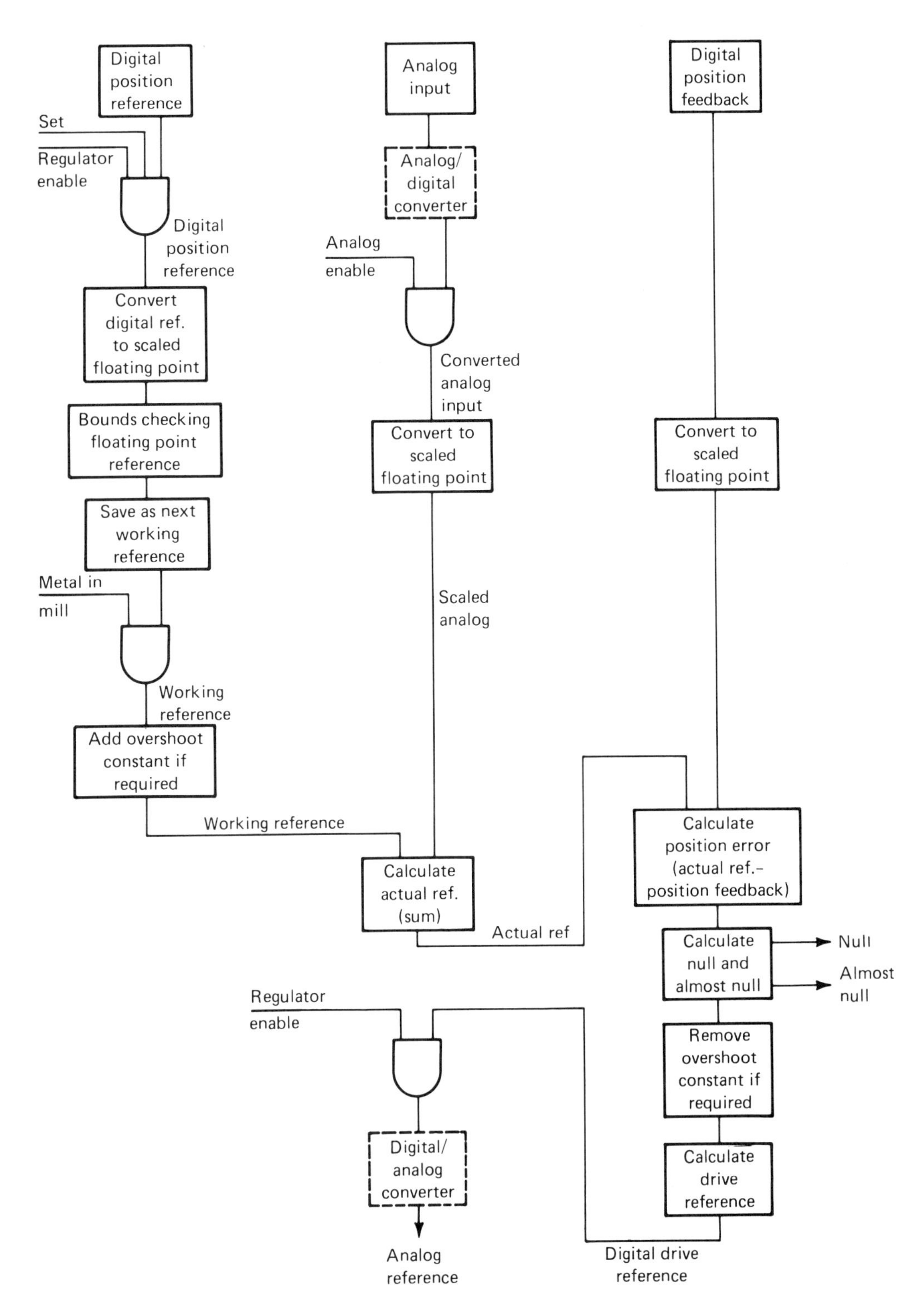

Figure 12-8 continued.

Assuming that the analog input has been enabled, that is, there is no metal in the stand then the analog-to-digital conversion is initiated. If the conversion is not completed within the prescribed time interval an error message is outputted and the scaled analog input is made equal to zero. The scaled analog input is also made equal to zero if the analog input was not enabled. If the conversion is completed in the time allocated the working reference is added to the scaled analog reference to give the actual reference. Provision has also been made to the regulator to enable external logic to permit a bipolar analog input (0 to ±10Vdc) giving a dc offset to be added to the digital reference. As a result

$$\text{actual reference} = \text{digital position reference} + \text{scaled analog offset} \qquad (12\text{-}1)$$

The digital position feedback is then read and converted to scaled floating point and then checked for any conversion error. If an error exists, the previous position feedback is retained and an error message is outputted; then the error is checked against allowed limits and if it is in excess of the limits the healthy digital output is disabled and the word is changed.

If a conversion error is not present, the regulator is checked to see if it is enabled; if it is not, the healthy word is changed. The outputs of the regulator enable and the error within prescribed limits are then combined to initiate the calculation of the position error. Then

$$\text{position error} = \text{actual reference} - \text{position feedback} \qquad (12\text{-}2)$$

In turn the resulting position error is checked to determine its magnitude. If not almost null, the null and almost null indicators are reset. If almost null, the overshoot word is checked and if not true the almost null indicator is set; then this output and the true output are combined and the position error is then checked to see if it is a null. If not a null, the null indicator is reset, and if true the overshoot word is checked and if true the dc offset is removed from the working reference and the overshoot word is changed to false. If not true, the null indicator is set.

At this point the regulator is again checked to see if it is enabled. If it is and provided that there is no metal in the mill or if there is and movement is permitted, the speed reference is calculated.

The speed reference is based upon the relationship

$$V^2 = 2as \qquad (12\text{-}3)$$

which relates the velocity (V) to the distance (s) covered under constant acceleration (a). The speed reference is developed from the position error as follows:

$$s = e \text{ (position error)},$$
$$a = d \text{ (deceleration rate), and}$$
$$v = e' \text{ (rate of change of position error).}$$

Then

$$e' = \sqrt{2de}. \tag{12-4}$$

This value of e' is directly proportional to the speed reference that must be applied to achieve the desired position as rapidly as possible without overshoot; that is, it is critically damped or in other words

$$\text{speed reference} = \sqrt{2de} \tag{12-5}$$

When small positional errors are involved there are some inherent problems with this technique. As the position error approaches zero, then de'/de (gain) approaches infinity. To overcome this problem a maximum gain clamp (K_1) is introduced when small positional errors exist, so that

$$\text{speed reference}_{\text{small errors}} = K_1 e \tag{12-6}$$

Since the drive can also achieve high speeds when there is a large position error a maximum speed clamp (K_2) is also used, so that

$$\text{speed reference}_{\text{large errors}} = K_2 e \tag{12-7}$$

A null band is also established so that the drive (s) will not respond if the position error is within the null band limits.

The single variable-voltage position regulator system is far more sophisticated than the multiple variable-voltage position regulator. Considering the single variable-voltage position regulator, a requirement exists to independently control the two screw-down motor drives during the leveling process. As a result, the system has been designed to average the two position feedback signals in order to calculate the position error. It should be noted that the two drives are synchronized during normal operation.

It is desirable to introduce a gain factor to the position error calculation; this gain factor is selected by a 4-bit binary code from any one of sixteen pre-programmed gain factors by either external logic or the microcomputer as desired.

Compensation for the drive reaction time may also be introduced as a modification to the speed reference. At low speeds the actual drive speed is determined during the interval between changes of the LSB of the position reference. In the case of high speed operation the actual drive speed is determined from the change in position feedback between updates. In either case the actual drive speed is multiplied by the drive reaction time and then added to the position error:

$$\text{position error} = \text{gain} \times (\text{actual reference} - \text{position feedback}) + (\text{drive speed} \times \text{reaction time}) \quad (12\text{-}8)$$

It is also desirable to be able to dynamically zero the drive position. This is achieved by producing an offset, termed the lock-on offset, that is equal to the current position error. A modified position error is then derived by subtracting the lock-on offset from the position error as follows:

$$\text{modified position error} = \text{position error} - \text{lock-on offset} \quad (12\text{-}9)$$

If dynamic zeroing is not desired the lock-on offset remains at zero.

The basic operation of the constant potential drive regulator is almost identical to that of the variable-voltage position regulator except for the positioning algorithm. The constant potential drive positioner is associated with a three-speed reversing drive.

The traditional constant potential drive regulator depends upon the use of breakpoints. At each breakpoint there is an associated position error. The selected speed is determined whenever the actual position error is greater than the position error associated with a particular breakpoint.

Since the constant potential drive has a high regulation and deceleration rate is proportional to the load, it is necessary to prevent overshoot. This is accomplished by selecting a breakpoint for lightly loaded conditions which achieves a minimum static drive accuracy. When starting the drive from standstill under small positional error conditions a low speed breakpoint that will produce a low drive torque is selected. As a result the system will take longer to position itself than if a large positional error were present.

A modified positioning algorithm was developed in which the actual drive speed is compared to the modified position error (Equation 12-9). This result is then used to select the most appropriate of three preprogrammed speeds. This method has resulted in significant improvements in the positioning accuracy and the speed of response of the drive as well as eliminating the overshoots inherent in the traditional approach.

A simplified logic representation of the complete drive is shown in Figure 12-8, Sheet 4.

GLOSSARY OF IMPORTANT TERMS

Programmable controller: A solid-state control system consisting of solid-state components that are easily programmed to control the performance of any repetitive process.

Uninterruptible power supply (UPS): An emergency electrical power supply system that provides an alternative source in the event of a power failure.

Switched-mode power supply: A regulated dc power supply that controls the output voltage by pulse width modulation control of a bipolar power transistor or MOSFET operating in the switching mode.

Robot: A programmable manipulator that performs useful work without the aid of a human operator.

REVIEW QUESTIONS

12-1 What are the advantages to using a programmable controller?

12-2 With the aid of a block diagram briefly explain the main elements of a programmable controller.

12-3 What is an uninterruptible power supply?

12-4 Explain the characteristics of a transfer and continuous or float type UPS.

12-5 Discuss the three types of continuous UPS systems.

12-6 Explain the terms nonredundant and parallel redundant as applied to UPS systems.

12-7 What are the advantages and disadvantages of using switched-mode power supplies?

12-8 With the aid of a schematic briefly describe the operation of a series-pass linear regulated power supply.

12-9 Repeat Question 12-8 for a switched-mode power supply.

12-10 Discuss the two types of industrial robots.

12-11 Briefly describe a PUMA robot.

Appendix A

Thyristor Protection

A-1 INTRODUCTION

Silicon controlled rectifiers and for that matter all semiconductor devices are limited in their capabilities. The correct application of thyristors depends upon ensuring that at all times the circuit operating conditions do not exceed their designed capabilities. This is assured by providing protection against overvoltage and overcurrent.

A-2 VOLTAGE TRANSIENTS

Voltage transients are primarily caused by switching disturbances, the source usually being the stored energy ($\frac{1}{2}LI^2$) in the inductive components of the system. The effect is to produce voltage transients whose peak may be as much as ten times the repetitive forward blocking voltage V_{FB}. In the forward direction anode, turn-on may occur, possibly causing the flow of large fault currents; in the reverse direction the transient can cause large currents to flow in small areas around the thyristor junction and can produce localized heating, unless transient suppression techniques are applied.

Voltage transients in a naturally commutated circuit, that is, a thyristor phase-controlled converter, can be broadly classified as ac side transients, dc side transients, and transients occurring within the converter.

A-2-1 Transients Originating on the AC Side of the Converter

These transients are caused by:

1. Line voltage switching resulting from fault clearance and switch-

ing in the ac supply system. The magnitude of the transient is dependent upon the speed of switching and the point on the voltage sinusoid at which switching occurs.

2. Lightning. It produces a line-to-ground surge with a steep-fronted waveform, building up to a maximum in a few microseconds, and it is completed in a few hundred microseconds.

3. Transformer primary inrush current. It produces oscillations in the resonant secondary winding circuit caused by the transformer leakage reactance and the distributed secondary winding capacitance. Applying the primary voltage at its peak produces a secondary voltage transient approximately twice the normal peak value. Even higher transients are produced if contact bounce is present.

4. Interruption of the transformer magnetization current. The rapid decay of core flux induces a transient at the secondary terminals which may approach ten times the peak secondary voltage. The transient is greatest at no load, when the connected rectifier is supplying an inductive load, for example, a choke input filter. The amplitude will be at its greatest if the interruption occurs when the primary current is passing through its zero value.

5. Energizing the primary of a step-down transformer. Because of the interwinding capacitance between the primary and secondary windings, when the primary is energized the primary voltage is momentarily coupled to the secondary. In high-ratio, step-down rectifier transformers the interwinding capacitance is of the order of 0.001 microfarads, and the charging voltage can give rise to transient secondary voltages several times the normal secondary voltage.

With reasonable care, the elimination of the sources of these voltage transients can be achieved. For example, switching the secondary of a transformer will overcome the problem of the oscillations caused by the primary inrush current. Transients caused by fault clearance in the ac supply system can be minimized by ensuring that the interrupting devices have a long arc time, thus dissipating the transient in the arc. Lightning arresters placed near the equipment to be protected will bypass overvoltages to ground. Connecting a capacitor across the secondary of a step-down transformer will cause voltage division and a reduction of the transient.

When it is not practical to suppress transients by these means, then R-C snubbers across the input ac lines may be used, or alternatively, voltage-dependent symmetrical resistors such as General Electric Thyrectors and GE-MOV varistors may be used.

A-2-2 Transients Originating on the DC Side of the Converter

When dc side switching takes place under load conditions, the stored magnetic energy in the ac supply system and the transformer leakage reactances produce a voltage $-Ldi/dt$. If there is no alternative leakage path, the induced voltage then appears across the SCRs and the switch, and it will be at its greatest under dc short-circuit conditions.

Inductive loads on the dc side will not cause voltage transients when switched, since by Lenz's Law the induced voltage will maintain the forward-biased SCRs in conduction.

Under overhauling load conditions the armature of a dc motor is accelerated, and the counter-emf is increased above the rectifier output level and applies a high reverse voltage across the SCRs. Similar conditions exist if the shunt field excitation is rapidly increased while the motor is driving a high inertia load.

The reduction of dc side generated transients is achieved by dissipation of the stored energy across the rectifier output terminals. Some methods are:

1. Adding additional inductance in series with the dc load.
2. Using a controlled rate of opening switch (or fuse for dc side fault clearance).
3. Connecting thyrectors or varistors across the rectifier output.
4. With regenerative loads, connecting a resistance in series with the armature when the voltage exceeds a safe value.

A-2-3 Transients Occurring Within the Rectifying Equipment

The principal causes of transients generated within the rectifier are the interruption of semiconductor fuses under fault conditions, and commutation transients.

1. Interruption of semiconductor fuses. Under fault conditions the rate of change of fault current is di/dt and is produced by the applied circuit voltage V, where $V = Ldi/dt$, L being the circuit inductance, at the end of the melting time $di/dt = 0$. The voltage across the arc is V. For di/dt to decay at the same rate as the rise, the voltage across the arc must be two times V. This voltage is applied across the semiconductor device, and unless the voltage rating of the device is sufficiently high, it will cause device failure.

2. Commutation transients. After turn-off has been initiated, time is required for the charge carriers to be swept away from the junctions, before the thyristor is capable of blocking a reverse voltage. During

this period the reverse recovery current is limited by the circuit impedance. When blocking is achieved, the sudden termination of the reverse recovery current causes a voltage transient to be generated by the commutating inductance.

A-3 OVERCURRENT PROTECTION

In overcurrent protection, the nature of the source impedance is very important. If the source impedance is "soft"—that is, there is sufficient reactance to limit the *di/dt* of the circuit—and its duration is unlikely to be in excess of a few cycles of the source frequency, then sufficient protection of the thyristor can be obtained by selecting a thyristor with an adequate surge current rating. If, on the other hand, the duration of the fault current is more than a few cycles, then it may be sufficient to rely upon a conventional short-time delay electromagnetic relay or thermal overloads, provided that prior to the clearance of the fault the thyristor does not become overheated.

If the source impedance is "stiff"—that is, it is a low-reactance source—then a rapid current buildup will occur under fault conditions. In this case, it is essential that the circuit be interrupted rapidly before permanent damage can occur to the thyristor. The cheapest and most effective interrupting device in these circumstances is the semiconductor fuse, which will clear the circuit when:

1. Another thyristor fails to block in the reverse direction, or
2. A short circuit occurs across the output terminals of the converter, or
3. A thyristor fails to turn off in time, or turns on at the incorrect time.

The action of a fuse under fault conditions is illustrated in Figure A-1. Initially, under fault conditions the current rises rapidly, being limited only by the circuit impedance. If there were no fuse action, it would build up to a peak value as shown and decay to zero, if it is assumed that a half-wave sinusoid voltage is being applied. In actual practice the current builds up to point A, at which point the fuse begins to melt. There will be a further slight increase in the current magnitude up to point B, the peak let-through current, before the energy is dissipated in the arc, with complete interruption of the protected circuit being achieved at point C.

The total clearing time, which is designed to occur in less than 8.3 ms, consists of two equal time segments, the melting time and the arcing time. The rate of decrease of current during the arcing time

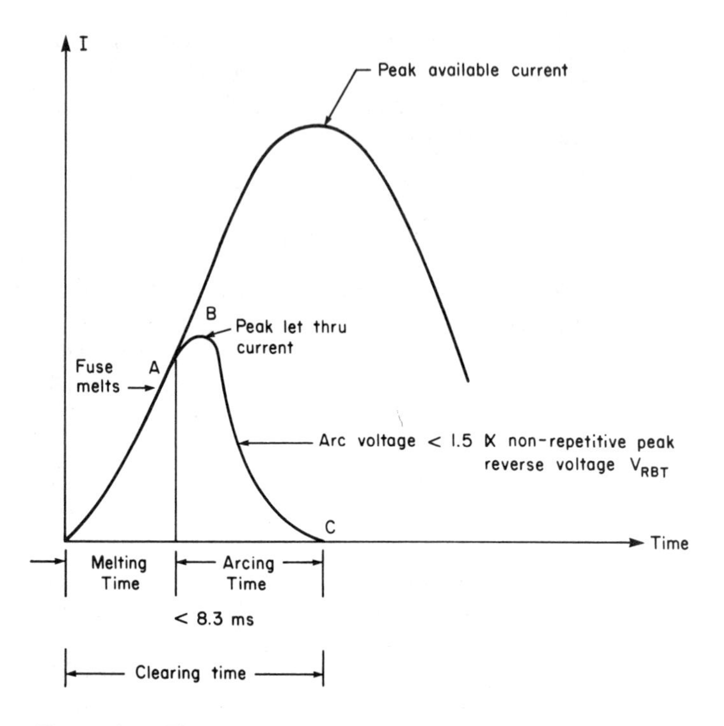

Figure A-1. The protective action of a semiconductor fuse.

must also be low enough that high induced voltages (Ldi/dt), which could destroy the thyristors, are not produced.

Semiconductors and fuses are designed to withstand specified I^2t values below a period of one-half cycle of the supply voltage.

A-3-1 I^2t Ratings

The energy that must be dissipated by a fuse during the clearing time comes from two sources: from the power supply and from the stored magnetic energy ($\frac{1}{2}LI^2$) in the inductive portions of the circuit. The heat energy to be dissipated by the fuse is $\int i^2Rdt$, where R represents the circuit resistance. Since the current is common to the whole circuit and the resistance is basically constant during the clearing time, the energy to be dissipated is proportional to $\int i^2dt$, that is, I^2t (amperes^2 $\times$ time).

By rating the fuses and the semiconductors on a common basis, namely their I^2t ratings, it is then possible to select fuses with a lower I^2t rating than the semiconductor, and thus protect the semi-conductor.

Fuse and thyristor manufacturers specify in their data the I^2t ratings of their devices, usually based on total clearing times of the fuses not exceeding 8.3 ms. The I^2t rating of the thyristor is based on the device operating at maximum rated current and maximum junction temperature; since this is not the normal practice, a factor of safety is introduced.

Since semiconductor fuses are much more expensive than normal renewable link fuses, it is normal practice to consider the relative costs of the fuse against the cost of the semiconductor and its replacement, and in the case of low cost thyristors to eliminate fusing.

A-4 GATE CIRCUIT PROTECTION

The gate circuits must also be provided with protection because of inductive and capacitive coupling between power and control circuits inducing voltage transients into the gating circuitry. Because of the high di/dt in the power circuits, it is essential that the power wiring and control wiring be as widely separated as possible, which also has the side effect of reducing the capacitive coupling.

Electrical isolation is standard practice in high-power converters and is usually provided by gate pulse transformers, which are ferrite cored, with a primary and usually multiple secondaries, or by optocouplers. Usually the leads connecting the gate and cathode of the SCR to either the pulse transformer or optocoupler are twisted with a minimum of two twists per foot, or are shielded, with the shield being connected to a common at one end only.

Figure A-2 illustrates a number of gate terminations designed to discriminate between signals and noise. In Figure A-2(a) the resistor decreases the gate sensitivity, increases the dv/dt of the device, and reduces the turn-off time, which at the same time increases the holding and latching currents. Connecting a capacitor between the gate and cathode leads (Figure A-2(b)) removes high-frequency noise components and increases the dv/dt capability of the thyristor; at the same time, it increases the gate-controlled delay time t_d and turn-on time and turn-off time, and reduces the gate-controlled rise time t_r. The circuit of Figure A-2(c) protects the gate supply during reverse transients in excess of V_{RBT}, and limits the negative bias applied to the gate to approximately 1 volt; alternatively, a zener diode may be used. Figure A-2(d) gives protection in low-power thyristor applications with dc gate signals. The capacitor C removes the ac component of a transient noise signal, and the resistor improves the dv/dt capability of the device. These gate terminations are only representative of the many combinations that can be used.

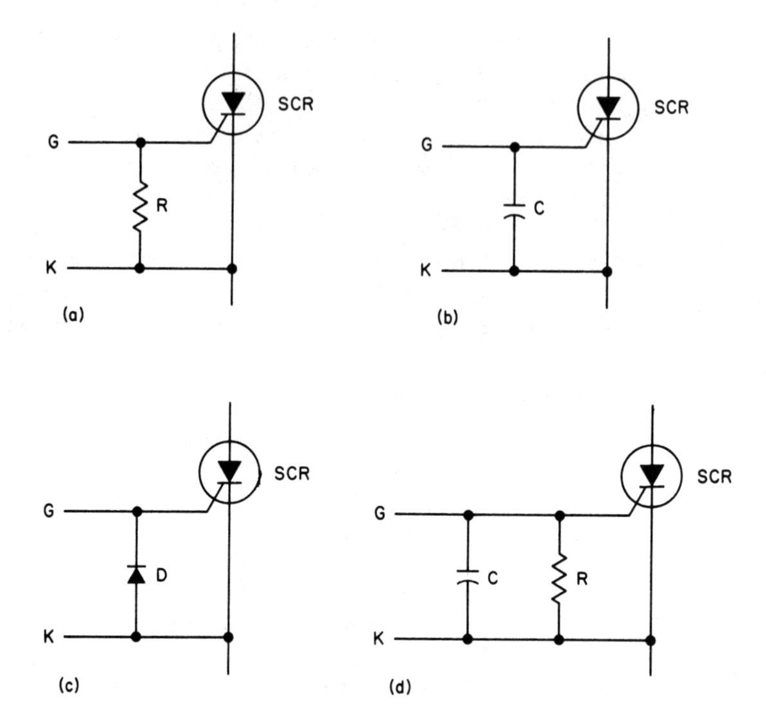

Figure A-2. Gate terminations.

Appendix B

Cooling

B-1 INTRODUCTION

The subject of efficient cooling of silicon devices (rectifier diodes, silicon controlled rectifiers, and triacs) is usually not afforded the attention that should be devoted to it. Since all semiconductor devices are temperature limited devices their current ratings are based upon thermal considerations. It is essential for the designer to understand the principles of thermal design in order to correctly select the device and the cooling method.

B-2 HEAT SOURCES IN THYRISTORS

The total forward current that an SCR can safely carry is dependent upon the maximum permissible junction temperature, the thermal impedances of the device, and the losses within the device which are primarily on-state losses, provided that the device is being operated at less than the specified di/dt. Normally these ratings are based on switching frequencies up to 400Hz and for specified conduction conditions.

The SCR is designed to operate with an operating junction temperature T_J within the specified range; for the 2N4361/2N4371 Series SCR this range is from $-40\,^{\circ}\mathrm{C}$ to $+125\,^{\circ}\mathrm{C}$. To maintain operation within this range it is necessary that the forward current not produce excessive junction temperatures. The maximum value of RMS current I_{RMS}, ($I_{T(RMS)}$) that the SCR may conduct is determined by the heating effects produced by:

1. The forward conduction loss, the major loss.

2. The forward or reverse leakage losses during blocking.

3. The switching losses produced during turn-on or turn-off.

4. The gate junction losses which are dependent upon the type and duration of the gate signal, that is, single-pulse, pulse trains, or dc signals.

Another important factor affecting the thyristor current rating is the efficiency of the heat sink or cooling system. Normally all calculations are based on the worst case situation, that is, maximum continuous load current for the maximum duty cycle with the highest thyristor forward voltage drop at the maximum ambient temperature of the cooling medium.

Figure B-1 defines the maximum ratings and characteristics of the 2N4361/2N4371 Series SCR produced by Westinghouse Electric Corporation.

Heat is transferred from the device by all three modes of heat transfer: conduction, convection, and radiation. However, usually one mode of heat transfer is more predominant and as a result the other modes may be ignored. In most industrial applications the predominant mode of heat transfer is by either natural or forced air convection cooling.

Since most of the heat is generated in the silicon wafer, the heat must flow from the silicon wafer to the case and then to the heat sink and into the cooling medium. The thermal path can be compared to an electric circuit consisting of a number of resistances connected in series. The electrical analog of a heat sink mounted SCR is illustrated in Figure B-2, where P_{AVE} is assumed to be a constant current source; then the junction temperature T_J is

$$T_J = P_{AVE}(R_{\theta JC} + R_{\theta CS} + R_{\theta SA}) + T_A \qquad \text{(B-1)}$$

where T_J = junction temperature, °C (125°C for 2N4361/2N4371 Series SCR)

P_{AVE} = average power dissipation in the *SCR*, W

$R_{\theta JC}$ = thermal resistance, junction to case, °C/W

$R_{\theta CS}$ = thermal resistance, case to sink, °C/W

$R_{\theta SA}$ = thermal resistance, sink to ambient, °C/W

T_A = ambient temperature, °C

Phase Control SCR 2N4361/2N4371 Series

70 A Avg. Up to 1400 Volts

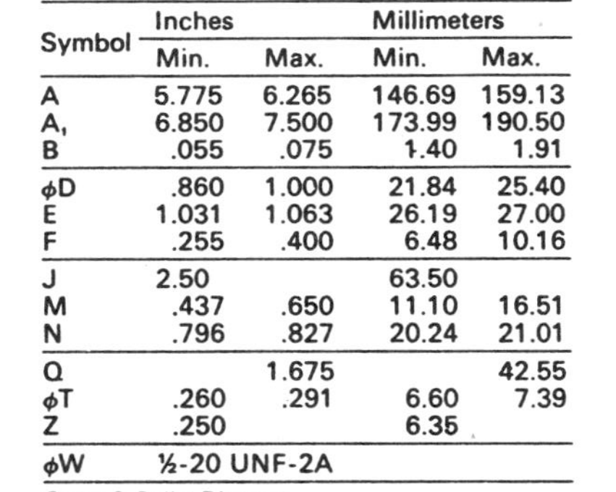

Conforms to TO-94 Outline

Features:
- All diffused design
- Low gate current
- Low V_{TM}
- Compression Bonded Encapsulation
- Low Thermal Impedance

Symbol	Inches Min.	Inches Max.	Millimeters Min.	Millimeters Max.
A	5.775	6.265	146.69	159.13
A,	6.850	7.500	173.99	190.50
B	.055	.075	1.40	1.91
φD	.860	1.000	21.84	25.40
E	1.031	1.063	26.19	27.00
F	.255	.400	6.48	10.16
J	2.50		63.50	
M	.437	.650	11.10	16.51
N	.796	.827	20.24	21.01
Q		1.675		42.55
φT	.260	.291	6.60	7.39
Z	.250		6.35	
φW	½-20 UNF-2A			

Creep & Strike Distance.
.50 in. min. (12.85 mm).
.10 in. min. (2.54 mm). **
(In accordance with NEMA standards.)
Finish—Nickel Plate.
Approx. Weight—5 oz. (142 g).

1. Complete threads to extend to within 2½ threads of seating plane.
2. Angular orientation of terminals is undefined.
3. Pitch diameter of ½-20 UNF-2A (coated) threads (ASA B1.1-1960).
4. Dimension "J" denotes seated height with leads bent at right angles.

Applications:
- Phase control
- Power supplies
- Motor control
- Light dimmers

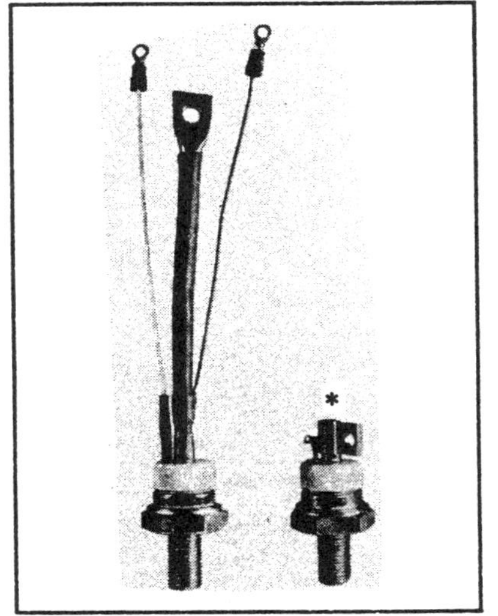

* For TO-83 Outline, see page S23.

Figure B-1. Westinghouse type 2N4361/2N4371 series, silicon controlled rectifier, maximum ratings and characteristics. (Courtesy of Westinghouse Corp.)

Voltage

Blocking State Maximums (2) (TJ = 125°C)	Symbol	2N4361 2N4371	2N4362 2N4372	2N4363 2N4373	2N4364 2N4374	2N4365 2N4375	2N4366 2N4376	2N4367 2N4377	2N4368* 2N4378
Repetitive peak forward blocking voltage, V	V_{DRM}	100	200	400	600	800	1000	1200	1400
Repetive peak reverse voltage, V	V_{RRM}	100	200	400	600	800	1000	1200	1400
Non-repetitive transient peak reverse voltage, t≤5 msec, V	V_{RSM}	200	300	500	700	950	1200	1450	1700
Forward leakage current, mA peak	I_{DRM}	←				10			→
Reverse leakage current, mA peak	I_{RRM}	←				10			→

Current

Conducting State Maximums (TJ = 125°C)	Symbol	
RMS forward current, A	$I_{T(rms)}$	110
Ave. forward current, A	$I_{T(av)}$	70
One-half cycle surge current, A (3)	I_{TSM}	1600
3 cycle surge current, A (3)	I_{TSM}	1250
10 cycle surge current, A (3)	I_{TSM}	1080
I^2t for fusing (for times 8.3 ms) A^2sec	I^2t	10,700
Forward voltage drop at I_{TM} = 500A and T_J = 25°C, V	I_{TM}	2.5

Thermal and Mechanical

	Symbol	
Min., Max. oper. junction temp., °C	T_J	—40 to +125
Min., Max. storage temp., °C	T_{stg}	—40 to +150
Max. mounting torque, (1) in lb.		130
Max. Thermal resistance (1) Junction to case, °C/Watt	$R_{\theta JC}$	.28
Case to sink, lubricated °C/Watt	$R_{\theta CS}$	.12

Switching

(TJ = 25°C)	Symbol	
Typical turn-off time, I_T = 50A, T_J = 125°C, di_R/dt = 5 A/µsec, reapplied dv/dt = 20V/µsec linear to 0.8 V_{DRM}, µsec	t_q	100
Typ. turn-on-time, I_T = 100A, V_D = 100V, µsec	t_{on}	4
Min. critical dv/dt, exponential to V_{DRM}, T_J = 125°C, V/µsec (5)	dv/dt	100
Min. di/dt non-repetitive, A/µsec	di/dt	800

(1) (4) (6)

Gate

Maximum Parameters (TJ = 25°C)	Symbol	
Gate current to trigger at V_D = 12V, mA	I_{GT}	250
Gate voltage to trigger at V_D = 12V, V	V_{GT}	3
Non-triggering gate voltage, T_J = 125°C, and rated V_{DRM}, V	V_{GDM}	0.15
Peak forward gate current, A	I_{GTM}	4
Peak reverse gate voltage, V	V_{GRM}	5
Peak gate power, Watts	P_{GM}	15
Average gate power, Watts	$P_{G(av)}$	3

(1) Consult recommended mounting procedures.
(2) Applies for zero or negative gate bias.
(3) Per JEDEC RS-397, 5.2.2.1.
(4) With recommended gate drive.
(5) Higher dv/dt ratings available, consult factory.
(6) Per JEDEC standard RS-397, 5.2.2.6.
***2N4361 Series in TO-94 PKG.**
2N4371 Series in TO-83 PKG.
**Glass-to-metal seal package.

Figure B-1 continued.

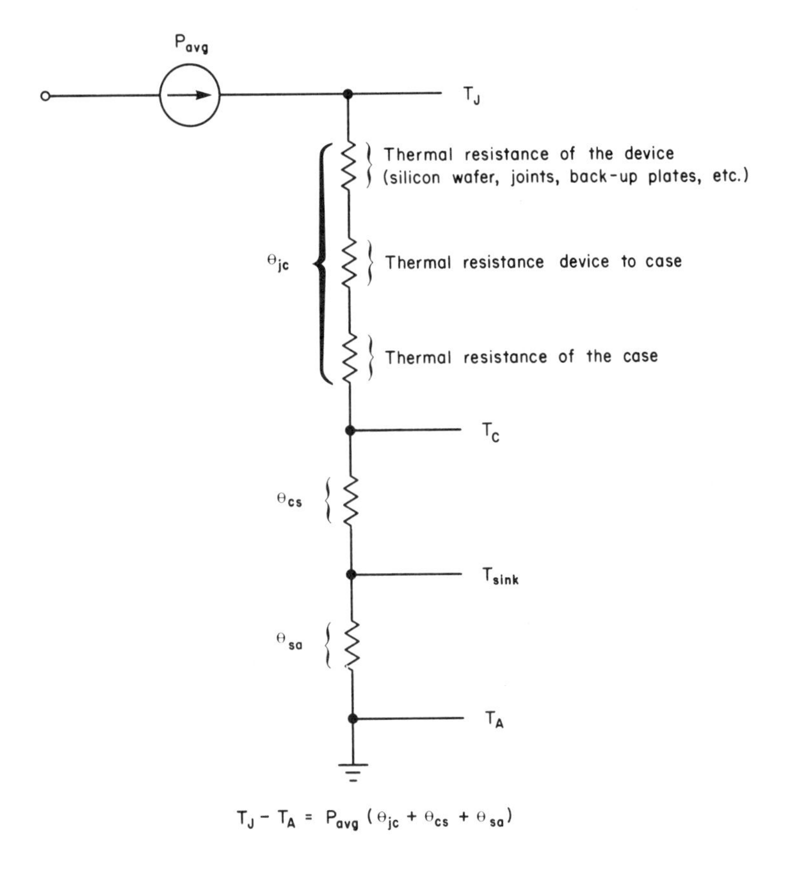

Figure B-2. Electrical analog describing the relationships between P_{AVE} and junction temperatures.

The thermal resistance $R_{\theta SA}$ of the heat sink to ambient depends upon the type of material, shape, treatment of the outside surface, and dimensions of the heat sink. Usually the information is presented in the form of graphs by the manufacturer, and relates the difference in temperature between the heat sink and the ambient temperature ΔT and the average power dissipated P_{AVE}. Typical graphs are shown in Figure B-3 where

$$\Delta T = T_{\text{SINK}} - T_A \,^{\circ}\text{C} \tag{B-2}$$

and

$$R_{\theta SA} = \frac{\Delta T}{P_{AVE}} \,^{\circ}\text{C/W} \tag{B-3}$$

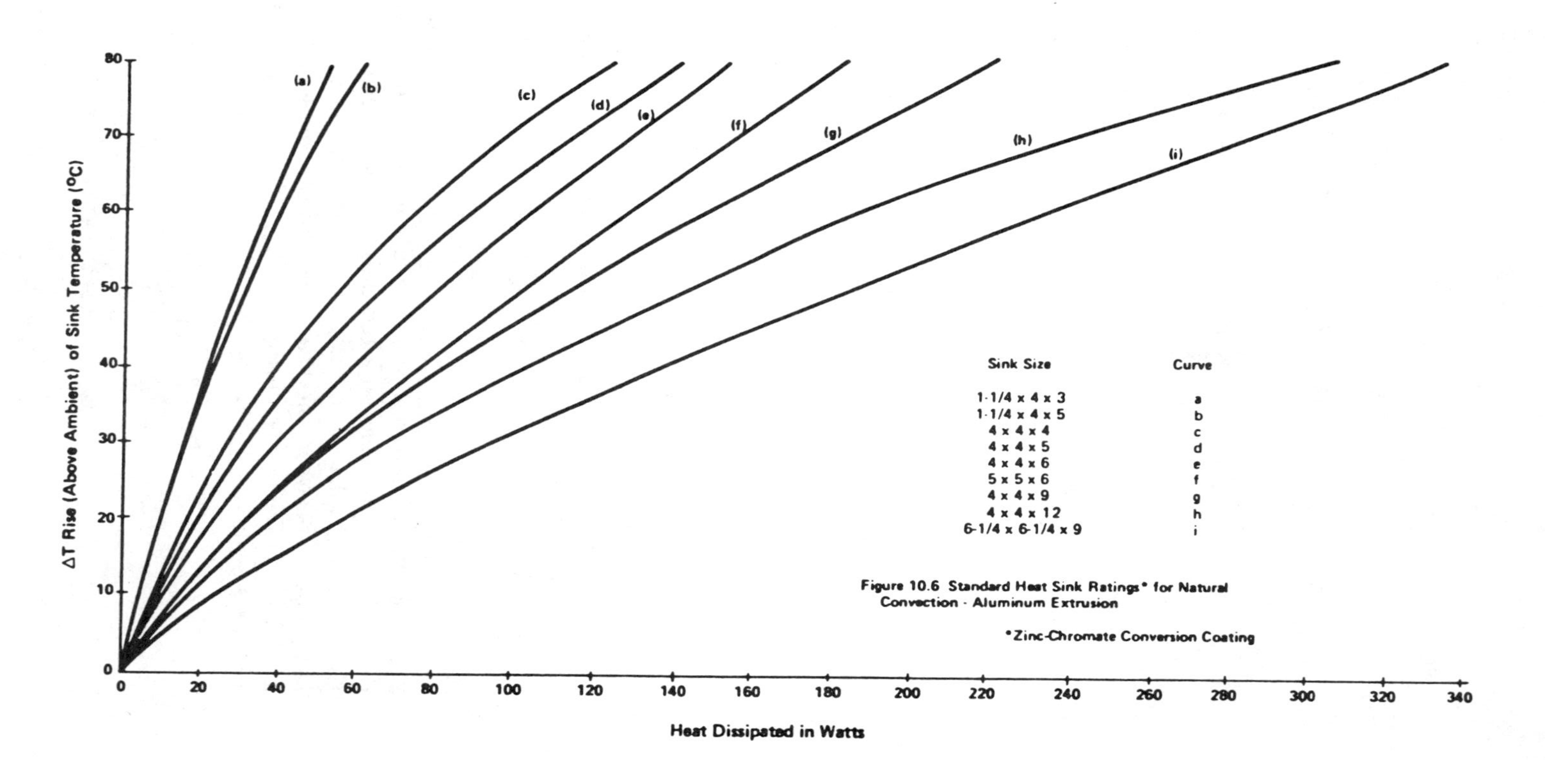

Figure B-3. Typical heat sink ratings—aluminum natural convection cooling. (Courtesy of Westinghouse Corp.)

To date the assumption has been that dc current is flowing and Equations (B-1), (B-2), and (B-3) are applicable to steady-state conditions. However, recognition must be made of the effects of rectangular current pulses as would be experienced in inverter and chopper applications as well as half-wave sinusoids as would be experienced in phase controlled applications.

All manufacturers provide data sheets; a typical example is shown in Figure B-4, for the Westinghouse 2N4361/2N4371 Series SCRs. The forward voltage drop V_F, (V_{TM}) which is the instantaneous voltage drop between the anode and cathode under load conditions with the SCR operating under pulsed conditions and properly heat sinked, must be equal to or less than the maximum value specified in the data sheet. A plot of V_F versus I_F, (V_{TM} versus I_{TM}) is shown in Figure B-4(e). From this plot, curves of power dissipation versus forward current for rectangular and half-wave sinusoid waveforms at various conduction angles are developed [see Figures B-4(a) and (c)]. In turn, these data combined with the thermal impedances, steady-state and transient, are used to present case temperature versus forward current for rectangular and half-wave sinusoid waveforms [see Figures B-4(b) and (d)].

The thermal capacity of the thyristor is very small compared to that of the load and source, and as a result the junction temperatures of the thyristor will vary rapidly and may exceed their designed maximum values, under cyclic load conditions.

Consider the application of a train of rectangular power pulses as shown in Figure B-5. The junction temperature T_J rises exponentially during the duration of the pulse and then decays exponentially during the off period. The difference of temperature ΔT *is*

$$\Delta T = PZ_{\theta(t)}\ ^\circ\mathrm{C} \qquad \text{(B-4)}$$

where $Z_{\Theta(t)}$ = transient thermal impedance, °C/W

A plot of the thermal impedance versus pulse duration is supplied by the manufacturer and forms part of the SCR data sheet [see Figure B-4(f)]. It can be seen that the thermal transient impedance will reach a steady state value, the dc thermal impedance, only after the application of a pulse of approximately 100 seconds in duration. Expressed in terms of the instantaneous forward current i_A, the average forward current $I_{AVE}(I_{T(AV)})$ is less than when a constant dc current is being carried [see Figure B-4 (b)].

When half-wave sinusoids are being considered, the dc rating of the thyristor is even greater because of the form factor of the pulse, resulting in a higher peak current than for a rectangular pulse. Curves illustrating this situation are shown in Figure B-4(c).

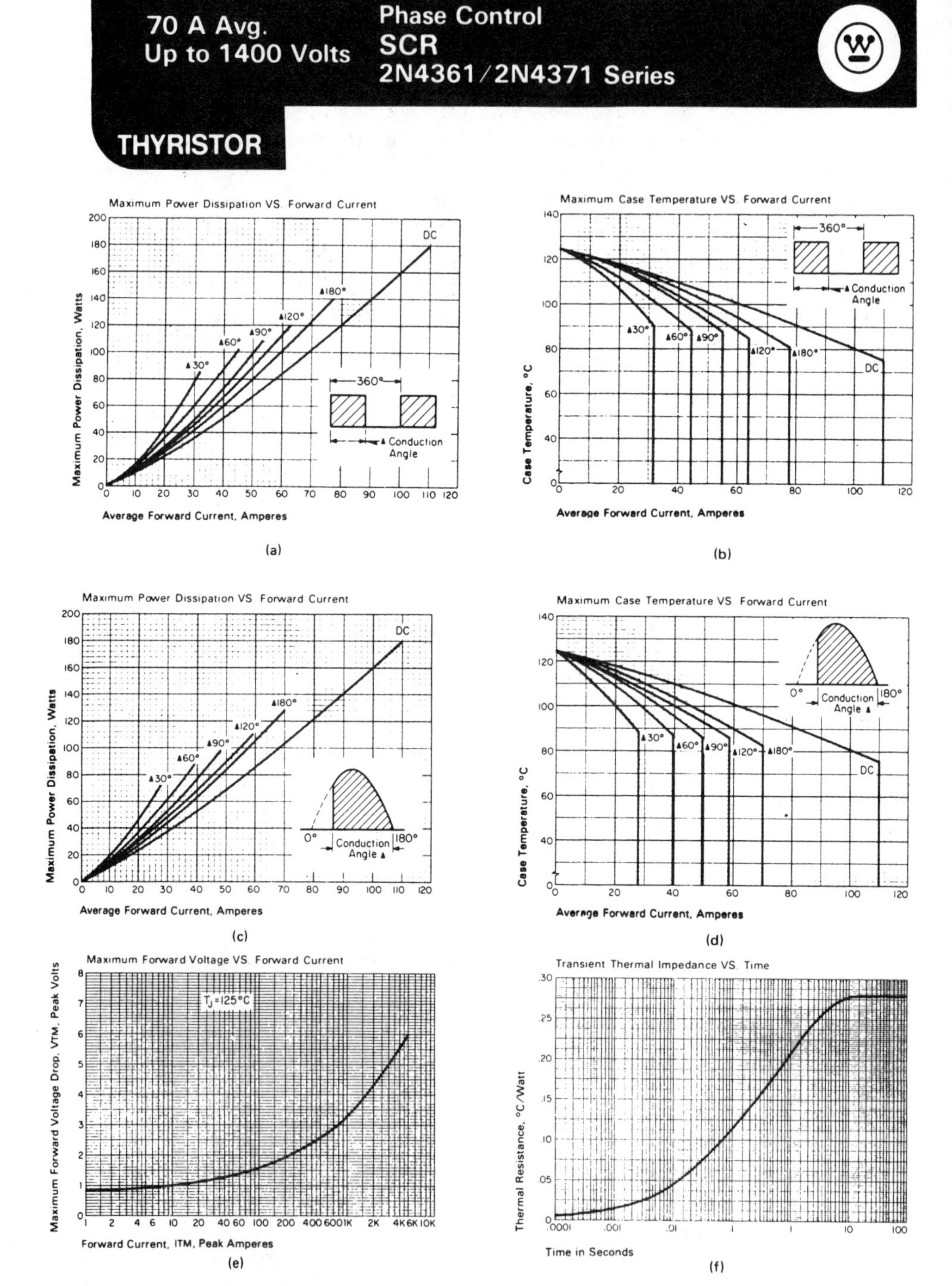

Figure B-4. Westinghouse phase control SCR, 2N4361/2N4371 series, electrical characteristics. (Courtesy of Westhinghouse Corp.)

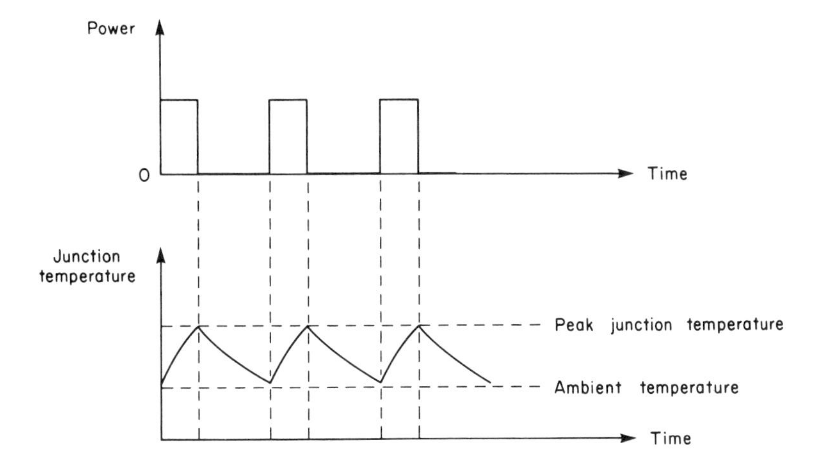

Figure B-5. Junction temperature variations resulting from a train of rectangular power pulses.

If the current pulses are irregular, they are approximated by a regular pulse train in order to simplify the calculations.

Example B-1

In Figure B-6, $V = 208\text{V}$, $R = 2\Omega$, and the ambient temperature $T_A = 40°\text{C}$. Using the Westinghouse 2N4361/2N4371 Series SCR, determine (1) the correct heat sink from Figure B-3 when the firing delay angle $\alpha = 0°$, and (2) the case and junction temperature when $\alpha = 120°$.

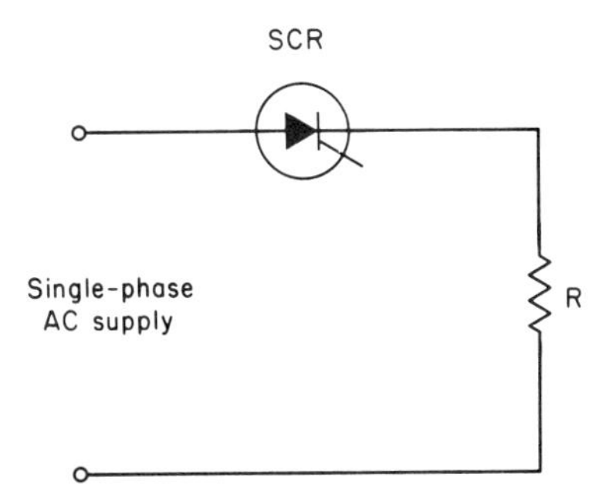

Figure B-6. Circuit for Example B-1.

Solution:

(1) When $\alpha = 0°$, the conduction angle $\gamma = 180°$ and the average current is

$$\frac{1}{2\pi}\int_0^{\pi} \frac{V_m}{R} \sin \omega t \, d(\omega t)$$
$$\frac{V_m}{R\pi} = \frac{\sqrt{2} \times 208}{2\pi} = 46.8\text{A}$$

and from Figure B-4(d), $T_{C(MAX)}$ °C = 108°C and from Figure B-4(c) $P_{AVE(MAX)} = 76$W. The heat sink temperature is

$$T_{\text{SINK}} = T_C - P_{AVE} \cdot \theta_{CS}$$

and assuming $\theta_{CS} = 0.075$, which is typical for Westinghouse SCRs, then

$$T_{\text{SINK}} = 108 - 76 \times 0.075 = 102 \cdot 43 \cong 103°\text{C}$$

and
$$\Delta T = T_{\text{SINK}} - T_A = 103 - 40 = 63°\text{C}$$

From Figure B-3, curve c defines the correct heat sink for natural convection cooling that will dissipate 76W when $\Delta T = 63°$C.

(2) When $\alpha = 120°$, the conduction angle $\gamma = 60°$ and the average current is

$$\frac{1}{2\pi}\int_{\alpha}^{\pi} \frac{V_m}{R} \sin \omega t \, d(\omega t)$$
$$= \frac{1}{2\pi}\int_{\alpha}^{\pi} \frac{\sqrt{2}V}{R} \sin \omega t \, d(\omega t)$$
$$= \frac{V}{\sqrt{2}\pi R}(1 + \cos\alpha)$$
$$= \frac{208}{\sqrt{2} \times 2 \times 2} = 11.7\text{A}$$

From Figure B-4(c), $P_{AVE} = 16.5$W
From Figure B-3, curve c for $P_{AVE} = 16.5$W, $\Delta T = 20°$C; therefore

$$T_{\text{SINK}} = T_A + \Delta T = 40 + 20 = 60°\text{C}$$

and the case temperature T_C is

$$T_C = T_{\text{SINK}} + P_{AVE}\,\theta_{CS} = 60 + 16.5 \times 0.075$$
$$\cong 61°\text{C}$$

and the temperature of the junction T_J is

$$T_J = T_C + P_{AVE}\theta_{JC} = 61 + 16.5 \times 0.28$$
$$= 64.6°\text{C}$$
$$\cong 65°\text{C}$$

B-3 AIR COOLING

A wide variety of extruded aluminum heat sinks are commercially available and all depend upon the use of cooling fins to increase the heat transfer capability. The limiting factor in the design of the extrusion is the extrusion process and its associated dies. A number of factors determine the performance of an air cooled heat sink; some of these factors are:

1. The extrusion design, that is, the ability to efficiently transfer heat from the semiconductor device be it a rectifier diode or thyristor to the exterior dissipating surface.
2. The type of material used in the construction of the heat sink.
3. The contact area between the semiconductor device and the heat sink.
4. The relative position of the semiconductor device on the heat sink.
5. The external surface area of the heat sink.
6. The ambient air temperature.
7. The volume of air flowing over the exterior surface of the heat sink.

The contact area between the rectifier/thyristor is extremely important in order to minimize the thermal impedance between the case and the heat sink. Both surfaces should be flat and smooth and free of dirt, corrosion, and surface oxides. It is normal practice to lubricate the mating surfaces with a substance to improve the thermal heat transfer. A commonly used material is Dow Corning silicone grease DC-200, which, in addition to improving the heat transfer capability, also minimizes the formation of oxides and corrosion products.

Another very important factor is the correct mounting of the semiconductor device to the heat sink to obtain the correct pressure between the mating surfaces. Correct installation procedures are usually detailed by the semiconductor manufacturer. However, in the case of stud mount devices it must be emphasized that excessive torquing will damage the threads and even cause mechanical damage to the silicon wafer. In addition, the threads of the stud and nut must never be lubricated since the tension on the stud may be as great as double that of the unlubricated device with the same torque loading.

Rectifier diodes and thyristors are usually plated with tin, cadmium, or nickel plating to protect them against corrosive conditions which are present in many industrial environments.

Natural air cooling is usually used in applications where the energy to be dissipated is less than 200W, since above this rating the required heat dissipating surface becomes excessive. Forced air cooling, where fans blow cooling air over the heat sink radiating surfaces at a high velocity, are used in applications where the heat to be dissipated is in excess of 200W. Heat sinks designed for forced air cooling also have their fins closer together since the predominant transfer of heat is by convection.

B-4 HEAT PIPES

A relative newcomer in the field of semiconductor device cooling is the heat pipe. The basic principle of the heat pipe is illustrated in Figure B-7. The heat pipe is a sealed metal pipe lined with a wick and partially filled with a low vapor pressure liquid. The application of heat to the pipe causes the liquid to vaporize at the heat source end. The vapor then flows to the condensing end where it condenses and the liquid returns to the heat source end by capillary action through the wick. The major advantage of this method of cooling semiconductor devices is that the heat sink or cooling fin arrangement may be at some distance from the device.

B-5 LIQUID COOLING

In high-power semiconductor rectifier and thyristor applications the devices are more effectively cooled by liquids. The two most commonly used liquids are water and oil. The major advantages of liquid cooling are compact design with the cooling of the heat transfer medium occurring remote to the semiconductors. There are several methods of liquid cooling; the most common are:

1. Mounting a number of semiconductor devices on one common liquid-cooled heat sink.

2. Sandwiching hockey-puck type devices between series connected liquid-cooled heat sinks.

3. Submerging the semiconductor device directly in the cooling medium. Obviously this method is only practical when oil liquid cooling techniques are employed.

Water cooling is very efficient and is approximately 2.5 times more effective in removing heat than oil. However, even though high flow rates may be achieved in a water cooled system, there are two major problems: corrosion and freezing. Corrosion is an electrolytic action caused by various parts of the cooling path being at different potentials. The use of distilled water and careful design to ensure that all metal components are compatible, that is, they will not initiate cathodic action, combined with the use of corrosion inhibitors will ensure satisfactory performance. Freezing can be overcome by using a glycol antifreeze.

Oil cooling even though not as effective as water cooling does eliminate the corrosion and freezing problems and provides good insulation. The major disadvantage of oil is that it is flammable which may restrict its use.

Both air-cooled and liquid-cooled heat sinks are normally equipped with a means of detecting cooling medium failure. This can be as simple as a bimetallic sensor (air-cooled) or a flow measuring device (liquid-cooled) connected to shut down the converter in the case of the heat sink overheating.

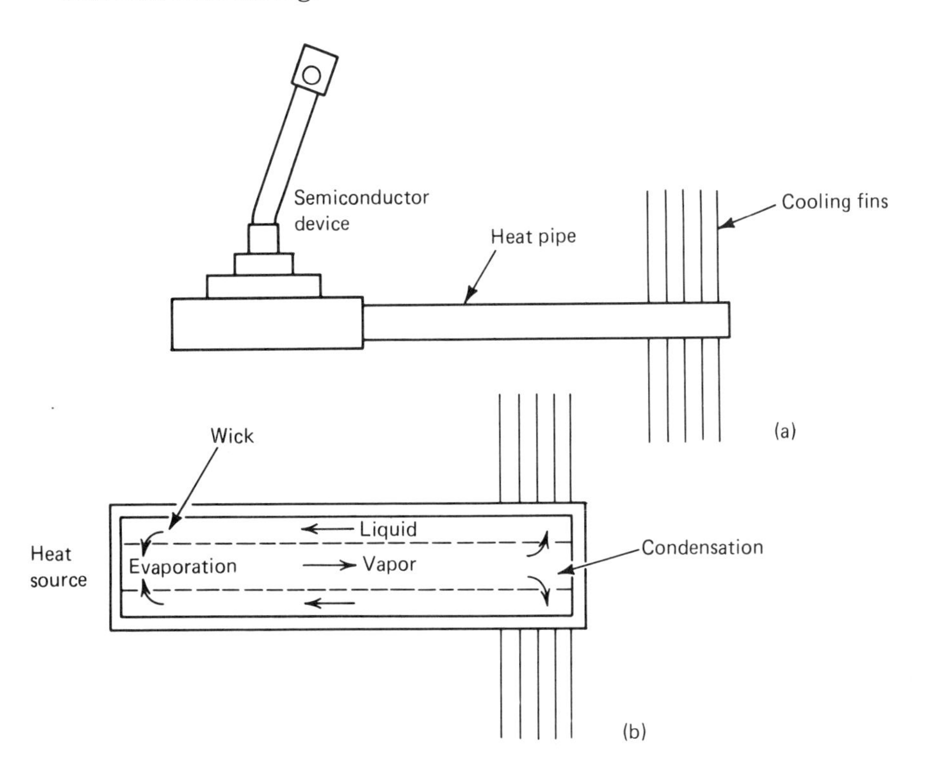

Figure B-7. Semiconductor cooling by means of a heat pipe: (a) application; (b) principle.

Bibliography

Bedford, B.D. and R.G. Hoft, *Principles of Inverter Circuits.* New York: John Wiley and Sons, Inc., 1964.

Bell, David A., *Electronic Devices and Circuits, 2nd ed.* Reston, Virginia: Reston Publishing Company, Inc., 1980.

Canadian General Electric, *Power Converter Handbook, Theory, Design and Application.* Peterborough, Ontario, Canada: Canadian General Electric Co., Ltd., 1976.

Carr, William N. and Jack P. Mize, *MOS/LSI Design and Application.* New York: Texas Instruments Electronics Series, McGraw-Hill Book Company, 1972.

Clayton, George B., *Linear Integrated Circuit Applications.* London: Macmillan Press Ltd., 1975.

Coughlin, Robert F. and Frederick F. Driscoll, *Operational Amplifiers and Linear Integrated Circuits.* Englewood Cliffs, N.J.: Prentice-Hall, Inc., 1977.

Fink, Donald G. and H. Wayne Beaty, *Standard Handbook for Electrical Engineers, 11th ed.* New York: McGraw-Hill Book Company, 1978.

Fitzgerald, A.E., Charles Kingsley, Jr., and Alexander Kusko, *Electrical Machinery, 3rd ed.* New York: McGraw-Hill Book Company, 1971.

Horowitz, Paul and Winfield Hill, *The Art of Electronics.* Cambridge, England: Cambridge University Press, 1980.

Hughes, Fredrick W., *Op Amp Handbook.* Englewood Cliffs, N.J.: Prentice-Hall Inc., 1981.

Hunter, Ronald P., *Automated Process Control Systems, Concepts and Hardware.* Englewood Cliffs, N.J.: Prentice-Hall, Inc., 1978.

Kosow, Irving L., *Control of Electric Machines.* Englewood Cliffs, N.J.: Prentice-Hall, Inc., 1973.

Kusko, A., *Solid-State D.C. Motor Drives.* Cambridge, Mass. and London: The M.I.T. Press, 1969.

Lancaster, Don, *CMOS Cookbook, 1st ed., 3rd Printing.* Indianapolis, Indiana: Howard W. Sons and Co., Inc., 1978.

Lancaster, Don, *TTL Cookbook, 1st ed, 6th Printing.* Indianapolis, Indiana: Howard W. Sons and Co., Inc., 1976.

Lenk, John D., *Handbook of Logic Circuits.* Reston, Virginia: Reston Publishing Company, Inc., 1972.

Lenk, John D., *Handbook of Modern Solid-State Amplifiers.* Englewood Cliffs, N.J.: Prentice-Hall, Inc., 1974.

Maloney, Timothy J., *Industrial Solid-State Electronics, Devices and Systems.* Englewood Cliffs, N.J.: Prentice-Hall, Inc., 1979.

Mazda, F.F., *Thyristor Control.* London: Newnes-Butterworth, 1973.

McDonald, A.C. and H. Lowe, *Feedback and Control Systems.* Reston, Virginia: Reston Publishing Company, Inc., 1981.

Metzger, Daniel L., *Electronic Components, Instruments, and Troubleshooting.* Englewood Cliffs, N.J.: Prentice-Hall, Inc., 1981.

Murphy, J.M.D., *Thyristor Control of A.C. Motors.* Oxford: Pergamon Press Ltd., 1973.

Norris, Bryan, *Digital Integrated Circuits, Operational-Amplifier, and Optoelectronic Circuit Design.* New York: Texas Instruments Electronic Series, McGraw-Hill Book Company, 1976.

Ogata, Katsuhiko, *Modern Control Engineering.* Englewood Cliffs, N.J.: Prentice-Hall, Inc., 1970.

Oliver, Frank J., *Practical Instrumentation Transducers.* New York: Hayden Book Co., Inc., 1971.

Pelly, B.R., *Thyristor Phase-Controlled Converters and Cycloconverters, Operation, Control and Performance.* New York: John Wiley and Sons, Inc., 1971.

Pressman, Abraham I., *Switching and Linear Power Supply, Power Converter Design.* Rochelle Park, N.J.: Hayden Book Company Inc., 1977.

Schaefer, Johannes, *Rectifier Circuits: Theory and Design.* New York: John Wiley and Sons, Inc., 1965.

SCR Applications Handbook. Dr. Richard G. Hoft, ed., International Rectifier Corp., Semiconductor Division, El Segundo, California, 1974.

SCR Manual, 5th ed. D.R. Graham and J.C. Hey, eds., General Electric Compay, 1972.

Seippel, Robert G., *Optoelectronics.* Reston, Virginia: Reston Publishing Company Inc., 1981.

Silicon Controlled Rectifier Designers Handbook, 2nd ed., Leslie R. Rice, ed., Westinghouse Electric Corporation, 1970.

Smeaton, Robert W., *Switchgear and Control Handbook.* New York: McGraw-Hill, Inc., 1977.

Soisson, Harold E., *Instrumentation in Industry.* New York: John Wiley and Sons, Inc., 1975.

Strangio, Christopher E., *Digital Electronics, Fundamental Concepts and Applications.* Englewood Cliffs, N.J.: Prentice-Hall, Inc., 1980.

Tyson, Forrest C., Jr., *Industrial Instrumentation.* Englewood Cliffs, N.J.: Prentice-Hall, Inc., 1961.

Watkin, R.V., *Computer Technology for Technicians and Technician Engineers, Vol. 1.* London: Longman Group Limited, 1976.

Westinghouse Electric Corp., *Introduction to Solid-State Power Electronics.* Youngwood, P.A.: Westinghouse Electric Corporation, 1977.

Woolvet, G.A., *Transducers in Digital Systems.* London: Peter Peregrinus Ltd., 1977.

ANSWERS TO PROBLEMS

Chapter 3

(1) 125.40V, 0.08 or 8%, 1.0032. **(2)** 99.03V, 110V, 0.4352, 1.04A, 0.33A, 311.13V. **(3)** 103.82V, 0.24 or 24%, 57.01W. **(4)** 93.63V, 93.6mA, 8.77W, 0.062 or 6.2%. **(5)** 12.43H. **(6)** 252V, 0.0962 or 9.62%, 64.09W.

Chapter 4

(1) 45.47°, 20.70A, 325.27V. **(2)** 64.23°, 20.70A, 644.00V. **(3)** 56.25° 72.76°. **(4)** 27.09°, 64.47°, 77.84°, 6.33A, 294.16V. **(5)** 96.32°, 124.68°, 142.13°, 6.33A, 294.16V. **(6)** 27.09°, 87.04°,5.12kVAR.

Chapter 6

(1) 11111111, 100001, 1111001, 10001, 101000101. **(2)** 10000.11, 1111011.101011, 1111.011, 101001100.100011, 1101.1001. **(3)** 10011, 1101000, 11000110, 100001, 111101. **(4)** 100000.011, 100100.111, 10101.1011, 10001.001, 1001.100. **(5)** 10, 111, 101, 1010, 110 **(6)** 1110, -1010, -1010, 1001, 1100. **(7)** 10101001, 10111101, 10001111, 1011011001, 1000.11. **(8)** 10.1001, 11.11011, 10.0011101, 1001.001011, 10.1101 **(9)** 377, 41, 505, 175.42, 1175.21634. **(10)** 185.609375, 125.40625, 1337, 213, 73. **(11)** 155.6, 11.3, 73, 753, 6736.66. **(12)** 7A.A147A, 2C, 23.2, 13.1, 1C.1A. **(13)** 9.D, D5.77, F5.1D, EF.D4, 5.5. **(14)** 101011100111.11000010, 1111000110.0010001011111, 110110100.011001011101, 10001111.01000001, 101110110.0100111. **(15)** 100010.01101000, 11.0001010000010110, 1.011100100110, 101000100, 1000100101.01000101. **(16)** 952, 7153, 182.7, 290.35, 699. **(18)** $A\bar{C}+\bar{A}B+\bar{B}C$, $AB+BC+AC$. **(19)** 1623.38 pulses/sec, 2164.5 pulses/sec. **(20)** 0.4 μs, 10MHz. **(21)** 100, 93ns. **(22)** Q_A=500kHz, Q_B=250kHz, Q_C=125kHz, Q_D=62.5kHz; 0000 to 1111, which is the decimal count from 0 to 15. **(23)** 15. **(24)** 128 μs

Chapter 8

(1) 2.0 x 10^3m, 1.5 x 10^{-1}m 7.143 x 10^{-7}m. **(2)** 2.0 x 10^9 μ, 1.5 x 10^5 μ, 0.7143 μ, 2 x 10^{12}m μ, 1.5 x 10^8m μ, 7.143 x 10^2m μ, 2.0 x 10^{13}Å, 1.5 x 10^9Å, 7.143 x 10^3Å. **(3)** 3770lm, 18.85 lm/W, 133.3lm/m² **(4)** 200 Ω, 100kΩ. **(5)** -0.312V, 0.9mA; -0.225V, 1.87mA; -0.125V, 2.8mA. **(6)** 347 Ω, 120 Ω, 45 Ω. **(7)** 15V, 7.5V, 11.5V.

Chapter 9

(1) -7.5, -0.75V, 1.6V, **(2)** 2.2, 0.44V, **(3)** -4.38V. **(4)** R_f=50kΩ, R_1=25kΩ, R_2=14.29kΩ, R_3=50kΩ. **(5)** -0.40V, 100V/s, 187ms. **(6)** 25.15%. **(7)** 3.92, 5100 Ω, 0.098 Ω. **(8)** 4287.5 Ω, 0.0333 Ω. **(9)** 0.0175 or 1.75%

Chapter 11

(1) 22.66°, 163V. **(2)** 164.89°. **(3)** 50° to 96°, 12,369 to 29,460 Ω. **(4)** 97° to 117°, 29,981 to 43,285 Ω.

Index